The Handbook of Attention

The Handbook of Attention

edited by Jonathan M. Fawcett, Evan F. Risko, and Alan Kingstone

The MIT Press
Cambridge, Massachusetts
London, England

This book was set in ITC Stone Serif Std 9/13pt by Toppan Best-set Premedia Limited. Printed and bound in the United States of America.

Library of Congress Cataloging-in-Publication Data is available.
ISBN: 978-0-262-02969-8

10 9 8 7 6 5 4 3 2 1

To our wives, children, and mentors

Contents

Attention: From the Laboratory to the Real World and Back Again

The scientific investigation of attention has advanced rapidly over the last decade as the contributions to the present volume clearly testify. This advance is evident in both the breadth of topics being tackled and the depth of their examination. And as one would expect when substantial progress is being made, new questions and issues emerge— many of which were unanticipated only a handful of years ago. As such, research in attention stands as one of the most exciting, diverse, and application-rich fields of inquiry in the psychological sciences today.

Our intention was for the present volume to bring together cutting-edge, authoritative reviews from the best and brightest researchers, writing on topics ranging from those that are traditional in the field of attention to those that are emerging areas of investigation. Even a cursory look at the table of contents should be enough to show the interested reader that we have more than achieved our goal. Here in this volume one can find definitive reviews, by well-known and rising stars, on foundational topics such as visual attention and cognitive control, underrepresented domains like auditory and temporal attention, and emerging areas of investigation including mind wandering and embodied attention. We also sought to sample broadly from the methods used by attention researchers, and accordingly, the chapters draw from an impressive scope of approaches including, but not limited to, psychophysics, mental chronometry, desktop eye tracking, mobile eye tracking, functional magnetic resonance imaging, event-related potentials, and field research.

Finally, and central to the present book, was our aim to establish a strong bridge between attention in everyday life and research on human attention. As such, we have combined in one volume, and on relatively equal footing, attention research that emphasizes traditional methodologies (part I) with research that focuses on attention in the context of people's natural everyday life (part II). These latter chapters review topics as diverse as driving, shopping, reading, multitasking, gambling, magic, lectures, and playing video games. One question that we had when we started this book was how tight the links were between laboratory-based and more real-world investigations. Therefore, in addition to choosing domains that fell clearly into one level of investigation or the other, we also challenged our authors to articulate how they saw laboratory-based studies in their domain connecting to everyday attention, or studies of everyday attention relating to laboratory-based studies.

Research in attention has often been conducted under conditions that, on the surface at least, little resemble—in terms of their complexity—situations in our everyday lives. This is, of course, intentional. The hope is that any cost to ecological validity will be offset by the discovery of principles of human attention and behavior that generalize to more complex situations. One of the many important functions served by research on everyday attention is that it allows researchers to evaluate the degree that their work is delivering on that greater promise. Specifically, do the principles derived from orthodox attention research (i.e., that which is conducted in the spartan conditions of the psychological laboratory) lead to predictions that are confirmed when tested in situations that more closely resemble people's day-to-day life? As the chapters in the second half of this volume (and many in the first as well) suggest, the answer to this question is complicated. While there are arguably some successes, there are also many remarkable failures and many other results that appear destined to fall short on their greater goals. This state of affairs, of course, raises important questions about the paths forward in attention research, and we hope this volume as a whole contributes meaningfully to that discussion.

Research on everyday attention of course makes a much deeper contribution than merely providing a check on orthodox studies. By first investigating attentional behavior or phenomena in complex, naturalistic settings, or in response to stimuli and tasks that better approximate those we encounter in our daily lives, researchers have the opportunity to ground attention research in the environment and behaviors that can subsequently be used to give shape to more orthodox investigations. The closer bond this latter approach brings between "lab" and "life" will surely increase the likelihood that the scientific investigation of attention yields generalizable principles. This "drilling down" approach, from life to lab, that we have referred to as cognitive ethology, can be juxtaposed with the more common "scaling up" approach that begins with the well-controlled approximation of some attentional behavior or phenomenon and progressively increases the complexity or naturalness of the setting, stimuli, or tasks. Both approaches have value, as evidenced by the contributions to this volume, and when one stands back and views the volume as a whole, it is clear that progress is being made along each of these paths. We believe that both paths promise future researchers many new and exciting discoveries.

—Jonathan M. Fawcett, Evan F. Risko, and Alan Kingstone

I

Attention in the Laboratory

1

Controlling Spatial Attention: Lessons from the Lab and Implications for Everyday Life

Charles L. Folk

Anyone who has spent several hours in a crowded shopping mall has no doubt returned home feeling not only physically exhausted but also mentally "drained." The visual sensory stimulation in such an environment is extreme, consisting of exposure to literally hundreds of thousands of stationary and moving objects. The fact that we feel drained provides anecdotal evidence that visual information processing requires cognitive "energy" or "resources" and that those resources are limited, in that they can be depleted. Thus, at any given point in time, limited attentional resources must be selectively allocated to a subset of the information available, which is presumably the information most relevant to current behavioral goals. A fundamental characteristic of visual information is that it is arrayed in space, and therefore the selective allocation of resources to objects or locations in space is referred to as *spatial attention*.

One obvious way in which the visual information processing system selectively "samples" the visual environment is with eye movements. Referred to as *overt orienting* or *overt attention,* eye movements provide a means of aligning objects of interest with the central fovea of the retina, which is capable of much finer spatial resolution than the peripheral retina. Thus, spatial acuity (or lack thereof in the peripheral retinal) is one form of processing limitation that the visual system overcomes by selectively aligning high-resolution sensory receptors. The concept of *spatial attention*, however, addresses limitations in processing that are distinct from spatial acuity and that require the selective allocation of resources to objects or locations in space, independent of (although certainly correlated with) eye movements. Referred to as *covert orienting* or *covert attention,* it is these shifts in spatial attention that are the focus of the current chapter.

Allocating resources to one source of information in the visual field can mean that important information at other locations might go unnoticed, a phenomenon known as *inattentional blindness* (Drew, Vo, & Wolfe, 2013; Mack & Rock, 1998; Simons & Chabris, 1999). The failure to shift attention to important environmental information can have dire consequences, such as a driver failing to notice that a traffic light has turned red, or a Transportation Security Administration officer failing to notice a weapon in a luggage x-ray scan. Thus, one of the most important issues with respect to spatial attention, in terms of both theory and application, is the nature of the mechanisms by which attentional shifts are controlled. With respect to the current chapter, the critical issue is the extent to which shifts of spatial attention are driven or "captured" by stimuli in the environment and whether such capture is subject to cognitive control. Understanding the interaction between top-down cognitive control and bottom-up stimulus-driven control of spatial attention has implications not only for attentional theory but also for behavior in everyday life.

Historical Context

Perhaps the earliest recorded demonstration of covert spatial attention was conducted by Von Helmholtz (1924), who sat in a dark room fixated on the center of wall display consisting of an array of many letters. During a series of electric sparks that were too brief to allow an eye movement, he found that although he was unable to process the entire array, he could selectively "attend" to the letters in one area and then move his attention (while remaining fixated) to another area prior to the next spark and attend to the letters there.

Systematic studies of spatial attention began in the 1950s and 1960s, with research on visual search for targets in cluttered displays (e.g., Green & Anderson, 1956) and also research on the degree to which the focus of attention is "narrowed" or "broadened" with changes in arousal (Easterbrook, 1959). This led to the influential metaphor of spatial attention as a "spotlight" whose focus or resolution can change under different circumstances (Wachtel, 1967). However, it wasn't until the 1970s that paradigms were developed that firmly established the existence of covert shifts of attention occurring independently of the eyes. One paradigm that was highly influential is the spatial cuing task developed by Posner and colleagues (see Posner, 1980). In its simplest form, the spatial cuing task involves the presentation of a target to the left or right of fixation, and the participant responds to its appearance (detection) or identity (discrimination). However, prior to the presentation of the target, a "cue" is presented that provides probabilistic information about the location of the subsequent target. For example, an arrow pointing to the left or right might appear at fixation indicating an 80% chance that the target will appear in that location (see figure 1.1, top). Response time or accuracy is measured as a function of whether the target appeared at the cued location (valid trials) or an uncued location (invalid trials). Importantly, the time interval between cue presentation and target presentation is typically less than 250 ms, which is the time it takes to initiate an eye movement. Thus, the presence of a cuing effect (e.g., faster response times to valid vs. invalid trials) is assumed to reflect a shift of covert spatial attention rather than a shift in eye position.

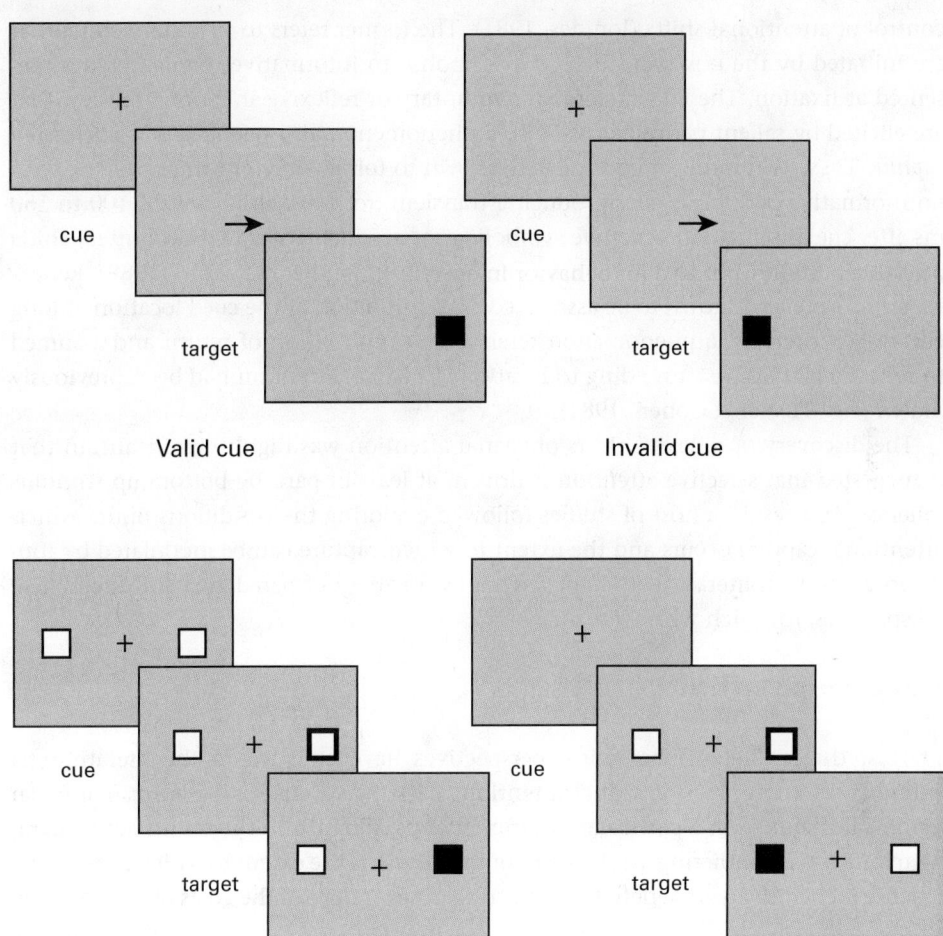

Figure 1.1
The spatial cuing paradigm with central, symbolic cues (top) and peripheral cues (bottom).

Research with the spatial cuing paradigm in the 1980s revealed robust cuing effects associated with both "central" cues, such as arrows presented at fixation, and "peripheral" cues, such as an abrupt flash or luminance increment appearing at one of the potential peripheral target locations (e.g., Jonides, 1981; see figure 1.1, bottom). Importantly, peripheral cues were found to produce cuing effects even when the cue was uninformative with respect to the subsequent target location (i.e., when the target was no more likely to appear at the cued location than any other location; Jonides, 1981). This led to the important theoretical distinction between *endogenous* and *exogenous* control of attentional shifts (Jonides, 1981). The former refers to voluntary shifts that are initiated by the observer, such as in response to informative, symbolic cues presented at fixation. The latter refers to involuntary or reflexive shifts of attention that are elicited by salient peripheral stimuli, a phenomenon also referred to as *attentional capture*. These two forms of control were shown to follow different time courses, with uninformative peripheral cues producing transient, reflexive shifts within 100 to 150 ms after cue onset and informative symbolic cues producing sustained voluntary shifts beginning around 200 to 300 ms after cue onset (e.g., Muller & Rabbitt, 1989). Reflexive shifts were also shown to be associated with inhibition at the cued location at long cue–target intervals, a phenomenon referred to as inhibition of return and assumed to reflect a bias against attending to locations to which attention had been previously drawn (e.g., Posner & Cohen, 1984).

The discovery of reflexive shifts of spatial attention was highly significant, in that it suggested that selective attention is driven, at least in part, by bottom-up stimulus salience. As a result, a host of studies followed exploring the conditions under which attentional capture occurs and the extent to which capture can be modulated by top-down goals. This literature is characterized by a variety of paradigms and theoretical perspectives, to which we now turn.

State-of-the-Art Review

At least three different theoretical perspectives have emerged in the literature on stimulus-driven control of spatial attention. Each perspective is based primarily on empirical results from a particular experimental paradigm, and the results across paradigms are often conflicting. At the heart of the debate is the extent to which attentional capture is cognitively impenetrable, occurring independent of the goals of the observer.

The Special Classes Perspective

The early discovery that cuing effects could be obtained for uninformative peripheral cues if those cues consisted of abrupt changes in luminance suggested that perhaps abrupt onset is a "special class" of stimuli to which the attention allocation system is particularly sensitive. According to this notion, abrupt onsets tend to be correlated with the appearance of behaviorally relevant events, and the system has therefore evolved to reflexively orient to such events. Support for this perspective came primarily from the *irrelevant singleton* paradigm developed by Yantis and Jonides (1984, 1990; Jonides & Yantis, 1988). In the typical task, participants engage in an attentionally demanding visual search for a specific letter among varying numbers of nontarget letters (display size), which leads to linear increases in response time. On each trial, one of the letters (the "singleton") is different from the rest in terms of some feature property, such as onset, color, luminance, and so forth (see figure 1.2, top). Critically, the

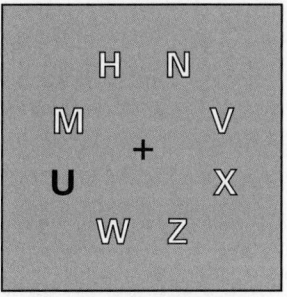

Target (H) is not the
color singleton

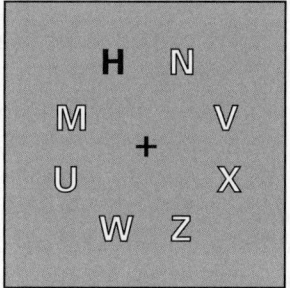

Target (H) happens to be
the color singleton

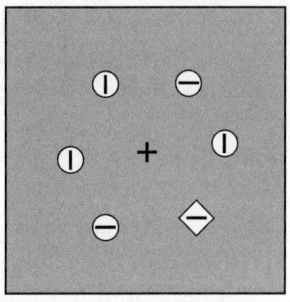

No distractor

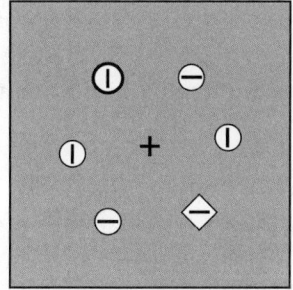

Color singleton distractor

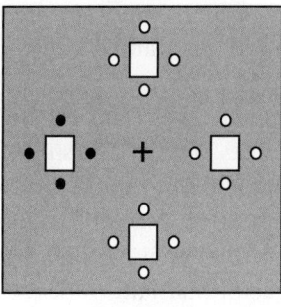

Cue display

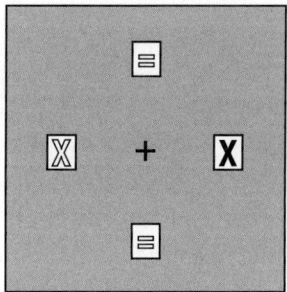

Target display

Figure 1.2
Paradigms for studying attentional capture: the irrelevant singleton paradigm (top), the additional
singleton paradigm (middle), and the modified spatial cuing paradigm (bottom).

target letter is no more likely to be the singleton than any other letter in the display (i.e., it is irrelevant to the search task). Thus, there is no incentive to voluntarily allocate attention to the singleton, allowing any effects of the singleton to be attributed to involuntary attention shifts. Yantis and Jonides (1984, 1990) found that when the singleton consisted of an abrupt onset of one of the letters among "no-onset" letters (revealed by removing masking line segments), response times were unaffected by display size on trials when the target happened to be the onset. These results suggest that attention was captured by the abrupt onset, obviating any additional search when the target happened to be the onset. In contrast, when the singleton consisted of a difference in color or luminance, there was no such search advantage for targets appearing as the singleton (Jonides & Yantis, 1988; Yantis & Egeth, 1999). This overall pattern is consistent with the special classes perspective, in that only onset singletons produced evidence of attentional capture. Subsequent studies provided evidence that other stimulus properties associated with change in information over time should be included in the special class, including the appearance of new objects (Yantis & Hillstrom, 1994) and looming or new motion (e.g., Al-Aidroos, Guo, & Pratt, 2010; Franconeri & Simons, 2003).

The Bottom-Up Salience Perspective

In contrast to the special classes approach, Theeuwes and colleagues have argued that attention shifts are driven entirely by the relative salience of objects in the visual field (see Theeuwes, 2010, for a review). According to this approach, any class of features has the ability to capture attention as long that feature produces the highest bottom-up salience signal in any given display. Evidence for this perspective comes primarily from the *additional singleton* paradigm. In the typical version, participants are presented with a circular array of shapes containing oriented line segments. One of the shapes (the target) is different from the others in terms of some feature property (e.g., a diamond among circles), and participants are required to search for this "singleton" and respond to the orientation of its resident line segment (see figure 1.2, middle). Critically, on half of the trials, one of the nontarget shapes appears as a singleton in a different dimension (e.g., color). Given that the additional singleton is completely irrelevant to the search task, it is assumed that any costs associated with its presence reflect the involuntary capture of attention. In a series of studies using this paradigm, Theeuwes and colleagues found that relative to no-distractor trials, the presence of the additional singleton produced significant costs in response time, but only when the distractor singleton was more salient than the target (Theeuwes, 1992, 1994). Moreover, response times to the target line segment are further influenced by whether the orientation of the line segment in the distractor singleton is the same or not (Theeuwes, 1995). These *response compatibility* effects are consistent with a shift of attention to the distractor. In summary, contrary to the special classes perspective, the results of the additional singleton paradigm seem to suggest that any preattentively processed singleton can elicit an involuntary shift of attention as long as it is the most salient item in the display.

The Contingent Attentional Capture Perspective

The third major perspective on the stimulus-driven control of spatial attention resides at the opposite end of the theoretical spectrum from the bottom-up salience approach. According to the contingent attentional capture perspective, the attention allocation system is "configurable" such that specific task requirements result in the offline instantiation of "attentional control settings." The ability of any given stimulus to generate an involuntary shift of attention is then contingent on whether the eliciting features of the stimulus match the current top-down control settings. Any stimulus carrying that property will elicit a shift of attention, even if the observer is aware that the stimulus is irrelevant (Folk, Remington, & Johnston, 1992).

Evidence for this perspective comes primarily from a modified spatial cuing task in which the feature property that defines the cue and target can be independently manipulated. For example, Folk et al. (1992) presented participants with a cue display consisting of clusters of four dots appearing around each of four boxes surrounding fixation. One set of dots (the cue) was different in color (red) than the rest (white), producing a color singleton. The target display, which appeared 150 ms after the cue display, consisted of the abrupt onset of a single "X" or "=" appearing in one of the four boxes, and participants responded to the identity of this target character (see figure 1.2, bottom). Importantly, across trials the target was no more likely to appear in the cued location than in any of the uncued locations, providing no incentive to voluntarily allocate attention to the color cue. Thus, any obtained cuing effects would be attributed to involuntary shifts of spatial attention. No such cuing effects were obtained for color singleton cues. However, when the cue was replaced by the abrupt onset of a single set of dots around one of the four boxes, significant cuing effects were obtained. In two other conditions both color and onset cues were also paired with target displays in which characters were presented in all four boxes, but one of them was red and the rest were white. In other words, the feature that defined the target was color. Under these conditions, the exact same color singleton cue that produced no evidence of capture when paired with an onset target now produced significant cuing effects when paired with a color-defined target. Moreover, the same onset cue that produced capture when paired with an onset target now had no effect on performance. This pattern has been replicated many times (e.g., Folk, Remington, & Wright, 1994; Folk & Remington, 1998, 2008) and provides strong evidence that attentional capture can be modulated by top-down set.

Accounting for Conflicting Results

Results from the additional singleton paradigm and the modified spatial cuing paradigm support very different accounts of attentional capture and, without further qualification, are mutually incompatible (see Theeuwes, 2010, for a critical review of the literature surrounding this debate). The viability of the salience-driven perspective is dependent on the ability to explain contingent cuing effects in the spatial cuing paradigm, and the viability of the contingent capture perspective is dependent on the ability to account for singleton distractor effects in the irrelevant singleton paradigm. Given the conflicting results of the various paradigms used to study attentional capture, an extended debate has ensued, with proponents of each perspective offering explanations for the results that appear inconsistent with their account. Here we will touch on some of the key proposals that have been offered.

Accounting for Additional Singleton Effects: Filtering Costs

From the contingent capture perspective, one explanation for the results of the additional singleton paradigm is that rather than indexing involuntary shifts of spatial attention, the costs associated with the salient, irrelevant singleton reflect a form of nonspatial distraction known as *filtering costs* (Kahneman, Treisman, & Burkell, 1983). It is well-established that there is a response time cost associated with the presence of an irrelevant object that is assumed to reflect the time needed to "filter out" the irrelevant object prior to a shift of attention to the target (Folk & Remington, 1998; Kahneman et al., 1983). In the case of the additional singleton paradigm, displays in which a salient distractor appears contain two singletons (the target and distractor) whereas displays on which the distractor is absent contain only one singleton (the target). Therefore, assuming that these two types of displays are functionally encoded in terms of one or two objects, the costs associated with the distractor may reflect filtering costs rather than shifts of spatial attention. Response compatibility effects associated with the match between the identity of the characters inside the distractor and target singletons have been attributed to the fact that the distractor and target both "pop out" from the background items, resulting in relatively low processing load (Lavie & Cox, 1997). There is evidence that under such low load conditions the identity of irrelevant items can be processed in parallel with the target (Folk, 2013; Lavie, 1995). In addition, a recent analysis of electrophysiological data from the additional singleton paradigm found no evidence for spatial capture by irrelevant singleton distractors (McDonald, Green, Jannati, & Di Lollo, 2013). Thus, the response compatibility effects associated with singleton distractors have been argued to reflect parallel processing of identity rather than a true shift of spatial attention (Folk, Remington, & Wu, 2009; but see Schreij, Theeuwes, & Olivers, 2010).

Accounting for Additional Singleton Effects: Search Mode

Another explanation for the results of the additional singleton paradigm centers on the notion that observers can adopt different types of top-down *search modes*. Bacon and Egeth (1994) have argued that in the standard additional singleton paradigm, observers tend to adopt a strategy of searching for singletons in general, rather than searching for the particular type of singleton defining the target (e.g., shape). In support of this claim, they found that distraction by the irrelevant color singleton was eliminated when the shape target was presented among heterogeneous nontarget shapes. They argued that under such conditions, observers are forced to search for a particular shape (i.e., the circle), thereby ensuring that observers adopt what they referred to as *feature search mode*. In contrast, when the target is consistently defined as a shape singleton, observers adopt *singleton detection mode*, in which the system is set to respond to singletons in general, regardless of the feature dimension in which the singletons are defined. Further evidence for the influence of search mode on performance in this task comes from studies showing that the costs associated with the irrelevant singleton can be predicted from past experience with the different search modes (Leber & Egeth, 2006) as well as variations in pretrial brain activity assumed to reflect the strength of the top-down control settings (Leber, 2010).

Accounting for Contingent Capture Effects: Speed of Disengagement

From the bottom-up salience perspective, the results of the modified spatial cuing paradigm are problematic, because at the time the cue appears, it is the most salient item in the display and should therefore capture attention regardless of top-down set. To account for this result, Theeuwes and colleagues have proposed the *speed of disengagement* hypothesis (Theeuwes, Atchley, & Kramer, 2000; Theeuwes, 2010). According to this notion, in the first sweep of visual information processing, a transient representation of visual salience is computed (see Van Zoest, Hunt, & Kingstone, 2010) and attention is always driven involuntarily to the object or location with the highest salience. However, the relationship between the features of the stimulus that capture attention and the top-down goals of the observer determines the speed with which attention is then disengaged from the irrelevant stimulus. In the spatial cuing paradigm, when the cue property does not match the target property, attention is able to quickly disengage and reorient back to fixation prior to the presentation of the target display, such that no cuing effects are observed. Thus, from this perspective, the lack of cuing effects in the spatial cuing paradigm reflects the influence of top-down attentional control settings on the speed of disengagement rather than the initial allocation of attention.

Evidence for the speed of disengagement account comes from studies showing that both the magnitude and the direction of cuing effects vary systematically with the similarity between the cue and the target (Anderson & Folk, 2012; Belopolsky, Schreij, & Theeuwes, 2010). Specifically, as the cue and the target become more dissimilar, cuing effects diminish and can even reverse, with slower response times on valid trials relative to invalid trials. This pattern is consistent with rapid disengagement followed by the application of top-down inhibition specific to the location of the cue (but see Anderson & Folk, 2012).

It should be noted that whether or not rapid disengagement is the mechanism underlying contingent attentional capture, there is clear evidence that top-down control settings do, in fact, influence the disengagement of attention. For example, Boot and Brockmole (2010) found that search for a peripherally presented target defined by color is significantly slowed by the presence of a fixation circle matching the target color relative to a fixation circle that does not match the target color, suggesting that the speed of attentional disengagement is contingent on top-down attentional control settings.

Accounting for Contingent Capture Effects: Intertrial, Feature-Based Priming

One other account of contingent capture effects that does not rely on top-down set is based on a large literature indicating that the processing of features on a given trial is influenced by whether that same feature was processed on previous trials (for a review, see Lamy & Kristjánsson, 2013). It has been suggested that since the property defining the target in the spatial cuing paradigm is typically held constant throughout a block of trials, feature priming may selectively enhance the processing of cues that match the target color relative to cues that do not. On this account, top-down effects are actually the result of bottom-up, trial-to-trial priming. In support of this claim, Belopolsky, Schreij, and Theeuwes (2010) found that when the property defining the target varied from trial to trial, evidence for capture was found regardless of the match between the cue property and the target property. Moreover, intertrial analyses found that the magnitude of the capture effect was influenced by whether the cue property on the current trial matched the target property on the previous trial (see also Folk & Remington, 2008), with large cuing effects for intertrial matches and small or nonexistent effects for intertrial mismatches (but see Lien, Ruthruff, & Johnston, 2010, for evidence that directly contradicts this pattern).

The Nature of Attentional Control Settings

As is evident from the research reviewed above, there is a continuing, vigorous debate over the influence of top-down set on attentional capture. The key issue is whether top-down attentional control settings prevent capture even during the initial sweep of visual information processing, or whether they simply influence the speed with which the system is able to disengage from the eliciting stimulus. What is clear, however, is that top-down control settings can have a substantial effect on attentional capture, even if the specific locus of that effect is unresolved. Thus, from a practical perspective, it is important to understand the nature of attentional control settings, including the kinds of properties for which the system can be set and how control settings are instantiated.

Attentional Set for Specific Feature Values

There is evidence that the attentional control system can be set for specific features values within a feature dimension. For example, Folk and Remington (1998) crossed red or green cues with red or green targets in the modified spatial cuing paradigm. They found the classic contingent capture pattern, with large cuing effects when the cue and target matched in color and very small effects when they did not, suggesting that the system can indeed be set for specific feature values.

Attentional Set for Dimensional Singletons

In addition to specific color values, there is evidence that the attention allocation system can be set for "singletons" within a given dimension. For example, using the modified spatial cuing paradigm, Folk and Anderson (2010) presented participants with targets that were color singletons appearing (unpredictably) in red or green. They found that both red and green cues produced significant cuing effects indicative of attentional capture. More importantly, they found that blue cues also produced evidence of capture, suggesting that the system was set for color singletons in general.

Attentional Set for Multiple Features

There is evidence to suggest that the attentional allocation system can also be set for more than one specific feature at a time. For example, Adamo, Pun, Pratt, and Ferber (2008) had participants search for targets defined by both color and location, such as blue targets on the right and green targets on the left. They found that irrelevant pre-cues produced evidence of capture only when the precues shared both features with the target, consistent with a simultaneous set for two defining features. Similar results were found for combinations of color and shape (Adamo, Wozny, Pratt, & Ferber, 2010). Irons, Folk, and Remington (2012) found similar effects for targets that could be defined by multiple colors (see also Moore & Weissman, 2010).

Attentional Set for Feature Relations

Recent research by Becker and colleagues suggests that it is possible for top-down attentional control settings to be based on the *relationship* between target and distractor colors rather than (or in addition to) particular color values (Becker, 2010; Becker, Folk, & Remington, 2010, 2013). For example, Becker et al. (2010) required participants to search for an orange target among gold nontargets. Notice that although a top-down set for the specific color orange could serve to enhance the activation of the target location relative to the nontarget locations, a top-down set for the relationship between target and nontarget colors (i.e., "redder") would achieve the same goal. Prior to the presentation of the search display, an irrelevant distractor was presented that was either red or orange. Surprisingly, the red distractors, which shared the relational property with the target, produced more disruption than the orange distractors, which shared the actual color of the target. These results suggest that it is possible to "set" the attentional guidance system for the relationship between target and nontarget colors when those relationships reliably distinguish targets from nontargets.

Attentional Control Settings and Working Memory

Some have argued that attentional control settings are manifested in terms of working memory "templates" for the relevant features of the target of search (e.g., Desimone & Duncan, 1995). However, an interesting question is whether the contents of working memory, even if irrelevant to a task and/or contrary to the relevant attentional set, can nonetheless involuntarily bias attention toward stimuli that match the working memory representation. There is growing evidence that this is indeed the case (see Olivers, 2008, for a review). For example, Olivers, Meijer, and Theeuwes (2006) found that color singleton distractors that match a color held in visual working memory produced more interference than other color singletons. It is important to note, however, that it is unclear whether *all* attentional control settings are manifested in terms of explicit visual working memory representations. It is possible, for example, that attentional control settings may be manifested in terms of top-down biasing or "weighting" of lower level feature channels independent of templates in working memory (e.g., Bundesen, 1990).

Attentional Control Settings and Semantic Memory

Finally, there is recent evidence that the activation of categories in semantic memory can also serve as an attentional control setting that results in capture by stimuli whose semantic content match the category. Wyble, Folk, and Potter (2013) presented participants with a category name (e.g., "amusement park rides") followed by a rapid sequence of natural scenes (100 ms per scene), one of which contained an exemplar of the category (e.g., bumper cars), which participants were required to identify at the end of the trial. The authors found that target report was significantly impaired by the presentation of an irrelevant distractor scene, appearing above or below the fixated sequence, containing another exemplar of the category for that trial (e.g., Ferris wheel). Importantly, the very same exemplar distractor produced no such effect when paired with a different category (e.g., "baby products"). This provides strong evidence that the activation of categories in semantic memory can serve as an attentional control setting that results in the capture of spatial attention by stimuli matching the semantic properties of the category.

Revisiting the "Special Classes" Perspective: The Behavioral Urgency Hypothesis

The bulk of this review has suggested that attentional capture is heavily modulated by top-down control settings, either by preventing capture or by influencing the speed of disengagement. There are, however, certain classes of stimuli that have been argued to be more resistant to top-down modulation than others. For example, as mentioned earlier, there is evidence that looming or "new" motion is one salient property that appears relatively insensitive to top-down modulation (e.g., Al-Aidroos et al., 2010; Franconeri & Simons, 2003). It has been suggested that looming motion is an example of a stimulus property associated with "behavioral urgency" in that such motion may indicate an impending collision that requires immediate action to avoid. According to this hypothesis, the attention allocation system may be "hardwired" to respond to such highly relevant events. In the following, several other candidates under the "behavioral urgency" rubric are considered.

Capture by Biological Motion/Animacy

There is evidence that in addition to looming/new motion, stimuli that exhibit motion consistent with a biological or animate entity can strongly attract attention. For example, Pratt, Radulescu, Guo, and Abrams (2010) compared the detection time for targets that appeared inside geometrical shapes that moved predictably because of collisions with other objects (i.e., nonbiological motion) or unpredictably without any such collisions (i.e., biological or animate motion). Targets were detected significantly faster when they appeared in objects exhibiting biological motion than when they appeared in objects exhibiting nonbiological motion. Such results are certainly consistent with the behavioral urgency hypothesis in that animate motion can be a harbinger of a potential threat in the environment. However, although these results clearly show capture by a property that is task irrelevant, what is needed is an experiment that tests whether similar effects would be observed when observers have adopted an explicit top-down set for some other feature property, such as color.

Capture by Unexpected/Rare Stimuli

Mounting evidence suggests that rare/unexpected stimuli are capable of overriding top-down control (e.g., Neo & Chua, 2006; Horstmann, 2002). For example, Neo and Chua (2006) found that when attention is highly focused, conditions that typically prevent capture by any irrelevant stimuli, irrelevant abrupt onsets produced evidence of capture if they appeared infrequently during the course of the experiment. Similar effects have been found for rare or unexpected color singletons. For example, Horstmann (2002) found that when a target letter in a visual search task appeared as an unexpected color singleton, performance was indistinguishable from conditions in which the target location was cued in advance, suggesting that attention was captured by the surprise singleton. Assuming that highly unexpected stimuli are correlated with events that require immediate behavioral actions, such results fit quite nicely with the behavioral urgency hypothesis.

Capture by Emotionally Valent Stimuli

There is a growing literature suggesting that emotionally valent stimuli may have special status with respect to the allocation of attention (e.g., Fox, Russo, & Dutton, 2002). For example, a number of studies have shown that angry or threatening faces are detected more rapidly and result in slower disengagement than happy faces (Fox et al., 2002). However, it is unclear the extent to which such effects reflect the emotional content of the stimuli or the presence of particular features (Cave & Batty, 2006). Moreover, several studies using the modified spatial cuing paradigm have shown no evidence for the capture of attention by emotional faces (Koster, Verschuere, Burssens, Custers, & Crombez, 2007; Lien, Taylor, & Ruthruff, 2013).

Capture by Stimuli Associated with Reward

Recent work by Anderson and colleagues suggests that stimuli associated with past reward can produce purely involuntary shifts of attention (Anderson, Laurent, & Yantis, 2011, 2012; Anderson & Yantis, 2012, 2013). Consider, for example, a study by Anderson, Laurent, and Yantis (2011) that involved a training phase and a test phase. During the training phase, participants searched for targets appearing in colors that were probabilistically associated with either high (5 cents) or low (1 cent) reward. During the test phase, distractors appearing in the previously rewarded color produced significant costs in response time even though participants searched for targets defined by shape. Subsequent studies have shown this effect is long lasting (Anderson & Yantis, 2013) and transfers across tasks (Anderson et al., 2012). The fact that these distraction effects occurred when color was completely irrelevant to the test task (targets were defined by shape) and when the colored distractor was not physically salient (it was not a color singleton) provides strong evidence for attentional selection by a reward-associated color independent of top-down set.

Summary

Research on special class stimuli suggests that some types of properties may indeed be able to override top-down set. However, more research is needed to more fully understand these phenomena and their implications. For example, do they truly reflect a failure of attentional control settings, or do they reflect limitations on our ability to manipulate attentional set in an experimental paradigm? Moreover, one might argue that the effects associated with unexpected and rewarded stimuli actually reflect a form of internal, top-down "tuning" of the system as a function of past experience.

Integration

As is evident in the above review, 30 years of research on the control of visual spatial attention has resulted in a wealth of information regarding the conditions under which spatial attention is involuntarily allocated to salient visual stimuli, as well as the development of several competing theoretical stances about the role of top-down set in modulating capture. What can these laboratory studies tell us about the allocation of spatial attention in everyday life? Three general points will be discussed.

Understanding the Successes and Failures of Attention Allocation

As mentioned in the introduction, the failure to allocate attention to important information in the visual field can have dramatic and tragic consequences. We hear about such failures almost daily on the news in the context of accidents (usually automobile accidents) caused by inattention. At the same time, when one considers the fact that throughout our waking day we are constantly engaged in shifting spatial attention from one object or location to another in a visual world composed of millions of objects that can change over space and time, perhaps the more surprising fact is that selective attention failures are relatively rare. In other words, one might argue that the attention allocation system is actually quite effective and efficient at allocating resources given the overwhelming amount of information it has to deal with.

The overarching theme of theories of selective attention reviewed above is the important interplay between top-down set and bottom-up salience in the allocation of attention (regardless of whether set affects capture or disengagement). It is precisely this interaction that provides a means of understanding the successes (and failures) of the system. It is rare that the most physically salient objects in the environment are the objects most relevant to our current goals. Thus, it would be quite inefficient to allow shifts of attention to be driven entirely by bottom-up salience. Instead, it makes much more sense to have a mechanism (i.e., top-down set) that increases the subjective salience of those objects that are most likely to be relevant to our current behavior. In doing so, the set of potential objects to shift attention to is dramatically reduced, resulting in a concomitant increase in efficiency and a high likelihood that attention will be shifted to the most important information. However, there will be cases where stimuli that are the most important for our current behavioral actions happen to carry properties that do not match the top-down set. It is those cases where selective attention is likely to fail. Analogous to the use of heuristics in decision making, conditionalizing the capture of attention on top-down set for feature properties results in a fast, efficient attention allocation system that "gets it right" most of the time.

Consider, for example, a study by Most and Astur (2007) in which participants in a driving simulator searched for road signs of a particular color (e.g., yellow) in order to determine which way to turn at each intersection. Unexpectedly, at one intersection a motorcycle turned in front of the vehicle. The critical result was that when the motorcycle was the same color as the road signs (e.g., yellow), participants stopped more rapidly and were significantly less likely to collide with the rider than if the motorcycle was in a different color (e.g., blue). This illustrates that for most of the "trip," the most relevant information was the colored road signs, to which attention was presumably allocated efficiently and appropriately. However, when a motorcycle appeared unexpectedly, it became the most important object in the environment. The success or failure of selective attention was then determined by whether the motorcycle event happened to carry the color for which the system was set.

Challenging Conventional Wisdom

Many information displays are designed with the explicit goal of capturing attention. For example, warnings and advertisements are designed to attract the attention of observers so that important information is processed (warnings) or familiarity with a product is enhanced (advertisements). Conventional wisdom typically dictates that the best way to attract consumers' attention is to enhance the bottom-up salience of the information (Wogalter & Vigilante, 2006). However, as the review of the laboratory research on attentional capture reveals, there is very little evidence that any of these stimulus properties are capable of capturing attention in a purely bottom-up fashion (see also Rauschenberger, 2003). Instead, what is critical is whether the information is displayed in a way that is likely to be part of the consumer's top-down set. As an extreme example, there is a considerable literature showing that addicts are more likely to be captured by stimuli associated with their drug of choice than are nonaddicts (see Field & Cox, 2008, for a review). Thus, ironically, warnings on cigarette packages are more likely to be noticed and remembered by those who are already addicted than by those who are just beginning to smoke (MacKinnon & Fenaughty, 1993).

Improving Information Display

Finally, laboratory research and theory on the interaction between top-down set and bottom-up salience in the allocation of spatial attention provides guiding principles for creating information displays that minimize distraction by irrelevant information and enhance the likelihood of attentional shifts to relevant information. Given the strong influence of top-down set on the allocation of attention, the most efficient and appropriate allocation of resources will occur when relevant information is presented in a way that is consistent with the current task set, and irrelevant information in a way that is not consistent with that set. The obvious problem for designing information displays, however, is that observers' task sets can and do change over time, and the particular set in place at any given point time is not directly observable. However, through careful task analysis, it may be possible to identify a limited range of potential top-down sets and to design information displays consistent with those sets.

As a simple, concrete example, consider the design of safety clothing for highway workers. Such clothing is typically designed to be as salient as possible, using neon colors such as orange or yellow. However, as pointed out above, conspicuity does not guarantee that spatial attention will be appropriately drawn to the conspicuous source. What is needed is a careful analysis of the kinds of attentional sets that drivers tend to favor. For example, one might argue that driving requires an attentional set for the color red, in that it is highly correlated with important events relevant to driving (e.g., a critical change in the status of a traffic light, the braking of the car in front of us, the detection of stop signs, etc.). Thus, one might argue that safety clothing for highway workers should not only be conspicuous but should include a hefty dose of the color red. Obviously, more complex and/or specific display conditions (e.g., the cockpit of an aircraft, safety warnings on medicine bottles, product advertisements) would require a much more detailed task analysis to determine what kinds of tasks sets are likely to be used in those specific environments.

One other principle of information display that falls out of the laboratory research on attentional control is the notion of consistency. Specifically, to the extent that task-relevant information is consistently presented in terms of a particular feature format (i.e., color, shape, size, etc.), it will be easier for observers to learn and use appropriate top-down sets to find information that is needed at any point in time. Finally, when it is imperative that attention be drawn to critical information that may not be part of the observer's current task set (e.g., an unexpected low fuel indicator), it may be possible to take advantage of some of the "special class" properties that tend to break through top-down set. For example, critical information might be presented in a rare/surprising format or might carry a feature property that, through prior training, has been previously associated with reward.

Future Directions

As is evident from this review, there remains a vigorous debate over the details of the mechanisms by which top-down set modulates the allocation of spatial attention. Clearly, additional research is needed to determine whether configuring the attention system results in the selective allocation of resources during the initial sweep of visual information processing or whether it influences postallocation disengagement processes. There is also disagreement over the nature of top-down control, with some arguing that top-down influence is entirely a function of bottom-up feature priming (Theeuwes, 2013) whereas others argue that top-down control is just one of many sources of information that can combine to determine the prioritization of resource allocation (Awh, Belopolsky, & Theeuwes, 2012). More research is needed to determine the appropriate conceptualization for the configurability of the attention allocation system. The recent work relevant to the special classes perspective has identified several stimulus properties that may be completely immune from top-down control. Additional research is needed to explore these properties, and their resistance to top-down set, more fully. Finally, with respect to spatial attention in everyday life, laboratory research suggests that top-down set may play a larger role than conspicuity in driving attention allocation to important environmental information. Thus, future research is needed that explores means of exploiting top-down task set to enhance the likelihood that relevant information will be attended and to decrease distraction by irrelevant information.

Box 1.1
Key Points

- Covert spatial attention is a cognitive resource that is selectively allocated to objects or locations in the visual array in the absence of eye movements.
- Shifts in spatial attention that occur against the intentions of the observer are referred to as attentional capture.
- The attention allocation system is "configurable" such that different task goals result in different attentional control settings.
- The capture of spatial attention is modulated by the match between the eliciting stimulus and current control settings.
- Although certain stimulus properties may have the ability to override attentional control settings, bottom-up conspicuity does not guarantee that a stimulus will elicit a shift of attention.

Box 1.2
Outstanding Issues

- Additional research is needed to determine whether the modulating effects of top-down set reflect the prevention of capture or the speed of attentional disengagement.
- There may be certain classes of stimuli that are capable of capturing attention independent of attention set, but additional research is needed to more fully explore this possibility.
- Designers of information displays, such as warnings, need to take into account the modulating effects of top-down attentional sets in addition to bottom-up salience.
- The efficiency of attention allocation to relevant information, as well as resistance to distraction by irrelevant information, can be enhanced by evaluating the nature of the likely top-down sets associated with a given real-world task.

References

Adamo, M., Pun, C., Pratt, J., & Ferber, S. (2008). Your divided attention, please! The maintenance of multiple attentional control sets over distinct regions of space. *Cognition, 107,* 295–303.

Adamo, M., Wozny, S., Pratt, J., & Ferber, S. (2010). Parallel, independent attentional control settings for colors and shapes. *Attention, Perception & Psychophysics, 72,* 1730–1735.

Al-Aidroos, N., Guo, R. M., & Pratt, J. (2010). You can't stop new motion: Attentional capture despite a control set for color. *Visual Cognition, 18,* 859–880.

Anderson, B. A., & Folk, C. L. (2012). Dissociating location-specific inhibition and attention shifts: Evidence against the disengagement account of contingent capture. *Attention, Perception & Psychophysics, 74,* 1183–1198.

Anderson, B. A., Laurent, P. A., & Yantis, S. (2011). Value-driven attentional capture. *Proceedings of the National Academy of Sciences of the United States of America, 108,* 10367–10371.

Anderson, B. A., Laurent, P. A., & Yantis, S. (2012). Generalization of value-based attentional priority. *Visual Cognition, 20,* 647–658.

Anderson, B. A., & Yantis, S. (2012). Value-driven attentional and oculomotor capture during goal-directed, naturalistic viewing. *Attention, Perception & Psychophysics, 74,* 1644–1653.

Anderson, B. A., & Yantis, S. (2013). Persistence of value-driven attentional capture. *Journal of Experimental Psychology: Human Perception and Performance, 39,* 6–9.

Awh, E., Belopolsky, A. V., & Theeuwes, J. (2012). Top-down versus bottom-up attentional control: A failed theoretical dichotomy. *Trends in Cognitive Sciences, 16,* 437–443.

Bacon, W. F., & Egeth, H. E. (1994). Overriding stimulus-driven attentional capture. *Perception & Psychophysics, 55,* 485–496.

Becker, S. I. (2010). The role of target–distractor relationships in guiding attention and the eyes in visual search. *Journal of Experimental Psychology. General, 139,* 247–265.

Becker, S. I., Folk, C. L., & Remington, R. W. (2010). The role of relational information in contingent capture. *Journal of Experimental Psychology: Human Perception and Performance, 36,* 1460–1476.

Becker, S. I., Folk, C. L., & Remington, R. W. (2013). Attentional capture does not depend on feature similarity, but on target–nontarget relations. *Psychological Science, 24,* 634–647.

Belopolsky, A. V., Schreij, D., & Theeuwes, J. (2010). What is top-down about contingent capture? *Attention, Perception & Psychophysics, 72,* 326–342.

Boot, W. R., & Brockmole, J. R. (2010). Irrelevant features at fixation modulate saccadic latency and direction in visual search. *Visual Cognition, 18,* 481–491.

Bundesen, C. (1990). A theory of visual attention. *Psychological Review, 97,* 523–547.

Cave, K. R., & Batty, M. J. (2006). From searching for features to searching for threat: Drawing the boundary between preattentive and attentive vision. *Visual Cognition, 14*, 629–646.

Desimone, R., & Duncan, J. (1995). Neural mechanism of selective visual attention. *Annual Review of Neuroscience, 18*, 193–222.

Drew, T., Vo, M., & Wolfe, J. (2013). The invisible gorilla strikes again: Sustained inattentional blindness in expert observers. *Psychological Science, 24*, 1848–1853.

Easterbrook, J. A. (1959). The effect of emotion on cue utilization and the organization of behavior. *Psychological Review, 66*, 183–201.

Field, M., & Cox, W. M. (2008). Attentional bias in addictive behaviors: A review of its development, causes, and consequences. *Drug and Alcohol Dependence, 97*, 1–20.

Folk, C. L. (2013). Dissociating search costs and compatibility effects in the additional singleton paradigm. *Frontiers in Cognitive Science, 4*, article 434.

Folk, C. L., & Anderson, B. A. (2010). Target uncertainty effects in attentional capture: Singleton detection mode or multiple attentional control settings? *Psychonomic Bulletin & Review, 17*, 421–426.

Folk, C. L., & Remington, R. W. (1998). Selectivity in distraction by irrelevant featural singletons: Evidence for two forms of attentional capture. *Journal of Experimental Psychology: Human Perception and Performance, 24*, 847–858.

Folk, C. L., & Remington, R. (2008). Bottom-up priming of top-down attentional control settings. *Visual Cognition, 16*, 215–231.

Folk, C. L., Remington, R. W., & Johnston, J. C. (1992). Involuntary covert orienting is contingent on attentional control settings. *Journal of Experimental Psychology: Human Perception and Performance, 18*, 1030–1044.

Folk, C. L., Remington, R. W., & Wright, J. H. (1994). The structure of attentional control: Contingent attentional capture by apparent motion, abrupt onset, and color. *Journal of Experimental Psychology: Human Perception and Performance, 20*, 317–329.

Folk, C. L., Remington, R. W., & Wu, S. C. (2009). Additivity of abrupt onset effects reflects non-spatial distraction, not the capture of spatial attention. *Attention, Perception & Psychophysics, 71*, 308–313.

Fox, E., Russo, R., & Dutton, K. (2002). Attentional bias for threat: Evidence for delayed disengagement from emotional faces. *Cognition and Emotion, 16*, 355–379.

Franconeri, S. L., & Simons, D. J. (2003). Moving and looming stimuli capture attention. *Perception & Psychophysics, 65*, 999–1010.

Green, B. F., & Anderson, L. K. (1956). Color coding in a visual search task. *Journal of Experimental Psychology, 51*, 19–24.

Horstmann, G. (2002). Evidence for attentional capture by a surprising color singleton in visual search. *Psychological Science, 13*, 499–505.

Irons, J. L., Folk, C. L., & Remington, R. W. (2012). All set! Evidence of simultaneous attentional control settings for multiple target colors. *Journal of Experimental Psychology: Human Perception and Performance, 38*, 758–775.

Jonides, J. (1981). Voluntary vs. automatic control over the mind's eye's movements. In J. B. Long & A. D. Baddeley (Eds.), *Attention and performance IX* (pp. 187–203). Hillsdale, NJ: Erlbaum.

Jonides, J., & Yantis, S. (1988). Uniqueness of abrupt visual onset in capturing attention. *Perception & Psychophysics, 43*, 346–354.

Kahneman, D., Treisman, A., & Burkell, J. (1983). The cost of visual filtering. *Journal of Experimental Psychology: Human Perception and Performance, 9*, 510–522.

Koster, E. H., Verschuere, B., Burssens, B., Custers, R., & Crombez, G. (2007). Attention for emotional faces under restricted awareness revisited: Do emotional faces automatically attract attention? *Emotion (Washington, D.C.), 7*, 285–295.

Lamy, D., & Kristjánsson, A. (2013). Is goal-directed attentional guidance just intertrial priming? A review. *Journal of Vision (Charlottesville, Va.), 14*, 1–19.

Lavie, N. (1995). Perceptual load as a necessary condition for selective attention. *Journal of Experimental Psychology: Human Perception and Performance, 21*, 451–468.

Lavie, N., & Cox, S. (1997). On the efficiency of visual selective attention: Efficient visual search leads to inefficient distractor rejection. *Psychological Science, 8*, 395–398.

Leber, A. B. (2010). Neural predictors of within-subject fluctuations in attentional control. *Journal of Neuroscience, 30*, 11458–11465.

Leber, A. B., & Egeth, H. E. (2006). It's under control: Top-down search strategies can override attentional capture. *Psychonomic Bulletin & Review, 13*, 132–138.

Lien, M.-C., Ruthruff, E., & Johnston, J. C. (2010). Attention capture with rapidly changing attentional control settings. *Journal of Experimental Psychology: Human Perception and Performance, 36*, 1–16.

Lien, M.-C., Taylor, R., & Ruthruff, E. (2013). Capture by fear revisited: An electrophysiological investigation. *Journal of Cognitive Psychology, 25*, 873–888.

Mack, A., & Rock, I. (1998). *Inattentional blindness*. Cambridge, MA: MIT Press.

MacKinnon, D. P., & Fenaughty, A. M. (1993). Substance use and memory for health warning labels. *Health Psychology, 12*, 147–150.

McDonald, J. J., Green, J. J., Jannati, A., & Di Lollo, V. (2013). On the electrophysiological evidence for the capture of visual attention. *Journal of Experimental Psychology: Human Perception and Performance, 39*, 849–860.

Moore, K. S., & Weissman, D. H. (2010). Involuntary transfer of a top-down attentional set into the focus of attention: Evidence from a contingent attentional capture paradigm. *Attention, Perception & Psychophysics, 72,* 1495–1509.

Most, S. B., & Astur, R. S. (2007). Feature-based attentional set as a cause of traffic accidents. *Visual Cognition, 15,* 125–132.

Muller, H. J., & Rabbitt, P. M. (1989). Reflexive and voluntary orienting of visual attention: Time course of activation and resistance to interruption. *Journal of Experimental Psychology: Human Perception and Performance, 15,* 315–330.

Neo, G., & Chua, F. (2006). Capturing focused attention. *Perception & Psychophysics, 68,* 1286–1296.

Olivers, C. (2008). Interactions between visual working memory and visual selection. *Frontiers in Bioscience, 13,* 1182–1191.

Olivers, C. N. L., Meijer, F., & Theeuwes, J. (2006). Feature-based memory-driven attentional capture: Visual working memory content affects visual attention. *Journal of Experimental Psychology: Human Perception and Performance, 32,* 1243–1265.

Posner, M. I. (1980). Orienting of attention. *Quarterly Journal of Experimental Psychology, 32,* 3–25.

Posner, M. I., & Cohen, Y. (1984). Components of attention. In H. Bouma & D. Bowhuis (Eds.), *Attention and performance X* (pp. 531–556). Hillsdale, NJ: Erlbaum.

Pratt, J., Radulescu, P., Guo, R., & Abrams, R. A. (2010). It's alive! Animate motion captures visual attention. *Psychological Science, 21,* 1724–1730.

Rauschenberger, R. (2003). Attentional capture by auto- and allo-cues. *Psychonomic Bulletin & Review, 10,* 814–842.

Schreij, D., Theeuwes, J., & Olivers, C. N. (2010). Abrupt onsets capture attention independent of top-down control settings: II. Additivity is no evidence for filtering. *Attention, Perception & Psychophysics, 72,* 672–682.

Simons, D. J., & Chabris, C. F. (1999). Gorillas in our midst: Sustained inattentional blindness for dynamic events. *Perception, 28,* 1059–1074.

Theeuwes, J. (1992). Perceptual selectivity for color and form. *Perception & Psychophysics, 51,* 599–606.

Theeuwes, J. (1994). Stimulus-driven capture and attentional set: Selective search for color and visual abrupt onsets. *Journal of Experimental Psychology: Human Perception and Performance, 20,* 799–806.

Theeuwes, J. (1995). Perceptual selectivity for color and form: On the nature of the interference effect. In A. F. Kramer, M. G. H. Coles, & G. D. Logan (Eds.), *Converging operations in the study of visual attention* (pp. 297–314). Washington, DC: American Psychological Association.

Theeuwes, J. (2010). Top-down and bottom-up control of visual selection. *Acta Psychologica, 123,* 77–99.

Theeuwes, J. (2013). Feature-based attention: It's all bottom-up priming. *Philosophical Transactions of the Royal Society. Series B, Biological Sciences, 368,* 1–11.

Theeuwes, J., Atchley, P., & Kramer, A. F. (2000). On the time course of top-down and bottom-up control of visual attention. In S. Monsell & J. Driver (Eds.), *Attention & performance XVIII* (pp. 105–125). Cambridge, MA: MIT Press.

van Zoest, W., Hunt, A. R., & Kingstone, A. (2010). Emerging representations in visual cognition: It's about time. *Current Directions in Psychological Science, 19,* 116–120.

Von Helmholtz, H. (1924). *Treatise on physiological optics.* Rochester, NY: Optical Society of America.

Wachtel, P. L. (1967). Conceptions of broad and narrow attention. *Psychological Bulletin, 68,* 417–429.

Wogalter, M. S., & Vigilante, W. J. (2006). Attention switch and maintenance. In M. S. Wogalter (Ed.), *The handbook of warnings* (pp. 245–266). Mahwah, NJ: Erlbaum.

Wyble, B., Folk, C. L., & Potter, M. (2013). Contingent attentional capture by conceptual relevant images. *Journal of Experimental Psychology: Human Perception and Performance, 39,* 861–871.

Yantis, S., & Egeth, H. (1999). On the distinction between visual salience and stimulus-driven attentional capture. *Journal of Experimental Psychology: Human Perception and Performance, 25,* 661–676.

Yantis, S., & Hillstrom, A. P. (1994). Stimulus-driven attentional capture: Evidence from equiluminant visual objects. *Journal of Experimental Psychology: Human Perception and Performance, 20,* 95–107.

Yantis, S., & Jonides, J. (1984). Abrupt visual onsets and selective attention: Evidence from visual search. *Journal of Experimental Psychology: Human Perception and Performance, 10,* 601–621.

Yantis, S., & Jonides, J. (1990). Abrupt visual onsets and selective attention: Voluntary vs. automatic allocation. *Journal of Experimental Psychology: Human Perception and Performance, 16,* 121–134.

2
Visual Search

Jeremy M. Wolfe

Visual search tasks are everywhere. Over and over, in the course of any day, we look for one thing in a visual world full of things that we are not looking for. The topic has been part of the field of experimental psychology since its inception and has been a particularly active area of research over the last quarter of a century. In this chapter, we will briefly review some of the historical roots of visual search research. The greater part of the chapter will be devoted to a review of laboratory search studies. In the final section, we will consider the current interest in studying real-world search tasks. We need to search because we are simply incapable of processing all visual stimuli at the same time. Even objects that are in clear view need to be attended to before they can be recognized. The central message of this chapter is that the deployment of attention in visual search is not random. It is "guided" in a manner that serves us well in most everyday settings. Attention is guided by the known properties of what we are looking for ("feature guidance") and, when possible, by the rules that govern the placement of that target in a scene ("scene guidance"). The result is a human search engine that works well most of the time but that can run into difficulty when civilization creates unusual search tasks like those in medical image perception or airport security.

In generalizing from laboratory search tasks to tasks in the real world, we need to be clever and careful. In a typical laboratory search experiment, observers (Os) with very little at stake typically do hundreds of trials of one type of search with each search lasting on the order of a second. The search stimuli and the response are generally quite artificial, and the search target is typically present on half of the trials or on all of them if the response is some sort of localization or identification task. These tasks are intended to tell us about search in the world where, in contrast to the lab, the stakes could be very high. (Where is my child? Is there a bomb here?) Unlike in the lab, most real tasks will not be repeated over and over again. How often do you need to find the baking powder on the supermarket shelf? Unlike a single search trial in the lab, many real search tasks take quite a while. For instance, searching a patient's chest CT scan for signs of lung cancer will take minutes, not seconds. Finally, the probability that a target is present varies widely, and your search behavior is governed by your estimate of the likelihood that the target is present. (How long should I spend looking for a good necktie on this bargain table?) Nevertheless, the rules of search, uncovered in the lab, will continue to constrain search performance in the world.

Historical Context

Key principles of visual attention were first scientifically described in the late nineteenth and early twentieth centuries. Consider, for example, that it is easier to report on a property of a briefly presented stimulus if you know that property in advance and can attend to it.

Egeth (1967) cites Kulpe (1904) as the beginning of experiments of this sort. Any standard search experiment in which the target is prespecified can be said to build on Kulpe's work on knowing what you are looking for. When you don't know what you are looking for or are misled, the result can be inattentional blindness (Rock, Linnett, Grant, & Mack, 1992) and missing gorillas (Drew, Vo, & Wolfe, 2013; Simons & Chabris, 1999).

Helmholtz (1924) probably gets credit for initiating the scientific study of where attention is deployed. He describes covert spatial attention, the ability to direct attention to one locus while keeping the eyes fixed at another, saying, "It is possible, simply by a conscious and voluntary effort, to focus the attention on some definite spot in an absolutely dark and featureless field" (*Treatise on Physiological Optics*, Vol. 3, p. 455 of the Southall translation). This is critical to our understanding of visual search because it provides the basis for theories of search based on limitations of attention and not simply on limits in peripheral vision. Certainly those perceptual limits are important. If the target of your search is not visible or not recognizable unless you are fixating on (or near) it, then search becomes largely a matter of the strategic deployment of the eyes (e.g., Najemnik & Geisler, 2005). The recognizability of items in the periphery will be based on limits imposed by the particular characteristics of peripheral vision (e.g., the effects of crowding; Levi, 2011). Indeed a "lossy" representation in peripheral vision can be used to explain the relative ease or difficulty of many search tasks (Rosenholtz, Huang, & Ehinger, 2012). From Helmholtz, we can derive the idea that attention, deployed covertly, is doing something different than eye movements in search.

Toward the end of this chapter, we will consider work on visual search in scenes, an important and growing piece of the current search literature. Some of the earliest work on visual search in scenes was done by Kingsley, who had his Os view "photographic reproductions of well-known paintings and suitable cuts from newspapers and magazines" (Kingsley, 1926, p. 21). He asked those subjects three types of questions that he labeled as perception questions, search questions, and thought questions, and he said of search that it "may include perception, but it also includes something more" (p. 55). He locates search intermediate between perception and thought, with aspects of both.

Kingsley's measures of search were largely observational. He noted that "the search was characterized by 'shifting of the eyes,' 'scanning of various objects' in the picture, numerous 'shifts of fixation' and 'glancing about' from one part of the picture to another" (Kingsley, 1932). With the development of eye-tracking methods (e.g., Ford, White, & Lichtenstein, 1959), these observations could be made more concrete as in the pioneering work of Yarbus (1967) and the scanpath work of Noton and Stark (1971). Early on, eye-movement recording was used to investigate more applied issues such as search of overhead surveillance imagery (Enoch, 1959) or medical x-ray images (Tuddenham, 1962). Kingsley also introspected with some success about the role of scene semantics (Biederman, Mezzanotte, & Rabinowitz, 1982; Henderson & Ferreira, 2004) in the guidance of attention in real-world search tasks: "I did not just let my eyes wander around. I looked for the table or furniture on which the ship model would be located and so avoided aimless eye-wandering" (Kingsley, 1932, p. 316).

Work on search in real scenes was a relatively small part of the search literature until fairly recently. Scenes are hard to control in a parametric manner, and, until relatively recently, they were hard to store and present to Os in the lab. In the mid-twentieth century, it was more practical to make use of displays of alphanumeric characters. What can you extract from a glimpse of a set of letters? One line of work, exemplified by Sperling (1960) and Sternberg (1966), developed into the study of iconic and short-term or working memory. Another line, exemplified by the work of Ulric Neisser (1967), became more interested in how information is extracted from the array of letters while it is visible.

When one was searching through a set of letters or words (as in Neisser's work), it was pretty clear that the search was proceeding word by word in a *serial* manner. In these experiments, one measured the reaction time or response time (same thing, same initials, RT), and, in the case of these serial searches, RTs increased linearly with the number of items in the display (known as the "set size"). In some searches, however, it did not seem necessary to search one item at a time. For instance, Egeth, Jonides, and Wall (1972) reported that, if you were searching for a *4* among *C*s or vice versa, the RT × set size function had a slope near zero. That is, it did not matter how many *C*s there were; a single *4* would just "pop out" of the display. In his book *Cognitive Psychology*, Neisser (1967) introduced the idea of a "preattentive" stage in which processing "must be genuinely 'global' and 'holistic'" (p. 89). These are terms we will see again.

Starting in the late 1970s, Anne Treisman (Treisman & Gelade, 1980) formulated her hugely influential feature integration theory (FIT). At the heart of FIT is the idea that a set of basic stimulus attributes can be processed in parallel but that "binding" features together into specific perceptual objects required attention and that this attention needed to be deployed serially from one object to the next. The sorts of features that could be processed in parallel included color, size, orientation, motion, curvature, and so forth. These came with the system. They were not learned. Treisman did not rule out the possibility of learned features, but when she did a series of experiments, training Os to search apparently "in parallel" for targets of novel shape, she found that this learning did not transfer to other tasks (Treisman, Vieira, & Hayes, 1992). FIT, therefore, situates the serial/parallel, attentive/preattentive distinctions in the structure and function of the visual system (or of perceptual systems more generally). This is somewhat different than models that see these distinctions as aspects of learning and memory.

Fundamental to feature *integration* theory (unsurprisingly) is the idea that features need to be integrated. Some tasks could get around this fundamental limitation—for instance, by having a target that required detecting the presence of only one feature. In other cases, the unbound co-occurrence of features might allow a task to be done (Vul & Rich, 2010). Otherwise, FIT holds that the need to bind is the critical bottleneck in processing. This insistence on a role for binding remains a debatable point (Di Lollo, 2012; Rosenholtz et al., 2012; Roskies, 1999; Treisman, 1999; Wolfe, 2012), but this review will favor a probinding position.

The "features" of FIT connected the study of attention more closely to the study of vision than had been the case previously. Unlike alphanumeric characters, Treisman's basic features like color, orientation, motion, and size were like the basic features being studied in single-unit physiology (e.g., Barlow, 1972) or in "channel" psychophysics (e.g., Braddick, Campbell, & Atkinson, 1978). The consequence was the emergence of a substantial field of research in visual attention, bringing together researchers whose primary interest had been attention with those whose focus was on vision. The following "state-of-the-art" section describes what has been learned at this intersection of the vision and attention literatures.

State-of-the-Art Review

The Basic Laboratory Visual Search Task

Treisman offered multiple lines of evidence in favor of FIT. For purposes of a chapter on visual search, the most relevant FIT evidence is the claim that searches for unique features are parallel, with RT × set size slopes near zero, while searches for anything involving a combination of features are serial self-terminating searches, with RTs that increased markedly and linearly with increasing set size.

In the classic visual search experiment, of the sort popularized by Treisman, Os search for a target item among a variable number of distractor items. Typically, Os press one response key if a target is present and another if it is absent and are asked to do so as quickly and accurately as possible. In this design, the primary measure of interest is the RT as a function of the number of items—the set size. Accuracy is monitored in order to assess the possibility of speed–accuracy trade-offs. It is typical for errors to rise as set size rises. Some of these errors occur when the O responds "absent" without having sufficiently examined the display. This generates a "miss" error and eliminates from the set of correct "hit" responses a response whose RT would have been on the longer side. As a consequence, it is not unusual for average RTs at larger set sizes to be lower than they would have been in the absence of such errors (Dukewich & Klein, 2009). One way to reduce such speed–accuracy trade-offs is to have targets present on every trial and to ask Os to localize the target. In other experimental designs, accuracy can be the dependent measure of primary interest (Bergen & Julesz, 1983). Typically, in such experiments, the stimulus duration (or interstimulus interval—ISI—between a stimulus and a mask) is brief, creating many errors by limiting the availability of the visual information. With a brief presentation, the rise in errors as set size increases is analogous to the rise in RTs with set size in free viewing conditions. More sophisticated use of error data involves systematically varying the exposure duration, the ISI, or the time until a mandatory response, and examining the shape of the resulting speed–accuracy trade-off functions (e.g., Dosher, Han, & Lu, 2004; McElree & Carrasco, 1999). In these experiments, the speed–accuracy trade-off is the method and not the problem.

Varying the set size in a display will vary the density and/or eccentricity of items in the display. Typically, items are placed in a semirandom array within a fixed area. In this situation, density must increase as set size increases. This is usually of little consequence, but it can be relevant (Cohen & Ivry, 1991), especially when crowding becomes a factor (Felisberti, Solomon, & Morgan, 2005). Eccentricity of targets is certainly a factor in search (Carrasco, Evert, Chang, & Katz, 1995; Wolfe, O'Neill, & Bennett, 1997). Eccentricity is often "controlled" by placing all items at a fixed radius around fixation. Of course, this does control eccentricity, but if the goal is to make all items equally visible, it is worth remembering that acuity isopters are actually roughly horizontal ovals. These measures of eccentricity assume that the O is holding fixation. In many search experiments (and in most real-world situations), this is not the case. Os are allowed to move their eyes freely. Interestingly, if the items in the search array do not need to be fixated in order to be identified, the pattern of RTs is essentially the same with and without eye movements (Zelinsky & Sheinberg, 1997) with a slight speed advantage to search when fixation is held. If items (e.g., small and/or crowded letters) need to be fixated, then search proceeds at the rate of voluntary saccadic eye movements: 3 to 4 items per second. The throughput of search is much higher when eye movements are not the limiting step. Depending on one's model of search (see below), the visual system is capable of handling on the order of 20 to 50 items per second in search. Note, again, that this is a throughput estimate and not a claim that 20 to 50 items are being processed in series, each taking only 20 to 50 ms to process fully (Wolfe, 2003).

The Continuum of Search Efficiency

The early work in visual search, enshrined in Treisman's FIT, described two types of search. For some search tasks, as in figure 2.1a, RT was essentially independent of set size. These tended to be searches where the target was defined by a unique basic feature

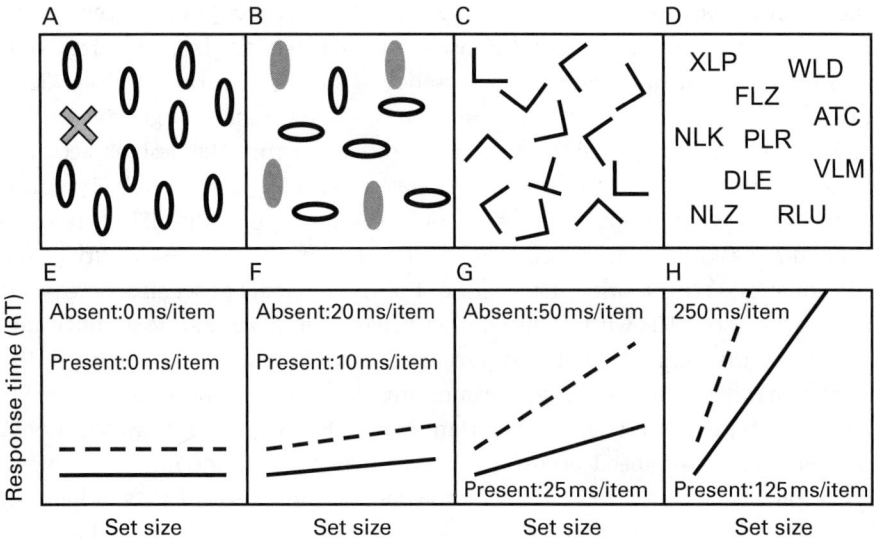

Figure 2.1
Examples of types of visual search tasks: feature (A), conjunction (B), spatial configuration (C), and a search requiring fixation of each trigram "item" (D). Hypothetical RT × set size functions are shown in E–H.

like color or size. In figure 2.1a, the target is defined by multiple, unique features. In these basic feature searches, slopes of RT × set size functions can actually become negative (Bravo & Nakayama, 1992), probably because feature singletons pop out more dramatically as density increases with the increase in set size. For other search tasks, like the search for a letter *T* among *L*s in figure 2.1c, RT increases as a roughly linear function of set size. Slopes for a wide range of such "serial" searches are in the range of 20–50 ms/item on target-present trials with slopes approximately twice that for target-absent trials (Wolfe, 1998). We will describe the differences in RT × set size slopes as differences in search "efficiency"—a theory-neutral term (Wolfe, 1998). Thus, feature searches would be said to be more efficient than a search for a *T* among *L*s, regardless of one's model of why that happened to be the case.

In Treisman's original conception, the feature searches were performed in parallel with the target being detected as a unique peak of activity in a feature map, or in a master map, created from pooling across feature maps. The absence of a feature would produce a low point in a feature map. This could lead to a "search asymmetry" in which search for *A* among *B* was more efficient than search for *B* among *A* (Treisman & Gormican, 1988; Treisman & Souther, 1985). For example, it is easier to find the one moving item among stationary items than it is to find the one stationary (motion

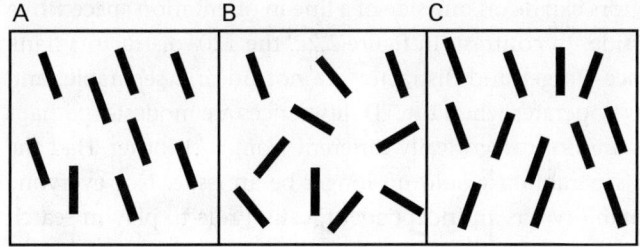

Figure 2.2
Examples of types of orientation search tasks. The target is vertical. This is easy among homogeneous, –20° distractors (A), harder among heterogeneous distractors (B), and hard when the ±20° distractors flank the target in orientation space (C).

absent) item among moving items (Dick, Ullman, & Sagi, 1987). For more extended discussion of search asymmetries, see the 2001 special issue of *Perception & Psychophysics* (Wolfe, 2001)—in particular, the cautionary note from Rosenholtz (2001).

Not every unique item will pop out in a search display. In the end, this is a type of signal detection problem. How big is the "signal" produced by the target item, relative to the "noise" produced by the distractors (Palmer, Verghese, & Pavel, 2000; Verghese, 2001)? Those interested in signal detection approaches to search should also consult the related work in medical image perception (e.g., Burgess, 2010; Chakraborty, 2006). For the sorts of search tasks shown in figure 2.1 and probably more generally, Duncan and Humphreys (1989) provide a couple of useful, qualitative rules. Search efficiency increases with increases in distractor–distractor (DD) similarity and *decreases* with increases in target–distractor (TD) similarity.

Thus, in figure 2.2a, it is reasonably easy to find the vertical target. DD similarity is high, and TD similarity is not too high to obscure the target. (It would be even easier to find a tilted target among vertical distractors—another search asymmetry; Vincent, 2011). If the 20° difference between target and distractor was reduced, search efficiency would decrease (Foster & Ward, 1991). In figure 2.2b, search would be less efficient because DD similarity is reduced. However, these simple rules do not fully explain the efficiency of feature search. Search will be quite inefficient in a search like that shown in figure 2.2c even though DD similarity is higher than in figure 2.2b and the TD differences are all 20° in both figure 2.2a and figure 2.2c (Wolfe, Friedman-Hill, Stewart, & O'Connell, 1992).

In figure 2.2c, the main issue is "linear separability." Described first in the context of color search (Bauer, Jolicoeur, & Cowan, 1996; D'Zmura, 1991), the rule is that a target is easy to find if a line (or a plane) can be drawn in the feature space placing distractors on one side and targets on the other. Thus, figure 2.2a is an efficient search because all of the –20° distractors can be on one side of a line in orientation space while the 0° target is on the other side. In contrast, in figure 2.2c, the ±20° distractors flank the target in orientation space. Target and distractor are not linearly separable, and search is inefficient. This effect operates when the TD differences are modest—perhaps when targets and distractors are not categorically different from each other. Had the distractors been ±60°, linear separability would no longer be an issue. Not everyone is convinced that linear separability has an independent, causal role to play in search (Vighneshvel & Arun, 2013). Real-world stimuli are defined in many dimensions. It is interesting to wonder whether the ease or difficulty of finding an object is related to the ability to separate that target from all distractors with some sort of hyperplane in some high-dimensional space. DiCarlo and Cox (2007) argue for something like this as a basis for object recognition. However, no such ability seems able to support efficient search for target real-world objects among arbitrary distractor objects (Vickery, King, & Jiang, 2005; Wolfe, Alvarez, Rosenholtz, Kuzmova, & Sherman, 2011).

Returning to figure 2.1, recall that the original FIT conception was of two types of search: efficient feature searches (figure 2.1a) and inefficient "serial" searches (figure 2.1b). The difficulty with that account is that there are intermediate cases. Egeth, Virzi, and Garbart (1984) produced an early, illustrative example. Using the Ts and Ls of figure 2.1c as an example, if you were looking for a red T among red and black Ls, the inefficient search would be restricted to the red items. As a consequence, the slope, measured as a function of all items, will be reduced by a factor very close to the percentage of red items in the display. In the figure 2.1c example, if half the items were red, the target-present slope would drop from 25 ms/item to about 12.5 ms, reflecting an inefficient search through the red items rather than through the whole display.

Wolfe argued that the preattentive features that support efficient feature search in FIT could "guide" the serial deployment of attention: hence, the name of the "guided search" (GS) model (Wolfe, Cave, & Franzel, 1989). GS borrowed FIT's two stages (see also Hoffman, 1979) and added a particular type of interaction between those stages. In later formulations, the representation doing the guiding in GS was conceived of as separate from but derived from the first stage (Wolfe, 1994, 2007). Conjunction searches like the example in figure 2.1b were important to the development of GS. Treisman had argued that searches like the search for a vertical 0 in figure 2.1b were "serial" and, thus, similar to the inefficient searches of figure 2.1c. However, evidence began to accumulate for conjunction searches that were too efficient for this account (e.g., McLeod, Driver, & Crisp, 1988; Nakayama & Silverman, 1986). Rather than treating each of these exceptions as special cases, GS offered a general account that said that Os could guide attention toward items with the target attributes (in figure 2.1b "vertical" and "black" or "holey") and that the intersection of those sources of guidance would prove to be a good place to deploy attention in the serial search for the conjunction.

In the GS view, all the searches in figure 2.1 lie on a continuum of guidance. In figure 2.1a, attention is guided to the target first time, every time, and, thus, the number of distractor items has no effect. In figure 2.1c, there is no guidance beyond guidance to the T and L objects in preference to attending to empty space, and search is serial/inefficient over the set of all items. In figure 2.1b, guidance can reduce the effective set size. The result is a reduced dependence on set size and more efficient search. Indeed, without some sort of noise, guidance should get attention to the target immediately in a conjunction search, and there are some very efficient conjunction searches (Theeuwes & Kooi, 1994). Finally, figure 2.1d shows that there are further constraints that can make search even less efficient. If each item needs to be fixated in order to identify it as target or distractor, then the slopes will be dependent on the rate of voluntary, saccadic eye movements at 3 to 4 per second.

An important conclusion of this analysis is that, while the search process may have two stages, preattentive and attentive, as described by Neisser and Treisman, it does not follow that search tasks fall into two categories that can be distinguished by their slopes/efficiency. In papers and textbooks, one still sees search tasks described as "serial" or "parallel" on the basis of their slopes, but, as the histogram in figure 2.3 illustrates, there is no point on the x-axis of slope efficiency that sensibly divides search tasks into two groups. Figure 2.3 shows a histogram of target-present and target-absent slopes, drawn from many search tasks of the sort illustrated in figure 2.1 (Wolfe, 1998). There is no hint of bimodality in these pooled data. Instead, the data are consistent with the idea of a continuum of search efficiencies, driven by the amount of preattentive guidance available.

Guidance and Guiding Attributes

In this chapter, we will not exhaustively discuss the candidate attributes that guide attention in visual search. That has been done elsewhere, repeatedly and fairly recently (Wolfe, 2014; Wolfe & Horowitz, 2004, 2007). There is little disagreement about the ability of some attributes (e.g., color, size, motion, orientation) to guide attention. Other candidates (e.g., faces or arbitrary categories like letters) are much more debatable. The attribute of "shape" poses an interesting problem. While it is possible to describe and parameterize features spaces for color, orientation, and other fundamental attributes of the visual stimulus, no one has ever proposed an entirely satisfactory description of shape space, though it seems very clear that shape guides search, as in a search for a circle among squares (Beck, 1966a, 1966b). In the realm of the guidance of visual attention, an interesting early effort was Julesz's texton theory (Julesz, 1981) in which he argued that it was possible to globally process first- and second-order statistics, by which he meant the distributions of dots and pairs of dots. Higher order statistics were not preattentively available. The most plausible approach may be to consider shape guidance to be guidance by several different properties including line termination (e.g.,

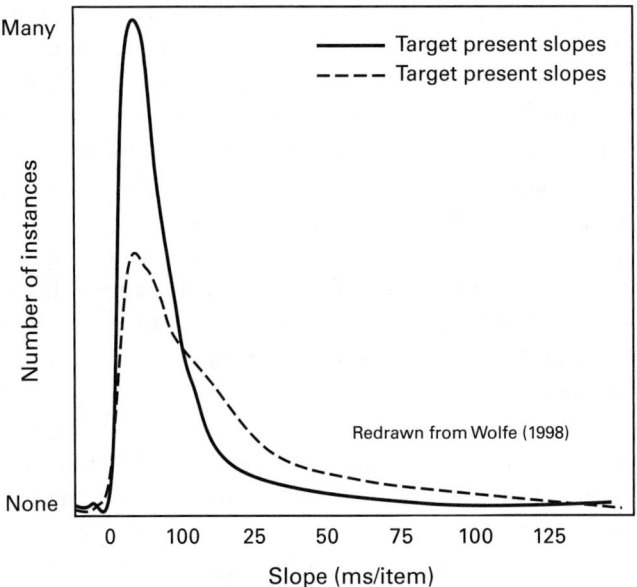

Figure 2.3
Graph (redrawn from Wolfe, 1998) showing that, when aggregated over many experiments and many search tasks, there is no bimodal distribution of search slopes into "parallel" and "serial" categories.

Cheal & Lyon, 1992), closure (Elder & Zucker, 1993), topological status (Chen, 2005), and curvature (Fahle, 1991; Wolfe, Yee, & Friedman-Hill, 1992). Still, this is a topic that will need more work before we truly understand how shape guides attention.

Guidance by basic features can be bottom-up or top-down. Bottom-up, stimulus-driven guidance is local in nature. An item will tend to attract attention if it is sufficiently different from its neighbors (Sagi & Julesz, 1987). Thus, in figure 2.4A, two vertical lines immediately pop out of this display because their neighbors are nearly horizontal in orientation. There are several other vertical lines in this image but they do not attract attention because they are embedded among other near-vertical items.

In figure 2.4B, there are many local feature gradients. They will not guide your attention to black horizontal targets. However, now that you have been told the targets are, in fact, black and horizontal, you will be able to use top-down, user-driven guidance to deploy attention to those targets. In standard conjunction search experiments, essentially all of the bottom-up guidance is misleading noise. This probably accounts for the difference between the efficiency of conjunction searches and feature searches. In a feature search for a known target among homogeneous distractors (e.g., find the

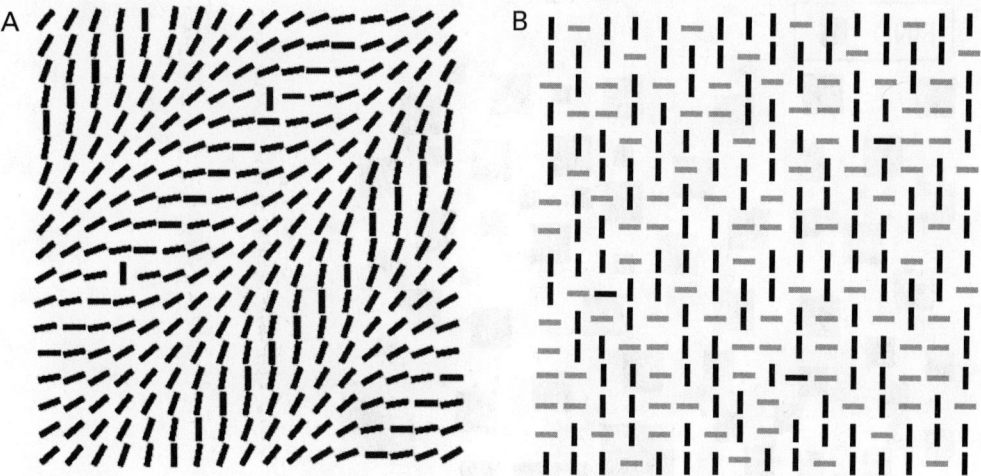

Figure 2.4
(A) Two vertical targets "pop out" in a bottom-up manner because they differ dramatically from their neighbors. Other verticals can be found, but they are much less salient, even in the same image. (B) It is easy to guide attention to black horizontal targets in a "top-down" manner, but these targets do not pop out.

horizontal among verticals), top-down and bottom-up guidance are working toward the same goal. In a conjunction search like that shown in figure 2.4B, bottom-up activity serves to hide the target, rather than to help find it.

In GS and related models, attention is deployed from item to item on the basis of a "priority map" (Serences & Yantis, 2006) that combines top-down and bottom-up signals. Different tasks can be implemented by changing the weights on the inputs to that map. Thus, returning to figure 2.4, if Os are looking for black horizontals, they will increase the weights on signals for "black" and "horizontal." Moreover, the dimensions of color and orientation will tend to be weighted over irrelevant dimensions like motion and size (Muller, Reimann, & Krummenacher, 2003; Weidner & Muller, 2009).

The "Language" of Guidance

The representation of the visual world that is used to guide attention is derived from the initial stages of visual processing, but it has its own distinct set of properties. Specifically, it is coarse, categorical, and, in many cases relational. These properties are illustrated in figure 2.5.

The figure is divided into four quadrants. Each contains one instance of the target (a 20 × 20 pixel square—in the original image). The target is the small item, except in

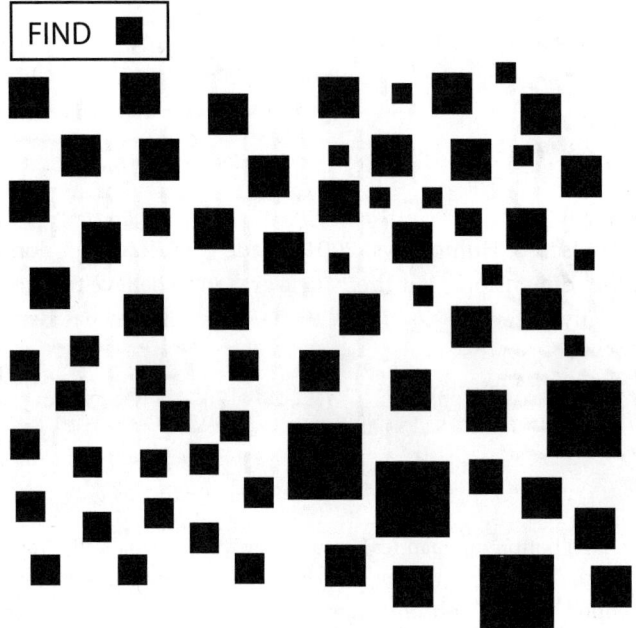

Figure 2.5
The coarse and categorical nature of guidance. The small target is easy to find in the upper left quadrant. Try to find it in the other three quadrants. See the text for details.

the upper right, where the same sized target item is the item of medium size. In the upper left, the target is easy to find with a combination of top-down and bottom-up guidance. In the lower left, it is hard to find. (All the targets are in roughly the same position in their respective quadrants.) This is predicted because we have reduced the TD difference. For present purposes, the point is that guidance cannot make use of a 20% size difference, present in that image, even though that difference can be registered by early visual processes and detected (perhaps depending on the reproduction of the image) once attention has landed on the item. The distinctions that guide attention are relatively coarse; much coarser than a just noticeable difference (Nagy & Sanchez, 1990).

The only difference between the upper left and upper right quadrants of figure 2.5 is the addition of some smaller items to the mix of distractors in the upper right. This makes the target harder to find. This particular effect can be accounted for in several ways. First, we have increased DD heterogeneity. As Duncan and Humphreys (1989) noted, this will decrease search efficiency. Secondly, the target is no longer linearly separable from the distractors. For present purposes, the important factor is that the target is no longer the small item. It is the medium item. The categorical status of search targets matters. Search is more efficient if the target is categorically unique and only a very small number of categorical terms seem to be available. In orientation, the categories appear to be "steep," "shallow," "left," and "right" (Hodsoll & Humphreys, 2005; Wolfe, Friedman-Hill, et al., 1992). The exact set of color categories is not clear, but color search is certainly influenced by the color categorical status of the targets and distractors (Yokoi & Uchikawa, 2005). In size, the effective categories appear to be big and small. While we can identify the item of medium size, we do not seem able to guide attention to medium (Hodsoll & Humphreys, 2001). Little work has been done on categorical guidance in other dimensions, but there is no reason to believe that the basic principle would be radically different. Note that, as Stephanie Becker has demonstrated, it is often the relationship between target and distractor that is the driving factor in a search task rather than the absolute value (Becker, 2010, 2013). That is, it is often more important that the target is "bigger" than the distractors and not that it is exactly "this big." This is illustrated in the lower right quadrant where it is quite easy to find the target, even though the distractors are still heterogeneous. The target is now the smallest item—both linearly separable and categorically distinct. However, it is not the precise target shown at the top left. In fact, it is 30% bigger than the actual targets—bigger than the distractors in the lower left. Even so, it is relatively easy to find this target because of its *relative* size. Relational effects, categorical effects, linear separability, and TD and DD similarity all modulate search efficiency.

There are other ways to think about what makes search less efficient in some cases than in others. One important line of theorizing holds that the real problem is with the representation of the visual world away from the point of fixation. As mentioned above, Rosenholtz et al. (2012) argue that crowding in peripheral vision creates a representation that is "lossy" in a manner that makes searches other than feature searches difficult. For instance, if one simulates a peripheral representation of *T*s and *L*s, the letters break up and reform in a manner that can make *T*s look like *L*s and vice versa. Obviously, this would make search for a *T* among *L*s problematic. In this view, under normal circumstances, eye movements would be used to disambiguate individual items. A similar lossy process in peripheral vision can take black verticals and white horizontals and create the appearance of white verticals and black horizontals—what Treisman would have called "illusory conjunctions" (Treisman & Schmidt, 1982). Indeed, this approach is, in a sense, an updated version of Treisman's description of the preattentive representation as an unbound soup of features that "float free" (Treisman & Gelade, 1980). An important difference is that Treisman's FIT and successors like GS hold that selective attention is required for the binding of the features into a stable object representation (or object file; Kahneman & Treisman, 1984; Kahneman, Treisman, & Gibbs, 1992) while accounts like that of Rosenholtz tend to dispense with binding and to give a larger role to the disambiguating role of eye movements (e.g., Geisler, Perry, & Najemnik, 2006; Zelinsky & Sheinberg, 1995).

This emphasis on the role of eye movements has the undeniable appeal of allowing the oculomotor system to do the work that otherwise requires a special "binding" mechanism. On the other hand, models of this sort need to explain why the efficiency of many search tasks is about the same under crowded and uncrowded, low-set-size conditions (Wolfe & Bennett,1997; Wolfe & DiMase, 2003; Wolfe et al., 2011). More generally, it is clear that selective attention is closely tied to the oculomotor system. As a general rule, attention goes where the eyes go (Kowler, Anderson, Dosher, & Blaser, 1995). Indeed, attention seems to be deployed to a saccade target before the eyes get there (Deubel & Schneider, 1996). However, fixations are not the same as deployments of attention. Volitional eye movements, in search and elsewhere, occur at a rate of 3 to 4 per second. The slope of RT × set size functions, on the other hand, shows Os processing items at a rate of 20 to 50 items per second, even for "inefficient" search, so long as identification of items does not require fixation. Apparently, Os can process multiple items on each fixation. There are multiple ways to model this, bracketed by serial and parallel processing extremes. On the parallel side, Os might be processing N items in parallel on each fixation. Alternatively, covert attention might be visiting N items in series on each fixation. The latter model is problematic if one imagines that one item must be fully identified before the next one can be selected. It is quite clear that object identification takes over 100 ms per item (Duncan, Ward, & Shapiro, 1994; Thorpe, Fize, & Marlot, 1996; Ward, Duncan, & Shapiro, 1996). Such a rate would not be compatible with throughput of 20 to 50 objects per second. A compromise imagines that visual search proceeds like a carwash (Moore & Wolfe, 2001; Wolfe, 2003) or a "pipeline," to use a more common computer science metaphor. Cars enter a carwash one at a time—let us say, once every minute. However, this does not mean it takes 1 minute to wash the car. Perhaps it takes 4 minutes, so four cars might be in different stages of the washing process at any moment. In the visual search analogy, items enter the visual search carwash in series; perhaps one every 50 ms. Maybe each object takes 200 to 300 ms to classify as target or distractor. Thus, at any moment, there would be about 4 to 6 "cars" in the visual search "carwash." This carwash is both serial and parallel, showing one face or the other, depending on the experiment. This could be one reason why, after decades, the serial/parallel discussion continues in the literature (e.g., Egeth, 1966; Sung, 2008; Townsend, 1971, Townsend & Wenger, 2004).

Search Termination

One of the enduring, if understudied problems in visual search is when to terminate the search. If it is known that there is one target and one target is found, search termination is relatively straightforward. However, if the target is not found or it is not known how many targets are present, the problem is trickier (for the problem of unknown numbers of targets, see the discussion of "satisfaction of search," later in the chapter). The simple answer would be to attend to each item and quit when every item has been either labeled as a target or rejected. This was the position of early versions of FIT and GS, but it can't be right. Horowitz and colleagues did a number of experiments that indicate that Os do not sample the visual display without replacement (Horowitz & Wolfe, 1998, 2003). The original claim that visual search had no memory for rejected distractors might have been too strong (Dickinson & Zelinsky, 2005; Korner & Gilchrist, 2007), but it is clear that Os are not successful at marking every rejected distractor during search. Inhibition of return (IOR; Posner & Cohen, 1984) was originally proposed as a mechanism for encouraging sampling without replacement in search (Klein, 1988). However, while there is debate about how many items can be subject to IOR at one time, the time course and persistence of IOR are not adequate if you need to inhibit any of the larger set sizes used in visual search experiments (e.g., Pratt & Abrams, 1995). Rather than being a mechanism of serial, exhaustive search, IOR is perhaps better thought of as a mechanism that encourages exploration and prevents perseveration (Klein & MacInnes, 1999; but see Hooge, Over, van Wezel, & Frens, 2005).

How should we explain search termination if not with some sort of exhaustive rejection of candidate targets? The other primary class of models can be called "quitting threshold" models. GS 2.0 (GS2) proposed two thresholds. Search ended when there were no items with attentional "activations" above some threshold or when the time reached a temporal quitting threshold. GS4 proposed a single quitting signal that accumulated during a trial and ended the trial if it reached its threshold. Chun and Wolfe (1996) argued that this threshold or these thresholds were set in an adaptive manner by the O over the course of a block of trials because, in their data, RTs tended to decrease after each correct response and increase after each error. Moreover, as was known in other literatures (e.g., vigilance; Parasuraman & Davies, 1976) RTs, as well as the pattern of errors, change with the prevalence of targets in a search task (Wolfe & VanWert, 2010). When targets are rare, the quitting threshold is made more liberal. Os quit sooner, and target-absent RTs become shorter. When targets are common, Os are much more reluctant to abandon the search with an "absent" response. Note that these quitting threshold accounts do not require "serial" search. Proponents of parallel models could incorporate a timing or an activation threshold, too.

While the idea of a quitting threshold is an improvement over a serial exhaustive account, it is not easy to satisfy all of the constraints provided by the data. A clever new approach is proposed by Moran et al. (2013). Their quitting threshold is probabilistic. After each deployment of attention, there is a probability that the trial will end with an absent response if it has not ended with a present response. That probability grows over the course of the trial and is governed by the following equation (using terminology a bit different from that in their paper):

P(quit) = QuitWeight/[sum(PriorityMap) + QuitWeight].

The QuitWeight accumulates over time in the trial. Placing the sum of activity in the priority map in the denominator means that the initial probability of quitting will be low if the display has a lot of attentionally interesting items. As the quit weight grows, P(quit) approaches 1.0. This model solves a variety of the problems of earlier quitting models. Interestingly, if we assume that a target-present display has a higher sum of activity in the priority map than an absent trial, then the probability of quitting a present trial will be a bit lower than the probability of quitting an absent trial for the same value of the quit weight. In a sense, the O "knows" that the target is present even though it has not yet been attended and found. Such an unlocalized sense of target presence can be documented in expert medical image searchers. They can discriminate positive from negative cases at above-chance levels, even when they cannot localize the lesion in the image (Evans, Georgian-Smith, Tambouret, Birdwell, & Wolfe, 2013; Evans, Wolfe, Tambouret, & Wilbur, 2010).

Interim Summary

The preceding has summarized some of the work on visual search for fairly simple targets, generally presented in random displays of items on a computer screen. While almost every point can be and has been argued over, a reasonable summary might be that the initial stages of visual processing are massively parallel in nature. Some products of that parallel processing appear to give rise to the visual experience of seeing something, everywhere, as soon as the eyes are opened. This "nonselective" visual processing also supports a limited amount of semantic processing of the "gist" of a scene (e.g., Greene & Oliva, 2009). Feature binding and the recognition of specific objects appear to be the products of a much more limited-capacity, "selective" pathway, capable of handling one or a very small number of items at one time. Thus, in search for a specific target object, Os use selective attention to deploy these limited resources (as well as the eyes) to candidate targets. This deployment is not random. Rather it is guided by knowledge of the basic features of the target. If Os were searching for pedestrians, they would look for vertical objects of human size and would not waste attentional resources on objects of markedly incorrect orientation and/or size. This guidance might actually slow search if the pedestrian happened to be horizontal for some reason. Failing to find a pedestrian, Os would terminate search when an internal quitting signal reached a quitting threshold (likely a probabilistic process, rather than a hard boundary). Feedback about the timing and success of the search would then change the quitting parameters for the next search.

Integration and Future Directions

After several decades of productive work on laboratory search tasks, the future of the field of visual search lies in asking how the rules that have been uncovered in studies of random arrays of items on a screen generalize to the search tasks of the real world. Of course, an interest in real-world search is not new. Recall the work of Kingsley as far back as the 1930s. These are "future directions" in the sense of being topics that are gaining ascendency in the field. Some of the more complex search tasks that are of current interest are "natural" tasks, like the search for pedestrians (see figure 2.6). Others are complex search tasks, invented by a complex civilization, like screening for cancer in mammograms or looking for threats at the airport. This is a future that is happening now with a substantial literature developing on topics like search in scenes (e.g., Biederman et al., 1982; Brockmole & Henderson, 2006; Eckstein, Drescher, & Shimozaki, 2006; Ehinger, Hidalgo-Sotelo, Torralba, & Oliva, 2009; Oliva, Wolfe, & Arsenio, 2004; Zelinsky & Schmidt, 2009) and medical image perception (e.g.,

Figure 2.6
What guides search in a real-world scene?

Berbaum, Caldwell, Schartz, Thompson, & Franken, 2007; Krupinski, Berger, Dallas, & Roehrig, 2003; Kundel & La Follette, 1972).

If we think a bit more about that search for a pedestrian, it will become clear that basic feature guidance will still play a role on the street. If you were looking for red lights, you would guide attention to red. However, other sources of guidance now come to play a critical role (Hollingworth, 2009; Vo & Henderson, 2009). Pedestrians are not typically presented to us in arrays of randomly placed objects. Pedestrians are found in outdoor, built environments. Such environments have rules. We know those rules, implicitly and/or explicitly, and use them to guide our search. If we track the eye movements of Os looking for people in natural scenes, we will find that those searches are strongly constrained to places where people are likely to be. For instance, pedestrians tend to be on the ground plane and not floating through the air. A model that includes such constraints will do better than models that do not (Ehinger, Hidalgo-Sotelo, Torralba, & Oliva, 2009).

It is useful to think about three types of guidance in search in scenes. Borrowing terms from elsewhere in psychology, these are *syntactic*, *semantic*, and *episodic* guidance (Biederman et al., 1982; Henderson & Ferreira, 2004). Syntax in linguistics refers to the grammar or structure of utterances while semantics refers to meaning. The definitions of scene syntax and semantics are subject to some ambiguity and disagreement. One way to think about the distinctions is that scene syntax refers to physical/structural constraints (cars belong on the ground plane, not on walls) while scene semantics refers to what is likely or plausible (cars *could* be found on railroad tracks, but they are more plausibly looked for on roads; Biederman et al., 1982). An alternative, overlapping definition uses syntax to refer to the local position of objects in scenes (a toaster on a kitchen table is syntactically appropriate, but a toaster on the kitchen floor would not be). Here semantics would refer to the more global relationship of an object with the meaning of the scene (a toaster would be semantically inappropriate in the bathroom, even if it were on a table; Vo & Henderson, 2009). In language processing, violations of semantic and syntactic rules produce different signals in scalp-recorded event-related potentials (ERPs). Interestingly, similar patterns of ERPs are seen when Os encounter semantic and syntactic violations in scenes (Vo & Wolfe, 2013a).

Episodic guidance borrows its name from episodic memory and refers to the fact that you will be faster to search for the toaster in your own kitchen because you remember where it was the last time you found it. Though this sort of guidance seems self-evident, there are circumstances under which visual search can appear to be quite amnesic—for instance, during some repeated searches of the same display. Search does improve when Os search repeatedly through the same displays in the contextual cuing paradigm (Chun & Jiang, 1998; Chun, 2000). In a standard contextual cuing experiment, Os might search through arrays of rotated *L*s, trying to determine the orientation of the one *T*. Os become faster to respond to repeated arrays than to unrepeated arrays even though they are at chance in identifying old from new displays.

It is often assumed that the pattern of distractors is guiding attention to the target in contextual cuing. If this were so, then, if a range of set sizes were used, the repeated arrays should produce shallower RT × set size slopes than unrepeated ones. However, this is not the case (Kunar, Flusberg, Horowitz, & Wolfe, 2007; but see Geyer, Zehetleitner, & Muller, 2010). Moreover, when Os search through the same unchanging small sets of three to six letters, even for several hundred trials, no improvement in search *efficiency* is seen (Wolfe, Klempen, & Dahlen, 2000). Even though Os clearly learn these simple displays, it appears that they search them de novo on each trial. This may not be a case of amnesic visual search. More plausibly, it is faster to redo a visual search through a handful of letters than it is to retrieve the target location from memory (Kunar, Flusberg, & Wolfe, 2008).

When scenes are used, the issue of memory guidance becomes more complex. When one searches for a succession of different items, search times do not improve even after many searches through an unchanging scene (find the pillow, find the vase, find the lamp, etc.). This might be due to the strength of semantic guidance. Semantic guidance might be so effective that initiating a new search is simply quicker and more reliable than using episodic memory to guide search. (Pillows are almost always on the bed. Why use memory to search for this, specific pillow?) The reliance on episodic memory can, however, be boosted when semantic guidance is made obsolete (e.g., by placing objects in uncommon locations; Vo & Wolfe, 2013b). A general finding in scene search is that, when Os are searching for a target for the second time, RTs become markedly faster (Wolfe et al., 2011). Eye tracking shows that Os seem to be searching less and less of the image on the second and third search for the same target, suggesting that, in this case, episodic guidance really is helping to speed search (Vo & Wolfe, 2012). Vo and Wolfe (2012) got this memory effect only for targets that had been found in search. Hollingworth (2012) obtained a memory effect for targets that had been previewed and not actively searched. A reasonable hypothesis is that multiple sources of guidance are available. Depending on the stimulus materials and experimental design, different sources of guidance may have an advantage. In particular, guidance based on memory may be relatively slow (Korner & Gilchrist, 2007) and may be seen most clearly when given a substantial head start on scene syntax and semantics and on feature guidance.

It is worth noting that there are situations in which memory for what you have found actually seems to interfere with subsequent search. The classic version of this problem is known as "satisfaction of search" (SoS) in the medical image perception literature (Berbaum, Franken, Caldwell, & Shartz, 2010; Tuddenham, 1962). SoS occurs in search tasks where there can be more than one target. The probability of finding the second target is lower than it would have been if the first had not been present in the image. Cain and Mitroff (2013) attribute at least part of the SoS phenomenon to the memory load created by the need to remember the first target.

SoS points to a class of extended search tasks that will receive more attention in the future. These are tasks with multiple targets. The bulk of the search literature in experimental psychology has dealt with displays containing one target or lacking targets altogether. The world, on the other hand, presents many tasks with multiple targets of either known number (find a matching pair of socks in the laundry) or unknown number (find all the metastases in this x-ray). As the number of targets becomes larger, these become foraging tasks (pick a quart of blueberries). Foraging has been extensively studied in the animal behavior literature (Stephens & Krebs, 1986) but much less as an example of human visual search (Wolfe, 2013). Beyond berry picking, it is relevant to fields as diverse as medical image perception and Web surfing (Pirolli, 2007).

At the time of the writing of this chapter, a Malaysian passenger plane has disappeared, probably into the Indian Ocean. Many nations have deployed highly sophisticated equipment to search for signs of the jet, but it is still possible to have an article in the March 28, 2014, *New York Times* headlined "Aircrew with Latest Gear Trusts Its Eyes to Find Jet." As the article reports, "'The human eye is the best way to search,' said Petty Officer Second Class Mike Burnett, one of the spotters." The task of the visual search researcher in the next 20 years will be to explain how humans perform search tasks in the real world and to help them to do those tasks more successfully.

In summary, when we search in the real world, in natural and artificial settings, our attention will be guided by multiple sources of information. There is classic feature guidance by a limited set of properties of the objects. Once search takes place in structured scenes, there is syntactic and semantic guidance derived from our expertise with those scenes. Domain experts like radiologists will have scene guidance for the scenes created by medical imaging devices. The rest of us will have guidance based on our knowledge of the natural and built environment. Finally, we will have specific episodic knowledge about the location of specific targets in specific environments. Under different circumstances, different sources of guidance will govern behavior, but under most circumstances, attention will be deployed in a rule-governed and adaptive manner.

Box 2.1

Key Points

- The data are consistent with the idea of a continuum of search efficiencies, rather than a dichotomy between serial and parallel search, driven by the amount of preattentive guidance available.

- The representation of the visual world that is used to guide attention is coarse, categorical, and in many cases relational.

- The search process appears to end when an internal quitting signal reaches a quitting threshold, and this threshold appears to be probabilistic rather than fixed.

Box 2.2

Outstanding Issues

- How do the principles of visual search generalize to complex search tasks (e.g., medical image perception)?
- How do semantic and episodic guidance interact in search through natural scenes?

References

Barlow, H. B. (1972). Single units and sensation: A neuron doctrine for perceptual psychology. *Perception, 1,* 371–394.

Bauer, B., Jolicoeur, P., & Cowan, W. B. (1996). Distractor heterogeneity versus linear separability in colour visual search. *Perception, 25,* 1281–1294.

Beck, J. (1966a). Effect of orientation and shape similarity on perceptual grouping. *Perception & Psychophysics, 1,* 300–302.

Beck, J. (1966b). Perceptual grouping produced by changes in orientation and shape. *Science, 154,* 538–540.

Becker, S. I. (2010). The role of target–distractor relationships in guiding attention and the eyes in visual search. *Journal of Experimental Psychology. General, 139,* 247–265.

Becker, S. I. (2013). Simply shapely: Relative, not absolute shapes are primed in pop-out search. *Attention, Perception & Psychophysics, 75,* 845–861.

Berbaum, K. S., Caldwell, R. T., Schartz, K. M., Thompson, B. H., & Franken, A. (2007). Does computer-aided diagnosis for lung tumors change satisfaction of search in chest radiography? *Academic Radiology, 14,* 1069–1076.

Berbaum, K. S., Franken, E. A., Caldwell, R. T., & Shartz, K. (2010). Satisfaction of search in traditional radiographic imaging. In E. A. Krupinski & E. Samei (Eds.), *The handbook of medical image perception and techniques* (pp. 107–138). Cambridge, UK: Cambridge University Press.

Bergen, J. R., & Julesz, B. (1983). Rapid discrimination of visual patterns. *IEEE Transactions on Systems, Man, and Cybernetics, SMC-13,* 857–863.

Biederman, I., Mezzanotte, R. J., & Rabinowitz, J. C. (1982). Scene perception: Detecting and judging objects undergoing relational violations. *Cognitive Psychology, 14,* 143–177.

Braddick, O., Campbell, F. W., & Atkinson, J. (1978). Channels in vision: Basic aspects. In R. Held, W. H. Leibowitz & H.-L. Teuber (Eds.), *Perception: Handbook of Sensory Physiology* (pp. 3–38). Berlin: Springer-Verlag.

Bravo, M., & Nakayama, K. (1992). The role of attention in different visual search tasks. *Perception & Psychophysics, 51*, 465–472.

Brockmole, J. R., & Henderson, J. M. (2006). Using real-world scenes as contextual cues for search. *Visual Cognition, 13*, 99–108.

Burgess, A. E. (2010). Signal detection in radiology. In E. Samei & E. A. Krupinski (Eds.), *The handbook of medical image perception and techniques* (pp. 26–46). Cambridge, UK: Cambridge University Press.

Cain, M. S., & Mitroff, S. R. (2013). Memory for found targets interferes with subsequent performance in multiple-target visual search. *Journal of Experimental Psychology: Human Perception and Performance, 39*, 1398–1408.

Carrasco, M., Evert, D. L., Chang, I., & Katz, S. M. (1995). The eccentricity effect: Target eccentricity affects performance on conjunction searches. *Perception & Psychophysics, 57*, 1241–1261.

Chakraborty, D. P. (2006). ROC curves predicted by a model of visual search. *Physics in Medicine and Biology, 51*, 3463–3482.

Cheal, M., & Lyon, D. (1992). Attention in visual search: Multiple search classes. *Perception & Psychophysics, 52*, 113–138.

Chen, L. (2005). The topological approach to perceptual organization. *Visual Cognition, 12*, 553–637.

Chun, M. M. (2000). Contextual cueing of visual attention. *Trends in Cognitive Sciences, 4*, 170–178.

Chun, M. M., & Jiang, Y. (1998). Contextual cuing: Implicit learning and memory of visual context guides spatial attention. *Cognitive Psychology, 36*, 28–71.

Chun, M. M., & Wolfe, J. M. (1996). Just say no: How are visual searches terminated when there is no target present? *Cognitive Psychology, 30*, 39–78.

Cohen, A., & Ivry, R. B. (1991). Density effects in conjunction search: Evidence for coarse location mechanism of feature integration. *Journal of Experimental Psychology: Human Perception and Performance, 17*, 891–901.

Deubel, H., & Schneider, W. X. (1996). Saccade target selection and object recognition: Evidence for a common attentional mechanism. *Vision Research, 36*, 1827–1837.

DiCarlo, J. J., & Cox, D. D. (2007). Untangling invariant object recognition. *Trends in Cognitive Sciences, 11*, 333–341.

Dick, M., Ullman, S., & Sagi, D. (1987). Parallel and serial processes in motion detection. *Science, 237*, 400–402.

Dickinson, C. A., & Zelinsky, G. J. (2005). Marking rejected distractors: A gaze-contingent technique for measuring memory during search. *Psychonomic Bulletin & Review, 12*, 1120–1126.

Di Lollo, V. (2012). The feature- binding problem is an ill-posed problem. *Trends in Cognitive Sciences, 16,* 317–321.

Dosher, B. A., Han, S., & Lu, Z. L. (2004). Parallel processing in visual search asymmetry. *Journal of Experimental Psychology: Human Perception and Performance, 30,* 3–27.

Drew, T., Vo, M. L.-H., & Wolfe, J. M. (2013). The invisible gorilla strikes again: Sustained inattentional blindness in expert observers. *Psychological Science, 24,* 1848–1853.

Dukewich, K. R., & Klein, R. M. (2009). Finding the target in search tasks using detection, localization, and identification responses. *Canadian Journal of Experimental Psychology, 63*(1), 1–7.

Duncan, J., & Humphreys, G. W. (1989). Visual search and stimulus similarity. *Psychological Review, 96,* 433–458.

Duncan, J., Ward, R., & Shapiro, K. (1994). Direct measurement of attention dwell time in human vision. *Nature, 369,* 313–315.

D'Zmura, M. (1991). Color in visual search. *Vision Research, 31,* 951–966.

Eckstein, M. P., Drescher, B. A., & Shimozaki, S. S. (2006). Attentional cues in real scenes, saccadic targeting, and Bayesian priors. *Psychological Science, 17,* 973–980.

Egeth, H. (1967). Selective attention. *Psychological Bulletin, 67,* 41–57.

Egeth, H., Jonides, J., & Wall, S. (1972). Parallel processing of multielement displays. *Cognitive Psychology, 3,* 674–698.

Egeth, H. E. (1966). Parallel versus serial processes in multidimensional stimulus discrimination. *Perception & Psychophysics, 1,* 245–252.

Egeth, H. E., Virzi, R. A., & Garbart, H. (1984). Searching for conjunctively defined targets. *Journal of Experimental Psychology: Human Perception and Performance, 10,* 32–39.

Ehinger, K. A., Hidalgo-Sotelo, B., Torralba, A., & Oliva, A. (2009). Modelling search for people in 900 scenes: A combined source model of eye guidance. *Visual Cognition, 17,* 945–978.

Elder, J., & Zucker, S. (1993). The effect of contour closure on the rapid discrimination of two-dimensional shapes. *Vision Research, 33,* 981–991.

Enoch, J. M. (1959). Effect of the size of a complex display upon visual search. *Journal of the Optical Society of America, 49,* 280–286.

Evans, K., Georgian-Smith, D., Tambouret, R., Birdwell, R., & Wolfe, J. M. (2013). The gist of the abnormal: Above-chance medical decision making in the blink of an eye. *Psychonomic Bulletin & Review, 20,* 1170–1175.

Evans, K., Wolfe, J. M., Tambouret, R. H., & Wilbur, D. C. (2010, November). In the blink of an eye: Discrimination and localization of abnormalities in cervical cytology screening from a global signal. *Cancer Cytopathology, 118*(6), 58th Annual Scientific Meeting of the American Society of Cytopathology.

Fahle, M. (1991). Parallel perception of vernier offsets, curvature, and chevrons in humans. *Vision Research, 31*, 2149–2184.

Felisberti, F. M., Solomon, J. A., & Morgan, M. J. (2005). The role of target salience in crowding. *Perception, 34*, 823–833.

Ford, A., White, C. T., & Lichtenstein, M. (1959). Analysis of eye movements during free search. *Journal of the Optical Society of America, 49*, 287–289.

Foster, D. H., & Ward, P. A. (1991). Asymmetries in oriented-line detection indicate two orthogonal filters in early vision. *Proceedings. Biological Sciences, 243*, 75–81.

Geisler, W. S., Perry, J. S., & Najemnik, J. (2006). Visual search: The role of peripheral information measured using gaze-contingent displays. *Journal of Vision, 6*, 858–873.

Geyer, T., Zehetleitner, M., & Muller, H. J. (2010). Contextual cueing of pop-out visual search: When context guides the deployment of attention. *Journal of Vision, 10*(5), article 20.

Greene, M. R., & Oliva, A. (2009). The briefest of glances: The time course of natural scene understanding. *Psychological Science, 20*, 464–472.

Helmholtz, H. v. (1924). *Treatise on physiological optics* (Southall, Trans. Trans. from 3rd German ed. of 1909, ed.). Rochester, NY: The Optical Society of America.

Henderson, J. M., & Ferreira, F. (2004). Scene perception for psycholinguists. In J. M. Henderson & F. Ferreira (Eds.), *The interface of language, vision, and action: Eye movements and the visual world* (pp. 1–58). New York: Psychology Press.

Hodsoll, J., & Humphreys, G. W. (2001). Driving attention with the top down: The relative contribution of target templates to the linear separability effect in the size dimension. *Perception & Psychophysics, 63*, 918–926.

Hodsoll, J., & Humphreys, G. (2005). The effect of target foreknowledge on visual search for categorically separable orientation targets. *Vision Research, 45*, 2346–2351.

Hoffman, J. E. (1979). A two-stage model of visual search. *Perception & Psychophysics, 25*, 319–327.

Hollingworth, A. (2009). Two forms of scene memory guide visual search: Memory for scene context and memory for the binding of target object to scene location. *Visual Cognition, 17*, 273–291.

Hollingworth, A. (2012). Task specificity and the influence of memory on visual search: Comment on Vo and Wolfe (2012). *Journal of Experimental Psychology: Human Perception and Performance, 38*, 1596–1603.

Hooge, I. T., Over, E. A., van Wezel, R. J., & Frens, M. A. (2005). Inhibition of return is not a foraging facilitator in saccadic search and free viewing. *Vision Research, 45*, 1901–1908.

Horowitz, T. S., & Wolfe, J. M. (1998). Visual search has no memory. *Nature, 394*, 575–577.

Horowitz, T. S., & Wolfe, J. M. (2003). Memory for rejected distractors in visual search? *Visual Cognition, 10*, 257–298.

Julesz, B. (1981). A theory of preattentive texture discrimination based on first order statistics of textons. *Biological Cybernetics, 41*, 131–138.

Kahneman, D., & Treisman, A. (1984). Changing views of attention and automaticity. In R. Parasuraman & D. R. Davies (Eds.), *Varieties of attention* (pp. 29–61). New York: Academic Press.

Kahneman, D., Treisman, A., & Gibbs, B. (1992). The reviewing of object files: Object-specific integration of information. *Cognitive Psychology, 24*, 179–219.

Kingsley, H. L. (1926). Search: A function intermediate between perception and thinking. *Psychological Monographs, 35*, 16–55.

Kingsley, H. L. (1932). An experimental study of "search." *American Journal of Psychology, 44*, 314–318.

Klein, R. (1988). Inhibitory tagging system facilitates visual search. *Nature, 334*, 430–431.

Klein, R. M., & MacInnes, W. J. (1999). Inhibition of return is a foraging facilitator in visual search. *Psychological Science, 10*, 346–352.

Korner, C., & Gilchrist, I. D. (2007). Finding a new target in an old display: Evidence for a memory recency effect in visual search. *Psychonomic Bulletin & Review, 144*, 846–851.

Kowler, E., Anderson, E., Dosher, B., & Blaser, E. (1995). The role of attention in the programming of saccades. *Vision Research, 35*, 1897–1916.

Krupinski, E. A., Berger, W. G., Dallas, W. J., & Roehrig, H. (2003). Searching for nodules: What features attract attention and influence detection? *Academic Radiology, 10*, 861–868.

Kulpe, O. (1904). *Versuche uber Abstraktion* [Experiments on abstraction]. Paper presented at the Bericht uber den le Kongresz fur Experimental Psychologie.

Kunar, M. A., Flusberg, S. J., Horowitz, T. S., & Wolfe, J. M. (2007). Does contextual cueing guide the deployment of attention? *Journal of Experimental Psychology: Human Perception and Performance, 33*, 816–828.

Kunar, M. A., Flusberg, S. J., & Wolfe, J. M. (2008). The role of memory and restricted context in repeated visual search. *Perception & Psychophysics, 70*, 314–328.

Kundel, H. L., & La Follette, P. S., Jr. (1972). Visual search patterns and experience with radiological images. *Radiology, 103*, 523–528.

Levi, D. M. (2011). Visual crowding. *Current Biology, 21*, R678–R679.

McElree, B., & Carrasco, M. (1999). The temporal dynamics of visual search: Evidence for parallel processing in feature and conjunction searches. *Journal of Experimental Psychology: Human Perception and Performance, 25*, 1517–1539.

McLeod, P., Driver, J., & Crisp, J. (1988). Visual search for conjunctions of movement and form is parallel. *Nature*, *332*, 154–155.

Moore, C. M., & Wolfe, J. M. (2001). Getting beyond the serial/parallel debate in visual search: A hybrid approach. In K. Shapiro (Ed.), *The limits of attention: Temporal constraints on human information processing* (pp. 178–198). Oxford, UK: Oxford University Press.

Moran, R., Zehetleitner, M. H., Mueller, H. J., & Usher, M. (2013). Competitive guided search: Meeting the challenge of benchmark RT distributions. *Journal of Vision*, *13*(8), article 24.

Muller, H. J., Reimann, B., & Krummenacher, J. (2003). Visual search for singleton feature targets across dimensions: Stimulus- and expectancy-driven effects in dimensional weighting. *Journal of Experimental Psychology: Human Perception and Performance*, *29*, 1021–1035.

Nagy, A. L., & Sanchez, R. R. (1990). Critical color differences determined with a visual search task. *Journal of the Optical Society of America. A, Optics and Image Science*, *7*, 1209–1217.

Najemnik, J., & Geisler, W. S. (2005). Optimal eye movement strategies in visual search. *Nature*, *434*, 387–391.

Nakayama, K., & Silverman, G. H. (1986). Serial and parallel processing of visual feature conjunctions. *Nature*, *320*, 264–265.

Neisser, U. (1967). *Cognitive psychology*. New York: Appleton, Century, Crofts.

Noton, D., & Stark, L. (1971). Eye movements and visual perception. In R. Held & W. Richards (Eds.), *Perception: Mechanisms and models—Readings from Scientific American* (pp. 218–227). San Francisco: Freeman.

Oliva, A., Wolfe, J. M., & Arsenio, H. C. (2004). Panoramic search: The interaction of memory and vision in search through a familiar scene. *Journal of Experimental Psychology: Human Perception and Performance*, *30*, 1132–1146.

Palmer, J., Verghese, P., & Pavel, M. (2000). The psychophysics of visual search. *Vision Research*, *40*, 1227–1268.

Parasuraman, R., & Davies, D. R. (1976). Decision theory analysis of response latencies in vigilance. *Journal of Experimental Psychology: Human Perception and Performance*, *2*, 578–590.

Pirolli, P. (2007). *Information foraging theory*. New York: Oxford University Press.

Posner, M. I., & Cohen, Y. (1984). Components of attention. In H. Bouma & D. G. Bouwhuis (Eds.), *Attention and performance X* (pp. 55–66). Hillside, NJ: Erlbaum.

Pratt, J., & Abrams, R. A. (1995). Inhibition of return to successively cued spatial locations. *Journal of Experimental Psychology: Human Perception and Performance*, *21*, 1343–1353.

Rock, I., Linnett, C. M., Grant, P., & Mack, A. (1992). Perception without attention: Results of a new method. *Cognitive Psychology*, *24*, 502–534.

Rosenholtz, R. (2001). Search asymmetries? What search asymmetries? *Perception & Psychophysics*, *63*, 476–489.

Rosenholtz, R., Huang, J., & Ehinger, K. A. (2012). Rethinking the role of top-down attention in vision: Effects attributable to a lossy representation in peripheral vision. *Frontiers in Psychology*, *3*, 13.

Roskies, A. (1999). The binding problem. *Neuron, 24*, 7–9.

Sagi, D., & Julesz, B. (1987). Short-range limitation on detection of feature differences. *Spatial Vision, 2*(1), 39–49.

Serences, J. T., & Yantis, S. (2006). Selective visual attention and perceptual coherence. *Trends in Cognitive Sciences, 10*, 38–45.

Simons, D. J., & Chabris, C. F. (1999). Gorillas in our midst: Sustained inattentional blindness for dynamic events. *Perception, 28*, 1059–1074.

Sperling, G. (1960). The information available in brief visual presentations. *Psychological Monographs, 15*, 201–293.

Stephens, D. W., & Krebs, J. R. (1986). *Foraging theory*. Princeton, NJ: Princeton University Press.

Sternberg, S. (1966). High-speed scanning in human memory. *Science, 153*, 652–654.

Sung, K. (2008). Serial and parallel attentive visual searches: Evidence from cumulative distribution functions of response times. *Journal of Experimental Psychology: Human Perception and Performance, 34*, 1372–1388.

Theeuwes, J., & Kooi, J. L. (1994). Parallel search for a conjunction of shape and contrast polarity. *Vision Research, 34*, 3013–3016.

Thorpe, S., Fize, D., & Marlot, C. (1996). Speed of processing in the human visual system. *Nature, 381*, 520–552.

Townsend, J. T. (1971). A note on the identification of parallel and serial processes. *Perception & Psychophysics, 10*, 161–163.

Townsend, J. T., & Wenger, M. J. (2004). The serial-parallel dilemma: A case study in a linkage of theory and method. *Psychonomic Bulletin & Review, 11*, 391–418.

Treisman, A. (1999). Solutions to the binding problem: Progress through controversy and convergence. *Neuron, 24*, 105–110, 111–125.

Treisman, A., & Gelade, G. (1980). A feature-integration theory of attention. *Cognitive Psychology, 12*, 97–136.

Treisman, A., & Gormican, S. (1988). Feature analysis in early vision: Evidence from search asymmetries. *Psychological Review, 95*, 15–48.

Treisman, A., & Souther, J. (1985). Search asymmetry: A diagnostic for preattentive processing of separable features. *Journal of Experimental Psychology: General, 114*, 285–310.

Treisman, A., Vieira, A., & Hayes, A. (1992). Automaticity and preattentive processing. *American Journal of Psychology, 105*, 341–362.

Treisman, A. M., & Schmidt, H. (1982). Illusory conjunctions in the perception of objects. *Cognitive Psychology*, *14*, 107–141.

Tuddenham, W. J. (1962). Visual search, image organization, and reader error in roentgen diagnosis: Studies of the psycho-physiology of roentgen image perception. *Radiology*, *78*, 694–704.

Verghese, P. (2001). Visual search and attention: A signal detection approach. *Neuron*, *31*, 523–535.

Vickery, T. J., King, L. W., & Jiang, Y. (2005). Setting up the target template in visual search. *Journal of Vision*, *5*, 81–92.

Vighneshvel, T., & Arun, S. P. (2013). Does linear separability really matter in visual search? Complex search is explained by its components. *Journal of Vision*, *13*(11).article 10

Vincent, B. T. (2011). Search asymmetries: Parallel processing of uncertain sensory information. *Vision Research*, *51*, 1741–1750.

Vo, M. L.-H., & Henderson, J. M. (2009). Does gravity matter? Effects of semantic and syntactic inconsistencies on the allocation of attention during scene perception. *Journal of Vision*, *9*, 1–15.

Vo, M. L.-H., & Wolfe, J. M. (2012). When does repeated search in scenes involve memory? Looking at versus looking for objects in scenes. *Journal of Experimental Psychology: Human Perception and Performance*, *38*, 23–41.

Vo, M. L.-H, & Wolfe, J. M. (2013a). Differential ERP signatures elicited by semantic and syntactic processing in scenes. *Psychological Science*, *24*, 1816–1823.

Vo, M. L.-H, & Wolfe, J. M. (2013b). The interplay of episodic and semantic memory in guiding repeated search in scenes. *Cognition*, *126*, 198–212.

Vul, E., & Rich, A. N. (2010). Independent sampling of features enables conscious perception of bound objects. *Psychological Science*, *21*, 1168–1175.

Ward, R., Duncan, J., & Shapiro, K. (1996). The slow time-course of visual attention. *Cognitive Psychology*, *30*, 79–109.

Weidner, R., & Muller, H. J. (2009). Dimensional weighting of primary and secondary target-defining dimensions in visual search for singleton conjunction targets. *Psychological Research*, *73*, 198–211.

Wolfe, J. M. (1994). Guided Search 2.0: A revised model of visual search. *Psychonomic Bulletin & Review*, *1*, 202–238.

Wolfe, J. M. (1998). What do 1,000,000 trials tell us about visual search? *Psychological Science*, *9*, 33–39.

Wolfe, J. M. (2001). Asymmetries in visual search: An introduction. *Perception & Psychophysics*, *63*, 381–389.

Wolfe, J. M. (2003). Moving towards solutions to some enduring controversies in visual search. *Trends in Cognitive Sciences, 7,* 70–76.

Wolfe, J. M. (2007). Guided Search 4.0: Current progress with a model of visual search. In W. D. Gray (Ed.), *Integrated models of cognitive systems* (pp. 99–119). New York: Oxford University Press.

Wolfe, J. M. (2012). The binding problem lives on: Comment on Di Lollo. *Trends in Cognitive Sciences, 16,* 307–308.

Wolfe, J. M. (2013). When is it time to move to the next raspberry bush? Foraging rules in human visual search. *Journal of Vision, 13*(3), article 10.

Wolfe, J. M. (2014). Approaches to visual search: Feature integration theory and guided search. In A. C. Nobre & S. Kastner (Eds.), *Oxford handbook of attention* (pp. 11–55). New York: Oxford University Press.

Wolfe, J. M., Alvarez, G. A., Rosenholtz, R., Kuzmova, Y. I., & Sherman, A. M. (2011). Visual search for arbitrary objects in real scenes. *Attention, Perception & Psychophysics, 73,* 1650–1671.

Wolfe, J. M., & Bennett, S. C. (1997). Preattentive object files: Shapeless bundles of basic features. *Vision Research, 37,* 25–43.

Wolfe, J. M., Cave, K. R., & Franzel, S. L. (1989). Guided search: An alternative to the feature integration model for visual search. *Journal of Experimental Psychology: Human Perception and Performance, 15,* 419–433.

Wolfe, J. M., & DiMase, J. S. (2003). Do intersections serve as basic features in visual search? *Perception, 32,* 645–656.

Wolfe, J. M., Friedman-Hill, S. R., Stewart, M. I., & O'Connell, K. M. (1992). The role of categorization in visual search for orientation. *Journal of Experimental Psychology: Human Perception and Performance, 18,* 34–49.

Wolfe, J. M., & Horowitz, T. S. (2004). What attributes guide the deployment of visual attention and how do they do it? *Nature Reviews. Neuroscience, 5,* 495–501.

Wolfe, J. M., & Horowitz, T. S. (2007). Visual search. *Scholaropedia, 3,* 3325.

Wolfe, J. M., Klempen, N., & Dahlen, K. (2000). Post-attentive vision. *Journal of Experimental Psychology: Human Perception and Performance, 26,* 693–716.

Wolfe, J. M., O'Neill, P. E., & Bennett, S. C. (1997). Why are there eccentricity effects in visual search? *Perception & Psychophysics, 60,* 140–156.

Wolfe, J. M., & VanWert, M. J. (2010). Varying target prevalence reveals two, dissociable decision criteria in visual search. *Current Biology, 20,* 121–124.

Wolfe, J. M., Yee, A., & Friedman-Hill, S. R. (1992). Curvature is a basic feature for visual search. *Perception, 21,* 465–480.

Yarbus, A. L. (1967). *Eye movements and vision*. New York: Plenum Press.

Yokoi, K., & Uchikawa, K. (2005). Color category influences heterogeneous visual search for color. *Journal of the Optical Society of America. A, Optics, Image Science, and Vision, 22*, 2309–2317.

Zelinsky, G., & Sheinberg, D. (1995). Why some search tasks take longer than others: Using eye movements to redefine reaction times. In J. M. Findlay, R. Walker, & R. W. Kentridge (Eds.), *Eye movement research: Mechanisms, processes and applications. Studies in visual information processing* (Vol. 6, pp. 325–336). Amsterdam, Netherlands: Elsevier Science.

Zelinsky, G. J., & Schmidt, J. (2009). An effect of referential scene constraint on search implies scene segmentation. *Visual Cognition, 17*, 1004–1028.

Zelinsky, G. J., & Sheinberg, D. L. (1997). Eye movements during parallel/serial visual search. *Journal of Experimental Psychology: Human Perception and Performance, 23*, 244–262.

3
Temporal Orienting of Attention

Anna C. Nobre and Simone G. Heideman

Attention is a highly selective and dynamic process that focuses our behavior and neural processing on the relevant aspects of our environment. While the topic of spatial selective attention is well described in the literature, the selective deployment of attention over time remains much less well investigated and understood. In this chapter we will argue that timing is essential for organizing and optimizing our interactions with the world. Being able to predict *when* something is going to happen is not very useful in itself, but this information becomes highly relevant when you combine it with predictions about *where* and *what* events will occur to guide perception and action (see also chapter 1 on spatial attention and chapter 14 on action and attention in the current volume). One of the main goals of the research in our lab is to establish how temporal predictions dynamically influence our perception and actions and to elucidate the neural bases of these attention-related mechanisms.

A classic example of a situation in which we orient our attention to a specific moment in time is when waiting at a red traffic light (see Coull & Nobre, 1998). When the light turns amber, we begin to anticipate the moment to "go" as the likelihood of a green light increases the longer we wait. In other situations timing is important because we need to keep to a rhythm—for example, when playing music or dancing.

Historical Context: Early Studies on Temporal Attention

Although the temporal attention literature is relatively sparse, the importance of the temporal structure of events was already recognized by some of the pioneers in psychology. Early investigations by Wundt (1874) and Woodrow (1914, 1916) show that the "foreperiod," that is, the interval between a warning stimulus and a target requiring a response, is an important determinant of response time. Wundt (1874) was the first to show that people respond faster when a target is preceded by a warning signal. In his experiment, participants had to release a key when they heard the sound of a steel ball hitting a metal plate. The mechanism releasing the ball could be adjusted to increase or decrease the foreperiod. Participants responded faster when they were able to view the release mechanism dropping the ball. Woodrow (1914, 1916) showed that reaction times are faster with regular foreperiods than with variable, unpredictable foreperiods. Later studies showed that the benefits of predictable foreperiods also appear in choice reaction-time tasks, where the nature of the stimulus and response is unknown in advance, although the effects are smaller in size (Bertelson & Boons, 1960). These foreperiod effects were usually interpreted as reflecting a general preparation or general increase in the readiness to respond (Niemi & Näätänen, 1981; Teichner, 1954).

Egan and colleagues (1961) showed that in addition to speeding up responses, temporal certainty also enhances perception. Their participants had to detect the occurrence of an auditory stimulus at an interval indicated by a light stimulus. The tone was presented on only half of the trials, and temporal uncertainty was manipulated by varying the duration of the observation interval. Tone detectability decreased progressively with interval lengthening and, thus, with temporal uncertainty. Later studies found similar results in the visual domain. Comparable results were obtained for lengthening a continuous interval and increasing the number of discrete temporal intervals in which the event of interest could occur (Earle & Lowe, 1971; Lasley & Cohn, 1981; Lowe, 1967; Westheimer & Ley, 1996). These studies all showed that we become not only faster but, importantly, more accurate when the occurrence of target events can be predicted in time. Temporal information thus improves not only reaction times but also perceptual processing.

Newhall (1923) was the first to test the effect of temporal expectations induced by rhythms. In his experiment, participants had to report changes in stimulus brightness under three different experimental conditions. Targets occurred 1 second after a series of nine regular auditory stimuli separated by 1-second intervals (i.e., on the next beat), 4 seconds after a series of five regular auditory stimuli separated by 1-second intervals, or after a 9-second empty interval. Perceiving the change in brightness was clearly facilitated in the first case, thus when the stimulus occurred immediately after the next temporal interval of the regular rhythm.

Although not always described in terms of "temporal" preparation, these early studies provided fundamental insights about how temporal orienting can be used to optimize our perception and action. Despite these early studies, temporal orienting of attention did not figure prominently in the research agenda until recently. This impacted the models that were developed; as a result of ignoring the dynamic, ever-changing nature of attentional processes, an important dimension was missing. The situation is changing. The importance of temporal information is being rediscovered and steadily reclaiming its place within the attention literature. New models are being developed, and old models are being adapted to include temporal aspects of attention. The neural instantiation of various types of temporal orienting remains an open question, attracting increasing investigation. In this chapter, we will review headway made in understanding three important types of temporal bias: hazard rates, rhythms, and temporal cues.

State-of-the-Art Review

Hazard Rates

Temporal expectations related to foreperiods are dependent on the passage of time, with the probability of the occurrence of an effect typically increasing over time. The "hazard function" is the conditional probability of an event's occurring at a given time, given that it has not yet occurred (Luce, 1986). Many cognitive experiments have a fixed temporal structure. In such tasks, the influence of hazard rates is visible in anticipatory saccades (Kingstone & Klein, 1993) and manual responses (Nickerson, 1965). Furthermore, hazard rates influence perceptual thresholds (e.g., Lasley & Cohn, 1981; Westheimer & Ley, 1996) and attenuate attentional-blink effects (Martens & Johnson, 2005; Shen & Alain, 2012).

Van Ede and colleagues (2011) studied the influence of hazard rates on contralateral sensorimotor alpha and beta desynchronization in anticipation of tactile stimuli. Desynchronization of alpha and beta activity over task-relevant areas is often linked to anticipation and preparation in the perceptual or motor domain (e.g., Gould et al., 2011; Pfurtscheller & Lopes da Silva, 1999; Wyart & Tallon-Baudry, 2008). Across experimental blocks, tactile hand stimulation was differentially distributed over the course of single experimental trials, occurring at one of three (1, 2, or 3 seconds) or only at two intervals (1 and 3 seconds). Beta-band desynchronization clearly adapted to the block-specific hazard rate, with desynchronization being strongest just before possible times of stimulus presentation and relaxing in between. Furthermore, this contralateral prestimulus beta suppression was associated with faster reaction times. Contralateral alpha desynchronization only relaxed in blocks with a two-point hazard rate, when relevant time points were spaced by 2 seconds. In this task, hazard rates thus influenced the time course of oscillatory activity related to spatiotemporal orienting in the tactile domain, which was associated with better performance.

Temporal expectations induced by hazard rates influences neural activity across many brain areas, from early perceptual input to late motor output stages. Animal studies have reported effects on spiking and oscillatory activity over many sensory areas, and as early as V1 (Ghose & Maunsell, 2002; Ghose & Bearl, 2010; Lima et al., 2011). At the other extreme of information processing, Schoffelen and colleagues (2005) showed that reaction times and motor cortex output, measured as corticospinal coherence in the gamma band, were both strongly affected by hazard-rate manipulations. A go/no-go study performed in our lab with electroencephalography (EEG) also showed hazard-rate effects on reaction times and indices of motor preparation (Cravo et al., 2011).

Rhythm and Entrainment Effects

Rhythms are everywhere around us, and they are important in many of our interactions with the environment. Rhythms are inherent to bodily processes like breathing and heart rate, and they are evident in speech and music, as well as in walking and performing other movements. We even sample the world in a rhythmic fashion since eye movements are primarily rhythmic (McAuley et al., 1999). Thus, rhythm is a powerful source of predictive temporal information (see also Schroeder et al., 2010), and, perhaps not surprisingly, rhythmicity is a hallmark of our brains (Buzsáki, 2006).

As Newhall demonstrated in 1923, stimuli coinciding with the beat of a rhythm have a clear perceptual advantage. Jones and colleagues (1976, 2006) launched the modern-day research on rhythmic orienting of attention, investigating the effects of rhythm on auditory discrimination. Auditory discrimination is maximal for targets occurring on the beat, and the bigger the interval between the predicted beat and the target, the worse performance becomes. This perceptual facilitation by rhythms appears to be largely automatic since the effects occur even when the rhythm is not predictive (Sanabria et al., 2011). Rhythmic benefits to perception are also found in the visual domain (e.g., de Graaf et al., 2013; Mathewson et al., 2010).

Jones (1976, 2010) proposed a "dynamic attending theory," in which entrainment to rhythmic regularities within our environment provided a mechanism for enhancing perception. Recently, neuroscientists have proposed that behavioral entrainment may be accompanied by neural entrainment. Oscillatory activity in the lower frequency ranges (e.g., delta and theta bands) has been shown to entrain to regular events in the environment by means of a phase reset (Lakatos et al., 2008, 2009, 2013; Schroeder & Lakatos, 2009). Entrainment of low-frequency oscillations in turn modulates excitability in local neuronal assembles, thereby synchronizing brain activity at higher frequencies (e.g., in the gamma band) coding information in these local ensembles (Canolty et al., 2006; Canolty & Knight, 2010).

The timing of slow, anticipatory brain potentials, like the contingent negative variation (CNV; Walter et al., 1964), also follows rhythms. Praamstra and colleagues (2006) showed rhythmic modulation of slow event-related potentials (ERPs), as well as preparatory desynchronization in the alpha and beta bands, when they presented series of targets with regular stimulus-onset asynchronies (SOAs). The time course of the CNV and oscillatory modulations varied with the length of the regular SOA in each series. Behavioral and neural responses to targets occurring after deviant intervals were compromised.

In our lab, we investigated effects of rhythmic temporal expectations in the presence or absence of spatial expectations to visual targets (Doherty et al., 2005). In this EEG experiment, a ball jumped across the screen in discrete steps before it disappeared behind an occluder. The participants' task was to discriminate the presence or absence of a small stimulus inside the target when it reappeared. Different types of expectation (spatiotemporal, temporal only, spatial only, and neutral) resulted from manipulating the spatial trajectory (linear or random) and/or rhythm (regular or random) of the ball. Both spatial and temporal expectations speeded reaction times, with an additive effect when both dimensions could be predicted. ERPs were recorded when the ball reappeared after occlusion—the target event in the task. Modulation of the earliest visual potential, P1, showed a striking interaction between temporal and spatial expectations. Whereas recordings showed no P1 enhancement by temporal expectation alone, the gain modulation of P1 by spatial expectation was significantly potentiated when temporal expectations were also present (spatiotemporal compared to spatial condition). Other, later, potentials (N1, P2, and P3) were also affected by temporal and spatial expectation.

These finding were subsequently replicated and extended by showing effects of combined spatiotemporal expectations on anticipatory alpha desynchronization during the occlusion period (Rohenkohl & Nobre, 2011). During occlusion, alpha-band desynchronization related to spatial expectation (e.g., Gould et al., 2011; Wyart & Tallon-Baudry, 2008) followed the time course of the preceding rhythm. Bursts of alpha foreshadowed the invisible jump of the ball during occlusion and its reappearance after occlusion. Temporal predictions thus influenced the time course of oscillatory activity related to anticipatory spatial attention. In a separate experiment, investigating temporal expectations in the absence of spatial expectations, responses to targets appearing after a predicted (50% valid) or unpredicted foreperiod were compared using a large set of possible isochronous (i.e., equally spaced) rhythms (Correa & Nobre, 2008). The results showed that temporal expectations were flexibly induced by multiple regular rhythms, and that the effects of temporal expectations strongly interacted with the foreperiod effects associated with the length of the occlusion period.

To probe the influences of rhythm on perception in greater depth, psychophysical experiments involving challenging perceptual discriminations were combined with modeling and EEG recordings. Behavioral accuracy and response times for target discrimination were compared when targets were embedded within a stimulus stream presented in a regular, rhythmic, or irregular, arrhythmic fashion (Cravo et al., 2013; Rohenkohl et al., 2012a). Stimuli consisted of Gaussian noise patches, in which a Gabor grating of different contrast levels could be embedded. Participants had to indicate the orientation of the grating in a forced-choice fashion. Importantly, the intervals immediately surrounding each target were identical for both the rhythmic and arrhythmic condition. Regular rhythm enhanced response times and lowered the threshold contrast levels for target discrimination (Rohenkohl et al., 2012a). Diffusion modeling (see Palmer et al., 2005) showed that temporal expectations influenced perceptual processing by enhancing the signal-to-noise gain of the visual targets. In a subsequent replication using EEG (Cravo et al., 2013), the same pattern of behavioral effects was observed, and these benefits followed entrainment of oscillatory activity in the delta band (the same range as the stimulus trains). Delta phase across the target events was better aligned for the rhythmic compared to the arrhythmic condition, and the correlation between delta phase and target discriminability was higher. These studies all suggest that rhythms are important sources of temporal expectations, and that entrainment to environmental rhythms has significant perceptual benefits.

Cuing

Foreperiods, hazard rates, and rhythms can all generate temporal expectations automatically. However, temporal information is, of course, not necessarily implicit. Explicit foreknowledge about intervals can also be used to orient attention in time voluntarily and, thereby, to improve our perception and actions.

The first study to use temporal cues (Coull & Nobre, 1998; see also Kingstone, 1992, experiment 4) adapted Posner's spatial orienting task (Posner et al., 1980, see also chapter 1 on spatial attention in the current volume) to test whether it is possible to orient attention voluntarily to anticipated moments of relevant events. The task required simple target detection of a left or right visual stimulus occurring after 300 or 1,500 ms. Symbolic cues preceding the target predicted (with 80% validity) its spatial location, time interval, or both location and interval. Validity benefits and invalidity costs occurred for both spatial and temporal cues and were even larger for temporal cues. The effects of temporal cues were restricted to targets at the short interval since after a short interval with no target, target appearance became predicted to occur at the long interval with absolute certainty.

The effects of temporal cuing have been extensively replicated and extended. The effects are flexible and robust; they occur using different tasks, sensory modalities, interval ranges, stimulus parameters, and response requirements (Correa, 2010; Griffin et al., 2001; Lange & Röder, 2010; Miniussi et al., 1999; Nobre, 2010; Nobre & Rohenkohl, 2014). ERP recordings during visual tasks have typically shown that, in the absence of spatial information, the visual P1 potential is unaffected by temporal cues, with effects occurring at later N2 and P3 potentials (Griffin et al., 2002; Miniussi et al., 1999). These effects are similar to those in experiments using rhythmically induced temporal and combined spatiotemporal expectations (Correa & Nobre, 2008; Doherty et al., 2005). Zanto and colleagues (2011) investigated the effects of temporal cuing using detection, discrimination, and go/no-go tasks in young and older adults. They replicated behavioral and neural effects in young participants but found a deficit in ability of elderly adults to use temporal cues.

Cuing tasks involving demanding perceptual discriminations for targets appearing at constant locations show that temporal cues may influence early stages of visual processing. Correa and colleagues (2006) used temporal cues to predict the occurrence of a target within a rapid serial visual presentation stream. They found that valid cues enhanced perceptual sensitivity (d') measures and modulated early P1 potentials.

A recent experiment in the lab (Rohenkohl et al., 2014) tested whether perceptual effects of temporal cuing depended on spatial attention. Participants made difficult perceptual discriminations on peripheral grating stimuli. These appeared after a cue predicting (80% validity) the side (left, right) and the interval (800, 2,000 ms) of the target. Results confirmed a synergistic interaction between temporal and spatial expectations. Temporal expectation boosted the effects of spatial expectations on perceptual variables, resulting in enhanced visual discriminations. However, valid temporal cues combined with invalid spatial cues had no significant effect on performance.

These studies show that cued temporal expectations can have a strong and flexible influence on behavior. Early perceptual effects may occur primarily by enhancing other types of top-down attention effects that map onto receptive-field properties of sensory areas.

Modes of Temporal Orienting: Commonalities and Dissociations

Hazard rates, rhythms, and cuing share some similar features in terms of behavioral as well as neural effects. Behaviorally, all these modes of temporal orienting can speed up reaction times and enhance accuracy of performance. The neural pattern of target modulation is similar, with a clear influence on later potentials (e.g., N2, P3 in vision) and flexible modulation of early potentials (P1) depending on whether spatial information is also present (Correa et al., 2006; Doherty et al., 2005; Griffin et al., 2002; Miniussi et al., 1999; Praamstra et al., 2006).

Modulation of anticipatory activity also follows similar patterns. The time course of anticipatory CNV-like brain potentials follows temporal expectations (e.g., Capizzi et al., 2013; Cravo et al., 2011; Miniussi et al., 1999; Praamstra et al., 2006; Trillenberg et al., 2000). Modulation of preparatory oscillatory activity is also observed. Alpha- and beta-band desynchronization over sensory and motor areas precedes temporally predictable targets depending on the specific task parameters and demands (e.g., Praamstra et al., 2006; Rohenkohl & Nobre, 2011; van Ede et al., 2011; Zanto et al., 2011). These anticipatory oscillatory effects often interact with other available information, like spatial expectation in visual tasks or effector expectation in motor tasks.

Despite these commonalities, there are some clear differences between different types of temporal biases. Sequential effects and rhythms seem to have more automatic influences than other types of temporal information. In a study from our lab, Rohenkohl and colleagues (2011) showed that rhythmic, regular motion patterns enhanced responses to visual targets automatically, but that symbolic color cues only enhanced behavior when participants were instructed to use cue predictions. Others showed that the automatic effects of rhythms occur even when they are unintentional and detrimental to the task that has to be performed (Breska & Deouell, 2014). Behavioral improvements have also been reported for auditory targets occurring on the beat of a rhythm, even when this rhythm is task-irrelevant and noninformative (Sanabria et al.,

2011). Furthermore, benefits associated with rhythms resist simultaneous performance of a demanding working memory task (de la Rosa et al., 2012). In a similar dual-task experiment, a competing working-memory task interfered with temporal cuing effects, but not with sequential effects (Capizzi et al., 2012, 2013).

Dissociations have also been reported in patient studies. Triviño and colleagues (2010) showed deficient temporal cuing effects in patients with right frontal lesions while foreperiod effects were affected in both left and right frontal patients. In contrast, sequential effects were preserved in both groups of patients. A third group of basal ganglia patients did not show deficits in any of the effects.

Computational modeling used to reveal functional mechanisms behind perceptual benefits also hint at different ways in which cuing, hazard rates, and rhythms affect performance variables in psychophysical tasks. Modeling of results in different tasks has yielded different explanations for the effects of temporal expectations. Vangkilde and colleagues (2012) applied computational modeling according to the "theory of visual attention" (Bundesen, 1990; Bundesen & Habekost, 2014) to non-speeded cued letter-recognition tasks with different hazard rates. They found that perceptual encoding became faster with increasing temporal expectations but did not lead to changes in perceptual threshold. Jepma and colleagues (2012) modeled effects in two experiments, using a fixed foreperiod and a temporal cuing paradigm. They found that temporal expectations induced by foreperiods and cues mainly influenced non-decision-related parameters like motor preparation or target encoding and had little influence on the rate of evidence accumulation or decision thresholds. Qualitatively different results were obtained in studies in the lab (Cravo et al., 2013; Rohenkohl et al., 2012a), in which we found the signal-to-noise gain of targets to be enhanced by rhythmic temporal expectations. It is possible that rhythms exert a stronger influence on early visual excitability than other types of temporal expectations, or that temporal orienting affects either perceptual or response-related parameters depending on other task parameters and demands (see Rohenkohl et al., 2012b; Nobre & Rohenkohl, 2014).

Of course, the distinctions we draw between different types of temporal expectation are somewhat arbitrary. In many tasks, and often in daily life, different types of temporal expectation are intermixed. An example is waiting for a traffic light to become green, with the cue given by the light turning amber signaling the upcoming change, and the hazard rate building during the wait. Furthermore, it remains unclear how multiple sources of temporal expectation interact and give rise to more complex patterns of expectations, for example, in the case of nonisochronous sequences, that have ordinal as well as temporal information embedded in them (see O'Reilly, McCarthy, et al., 2008). In such tasks, a clear interaction between implicitly learned spatial and temporal information is generally found, but how these effects interact on a neural level is still unclear.

Models and Mechanisms of Timing

There is no agreement about how temporal information is represented in the brain. Models tend to fall into two classes (see Ivry & Schlerf, 2008, for a review). One class proposes a general timing control network regardless of the goals and domains involved. Within such models, temporal expectations would be generated within this control network, or at least in interaction with this dedicated neural clock (e.g., Merchant et al., 2013). The other class of models questions the existence of a central-clock mechanism and proposes that timing is embedded within brain areas involved in the task at hand.

A classic "clock" model is the pacemaker-accumulator model (Gibbon et al., 1984; Treisman, 1963). In this model, an internal pacemaker emits regular pulses that can be recorded by an accumulator or counter on the basis of sensory signals, for example, a warning cue or a target duration that has to be remembered. The number of pulses counted represents the interval duration. Series of pulses can compared with previously stored pulse series to compare intervals. In contrast, in state-dependent network models (Mauk & Buonomano, 2004) no clocks or brain areas are dedicated to temporal processing. In such models, timing arises from computations that include temporal processing but are not mediated by central mechanisms. In such models timing is an intrinsic property of areas dedicated to the processing of other features, like space, and involves state-dependent changes in network dynamics. Thus, timing is distributed across the brain. An alternative distributed view of temporal processing proposes "climbing neural activity" as the process by which we track time (Wittmann, 2013). This model suggests that neural activity ramps up until it reaches a peak at the end of a tracked interval. Others argue that our representation of durations is dependent on memories, whereby the decaying strength of memory traces is used to keep track of time (Staddon, 2005).

The dissociations across temporal orienting effects tend to argue against there being one single mechanism for organizing all aspects of interval timing in the brain. Studies showing temporal modulations of the receptive field properties of sensory neurons make it clear that models proposing one central dedicated neural clock are unlikely. However, central timing mechanisms may interact with distributed neural populations to cause highly localized timing effects. At the moment it remains difficult to arbitrate between the competing accounts. What is becoming evident is that oscillatory activity likely plays an important role in at least some of these processes.

Temporal Orienting and the Brain

Echoing the debate concerning central versus distributed models of timing, the neural correlates of temporal expectations are not fully resolved. The EEG and magnetoencephalography studies discussed in this chapter indicate that temporal information can interact with receptive-field properties of relevant sensory and motor areas to improve our perception and actions. However, they are moot on the possible contribution of a central, dedicated timing system. Despite the obvious disadvantages of using functional magnetic resonance imaging to investigate temporal processing, it is clear that some brain areas seem to be more involved in timing tasks than others. Neuroimaging, lesion, and patient studies have consistently implicated the cerebellum, basal ganglia, supplementary motor area, and inferior frontal cortex in explicit temporal processing while parietal and left premotor areas seem more involved in temporal orienting (Buhusi & Meck, 2005; Coull & Nobre, 2008; Coull et al., 2004; Ivry, 1996; Ivry & Spencer, 2004). However, the precise combination of areas that are activated together seems to be highly dependent on task demands (Coull & Nobre, 2008). Furthermore, different brain structures may participate at different timescales; for example, the cerebellum has been implicated in temporal processing in the millisecond range (Ivry & Spencer, 2004) while the basal ganglia have been implicated in timing ranging from hundreds of milliseconds to minutes (Buhusi & Meck, 2005).

A frontoparietal network containing the intraparietal sulcus, the anterior parietal lobule, and the inferior premotor cortex is often found to be activated in studies looking at temporal orienting (Cotti et al., 2011; Coull et al., 2000; Coull & Nobre, 1998; Coull et al., 2013; Davranche et al., 2011). A meta-analysis conducted in our lab (Nobre & Rohenkohl, 2014), including eight studies on temporal cuing and related tasks, confirmed reliable activations in posterior portion of the intraparietal sulcus, anterior inferior parietal lobule, and premotor cortex. In addition, activations in the cerebellum were found although this area is not consistently reported in the temporal orienting literature. The left parietal cortex seems selectively involved in temporal orienting compared to other types of orienting. It is more active for temporal cuing compared to spatial (Coull & Nobre, 1998) or motor-effector cuing (Cotti et al., 2011). Furthermore, activity in parietal areas was more strongly correlated with task-relevant areas during a perceptual as well as a motor task (Davranche et al., 2011). The functional overlap between brain areas involved in temporal orienting and in manual sensorimotor control has often been noted (e.g., Coull & Nobre, 1998; Nobre, 2001; Nobre & Rohenkohl, 2014). One intuitive possibility is that temporal orienting relies on computations available in manual-control circuits in an analogous way to how spatial orienting relies on computations available through the oculomotor circuitry (O'Reilly, Mesulam, & Nobre, 2008).

Timing in Aging

Temporal processing breaks down in certain neurodegenerative disorders, such as Parkinson's disease (PD) and Huntington's disease (e.g., Grahn & Brett, 2009; Harrington et al., 1998; Pastor et al., 1992; Paulsen et al., 2004), but may also degrade during aging (see Balci et al., 2009). The explicit timing literature shows that temporal precision decreases with age. Older adults give larger verbal estimates of duration and shorter productions of durations than younger participants, and timing effects in elderly adults are significantly influenced by task complexity (Block et al., 1998). Deterioration of temporal processing may have substantial deleterious consequences. For instance, in order to navigate through traffic, it is important to integrate temporal and spatial information about one's own location and movement speed, but also with respect to other vehicles. Elderly people seem less able to use and integrate temporal information, which causes problems in highly dynamic contexts like traffic (see, e.g., Andersen & Cisneros, 2000).

Although the relation between aging and the types of temporal expectation described in this chapter has not been fully investigated, there are studies that suggest that temporal orienting deteriorates with age. Zanto and colleagues (2011) studied the effects of temporal cuing in young and elderly participants using detection, forced-choice discrimination, and go/no-go discrimination tasks. They showed that young participants benefited from temporal cues and thereby improved their performance in all tasks. The time course of CNV potential and alpha-power desynchronization also followed the temporal predictions by the cues. In contrast, elderly adults did not use the temporal cues to improve performance, and the neural markers of temporal orienting were also diminished.

These results suggest that the effects of temporal expectations diminish with age. Note that it is not clear yet how these effects come about, or how effects of explicit timing relate to the different types of temporal expectations. These age-related deficits might imply a decline of temporal processing in itself or instead be related to other factors, like task demands. Thus, it might be the case that when resources are needed for the processing of other information, temporal accuracy suffers. We are currently investigating these questions in the lab.

Integration: Temporal Orienting in Our Daily Life
Temporal expectations mediate most of our interactions with the environment. Examples are not hard to find. We already mentioned the example of traffic lights, with the different colors signaling upcoming changes. Another example is waiting for the start of a race, where the "ready, set, ..." said by the referee allows you to prepare optimally for the crucial moment of the "... go." Rhythms are also important during more continuous types of movement, for example, when walking, learning how to cycle, dancing, or when performing more complex behaviors like juggling or playing an instrument. During complex, nonisochronous sequences of behaviors different types of temporal expectations and temporal processes likely interact to allow for optimal performance.

Demonstrating how temporal information improves our perception is more difficult. One example might be human speech. Speech has very distinct temporal properties, and if those characteristics are altered, performance suffers (Jones, 1976). Complementarily, emphasizing the rhythmic structure of speech can boost performance. Experiments have shown that rhythmic priming enhances speech perception (Cason & Schön, 2012) and comprehension (Rothermich et al., 2012), and that it can improve speech production (Tilsen, 2011). In many situations temporal information will be especially perceptually advantageous when combined with spatial information. For example, when driving, you have to take into account spatial as well as temporal information related to your environment and the vehicles around you to be able to anticipate changes and drive safely.

Most of the research into temporal expectations has looked at basic, sensory processes. However, there are some studies that have investigated timing related to higher level processing (e.g., Chapin et al., 2010; Fujioka et al., 2012; Gordon et al., 2011; Large et al., 2002; Large & Snyder, 2009; Nozaradan et al., 2011). One of these studies by Nozaradan and colleagues (2011) looked at neuronal entrainment to a musical beat. They presented their participants with a 33-second-long auditory stimulus consisting of a pure tone with a 2.4-Hz auditory beat. There were three different conditions: a control task, a binary meter imagery task, and a ternary meter imagery task. The tone participants heard was the same for all conditions, but the imagined meter, that is, the rhythm that was imagined on top of the beat, was different. Participants' EEG results showed entrainment to the 2.4-Hz beat, which was similar for all conditions. However, interestingly, imagining a binary meter caused an additional response at 1.2 Hz, while imagining the ternary meter caused an additional response at 0.8 Hz. Thus, both conditions showed entrainment related to the psychological interpretation of the beat. These results show that entrainment may play an important role in music perception and, furthermore, that this entrainment can be influenced dynamically, depending on subjective factors.

Besides facilitating perception and action toward relevant events, timing is also likely to play a role in canceling out irrelevant input. In our daily lives we are bombarded with distracting sensory input. It has been proposed that oscillatory brain activity is important for selecting relevant "streams" of information (Lakatos et al., 2008; Schroeder & Lakatos, 2009), but perhaps just as interesting is the idea that temporal expectations can also be used to suppress irrelevant input, a phenomenon well known to happen for selective spatial attention. For unattended task-irrelevant spatial locations, alpha band synchronization over corresponding visual areas is often found (Rihs et al., 2007; Worden et al., 2000), which may involve suppression of distracting, irrelevant input.

Relatively rhythmic, predictable input, like the ticking of a clock in the room, people talking in the background, or the sound of our own footsteps, could be canceled out by processes involving rhythmic oscillatory brain activity. In a recent study, Lakatos and colleagues (2013) showed opposite phase entrainment in monkey auditory area

A1 for neural ensembles that coded for task-relevant versus irrelevant frequency content when monkeys had to select between competing auditory streams. This opposite entrainment effect was aligned with the rhythm of the attended auditory stream with a preferred phase for task-relevant frequencies and opposite phase for task-irrelevant frequencies at task-relevant time points. These opposite entrainment effects seemed to sharpen the contrast between the processing of targets versus distractors. Horton and colleagues (2013) found opposite entrainment effects for speech suppression in a "cocktail party" situation. They measured EEG while participants heard multiple spoken sentences. Participants had to attend to either the left or the right speaker and after each trial had to indicate whether a prompted sentence had originated from the cued or uncued speaker. The results showed not only that oscillatory activity entrained to the attended speech stream, but furthermore that entrainment to the unattended stream was also present and that this entrainment had an opposite phase such that the speech occurred at moments of smallest neural excitability. These studies show that entrainment to rhythmic streams can be used as a mechanism of enhancement, as well as suppression. Entrainment-based suppression could provide a highly useful general mechanism for inhibiting irrelevant rhythmic input.

The process by which we cancel out our own sensory input has for a long time been considered under the notion of the "efference copy" (Cullen, 2004; Von Holst & Mittelstaedt, 1950). Efference copies are involved whenever we perform a motor act, to account for or cancel out the sensory consequences of our actions. They help us to distinguish between internally and externally produced sensations and are, for example, thought to be the reason we cannot tickle ourselves (Blakemore et al., 2000). Timing is critical to the efference copy mechanism; by artificially increasing the interval between the motor act and the subsequent sensory experience, self-produced tickling feels more and more tickly (Blakemore et al., 1999). Some scholars have suggested that deficits in cancellation of efference copies may contribute to symptoms in psychiatric conditions such as schizophrenia (see Frith, 1992, and Pynn & DeSouza, 2013, for a review). For example, auditory hallucinations are proposed to be induced when efference copies are altered and do not fully predict upcoming sensations. In such a case internally generated thoughts are not canceled out and auditory hallucinations can occur.

Future Directions

In this chapter, we argued that temporal expectations are important for shaping our perception and actions. Temporal expectations, whether caused by hazard rates, rhythms, or cues, influence behavior by interacting with receptive-field properties of relevant sensory, sensorimotor, and motor neurons. In this way, temporal expectations operate synergistically with other anticipatory top-down expectations related to space or upcoming responses to enhance their effects. This "boosting" of relevant features by temporal expectations not only results in speeded responses but also improves the quality of our perception. Oscillations are increasingly implicated in this process. Oscillatory activity can become entrained to behaviorally important streams of information. Although some frequency bands seem more relevant than others, at the moment the precise role the different frequencies play is not clear.

It is clear that temporal information can influence different stages of processing. Influences on early sensory areas, as well as late motor output stages, have been described, but we do not know yet what happens at higher stages of processing. The precise stage at which temporal processing interacts with neural activity is likely highly dependent on task demands. Faced with competing sensory stimuli, it might be beneficial to boost early perceptual analysis, while in speeded response tasks with a primarily motor bottleneck, temporal expectations may focus on late stages of processing.

Studies in the lab have shown that temporal information interacts with spatial bias to boost performance. However, it is still unclear how temporal processing interacts with other types of top-down bias. In addition, it is unclear how the different types of temporal expectations work together in cases of predictable but nonisochronous repeating patterns.

With respect to the different types of temporal bias described in this chapter, it is still unclear whether those are supported by coextensive or distinct neural mechanisms. It is probably important to distinguish between more automatic, rhythmic types of temporal processing and less automatic types, like those triggered by temporal cues. A related question is how explicit and implicit temporal information interact. The literature suggests that these are supported by distinct neural systems, but some overlap might be present. In this respect it is also important to investigate how we integrate temporal information from different sources and sensors, and how we deal with temporal information over different timescales. Furthermore it will be relevant to investigate how temporal expectations in perceptual and motor areas interact and constrain each other.

We have suggested that temporal information may be just as important for canceling out irrelevant input as it is for prioritizing relevant stimuli. We described two mechanisms that may be important in this respect, opposite entrainment and efference copy. It is not yet clear how widespread these effects are and how they interact with each other and with other temporal expectation effects to contribute to the optimal use of temporal information.

Because deficient temporal processing is an important characteristic of disorders like PD and schizophrenia but may also be compromised during normal healthy aging, it is important to investigate the aforementioned effects in these populations. This might also provide us with more insight into the nature of temporal expectations in the brain.

Box 3.1

Key Points

- Predictive temporal information carried by hazard rates, rhythms, and cues confers sizeable and reliable benefits to behavioral performance. Mechanisms are likely to include prioritization of relevant events as well as suppression of distraction.

- Temporal expectations modulate several stages of stimulus analysis, from early sensory processing to response execution.

- Temporal expectations interact significantly with predictions about spatial locations to enhance perception and action.

- Entrainment and timing of oscillatory activity are implicated in these processes although the precise mechanisms and the boundary conditions are not yet fully understood.

Box 3.2

Outstanding Issues

- How do the various sources of temporal expectation (hazard rates, rhythms, cues) relate to and interact with one another, and how do these processes interact with predictive information about other stimulus dimensions to enhance processing of targets and suppression of distraction?

- How do temporal expectations develop throughout the life span, and how are they compromised by aging, brain lesions, and neurodegenerative diseases?

References

Andersen, G., & Cisneros, J. (2000). Age-related differences in collision detection during deceleration. *Psychology and Aging, 15*, 241–252.

Balci, F., Meck, W., Moore, H., & Brunner, D. (2009). Timing deficits in aging and neuropathology. In J. L. Bizon & A. G. Woods (Eds.), *Animal models of human cognitive aging* (pp. 161–201). Totowa, NJ: Humana Press.

Bertelson, P., & Boons, J. P. (1960). Time uncertainty and choice reaction time. *Nature, 187*, 531–532.

Blakemore, S. J., Frith, C. D., & Wolpert, D. M. (1999). Spatio-temporal prediction modulates the perception of self-produced stimuli. *Journal of Cognitive Neuroscience, 11*, 551–559.

Blakemore, S. J., Wolpert, D., & Frith, C. (2000). Why can't you tickle yourself? *Neuroreport, 11*, 11–16.

Block, R. A., Zakay, D., & Hancock, P. A. (1998). Human aging and duration judgments: A meta-analytic review. *Psychology and Aging, 13*, 584–596.

Breska, A., & Deouell, L. (2014). Automatic bias of temporal expectations following temporally regular input independently of high-level temporal expectation. *Journal of Cognitive Neuroscience, 26*, 1555–1571.

Buhusi, C. V., & Meck, W. H. (2005). What makes us tick? Functional and neural mechanisms of interval timing. *Nature Reviews. Neuroscience, 6*, 755–765.

Bundesen, C. (1990). A theory of visual attention. *Psychological Review, 97*, 523–547.

Bundesen, C., & Habekost, T. (2014). Theory of visual attention (TVA). In A. C. Nobre & S. Kastner (Eds.), *The Oxford handbook of attention* (pp. 1095–1121). New York: Oxford University Press.

Buzsáki, G. (2006). *Rhythms of the brain*. Oxford, UK: Oxford University Press.

Canolty, R., Edwards, E., & Dalal, S. (2006). High gamma power is phase-locked to theta oscillations in human neocortex. *Science, 313*, 1626–1628.

Canolty, R. T., & Knight, R. T. (2010). The functional role of cross-frequency coupling. *Trends in Cognitive Sciences, 14*, 506–515.

Capizzi, M., Correa, A., & Sanabria, D. (2013). Temporal orienting of attention is interfered by concurrent working memory updating. *Neuropsychologia, 51*, 326–339.

Capizzi, M., Sanabria, D., & Correa, Á. (2012). Dissociating controlled from automatic processing in temporal preparation. *Cognition, 123*, 293–302.

Cason, N., & Schön, D. (2012). Rhythmic priming enhances the phonological processing of speech. *Neuropsychologia, 50*, 2652–2658.

Chapin, H. L., Zanto, T., Jantzen, K. J., Kelso, S. J. A., Steinberg, F., & Large, E. W. (2010). Neural responses to complex auditory rhythms: The role of attending. *Frontiers in Psychology, 1*, 224.

Correa, A. (2010). Enhancing behavioural performance by visual temporal orienting. In A. Nobre & J. Coull (Eds.), *Attention and time* (pp. 359–370). New York: Oxford University Press.

Correa, A., Lupiáñez, J., Madrid, E., & Tudela, P. (2006). Temporal attention enhances early visual processing: A review and new evidence from event-related potentials. *Brain Research, 1076*, 116–128.

Correa, A., & Nobre, A. C. (2008). Neural modulation by regularity and passage of time. *Journal of Neurophysiology, 100*, 1649–1655.

Cotti, J., Rohenkohl, G., Stokes, M., Nobre, A. C., & Coull, J. T. (2011). Functionally dissociating temporal and motor components of response preparation in left intraparietal sulcus. *NeuroImage, 54*, 1221–1230.

Coull, J., & Nobre, A. (2008). Dissociating explicit timing from temporal expectation with fMRI. *Current Opinion in Neurobiology, 18*, 137–144.

Coull, J. T., Davranche, K., Nazarian, B., & Vidal, F. (2013). Functional anatomy of timing differs for production versus prediction of time intervals. *Neuropsychologia, 51,* 309–319.

Coull, J. T., Frith, C. D., Büchel, C., & Nobre, A. C. (2000). Orienting attention in time: Behavioural and neuroanatomical distinction between exogenous and endogenous shifts. *Neuropsychologia, 38,* 808–819.

Coull, J. T., & Nobre, A. C. (1998). Where and when to pay attention: The neural systems for directing attention to spatial locations and to time intervals as revealed by both PET and fMRI. *Journal of Neuroscience, 18,* 7426–7435.

Coull, J. T., Vidal, F., Nazarian, B., & Macar, F. (2004). Functional anatomy of the attentional modulation of time estimation. *Science, 303,* 1506–1508. doi:10.1126/science.1091573.

Cravo, A. M., Rohenkohl, G., Wyart, V., & Nobre, A. C. (2011). Endogenous modulation of low frequency oscillations by temporal expectations. *Journal of Neurophysiology, 106,* 2964–2972.

Cravo, A. M., Rohenkohl, G., Wyart, V., & Nobre, A. C. (2013). Temporal expectation enhances contrast sensitivity by phase entrainment of low-frequency oscillations in visual cortex. *Journal of Neuroscience, 33,* 4002–4010.

Cullen, K. E. (2004). Sensory signals during active versus passive movement. *Current Opinion in Neurobiology, 14,* 698–706.

Davranche, K., Nazarian, B., Vidal, F., & Coull, J. (2011). Orienting attention in time activates left intraparietal sulcus for both perceptual and motor task goals. *Journal of Cognitive Neuroscience, 23,* 3318–3330.

de Graaf, T. A., Gross, J., Paterson, G., Rusch, T., Sack, A. T., & Thut, G. (2013). Alpha-band rhythms in visual task performance: Phase-locking by rhythmic sensory stimulation. *PLoS ONE, 8*(3), e60035.

de la Rosa, M. D., Sanabria, D., Capizzi, M., & Correa, A. (2012). Temporal preparation driven by rhythms is resistant to working memory interference. *Frontiers in Psychology, 3,* 308.

Doherty, J. R., Rao, A., Mesulam, M. M., & Nobre, A. C. (2005). Synergistic effect of combined temporal and spatial expectations on visual attention. *Journal of Neuroscience, 25,* 8259–8266.

Earle, D., & Lowe, G. (1971). Channel, temporal, and composite uncertainty in the detection and recognition of auditory and visual signals. *Perception & Psychophysics, 9*(2A), 177–181.

Egan, J., Greenberg, G., & Schulman, A. (1961). Interval of time uncertainty in auditory detection. *Journal of the Acoustical Society of America, 33,* 771–778.

Frith, C. D. (1992). *The cognitive neuropsychology of schizophrenia.* Hove, UK: Erlbaum.

Fujioka, T., Trainor, L. J., Large, E. W., & Ross, B. (2012). Internalized timing of isochronous sounds is represented in neuromagnetic β oscillations. *Journal of Neuroscience, 32,* 1791–1802.

Ghose, G., & Bearl, D. (2010). Attention directed by expectations enhances receptive fields in cortical area MT. *Vision Research, 50,* 441–451.

Ghose, G. M., & Maunsell, J. H. R. (2002). Attentional modulation in visual cortex depends on task timing. *Nature*, *419*, 616–620.

Gibbon, J., Church, R. M., & Meck, W. H. (1984). Scalar timing in memory. *Annals of the New York Academy of Sciences*, *423*, 52–77.

Gordon, R. L., Magne, C. L., & Large, E. W. (2011). EEG correlates of song prosody: A new look at the relationship between linguistic and musical rhythm. *Frontiers in Psychology*, *2*, 352.

Gould, I. C., Rushworth, M. F., & Nobre, A. C. (2011). Indexing the graded allocation of visuospatial attention using anticipatory alpha oscillations. *Journal of Neurophysiology*, *105*, 1318–1326.

Grahn, J. A., & Brett, M. (2009). Impairment of beat-based rhythm discrimination in Parkinson's disease. *Cortex*, *45*(1), 54–61.

Griffin, I. C., Miniussi, C., & Nobre, A. C. (2001). Orienting attention in time. *Frontiers in Bioscience*, *6*, 660–671.

Griffin, I. C., Miniussi, C., & Nobre, A. C. (2002). Multiple mechanisms of selective attention: Differential modulation of stimulus processing by attention to space or time. *Neuropsychologia*, *40*, 2325–2340.

Harrington, D. L., Haaland, K. Y., & Hermanowicz, N. (1998). Temporal processing in the basal ganglia. *Neuropsychology*, *12*, 3–12.

Horton, C., D'Zmura, M., & Srinivasan, R. (2013). Suppression of competing speech through entrainment of cortical oscillations. *Journal of Neurophysiology*, *109*, 3082–3093.

Ivry, R. B. (1996). The representation of temporal information in perception and motor control. *Current Opinion in Neurobiology*, *6*, 851–857.

Ivry, R. B., & Schlerf, J. E. (2008). Dedicated and intrinsic models of time perception. *Trends in Cognitive Sciences*, *12*, 273–280.

Ivry, R. B., & Spencer, R. M. C. (2004). The neural representation of time. *Current Opinion in Neurobiology*, *14*, 225–232.

Jepma, M., Wagenmakers, E.-J., & Nieuwenhuis, S. (2012). Temporal expectation and information processing: A model-based analysis. *Cognition*, *122*, 426–441.

Jones, M. R. (1976). Time, our lost dimension: Toward a new theory of perception, attention and memory. *Psychological Review*, *83*, 323–355.

Jones, M. R. (2010). Attending to sound patterns and the role of entrainment. In A. Nobre & J. Coull (Eds.), *Attention and time* (pp. 317–330). New York: Oxford University Press.

Jones, M. R., Johnston, H. M., & Puente, J. (2006). Effects of auditory pattern structure on anticipatory and reactive attending. *Cognitive Psychology*, *53*, 59–96.

Kingstone, A. (1992). Combining expectancies. *The Quarterly Journal of Experimental Psychology, A*, *44*, 69–104.

Kingstone, A., & Klein, R. M. (1993). What are human express saccades? *Perception & Psychophysics, 54*, 260–273.

Lakatos, P., Karmos, G., Mehta, A. D., Ulbert, I., & Schroeder, C. E. (2008). Entrainment of neuronal oscillations as a mechanism of attentional selection. *Science, 320*, 110–113.

Lakatos, P., Musacchia, G., O'Connel, M. N., Falchier, A. Y., Javitt, D. C., & Schroeder, C. E. (2013). The spectrotemporal filter mechanism of auditory selective attention. *Neuron, 77*, 750–761.

Lakatos, P., O'Connell, M. N., Barczak, A., Mills, A., Javitt, D. C., & Schroeder, C. E. (2009). The leading sense: Supramodal control of neurophysiological context by attention. *Neuron, 64*, 419–430.

Lange, K., & Röder, B. (2010). Temporal orienting in audition, touch and across modalities. In A. Nobre & J. Coull (Eds.), *Attention and time* (pp. 393–405). New York: Oxford University Press.

Large, E. W., Fink, P., & Kelso, S. J. (2002). Tracking simple and complex sequences. *Psychological Research, 66*, 3–17.

Large, E. W., & Snyder, J. S. (2009). Pulse and meter as neural resonance. *Annals of the New York Academy of Sciences, 1169*, 46–57.

Lasley, D. J., & Cohn, T. (1981). Detection of a luminance increment: Effect of temporal uncertainty. *Journal of the Optical Society of America, 71*, 845–850.

Lima, B., Singer, W., & Neuenschwander, S. (2011). Gamma responses correlate with temporal expectation in monkey primary visual cortex. *Journal of Neuroscience, 31*, 15919–15931.

Lowe, G. (1967). Interval of time uncertainty in visual detection. *Perception & Psychophysics, 2*, 278–280.

Luce, R. (1986). *Response times: Their role in inferring elementary mental organization.* New York: Oxford University Press.

Martens, S., & Johnson, A. (2005). Timing attention: Cuing target onset interval attenuates the attentional blink. *Memory & Cognition, 33*, 234–240.

Mathewson, K. E., Fabiani, M., Gratton, G., Beck, D. M., & Lleras, A. (2010). Rescuing stimuli from invisibility: Inducing a momentary release from visual masking with pre-target entrainment. *Cognition, 115*, 186–191.

Mauk, M. D., & Buonomano, D. V. (2004). The neural basis of temporal processing. *Annual Review of Neuroscience, 27*, 307–340.

McAuley, J., Rothwell, J., & Marsden, C. (1999). Human anticipatory eye movements may reflect rhythmic central nervous activity. *Neuroscience, 94*, 339–350.

Merchant, H., Harrington, D. L., & Mech, W. H. (2013). Neural basis of the perception and estimation of time. *Annual Review of Neuroscience, 36*, 313–336.

Miniussi, C., Wilding, E. L., Coull, J. T., & Nobre, A. C. (1999). Orienting attention in time: Modulation of brain potentials. *Brain, 122*, 1507–1518.

Newhall, S. N. (1923). Effects of attention on the intensity of cutaneous pressure and on visual brightness. *Archives de Psychologie, 61*, 5–75.

Nickerson, R. (1965). Response time to the second of two successive signals as a function of absolute and relative duration of intersignal interval. *Perceptual and Motor Skills, 21*, 3–10.

Niemi, P., & Näätänen, R. (1981). Foreperiod and simple reaction time. *Psychological Bulletin, 89*, 133–162.

Nobre, A. C. (2001). Orienting attention to instants in time. *Neuropsychologia, 39*, 1317–1328.

Nobre, A. C. (2010). How can temporal expectations bias perception and action? In A. Nobre & J. Coull (Eds.), *Attention and time* (pp. 371–391). New York: Oxford University Press.

Nobre, A. C., & Rohenkohl, G. (2014). Time for the fourth dimension in attention. In A. C. Nobre & S. Kastner (Eds.), *The Oxford handbook of attention* (pp. 676–721). New York: Oxford University Press.

Nozaradan, S., Peretz, I., Missal, M., & Mouraux, A. (2011). Tagging the neuronal entrainment to beat and meter. *Journal of Neuroscience, 31*, 10234–10240.

O'Reilly, J. X., McCarthy, K. J., Capizzi, M., & Nobre, A. C. (2008). Acquisition of the temporal and ordinal structure of movement sequences in incidental learning. *Journal of Neurophysiology, 99*, 2731–2735.

O'Reilly, J. X., Mesulam, M. M., & Nobre, A. C. (2008). The cerebellum predicts the timing of perceptual events. *Journal of Neuroscience, 28*, 2252–2260.

Palmer, J., Huk, A., & Shadlen, M. (2005). The effect of stimulus strength on the speed and accuracy of a perceptual decision. *Journal of Vision, 5*, 376–404.

Pastor, M. A., Artieda, J., Jahanshahi, M., & Obeso, J. A. (1992). Time estimation and reproduction is abnormal in Parkinson's disease. *Brain, 115*, 211–225.

Paulsen, J. S., Zimbelman, J. L., Hinton, S. C., Langbehn, D. R., Leveroni, C. L., Benjamin, M. L., et al. (2004). fMRI biomarker of early neuronal dysfunction in presymptomatic Huntington's disease. *AJNR. American Journal of Neuroradiology, 25*, 1715–1721.

Pfurtscheller, G., & Lopes da Silva, F. H. (1999). Event-related EEG/MEG synchronization and desynchronization: Basic principles. *Clinical Neurophysiology, 110*, 1842–1857.

Posner, M. I., Snyder, C. R., & Davidson, B. J. (1980). Attention and the detection of signals. *Journal of Experimental Psychology, 109*, 160–174.

Praamstra, P., Kourtis, D., Kwok, H. F., & Oostenveld, R. (2006). Neurophysiology of implicit timing in serial choice reaction-time performance. *Journal of Neuroscience, 26*, 5448–5455.

Pynn, L. K., & DeSouza, J. F. X. (2013). The function of efference copy signals: Implications for symptoms of schizophrenia. *Vision Research, 76,* 124–133.

Rihs, T. A., Michel, C. M., & Thut, G. (2007). Mechanisms of selective inhibition in visual spatial attention are indexed by alpha-band EEG synchronization. *European Journal of Neuroscience, 25,* 603–610.

Rohenkohl, G., Coull, J. T., & Nobre, A. C. (2011). Behavioural dissociation between exogenous and endogenous temporal orienting of attention. *PLoS ONE, 6*(1), e14620.

Rohenkohl, G., Cravo, A. M., Wyart, V., & Nobre, A. C. (2012a). Re: Temporal expectation may affect the onset, not the rate, of evidence accumulation. Reply to Nieuwenhuis, S., Jepma, M., and Wagenmakers, E.-J. Temporal expectation may affect the onset, not the rate, of evidence accumulation. *Journal of Neuroscience, 32,* 8424–8428.

Rohenkohl, G., Cravo, A. M., Wyart, V., & Nobre, A. C. (2012b). Temporal expectation improves the quality of sensory information. *Journal of Neuroscience, 32,* 8424–8428.

Rohenkohl, G., Gould, I. C., Pessoa, J., & Nobre, A. C. (2014). Combining spatial and temporal expectations to improve visual perception. *Journal of Vision, 14,* 1–13.

Rohenkohl, G., & Nobre, A. C. (2011). α oscillations related to anticipatory attention follow temporal expectations. *Journal of Neuroscience, 31,* 14076–14084.

Rothermich, K., Schmidt-Kassow, M., & Kotz, S. A. (2012). Rhythm's gonna get you: Regular meter facilitates semantic sentence processing. *Neuropsychologia, 50,* 232–244.

Sanabria, D., Capizzi, M., & Correa, A. (2011). Rhythms that speed you up. *Journal of Experimental Psychology: Human Perception and Performance, 37,* 236–244.

Schoffelen, J.-M., Oostenveld, R., & Fries, P. (2005). Neuronal coherence as a mechanism of effective corticospinal interaction. *Science, 308,* 111–113.

Schroeder, C. E., & Lakatos, P. (2009). Low-frequency neuronal oscillations as instruments of sensory selection. *Trends in Neurosciences, 32,* 1–16. doi:10.1016/j.tins.2008.09.012.Low-frequency.

Schroeder, C. E., Wilson, D. A., Radman, T., Scharfman, H., & Lakatos, P. (2010). Dynamics of active sensing and perceptual selection. *Current Opinion in Neurobiology, 20,* 172–176.

Shen, D., & Alain, C. (2012). Implicit temporal expectation attenuates auditory attentional blink. *PLoS ONE, 7*(4), e36031.

Staddon, J. E. R. (2005). Interval timing: Memory, not a clock. *Trends in Cognitive Sciences, 9,* 312–314.

Teichner, W. H. (1954). Recent studies of simple reaction time. *Psychological Bulletin, 51,* 128–149.

Tilsen, S. (2011). Metrical regularity facilitates speech planning and production. *Laboratory Phonology, 2,* 185–218.

Treisman, M. (1963). The temporal discrimination and indifference interval: Implications for a model of the internal clock. *Psychological Monographs, 77*(13), 1–31.

Trillenberg, P., Verleger, R., Wascher, E., Wauschkuhn, B., & Wessel, K. (2000). CNV and temporal uncertainty with "ageing" and "non-ageing" S1–S2 intervals. *Clinical Neurophysiology, 111*, 1216–1226.

Triviño, M., Correa, A., Arnedo, M., & Lupiáñez, J. (2010). Temporal orienting deficit after prefrontal damage. *Brain, 133*, 1173–1185.

van Ede, F., de Lange, F., Jensen, O., & Maris, E. (2011). Orienting attention to an upcoming tactile event involves a spatially and temporally specific modulation of sensorimotor alpha- and beta-band oscillations. *Journal of Neuroscience, 31*, 2016–2024.

Vangkilde, S., Coull, J. T., & Bundesen, C. (2012). Great expectations: Temporal expectation modulates perceptual processing speed. *Journal of Experimental Psychology, 38*, 1183–1191.

Von Holst, E., & Mittelstaedt, H. (1950). Das reafferenzprinzip—Wechselwirkungen zwischen Zentrainervensystem und Peripherie. *Naturwissenschaften, 37*, 464–476.

Walter, W. G., Cooper, R., Aldridge, V. J., McCalum, W. C., & Winter, A. L. (1964). Contingent negative variation: An electric sign of sensorimotor association and expectancy in the human brain. *Nature, 203*, 380–384.

Westheimer, G., & Ley, E. (1996). Temporal uncertainty effects on orientation discrimination and stereoscopic thresholds. *Journal of the Optical Society of America, 13*, 884–886.

Wittmann, M. (2013). The inner sense of time: How the brain creates a representation of duration. *Nature Reviews. Neuroscience, 14*, 217–223.

Woodrow, H. (1914). *The measurement of attention*. Whitefish, MT: Kessinger.

Woodrow, H. (1916). The faculty of attention. *Journal of Experimental Psychology, 1*, 285–318.

Worden, M. S., Foxe, J. J., Wang, N., & Simpson, G. V. (2000). Anticipatory biasing of visuospatial attention indexed by retinotopically specific alpha-band electroencephalography increases over occipital cortex. *Journal of Neuroscience, 20*, RC63, 1–6.

Wundt, W. M. (1874). *Grundzüge der physiologischen Psychologie*. Leipzig, Germany: W. Engelmann.

Wyart, V., & Tallon-Baudry, C. (2008). Neural dissociation between visual awareness and spatial attention. *Journal of Neuroscience, 28*, 2667–2679.

Zanto, T. P., Pan, P., Liu, H., Bollinger, J., Nobre, A. C., & Gazzaley, A. (2011). Age-related changes in orienting attention in time. *Journal of Neuroscience, 31*, 12461–12470.

4

What Laboratory Studies of Symbolic Spatial Cues Reveal about the Control of Attention in Everyday Life

Bradley S. Gibson and Pedro Sztybel

Historical Context

Humans use a variety of linguistic and iconic spatial symbols in everyday life to communicate where others should focus their attention. For instance, we may utter words such as "above," point our fingers, or direct our gaze to signal where others should focus their attention. For the past 30 years, the cognitive and neurobiological mechanisms that enable us to shift the focus of our attention from one location (or object) to another have been studied extensively in the laboratory (see, e.g., Corbetta, Kincade, Ollinger, McAvoy, & Shulman, 2000; Eriksen & Hoffman, 1972; Gibson & Bryant, 2005; Goldberg, Maurer, & Lewis, 2001; Hommel, Pratt, Colzato, & Godijn, 2001; Hopfinger, Buonocore, & Mangun, 2000; Klein & Hanson, 1990; Logan, 1995; Mayer & Kosson, 2004; Nobre, Sebestyen, & Miniussi, 1995; Posner & Petersen, 1990; Posner, Snyder, & Davidson, 1980; Ristic & Kingstone, 2006; Tipples, 2002). However, despite this extensive study, very little is known about the role that different spatial symbols may play in guiding attention in space and whether some symbols might be more effective than others. In large part, this lack of knowledge about how different symbols distribute attention in space derives from a lack of theory that is capable of defining the relevant spatial dimensions along which these symbols can differ.

Early research focused on potential differences between spatial symbols and other cues that could convey information about space more directly such as "onset" cues (i.e., cues that appear abruptly at the target's location) by distinguishing between so-called "central" and "peripheral" cues, respectively (Jonides, 1981; Posner, 1980; see Gibson & Bryant, 2005, and, Yantis, 1996, for reviews). However, this distinction did little more than describe the physical location of the cues on the screen, with symbolic cues appearing centrally at fixation and onset cues appearing in the periphery. Nevertheless, as a result of this taxonomy, there has been a tendency to think that all spatial symbols control attention equally (see Gibson & Kingstone, 2006, for a fuller discussion of this issue).

More recently, we have attempted to provide a more adequate theory of symbolic control that is based on the semantics of space, which has the potential to make more principled distinctions between symbolic cues. Over a decade ago, Logan (1995) suggested that theories of spatial language may provide a useful foundation for understanding how words may control the distribution of attention in space (see also Kemmerer, 2006; Levinson, 2003). Logan suggested that the comprehension of spatial linguistic terms such as "above," "below," "left," and "right" requires the use of a spatial frame of reference that defines the location of one object (the located object) relative to another (the reference object). A reference frame is like a Cartesian coordinate system used in algebra, and it consists of a set of orthogonal axes together with a set of parameters that define the origin, orientation, and endpoints (direction) of the axes and that also provide some specification of distance along the axes (Logan & Sadler, 1996). Moreover, these parameters are considered to be separable and not necessarily set in an all-or-none fashion.

The analysis of spatial semantics in terms of a reference system with separable parameters provides a promising foundation for advancing our understanding of symbolic control because these representations define the dimensions along which spatial symbols can differ. The main goal of this chapter is to review recent evidence showing that spatial symbols do indeed differ according to how the parameters of a reference frame are set. In addition, we will also demonstrate the utility of this theory by discussing how it has exposed at least one other blind spot in our understanding of symbolic control.

Before beginning, it is important to make two additional remarks about our laboratory studies of symbolic control. Like most laboratory studies of attention, our studies are based on a variety of different spatial processes and representations that have developed outside the lab. Because these spatial processes and representations have developed outside the lab, they have been subject to the statistical regularities and biases that characterize the communication patterns between those who share a common symbol system. Our goal has been to use a theory of spatial semantics to develop hypotheses about the nature of these processes and representations and then to use tightly controlled experimental manipulations and standardized tasks in the lab to test these hypotheses and reveal how these statistical regularities and biases influence the nature of these processes and representations. We will return to this issue in the Integration section of this chapter.

The second remark concerns the standardized tasks that we have used to study the control of attention in the laboratory. Generally speaking, we believe that spatial symbols are used to control the orientation of spatial attention in two basic contexts outside the laboratory, and we have tried to mimic these contexts inside the laboratory. In mandatory contexts, spatial symbols are required to find the object of interest because there is no other way to locate the target whereas, in nonmandatory contexts, spatial symbols are not strictly necessary because the object of interest is unique in some other way (e.g., in virtue of its visual features or identity) and can be located by other means such as visual search. Notice that the distinction between mandatory and nonmandatory contexts is based on the extent to which spatial symbols are required to locate the target and not on how informative (or valid) the cue is about the location of the target. Thus, mandatory and nonmandatory spatial symbols could both be 100% valid about the location of the target, but observers would only be obligated to use the symbol in the mandatory context whereas their use of the symbol in the nonmandatory context would be considered voluntary because the observer could have used visual search to find the target.

The distinction between mandatory and nonmandatory contexts is important because we have found that observers do not always use spatial symbols to control their attention in nonmandatory contexts, even when the symbols are 100% valid (see Davis & Gibson, 2012, and Gibson & Sztybel, 2014, for a more detailed discussion of this issue as it relates to both voluntary and involuntary control mechanisms). Hence, unless otherwise noted, most of the studies that were designed to compare different kinds of spatial symbols used variants of the mandatory spatial cuing paradigm shown in figure 4.1. In this paradigm, one colored circle (red or green) is typically presented in each of the four cardinal directions such that the target display contains two red circles and two green circles, and observers' task is to discriminate the color of the circle that is indicated by the central symbolic cue. Of critical importance, cue processing is considered to be mandatory in this paradigm because the correct target cannot be identified without the aid of the cue (Davis & Gibson, 2012). Such mandatory cue processing is important because it ensures that different spatial symbols are processed fully and therefore equally. However, such mandatory cue processing also requires the use of 100%-valid cues. Hence, in this paradigm, the correct discrimination of the target can be taken as evidence that the cue was comprehended and attention was shifted in accordance with the meaning of the cue.

State-of-the-Art Review

The Specification of Orientation and Direction

Gibson and Kingstone (2006) provided initial evidence that different spatial symbols may depend on different reference systems. In particular, they reported a "cued axis effect" when spatial word cues were shown, observing that reaction times (RTs) were significantly faster when "above/below" word cues were shown than when "left/right" word cues were shown (see also Logan, 1995). In contrast, no such RT differences were observed when arrow (or eye gaze) cues were shown. This cued axis effect has been interpreted as showing that word cues access the vertical orientation more efficiently than the horizontal orientation. The functional consequence of this vertical primacy

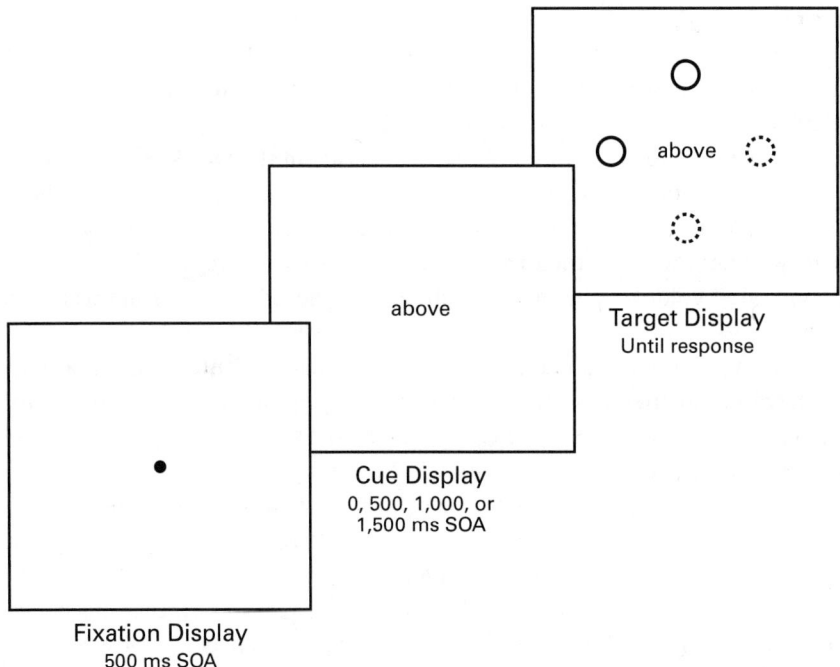

Figure 4.1
Illustration of a typical display sequence shown in the mandatory spatial cuing paradigm when direction cues are used. The solid circle appeared red and the dotted circle appeared green in the actual experiments, and the displays are not drawn to scale. SOA, stimulus-onset asynchrony.

effect is that shifts of attention can be initiated sooner along the vertical axis in response to words such as "above" and "below" than along the horizontal axis in response to words such as "left" and "right." In contrast, corresponding arrow cues access both the horizontal and vertical orientations equally well. Consequently, shifts of attention can be initiated equally fast along both axes in response to arrow cues.

In addition to the cued axis effect, Gibson et al. (2009) also reported a "cued end-point effect" when spatial word cues were shown, signifying a functional difference in the extent to which direction was specified. For instance, consider the color of the circle that appeared on the same axis but opposite the cued circle in the spatial cuing paradigm shown in figure 4.1. On some trials, the colors of the cued circle and the opposite circle were response compatible (e.g., both were red), but on other trials the colors of the cued circle and the opposite circle were response incompatible (e.g., one was red and one was green). Gibson et al. (2009) showed that the incompatibility of the opposite circle slowed RTs when "left/right" word cues were shown, but not when leftward/rightward-pointing arrow cues or "above/below" word cues were shown, suggesting that the endpoints of the horizontal axis were less differentiated when "left/right" word cues were shown (see also Gibson & Davis, 2011).

Furthermore, Gibson et al. (2009) showed that the cued endpoint effect associated with spatial terms such as "left" and "right" could not be attributed to *left/right* scanning patterns associated with the processing of linguistic cues, nor could it be attributed to *left/right* confusions associated with the misapprehension of "left" and "right." In addition, the cued endpoint effect persisted across the duration of the entire experiment, and it was observed across the entire distribution of response times. Thus, the cued endpoint effect associated with "left" and "right" did not dissipate in magnitude as observers became more familiar with the experimental context, nor could this effect be attributed solely to the fastest or slowest RTs. We will discuss the nature of this cued endpoint effect in greater detail in the Integration section of this chapter.

The Specification of Distance

Thus far, the utility of this semantic-based theory of symbolic control has been validated by findings suggesting that linguistic and iconic cues can be distinguished based on how the orientation and direction parameters are set. However, perhaps the greatest contribution of this theory has been its ability to reveal an important empirical and theoretical blind spot in our understanding of symbolic control. In particular, it is typically assumed that spatial symbols guide attention to specific locations, but a spatial symbol can guide spatial attention to a specific location in space only when it is capable of fully specifying both the direction and distance parameters of a spatial frame of reference (Landau & Jackendoff, 1993).

For instance, consider the four boxes depicted in the spatial cuing paradigm shown in figure 4.2. Notice that two of the boxes appear *above* the central fixation point (and the cue, when present) and two of the boxes appear *below* the central fixation point. In addition, notice that two of the boxes appear *far* from the central fixation point and two of the boxes appear *near* to the central fixation point. Therefore, spatial symbols that convey only direction (e.g., *above*) or distance (e.g., *near*) cannot specify the unique location of any single box; rather, this can be accomplished only by spatial symbols that convey both direction and distance (e.g., *above* and *near*). However, despite this necessity, the spatial symbols used in the vast majority of previous cuing studies conducted over the past 30 years have only conveyed information about direction, not about distance. Thus, there is a potential discrepancy between what observers are thought to be doing in these studies (i.e., shifting attention to specific locations) and the nature of the symbolic information they have been provided to do so.

One potentially important factor concerns the fact that targets typically appear at a single, fixed distance in previous symbolic cuing studies that have used directional symbols to investigate the nature of symbolic control. For instance, this could be accomplished in figure 4.2 by eliminating the two near boxes or the two far boxes

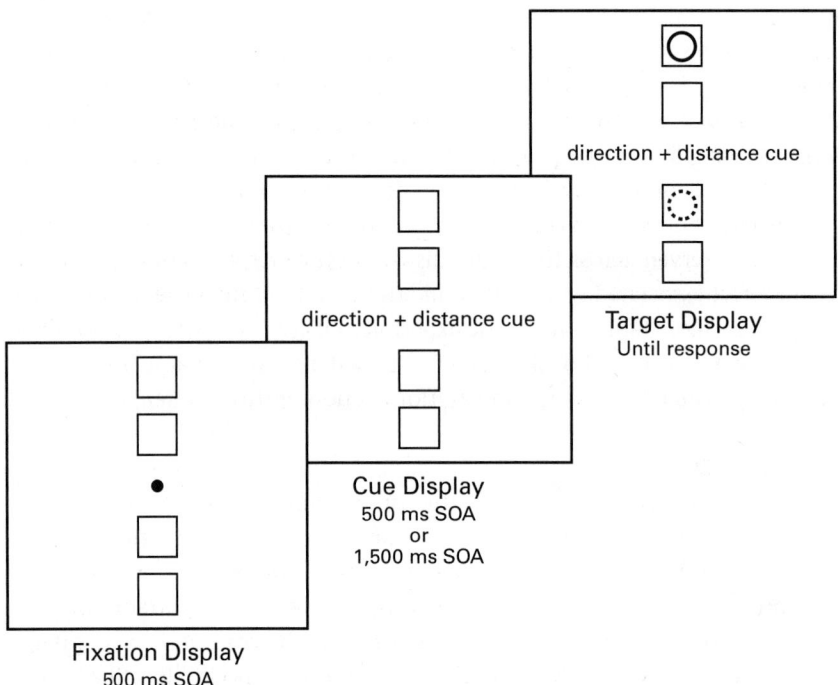

Figure 4.2

Illustration of a typical display sequence shown in Gibson and Sztybel's (2014) study. The term "direction + distance cue" was not actually shown but is used as an abstract marker for the various cues that were shown in their experiments. The solid circle appeared red and the dotted circle appeared green in the actual experiments, and the displays are not drawn to scale. SOA, stimulus-onset asynchrony. Reprinted with permission from Gibson, B. S., & Sztybel, P. (2014). Guiding attention to specific locations by combining symbolic information about direction and distance: Are human observers direction experts? *Journal of Experimental Psychology: Human Perception and Performance, 40,* 731–751. Copyright © 2014 by the American Psychological Association.

so that there is only one box on each side of the central fixation point. Notice that in this situation, the boxes do more than simply define the two potential target locations as they now provide 100%-valid spatial information about the distance of the target.

Using a slightly different setup, Gibson, Thompson, Davis, and Biggs (2011) have recently shown that observers can readily learn the distance of the target when it is most likely to appear at either a *near, middle,* or *far* distance, and observers can use this knowledge to supplement the partial spatial information that is conveyed by the directional cue to guide attention. Gibson et al. (2011) referred to this learned spatial information about distance as "extrasymbolic" knowledge in order to distinguish it from the symbolic spatial information about direction that was conveyed by the cue.

Of greatest importance, they showed that observers could shift their attention to specific locations by combining symbolic information about direction (derived from their comprehension of a symbolic direction cue) with extrasymbolic information about distance (derived from their knowledge of the distance of the target).

The findings reported by Gibson et al. (2011) are important because they reveal a lack of theoretical clarity in previous studies of symbolic control. In particular, these findings demonstrate that observers do shift their attention to specific locations, but they do so by combining both symbolic and extrasymbolic information. Hence, the spatial orienting effects that have been observed in previous symbolic cuing studies do not all derive from symbolic processing mechanisms and cannot all be explained by theories of symbolic control.

Furthermore, regardless of whether the information about distance is conveyed symbolically or extrasymbolically, previous symbolic cuing studies have typically shown only that observers can maintain a constant setting for the distance parameter across the entire duration of the study. However, this is a rather limited context, and the extent to which observers can specify the distance parameter in more realistic contexts in which it varies on a trial-by-trial basis remains unknown. Thus, there is a critical need to examine the extent to which the specification of the direction parameter and the distance parameter can both be brought under symbolic control by including symbolic cues that convey information about both direction and distance, under conditions in which this information varies on a trial-by-trial basis.

In addition, although observers may routinely combine symbolic information about direction and distance when both spatial dimensions are available symbolically, it is still possible that observers might be more proficient at using one spatial dimension than the other. For instance, the predominant use of directional symbols in the spatial cuing literature may not simply represent a neglected topic but rather may be indicative of a more fundamental spatial bias in which symbolic information about direction is preferred, and therefore used more frequently than symbolic information about distance, to control the orientation of spatial attention in the outside world. Indeed, the predominant use of directional symbols in the spatial cuing literature over the past 30 years appears to reflect the unanimous scientific practices of attention researchers, and unanimous scientific practices that go unquestioned often reflect the intrinsic cognitive biases of the individuals who engage in those practices (Kuhn, 1970; see also Gibson, 1984). As a result of this preference, observers may become experts at using direction symbols to orient their attention, but only novices at using distance symbols. Evidence in support of this direction bias would therefore be important because it would explain why the distance parameter has been excluded from the scientific literature on symbolic control. However, it is also possible that symbolic information about direction is on an equal footing with symbolic information about distance. This finding would also be significant because it would highlight the importance of including both spatial dimensions in the symbolic cuing literature.

We have recently attempted to shed light on these issues by presenting symbolic cues that conveyed information about both direction (*above* vs. *below*) and distance (*near* vs. *far*) within the context of the spatial cuing paradigm shown in figure 4.2 (Gibson & Sztybel, 2014). Across four experiments we used a variety of different direction + distance cues to refer to the four specific locations shown in figure 4.2: *above/far, above/ near, below/far,* and *below/near.* For instance, in one experiment, eye gaze cues that varied both the direction and amplitude (distance) of the pupils were used to refer to these four specific locations. In other experiments, upper- and lowercase spatial words in which the word conveyed one spatial dimension in a familiar fashion and the letter case conveyed the other spatial dimension in an unfamiliar fashion were used to refer to these four specific locations. Finally, signed numbers such as "+3," "+1," "–3," and "–1" were used to refer to these four specific locations.

In our view, the primary challenge of using direction + distance cues was to isolate the effect of one spatial dimension while holding the effect of the other spatial dimension constant. We held the effect of one spatial dimension constant by making this dimension mandatory for performance (Davis & Gibson, 2012). In other words, the primary task was designed so that observers could not reliably identify the correct target without the aid of the mandatory, 100%-valid spatial dimension (as in the spatial cuing paradigm shown in figure 4.1). Although the mandatory, 100%-valid spatial dimension was sufficient to identify the target, it only provided partial spatial information about the specific location of the target. Complete spatial information about specific locations could be obtained by combining the setting specified by the 100%-valid spatial dimension with the setting specified by the other spatial dimension, but this latter information was valid only 75% of the time, resulting in both "valid" and "invalid" trials. Because the use of the 75%-valid spatial dimension was not strictly necessary to perform the task, an observer's willingness to use this information can be considered voluntary (Davis & Gibson, 2012).

In addition, we also manipulated which spatial dimension was mandatory (100% valid) and which spatial dimension was nonmandatory (75% valid) to examine whether there were asymmetries in the voluntary use of symbolic information about direction and distance to orient attention. This manipulation resulted in two cue conditions: one cue condition in which symbolic information about distance was 100% valid and symbolic information about direction was 75% valid—which we referred to as the "nonmandatory direction condition"—and another cue condition in which symbolic information about direction was 100% valid and symbolic information about distance was 75% valid—which we referred to as the "nonmandatory distance condition."

If there is a preference to use direction symbols to orient attention in the outside world, then symbolic information about direction should have a larger effect on performance (response times and percent error rates) than symbolic information about distance. Although this preference may be reflected in a variety of ways, we focused on the "costs plus benefits" of orienting attention (Posner, 1980; Posner, Snyder, & Davidson, 1980; Jonides, 1981; Jonides & Mack, 1984), as measured by the magnitude of the spatial cuing effect (Invalid condition—Valid condition) observed when each spatial dimension was nonmandatory (75% valid). For instance, if there is a bias for direction, then observers may not prioritize symbolic information about distance when this information is not mandatory, and the additional spatial information it provides about the specific location of the target may not be deemed useful. As a result, the spatial cuing effect should be relatively small in the nonmandatory distance condition. In contrast, if there is a bias for direction, observers should consistently use symbolic information about direction even when this information is not mandatory. As a result, the spatial cuing effect should be relatively large in the nonmandatory direction condition.

The main results obtained in each of our four experiments showed that observers routinely combined symbolic information about direction and distance in that significant spatial cuing effects were found in each of the different cue conditions regardless of which spatial dimension was nonmandatory. These findings are important because they suggest that observers do not ignore nonmandatory information about distance, and they provide the first evidence that current semantic-based theories of symbolic control can be extended beyond the orientation and direction parameters to include the distance parameter as well. As such, the findings reported by Gibson and Sztybel (2014) suggest that spatial symbols can provide symbolic control over the guidance of attention to specific locations, even when the symbolic information about distance is nonmandatory and varies on a trial-by-trial basis. Perhaps more importantly, these findings also suggested that the spatial cuing effects were larger when the direction symbol was nonmandatory than when the distance symbol was nonmandatory, at least under some conditions. We will discuss the nature of this direction bias in greater detail in the Integration section of this chapter.

Integration

In this section, we attempt to reveal the statistical regularities and biases in the outside world that led to the cued endpoint effect and direction bias.

What the Cued Endpoint Effect Reveals about the Control of Attention in Everyday Life

The cued endpoint effect has been interpreted to suggest that attention is distributed more broadly across space when "left/right" word cues are shown relative to when "above/below" word cues are shown because of learned differences in *cue validity*—the extent to which a spatial cue provides consistent and accurate information about the location of a target object. Furthermore, these learned differences in cue validity are thought to arise because spatial terms can be defined with respect to a variety of different frames of reference, not all of which are compatible. For instance, the spatial perspective of a speaker is often different from the spatial perspective of a listener in everyday life, which tends to have a greater impact on spatial terms such as "left" and "right." In the extreme, the spatial term "left" may actually mean *right* (and vice versa) when the word is uttered from the speaker's egocentric perspective and interpreted from the listener's egocentric perspective and the two are offset by 180° (Schober, 1993, 1995). In contrast, no such ambiguity arises for the spatial terms "above" and "below" in this context so long as the two individuals remain in their upright, canonical orientations. The consequence of this learned difference in consistency is that spatial terms such as "above" and "below" function like more valid spatial cues than spatial terms such as "left" and "right" even though both may be 100% valid in the laboratory. Indeed, it is well-known that experimentally manipulated differences in cue validity can influence the focus of attention, with less valid cues' leading to more broadly distributed attention (Johnson & Yantis, 1995).

In contrast, leftward- and rightward-pointing arrows may specify direction in a more consistent fashion along the horizontal axis than spatial terms such as "left" and "right," despite changes in perspective, by trading *spatial* ambiguity for *perceptual* ambiguity. For instance, consider a situation in which two individuals who are offset by 180° are looking at a horizontally oriented arrow. In this situation, the arrow can appear to be pointing in the leftward direction from one perspective at the same time that it can appear to be pointing in the opposite, rightward direction from the other perspective. In other words, the arrow remains equally valid for both individuals because it can simultaneously convey a location on the *left* to one individual and a location on the *right* to the other individual. Consequently, leftward- and rightward-pointing arrows function like more valid spatial cues than spatial terms such as "left" and "right."

The conclusion that the distribution of visual attention can be influenced by the statistical structure of spatial semantics is important because it suggests that previously learned linguistic meanings can exert powerful control over visual attention. However, the persistent occurrence of this cued endpoint effect is also somewhat surprising given that semantic ambiguities are routinely encountered and resolved, if only momentarily, during the comprehension of everyday discourse (see Spivey, 2006, for a review). For instance, the precise meaning of the word "bank" is momentarily ambiguous in the sentence "Bill went to the bank to deposit his money" until we learn what he went there to do. One reason the preestablished semantic ambiguities associated with "left" and "right" may not have been fully resolved in the spatial cuing experiments reported by Gibson et al. (2009) is because the distribution of visual attention does not demand disambiguation. In other words, spatially directed shifts of visual attention may tolerate directional ambiguity because the distribution of attention can itself be spatially ambiguous in the sense that it can be divided across multiple locations simultaneously.

In contrast, other spatial-directed movements such as manual reaching are more discrete and cannot be divided in quite the same manner: The arm cannot be directed to multiple locations simultaneously. Consequently, spatially directed reaching responses may require stronger disambiguation of the endpoints of the horizontal axis than spatially directed shifts of visual attention before they are generated in response to "left/right" cues. Hence, the precise meaning of "left" and "right" in any given context may depend not only on the computation of axes and endpoints but also on whether these terms are "grounded" in the attention system or in the motor system (Barsalou, 2008).

Gibson and Davis (2011) investigated the possibility that the dynamics of semantic disambiguation may interact with the type of spatially directed movement within the context of a repetition priming paradigm. In this paradigm, prime and probe displays were paired on each trial, and two spatially directed movements were performed in rapid succession. The first movement (if required) was generated during the prime phase of the trial in response to spatial word cues. Two different types of spatially directed movements served as prime responses in this study: attentional orienting and manual reaching. Regardless of the spatially directed movement generated during the prime phase, the second movement always involved a shift of visual attention that was generated during the probe phase of the trial. By design, this second movement was always directed by the same spatial cue that appeared during the prime phase of the trial, though it was always initiated relative to a different reference object than the prime.

Consistent with expectations, the results showed that the magnitude of the cued endpoint effect was significantly reduced when it was preceded by a spatially directed reaching response relative to when it was preceded by a spatially directed shift of visual attention. Thus, these findings extend our conclusion that spatial semantics can influence the spatial distribution of attention by demonstrating that different types of spatially directed movements can also influence the semantics of space.

What the Direction Bias Reveals about the Control of Attention in Everyday Life

Although the precise explanation of the observed direction bias is still unknown, we tentatively interpreted our findings to suggest that human observers are experts at using direction symbols to orient their attention, but only novices at using distance symbols, presumably because they have had greater experience shifting their attention in response to direction symbols than in response to distance symbols in everyday life (Gibson & Sztybel, 2014). Initial evidence for this direction expertise hypothesis was obtained using eye gaze cues (Gibson & Sztybel, 2014; see Experiment 1), but we also attempted to test this hypothesis in subsequent experiments by distinguishing between the familiarity of a symbol and its purported use as an attention cue in the outside world. According to this distinction, spatial symbols can either be commonly available for communication or not in everyday life, and they can either be commonly used to orient attention or not in everyday life. By hypothesis, human observers are thought to be experts at using familiar direction symbols because these symbols are both commonly available for communication and are commonly used to orient attention in the outside world. Conversely, human observers are thought to be novices at using familiar distance symbols because although these symbols are commonly available for communication, they are not commonly used to orient attention in the outside world. As a result, the spatial cuing effect associated with familiar direction symbols should be larger than the spatial cuing effect associated with familiar distance symbols, a finding that was purportedly consistent with the results obtained when eye gaze cues were shown (Gibson & Sztybel, 2014; see Experiment 1).

Several more novel predictions arise when one considers the effects of both familiar and unfamiliar (or novel) symbols on spatial orienting effects. For instance, a novel direction symbol could be created that expresses a spatial dimension that is commonly used to orient attention but that is not commonly available for communication. In this situation, it is reasonable to expect that any orienting expertise that is acquired in the outside world would be specific to familiar spatial symbols and should not generalize to novel spatial symbols without additional training. As a result, the spatial cuing effect associated with novel direction symbols should be smaller than the spatial cuing effect associated with familiar direction symbols.

Similarly, a novel distance symbol could also be created that expresses a spatial dimension that is neither commonly used to orient attention nor commonly used for communication. In this situation, it is reasonable to expect that the spatial cuing effect associated with novel distance symbols should be equal to the spatial cuing effect associated with familiar distance symbols because all distance symbols should function like novel spatial symbols without proper training, at least when it comes to orienting attention. Likewise, the spatial cuing effect associated with familiar distance symbols should also be equal to the spatial cuing effect associated with novel direction symbols because both are functionally novel in this context.

In order to test these predictions, we used direction + distance cues in which either the symbolic information about direction was more familiar than the symbolic information about distance or the symbolic information about distance was more familiar than the symbolic information about direction (Gibson & Sztybel, 2014; see Experiments 2a and 2b). Spatial word cues were used instead of eye gaze cues in these experiments because the familiarity of these cues could be objectively assessed using established databases that can estimate the frequency of occurrence in the outside world. In addition, other physical features of these words could also be used in a novel way to convey the other spatial dimension.

More specifically, in one experiment, symbolic information about direction was conveyed in a familiar fashion by using the spatial words "above" and "below" whereas symbolic information about distance was conveyed in a novel fashion by arbitrarily using uppercase letters to convey *far* and lowercase letters to convey *near*. For instance, the cue "ABOVE" was used to refer to the *above/far* location whereas the cue "above" was used to refer to the *above/near* location. Conversely, in another experiment, symbolic information about distance was conveyed in a familiar fashion by using the spatial words "near" and "far" whereas symbolic information about direction was conveyed in a novel fashion by arbitrarily using uppercase letters to convey *above* and lowercase letters to convey *below*. For instance, the cue "NEAR" was used to refer to the *near/above* location whereas the cue "near" was used to refer to the *near/below* location.

If the direction bias reflects a difference in orienting expertise, the magnitude of the spatial cuing effects elicited by direction symbols should depend on the familiarity of those individual symbols whereas the magnitude of the spatial cuing effects elicited by distance symbols should not depend on the familiarity of those individual symbols. Specifically, the spatial cuing effect observed in the nonmandatory direction condition should be significantly larger than the spatial cuing effect observed in the nonmandatory distance condition when the direction symbol is more familiar than the distance symbol. However, the spatial cuing effect observed in the nonmandatory direction condition may not differ from the spatial cuing effect observed in the nonmandatory distance condition when the distance symbol is more familiar than the direction symbol.

The main results of these two experiments were consistent with the direction expertise hypothesis (Gibson & Sztybel, 2014; see Experiments 2a and 2b). In particular, the spatial cuing effect observed in the nonmandatory direction condition was significantly larger than the spatial cuing effect observed in the nonmandatory distance condition when the direction symbol was more familiar than the distance symbol. In contrast, the spatial cuing effect observed in the nonmandatory distance condition was equal to the spatial cuing effect observed in the nonmandatory direction condition when the distance symbol was more familiar than the direction symbol. In this latter situation, the findings showed that the higher familiarity of the distance symbols could offset, but not reverse, the asymmetrical influence of the more novel direction symbols. Moreover, when the effects of familiarity were compared within the same spatial dimension, the results suggested that the familiarity of the individual symbols altered the spatial cuing effect associated with the direction symbols, but not the spatial cuing effect associated with the distance symbols. These findings were therefore interpreted to suggest that the direction bias derives from observers' greater experience using familiar direction symbols to orient their attention. In contrast, observers generally do not have the same experience using familiar distance symbols to orient their attention. As a result, all distance symbols function like unfamiliar symbols within the context of symbolic cuing experiments.

The hypothesis under consideration is that human observers are experts at using familiar direction symbols to orient their attention, but only when these symbols are both commonly available for communication and commonly used to orient attention in the outside world. Based on the evidence reported thus far, it may be reasonable to assume that if the direction symbol is familiar, then it also has been commonly used to orient attention in the outside world. However, this assumption may be incorrect because there appear to be direction symbols that are commonly used in some contexts, yet they are not commonly used to orient attention in the outside world.

For instance, positively and negatively signed numbers are commonly used in algebra to define the specific location of points within a Cartesian coordinate axis system, with the sign conveying the direction of the point along an axis and the number conveying the distance of the point along an axis. However, although these spatial symbols may be commonly used in algebra, they do not seem to be commonly used to orient attention in the outside world. For instance, we do not typically say that "the keys are hanging *plus* the coats" to communicate where another should look for the keys (even assuming a context in which the vertical axis might be prespecified). Rather, we typically say that "the keys are hanging *above* the coats" to communicate this spatial information. Consequently, if the present analysis is correct, then although observers may routinely combine these two spatial dimensions when signed numbers such as "+3," "+1," "−3," and "−1" are used to convey information about the direction and distance of a target, they should not exhibit any directional expertise. The main results were consistent with this expectation by showing that the spatial cuing effect observed in the nonmandatory direction condition was equal to (if not smaller than) the spatial cuing effect observed in the nonmandatory distance condition (Gibson & Sztybel, 2014; see Experiment 3).

Future Directions

The cuing effects reviewed in this chapter are thought to reflect semantic constraints that are established in everyday life. However, because these different cuing effects were not explicitly manipulated within the context of the experiment, there may be lingering doubt that some other, uncontrolled, difference between symbolic cues may have caused the observed functional differences. For this reason, future studies should continue to examine alternative explanations and/or attempt to simulate the effects of the outside world inside the lab. For instance, future studies of the cued endpoint effect could provide some reassurance that the inefficient selection observed in response to "left/right" cues is not a simple artifact by training inconsistent mappings between number word cues and cued directions. Of particular interest will be whether such ambiguity can be induced simply by stipulating different mappings (e.g., a novel symbol means *left* for the first 100 trials, but it means *right* for the next 100 trials) or whether such ambiguity depends on the presence of opposing frames of reference.

In addition, future studies should also clarify the nature of the observed direction bias. For instance, the findings reported by Gibson and Sztybel (2014) were interpreted to support the direction expertise hypothesis by suggesting that observers usually depended more on symbolic information about direction than on symbolic information about distance to guide their attention to specific spatial locations. However, one limitation of the evidence reported thus far is that the spatial cuing effects that led to this interpretation reflected only the combined costs plus benefits of orienting attention to a specific location. Thus, it is unclear whether the asymmetrical influence of direction symbols on the control of attention to specific locations reflects larger benefits, larger costs, or a combination of both.

This limitation is potentially important because the direction expertise hypothesis contends that direction symbols should at least produce larger benefits than distance symbols. This is because observers should be more likely to combine the nonmandatory symbolic information about direction with the mandatory symbolic information about distance (in the nonmandatory direction condition) than they are to combine the nonmandatory symbolic information about distance with the mandatory symbolic information about direction (in the nonmandatory distance condition). Hence, if the direction expertise hypothesis is correct, then observers should be able to discriminate the color of the target significantly faster in the nonmandatory direction condition than in the nonmandatory distance condition when the target actually appears at the validly cued location. However, if this result is not obtained, then alternative explanations of the direction bias will be required.

Conclusion

In conclusion, the evidence reviewed in this chapter was interpreted to support a semantic-based theory of symbolic attention control. According to this theory, linguistic and iconic spatial symbols are comprehended by specifying one or more of the orientation, direction, and distance parameters of a spatial frame of reference, which in turn can influence the manner in which attention is distributed in space. Contrary to early theories which did not draw any functional distinctions between different spatial symbols, the evidence reviewed in this article suggested that linguistic and iconic spatial symbols can in fact be distinguished based on how the orientation and direction parameters are set. These findings have suggested that there is a cost associated with using spatial linguistic symbols to control attention, relative to other iconic symbols, especially along the horizontal axis—a cost that includes not only the initiation of spatial shifts of attention but also the extent to which attention can be selectively focused at a location. Perhaps more importantly, this theory has also revealed that the distance parameter has been inadvertently ignored by researchers over the past 30 years, perhaps because distance symbols are not commonly used to orient attention in the outside world. Consistent with this notion, observers demonstrated greater proficiency orienting their attention in response to direction symbols than distance symbols, at least under some conditions. In so doing, this theory has begun to extend the boundaries of our understanding of symbolic attention control.

Box 4.1
Key Points

- Symbols convey spatial information by specifying one or more of the various parameters of a spatial frame of reference that define orientation, direction, and distance relative to the observer (or the environment or an object).

- Using this framework, we have shown that spatial word cues specify the orientation and direction parameters differently than arrow and eye gaze cues, especially along the horizontal axis (the cued axis and cued endpoint effects, respectively). Thus, spatial word cues control the spatial distribution of attention differently than arrow and eye gaze cues. Spatial word cues that refer to locations along the horizontal axis are construed as functioning like less valid spatial cues than arrow and eye gaze cues because of statistical patterns that develop between individuals with potentially opposing frames of reference in the outside world.

- Furthermore, using this framework, we have shown that symbolic direction cues generate larger spatial cuing effects than symbolic distance cues (the direction bias effect). This finding is interpreted to reflect greater expertise using direction symbols to control attention than using distance symbols, presumably because individuals prefer to use direction symbols to control attention in the outside world.

Box 4.2
Outstanding Issues

- The different cuing effects reviewed in this chapter are thought to be based on observers' prior experiences with spatial symbols outside the lab. Because these experiences were not explicitly manipulated within the context of the experiment, there may be lingering doubt that some other, uncontrolled, difference between symbolic cues may have caused the observed functional differences.

- Future studies of the cued endpoint effect could provide some reassurance that the inefficient selection observed in response to "left/right" cues is not a simple artifact by training inconsistent mappings between number word cues and cued directions. Of particular interest will be whether such ambiguity can be induced simply by stipulating different mappings (e.g., a novel symbol means *left* for the first 100 trials, but it means *right* for the next 100 trials) or whether such ambiguity depends on the presence of opposing frames of reference.

- Future studies of the direction bias should clarify whether the asymmetrical influence of direction symbols on the control of attention to specific locations reflects larger benefits, larger costs, or a combination of both. The direction expertise account of this bias specifically predicts greater benefits associated with direction symbols than distance symbols.

References

Barsalou, L. W. (2008). Grounded cognition. *Annual Review of Psychology, 59*, 617–645.

Corbetta, M., Kincade, J., Ollinger, J., McAvoy, M. P., & Shulman, M. M. (2000). Voluntary orienting is dissociated from target detection in human posterior parietal cortex. *Nature Neuroscience, 3*, 292–297.

Davis, G. J., & Gibson, B. S. (2012). Going rogue in the spatial cuing paradigm: High spatial validity is insufficient to elicit voluntary shifts of attention. *Journal of Experimental Psychology: Human Perception and Performance, 38*, 1192–1201.

Eriksen, C. W., & Hoffman, J. E. (1972). Temporal and spatial characteristics of selective encoding from visual displays. *Perception & Psychophysics, 12*, 201–204.

Gibson, B. S. (1984). The convergence of Kuhn and cognitive psychology. *New Ideas in Psychology, 2*, 211–221.

Gibson, B. S., & Bryant, T. (2005). Variation in cue duration reveals top-down modulation of involuntary orienting to uninformative symbolic cues. *Perception & Psychophysics, 67*, 749–758.

Gibson, B. S., & Davis, G. J. (2011). Grounding spatial language in the motor system: Reciprocal interactions between conceptual control and spatial orienting. *Visual Cognition, 19*, 79–116.

Gibson, B. S., & Kingstone, A. (2006). Visual attention and the semantics of space: Beyond central and peripheral cues. *Psychological Science, 17*, 622–627.

Gibson, B. S., Scheutz, M., & Davis, G. J. (2009). Symbolic control of visual attention: Semantic constraints on the spatial distribution of attention. *Attention, Perception & Psychophysics, 71*, 363–374.

Gibson, B. S., & Sztybel, P. (2014). Guiding attention to specific locations by combining symbolic information about direction and distance: Are human observers direction experts? *Journal of Experimental Psychology: Human Perception and Performance, 40*, 731–751.

Gibson, B. S., Thompson, A. N., Davis, G. J., & Biggs, A. T. (2011). Going the distance: Extra-symbolic contributions to the symbolic control of spatial attention. *Visual Cognition*, *19*, 1237–1261.

Goldberg, M., Maurer, D., & Lewis, T. (2001). Developmental changes in attention: The effects of endogenous cueing and of distractors. *Developmental Science*, *4*, 209–219.

Hommel, B., Pratt, J., Colzato, L., & Godijn, R. (2001). Symbolic control of visual attention. *Psychological Science*, *12*, 360–365.

Hopfinger, J., Buonocore, M., & Mangun, G. (2000). The neural mechanisms of top-down attentional control. *Nature Neuroscience*, *3*, 284–291.

Johnson, D. N., & Yantis, S. (1995). Allocating visual attention: Tests of a two-process model. *Journal of Experimental Psychology: Human Perception and Performance*, *21*, 1376–1390.

Jonides, J. (1981). Voluntary versus automatic control over the mind's eye movement. In J. B. Long & A. D. Baddeley (Eds.), *Attention and performance IX* (pp. 187–203). Hillsdale, NJ: Erlbaum.

Jonides, J., & Mack, R. (1984). On the cost and benefit of cost and benefit. *Psychological Bulletin*, *96*, 29–44.

Kemmerer, D. (2006). The semantics of space: Integrating linguistic typology and cognitive neuroscience. *Neuropsychologia*, *44*, 1607–1621.

Klein, R. M., & Hanson, E. (1990). Chronometric analysis of spotlight failure in endogenous visual orienting. *Journal of Experimental Psychology: Human Perception and Performance*, *16*, 790–801.

Kuhn, T. S. (1970). *The structure of scientific revolutions*. Chicago: University of Chicago Press.

Landau, B., & Jackendoff, R. (1993). What and where in spatial language and spatial cognition. *Behavioral and Brain Sciences*, *16*, 217–265.

Levinson, S. C. (2003). *Space in language and cognition*. Cambridge, UK: Cambridge University Press.

Logan, G. D. (1995). Linguistic and conceptual control of visual spatial attention. *Cognitive Psychology*, *28*, 103–174.

Logan, G. D., & Sadler, D. D. (1996). A computational analysis of the apprehension of spatial relations. In P. Bloom, M. A. Peterson, L. Nadel, & M. Garrett (Eds.), *Language and space* (pp. 493–529). Cambridge, MA: MIT Press.

Mayer, A. R., & Kosson, D. S. (2004). The effects of auditory and visual linguistic distractors on target localization. *Neuropsychology*, *18*, 248–257.

Nobre, A., Sebestyen, G., & Miniussi, C. (1995). The dynamics of shifting visuospatial attention revealed by event-related potentials. *Neuropsychologia*, *38*, 964–974.

Posner, M. I. (1980). Orienting of attention. *Quarterly Journal of Experimental Psychology*, *32*, 3–25.

Posner, M. I., & Petersen, S. E. (1990). The attention system of the human brain. *Annual Review of Neuroscience, 13*, 25–42.

Posner, M. I., Snyder, C. R., & Davidson, B. J. (1980). Attention and the detection of signals. *Journal of Experimental Psychology. General, 109*, 160–174.

Ristic, J., & Kingstone, A. (2006). Attention to arrows: Pointing to a new direction. *Quarterly Journal of Experimental Psychology, 59*, 1921–1930.

Schober, M. F. (1993). Spatial perspective-taking in conversation. *Cognition, 47*, 1–24.

Schober, M. F. (1995). Speakers, addressees, and frames of reference: Whose effort is minimized in conversations about locations? *Discourse Processes, 20*, 219–247.

Spivey, M. J. (2006). *The continuity of mind*. Oxford, UK: Oxford University Press.

Tipples, J. (2002). Eye gaze is not unique: Automatic orienting in response to uninformative arrows. *Psychonomic Bulletin & Review, 9*, 314–318.

Yantis, S. (1996). Attentional capture in vision. In A. F. Kramer, M. G. H. Coles, & G. D. Logan (Eds.), *Converging operations in the study of visual selective attention* (pp. 45–76). Washington, DC: American Psychological Association.

5
Auditory Selective Attention

Barbara Shinn-Cunningham and Virginia Best

Historical Context

Some of the earliest work examining human selective attention considered auditory communication signals, focusing on the issue of how we choose what to listen to in a mixture of competing speech sounds (commonly referred to as the "cocktail party problem"; Cherry, 1953). The earliest selective attention models were bottleneck theories, which propose that a filtering step restricts how much sensory information is passed on to a limited-capacity processing system. Broadbent initially proposed a strict filter that rejects unwanted information based on physical parameters such as location, pitch, loudness, and timbre (Broadbent, 1958). In Broadbent's theory, listeners can access only very basic properties about a rejected source and cannot extract detailed information about its content. While Broadbent argued that listeners can direct volitional attention to physical sound features to determine what information is filtered out, he also recognized that certain stimuli may capture attention involuntarily, overriding listener goals; that is, what is attended and what is rejected depends on both volitional attention and inherent stimulus salience.

While many of the basic tenets of Broadbent's filter theory guide modern thinking, subsequent observations prompted modifications of the theory. For example, some listeners notice their own name when it occurs in an unattended channel, suggesting that some information from filtered-out signals "gets through" (e.g., see Moray, 1959; Wood & Cowan, 1995; Colflesh & Conway, 2007). Similarly, even when listeners are not consciously aware of the content of an unattended stimulus, they can show priming effects where the ignored stimuli influence perception of subsequent items, demonstrating that some meaning from to-be-ignored streams is extracted (e.g., see Gray & Wedderburn, 1960; Treisman, 1960; Corteen & Wood, 1972; Mackay, 1973; however, see also Lachter et al., 2004). To account for such effects, Treisman (1960) proposed that the filter attenuates unwanted stimuli rather than filtering them out completely.

In the late 1960s and early 1970s, many seminal studies of selective attention explored auditory perception. Following this early work, though, visual studies overshadowed work in the auditory domain. Indeed, Treisman herself developed the influential feature integration theory (FIT) to describe attention in the visual domain (Treisman & Gelade, 1980). FIT argued that features of different objects in a visual scene are processed in parallel, but only those features that are at an attended location are bound together (integrated) and processed in detail as one object. Work testing such ideas about visual attention flourished in the 1970s and 1980s, while most auditory behavioral studies focused on how information is coded in the auditory periphery, with little consideration of limits related to attention. While some key electroencephalography studies showed that selective attention modulates the strength of early cortical responses to sound (e.g., see Picton & Hillyard, 1974; Hansen & Hillyard, 1980), these results did not strongly influence psychoacoustic studies at that time. A handful of studies published in the 1990s used cuing to discriminate between endogenous and exogenous components of auditory attention, largely inspired by previous studies in the visual domain (e.g., Spence & Driver, 1994; Mondor & Zatorre, 1995; Quinlan & Bailey, 1995), but these again used relatively simple stimuli and tasks with low attentional demands.

When auditory researchers began once again to explore auditory perception where central bottlenecks rather than sensory limitations determined performance, the work was rarely related to modern theories of attention and memory. Instead, the term "informational masking" (IM) was coined (see Kidd et al., 2008). IM is a catchall phrase, encompassing any perceptual interference between sounds that was not explained by "energetic masking" (EM), where EM was defined as interference explained by masking within the auditory nerve (e.g., see Kidd et al., 1994; Oh & Lutfi, 1998; Freyman et al., 1999; Watson, 2005). That is, IM encompassed any suboptimal processing of information, including loss of information due to attentional filtering. Still, key components of modern theories of selective attention, both in visual and auditory science, trace back to early auditory work by Cherry, Broadbent, Treisman, and their contemporaries.

State-of-the-Art Literature Review

Object-Based Attention and Auditory Streams

Recent work on auditory attention draws heavily on key findings from vision science (e.g., see Spence et al., 2001; Cusack & Carlyon, 2003; Shinn-Cunningham, 2008; Snyder et al., 2012). In vision, attention is argued to operate as a "biased competition" between the neural representations of perceptual objects (e.g., see Desimone & Duncan, 1995; Kastner & Ungerleider, 2001). Like Broadbent's early theories, the biased-competition view argues the focus of attention is determined by the interplay between the salience of stimuli (exogenously guided attention) and observer goals (endogenously guided attention). However, biased competition arises specifically between objects, each of which is a collection of attributes. At any one time, one object is the focus of attention and is processed in greater detail than other objects in the scene. Recently, evidence for such effects in auditory processing has started to emerge from physiological studies (e.g., Chait et al., 2010; Maddox et al., 2012; Mesgarani & Chang, 2012; Middlebrooks & Bremen, 2013).

If selective auditory attention is object based, it is important to define what constitutes an auditory object. Yet it is challenging to come up with a clear, irrefutable definition. Seminal work by Al Bregman (1990) laid out some of the rules governing the perceptual organization of sound mixtures. Later researchers explored these rules to reveal the sound features that lead to object formation (see reviews by Carlyon, 2004; Griffiths & Warren, 2004). For instance, sounds tend to group together if they turn on and off together (i.e., are comodulated), are harmonically related, and are continuous in time and frequency. Such sound features are relatively "local" in time, operating at the timescale of, for example, a speech syllable. However, we perceive ongoing speech as one stream even though there are silent gaps across which "local" features cannot operate. Grouping across acoustic discontinuities is driven by higher order perceptual features, such as location, pitch, and timbre, as well as signal structure learned through experience (e.g., phonetic, semantic, and lexical structure).

The relationship between object formation and auditory selective attention remains a subject of debate. Some argue that objects form only when a stream (an auditory object extending through time) is attended (e.g., Jones, 1976; Alain & Woods, 1997; Cusack et al., 2004). However, other studies suggest that auditory streams form automatically and preattentively (e.g., Bregman, 1990; Macken et al., 2003; Sussman et al., 2007). Most likely, both automatic and attention-driven processes influence stream formation. In cases where low-level attributes are sufficiently distinct to define a stream unambiguously, it will be segregated from a sound mixture even without attention. However, sound mixtures are often ambiguous, in which case attention to a particular perceptual feature may help "pull out" the stream that is attended (e.g., see Alain et al., 2001; Macken et al., 2003). This view is supported by studies that show that perception of a complex auditory scene is refined through time (e.g., see Carlyon et al., 2003; Teki et al., 2013).

Our own work supports the idea that object formation and attention interact. Specifically, when listeners are instructed to attend to one sound feature (e.g., pitch or timbre) and report a stream of words, they make more errors when a task-irrelevant sound feature changes between words, breaking down the perceived continuity of the stream (Maddox & Shinn-Cunningham, 2012). Conversely, when the talker of a target word stream is consistent, continuity of talker identity enhances performance even when talker identity is task irrelevant, such as when listeners are trying to attend to words from a particular location (Best et al., 2008; Bressler et al., 2014). These results demonstrate that even irrelevant features influence selective attention performance, presumably by influencing object formation. On the other hand, listeners weight a given acoustic cue more when it is task relevant than when it is task irrelevant (Maddox & Shinn-Cunningham, 2012). Together, these results suggest that attention is object based, but that attention itself affects object formation (Fritz et al., 2007; Shinn-Cunningham, 2008; Shinn-Cunningham & Best, 2008; Shamma et al., 2011).

Although the idea of object-based attention came from vision, there is relatively little discussion of the relationship between visual object formation and visual attention. We speculate that this is because auditory objects emerge only through time. Consider that a static two-dimensional picture of a natural scene generally contains enough information for visual objects to emerge without any further information. In contrast, auditory information is typically conveyed by changes in sounds as a function of time. Similarly, only by analyzing sound's time-frequency content can the features and structure that drive auditory stream formation be extracted. "Local" grouping features emerge over 10s of milliseconds, but higher order features can require on the order of seconds to be perceived (e.g., see Cusack et al., 2004; Chait et al., 2010). Object formation can take time (e.g., see Cusack et al., 2004) and can be unstable (e.g., see Hupe et al., 2008). Thus, in natural auditory scenes, it is nearly impossible to discuss selective attention without simultaneously considering object formation. Current theories of auditory object formation and attention deal directly with the fact that auditory objects emerge through time, and that this process may both influence and be influenced by attention (Elhilali et al., 2009; Shamma et al., 2011).

Dimensions of Auditory Selective Attention

Although the "unit" of auditory attention seems to be an auditory object, listeners can direct top-down selective attention by focusing on different acoustic dimensions, many of which also influence object and stream formation (e.g., frequency, location, pitch, timbre, etc.). Other dimensions are even more basic; perhaps the most fundamental feature, given the tonotopic arrangement of the auditory system, is frequency. Some of the earliest experiments exploring auditory selective attention used the probe-signal method to demonstrate that listeners can focus attention on a certain frequency region, which enhances detection of a quiet tone at or near that frequency (e.g., Greenberg & Larkin, 1968; Scharf et al., 1987). Perceived location is another powerful cue for directing selective auditory attention, improving the ability to extract information about a target stream (e.g., see Arbogast & Kidd, 2000; Kidd et al., 2005). There are also examples demonstrating that attention can be directed to pitch (Maddox & Shinn-Cunningham, 2012), level (e.g., attending to the softer of two voices; Brungart, 2001; Kitterick et al., 2013), and talker characteristics such as timbre and gender (e.g., Culling et al., 2003; Darwin et al., 2003). Auditory attention can also be focused in time, such that sounds occurring at expected times are better detected than those occurring at unpredictable times (Wright & Fitzgerald, 2004). This idea has been elaborated to describe attention that is distributed in time either to enhance sensitivity to target sequences ("rhythmic attention"; e.g., Jones et al., 1981) or to cancel irrelevant sounds (Devergie et al., 2010).

Evidence from the neuroimaging literature suggests that selective spatial auditory attention engages some of the same frontoparietal circuitry that controls spatial visual attention (e.g., see Tark & Curtis, 2013; Lee et al., 2014). These brain regions engaged during auditory selective spatial attention are also engaged to some degree during non-spatial attention. However, there are also differences in activity, depending on exactly what acoustic feature guides attentional focus (e.g., see Hill & Miller, 2010; Lee et al., 2013). Still, there is no definitive list of which sound attributes or statistics can be used to focus auditory attention, or exactly which cortical regions are engaged by attention to specific acoustic features.

Failures of Selective Attention

There are several contextual factors that interfere with selective attention. First, the presence of acoustically similar distractors interferes with a variety of tasks that depend on selective attention (e.g., see the discussion in Durlach et al., 2003; Kidd et al., 2008). Second, uncertainty about what sound is a target and what is a distractor can lead to suboptimal selective attention (e.g., because the target or distractor is random and unpredictable, it can be difficult to filter out the distractor). Failures of attention due to these factors ("similarity" and "uncertainty") have been intensively studied and are discussed as different forms of IM (Durlach et al., 2003; Kidd et al., 2008).

For example, detection and discrimination of a target tone is much more difficult in the presence of simultaneous tones that are remote in frequency (Neff & Green, 1987) and/or tones that may overlap with the target in frequency but not in time (Watson et al., 1975, 1976), even though such interfering tones do not alter the target's representation at the level of the auditory nerve. In addition, performance suffers if uncertainty is increased by roving the characteristics of the target or distractors from trial to trial (e.g., Green, 1961; Spiegel et al., 1981; Neff & Dethlefs, 1995; see the review in Kidd et al., 2008). In the case of spatial attention, it seems intuitive that attending to one location might fail when competing sounds are added at nearby locations (akin to visual "crowding"), but few past experiments have addressed this issue directly. Spatial uncertainty (i.e., not knowing where a target stream will come from) disrupts selective attention; for example, the ability to focus on and analyze a stream of speech is poorer when its spatial location changes compared to when its location is fixed and known (Kidd et al., 2005; Brungart & Simpson, 2007; Best et al., 2008). In addition, listeners make fewer errors when attending to a fixed and known voice than when the target voice characteristics change regularly. In such cases, performance can be improved by providing cues to prime the listener to focus on target features such as frequency, spatial location, or voice (Darwin et al., 2003; Richards & Neff, 2004; Kidd et al., 2005; Brungart & Simpson, 2007).

The failures of attention described above often show up in performance measures as target–masker confusions or substitutions, indicating that the listener extracted meaning from the sound mixture he or she heard but analyzed the wrong auditory object (e.g., Brungart, 2001; Brungart et al., 2001; Ihlefeld & Shinn-Cunningham, 2008b). However, there is another stage at which failures of attention can occur. As mentioned above, attention relies on the appropriate segregation of acoustic mixtures into well-formed objects. Perceptual segregation requires fine spectrotemporal details; if the sensory representation of sound is too muddled to support sound segregation, then selective attention can fail even when listeners know what to attend to (Shinn-Cunningham, 2008). As an example, listeners with hearing loss, especially those hearing through cochlear implants, receive highly distorted auditory inputs and are poor at segregating acoustic mixtures; consistent with this, the main complaint of such listeners is that they have trouble understanding sound in noisy backgrounds, where selective attention is necessary (see Shinn-Cunningham & Best, 2008, for a more detailed discussion of these ideas). Even listeners with good peripheral hearing may hear a sound mixture whose content is too chaotic to support source segregation, for instance, if there are too many sound sources in the scene or if there is significant reverberant energy distorting the spectrotemporal structure important for segregating sound sources (e.g., see Lee & Shinn-Cunningham, 2008; Mandel et al., 2010).

Dividing and Shifting Attention

While the basic filter theory of attention is built around the premise that attention operates to select one source and exclude others, there have been several attempts over the years to investigate the extent to which this selectivity is obligatory. In other words, to what extent can attention be divided between two or more auditory objects if that is the goal?

For relatively simple detection tasks, it seems that listeners can divide attention between two sound streams, with performance comparable to that achieved when monitoring a single stream. For example, when listeners monitor either two frequencies or two ears for target stimuli, detection in one channel seems to suffer only when a target occurs simultaneously in the other channel (Pashler, 1998). It is worth noting, however, that the ability to monitor multiple streams successfully may require more "effort" (a notoriously difficult thing to measure).

In the extensive literature on competing speech mixtures, there is little evidence that listeners actually divide attention. When listeners follow one talker, they appear to recall little about unattended talkers (Cherry, 1953). It is true that when listeners are instructed in advance to report back both of two competing messages, listeners can perform relatively well (Best et al., 2006; Gallun et al., 2007; Ihlefeld & Shinn-Cunningham, 2008a). However, it is not clear that this good performance indicates a true sharing of attention across streams. One possibility is that attention can be divided to a point, when the stimuli are brief, when the two tasks are not demanding, and/or when the two tasks do not compete for a limited pool of processing resources (Gallun et al., 2007). Another possibility is that listeners process simultaneous inputs to the auditory system serially (Broadbent, 1954, 1956). When two sequences of digits are presented simultaneously to the two ears (or in two voices), listeners can recall all digits, but they first report the digits presented to one ear (or spoken by one voice) and then recall the content of the other stream (Broadbent, 1954, 1956). Broadbent postulated that simultaneous sensory inputs are stored temporarily via immediate auditory memory and then processed serially by a limited-capacity mechanism (Broadbent, 1957; see chapter 9 in Broadbent, 1958, and Lachter et al., 2004). A consequence of such a scheme is that the secondary message in the pair must be held in a volatile memory store while the primary message is processed.

Another question that has interested researchers is how easily and rapidly selective attention can be switched from one object to another when the focus of interest changes. There are many examples showing that there is a cost associated with switching auditory attention. Early experiments demonstrated deficits in recall of speech items when presented alternately to the two ears (Broadbent, 1954, 1956; Cherry & Taylor, 1954; Treisman, 1971). This cost is also apparent in more complex scenarios where listeners must switch attention on cue between multiple simultaneous streams of speech (Best et al., 2008). Costs of switching attention have also been demonstrated when the switch is from one voice to another (Larson & Lee, 2013; Lawo & Koch, 2014). The cost of switching attention is associated with the time required to disengage and reengage attention but may also come from an improvement in performance over time when listeners are able to hone the attentional filter more finely when they maintain focus on a single stream (e.g., see Best et al., 2008; Bressler et al., 2014).

Selective Attention in Complex Listening Scenarios

Over the last decade or so, there has been a surge of interest in studying attention in more natural, complex listening scenarios to try to strengthen the link between laboratory results and real-world behavior. Earlier improvements in technology facilitated this new research by providing sophisticated new ways to create complex auditory scenes like those encountered in everyday settings (e.g., see Carlile, 1996).

In most controlled experiments, listeners are asked to repeat back, verbatim, the contents of a target sentence or digit sequence. In contrast, in most verbal exchanges outside the lab, the exact wording of a message is irrelevant; only the meaning of the words is critical (albeit with some notable exceptions, such as understanding telephone numbers, etc.). Moreover, in typical conversations, that meaning must be tracked continuously as the conversation goes on. In the lab, however, messages are brief and often organized into trials where the subject has ample time to report back the content in between trials. Some recent experiments have attempted to bridge this gap, asking listeners to maintain attention on speech that flows rapidly and continuously (Hafter, Xia, & Kalluri, 2013; Hafter, Xia, Kalluri, Poggesi, et al., 2013). Listeners were presented with competing stories and asked interpretive questions (i.e., requiring semantic, not just phonetic processing); the questions were most often about one of the stories (the target) but occasionally were about the competing story. Results revealed different kinds of limits on the processing of simultaneous speech from multiple talkers. For example, listeners only had the capacity to partially process one (but not two) competitors, and spatial attention limited that capacity to nearby competitors. Studies like this, which better recreate the pressures associated with continuously extracting meaning from ongoing conversations, are likely to produce new insights into how attention influences everyday communication.

In an attempt to understand the role of selective and divided attention in busy, natural listening scenarios, Eramudugolla and colleagues (2005) designed a novel "change deafness" paradigm. In a scene consisting of four to eight spatially separated natural sounds, they showed that when selective attention was directed in advance to one object, listeners were remarkably good at monitoring that object and detecting its disappearance in a subsequent exposure to the scene. However, in the absence of directed attention (i.e., when relying on divided attention) listeners were unable to reliably detect the disappearance of one of the objects.

Conversely, when listeners do focus attention selectively within a complex scene, it can leave them completely unaware of unusual or unexpected auditory events. For instance, one demonstration of "inattentional deafness" used an auditory analogue of the famous visual gorilla experiment. In the auditory version, listeners sustained their attention on one of two conversations in a simulated cocktail party. Under these conditions, many of the listeners failed to notice the presence of a man walking around the simulated scene repeatedly saying "I am a gorilla," despite both the prolonged duration of his message (19 seconds) and the fact that the message was audible and intelligible when attended (Dalton & Fraenkel, 2012). Inattentional deafness has also been demonstrated in complex musical scenes for nonmusicians and musicians alike (Koreimann et al., 2014). That is, focusing on one object can leave people unaware of the content of a competing object. In addition, focusing on one aspect of one object can leave people unaware of changes in other aspects of that object. When listeners are selectively focused on the content of an important message, they are often not even aware of a switch of the talker identity midway through the message (Vitevitch, 2003); moreover, the likelihood of listeners' noting such talker changes decreases as the demands of processing the lexical content of the message increases (Vitevitch & Donoso, 2011). While the first form of inattentional deafness (missing the presence of an unattended object) probably reflects the filtering out of unwanted information, so that its content is never processed, the second (missing discontinuities in one feature of a stream of information that is being processed) may be due to a failure to extract and store in memory high-order aspects of an attended object (such as talker identity) when they are not critical to performance and other task demands are high. Further work is needed to explore how auditory memory interacts with perception in complex, demanding scenes (see Snyder & Gregg, 2011, for a recent review of such issues).

Integration

The expansive literature on auditory selective attention has come about because of an intuitive recognition of its importance in how we function in the world. Competing sounds are a feature of most everyday environments, and selective attention is critical for enabling us to navigate and communicate in common situations. Yet, there are some major distinctions between what we typically test in laboratory experiments and how listeners normally operate in the real world. One major difference is the kind of information listeners typically extract from auditory scenes. Almost no laboratory experiments test listeners' awareness of the ambience of a setting (busy indoor café or a forest in the height of spring?), yet such information is something we are constantly monitoring, even when we are not consciously aware that we are doing so. Similarly, as noted in the previous section, in the laboratory we often test how well listeners can report back the exact content of brief spoken messages, even though it is really the meaning of ongoing conversations that must be conveyed in the boardroom, the football field, the coffee shop, and so on. Of course, it is relatively easy to come up with objective scores for how well listeners can report back an exact sequence of words compared to testing how well they understood the meaning of a long verbal exchange. Yet to truly understand how selective attention affects our everyday interactions, we must quantify performance in more natural scenarios.

While there has been a lot of experimental work on failures of attention related to selection (e.g., studies of IM), there is surprisingly little known about how often these kinds of failures occur in the real world. In many laboratory paradigms, stimuli are manipulated to create scenes in which competing streams are unnaturally similar or unnaturally correlated in their timing. Yet naturally occurring sounds in the real world come from independent physical sources, so are generally distinct, temporally uncorrelated, from different locations, and unrelated in all other dimensions. Moreover, other nonacoustic information, such as visual cues, reinforces these acoustic cues and reduces uncertainty about where and when to attend to extract a particular source. These (and other) properties of real-world listening scenarios reduce the likelihood of confusion compared to what we researchers often create in a "good" experimental design. Despite this, introspectively, one can think of occasions where a competing conversation "intruded" into the conversation of interest. Imagine trying to listen to the song of one bird in the dawn chorus (where the similarity of the individual bird songs makes it difficult to segregate any one physical source from the rest). Moreover, there is evidence that certain populations, such as children and the elderly, have particular difficulties in suppressing unwanted sounds (e.g., see Tun et al., 2002; Elliott, 2002). A recent study showed that performance on a laboratory test that measures people's ability to maintain attention on an auditory task in the presence of a distractor can be predicted by performance on a questionnaire that rates everyday distractibility (the Cognitive Failures Questionnaire; Murphy & Dalton, 2014). These studies hint that even though laboratory experiments rarely reflect the kinds of challenges facing ordinary listeners in ordinary settings, the factors studied in the laboratory nonetheless affect our ability to communicate in everyday life.

We touched above on object formation and its role in selective attention. In the real world it is very likely that failures of object formation are a significant contributor to failures of selective attention. In real-world listening environments, the defining features of sound objects are often degraded by interactions with other sounds and reflections from walls and other surfaces. Moreover, it is common for sounds with similar spectra to occur simultaneously, causing masking in the peripheral representation of sound (EM) to affect everyday perception. These factors are exacerbated in listeners with hearing loss and in cochlear implant users because of their reduced peripheral resolution (Shinn-Cunningham & Best, 2008).

The challenges associated with dividing and switching attention discussed above are crucially important for understanding real-world behavior. To return to our starting point of the cocktail party problem, it is clear that most social interactions involve not just paying attention to one talker in the presence of unwanted distractors but rather dynamically redirecting attention to different talkers with different vocal characteristics, who are generally at different locations or even moving around, all in unpredictable and unexpected ways. Indeed, the more boisterous, and "social" a conversation is, the more likely it will involve unexpected interruptions by some animated, engaged participant responding to a previous talker's point. Thus, attention must constantly be divided and shifted in order to follow the flow of conversation and participate meaningfully in social settings, stressing our perceptual skills in ways rarely tested in the laboratory.

A final issue worth considering is the possible interplay between technology and auditory selective attention. Recent advances in hearing devices provide a number of new opportunities for aiding listeners who have difficulty communicating in normal social scenes (e.g., see Edwards, 2007; Kidd et al., 2013). Our scientific understanding of auditory attention should guide the development of such new devices and focus resources on those aspects of the problem that lead to failures of communication—and to social isolation. On the other side of the spectrum, new technologies can also bring new challenges related to auditory selective attention that we are only just beginning to understand. For example, it is clear that focused attention to portable music players and cell phones has consequences for awareness of other critical sounds in the environment (sirens, alarms, approaching cars) that can interfere with fundamental tasks such as walking and driving.

Summary and Future Directions

Some of the earliest research on attention came out of work on auditory selective attention. While visual researchers embraced this early work, many psychoacousticians instead focused on bottom-up processing limitations of the auditory system. Over the last 15 to 20 years, auditory researchers have turned back to studying the importance of selective attention in auditory perception, basing much of their work on the breakthroughs from visual science.

Many aspects of auditory selective attention seem to operate analogously to how selective attention operates in vision. For instance, there is a close relationship between object formation and attention in both modalities. In line with the idea that the "unit" of selective attention is an object, listeners show little evidence for being able to truly divide attention; instead, they appear to switch attention rapidly between sources and to use memory to fill in the gaps. Many other perceptual "failures" in complex listening scenarios are also consistent with auditory attention's analyzing one, and only one, object at a time. For instance, the idea that listeners are unable to monitor multiple sources simultaneously in a complex scene explains both change deafness and inattentional deafness.

Various auditory features both support auditory scene analysis (source segregation) and serve as perceptual dimensions that can be used to direct attention, selecting out a target object from a complex acoustic scene. These dimensions include frequency content, location, timbre, pitch, and rhythm; however, there is no well-defined list of discrete features that can be used to direct attention. This is one area where much work remains.

While many of the basic principles of selective attention are similar in auditory and visual perception, one striking difference is in the importance of time in auditory perception. Auditory sources are inherently temporal; auditory information can be extracted only by listening to a sound through time. In line with this, an auditory object can be resolved from an auditory scene only by analyzing the spectrotemporal content of sound across a range of timescales, from milliseconds to tens of seconds. Because of this, and because acoustically, sound combines additively before entering the ears, auditory selective attention is often limited by difficulties with *segregating* an object of interest from a scene. Source segregation is even harder in the presence of reverberant energy and background noise; any degradation of the sensory representation of sound in the auditory periphery, such as from hearing loss, exacerbates the problems of source segregation. Thus, in everyday settings, selective auditory attention often fails because of failures of object formation. Given the important role that time plays in auditory information, the dynamics of attention (e.g., the time course for attention to be focused, reoriented, maintained, etc.) is, if anything, more critical than in other sensory dimensions. Yet, relatively few studies have tackled the problem of exploring the dynamics of selective auditory attention.

The demands we face every day are very different than those tested in most laboratories. In the real world, auditory scenes are unpredictable and ongoing. Real-world social settings require listeners to keep up and constantly extract meaning from ongoing conversations that are full of unpredictable interruptions and shifts and that often take place in noisy, reverberant settings. In addition, competition from nonauditory modalities is inevitable in busy everyday environments (think about how distracting a TV screen can be in a pub or a restaurant when you are trying to listen to a conversation). New insights will come from expanding our research to better match the complexities of everyday settings. Insights from such studies will be especially helpful in understanding the difficulties faced by various special populations, from listeners with hearing impairment to children to veterans with traumatic brain injury to listeners with deficits in cognitive function.

Box 5.1
Key Points

- Selective auditory attention cannot operate unless listeners can segregate the acoustic mixture reaching the ears into constituent sound sources.
- Listeners can focus selective auditory attention on a sound object based on any number of features, from location to timbre to talker characteristics.
- Because sound conveys information through time, it is critical to study the dynamics of auditory attention.
- While laboratory experiments have demonstrated many of the key factors important for selective auditory attention, real-world settings require listeners to maintain awareness in dynamic and complicated sound mixtures and to keep up with ongoing, unpredictable conversations, unlike typical controlled experimental conditions.

Box 5.2
Outstanding Issues

- Further research is needed to explore and identify the discrete auditory features that can be used to direct attention.
- What is the time course for attention in the auditory domain to be focused, reoriented, maintained?
- How do the principles of auditory attention identified in the laboratory scale up to the more complex environments characteristic of our everyday lives?

References

Alain, C., Arnott, S. R., & Picton, T. W. (2001). Bottom-up and top-down influences on auditory scene analysis: Evidence from event-related brain potentials. *Journal of Experimental Psychology: Human Perception and Performance, 27,* 1072–1089.

Alain, C., & Woods, D. L. (1997). Attention modulates auditory pattern memory as indexed by event-related brain potentials. *Psychophysiology, 34,* 534–546.

Arbogast, T. L., & Kidd, G., Jr. (2000). Evidence for spatial tuning in informational masking using the probe-signal method. *Journal of the Acoustical Society of America, 108,* 1803–1810.

Best, V., Gallun, F. J., Ihlefeld, A., & Shinn-Cunningham, B. G. (2006). The influence of spatial separation on divided listening. *Journal of the Acoustical Society of America, 120,* 1506–1516.

Best, V., Ozmeral, E. J., Kopčo, N., & Shinn-Cunningham, B. G. (2008). Object continuity enhances selective auditory attention. *Proceedings of the National Academy of Sciences of the United States of America, 105,* 13174–13178.

Bregman, A. S. (1990). *Auditory scene analysis: The perceptual organization of sound.* Cambridge, MA: MIT Press.

Bressler, S., Masud, S., Bharadwaj, H., & Shinn-Cunningham, B. (2014). Bottom-up influences of voice continuity in focusing selective auditory attention. *Psychological Research, 78,* 349–360.

Broadbent, D. E. (1954). The role of auditory localization in attention and memory span. *Journal of Experimental Psychology, 47,* 191–196.

Broadbent, D. E. (1956). Successive responses to simultaneous stimuli. *Quarterly Journal of Experimental Psychology, 8,* 145–152.

Broadbent, D. E. (1957). Immediate memory and simultaneous stimuli. *Quarterly Journal of Experimental Psychology, 9,* 1–11.

Broadbent, D. E. (1958). *Perception and communication.* New York: Pergamon Press.

Brungart, D. S. (2001). Informational and energetic masking effects in the perception of two simultaneous talkers. *Journal of the Acoustical Society of America, 109,* 1101–1109.

Brungart, D. S., & Simpson, B. D. (2007). Cocktail party listening in a dynamic multitalker environment. *Perception & Psychophysics, 69,* 79–91.

Brungart, D. S., Simpson, B. D., Ericson, M. A., & Scott, K. R. (2001). Informational and energetic masking effects in the perception of multiple simultaneous talkers. *Journal of the Acoustical Society of America, 110,* 2527–2538.

Carlile, S. (1996). *Virtual auditory space: Generation and applications.* New York: R.G. Landes.

Carlyon, R. P. (2004). How the brain separates sounds. *Trends in Cognitive Sciences, 8,* 465–471.

Carlyon, R. P., Plack, C. J., Fantini, D. A., & Cusack, R. (2003). Cross-modal and non-sensory influences on auditory streaming. *Perception, 32*, 1393–1402.

Chait, M., de Cheveigne, A., Poeppel, D., & Simon, J. Z. (2010). Neural dynamics of attending and ignoring in human auditory cortex. *Neuropsychologia, 48*, 3262–3271.

Cherry, E. C. (1953). Some experiments on the recognition of speech, with one and with two ears. *Journal of the Acoustical Society of America, 25*, 975–979.

Cherry, E. C., & Taylor, W. K. (1954). Some further experiments upon the recognition of speech, with one and with two ears. *Journal of the Acoustical Society of America, 26*, 554–559.

Colflesh, G. J., & Conway, A. R. (2007). Individual differences in working memory capacity and divided attention in dichotic listening. *Psychonomic Bulletin & Review, 14*, 699–703.

Corteen, R. S., & Wood, B. (1972). Autonomic responses to shock-associated words in an unattended channel. *Journal of Experimental Psychology, 94*, 308–313.

Culling, J. F., Hodder, K. I., & Toh, C. Y. (2003). Effects of reverberation on perceptual segregation of competing voices. *Journal of the Acoustical Society of America, 114*, 2871–2876.

Cusack, R., & Carlyon, R. P. (2003). Perceptual asymmetries in audition. *Journal of Experimental Psychology: Human Perception and Performance, 29*, 713–725.

Cusack, R., Deeks, J., Aikman, G., & Carlyon, R. P. (2004). Effects of location, frequency region, and time course of selective attention on auditory scene analysis. *Journal of Experimental Psychology: Human Perception and Performance, 30*, 643–656.

Dalton, P., & Fraenkel, N. (2012). Gorillas we have missed: Sustained inattentional deafness for dynamic events. *Cognition, 124*, 367–372.

Darwin, C. J., Brungart, D. S., & Simpson, B. D. (2003). Effects of fundamental frequency and vocal-tract length changes on attention to one of two simultaneous talkers. *Journal of the Acoustical Society of America, 114*, 2913–2922.

Desimone, R., & Duncan, J. (1995). Neural mechanisms of selective visual attention. *Annual Review of Neuroscience, 18*, 193–222.

Devergie, A., Grimault, N., Tillmann, B., & Berthommier, F. (2010). Effect of rhythmic attention on the segregation of interleaved melodies. *Journal of the Acoustical Society of America, 128*, EL1–EL7.

Durlach, N. I., Mason, C. R., Shinn-Cunningham, B. G., Arbogast, T. L., Colburn, H. S., & Kidd, G., Jr. (2003). Informational masking: Counteracting the effects of stimulus uncertainty by decreasing target–masker similarity. *Journal of the Acoustical Society of America, 114*, 368–379.

Edwards, B. (2007). The future of hearing aid technology. *Trends in Amplification, 11*, 31–45.

Elhilali, M., Ma, L., Micheyl, C., Oxenham, A. J., & Shamma, S. A. (2009). Temporal coherence in the perceptual organization and cortical representation of auditory scenes. *Neuron, 61*, 317–329.

Elliott, E. M. (2002). The irrelevant-speech effect and children: Theoretical implications of developmental change. *Memory & Cognition, 30,* 478–487.

Eramudugolla, R., Irvine, D. R., McAnally, K. I., Martin, R. L., & Mattingley, J. B. (2005). Directed attention eliminates "change deafness" in complex auditory scenes. *Current Biology, 15,* 1108–1113.

Freyman, R. L., Helfer, K. S., McCall, D. D., & Clifton, R. K. (1999). The role of perceived spatial separation in the unmasking of speech. *Journal of the Acoustical Society of America, 106,* 3578–3588.

Fritz, J. B., Elhilali, M., David, S. V., & Shamma, S. A. (2007). Auditory attention: Focusing the searchlight on sound. *Current Opinion in Neurobiology, 17,* 437–455.

Gallun, F. J., Mason, C. R., & Kidd, G., Jr. (2007). Task-dependent costs in processing two simultaneous auditory stimuli. *Perception & Psychophysics, 69,* 757–771.

Gray, G., & Wedderburn, A. (1960). Grouping strategies with simultaneous stimuli. *Quarterly Journal of Experimental Psychology, 12,* 180–185.

Green, D. M. (1961). Detection of auditory sinusoids of uncertain frequency. *Journal of the Acoustical Society of America, 33,* 897–903.

Greenberg, G. Z., & Larkin, W. D. (1968). Frequency-response characteristic of auditory observers detecting signals of a single frequency in noise: The probe-signal method. *Journal of the Acoustical Society of America, 44,* 1513–1523.

Griffiths, T. D., & Warren, J. D. (2004). What is an auditory object? *Nature Reviews. Neuroscience, 5,* 887–892.

Hafter, E. R., Xia, J., & Kalluri, S. (2013). A naturalistic approach to the cocktail party problem. *Advances in Experimental Medicine and Biology, 787,* 527–534.

Hafter, E. R., Xia, J., Kalluri, S., Poggesi, R., Hansen, C., & Whiteford, K. (2013). Attentional switching when listeners respond to semantic meaning expressed by multiple talkers. *Journal of the Acoustical Society of America, 133,* 3381.

Hansen, J. C., & Hillyard, S. A. (1980). Endogenous brain potentials associated with selective auditory attention. *Electroencephalography and Clinical Neurophysiology, 49,* 277–290.

Hill, K. T., & Miller, L. M. (2010). Auditory attentional control and selection during cocktail party listening. *Cerebral Cortex, 20,* 583–590.

Hupe, J. M., Joffo, L. M., & Pressnitzer, D. (2008). Bistability for audiovisual stimuli: Perceptual decision is modality specific. *Journal of Vision, 8*(7), 1, 1–15.

Ihlefeld, A., & Shinn-Cunningham, B. (2008a). Spatial release from energetic and informational masking in a divided speech identification task. *Journal of the Acoustical Society of America, 123,* 4380–4392.

Ihlefeld, A., & Shinn-Cunningham, B. (2008b). Spatial release from energetic and informational masking in a selective speech identification task. *Journal of the Acoustical Society of America, 123*, 4369–4379.

Jones, M. R. (1976). Time, our lost dimension: Toward a new theory of perception, attention, and memory. *Psychological Review, 83*, 323–355.

Jones, M. R., Kidd, G., & Wetzel, R. (1981). Evidence for rhythmic attention. *Journal of Experimental Psychology: Human Perception and Performance, 7*, 1059–1073.

Kastner, S., & Ungerleider, L. G. (2001). The neural basis of biased competition in human visual cortex. *Neuropsychologia, 39*, 1263–1276.

Kidd, G., Jr., Arbogast, T. L., Mason, C. R., & Gallun, F. J. (2005). The advantage of knowing where to listen. *Journal of the Acoustical Society of America, 118*, 3804–3815.

Kidd, G., Jr., Favrot, S., Desloge, J. G., Streeter, T. M., & Mason, C. R. (2013). Design and preliminary testing of a visually guided hearing aid. *Journal of the Acoustical Society of America, 133*, EL202–EL207.

Kidd, G., Jr., Mason, C. R., Deliwala, P. S., Woods, W. S., & Colburn, H. S. (1994). Reducing informational masking by sound segregation. *Journal of the Acoustical Society of America, 95*, 3475–3480.

Kidd, G., Jr., Mason, C. R., Richards, V. M., Gallun, F. J., & Durlach, N. I. (2008). Informational masking. In W. Yost, A. N. Popper, & R. R. Fay (Eds.), *Springer handbook of auditory research: Vol. 29. Auditory perception of sound sources* (pp. 143–189). New York: Springer.

Kitterick, P. T., Clarke, E., O'Shea, C., Seymour, J., & Summerfield, A. Q. (2013). Target identification using relative level in multi-talker listening. *Journal of the Acoustical Society of America, 133*, 2899–2909.

Koreimann, S., Gula, B., & Vitouch, O. (2014). Inattentional deafness in music. *Psychological Research, 78*, 304–312.

Lachter, J., Forster, K. I., & Ruthruff, E. (2004). Forty-five years after Broadbent (1958): Still no identification without attention. *Psychological Review, 111*, 880–913.

Larson, E., & Lee, A. K. C. (2013). Influence of preparation time and pitch separation in switching of auditory attention between streams. *Journal of the Acoustical Society of America, 134*, EL165–EL171.

Lawo, V., & Koch, I. (2014). Dissociable effects of auditory attention switching and stimulus–response compatibility. *Psychological Research, 78*, 379–386.

Lee, A. K. C., Larson, E., Maddox, R. K., & Shinn-Cunningham, B. G. (2014). Using neuroimaging to understand the cortical mechanisms of auditory selective attention. *Hearing Research, 307*, 111–120.

Lee, A. K. C., Rajaram, S., Xia, J., Bharadwaj, H., Larson, E., Hamalainen, M., et al. (2013). Auditory selective attention reveals preparatory activity in different cortical regions for selection based on source location and source pitch. *Frontiers in Neuroscience, 6*, 190.

Lee, A. K. C., & Shinn-Cunningham, B. G. (2008). Effects of reverberant spatial cues on attention-dependent object formation. *Journal of the Association for Research in Otolaryngology, 9*, 150–160.

Mackay, D. G. (1973). Aspects of the theory of comprehension, memory and attention. *Quarterly Journal of Experimental Psychology, 25*, 22–40.

Macken, W. J., Tremblay, S., Houghton, R. J., Nicholls, A. P., & Jones, D. M. (2003). Does auditory streaming require attention? Evidence from attentional selectivity in short-term memory. *Journal of Experimental Psychology: Human Perception and Performance, 29*, 43–51.

Maddox, R. K., Billimoria, C. P., Perrone, B. P., Shinn-Cunningham, B. G., & Sen, K. (2012). Competing sound sources reveal spatial effects in cortical processing. *PLoS Biology, 10*(5), e1001319.

Maddox, R. K., & Shinn-Cunningham, B. G. (2012). Influence of task-relevant and task-irrelevant feature continuity on selective auditory attention. *Journal of the Association for Research in Otolaryngology, 13*, 119–129.

Mandel, M. I., Bressler, S., Shinn-Cunningham, B., & Ellis, D. P. W. (2010). Evaluating source separation algorithms with reverberant speech. *IEEE Transactions on Audio Speech and Language Processing, 18*, 1872–1883.

Mesgarani, N., & Chang, E. F. (2012). Selective cortical representation of attended speaker in multi-talker speech perception. *Nature, 485*, 233–236.

Middlebrooks, J. C., & Bremen, P. (2013). Spatial stream segregation by auditory cortical neurons. *Journal of Neuroscience, 33*, 10986–11001.

Mondor, T. A., & Zatorre, R. J. (1995). Shifting and focusing auditory spatial attention. *Journal of Experimental Psychology: Human Perception and Performance, 21*, 387–409.

Moray, N. (1959). Attention in dichotic listening: Affective cues and the influence of instructions. *Quarterly Journal of Experimental Psychology, 11*, 56–60.

Murphy, S., & Dalton, P. (2014). Ear-catching? Real-world distractibility scores predict susceptibility to auditory attentional capture. *Psychonomic Bulletin & Review, 21*, 1209–1213.

Neff, D. L., & Dethlefs, T. M. (1995). Individual differences in simultaneous masking with random-frequency, multicomponent maskers. *Journal of the Acoustical Society of America, 98*, 125–134.

Neff, D. L., & Green, D. M. (1987). Masking produced by spectral uncertainty with multicomponent maskers. *Perception & Psychophysics, 41*, 409–415.

Oh, E. L., & Lutfi, R. A. (1998). Nonmonotonicity of informational masking. *Journal of the Acoustical Society of America, 104*, 3489–3499.

Pashler, H. E. (1998). *The psychology of attention*. Cambridge, MA: MIT Press.

Picton, T. W., & Hillyard, S. A. (1974). Human auditory evoked potentials: II. Effects of attention. *Electroencephalography and Clinical Neurophysiology*, *36*, 191–199.

Quinlan, P. T., & Bailey, P. J. (1995). An examination of attentional control in the auditory modality: Further evidence for auditory orienting. *Perception & Psychophysics*, *57*, 614–628.

Richards, V. M., & Neff, D. L. (2004). Cuing effects for informational masking. *Journal of the Acoustical Society of America*, *115*, 289–300.

Scharf, B., Quigley, S., Aoki, C., Peachey, N., & Reeves, A. (1987). Focused auditory attention and frequency selectivity. *Perception & Psychophysics*, *42*, 215–223.

Shamma, S. A., Elhilali, M., & Micheyl, C. (2011). Temporal coherence and attention in auditory scene analysis. *Trends in Neurosciences*, *34*, 114–123.

Shinn-Cunningham, B. G. (2008). Object-based auditory and visual attention. *Trends in Cognitive Sciences*, *12*, 182–186.

Shinn-Cunningham, B. G., & Best, V. (2008). Selective attention in normal and impaired hearing. *Trends in Amplification*, *12*, 283–299.

Snyder, J. S., & Gregg, M. K. (2011). Memory for sound, with an ear toward hearing in complex auditory scenes. *Attention, Perception & Psychophysics*, *73*, 1993–2007.

Snyder, J. S., Gregg, M. K., Weintraub, D. M., & Alain, C. (2012). Attention, awareness, and the perception of auditory scenes. *Frontiers in Psychology*, *3*, 15.

Spence, C., Nicholls, M. E. R., & Driver, J. (2001). The cost of expecting events in the wrong sensory modality. *Perception & Psychophysics*, *63*, 330–336.

Spence, C. J., & Driver, J. (1994). Covert spatial orienting in audition: Exogenous and endogenous mechanisms. *Journal of Experimental Psychology: Human Perception and Performance*, *20*, 555–574.

Spiegel, M. F., Picardi, M. C., & Green, D. M. (1981). Signal and masker uncertainty in intensity discrimination. *Journal of the Acoustical Society of America*, *70*, 1015–1019.

Sussman, E. S., Horvath, J., Winkler, I., & Orr, M. (2007). The role of attention in the formation of auditory streams. *Perception & Psychophysics*, *69*, 136–152.

Tark, K. J., & Curtis, C. E. (2013). Deciding where to look based on visual, auditory, and semantic information. *Brain Research*, *1525*, 26–38.

Teki, S., Chait, M., Kumar, S., Shamma, S., & Griffiths, T. D. (2013). Segregation of complex acoustic scenes based on temporal coherence. *eLife*, *2*, e00699.

Treisman, A. M. (1960). Contextual cues in selective listening. *Quarterly Journal of Experimental Psychology*, *12*, 157–167.

Treisman, A. M. (1971). Shifting attention between the ears. *Quarterly Journal of Experimental Psychology, 23*, 157–167.

Treisman, A. M., & Gelade, G. (1980). A feature-integration theory of attention. *Cognitive Psychology, 12*, 97–136.

Tun, P. A., O'Kane, G., & Wingfield, A. (2002). Distraction by competing speech in young and older adult listeners. *Psychology and Aging, 17*, 453–467.

Vitevitch, M. S. (2003). Change deafness: The inability to detect changes between two voices. *Journal of Experimental Psychology: Human Perception and Performance, 29*, 333–342.

Vitevitch, M. S., & Donoso, A. (2011). Processing of indexical information requires time: Evidence from change deafness. *Quarterly Journal of Experimental Psychology, 64*, 1484–1493.

Watson, C. S. (2005). Some comments on informational masking. *Acta Acustica united with Acustica, 91*, 502–512.

Watson, C. S., Kelly, W. J., & Wroton, H. W. (1976). Factors in the discrimination of tonal patterns. II. Selective attention and learning under various levels of stimulus uncertainty. *Journal of the Acoustical Society of America, 60*, 1176–1186.

Watson, C. S., Wroton, H. W., Kelly, W. J., & Benbassat, C. A. (1975). Factors in the discrimination of tonal patterns. I. Component frequency, temporal position, and silent intervals. *Journal of the Acoustical Society of America, 57*, 1175–1185.

Wood, N., & Cowan, N. (1995). The cocktail party phenomenon revisited: How frequent are attention shifts to one's name in an irrelevant auditory channel? *Journal of Experimental Psychology: Learning, Memory, and Cognition, 21*, 255–260.

Wright, B. A., & Fitzgerald, M. B. (2004). The time course of attention in a simple auditory detection task. *Perception & Psychophysics, 66*, 508–516.

6

Crossmodal Attention: From the Laboratory to the Real World (and Back Again)

Charles Spence and Cristy Ho

Historical Context

Research on the topic of attention really got going as a result of the problems associated with the processing of multiple sources of information out there in the real world. Specifically, psychologists were tasked with trying to understand why it was that pilots were missing important messages presented over their headsets while in flight during the Second World War. Meanwhile, during the 1960s, scientists (e.g., at the Applied Psychology Unit in Cambridge, UK) started to investigate the cognitive/ perceptual challenges associated with using a mobile phone or car radio while driving (e.g., Brown, 1962, 1965; Brown & Poulton, 1961; Brown, Simmonds, & Tickner, 1967; Brown, Tickner, & Simmonds, 1969). While such research touched on the topic of crossmodal limits on the allocation of auditory and visual attention, it is undoubtedly true to say that the vast majority of the early research involved studying attention from a purely unisensory perspective—that is, focusing on the selection of one voice among others or one visual stimulus among a host of other competing visual inputs (see Driver, 2001, for a comprehensive review). Psychologists working in the field of attention rapidly moved on from studying such real-world selection problems to concentrating on much more theoretical and laboratory-based questions instead. In fact, during the early decades, say from the 1950s through the 1980s, much of the focus in the research laboratories was on trying to determine whether attentional selection took place "early" or "late" in human information processing (Allport, 1993).[1]

Undoubtedly, a fundamental switch in the focus of research occurred with the arrival of the personal computer in the 1970s (Abbate, 1999). Almost overnight, this technology replaced the reel-to-reel tape recorder as the "must have" device for psychologists and in so doing led to a much greater interest in the topic of *visual* attention (and a concomitant loss of interest among researchers in problems associated with auditory attentional selection). Visual search, the flanker task, and object-based attention research (e.g., Duncan, 1980; Egly, Driver, & Rafal, 1995; Eriksen & Eriksen, 1974) soon became the fundamental paradigms motivating many of the researchers working in the field (e.g., Quinlan, 2003; Treisman & Gelade, 1980). Mike Posner, with his iconic spatial cuing task, also helped to usher in a new experimental paradigm for attention researchers to work with (e.g., Posner, 1978). In a typical Posnerian cuing study, participants would be presented with a peripheral visual cue (usually from the left or right of central fixation) prior to the onset of a visual target requiring a speeded manual detection or discrimination response. The results of many such studies showed that people could not ignore the cue, and their spatial attention was briefly drawn to the location of the cue, thus facilitating responses to targets subsequently presented from the same (as compared to the opposite) location. In terms of the role of selective attention on tactile information processing, however, the psychologists really had to wait for the arrival of cheap actuators (i.e., tactile stimulators), and hence this has only really been an active research area for the last decade or two (see Gallace & Spence, 2014; Spence & Gallace, 2007, for reviews).

In recent years, much of the cognitive neuroscience research interest has shifted to studying where exactly in the human brain selection takes place—that is, how early could an attentional modulation of information processing be observed in humans (e.g., Lavie, 2005; Reynolds & Chelazzi, 2004; Shulman, 1990)? While the focus of research in the field of attention has traditionally always been on spatial selection and orienting, temporal selection (see Nobre & Coull, 2010, for a review of the literature) and memory-based selection (see Gazzaley & Nobre, 2012, for a review) are now becoming increasingly popular topics of research interest.

State-of-the-Art Review

What is striking, in hindsight, is how much of the early attention research (no matter whether it be visual, auditory, or tactile) had a purely unisensory focus (Driver, 2001). The shift toward the study of crossmodal attention and multisensory perception really emerged later, prompted in part by the findings coming out of neuroscience laboratories (see Spence & Driver, 2004). In particular, neuroscience research findings documenting the multisensory integration taking place in single cells in subcortical orienting sites such as the superior colliculus (SC; Stein & Meredith, 1993; Stein & Stanford, 2008) were particularly influential here. The results of such laboratory-based research came to a number of generalizations concerning the nature of multisensory integration. Three of the fundamental rules that have inspired much of your current authors' own research interest over the years are related to the notions of superadditivity, subadditivity, and the law of inverse effectiveness (Holmes & Spence, 2005). These concepts all emerged from animal neurophysiology. Superadditivity is said to occur when the response of a cell, or human observer, to a combined multisensory input is significantly greater than the response that would have been predicted from simply summing the component unisensory responses. Subadditivity is said to occur when the response of the cell, or observer, to the multisensory input is less that that seen to the best of the unisensory responses. According to the principle of inverse effectiveness, multisensory interactions tend to be more pronounced when the cell, or organism's, responses to the component unisensory stimuli are weak (e.g., Holmes & Spence, 2005; Stein & Stanford, 2008).

The claim coming out of this neurophysiological research was very much that combined sensory inputs, if congruent in both space and time (Holmes & Spence, 2005), would potentially give rise to enhanced multisensory integration (i.e., super-additivity) and presumably also enhanced spatial (i.e., overt) orienting. There was a tempting link between the multisensory interactions documented at the cellular level in the anesthetized, and then awake, preparation (e.g., Stein, Huneycutt, & Meredith, 1988; Stein, Meredith, Honeycutt, & McDade, 1989) and the overt (and possibly also covert) orienting of an animal's (and hence likely also a human's) spatial attention (Müller, Philiastides, & Newsome, 2005) in, say, a Posnerian cuing paradigm (Spence & Driver, 1997a).

To the extent that we perceive more or less that which we attend to (see Luck & Ford, 1998), it seemed reasonable to assume that one would need to know how attention was controlled and distributed, both spatially (and, latterly, temporally) in order to really appreciate how people would process and perceive the variety of sensory inputs that were available to them. However, that said, despite our best early attempts (e.g., see Spence & Driver, 1999), it turned out to be impossible to demonstrate that multisensory cues captured spatial attention any more effectively than the best of the component unisensory cues. That is, the rules of multisensory integration derived from animal neurophysiology did not necessarily seem to align all that closely with the results of crossmodal exogenous spatial cuing studies published in awake humans (see Spence & Driver, 2004). In fact, after several years of trying, it really seemed as though our group, as well as many others, had simply given up on trying to demonstrate any kind of special attentional enhancement following the presentation of multisensory (as opposed to unisensory) cues (see Spence, 2010).

To be clear, a large body of research was able to extend Posner's spatial cuing paradigm to a crossmodal setting and, in so doing, show that the sudden presentation of a spatially nonpredictive auditory, visual, or tactile cue would give rise to a short-lasting shift of an observer's spatial attention to the cued location. At short cue–target stimulus-onset asynchronies (SOAs), a facilitation of responding to visual, auditory, and tactile targets was reported (Spence & Driver, 2004; Wright & Ward, 2008), while at slightly longer SOAs (and especially in simple speeded target detection paradigms), this exogenous attentional facilitation would often be replaced by a slightly longer lasting inhibition of return (Klein, 2000; Spence, Lloyd, McGlone, Nicholls, & Driver, 2000). However, despite the fact that multisensory spatial cues were no more effective than unisensory cues in capturing an observer's spatial attention, there was still an immediate applied relevance for the crossmodal attentional findings coming out of the laboratory research (see Spence & Driver, 1997b). For example, such research clearly demonstrated that the sudden presentation of auditory, visual, or tactile cues could potentially be used to drive spatial attention in a particular direction (or to a specific location). Such findings were highly relevant for real-world applications such as in terms of thinking about warning signal design in the case of driving (see Ho & Spence, 2005, 2008; Spence & Driver, 1997b). Such crossmodal attentional capture also appeared to be fairly automatic, given that such effects were observed even when the cue was entirely nonpredictive with regard to the likely position from which the visual target, say, was subsequently presented. Of course, it makes little sense to think about presenting spatially nonpredictive cues in the real-world design of warning signals (what, after all, would be the point?). However, such effects are worth studying, from the vantage point of a laboratory-based psychologist, in terms of the light they were thought to shine on the automaticity of the underlying process of attentional capture. That said, across a number of studies, it has been shown that spatial cuing effects are larger following the presentation of spatially predictive than following nonpredictive cues (Ho & Spence, 2005; Ho, Tan, & Spence, 2005). Here, though, it should be noted that research over the last decade has demonstrated that such crossmodal cuing effects, following the presentation of a unimodal cue, are, in fact, anything but automatic (see Santangelo & Spence, 2008, for a review).

The fundamental question that we addressed in our applied research was how best to present warning signals to a distracted driver. The problem with the use of visual warning signals is that the majority of the information that the driver has to process is visual (e.g., Senders, Kristofferson, Levison, Dietrich, & Ward, 1967; Sivak, 1996; Spence & Ho, 2008a). Using the auditory modality to present relevant information to a distracted driver therefore seemed like a sensible strategy. However, it is important to note that there is also a danger here, namely, that a driver's auditory attention might be distracted (Mohebbi, Gray, & Tan, 2009), not to mention the competition from the car radio, which can, on occasion, reach surprisingly high levels (Ramsey & Simmons, 1993). Indeed, designing warning signals that can compete effectively with the emerging in-car technologies may become an increasingly tough challenge in the coming years (see also McKeown & Isherwood, 2007).

It is our contention that spatial colocation is an important rule of thumb when it comes to the design of multisensory safety and information systems for drivers (or for that matter, for any other interface operators; e.g., Baldwin et al., 2012). A variety of emerging technologies and intelligent transport systems are now available, and many of them are capable of providing information to drivers, warning them about potential dangers, and hopefully providing the necessary signal to reorient their attention if they should happen to be distracted by a roadside hoarding, the Satnav, the Internet, or any other stimuli that might attract their attention while they are driving (e.g., Ashley, 2001; Sarter, 2013).

In our own applied research, we have adapted the spatial cuing paradigm (see Spence & Driver, 1994, 1997a) in order to assess the potential benefits of spatial warning signals in terms of facilitating a driver's responses to a restricted range of potential emergency driving scenarios. Subsequently, we investigated which cues (or warning signals) worked best in terms of capturing and thence redirecting a driver's spatial attention back to the roadway (e.g., Ho & Spence, 2005, 2008). We simply took the standard Posnerian spatial cuing study from laboratory research (see Posner, 1978) and turned the setup 90°, so that now we cued drivers to either the front or rear (in order to direct their attention in one direction or the other). Note here that drivers typically look in the rearview mirror in order to determine what is going on behind them (i.e., they look forward to, in some sense, attend backward). The participants in our studies viewed videos of real driving scenes (recorded on the roadway in rural Oxfordshire) while we measured their reaction times (RTs). Importantly, these were not to manual button-press responses, as is typically the case in the majority of laboratory-based studies of attention, but rather RTs to make foot-pedal responses instead (cf. Levy, Pashler, & Boer, 2006). Specifically, our participants would typically either have to hit the brake pedal whenever the car seen in front slowed suddenly or else to depress the accelerator whenever the car behind (as seen in rearview mirror) accelerated suddenly.

The general findings to emerge from our own laboratory-based driving research were that the effects of spatial cuing using informative cues gave rise to a significantly larger performance advantage than did noninformative cues, and that iconic sounds (such as the sound of a car horn) were much more effective than meaningless sounds such as pure tones (see Ho & Spence, 2005, 2008; though see also Ho, Santangelo, & Spence, 2009). There was also some interest in the use of verbal cues, that is, warning signals consisting of verbal instructions (or suggestions) such as "brake" or "accelerate" and so forth (Ho & Spence, 2006). Here, though, there is of course always going to be a problem of knowing whether the driver will necessarily understand the language in which the signal is uttered (just imagine oneself with a hire car on a foreign holiday).

Auditory cues were always the most effective (in terms of reducing braking RTs), with tactile cues being somewhat less effective. This pattern of results is one that we have now replicated across a range of different situations (e.g., see Ngo, Pierce, & Spence, 2012; see also Scott & Gray, 2008). There is an immediate relevance of these research findings that spans all the way from the design of warning signals for the commercial vehicle sector out to a variety of military applications (e.g., Oskarsson, Eriksson, & Carlander, 2012; Sklar & Sarter, 1999; Van Erp, Eriksson, Levin, Carlander, Veltman, & Vos, 2007). Furthermore, there would also appear to be fruitful insights for a number of other domains such as air traffic control (Ngo et al., 2012) or, in fact, any other situation that requires lightning-fast responses (e.g., think of financial traders working on the trading floor in an investment bank—see http://www.informationweek.com/news/199200297—all the way through to those working in the setting of the operating theater; see also Baldwin et al., 2012; Haas & van Erp, 2014, for reviews).

In the context of driving, it is worth noting that car companies typically tend to be much more interested in knowing how much of a time saving they can hope to achieve by implementing a given warning signal than necessarily wanting to learn about the precise neural mechanisms (such as those underpinning spatial attention or response priming) that may underlie it. However, laboratory-based researchers (or if not the researchers themselves, then certainly the reviewers of their papers) often worry about whether the spatial cuing effects that they have documented in the laboratory or simulator result from some form of attentional facilitation versus from response priming. These factors are, of course, going to be somewhat confounded in the driving situation since a sudden event happening on the roadway ahead is more likely going to require a braking, rather than an acceleration, response. In other words, the warning signal may simply prime the likely response. Nevertheless, in order to address this theoretical question empirically, Ho, Tan, and Spence (2006) introduced a colored number plate change discrimination task. The participants in their study had to discriminate whether the number plate of the car in front or behind changed color (to either red or blue). Since the location of the cued car (front vs. back) was orthogonal to the dimension along which participants had to make their speeded discrimination responses (red vs. blue), any spatial cuing effects that were observed could unambiguously be attributed to the attentional facilitation of the participant's perception of the car (and the color of its number plate) rather than to the priming of a particular response (e.g., hitting the brake or the accelerator pedal).

The results of Ho et al.'s (2006) research revealed reduced cuing effects following the spatially nonpredictive auditory car horn cue, and an elimination of cuing benefits following the presentation of the spatially nonpredictive vibrotactile cue to the participant's stomach or back. When taken together with the other findings coming out of the Crossmodal Research Laboratory, these results therefore suggested that vibrotactile cues presented to the body surface prime the appropriate (or at least the required) driver response. By contrast, the presentation of a car horn sound, when presented from in front or behind of the driver, appears capable both of inducing a response priming effect and also giving rise to a small but significant attentional facilitation effect as well (see Ho & Spence, 2008, chapter 6).[2] The fact that auditory cues result in both attentional facilitation *and* response priming whereas tactile cues only give rise to response priming may, then, help to explain why it is that the latter always seem to be somewhat less effective in facilitating an operator's performance than are the former (though see also Ngo et al., 2012).

Integration

Why, and When, Do Multisensory Cues Work Better?

One of the important differences between laboratory-based studies of spatial attention and applied studies of driving and warning signal design concerns the question of perceptual/attentional load (e.g., Lavie, 1995, 2005). Typically, there is simply much more going on in the roadway (or in the driving simulator) than in most dark and silent psychophysics laboratories. The increased perceptual load faced by the drivers in our simulator study (Ho, Reed, & Spence, 2007) where multisensory cues were, for the first time, shown to be more effective than the best of the unisensory (either auditory or tactile) warning signals turned out to be the key factor modulating the relative effectiveness of multisensory, as compared to the component unisensory, cues. That perceptual load was key to eliciting the multisensory cue benefit was something that we were able to confirm by returning from the driving simulator to the psychophysics laboratory. In particular, we conducted a series of experimental studies in which we gave the participants a straightforward orthogonal spatial cuing task to perform while varying the perceptual load that they were under, by presenting a rapid serial visual presentation (RSVP) stream of characters at fixation. In the low load condition, the participants could ignore whatever was going on at fixation and only had to respond to the peripheral targets requiring an elevation discrimination response. Under the high load condition, the participants had to detect the target characters that were sometimes presented in the central visual stream. On other trials where no central target appeared in the RSVP stream, the participants simply had to respond to the elevation of the peripheral target instead (e.g., as in a traditional orthogonal spatial cuing study). What kind of target might be presented on an upcoming trial was entirely unpredictable.

The results of this research were completely unambiguous (see Santangelo, Ho, & Spence, 2008; Santangelo & Spence, 2007). In particular, under the no/low load conditions, multisensory (in this case, audiovisual) peripheral cues were found to be no more effective than unisensory cues (thus replicating the results of our earlier research; see Spence & Driver, 1999). However, as soon as the load of the participant's task was increased (i.e., when the participant had to monitor the central RSVP stream), multisensory cues were found to retain their effectiveness while unisensory cues suddenly lost their ability to capture an operator's spatial attention (see Spence, 2010, for a review). That is, unimodal auditory, visual, or tactile cues presented unpredictably to the left or right of fixation no longer exogenously captured participants' spatial attention (Santangelo & Spence, 2007). However, both audiovisual and audiotactile cues still did, even under the most demanding of dual-task conditions (Santangelo, Ho, & Spence, 2008). It turned out, then, that the lack of perceptual load was the key reason why multisensory cues were no more effective than unimodal cues in the earlier research (e.g., see Spence & Driver, 1999). One unanswered question currently is whether all multi-sensory cues, for example, audiovisual, visuotactile, audiotactile, are equally effective or not under load. Future comparative studies will be needed to deliver a meaningful answer to this question. A related question here is whether trimodal cues are even more effective than bimodal cues. Once again, though, more research will be needed to address this intriguing question.

Now, whether such results should really be considered in terms of the notion of superadditivity (that inspired this line of research in the first place) remains something of an open question. Here, it should be noted that while superadditivity can be demonstrated at the level of a single neuron, at a population level in the SC, the distribution of neural responses actually appears to be fairly normally distributed around straight additivity (Stein, Stanford, Wallace, Vaughan, & Jiang, 2004).

Many of the studies that were conducted over the last few years have demonstrated the same pattern of results: namely, that multisensory cues appear to have an automatic interrupt function under load (see Spence, 2010, for a review) whereas unisensory cues do not (see Koelewijn, Bronkhorst, & Theeuwes, 2009). This example therefore represents just one of the ways in which laboratory-based and real-world research can mutually inform one another, especially as part of an interactive and iterative "... lab to real world ... to lab ... and back again" process. In the case just mentioned, the fundamental insights (or rather the empirical observations that stimulated the relevant laboratory research) emerged from the real world, but further insight/explanation of the constraints and robustness of the underlying phenomenon came from psychophysical research conducted in the psychophysics laboratory.

Spatial Coincidence and the Benefits of Multisensory Warning Signals

One important example of a research finding that moved in the opposite direction (i.e., from the laboratory to the real world) comes from single-cell neurophysiological studies demonstrating the importance of spatial coincidence to multisensory integration: If two stimuli excite the overlapping neuronal receptive fields, then superadditive responses may well be observed (especially if the responses to the individual stimuli are weak—this is "the law of inverse effectiveness" mentioned earlier; e.g., see Holmes & Spence, 2005). However, subadditive responses are more likely to be observed if the two stimuli do not match up (i.e., if they are incongruent in either space or time). Now, as the engineers working in the real world have slowly come to realize, when it comes to trying to elicit multisensory benefits in their multisensory warning signal design, what they have typically failed to appreciate is the importance of spatial colocation (even if not precise colocation then at least spatial correspondence). One sees many human-factors research papers in which the authors report on their attempts to compare unisensory to multisensory cues. The results are, however, often disappointing—in the sense that multisensory cues provide no additional benefit relative to the best of the unisensory cues (e.g., see Fitch, Kiefer, Hankey, & Kleiner, 2007; Selcon, Taylor, & McKenna, 1995, for representative examples). If one looks carefully at the methods that were utilized in these various studies, one immediately sees that the cues that were presented to the participants came from markedly *different* spatial locations. Thus, for example, in Fitch et al.'s study, vibrotactile signals were provided by stimulators mounted in the bottom of the driver's seat while auditory cues were presented from one of eight loudspeakers mounted around the vehicle's interior.

We have now conducted a number of laboratory studies that have helped to emphasize the importance of spatial colocation for both the exogenous and endogenous driving of a driver's spatial attention (Ho, Santangelo, & Spence, 2009; Ho & Spence, 2008; Spence & Read, 2003). We have, for example, been able to demonstrate that people find it easier to combine speech shadowing and driving when the speech stream comes from directly in front of drivers rather than from their side (Spence & Read, 2003). Meanwhile, attentional capture simply does not occur under those conditions in which the auditory and tactile cues are spatially misaligned (e.g., when the tactile cue is presented on the middle of a person's stomach while the auditory cues are delivered to one or other side; Ho et al., 2009). However, as soon as the various component cues are brought into directional alignment, then the robust spatial cuing effect returns (see Ho et al., 2009, figures 2 and 3).[3]

This is, of course, all great news for the human-factors researcher. However, there is certainly still a lot more research to be done. For example, one of the current interests for a number of researchers relates to the fact that slightly desynchronizing auditory and visual signals can sometimes lead to SC neurons' being driven more effectively than when the component signals are presented in exact synchrony, especially if those signals happen to be presented from close to the observer's head (King & Palmer, 1985; Meredith, Nemitz, & Stein, 1987; Spence & Squire, 2003), as is, of course, typically the case in the driving situation (where any warning signals will likely be presented within the confines of the car interior). The question therefore becomes one of whether slightly desynchronizing the component parts of an audiovisual signal might give rise to warning signals that are even more effective than synchronous ones (Spence & Driver, 1999; see also Chan & Chan, 2006). Here, what we see is that a number of the larger car companies are going straight to very expensive and time-consuming simulator studies in order to test this idea. We would, however, like to argue that a better (or at least more efficient) option here might well be to start out in the laboratory (Ho, Gray, & Spence, 2013). There, one could investigate a range of stimulus types and asynchronies before moving over to real-world testing and implementation. In passing, it is perhaps also worth noting that slightly desynchronizing multisensory (i.e., audiovisual) signals might also serve to provide the driver's brain with a subtle (perhaps subconscious) cue as to the distance of the event of interest (cf. Jaekl, Soto-Faraco, & Harris, 2011).

The Discovery and Development of Near-Rear Warning Signals

As a final example of how the findings of laboratory-based studies can have an impact on real-world attention research, we would like to mention the relatively recent discovery of neural circuits in the brain that appear to be dedicated to the monitoring of near-rear peripersonal space: that part of space that we never see (and most certainly rarely consciously think about; see Spence & Driver, 2004), but which may have been crucial for our ancestors to monitor in terms of their evolutionary survival. After all, if a predator were to have grabbed us by the back of the neck, then the consequences would most likely be serious to say the least! Graziano and his colleagues originally demonstrated the existence of such neural circuits in monkeys (e.g., see Graziano & Cooke, 2006). Other researchers have subsequently shown the existence of similar circuits in normal and brain-damaged human participants as well (see Occelli, Spence, & Zampini, 2011, for a review). Ho and Spence (2009) took this insight and tested whether a near-rear auditory warning signal (perhaps mounted in the driver's headrest) could be used to trigger a rapid head-turning response when a driver is distracted from monitoring the roadway. Once again, this is something that does not appear to have been intuitive to the engineer (after all, how often have you come across a headrest-mounted warning signal before?). Ho and Spence's results demonstrated the potential benefits of presenting warning signals in near-rear space as compared to the same auditory warning signal when presented from in front of the driver or a warning signal presented in another modality (though see also Sambo & Iannetti, 2013). We know of at least one car company that will launch a headrest-mounted display for their vehicles in the coming year.

Future Directions

Now, while there is clearly a benefit to be had in terms of understanding attention and hence the perception of drivers in the real world (from the results of laboratory studies), we would like to argue that thinking about real-world research also provides some intriguing and important directions for (or constraints on the design of) future laboratory research on attention. For example, an experimentalist working in the laboratory typically wants to get as many data points as possible. Hence, in a standard exogenous spatial cuing study, there might be as many as 1,000 "unexpected" cues over the course of an hour-long testing session. By contrast, real-world car warning signals may only occur once a week, once a month, or even less often. That means that they may really be unexpected. Addressing the question of how to make sure that such unexpected warnings are effective without being unnerving is a challenge currently awaiting the research community (Ho & Spence, 2008). In our own work, we have been experimenting with the idea of using looming auditory, tactile, and audiotactile warning signals since they avoid the sudden onset while also potentially providing useful information about the speed of approach of another vehicle (Gray, 2011; Ho, Spence, & Gray, 2013; Meng, Gray, Ho, Ahtamad, & Spence, 2015; see also Ho & Spence, 2013).

We would argue that five issues really stand out as needing to be tackled in future research: first, studying spatial attention in the laboratory under conditions where attention-capturing events are presented much less frequently than they are in typical laboratory-based studies of attention. Second, more thought needs to be given to the design of warning signals that avoid the shock of sudden onset but that nevertheless appear to provide especially effective attention-capturing signals (Gray, 2011; Ho, Spence, & Gray, 2013; Leo, Romei, Freeman, Ladavas, & Driver, 2011).

Third, we would like to suggest that too much of the research in laboratory studies that has been conducted to date has tended to focus on the use of pure tones and white noise bursts as attention-capturing cues. Note that these are essentially meaningless stimuli that are rarely encountered in the real world. Thinking about the real-word context in which meaningful icons, verbal cues ("brake!"; Edworthy & Hellier, 2006; Ho & Spence, 2006), and iconic sounds (car horn sounds, screeching car tires, etc.) are heard (Graham 1999; McKeown & Isherwood, 2007; Oyer & Hardick, 1963), in the future it might be better to start using these stimuli in those laboratory studies that have any pretensions of real-world relevance/application. Of course, given that drivers are unlikely to accept training in order to interpret warning signals, wherever possible, it may be good if the stimuli, especially sounds, used are iconic, in that the relation between signal and referent is as clear as possible (see Petocz, Keller, & Stevens, 2008).

Fourth, there seems to be too much focus in the laboratory on button pressing as the preferred response from participants. Indeed, many researchers appear to believe that response modality such as manual/vocal does not affect the results that are observed. However, perhaps the mode of responding does have the potential to exert more of an impact than many researchers realize (Gallace, Soto-Faraco, Dalton, Kreukniet, & Spence, 2008). Hence, the new approach is to think about which response one wishes to influence—for example, is it getting the driver to turn the steering wheel (Fitch, Hankey, Kleiner, & Dingus, 2011), to press the brake pedal (Ho & Spence, 2008), or to turn his or her head back to the roadway (Ho & Spence, 2009)?—and then to test the specific response modality concerned (see Ho & Spence, 2014). It is certainly worth bearing in mind that different signals may be preferentially linked to specific behavioral responses. Once again, such a conclusion is not one that naturally emerges from laboratory-based attention research.

Fifth, a final challenge for future research that is worth mentioning here relates to the topic of aging drivers. Here, at least, there is some hope that multisensory cues may prove to make up for sensory loss (Laurienti & Hugenschmidt, 2012; Laurienti, Burdette, Maldjian, & Wallace, 2006). Laurienti and his colleagues demonstrated in the laboratory that older participants were able to use multisensory integration in order to make up for their unisensory decline in the laboratory setting. One can only imagine how important such a finding would be were a similar pattern of results to be obtained in a real-world driving scenario, for example (Lees et al., 2012; Spence & Ho, 2008b).

Conclusions

Ultimately, our sense is that while applied research has always seemed to fall some years behind laboratory research in terms of its theoretical underpinnings, the situation has changed rapidly. In the area of attention capture and driving research, as in many other areas, it turns out that there is a rich interplay to be had between the laboratory and the real world. It is also becoming increasingly apparent how many laboratory-based researchers need to move beyond the WEIRDo's (i.e., the Western, Educated, Intelligent, Rich, and Democratic North American psychology students) who constitute the subject pool in 95% of published psychology research (see Henrich, Heine, & Norenzayan, 2010). In the above discussion, we have chosen to focus on the situation of driver warning systems. However, one might want to ask where else time-critical responding by a human operator would be important? Here, one might think about military and aviation scenarios (e.g., Baldwin et al., 2012; Haas & van Erp, 2014; Oskarsson et al., 2012; Sklar & Sarter, 1999). Perhaps even more interesting is to consider air traffic control and financial trading (Ngo et al., 2012; http://www.informationweek.com/news/199200297).

To conclude, in our own research we have found it extremely fruitful to switch back and forth between laboratory-based studies of attention and real-world research on the design of warning signals and information systems (albeit with most of our research having been conducted in a driving simulator rather than on-road; Reed & Green, 1999). We have frequently found that the basic research provides relevant ideas for real-world research, and that conducting research in the real world can sometimes get one to think differently about the basic research paradigms that we typically use. The benefits, it has to be said, have often been unexpected/surprising (and most certainly were not predicted as outcomes of the research). Of course, in the era of impactful research, there are few places better to be than in the study of crossmodal attention and multisensory integration and its application in interface design (see Ho, Gray, & Spence, 2014; Spence, 2012, for reviews).

Box 6.1
Key Points

- Perception is fundamentally multisensory, and attention is controlled crossmodally.
- Spatially colocated multisensory warning signals capture an interface operator's attention more effectively than unisensory warning signals when the operator is working under conditions of high perceptual load.

Box 6.2

Outstanding Issues

- Future research should study spatial attention under those conditions in which the attention-capturing events are presented much less frequently than they are in typical laboratory research (in order to better approximate real-world conditions).

- Laboratory-based attention researchers should consider presenting more meaningful cues (that have real-world relevance) in future laboratory studies of attentional orienting.

- Researchers should think about the specific response modality relevant to the research question and test it accordingly, given that different signals may be preferentially linked to specific behavioral responses.

Notes

1. Though "early or late relative to what?" was rarely explicitly specified.

2. Interestingly, when the loudspeakers used to present the auditory cues were brought close to the front/back of the participant's head, auditory spatial cuing effects were suddenly eliminated in the colored number plate discrimination task, despite the fact that the participants could still discriminate the direction of the warning signal without any trouble (Ho & Spence, 2008). This intriguing result therefore suggests that the fundamental difference between the effectiveness of tactile and auditory warning signals is really attributable to the fact that tactile signals must, of necessity, be presented to the driver's body itself whereas auditory signals are normally presented from further away (cf. Previc, 1998, 2000), rather than in terms of there being any fundamental difference between the senses themselves (Spence & Ho, 2008c).

3. It is perhaps worth noting here that if one were to present a relatively unlocalizable auditory cue at the same time as the onset of the brake lights on the lead car, say, a ventriloquism effect might well result in a driver's mislocalizing the sound in the direction of the brake lights (e.g., Alais & Burr, 2004; see also Spence, 2013).

References

Abbate, J. (1999). Getting small: A short history of the personal computer. *Proceedings of the IEEE, 87*, 1695–1698.

Alais, D., & Burr, D. (2004). The ventriloquist effect results from near-optimal bimodal integration. *Current Biology, 14*, 257–262.

Allport, A. (1993). Attention and control: Have we been asking the wrong questions? A critical review of twenty-five years. In D. E. Meyer & S. Kornblum (Eds.), *Attention and performance XIV:*

Synergies in experimental psychology, artificial intelligence, and cognitive neuroscience (pp. 183–218). Cambridge, MA: MIT Press.

Ashley, S. (2001). Driving the info highway. *Scientific American, 285*(4), 44–50.

Baldwin, C. L., Spence, C., Bliss, J. P., Brill, J. C., Wogalter, M. S., Mayhorn, C. B., & Ferris, T. (2012). Multimodal cueing: The relative benefits of the auditory, visual, and tactile channels in complex environments. *Proceedings of the 56th Human Factors and Ergonomics Society Meeting, 56,* 1431–1435.

Brown, I. D. (1962). Measuring the "spare mental capacity" of car drivers by a subsidiary auditory task. *Ergonomics, 5,* 247–250.

Brown, I. D. (1965). Effect of a car radio on driving in traffic. *Ergonomics, 8,* 475–479.

Brown, I. D., & Poulton, E. C. (1961). Measuring the spare "mental capacity" of car drivers by a subsidiary task. *Ergonomics, 4,* 35–40.

Brown, I. D., Simmonds, D. C. V., & Tickner, A. H. (1967). Measurement of control skills, vigilance, and performance on a subsidiary task during 12 hours of car driving. *Ergonomics, 10,* 665–673.

Brown, I. D., Tickner, A. H., & Simmonds, D. C. V. (1969). Interference between concurrent tasks of driving and telephoning. *Journal of Applied Psychology, 53,* 419–424.

Chan, A. H. S., & Chan, K. W. L. (2006). Synchronous and asynchronous presentations of auditory and visual signals: Implications for control console design. *Applied Ergonomics, 37,* 131–140.

Driver, J. (2001). A selective review of selective attention research from the past century. *British Journal of Psychology, 92,* 53–78.

Duncan, J. (1980). The demonstration of capacity limitation. *Cognitive Psychology, 12,* 75–96.

Edworthy, J., & Hellier, E. (2006). Complex nonverbal auditory signals and speech warnings. In M. S. Wogalter (Ed.), *Handbook of warnings* (pp. 199–220). Mahwah, NJ: Erlbaum.

Egly, R., Driver, J., & Rafal, R. D. (1995). Shifting visual attention between objects and locations: Evidence from normal and parietal lesion subjects. *Journal of Experimental Psychology. General, 123,* 161–177.

Eriksen, B. A., & Eriksen, C. W. (1974). Effects of noise letters upon the identification of a target letter in a nonsearch task. *Perception & Psychophysics, 16,* 143–149.

Fitch, G. M., Hankey, J. M., Kleiner, B. M., & Dingus, T. A. (2011). Driver comprehension of multiple haptic seat alerts intended for use in an integrated collision avoidance system. *Transportation Research Part F: Traffic Psychology and Behaviour, 14,* 278–290.

Fitch, G. M., Kiefer, R. J., Hankey, J. M., & Kleiner, B. M. (2007). Toward developing an approach for alerting drivers to the direction of a crash threat. *Human Factors, 49,* 710–720.

Gallace, A., Soto-Faraco, S., Dalton, P., Kreukniet, B., & Spence, C. (2008). Response requirements modulate tactile spatial congruency effects. *Experimental Brain Research, 191*, 171–186.

Gallace, A., & Spence, C. (2014). *In touch with the future: The sense of touch from cognitive neuroscience to virtual reality.* Oxford, UK: Oxford University Press.

Gazzaley, A., & Nobre, A. C. (2012). Top-down modulation: Bridging selective attention and working memory. *Trends in Cognitive Sciences, 16*, 129–135.

Graham, R. (1999). Use of auditory icons as emergency warnings: Evaluation within a vehicle collision avoidance application. *Ergonomics, 42*, 1233–1248.

Gray, R. (2011). Looming auditory collision warnings for driving. *Human Factors, 53*, 63–74.

Graziano, M. S. A., & Cooke, D. F. (2006). Parieto-frontal interactions, personal space, and defensive behavior. *Neuropsychologia, 44*, 845–859.

Haas, E. C., & van Erp, J. B. F. (2014). Multimodal warnings to enhance risk communication and safety. *Safety Science, 61*, 29–35.

Henrich, J., Heine, S. J., & Norenzayan, A. (2010). The weirdest people in the world? *Behavioral and Brain Sciences, 33*, 61–135.

Ho, C., Gray, R., & Spence, C. (2013). Role of audiovisual synchrony in driving head orienting responses. *Experimental Brain Research, 227*, 467–476.

Ho, C., Gray, R., & Spence, C. (2014). To what extent do the findings of laboratory-based spatial attention research apply to the real-world setting of driving? *IEEE Transactions on Human–Machine Systems, 44*, 524–530.

Ho, C., Reed, N., & Spence, C. (2007). Multisensory in-car warning signals for collision avoidance. *Human Factors, 49*, 1107–1114.

Ho, C., Santangelo, V., & Spence, C. (2009). Multisensory warning signals: When spatial correspondence matters. *Experimental Brain Research, 195*, 261–272.

Ho, C., & Spence, C. (2005). Assessing the effectiveness of various auditory cues in capturing a driver's visual attention. *Journal of Experimental Psychology. Applied, 11*, 157–174.

Ho, C., & Spence, C. (2006). Verbal interface design: Do verbal directional cues automatically orient visual spatial attention? *Computers in Human Behavior, 22*, 733–748.

Ho, C., & Spence, C. (2008). *The multisensory driver: Implications for ergonomic car interface design.* Aldershot, UK: Ashgate.

Ho, C., & Spence, C. (2009). Using peripersonal warning signals to orient a driver's gaze. *Human Factors, 51*, 539–556.

Ho, C., & Spence, C. (2013). Affective multisensory driver interface design. [Special Issue on Human Emotional Responses to Sound and Vibration in Automobiles]. *International Journal of Vehicle Noise and Vibration, 9*, 61–74.

Ho, C., & Spence, C. (2014). Effectively responding to tactile stimulation: Do homologous cue and effector locations really matter? *Acta Psychologica, 151*, 32–39.

Ho, C., Spence, C., & Gray, R. (2013). Looming auditory and vibrotactile collision warnings for safe driving. In *Proceedings of the 7th International Driving Symposium on Human Factors in Driver Assessment, Training, and Vehicle Design* (pp. 551–557). Iowa City: University of Iowa, Public Policy Center.

Ho, C., Tan, H. Z., & Spence, C. (2005). Using spatial vibrotactile cues to direct visual attention in driving scenes. *Transportation Research Part F: Traffic Psychology and Behaviour, 8*, 397–412.

Ho, C., Tan, H. Z., & Spence, C. (2006). The differential effect of vibrotactile and auditory cues on visual spatial attention. *Ergonomics, 49*, 724–738.

Holmes, N. P., & Spence, C. (2005). Multisensory integration: Space, time, and superadditivity. *Current Biology, 15*, R762–R764.

Jaekl, P., Soto-Faraco, S., & Harris, L. (2011). Perceived size change induced by audio-visual temporal delays. *Experimental Brain Research, 216*, 457–462.

King, A. J., & Palmer, A. R. (1985). Integration of visual and auditory information in bimodal neurones in the guinea-pig superior colliculus. *Experimental Brain Research, 60*, 492–500.

Klein, R. (2000). Inhibition of return. *Trends in Cognitive Sciences, 4*, 138–147.

Koelewijn, T., Bronkhorst, A., & Theeuwes, J. (2009). Auditory and visual capture during focused visual attention. *Journal of Experimental Psychology: Human Perception and Performance, 35*, 1303–1315.

Laurienti, P. J., Burdette, J. H., Maldjian, J. A., & Wallace, M. T. (2006). Enhanced multisensory integration in older adults. *Neurobiology of Aging, 27*, 1155–1163.

Laurienti, P. J., & Hugenschmidt, C. E. (2012). Multisensory processes in old age. In A. Bremner, D. Lewkowicz, & C. Spence (Eds.), *Multisensory development* (pp. 251–270). Oxford, UK: Oxford University Press.

Lavie, N. (1995). Perceptual load as a necessary condition for selective attention. *Journal of Experimental Psychology: Human Perception and Performance, 21*, 451–468.

Lavie, N. (2005). Distracted and confused? Selective attention under load. *Trends in Cognitive Sciences, 9*, 75–82.

Lees, M. N., Cosman, J., Lee, J. D., Vecera, S. P., Dawson, J. D., & Rizzo, M. (2012). Crossmodal warnings for orienting attention in older drivers with and without attention impairment. *Applied Ergonomics, 43*, 768–776.

Leo, F., Romei, V., Freeman, E., Ladavas, E., & Driver, J. (2011). Looming sounds enhance orientation sensitivity for visual stimuli on the same side as such sounds. *Experimental Brain Research, 213*, 193–201.

Levy, J., Pashler, H., & Boer, E. (2006). Central interference in driving: Is there any stopping the psychological refractory period? *Psychological Science, 17*, 228–235.

Luck, S. J., & Ford, M. A. (1998). On the role of selective attention in visual perception. *Proceedings of the National Academy of Sciences of the United States of America, 95*, 825–830.

McKeown, J. D., & Isherwood, S. (2007). Mapping the urgency and pleasantness of speech, auditory icons, and abstract alarms to their referents within the vehicle. *Human Factors, 49*, 417–428.

Meng, F., Gray, R., Ho, C., Ahtamad, M., & Spence, C. (2015). Dynamic vibrotactile signals for forward collision avoidance warning systems. *Human Factors, 57*, 329–346.

Meredith, M. A., Nemitz, J. W., & Stein, B. E. (1987). Determinants of multisensory integration in superior colliculus neurons: I. Temporal factors. *Journal of Neuroscience, 7*, 3215–3229.

Mohebbi, R., Gray, R., & Tan, H. Z. (2009). Driver reaction time to tactile and auditory rear-end collision warnings while talking on a cell phone. *Human Factors, 51*, 102–110.

Müller, J. R., Philiastides, M. G., & Newsome, W. T. (2005). Microstimulation of superior colliculus focuses attention without moving the eyes. *Proceedings of the National Academy of Sciences of the United States of America, 102*, 524–529.

Ngo, M. K., Pierce, R., & Spence, C. (2012). Utilizing multisensory cues to facilitate air traffic management. *Human Factors, 54*, 1093–1103.

Nobre, A. C., & Coull, J. T. (Eds.). (2010). *Attention and time.* Oxford, UK: Oxford University Press.

Occelli, V., Spence, C., & Zampini, M. (2011). Audiotactile interactions in front and rear space. *Neuroscience and Biobehavioral Reviews, 35*, 589–598.

Oskarsson, P.-A., Eriksson, L., & Carlander, O. (2012). Enhanced perception and performance by multimodal threat cueing in simulated combat vehicle. *Human Factors, 54*, 122–137.

Oyer, J., & Hardick, J. (1963). *Response of population to optimum warning signal* (Final Report No. SHSLR163, Contract No. OCK-OS-62–182). Washington, DC: Office of Civil Defense.

Petocz, A., Keller, P. E., & Stevens, C. J. (2008). Auditory warnings, signal-referent relations, and natural indicators: Re-thinking theory and application. *Journal of Experimental Psychology. Applied, 14*, 165–178.

Posner, M. I. (1978). *Chronometric explorations of mind.* Hillsdale, NJ: Erlbaum.

Previc, F. H. (1998). The neuropsychology of 3-D space. *Psychological Bulletin, 124*, 123–164.

Previc, F. H. (2000). Neuropsychological guidelines for aircraft control stations. *IEEE Engineering in Medicine and Biology Magazine, 19*, 81–88.

Quinlan, P. T. (2003). Visual feature integration theory: Past, present, and future. *Psychological Bulletin, 129*, 643–673.

Ramsey, K. L., & Simmons, F. B. (1993). High-powered automobile stereos. *Otolaryngology—Head and Neck Surgery, 109*, 108–110.

Reed, M. P., & Green, P. A. (1999). Comparison of driving performance on-road and in a low-cost simulator using a concurrent telephone dialling task. *Ergonomics, 42*, 1015–1037.

Reynolds, J. H., & Chelazzi, L. (2004). Attentional modulation of visual processing. *Annual Review of Neuroscience, 27*, 611–647.

Sambo, C. F., & Iannetti, G. D. (2013). Better safe than sorry? The safety margin surrounding the body is increased by anxiety. *Journal of Neuroscience, 33*, 14225–14230.

Santangelo, V., Ho, C., & Spence, C. (2008). Capturing spatial attention with multisensory cues. *Psychonomic Bulletin & Review, 15*, 398–403.

Santangelo, V., & Spence, C. (2007). Multisensory cues capture spatial attention regardless of perceptual load. *Journal of Experimental Psychology: Human Perception and Performance, 33*, 1311–1321.

Santangelo, V., & Spence, C. (2008). Is the exogenous orienting of spatial attention truly automatic? Evidence from unimodal and multisensory studies. *Consciousness and Cognition, 17*, 989–1015.

Sarter, N. (2013). Multimodal displays: Conceptual basis, design guidance, and research needs. In J. D. Lee & A. Kirklick (Eds.), *The Oxford handbook of cognitive engineering* (pp. 556–565). Oxford, UK: Oxford University Press.

Scott, J. J., & Gray, R. (2008). A comparison of tactile, visual, and auditory warnings for rear-end collision prevention in simulated driving. *Human Factors, 50*, 264–275.

Selcon, S. J., Taylor, R. M., & McKenna, F. P. (1995). Integrating multiple information sources: Using redundancy in the design of warnings. *Ergonomics, 38*, 2362–2370.

Senders, J. W., Kristofferson, A. B., Levison, W. H., Dietrich, C. W., & Ward, J. L. (1967). The attentional demand of automobile driving. *Highway Research Record, 195*, 15–33.

Shulman, G. L. (1990). Relating attention to visual mechanisms. *Perception & Psychophysics, 47*, 199–203.

Sivak, M. (1996). The information that drivers use: Is it indeed 90% visual? *Perception, 25*, 1081–1089.

Sklar, A. E., & Sarter, N. B. (1999). Good vibrations: Tactile feedback in support of attention allocation and human–automation coordination in event-driven domains. *Human Factors, 41*, 543–552.

Spence, C. (2010). Crossmodal spatial attention. *Annals of the New York Academy of Sciences, 1191*, 182–200.

Spence, C. (2012). Drive safely with neuroergonomics. *Psychologist, 25*, 664–667.

Spence, C. (2013). Just how important is spatial coincidence to multisensory integration? Evaluating the spatial rule. *Annals of the New York Academy of Sciences, 1296*, 31–49.

Spence, C. J., & Driver, J. (1994). Covert spatial orienting in audition: Exogenous and endogenous mechanisms. *Journal of Experimental Psychology: Human Perception and Performance, 20,* 555–574.

Spence, C., & Driver, J. (1997a). Audiovisual links in exogenous covert spatial orienting. *Perception & Psychophysics, 59,* 1–22.

Spence, C., & Driver, J. (1997b). Cross-modal links in attention between audition, vision, and touch: Implications for interface design. *International Journal of Cognitive Ergonomics, 1,* 351–373.

Spence, C., & Driver, J. (1999). A new approach to the design of multimodal warning signals. In D. Harris (Ed.), *Engineering psychology and cognitive ergonomics: Vol. 4. Job design, product design and human–computer interaction* (pp. 455–461). Aldershot, UK: Ashgate.

Spence, C., & Driver, J. (Eds.). (2004). *Crossmodal space and crossmodal attention.* Oxford, UK: Oxford University Press.

Spence, C., & Gallace, A. (2007). Recent developments in the study of tactile attention. *Canadian Journal of Experimental Psychology, 61,* 196–207.

Spence, C., & Ho, C. (2008a). Crossmodal information processing in driving. In C. Castro (Ed.), *Human factors of visual performance in driving* (pp. 187–200). Boca Raton, FL: CRC Press.

Spence, C., & Ho, C. (2008b). Multisensory driver interface design: Past, present, and future. *Ergonomics, 51,* 65–70.

Spence, C., & Ho, C. (2008c). Multisensory warning signals for event perception and safe driving. *Theoretical Issues in Ergonomics Science, 9,* 523–554.

Spence, C., Lloyd, D., McGlone, F., Nicholls, M. E. R., & Driver, J. (2000). Inhibition of return is supramodal: A demonstration between all possible pairings of vision, touch and audition. *Experimental Brain Research, 134,* 42–48.

Spence, C., & Read, L. (2003). Speech shadowing while driving: On the difficulty of splitting attention between eye and ear. *Psychological Science, 14,* 251–256.

Spence, C., & Squire, S. B. (2003). Multisensory integration: Maintaining the perception of synchrony. *Current Biology, 13,* R519–R521.

Stein, B. E., Huneycutt, W. S., & Meredith, M. A. (1988). Neurons and behavior: The same rules of multisensory integration apply. *Brain Research, 448,* 355–358.

Stein, B. E., & Meredith, M. A. (1993). *The merging of the senses.* Cambridge, MA: MIT Press.

Stein, B. E., Meredith, M. A., Honeycutt, W. S., & McDade, L. (1989). Behavioral indices of multisensory integration: Orientation to visual cues is affected by auditory stimuli. *Journal of Cognitive Neuroscience, 1,* 12–24.

Stein, B. E., & Stanford, T. R. (2008). Multisensory integration: Current issues from the perspective of the single neuron. *Nature Reviews. Neuroscience, 9,* 255–267.

Stein, B. E., Stanford, T. R., Wallace, M. T., Vaughan, W. J., & Jiang, W. (2004). Crossmodal spatial interactions in subcortical and cortical circuits. In C. Spence & J. Driver (Eds.), *Crossmodal space and crossmodal attention* (pp. 25–50). Oxford, UK: Oxford University Press.

Treisman, A. M., & Gelade, G. (1980). A feature-integration theory of attention. *Cognitive Psychology*, *12*, 97–136.

Van Erp, J. B. F., Eriksson, L., Levin, B., Carlander, O., Veltman, J. A., & Vos, W. K. (2007). Tactile cueing effects on performance in simulated aerial combat with high acceleration. *Aviation, Space, and Environmental Medicine*, *78*, 1128–1134.

Wright, R. D., & Ward, L. M. (2008). *Orienting of attention*. Oxford, UK: Oxford University Press.

7

Task-Set Control and Task Switching

Stephen Monsell

The brain contains a repertoire of perceptual, motor, linguistic, spatial, and many other representations and processes. To get any cognitive task accomplished, we have to connect and configure a subset of these components so that they will select relevant perceptual inputs, perform operations upon those inputs that will yield the required type of decision, and generate appropriate overt actions or information to store in memory. This connecting and configuring is described as adopting a "task set." To investigate the processes which control task set, we need to exercise those processes in the lab and measure and manipulate their behavioral consequences and associated brain activity. The term "task switching" has come to label a set of experimental paradigms that exercise task-set control by requiring participants to switch frequently among tasks specified by the experimenter (Grange & Houghton, 2014a; Kiesel et al., 2010; Monsell, 2003; Vandierendonck, Liefooghe, & Verbruggen, 2010).

In such an experiment, the participant is typically introduced to, and practices, each of a small set of choice–reaction time tasks (typically two or three). The tasks require the participant to classify a stimulus in different ways, identify different properties of the stimulus, or perform different operations, such as addition versus subtraction, on combinations of stimuli. (For a somewhat arbitrary collection of examples, see table 7.1.) Participants then progress to blocks of trials on each of which any one of the tasks may have to be performed, sometimes the same task as on the previous trial (a *task-repeat* trial) and sometimes a different task (a *task-switch* trial). Averaging performance over many trials, we usually find a substantially longer mean reaction time (RT), and often more errors, on task-switch than on task-repeat trials. These "switch costs"[1] and other phenomena revealed by these experiments supply clues about task-set control: how the brain reconfigures itself to achieve a different processing goal, and limits on this control. This chapter samples what we have learned from such experiments, focusing largely on behavioral data. I should acknowledge from the outset that although we switch tasks frequently in everyday life, and although there is an applied literature (see the Integration section), I focus mainly on attempts to probe basic mechanisms of task-set control in the laboratory.

Table 7.1

Examples of Task-Switching Studies, Chosen to Illustrate the Range of Stimuli, Tasks, Paradigms and Issues Examined

Stimulus Set	Tasks	Paradigm	Citation	Comment
8 digits	Classify digit as high/low or odd/even with L/R key press	Task cuing, two verbal cues per task (e.g., PARITY or ODD/EVEN)	Logan & Bundesen (2003)	The use of two cues per task in this case allows unconfounding of task and cue change. These digit-classification tasks have been used by numerous authors and with other paradigms (e.g., Altmann, 2002).
8 digits	Classify digit as high/low, odd/even, or "inner/outer" with L/R key press	Task cuing, cue is background shape	Schuch & Koch (2003)	Use of 3 tasks allows assessment of $n-2$ task repetition.
144 object names	Classify as (1) mono/disyllable, (2) big/small object with L/R key press	Task cuing, two transparent pictorial cues, 2 per task	Monsell & Mizon (2006, Experiment 3)	An example of a large stimulus set; for much larger sets see Arrington & Logan (2004b) and Elchlepp et al. (2015). Also illustrates use of several CSIs to plot preparation function: see figure 7.1.
4 shapes in 4 colors	Identify shape or color with one of 4 keys	Task cuing, one verbal and one pictorial cue per task	Lavric et al. (2008)	An example of four-choice identification rather than the usual binary classification. Looks at ERP correlates of task-set preparation; includes comparison of two types of task cue.
Digit + Letter pair (e.g., "4L") or (as univalent stimuli) pairs such as "4&" and "G#"	Classify digit as odd/even or letter as consonant/vowel	Alternating runs of 2 or 4, cued by position in frame	Rogers & Monsell (1995)	Allows comparison of univalent (irrelevant character neither letter nor digit) to bivalent (letter and digit mapped to congruent or incongruent responses). Popular in neuroscience and applied research (in spite of the complexity of having to locate the relevant character).
Bar with/without L/R arrowhead contains "LEFT," "RIGHT," or "XXXX"	Respond with L/R key press to direction of the arrow or the word	Alternating runs of 3, task and run position cued by background "clock hand"	Aron et al. (2004)	S-R mapping easy to acquire because of preexisting meanings of arrows and words, and remains constant: Only the relevant attribute switches. Also allows univalent/multivalent comparison.
108 object pictures with superimposed object name	Name object or word	Intermittent instruction with runs of 3	Waszak et al. (2003, Experiment 1)	Identification/naming responses for each task. Use of many stimuli also allows manipulation of stimulus history (in this case the task previously performed on it).

Note. CSI, cue–stimulus interval.

Historical Context

From the beginnings of experimental psychology in the nineteenth century, it has been recognized as a core property of our mental architecture that in different contexts the same input can trigger different mental operations and actions. Changing ones mental "set" (Einstellung), and hence how the stimulus is processed, as a function of the context, involves a delicate balance between deliberate choice or "will," on the one hand, and on the other hand processing being modulated by environmental context in a relatively automatic way (see Goschke, 2000, on dimensions of this balance). Current terminology refers to control being "endogenous" or "exogenous." Psychology loves a categorical dichotomy, but it is important to stress that these terms describe endpoints on a continuum of voluntariness, and it is the interaction between top-down and bottom-up control (as also discussed in other chapters in this volume) that determines the task set in play. Someone with long experience of driving a car without automatic transmission usually shifts gear "automatically" when the engine pitch reaches a certain level, though they can also do so with more deliberation; a novice has to exert deliberate control.

The "set" part of the term "task set" captures another important idea. Voluntary action is often not a matter of perceiving the stimulus, then deliberating what to do about it, and then choosing an action. Instead one voluntarily "sets" oneself *in advance* (e.g., one gets oneself into tennis, or driving, or reading mode—or, in the lab, into odd–even or high–low digit-classification mode) so that when the relevant stimulus appears, it drives a chain of processes that run through to initiation of an action without further deliberation or exercise of volition (see Hommel, 2000, on Exner's 1879 "prepared reflex"); that is not to say that we cannot deliberately interrupt the chain if needed.

Another idea with a long lineage (e.g., Fitts & Posner, 1967) has to do with how a task set is acquired. Many tasks are initially specified to us via verbal instructions, or we figure out a set of instructions for ourselves. These instructions are initially represented as such in *declarative* memory, and we perform the task by following instructions actively maintained in declarative *working memory.* Then, through practice, we rapidly develop and "compile" into *procedural* memory a fluent procedure for performing the task. This integrated task-set package can then be reactivated—possibly through being retrieved into *procedural working memory* (as discussed in more detail later)—when needed to control current performance. Both the operation of a task set and its reinstatement become more fluent with further practice so that it is retrieved more and more involuntarily and with less and less conscious thought when the environment affords it (as when text involuntarily activates "reading mode" in the literate adult, hence the famous Stroop effect: difficulty naming the color in which conflicting color names are written).

The idea of capturing task-set control in the lab by getting people to switch between tasks is usually credited to Jersild (1927). He compared the time taken to read through a list of stimuli responding to each one according to a single rule (e.g., add 3 to each number) to the time taken to alternate between two tasks (add 3, subtract 3, add 3 ...). Although still occasionally used (e.g., Emerson & Miyake, 2003), this comparison of alternating and pure lists confounds two different kinds of switch cost: transient slowing when the task changes and a more global difference observed between performance on task-repeat trials in a block of trials where one must be ready to switch and performance in a single task block. The global contrast, usually called the "mixing cost," and the transient switch cost are influenced by different factors. The mixing cost seems to derive principally from uncertainty about which task will be required (Rubin & Meiran, 2005) and may reflect sustained rather than (or as well as) transient control settings to deal with the high level of anticipated task competition (Braver, Reynolds, & Donaldson, 2003; Mari-Beffa, Cooper, & Houghton, 2012).

In what follows, I will focus largely on within-block transient effects of a task change, hereinafter simply called the "switch cost." (For a recent survey of the mixing cost, see Mari-Beffa & Kirkham, 2014). In the 1990s, development or refinement of several ways of measuring task-switch costs, together with contrasting interpretations of the phenomena then observed, stimulated the growth of a large literature. Among the more important observations (Meiran, 1996; Rogers & Monsell, 1995) was that allowing participants a second or so to prepare for a change of task resulted in a substantial *reduction in switch cost* (RISC) but did not eliminate it: There was an asymptotic *residual cost* even with several seconds to prepare (see figure 7.1).

State-of-the-Art Review

Paradigms

To reveal these and other phenomena, we must get participants to switch tasks on some trials and not others (other things being equal) while controlling time available for preparation. I start by briefly introducing some of the paradigms used, and their virtues and limitations:

Task Cuing   On each trial the stimulus is preceded (or accompanied) by a cue that specifies the task, sometimes the same task as on the previous trial, sometimes one of the other tasks in play. The pioneers of this paradigm were Shaffer (1965), Biederman (1972), and Sudevan and Taylor (1987), but refinements were needed (see Monsell & Mizon, 2006, and Meiran, 2014, for recommendations). First, as Meiran (1996) was the first to appreciate, one needs to separate the time available for active preparation for the cued task from the time available for any passive dissipation of a task set after the previous trial, by manipulating the cue–stimulus interval (CSI) separately from the response–stimulus interval (RSI). For example, one can keep the RSI constant as the

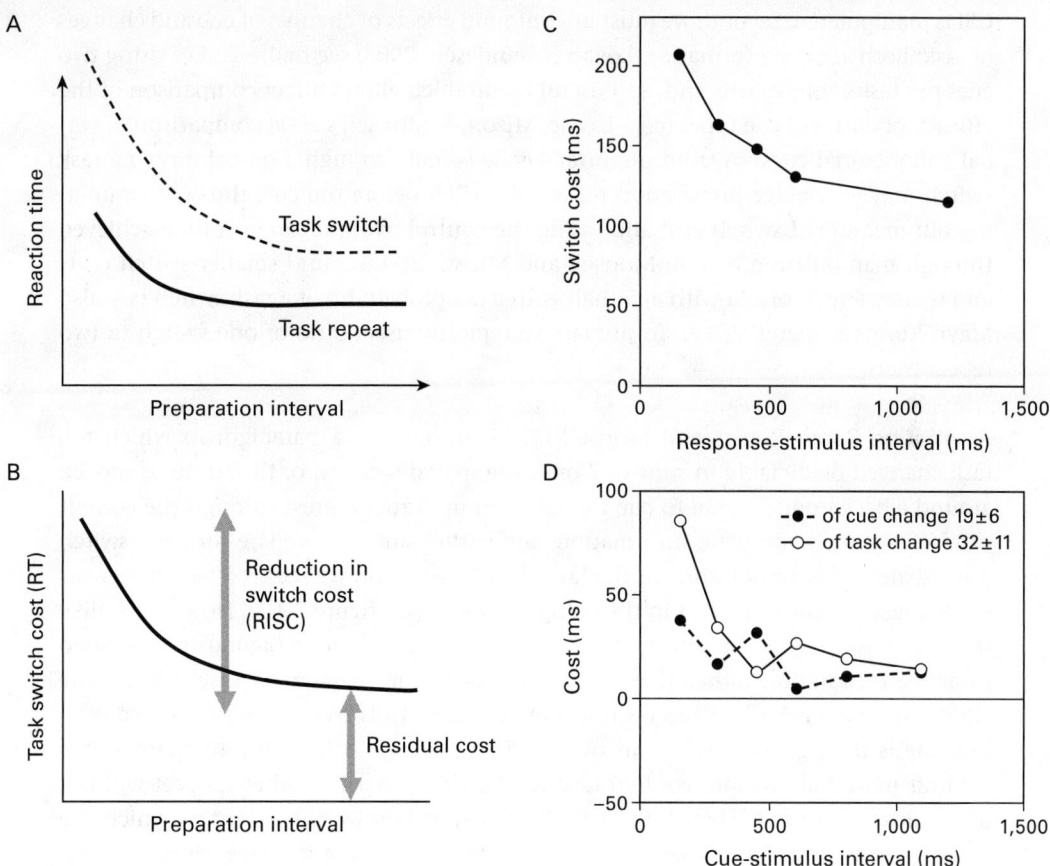

Figure 7.1

(A) Reaction time (RT) for task-switch and -repeat trials as a function of preparation time: idealized data. (B) Task-switch cost (the difference between task-switch and task-repeat RT) as a function of preparation time. Illustrative data from (C) an experiment with runs of 2 alternating between classifying the digit and classifying the letter of a pair of characters (redrawn from Rogers & Monsell, 1995, Experiment 3) and (D) a task-cuing experiment (from figure 5 of Monsell & Mizon, 2006, Experiment 3): The stimulus was an object name, classified by either number of syllables or size of object; the empty symbols indicate the difference between task-switch and task-repeat trials (the cue changes in each case); the filled symbols indicate the difference between cue-change and cue-repeat trials (the task stays the same in each case); the mean and standard error of each cost, averaged over CSI, is also shown.

CSI is manipulated. Second, we must unconfound effects of changes of cue and changes of task: Both affect performance (Logan & Bundesen, 2003; see figure 7.1D). Using two cues per task achieves this and, as a useful by-product, allows direct comparison of the efficacy of different cue types (e.g., Lavric, Mizon, & Monsell's 2008 comparison of verbal and pictorial cues). A third important issue is that too high a probability of a task switch may encourage preparation for a task switch before the cue, thus contaminating our measure of switch cost and losing the control over preparation time achieved through manipulation of CSI; Monsell and Mizon (2006) found smaller switch costs and weaker effects of CSI with a .5 than with a .25 probability of a task switch (see also Mayr, Kuhns & Rieter, 2013). In my lab we typically use a ratio of one switch to two repeat trials.

Alternating Runs   Rogers and Monsell (1995) introduced a paradigm in which the task changed predictably in runs of 2 or 4; the spatial location of the stimulus moved around a background frame to cue the task and indicate progress through the current run.[2] Given such complete information, the participant can prepare for each switch immediately after responding on the last trial of a run, and we see an effect of increasing RSI just like the CSI effect in the cuing paradigm (see figure 7.1D). However, unlike the cuing paradigm, the effect of RSI cannot always be unambiguously attributed to active preparation rather than passive decay of the previous task set.[3] Thus, the alternating runs paradigm has been less used of late. However, it does have two merits. One is that, guaranteed a run of (say) four trials on a task, the participant can commit more fully to the required task set, knowing that the other task set will not be needed for a while. Hence, the paradigm may maximize the extent to which the participant changes his or her task set at the beginning of a run (see Monsell, Sumner, & Waters, 2003, for evidence). In addition, short alternating runs with task and position in run indicated by location seems an especially transparent way of signaling the task required—useful if one wants to minimize the difficulty of interpreting task cues—for example with brain-damaged participants (e.g., Aron, Monsell, Sahakian, & Robbins, 2004).

Predictable Short Task Sequences   To avoid external cuing between trials, one can give participants a short sequence of trials in which the task sequence is prespecified, either by a sequence cue or by repeating the same sequence for a block of trials. When the sequence is only two trials (e.g., Allport, Styles, & Hsieh, 1994; Goschke, 2000) the overhead of having to remember the identity of the second task will be modest. However, for longer sequences, the issue becomes how the sequence of trials is retrieved from memory as much as how the shift of task sets is achieved. See Logan (2004), Koch, Philipp, and Gade (2006), and Mayr (2009) on "task span": performance of longer task sequences held in memory.

Intermittent Instruction   The paradigms mentioned so far may seem distant from most "natural" task switching. In everyday life some event (phone ringing, door opening, baby crying) cues a switch to another task, which we often perform for a while until another salient event summons us to pursue another goal. In a paradigm developed by Gopher and colleagues (Gopher, 1996; Gopher, Armony, & Greenshpan, 2000), and recently reviewed by Altmann (2014), who calls it the "extended runs" procedure, a series of trials, on each of which just the stimulus is presented, is interrupted from time to time by an instruction, which specifies which task is to be performed on subsequent trials until the next instruction. Breaking the sequence of trials with an instruction results in a "restart cost" (longer RT on the first trial of a run) even when the instruction is to continue performing the same task, but RT is even longer on the first trial when the task changes, and this excess provides a measure of the task-switch cost. For instance Kramer, Hahn, and Gopher (1999) used this method to test the effects of age on switch costs, finding that, although it is often claimed that the elderly are impaired in executive function, switch costs were no larger in elderly participants provided that they were allowed sufficient practice. Altmann has also favored this method for developing a theory of task switching based on the memory strength of a code representing the most recently cued task (Altmann & Gray, 2008).

That there is a "restart cost" even when the task repeats (see also Allport & Wylie, 2000) raises the interesting question of the relation between switch costs observed in the other paradigms and "restart costs." Mayr and colleagues (Bryck & Mayr, 2008; Mayr, Kuhns, & Hubbard, 2014) have argued (see below) that, even without a task switch, an interruption of sufficient duration may have the same effect as a task switch in requiring (re-) retrieval of a task set.

The greater naturalness of the intermittent instruction paradigm is bought at some price in efficiency: The switch cost must be estimated from a relatively small subset of the data: RTs to the first trials of long runs. On the other hand, phenomena may emerge from examination of long runs that are not obvious in the shorter runs of other paradigms. Altmann and Gray's (2008) theory relies on their observation (e.g., Altmann, 2002) that RT gets slightly longer over the course of a long run in this paradigm ("within-run slowing").[4]

Voluntary Task Switching   There has been a persistent concern with the paradigms described so far (cf. Wylie, Sumowski, & Murray, 2011). We do task-switching experiments primarily to capture "endogenous control" in the lab. And when participants perform their first switching blocks after practicing the tasks separately, they do experience a voluntary effort to get organized for the task on switch trials. However, after several hundred trials, performance feels more fluent, if not exactly "automatic." Even if we are sure participants are still shifting between task sets (I consider later the possibility that they are not), is this still the intentional application of endogenous voluntary control we hoped to capture, or it more like exogenous driving of task-set selection by the cue? After all, given 10,000 trials, even a rhesus monkey can master a version of cued task switching (Stoet & Snyder, 2003); does it engage in an act of will on each (switch) trial? There is also evidence that task-set selection (in people) can be primed by subliminal task cues (Lau & Passingham, 2007; Reuss, Kiesel, Kunde, & Hommel, 2011); again this hints at automatic activation of task set via a learned cue–task association (Gade & Koch, 2007) rather than a deliberate act of will.

Discomfort with practiced task cuing as a measure of "real" top-down control inspired some researchers (e.g., Arrington & Logan, 2004a, 2005, see Arrington, Reiman & Weaver, 2014, for review) to ask participants to decide for themselves which of two or three tasks to perform on the stimulus presented on each trial, while keeping the task frequencies roughly equal, and the task sequence approximately random. (Different response sets are assigned to each task to enable the experimenter to know which task has been chosen.) Switch costs are obtained, but it is difficult to disentangle effects of the process of choosing which task to do (random sequence generation being a complex task in its own right) from effects of the control operations that implement the chosen task set. It is also necessary to unconfound effects of switching and the length of the preceding run. Nor does voluntary task switching escape exogenous influences: Mayr and Bell (2006) found that stimulus repetition strongly determines the likelihood of a switch, Lien and Ruthruff (2008) that people are more likely to switch back to a task from which they recently switched, and Yeung (2010) that voluntary task switching exhibits an asymmetry of switch costs that (see below) has been interpreted as a consequence of involuntary competition from a recently activated task.

Explaining the Switch Cost

All the paradigms yield a switch cost, and most enable us to partition it into a part that, with a suitably motivated participant, can be eliminated by preparation (the RISC effect) and the part that cannot (the residual cost)—see figure 7.1B. It is tempting to attribute these empirical components to different mechanisms. Many authors attribute the RISC effect to active *task-set reconfiguration* (TSR). Below I will discuss components of task set that might be reconfigured, but there are several ways in which some sort of reconfiguration process might lead to a RISC effect:

• TSR operations take time. If they are not completed before stimulus onset, they must be completed (or nearly so) after the onset and before the domain-specific processes which they link or reconfigure get under way. For example, if S-R rules for a task are loaded into procedural working memory, this must be done before they can be used to select a response (Mayr & Kliegl, 2000). The impact of TSR on RT might be o due to the addition of TSR to poststimulus processing as an extra Sternbergian processing stage, delaying processes that follow it (Rogers & Monsell, 1995) or

o due to TSR operations being performed concurrently with task-specific processes and prolonging them (e.g., by competing with them for resources).

- Inasmuch as TSR operations are not completed before initiation of the domain-specific processes they reconfigure, the latter will run, if at all, less efficiently. For example, theorists of a connectionist bent (Brown, Reynolds, & Braver, 2007; Gilbert & Shallice, 2002; Miller & Cohen, 2001) conceive of the cue as activating task or context units which in turn activate other units so as to bias the processing of particular stimulus attributes or favor particular stimulus–response connections, in an incremental rather than an all-or-none way. The weaker the biasing achieved by the time the relevant information is processed, the less efficient (and more subject to cross-task interference) will be the appropriate encoding of the stimulus or its mapping to action.

To attribute the RISC effect to task-set reconfiguration is not to claim that on task-repeat trials participants do nothing to prepare. In the cuing paradigm we see a substantial reduction in mean RT with preparation on repeat as well as on switch trials (as in figure 7.1A). A century of RT research tells us that warning and alerting signals enable generic preparation for stimulus processing and responding (see chapter 3 in the current volume). On task-repeat trials participants may in addition reinforce their task-specific preparation, but, if necessary at all, maintenance of the existing task-set configuration surely involves less processing effort than *re*configuration.

Theories of the residual cost include the following:

- Competition (at the levels of both tasks and responses; see below) resulting from "task-set inertia"—passive carryover of the task-set state (whether activation, inhibition, or both) from the previous trial (Allport et al., 1994; Yeung & Monsell, 2003b; Yeung, Nystrom, Aronson, & Cohen, 2006). For some reason, this inertia cannot be fully overcome in advance by TSR.

- Competition due to reactivation of a competing task set previously associated with the stimulus (Waszak, Hommel, & Allport, 2003). Again, TSR cannot fully preempt such reactivation, perhaps because it cannot be countered until it happens.

- A component of TSR cannot happen until after some stimulus processing (Hübner, Futterer, & Steinhauser, 2001; Rogers & Monsell, 1995; Rubinstein, Meyer, & Evans, 2001). For example, encoding a stimulus of the relevant category might be needed to fully re-enable a set of S-R rules for dealing with that category.

- A component of TSR cannot happen until after a task-specific process. For example, Schuch and Koch (2003) suggest that the previous task's category-response rules cannot be inhibited until after the conflict they induce has been detected, which is why aborting the response on trial $n - 1$ with a "no-go" signal abolishes the switch cost on trial n (see Los & Van der Burg, 2010, for a review of this effect).

- A component of TSR cannot happen until the changed task set has been used to generate one response. For example, Meiran (2000a, 2000b) argued that while attention could be reoriented to the appropriate stimulus attribute in advance of stimulus onset, biasing the response set (changing the meanings of the responses) to optimize the task set could be accomplished only after the generation of one response.

- There is a failure, on a proportion of trials, to successfully "engage" a discrete TSR process. De Jong (2000) argued that the residual cost results from a mixture of full preparation (as good as on a task-repeat trial) and zero preparation (as bad as at zero CSI) trials. A prediction this makes about RT distributions (under circumstances where there is little effect of preparation interval on task-repeat RT) has been broadly confirmed (Brown, Lehmann, & Poboka, 2006; De Jong, 2000; Nieuwenhuis & Monsell, 2002). However, it remains unexplained why the switch cost should so soon reach an asymptote as the CSI is extended, which should afford further opportunities to "engage."

Other Basic Phenomena

I have so far discussed the switch cost and the reduction in that switch cost with preparation. Next I consider briefly some other robustly established phenomena observed using the paradigms outlined above.

Effects of Response Congruence In many experiments the same response set is used for two tasks (e.g., the same two keys for odd/even and low/high classification of a digit). RT is longer and errors more frequent for response-incongruent stimuli (for which the two tasks assign different responses) than for response-congruent stimuli (see Kiesel et al., 2010, for a review). This Stroop-like congruence effect clearly indicates that the currently irrelevant task set is not entirely suppressed: To some degree the irrelevant attribute is still processed through to response activation according to the irrelevant task-set rule. The congruence effect is often somewhat stronger on switch than on repeat trials (Kiesel et al., 2010; Rogers & Monsell, 1995), but any such modulation of the congruence effect by switching is modest compared to the robust congruence effect on task-repeat trials still seen several trials after a task switch (e.g., Monsell et al., 2003). Hence, the partial processing of the stimulus via the wrong task set's S-R rules long outlives the transient task cost and looks more like a source of the mixing cost.

Response-conflict effects observed in task switching differ from response-conflict effects in single-task paradigms (e.g., Stroop, Simon, and flanker effects) in two interesting ways. First if, in the single-task paradigms, there are "neutral" or "univalent" stimuli which afford neither of the responses (e.g., a string of black *X*s for color-naming, or irrelevant flankers not associated with either response), these yield RTs intermediate between congruent and incongruent conditions. In contrast, in task-switching experiments, we observe mean RT shorter for univalent than for congruent stimuli (Aron et al., 2004; Rogers & Monsell, 1995; Steinhauser & Hübner, 2007). The implication is that, in addition to response conflict, there is competition between task representations: Univalent stimuli never occur in the context of the other task and hence do not cause competition at the task level, whereas congruent stimuli, being multivalent, do, and this can outweigh the benefit they gain from facilitation at the response level.

A second difference arises in the case where, in single-task experiments, the interference is radically asymmetrical. For Stroop stimuli, for example, an incongruent color name interferes much more with color naming than an incongruent color interferes with reading of the color name (in single-task blocks), reflecting the huge difference in practice between the two tasks (MacLeod & Dunbar, 1988). However, when participants switch between two such tasks, the interference becomes much more symmetrical (Yeung & Monsell, 2003b). Recent practice makes both task sets highly available, attenuating the effect of the long-term difference in task strength.

The Paradoxical Asymmetry of Switch Costs   Allport et al. (1994) were the first to report the counterintuitive observation that, when switching between task pairs of unequal strength, such as reading versus color-naming Stroop stimuli, it is easier to switch from the dominant to the subordinate task than vice versa, to judge by the smaller switch cost in the former case. Allport et al. interpreted this by appeal to inertia of task-set inhibition: To perform the subordinate task requires stronger suppression of the dominant task set than vice versa, and this suppression carries over into the next trial. Yeung and Monsell (2003a, 2003b) modified details of this account (with two tasks we cannot tell whether it is extra inhibition of the dominant task, or extra activation of the subordinate task, that persists from trial $n - 1$) and demonstrated that the asymmetry of costs could be reversed by manipulations that reduced, but did not eliminate, the dominance relation indicated by the response-conflict effects. They were able to model both the asymmetry of costs, and its reversal, with a simple quantitative model assuming both task-set inertia—carryover of task-set activation[5] from trial to trial—and application of top-down biasing of task activation just enough to get the appropriate task done with an acceptable error rate in the face of the built-in bias toward the dominant task. Yeung et al. (2006) reported what appears to be a neurophysiological correlate of task-set inertia; when participants switched between naming the face or word in a composite of the two stimuli, the level of activation in brain regions selectively activated by face- (or word-) processing predicted the behavioral cost of switching to the word- (or face-) task.

However, Bryck and Mayr (2008) and Mayr et al. (2014) have reported that even when participants do not perform both tasks within a block but alternate between pure-task blocks, the "restart" cost induced by an interruption (e.g., just a 5-second pause) shows the same asymmetry. They argue that the asymmetry is due not to task-set inertia (as the competing task has not been performed since the previous block of trials) but to the interruption requiring updating (re-retrieval) of task-set parameters from long-term memory; the dominant task set is more prone to retrieval of associated interfering memory traces during updating.

N – 2 Repetition Effect   As already noted, switching between tasks can be seen as biasing the system to favor one of the tasks afforded by the stimulus set. Is this bias achieved by activating the desired task set or by inhibiting the currently irrelevant task set? With only two tasks in play, we cannot tell. However, with three, we can examine performance on the last trial of task sequences such as ABA and CBA. It is frequently reported that performance is worse for ABA (Mayr, 2002; Mayr & Keele, 2000), and this is attributed to the need to overcome the residual inhibition of task set A incurred during the transition to task B on trial $n - 1$ (see Koch, Gade, Schuch, & Philipp, 2010, for a review).

Does Task Cuing Capture Task-Set Control Processes?

The theoretical interpretations I have reviewed so far assume that it does, provided that we acknowledge that the TSR processes posited may during extended practice move along a continuum from effortful to relatively automatic. A more fundamental challenge came from Logan and Bundesen (2003), who proposed that (presumably after a little practice) the participant in a task-cuing experiment does no more than apply, on every trial, just *one* task set: Participants retrieve the response that they have learned goes with each combination of cue and stimulus presented—a "compound retrieval" account. Where then do "switch" costs, the RISC effect, and so forth, come from? Based initially on the observation that RTs on task-repeat trials are faster when the task cue also repeats than when it does not (see, e.g., figure 7.1D), Logan and Bundesen (2003) proposed that (1) the apparent "task" switch cost was due to priming of cue encoding by cue repetition and (2) the RISC effect was due to the diminishing contribution of cue encoding time to the initiation of compound retrieval as the CSI increases. However, robust task switch costs are seen even when cue repetition is controlled for or prevented (e.g., Mayr & Kliegl, 2003; Monsell & Mizon, 2006). In response, Logan and colleagues (Logan & Bundesen, 2004; Schneider & Logan, 2005) have argued that even when the cue changes on a task-repeat trial, cue encoding is partially primed when a *related* prime (related by signaling the same task) was presented on the previous trial.

I mention here just two pieces of evidence[6] particularly difficult for this proposal as a general account of task-cuing performance. Forrest, Monsell, and McLaren (2014) gave participants about 700 standard trials cuing odd/even and high/low digit classifications, with two cues per task and only four digits, which should make the compound cue + stimulus → response (CSR) rules especially easy to acquire. Half the participants were given standard task-cuing instructions. The other half were given instructions or training designed to induce compound retrieval: that is, to learn and use the compound CSR mappings (or—for the two congruent digits—just S-R mappings). If Logan et al.'s account is correct, Forrest et al. were just encouraging the CSR-instructed group to do what the tasks-instructed group was already doing: compound retrieval. However, the two groups behaved quite differently: The CSR group showed enormous congruence effects (because they learned responses for congruent stimuli quickly and incongruent stimuli slowly), only a small "task-switch" cost, on incongruent trials only, and no RISC effect. In contrast, the standardly instructed participants showed the usual pattern: modest congruence effects, larger switch costs, and a substantial RISC effect. They also transferred easily after 700 trials to a new set of four digits, as one would expect if they were applying the same task rules to new stimuli, whereas the CSR-instructed group behaved as we would expect if they were having to learn a new set of associations. We concluded that, although learning of CSR associations is a possible strategy when the stimulus set is very small, the natural strategy, even with a small stimulus set, is to select and activate the relevant cued classification task set and use it to select the response.

Another result problematic for the idea that the apparent task-switch cost is really a cue-encoding benefit came from a task-cuing study reported by Arrington, Logan, and Schneider (2007). Participants had to respond overtly to indicate they had identified the cue category (i.e., the task) and only then saw the stimulus. RT for the response to the cue ought to reflect cue-encoding time, excluding its contribution from poststimulus processing. However, there was no effect of a task switch on RT to the cue, only on RT to the stimulus.

The task-cuing paradigm used in these studies was the standard one with bivalent stimuli affording both tasks. Dreisbach and colleagues (see Dreisbach, 2012, for a summary) conducted experiments in which four red and four green words were mapped to two responses by eight simple S-R rules. Nonobviously, the response could also be derived by two classification rules of the form: If green, respond to consonant-initial word with a left response, and a vowel-initial word with a right response; if red, respond left for animal, right for nonanimal (i.e., the two-level contingencies of a task-cuing experiment). Participants who did not notice these task rules performed well; learning eight S-R rules is an efficient way to perform this task. Participants informed of the hidden tasks structure actually performed worse but, in spite of this, adopted the task-rule strategy, as indicated by costs appearing when the color changed from red to green.

A general moral seems to be that, when the contingencies afford a two-level control strategy—that is, choose the relevant task set to activate, and then apply it to the stimulus to retrieve a response—participants tend to adopt this strategy, even in circumstances in which it is disadvantageous in terms of overall performance. Such a strategy is usually adaptive, conferring several crucial advantages over brute-force S-R and CSR learning: It generalizes easily to new sets of stimuli (so that new S-R or CSR rules do not need to be learned), it allows advance preparation with a long CSI (the task set can be readied before the stimulus onset), and, as other data from Dreisbach's lab demonstrate, selectively activating a task set shields against interference from irrelevant information. Collins and Frank (2013) discuss other evidence for, and model, the emergence of task-level restructuring of S-R contingencies.

I conclude that, while "direct" associative learning of one-step environment-to-action links is possible, and sometimes occurs, a hierarchical strategy which parses the contingencies into tasks, then S-R correspondences within tasks, may be the brain's preferred and natural strategy. It is this control strategy that the task-cuing paradigm usually elicits and allows us to investigate.[7]

Components of Task Set

A task set, even for the rather simple "one-step" tasks typically used in task-switching experiments, is a complex. We can distinguish a number of "parameters" (Logan & Gordon, 2001) or components that may need to be reset to perform such a task: which sensory modality to attend to, where to orient attention in space and when in time, which attribute(s) of the attended object(s) to process, what the S-R rules are or what operation to perform, which effectors to express the abstract response category with, how to set decision criteria (e.g., fast and careless vs. slow and careful). More broadly, we can distinguish (Meiran, 2000a, 2000b) between "stimulus set" (what attribute of the stimulus to evaluate) and "response set" (how to express the decision in action) or, equivalently, between "attentional set" and "intentional set" (Rushworth, Passingham, & Nobre, 2002, 2005). I illustrate next three kinds of evidence that may help us pick apart the contribution of different components to switch costs.

One-Component Shifts   Some experiments badged as "task switching" use task pairs which (a priori) would appear to require resetting of just one component. Examples include the following:

- *Only the relevant stimulus dimension changes.* For example, Meiran and Marciano (2002) presented a pair of objects varying in several dimensions (color, shape, tilt, etc.). Participants were cued on each trial to decide whether the objects matched on one of the dimensions. The S-R rules for indicating "same" and "different" remained constant. Shift costs were observed. As another example, to localize brain mechanisms associated with the costs of a change of "attentional set," Rushworth, Passingham, and Nobre (2005) used shifts between attending to targets defined by shape and color, to classify a superimposed character according to a constant S-R rule.

- *Only the S-R rules change.* In a parallel neuroimaging study of "intentional set," Rushworth, Passingham, and Nobre (2002) conducted an intermittent instruction experiment in which participants classified a shape as triangle or square: Sometimes the instruction reversed the assignment of shapes to left and right responses. Again, switch costs were observed.

- *Only the effector changes.* Philipp and Koch (2011) had participants classify a digit as (e.g.) odd or even on every trial, but they were cued to make their left/right response with hands, feet, or voice. A substantial shift cost was obtained, reduced but not eliminated by preparation. An $n - 2$ effector repetition cost can also be observed (Philipp & Koch, 2005).

Hence, it appears that changing any one parameter or component of the task can yield at least some of the standard phenomenology of task-switch costs. However, stimulus set and response set may differ in their amenability to advance preparation. I mentioned earlier Meiran's (2000a, 2000b) assumption that participants could prepare stimulus set (e.g., attend to the relevant dimension) in advance but were able to adjust their response set optimally only after responding once: hence, the residual switch cost. However, in the Meiran and Marciano (2002) study just described, they also found that the cost of shifting attention between dimensions was not attenuated by increasing the CSI. In contrast, in a similar condition in which the relevant dimension remained constant while the assignment of responses for "same" and "different" was cued, they found that the cost of this "response-set" shift did reduce with preparation. Thus, in their revised model Meiran, Kessler, and Adi-Japha (2008) reversed the previous assumption about preparation: Response set can be reconfigured before the stimulus while stimulus set cannot.

This may seem quite a counterintuitive claim. As represented elsewhere in this volume, there are well-established traditions of research on shifting attention between visual locations (Posner, 1980), ears (Treisman, 1971), modalities (Spence & Driver, 1997), and (in visual search) between dimensions such as color and form and between features such as red versus blue objects (see Krummenacher & Müller, 2012, and chapters 1 and 2 in the current volume). Let's call all these various potential "objects of attention" or "filter settings" *channels*. In each case, an unpredictable switch between channels incurs costs, we can choose which channel to attend to, and it seems that precuing the relevant channel before stimulus onset reduces the cost of a switch. For example, Müller, Reimann, and Krummenacher (2003) found that in visual search for a color or form singleton, verbally cuing the dimension on which there would be a singleton reduced the RT cost of a dimension switch. The general assumption seems to be that we have top-down control of which channel we attend to, and we can set ourselves to attend it in advance.[8]

However, these literatures have not really looked for limits on intentional switching between attentional channels using the precautions of the task-cuing paradigm. Recently Koch and colleagues have begun to do this for switching between auditory channels defined by speaker properties. Koch, Lawo, Fels, and Vorlander (2011) presented two digit names, one in a female, one in a male voice, one randomly on the left and the other on the right, preceded by a gender cue. The task was to classify as odd/ even the digit spoken in the voice type cued. They observed substantial costs of switching between genders but (provided two cues per gender were used to partial out effects of a cue change) obtained no reduction in switch cost when 1,000 rather than 100 ms was allowed for preparation. Lawo, Fels, Oberem, and Koch (2014) report a benefit of preparation for a switch of ear but not a switch of gender.

Partitioning the Switch Cost   Another way to address the contribution of the components of task set is to ask whether, in the typical task-switching experiment in which both stimulus set and response set must be reconfigured, the RT switch cost can be partitioned among components. Elchlepp, Lavric, and Monsell (2015) used event-related potentials (ERPs) to do this. They presented on each trial a string of red and blue letters. In one experiment the letter string was a word, and participants were cued, with an 800-ms CSI, either to classify the word's meaning (living/nonliving) or to decide whether its color pattern was symmetrical. RTs showed the sort of residual switch cost (50 ms) one would expect (much smaller than with a 200-ms CSI), so preparation was clearly effective. On task-repeat trials, for which mean RT was ~800 ms, ERPs for high- and low-frequency words began to diverge around 250 ms, a marker for early stages of lexical access. On task switch trials, the latency of this evolution of the ERP frequency effect was delayed by a substantial fraction of the RT switch cost. In a second experiment, the reading task was lexical decision, and the onset of the ERP word/nonword difference at around 180 ms was also delayed by an amount similar to the (59-ms) RT switch cost. Thus, in this case a large part of the residual switch cost appears to be due to early processing being prolonged. We speculate that, even after preparation to read a letter string, attention remains partially oriented to the (color pattern) stimulus attributes relevant on the previous trial until one reading trial has been completed. We term this "attentional inertia."

Another way to detect attentional inertia is with eye tracking. Longman, Lavric, Munteanu, and Monsell (2014) presented a triangular array of three digits, and cued one of three classification tasks, each linked consistently to a digit location (e.g., if odd/even was cued, a participant might have to classify the top digit). Eye fixations provided an index of the orientation of spatial attention during preparation and performance. On switch trials orientation to the relevant location was delayed, and the delay predicted the RT switch cost, both between and within participants. There was also a tendency to fixate the previously relevant location, which was reduced but not eliminated at the longer CSIs. (A control experiment with three locations but only one task showed that, without task switches, any tendency to fixate the previously relevant location and the delay in fixating the relevant one were very small and eliminated at long CSIs.) Hence, when the orientation of spatial attention is a task parameter that needs resetting, it shows substantial inertia and contributes to the residual cost. Mayr et al. (2013) also used eye tracking, this time to measure attention to a dimension. Participants were cued to detect a color or form singleton among three objects in order to discriminate its value (e.g., light or dark blue). Again, a substantial residual switch cost in orienting attention to the relevant dimension-defined target was found.

In summary, when a change in task set requires us to reorient attention to a different location, dimension, voice channel, or attribute, our ability to do this in advance of the stimulus appears to be, if not negligible, at least limited: Attentional inertia appears to be a significant component of the residual cost for at least some objects of attention.

Switching One versus Two Components   When a task set comprises components that could logically be shifted independently, we can ask if in practice they are. Kieffaber, Kruschke, Cho, Walker, and Hetrick (2013) presented two objects and cued one of three tasks with a dimension cue ("shape," "size," or "color"). For two of these dimensions, a same/different judgment was required (so that switches between them involved only a one-component shift), while for the third, the participant had to response to the value of the dimension (so that for switches involving this task, both dimension and response rules changed). The one-component shifts showed shorter RTs and less of an impact on ERP amplitudes, suggesting that switching more components takes more time, as we might expect if they are reconfigured separately.

On the other hand, several authors have devised sets of four tasks in which the switches/repeats of one component can be factorially combined with switches/repeats of the other. For example, Kleinsorge (2004) presented a string of identical digits (e.g., "7777") whose color cued the relevant dimension (numerosity vs. value) and whose location cued the judgment to be made (odd/even or high/low). Responses for switches of both dimension and judgment were slower than switches of dimension alone, but not detectably slower than switches of judgment alone. They concluded that the components were structured hierarchically so that shifting the higher feature (judgment) facilitated shifting of the lower (dimension) but not vice versa. Vandierendonck, Christiaens, and Liefooghe (2008) conducted a study very similar except that the cue was verbal (e.g., the Dutch equivalent of "V HI-LO" for a high/low judgment on the digit's value) and found a different pattern: The switch cost was the same regardless of whether either or both components changed. They concluded the two components are reconfigured as an integral package.

Clearly the jury is still out on this issue. We should perhaps entertain the possibility that whether the components of task set are treated as an integrated "package," reconfigured as one, or as a collective, reconfigured separately, may depend on factors such as the amount of practice with the whole package, the ease of figuring out what needs to be configured, and the costs and benefits of separately reconfiguring the components. For example, in their eye-tracking studies described above, Longman et al. (2014) observed attentional inertia and delay with verbal cues that labeled the task (e.g., "Odd?" "Low?"). But when Longman, Lavric, and Monsell (2015) repeated the experiment with instructions that emphasized the relevant location over the classification to be performed, and used arrows or verbal cues (e.g., "up," "left") to cue the task via the relevant location, the attentional delays and evidence of inertia disappeared; however, the RT switch cost was now much larger. We think using explicit location cues encouraged participants to decouple the location from other parameters of the task set, shift attention to the relevant location first, and worry about how to classify the digit later. The downside of this decoupling was that the CSI was not used efficiently to prepare the decision component.

Response Selection and Task-Switch Costs In this section, I have stressed the contribution of attentional selection to residual task-switch costs, partly because this is a Handbook of Attention, and partly as a corrective to a general tendency in the task-switching literature to think that the main problem in switching tasks is the clash of S-R rules between competing task sets. Of course, the occurrence of response congruence effects with bivalent stimuli indisputably demonstrates involuntary application, to some degree, of the irrelevant task set's S-R rules. However, the extent to which response conflict is a source of the RT switch cost, per se, is much less clear.

First, as noted earlier, the effect of response conflict way outlasts the transient cost of a task switch. Second, although the congruence effect is usually larger on switch than on repeat trials, it is often not very much larger (e.g., Rogers & Monsell, 1995) and sometimes not larger at all (e.g., Monsell et al., 2003). A third point is that in experiments demonstrating a robust RISC effect, there is often no corresponding reduction in congruence effects with preparation, though under other conditions there is (see Monsell & Mizon, 2006, for a discussion). Response congruence effects seem to stem more from the use of the competing task rules in the extended recent past, not on the last trial; they may be more important for the mixing cost than for the switch cost. To put it another way, they are due to the competing task's rules in long-term memory being primed by earlier blocks, not because they linger in working memory from the previous trial. Consistent with this, S-R congruence effects are strongly influenced by the frequency with which competing S-R mappings have been used, but not by a concurrent load (Kiesel, Wendt, & Peters, 2007).

Procedural Working Memory and Task-Set Control

As I have already mentioned, one conception of task-set control is of task units (possibly in prefrontal cortex) incrementally biasing other components of the system to favor the processing of particular sensory objects, attributes, and so forth (cf. the "biased competition" of Desimone & Duncan, 1995) and/or particular stimulus–response or category–response mappings, over others afforded by the same input (Brown et al., 2007; Miller & Cohen, 2001; Gilbert & Shallice, 2002). Task representations differing in their relative degree of activation are also a feature of Altmann and Gray's (2008) model.

An alternative conception of task-set control has the cued task set being retrieved into a component of working memory I call *procedural working memory* (PWM; cf. Kieras, Meyer, Ballas, & Lauber, 2000; Mayr & Kliegl, 2000; Oberauer, 2009; Rubinstein et al., 2001; Vandierendonck, 2012). Oberauer (2009) assumes that PWM contains only the S-R rules for the current task; this is presumably how "goal-shielding" (Dreisbach, 2012) is accomplished. However, S-R rules for other tasks recently in play (as in a task-switching experiment) are still active in procedural long-term memory and may partially activate responses if the stimulus matches them: hence, response congruence effects. The idea that working memory toggles rather discretely between maintenance mode and updating mode is a feature of some neurocomputational models (Durstewitz, Kelc, & Güntürkün, 1999; O'Reilly, 2006). Applying this assumption to PWM, we might assume (cf. Mayr et al., 2014) that a task switch (usually) initiates PWM updating with the set of rules required for the upcoming task, and a task repeat (usually) leaves the system in maintenance mode. Circumstances other than a task switch (e.g., interruptions, as proposed by Mayr et al., 2014) may also prompt updating.

Given this conception of a PWM, questions arise about the process of promoting the S-R rules and other task-set parameters of the current task into PWM from the semi-active state in procedural long-term memory enjoyed by the several tasks currently "in play." For example, does the number of task sets currently in play influence the difficulty of retrieving one? Simply increasing the number of tasks among which the participant is asked to switch will also reduce the frequency and average recency with which each task occurs. Van 't Wout, Lavric, and Monsell (2015) unconfounded these factors by keeping the frequency and recency of two tasks constant while varying the number of other tasks in play between one and three. They found no effect of number of tasks, thus controlled, on performance, on switch costs, or on the preparation of task set as measured by the RISC effect. Hence, it appears that promoting a task's parameters into PWM does not depend on the number of tasks in contention for such promotion. In contrast, the number of S-R rules which comprise an active task set is well-known to influence the difficulty of activating one of them, as indicated by Hick's law.[9]

The Role of Language

To what extent is task-set control facilitated, or even implemented (Goschke, 2000), by verbal self-instruction? A number of studies have explored the role of verbalization of elements of task set by combining task switching with concurrent "articulatory suppression," on the grounds that this should preempt the use of phonological working memory to maintain verbal self-instructions. Verbalization could support task-set control in several ways. In situations where the participant has to hold on to the sequence of tasks for themselves, phonological working memory clearly aids retention of the necessary sequence (Baddeley, Chincotta, & Adlam, 2001; Bryck & Mayr, 2005; Emerson & Miyake, 2003; Saeki & Saito, 2004, 2009). In a task-cuing experiment Miyake, Emerson, Padilla, and Ahn (2004) found that articulatory suppression during the CSI impaired preparation unless the cue explicitly named the relevant attribute (color or shape); Goschke (2000), using predictable two-trial sequences, similarly found that articulatory suppression impaired preparation during the RSI. In these cases covert verbalization could be helping to activate the overall task goal or to direct attention to the relevant attribute. Are verbal representations in phonological working memory also used to represent the S-R rules then activated (as proposed by Liefooghe, Vandierendonck, Muyllaert, Verbruggen, & Vanneste, 2005)? Van 't Wout, Lavric, and Monsell (2013) manipulated the phonological similarity and complexity of the stimulus terms in the S-R rules for the tasks but could detect no effect of these variables on performance, switch costs, or the RISC effect in the task-switching blocks. However, phonological similarity did impair performance early in single-task practice, consistent with the idea that declarative working memory represents S-R rules when they are first introduced, but the rules are then rapidly compiled by practice into a nonlinguistic code, after which this compiled representation can be retrieved directly into PWM when the task changes. Typical task switching experiments capture performance only after this transformation.

Integration

What has all this to do with task switching in ordinary life? It is not a necessary requirement of an experimental paradigm that it have direct analogues in the real world, provided that it engages and/or measures mechanisms that do contribute to performance in the real world.[10] Nevertheless, task switching surely does happen frequently in the real world. It may justly be argued that real-world task switching rarely involves multivalent stimuli, as used in most task-switching experiments. If I switch between using the phone and cooking, the cooking pots and hob do not afford conversation, and the phone does not afford cooking operations. However, while task switch costs are often reduced in laboratory experiments when the stimuli and the responses are univalent, they are not eliminated (e.g., Rogers & Monsell, 1995). Moreover, IT companies compete to introduce devices into our lives that are task multivalent, such as smartphones and computers. Task errors that involve acting upon the wrong window, or upon one window according to the rules of another application, are common experiences. Nor are the costs of switching among these tasks ameliorated by the tendency of companies to change task parameters (such as the location of control icons) with each upgrade.

More generally, because most "multitasking" situations are dealt with by frequent and rapid switching among the various tasks to be performed—that is, "serial" multitasking as distinguished and discussed in chapter 19 of this volume—task-set reconfiguration and resistance to cross-task interference are surely important contributors to successful multitasking, and lab task-switching studies should help isolate properties of that contribution. However, large though task-switch costs are (typically well over 100 ms without an opportunity to prepare) in relation to many RT effects studied in experimental psychology, they reflect but one component of the huge overhead required for real-world serial multitasking. The overhead includes reconstruction from memory of the pre-interruption context, the setting and use of prospective memory triggers to schedule task switches appropriately, the requirement to prioritize tasks, plan task scheduling, troubleshoot, and more, as discussed in chapter 19.

One link between real-world multitasking and lab task switching is research that asks whether the effect of "real-world" practice at multitasking can be detected in performance on lab tests. In chapter 23, Bavelier and Föcker review studies on the impact of video-game practice on cognitive skills assessed with laboratory measures that include switch costs, flanker effects, and so on. As they explain, there are numerous methodological pitfalls. Cross-sectional comparisons between experienced gamers and novices on performance in lab tests are particularly problematic. More convincing is a demonstration that practice on a video game that (plausibly) exercises task-set control selectively improves lab measures of set shifting while equivalent and equally motivating practice on a game requiring continuous focus on a single task (e.g., Tetris) does not. Boot, Kramer, Simons, Fabiani, and Gratton (2008) tested experienced gamers on cued switching between odd/even and high/low classification of digits. They found that expert gamers had a smaller switch cost than nongamers. However, in a longitudinal study, they obtained no evidence that 20 hours of practice of an action game changed switch costs more than equivalent practice on a puzzle game or a strategy game. On the other hand, Strobach, Frensch, and Schubert (2012) did find that in novice gamers, 15 hours of practice on an action video game (Medal of Honor) reduced switch costs (using Rogers & Monsell's, 1995, digit/letter pairs—see table 7.1) while equivalent practice on Tetris, or no practice, did not; critically, the effect of practice was selective to task-switch trials.

A modern form of real-world multitasking that causes concern, especially to educators and parents, is "media multitasking": consuming more than one stream of video, audio, and text content at a time. Ophir, Nass, and Wagner (2009) compared performance of habitual and infrequent media multitaskers in an experiment cuing the letter/digit classification tasks and found a much larger switch cost in the habitual multitaskers: an apparent conflict with the results with gamers mentioned above. However, in a direct replication of the Ophir et al. study, Alzahabi and Becker (2013) found the opposite result—a smaller switch cost in the habitual multitaskers. Conceivably, the two studies sampled rather different populations of media multitaskers: Some may specialize in monitoring multiple media in parallel and become more open to interference from irrelevant attributes whereas others learn to shift their focus between media efficiently. These studies illustrate the difficulty of a cross-sectional comparison dependent on natural experience.

One reason for interest in such studies is the hope that any transfer from practice on such complex "real" (or simulated) world skills to effects measured in the lab could be reversed, that is, that training on lab tasks chosen to develop particular cognitive skills, such as task-set control, could be used to remediate developmental or acquired deficits in the real world, or to develop such skills in people working in pressured multitasking environments, such as fighter pilots or air-traffic controllers. There are cognitive training programs of this general kind, some delivering training via the Web, but also quite widespread skepticism in the psychological community about whether the improvements in performance on the small set of lab tasks used generalize beyond the specific skills practiced in the training (see, e.g., Redick et al., 2013).

The more general relevance of task-switching research to the real world is the potential application of our increasing understanding of task-set control and its limitations to the analysis and design of built environments, tools, and systems in which such limitations are potential sources of error, danger, inefficiency, and stress. If attentional inertia, for example, is an important source of task-switch costs, then how can we design the interface such that necessary shifts of attention to other attributes are minimized or facilitated? Which cues or display arrangements are optimal for facilitating task-set reconfiguration and reducing cross talk between the task sets in play?

Summary and Future Directions

The last two decades have seen the development of several task-switching paradigms that attempt to capture task-set control in the lab and enable us to manipulate and measure its efficiency. None is without its limitations, but I have argued they do indeed capture aspects of task-set control in the laboratory (in spite of some claims to the contrary) and reveal its properties. Although there remains considerable controversy about the interpretation of the phenomena these paradigms reveal, the evolving richness of the database increasingly constrains theory.

At the same time it should be acknowledged that the typical experiment in this genre, designed to average performance over hundreds of trials, captures the control of a task set at an intermediate point in its history, when it has been securely compiled into and can be retrieved from procedural memory but has not yet become a habit. Such experiments do not get at the critical early transition from following verbal instructions to executing a compiled procedure. To study task-set formation rather than activation/retrieval, we need paradigms that issue new task instructions frequently—such as the "rapid instructed task learning" paradigm recently reviewed by Cole, Laurent, and Stocco (2013). Another potential entry point to this early stage of task-set control is the dissociations sometimes observed between knowing the task set, in the sense of being able to repeat instructions, and executing it when required, as in the "goal-neglect" phenomenon studied by Duncan and colleagues (Duncan et al., 2008). At the other end of a task set's life history, we need heroic studies of how a task set evolves, over many hours of practice, from a procedure that is voluntarily retrievable with a modicum of effort to one that is hard to suppress. Also, most task-switching research has been limited to simple one-step operations such as classification, identification, matching, or addition; we should be trying to extend our scope to more complex tasks.

Space limitations have restricted this review largely to (an idiosyncratic selection of) research using behavioral measures in the average normal adult. There is also a large literature using functional magnetic resonance imaging (fMRI) with these paradigms to explore the network of brain regions in prefrontal cortex, parietal cortex, and basal ganglia which contribute components of task-set control (for reviews see: Richter & Yeung, 2014; Ruge, Jamadar, Zimmermann, & Karayanidis, 2013; Wager, Jonides, Smith, & Nichols, 2005), neuropsychological studies such as Shallice, Stuss, Picton, Alexander, & Gillingham (2008) exploring effects of damage to the network (see Das & Wylie, 2014, for review), and studies using transcranial magnetic stimulation to suppress components of the network (e.g., Rushworth, Hadland, Paus, & Sipila, 2002). Another large literature examines electrophysiological correlates of prestimulus task-set preparation and poststimulus effects of task switching and competition (for reviews see Karayanidis et al., 2010; Karayanidis & Jamadar, 2014), which are hard to separate with fMRI (though see Ruge et al., 2013). For a survey of computational and mathematical models of control in task switching, see Grange and Houghton (2014b). There are numerous studies of how task switching changes across the life span (see Kray & Ferdinand, 2014; Peters & Crone, 2014; Wasylyshyn, Verhaeghen, & Sliwinski, 2011) and some using switching paradigms to assess potential abnormalities of task-set control in psychopathology (Ravizza & Salo, 2014). Other researchers have exploited naturally occurring polymorphisms of the genes controlling dopamine production to examine the role of neurotransmitter systems in cognitive flexibility, as assessed by task switching (Colzato, Waszak, Nieuwenhuis, Posthuma, & Hommel, 2010). Finally, there are important factor-analytic studies of individual differences in task-set control, their association with and dissociation from other aspects of executive function, and their heritability (see Mikaye & Friedman, 2012, for a review).

Box 7.1
Key Points

- The chapter describes several "task-switching" paradigms that have been developed to capture task-set control and the principal phenomena observed.

- The phenomena index both an active process of task-set reconfiguration and limits on control revealed by "residual" switch costs and cross talk between tasks, even after ample opportunity for preparation. Variants of and challenges to this characterization are summarized.

- The review is limited largely to behavioral evidence from neurotypical adults, but links are provided to literatures on fMRI, electrophysiological, neuropsychological, life-cycle, and individual-differences investigations using these paradigms, as well as computational modeling approaches.

- In contrast to the more typical focus on the difficulty of shifting between conflicting sets of S-R rules, this chapter emphasizes evidence for "attentional inertia" as an important source of the "residual" switch cost.

- Views of the role of procedural working memory and language in task-set control are sketched.

- The relation between lab task-switching experiments and "real-world" multitasking is reviewed.

Box 7.2
Outstanding Issues

- These paradigms typically examine switches between task sets during a few hundred trials after their compilation from verbal instructions into procedural memory. Explorations are needed both of the initial translation from verbal instruction to a compiled task set, and of its later evolution from this procedural "package," that can be voluntarily activated, into a habit that may need to be suppressed.

- Numerous theoretical conceptions of task-set control and its limitations, and of the memory systems supporting the acquisition and representation of task set, remain in contention. Analysis of the components of task set and their integration, and the role of language, is at an early stage. Further development of computational models addressing these issues is needed.

- The limitations in, and factors determining, our ability to shift attentional and intentional set revealed by lab experiments need to be translated into design principles for the multitasking workspace.

Notes

1. This is just a label: As is sometimes pointed out, we could equally conceptualize these as task-repetition benefits.

2. For other run lengths and cuing strategies see Aron, Monsell, Sahakian, and Robbins (2004) and Monsell, Sumner, and Waters (2003).

3. As it happens, in the Rogers and Monsell study, randomly varying RSI within a block in Experiment 2 abolished the RISC observed when RSI varied between blocks in Experiment 3 (see figure 7.1C), indicating that the latter reflected voluntary preparation, not passive decay.

4. I note however that though a similar small decline in performance could be seen in the alternating runs of 4 used by Rogers and Monsell (1995, Experiment 6), it was not clearly evident in the runs of 8 used by Monsell et al. (2003), especially if errors are examined as well as RTs.

5. They assumed carryover of activation rather than inhibition—but, with only two tasks, either can work.

6. For other difficulties see Forrest, Monsell, and McLaren (2014).

7. A few participants in Forrest et al.'s (2014) task-instructed group reported that they discovered and used S-R and CSR associations. They were replaced for the analyses described, but they suggest a methodological recommendation: To avoid the possibility of (some) participants (sometimes) using compound retrieval in the task-cuing paradigm, thus contaminating our measures of task-set control, we should use relatively large stimulus sets.

8. I gloss here over the fierce debate about how early in stimulus processing this voluntary selection penetrates: See chapter 1 in the current volume.

9. Van 't Wout (2013, PhD thesis) showed that the Hick's law effect of number of S-R rules in single-task performance survives the controls for frequency and recency applied in the task-switching study.

10. In the era of steam trains, railroad engineers used to walk down a train hitting each wheel with a hammer listening for the "ding" of an intact wheel versus the "thud" made by a cracked one. The sound made was usefully diagnostic of properties relevant to, but not part of, the wheel's "real-world" function.

References

Allport, D. A., Styles, E. A., & Hsieh, S. (1994). Shifting intentional set: Exploring the dynamic control of tasks. In C. Umiltà & M. Moscovitch (Eds.), *Attention and Performance XV: Conscious and nonconscious information processing* (pp. 421–452). Cambridge, MA: MIT Press.

Allport, D. A., & Wylie, G. (2000). Task-switching, stimulus–response bindings and negative priming. In S. Monsell & J. Driver (Eds.), *Control of cognitive processes: Attention and performance XVIII* (pp. 35–70). Cambridge, MA: MIT Press.

Altmann, E. M. (2002). Functional decay of memory for tasks. *Psychological Research–Psychologische Forschung, 66*, 287–297.

Altmann, E. M. (2014).The extended-runs procedure and restart cost. In J. A. Grange & G. Houghton (Eds.), *Task switching and cognitive control* (pp. 101–116). Oxford: Oxford University Press.

Altmann, E. M., & Gray, W. D. (2008). An integrated model of cognitive control in task switching. *Psychological Review, 115*, 602–639.

Alzahabi, R., & Becker, M. W. (2013). The association between media multitasking, task-switching, and dual-task performance. *Journal of Experimental Psychology: Human Perception and Performance, 39*, 1485–1495.

Aron, A. R., Monsell, S., Sahakian, B. J., & Robbins, T. W. (2004). A componential analysis of task-switching deficits associated with lesions of left and right frontal cortex. *Brain, 127*, 1561–1573.

Arrington, C. M., & Logan, G. D. (2004a). The cost of a voluntary task switch. *Psychological Science, 15*, 610–615.

Arrington, C. M., & Logan, G. D. (2004b). Episodic and semantic components of the compound stimulus strategy in the explicit task-cuing procedure. *Memory & Cognition, 32*, 965–978.

Arrington, C. M., & Logan, G. D. (2005). Voluntary task switching: Chasing the elusive homunculus. *Journal of Experimental Psychology: Learning, Memory, and Cognition, 31*, 683–702.

Arrington, C. M., Logan, G. D., & Schneider, D. W. (2007). Separating cue encoding from target processing in the explicit task-cuing procedure: Are there "true" task switch effects? *Journal of Experimental Psychology: Learning, Memory, and Cognition, 33*, 484–502.

Arrington, C. M., Reiman, K. M., & Weaver, S. M. (2014). Voluntary task switching. In J. A. Grange & G. Houghton (Eds.), *Task switching and cognitive control* (pp. 117–136). Oxford: Oxford University Press.

Baddeley, A. D., Chincotta, D., & Adlam, A. (2001). Working memory and the control of action: Evidence from task switching. *Journal of Experimental Psychology: General, 130*, 641–657.

Biederman, I. (1972). Human performance in contingent information processing tasks. *Journal of Experimental Psychology, 93*, 219–238.

Boot, W. R., Kramer, A. F., Simons, D. J., Fabiani, M., & Gratton, G. (2008). The effects of video game playing on attention, memory, and executive control. *Acta Psychologica, 129*, 387–398.

Braver, T. S., Reynolds, J. R., & Donaldson, D. I. (2003). Neural mechanisms of transient and sustained cognitive control during task switching. *Neuron, 39*, 713–726.

Brown, J. W., Reynolds, J. R., & Braver, T. S. (2007). A computational model of fractionated conflict-control mechanisms in task-switching. *Cognitive Psychology, 55*, 37–85.

Brown, S., Lehmann, C., & Poboka, D. (2006). A critical test of the failure-to-engage theory of task switching. *Psychonomic Bulletin & Review, 13*, 152–159.

Bryck, R. L., & Mayr, U. (2005). On the role of verbalization during task set selection: Switching or serial order control? *Memory & Cognition, 33*, 611–623.

Bryck, R. L., & Mayr, U. (2008). Task selection cost asymmetry without task switching. *Psychonomic Bulletin & Review, 15*, 128–134.

Cole, M. W., Laurent, P., & Stocco, A. (2013). Rapid instructed task learning: A new window into the human brain's unique capacity for flexible cognitive control. *Cognitive, Affective & Behavioral Neuroscience, 13*, 1–22.

Collins, A. G. E., & Frank, M. J. (2013). Cognitive control over learning: Creating, clustering, and generalizing task-set structure. *Psychological Review, 120*, 190–229.

Colzato, L. S., Waszak, F., Nieuwenhuis, S., Posthuma, D., & Hommel, B. (2010). The flexible mind is associated with the catechol-O-methyltransferase (COMT) Val(158)Met polymorphism: Evidence for a role of dopamine in the control of task-switching. *Neuropsychologia, 48*, 2764–2768.

Das, A., & Wylie, G. R. (2014). Task switching and executive dysfunction. In J. A. Grange & G. Houghton (Eds.), *Task switching and cognitive control* (pp. 272–299). Oxford: Oxford University Press.

De Jong, R. (2000). An intention-activation account of residual switch costs. In S. Monsell & J. Driver (Eds.), *Control of cognitive processes: Attention and performance XVIII* (pp. 357–376). Cambridge, MA: MIT Press.

Desimone, R., & Duncan, J. (1995). Neural mechanisms of selective attention. *Annual Review of Neuroscience, 18*, 193–222.

Dreisbach, G. (2012). Mechanisms of cognitive control: The functional role of task rules. *Current Directions in Psychological Science, 21*, 227–231.

Duncan, J., Parr, A., Woolgar, A., Thompson, R., Bright, P., Cox, S., et al. (2008). Goal neglect and Spearman's g: Competing parts of a complex task. *Journal of Experimental Psychology: General, 137*, 131–148.

Durstewitz, D., Kelc, M., & Güntürkün, O. (1999). A neurocomputational theory of dopaminergic modulation of working memory functions. *Journal of Neuroscience, 19*, 2807–2822.

Elchlepp, H., Lavric, A., & Monsell, S. (2015). A change of tasks prolongs early processes: Evidence from ERPs in lexical tasks. *Journal of Experimental Psychology: General, 144*, 299–325.

Emerson, M. J., & Miyake, A. (2003). The role of inner speech in task switching: A dual-task investigation. *Journal of Memory and Language, 48*, 148–168.

Fitts, P. M., & Posner, M. I. (1967). *Human performance*. Belmont, CA: Brooks-Cole.

Forrest, C. L. D., Monsell, S., & McLaren, I. P. L. (2014). Is performance in task-cuing experiments mediated by task set selection or associative compound retrieval? *Journal of Experimental Psychology: Learning, Memory, and Cognition, 40*, 1002–1024.

Gade, M., & Koch, I. (2007). Cue–task associations in task switching. *Quarterly Journal of Experimental Psychology, 60,* 762–769.

Gilbert, S. J., & Shallice, T. (2002). Task switching: A PDP model. *Cognitive Psychology, 44,* 297–337.

Gopher, D. (1996). Attention control: Explorations of the work of an executive controller. *Brain Research. Cognitive Brain Research, 5,* 23–38.

Gopher, D., Armony, L., & Greenshpan, Y. (2000). Switching tasks and attention policies. *Journal of Experimental Psychology. General, 129,* 308–339.

Goschke, T. (2000). Intentional reconfiguration and involuntary persistence in task set switching. In S. Monsell & J. Driver (Eds.), *Control of cognitive processes: Attention and performance XVIII* (pp. 331–355). Cambridge, MA: MIT Press.

Grange, J. A., & Houghton, G., Eds. (2014a). *Task switching and cognitive control.* Oxford: Oxford University Press.

Grange, J. A., & Houghton, G. (2014b). Models of cognitive control in task switching. In J. A. Grange & G. Houghton (Eds.), *Task switching and cognitive control* (pp. 160–199). Oxford: Oxford University Press.

Hommel, B. (2000). The prepared reflex: Automaticity and control in stimulus–response translation. In S. Monsell & J. Driver (Eds.), *Control of cognitive processes: Attention and Performance XVIII* (pp. 247–273). Cambridge, MA: MIT Press.

Hübner, R., Futterer, T., & Steinhauser, M. (2001). On attentional control as a source of residual shift costs: Evidence from two-component task shifts. *Journal of Experimental Psychology: Learning, Memory, and Cognition, 27,* 640–653.

Jersild, A. T. (1927). Mental set and shift. *Archives of Psychology, Whole No. 89.*

Karayanidis, F., Jamadar, S., Ruge, H., Phillips, N., Heathcote, A. J., & Forstmann, B. U. (2010). Advance preparation in task-switching: Converging evidence from behavioral, brain activation, and model-based approaches. *Frontiers in Psychology, 25,* 1–13.

Karayanidis, F., & Jamadar, S. D. (2014). Event-related potentials reveal multiple components of proactive and reactive control in task switching. In J. A. Grange & G. Houghton (Eds.), *Task switching and cognitive control* (pp. 200–236). Oxford: Oxford University Press.

Kieffaber, P. D., Kruschke, J. K., Cho, R. Y., Walker, P. M., & Hetrick, W. P. (2013). Dissociating stimulus-set and response-set in the context of task-set switching. *Journal of Experimental Psychology: Human Perception and Performance, 39,* 700–719.

Kieras, D. E., Meyer, D. E., Ballas, J. A., & Lauber, E. J. (2000). Modern computational perspectives on executive mental processes and cognitive control: Where to from here? In S. Monsell & J. S. Driver (Eds.), *Control of cognitive processes: Attention and Performance XVIII* (pp. 681–712). Cambridge, MA: MIT press.

Kiesel, A., Steinhauser, M., Wendt, M., Falkenstein, M., Jost, K., Philipp, A. M., et al. (2010). Control and interference in task switching—A review. *Psychological Bulletin, 136*, 849–874.

Kiesel, A., Wendt, M., & Peters, A. (2007). Task switching: On the origin of response congruency effects. *Psychological Research, 71*, 117–125.

Kleinsorge, T. (2004). Hierarchical switching with two types of judgment and two stimulus dimensions. *Experimental Psychology, 51*, 145–149.

Koch, I., Gade, M., Schuch, S., & Philipp, A. M. (2010). The role of inhibition in task switching: A review. *Psychonomic Bulletin & Review, 17*, 1–14.

Koch, I., Lawo, V., Fels, J., & Vorlander, M. (2011). Switching in the cocktail party: Exploring intentional control of auditory selective attention. *Journal of Experimental Psychology: Human Perception and Performance, 37*, 1140–1147.

Koch, I., Philipp, A. M., & Gade, M. (2006). Chunking in task sequences modulates task inhibition. *Psychological Science, 17*, 346–350.

Kramer, A. F., Hahn, S., & Gopher, D. (1999). Task coordination and aging: Explorations of executive control processes in the task switching paradigm. *Acta Psychologica, 101*, 339–378.

Kray, J., & Ferdinand, N. K. (2014). Task-switching and aging. In J. A. Grange & G. Houghton (Eds.), *Task switching and cognitive control* (pp. 350–371). Oxford: Oxford University Press.

Krummenacher, J., & Müller, H. J. (2012). Dynamic weighting of feature dimensions in visual search: Behavioral and psychophysiological evidence. *Frontiers in Psychology, 3*.

Lau, H. C., & Passingham, R. E. (2007). Unconscious activation of the cognitive control system in the human prefrontal cortex. *Journal of Neuroscience, 27*, 5805–5811.

Lavric, A., Mizon, G. A., & Monsell, S. (2008). Neurophysiological signature of effective anticipatory task-set control: A task-switching investigation. *European Journal of Neuroscience, 28*, 1016–1029.

Lawo, V., Fels, J., Oberem, J., & Koch, I. (2014). Intentional attention switching in dichotic listening: Exploring the efficiency of nonspatial and spatial selection. *Quarterly Journal of Experimental Psychology, 67*, 2010–2024.

Liefooghe, B., Vandierendonck, A., Muyllaert, I., Verbruggen, F., & Vanneste, W. (2005). The phonological loop in task alternation and task repetition. *Memory, 13*, 550–560.

Lien, M.-C., & Ruthruff, E. (2008). Inhibition of task set: Converging evidence from task choice in the voluntary task-switching paradigm. *Psychonomic Bulletin & Review, 15*, 1111–1116.

Logan, G. D. (2004). Working memory, task switching, and executive control in the task span procedure. *Journal of Experimental Psychology: General, 133*, 218–236.

Logan, G. D., & Bundesen, C. (2003). Clever homunculus: Is there an endogenous act of control in the explicit task-cuing procedure? *Journal of Experimental Psychology: Human Perception and Performance, 29*, 575–599.

Logan, G. D., & Bundesen, C. (2004). Very clever homunculus: Compound stimulus strategies for the explicit task-cuing procedure. *Psychonomic Bulletin & Review, 11*, 832–840.

Logan, G. D., & Gordon, R. D. (2001). Executive control of visual attention in dual-task situations. *Psychological Review, 108*, 393–434.

Longman, C. S., Lavric, A., Munteanu, C., & Monsell, S. (2014). Attentional inertia and delayed orienting of spatial attention in task-switching. *Journal of Experimental Psychology: Human Perception and Performance, 40*, 1580–1602.

Longman, C. S., Lavric, A., & Monsell, S. (2015). The coupling between spatial attention and other components of task set. Manuscript submitted.

Los, S. A., & Van der Burg, E. (2010). The origin of switch costs: Task preparation or task application? *Quarterly Journal of Experimental Psychology, 63*, 1895–1915.

MacLeod, C. M., & Dunbar, K. (1988). Training and Stroop-like interference: Evidence for a continuum of automaticity. *Journal of Experimental Psychology: Learning, Memory, and Cognition, 14*, 126–135.

Mari-Beffa, P., Cooper, S., & Houghton, G. (2012). Unmixing the mixing cost: Contributions from dimensional relevance and stimulus–response suppression. *Journal of Experimental Psychology: Human Perception and Performance, 38*, 478–488.

Mari-Beffa, P., & Kirkham, A. (2014). The mixing cost as a measure of cognitive control. In Grange J. A. & Houghton, G. (Eds.), *Task switching and cognitive control* (pp. 74–100). Oxford: Oxford University Press.

Mayr, U. (2002). Inhibition of action rules. *Psychonomic Bulletin & Review, 9*, 93–99.

Mayr, U. (2009). Sticky plans: Inhibition and binding during serial-task control. *Cognitive Psychology, 59*, 123–153.

Mayr, U., & Bell, T. (2006). On how to be unpredictable: Evidence from the voluntary task-switching paradigm. *Psychological Science, 17*, 774–780.

Mayr, U., & Keele, S. W. (2000). Changing internal constraints on action: The role of backward inhibition. *Journal of Experimental Psychology: General, 129*, 4–26.

Mayr, U., & Kliegl, R. (2000). Task-set switching and long-term memory retrieval. *Journal of Experimental Psychology: Learning, Memory, and Cognition, 26*, 1124–1140.

Mayr, U., Kuhns, D., & Hubbard, J. (2014). Long-term memory and the control of attentional control. *Cognitive Psychology, 72*, 1–26.

Mayr, U., Kuhns, D., & Rieter, M. (2013). Eye movements reveal dynamics of task control. *Journal of Experimental Psychology: General, 142*, 489–509.

Meiran, N. (1996). Reconfiguration of processing mode prior to task performance. *Journal of Experimental Psychology: Learning, Memory, and Cognition, 22*, 1423–1442.

Meiran, N. (2000a). Modeling cognitive control in task-switching. *Psychological Research–Psychologische Forschung, 63*, 234–249.

Meiran, N. (2000b). Reconfiguration of stimulus task sets and response task sets during task switching. In J. Driver & S. Monsell (Eds.), *Control of cognitive processes: Attention and performance XVIII* (pp. 377–399). Cambridge, MA: MIT Press.

Meiran, N. (2014). The task-cuing paradigm: A user's guide. In J. A. Grange & G. Houghton (Eds.), *Task switching and cognitive control* (pp. 45–73). Oxford: Oxford University Press.

Meiran, N., Kessler, Y., & Adi-Japha, E. (2008). Control by action representation and input selection (CARIS): A theoretical framework for task switching. *Psychological Research, 72*, 473–500.

Meiran, N., & Marciano, H. (2002). Limitations in advance task preparation: Switching the relevant stimulus dimension in speeded same–different comparisons. *Memory & Cognition, 30*, 540–550.

Mikaye, A., & Friedman, N. P. (2012). The nature and organization of individual differences in executive functions: Four general conclusions. *Current Directions in Psychological Science, 21*, 8–14.

Miller, E. K., & Cohen, J. D. (2001). An integrative theory of prefrontal cortex function. *Annual Review of Neuroscience, 24*, 167–202.

Miyake, A., Emerson, M. J., Padilla, F., & Ahn, J. C. (2004). Inner speech as a retrieval aid for task goals: The effects of cue type and articulatory suppression in the random task cuing paradigm. *Acta Psychologica, 115*, 123–142.

Monsell, S. (2003). Task switching. *Trends in Cognitive Sciences, 7*, 134–140.

Monsell, S., & Mizon, G. A. (2006). Can the task-cuing paradigm measure an endogenous task-set reconfiguration process? *Journal of Experimental Psychology: Human Perception and Performance, 32*, 493–516.

Monsell, S., Sumner, P., & Waters, H. (2003). Task-set reconfiguration with predictable and unpredictable task switches. *Memory & Cognition, 31*, 327–342.

Müller, H. J., Reimann, B., & Krummenacher, J. (2003). Visual search for singleton feature targets across dimensions: Stimulus- and expectancy-driven effects in dimensional weighting. *Journal of Experimental Psychology: Human Perception and Performance, 2003*, 1021–1035.

Nieuwenhuis, S., & Monsell, S. (2002). Residual costs in task switching: Testing the failure-to-engage hypothesis. *Psychonomic Bulletin & Review, 9*, 86–92.

Oberauer, K. (2009). Design for a working memory. In B. H. Ross (Ed.), *Psychology of learning and motivation: Advances in research and theory* (Vol. 51, pp. 45–100). San Diego: Academic Press.

Ophir, E., Nass, C., & Wagner, A. D. (2009). Cognitive control in media multitaskers. *Proceedings of the National Academy of Sciences of the United States of America, 106*, 15583–15587.

O'Reilly, R. C. (2006). Biologically based computational models of high-level cognition. *Science, 314*, 91–94.

Peters, S., & Crone, E. (2014). Cognitive flexibility in childhood and adolescence. In J. A. Grange & G. Houghton (Eds.), *Task switching and cognitive control* (pp. 332–349). Oxford: Oxford University Press.

Philipp, A. M., & Koch, I. (2005). Switching of response modalities. *Quarterly Journal of Experimental Psychology, A, 58*, 1325–1338.

Philipp, A. M., & Koch, I. (2011). The role of response modalities in cognitive task representations. *Advances in Cognitive Psychology, 7*, 31–38.

Posner, M. I. (1980). Orienting of attention. *Quarterly Journal of Experimental Psychology, 32*, 3–25.

Ravizza, S. M. & Salo, R. E. (2014). Task switching in psychiatric disorders. In J. A. Grange & G. Houghton (Eds.), *Task switching and cognitive control* (pp. 300–331). Oxford: Oxford University Press.

Redick, T. S., Shipstead, Z., Harrison, T. L., Hicks, K. L., Fried, D. E., Hambrick, D. Z., et al. (2013). No evidence of intelligence improvement after working memory training: A randomized, placebo-controlled study. *Journal of Experimental Psychology: General, 142*, 359–379.

Reuss, H., Kiesel, A., Kunde, W., & Hommel, B. (2011). Unconscious activation of task sets. *Consciousness and Cognition, 20*, 556–567.

Richter, F. R. & Yeung, N. (2014). Neuroimaging studies of task-switching. In J. A. Grange & G. Houghton (Eds.), *Task switching and cognitive control* (pp. 237–271). Oxford: Oxford University Press.

Rogers, R. D., & Monsell, S. (1995). The costs of a predictable switch between simple cognitive tasks. *Journal of Experimental Psychology: General, 124*, 207–231.

Rubin, O., & Meiran, N. (2005). On the origins of the task mixing cost in the cuing task-switching paradigm. *Journal of Experimental Psychology: Learning, Memory, and Cognition, 31*, 1477–1491.

Rubinstein, J. S., Meyer, D. E., & Evans, J. E. (2001). Executive control of cognitive processes in task switching. *Journal of Experimental Psychology: Human Perception and Performance, 27*, 763–797.

Ruge, H., Jamadar, S., Zimmermann, U., & Karayanidis, F. (2013). The many faces of preparatory control in task switching: Reviewing a decade of fMRI research. *Human Brain Mapping, 34*, 12–35.

Rushworth, M. F. S., Hadland, K. A., Paus, T., & Sipila, P. K. (2002). Role of the human medial frontal cortex in task switching: A combined fMRI and TMS study. *Journal of Neurophysiology, 87*, 2577–2592.

Rushworth, M. F. S., Passingham, R. E., & Nobre, A. C. (2002). Components of switching intentional set. *Journal of Cognitive Neuroscience, 14*, 1139–1150.

Rushworth, M. F. S., Passingham, R. E., & Nobre, A. C. (2005). Components of attentional set-switching. *Experimental Psychology, 52*, 83–98.

Saeki, E., & Saito, S. (2004). Effect of articulatory suppression on task-switching performance: Implications for models of working memory. *Memory, 12*, 257–271.

Saeki, E., & Saito, S. (2009). Verbal representation in task order control: An examination with transition and task cues in random task switching. *Memory & Cognition, 37*, 1040–1050.

Schneider, D. W., & Logan, G. D. (2005). Modeling task switching without switching tasks: A short-term priming account of explicitly-cued performance. *Journal of Experimental Psychology: General, 134*, 343–367.

Schuch, S., & Koch, I. (2003). The role of response selection for inhibition of task sets in task shifting. *Journal of Experimental Psychology: Human Perception and Performance, 29*, 92–105.

Shaffer, L. H. (1965). Choice reaction with variable S-R mapping. *Journal of Experimental Psychology, 70*, 284–288.

Shallice, T., Stuss, D. T., Picton, T. W., Alexander, M. P., & Gillingham, S. (2008). Multiple effects of prefrontal lesions on task-switching. *Frontiers in Human Neuroscience, 1.*

Spence, C., & Driver, J. (1997). On measuring selective attention to a specific sensory modality. *Perception & Psychophysics, 59*, 389–403.

Steinhauser, M., & Hübner, R. (2007). Automatic activation of task-related representations in task shifting. *Memory & Cognition, 35*, 138–155.

Stoet, G., & Snyder, L. H. (2003). Executive control and task-switching in monkeys. *Neuropsychologia, 41*, 1357–1364.

Strobach, T., Frensch, P. A., & Schubert, T. (2012). Video game practice optimizes executive control skills in dual-task and task switching situations. *Acta Psychologica, 140*, 13–24.

Sudevan, P., & Taylor, D. A. (1987). The cuing and priming of cognitive operations. *Journal of Experimental Psychology: Human Perception and Performance, 13*, 89–103.

Treisman, A. M. (1971). Shifting attention between the ears. *Quarterly Journal of Experimental Psychology, 23*, 157–167.

Van 't Wout, F., Lavric, A., & Monsell, S. (2013). Are stimulus–response rules represented phonologically for task-set preparation and maintenance? *Journal of Experimental Psychology: Learning, Memory, and Cognition, 39*, 1538–1551.

Van 't Wout, F., Lavric, A., & Monsell, S. (2015). Is it harder to switch among a larger set of tasks? *Journal of Experimental Psychology: Learning, Memory, and Cognition, 41*, 363–376.

Vandierendonck, A. (2012). Role of working memory in task-switching. *Psychologica Belgica, 52*, 229–253.

Vandierendonck, A., Christiaens, E., & Liefooghe, B. (2008). On the representation of task information in task switching: Evidence from task and dimension switching. *Memory & Cognition, 36*, 1248–1261.

Vandierendonck, A., Liefooghe, B., & Verbruggen, F. (2010). Task switching: Interplay of reconfiguration and interference control. *Psychological Bulletin, 136*, 601–626.

Wager, T. D., Jonides, J., Smith, E. E., & Nichols, T. E. (2005). Toward a taxonomy of attention shifting: Individual differences in fMRI during multiple shift types. *Cognitive, Affective & Behavioral Neuroscience, 5,* 127–143.

Wasylyshyn, C., Verhaeghen, P., & Sliwinski, M. J. (2011). Aging and task switching: A meta-analysis. *Psychology and Aging, 26,* 15–20.

Waszak, F., Hommel, B., & Allport, A. (2003). Task-switching and long-term priming: Role of episodic stimulus–task bindings in task-shift costs. *Cognitive Psychology, 46,* 361–413.

Wylie, G. R., Sumowski, J. F., & Murray, M. (2011). Are there control processes, and (if so) can they be studied? *Psychological Research–Psychologische Forschung, 75,* 535–543.

Yeung, N. (2010). Bottom-up influences on voluntary task switching: The elusive homunculus escapes. *Journal of Experimental Psychology: Learning, Memory, and Cognition, 2010,* 348–362.

Yeung, N., & Monsell, S. (2003a). The effects of recent practice on task switching. *Journal of Experimental Psychology: Human Perception and Performance, 29,* 919–936.

Yeung, N., & Monsell, S. (2003b). Switching between tasks of unequal familiarity: The role of stimulus-attribute and response-set selection. *Journal of Experimental Psychology: Human Perception and Performance, 29,* 455–469.

Yeung, N., Nystrom, L. E., Aronson, J. A., & Cohen, J. D. (2006). Between-task competition and cognitive control in task switching. *Journal of Neuroscience, 26,* 1429–1438.

8

Using Working Memory to Control Attentional Deployment to Items in Complex Scenes

Geoffrey F. Woodman and Steven J. Luck

When we look out on the world, we are often looking for something in particular. For example, when we approach a stop sign we are looking for cross-traffic. When we look out on the soccer field, we are looking for our children. When we open the refrigerator, we are looking for the milk to pour over the cereal. These acts of searching for a particular target require us to control attention in a top-down manner, as opposed to focusing processing in a bottom-up manner on the largest, brightest, or suddenly appearing objects. A dominant view in theories and empirical studies of attention is that top-down control over attention is made possible by holding representations of the searched for items in working memory (for reviews, see Olivers, Peters, Houtkamp, & Roelfsema, 2011; Soto, Hodsoll, Rotshtein, & Humphreys, 2008; Woodman, Carlisle, & Reinhart, 2013). That is, when we want to find Alison, Carter, Henry, Hunter, Paige, or Sam on the soccer field, we maintain representations of them in visual working memory. This mechanistic explanation for how we implement top-down attentional control was proposed in the earliest theories of attention and has been a hot topic of debate in recent empirical studies.

Our goal in this chapter is to discuss the origins of this idea, review recent findings, and lay out the unresolved issues about how working memory is involved in controlling attention. We begin by discussing the historical origins of the idea that the representations in memory determine what attention selects.

Historical Context

The idea that attention and working memory are intimately intertwined is as old as the field of psychology itself. James (1890) proposed that our ability to select one of the multitude of possible inputs to process was as much determined by the active thoughts in our minds as the stimuli that impinge upon us: "While the object excites it [attention] from without, other brain-cells ... arouse it from within" (p. 441). This quote captures the idea that attentional selection depends on both the bottom-up input (i.e., "without") and the top-down representations (i.e., "within") in mind.

This general idea was expanded upon in considerable detail by Pillsbury (1908) in his book *Attention*. He proposed that attentional selection was largely determined by the nature of our memory representations. Like James, he discussed at length the idea that attention is controlled by the representations that we kept active in primary memory (i.e., short-term memory or working memory in modern terminology). Pillsbury (1908) provides us with this concise description of the top-down control of attention during search:

> It is much easier to see any lost article if you have a definite picture of what is sought. In fact, searching for anything consists ordinarily of nothing more than walking about the place where the object is supposed to be, with the idea of the object kept prominently in mind, and thereby standing ready to facilitate the entrance of the perception when it offers itself. (p. 36)

As we will discuss below, the idea that representations stored in working memory allow us to select matching perceptual inputs has dominated the thinking of cognitive scientists and theoreticians for at least the last century. To substantiate that statement, we need look no further than the first major theory of attention proposed in the middle of the last century. Broadbent (1957) proposed that attention works like a filter to determine which perceptual input channels receive the benefit of detailed analysis and can therefore be recognized. Stimuli presented in other input channels went unrecognized. His elegant mechanistic analogy for the operation of attention was a system of tubes that converged on a bottleneck governed by the attentional filter that worked like a flap to gate which inputs were allowed access to the bottleneck. However, Broadbent also proposed that this system had temporary storage of information in short-term memory via recirculation tubes that could keep inputs from an unselected channel (or tube) circulating through the system for a brief period. This allowed the model to account for findings from immediate memory tasks that were recent at that time (Brown, 1954).

The idea that memory representations are integral to the selection process carried out by attention was carried to its most extreme theoretical position by Norman (1968). This theory proposed that what we call attention is a consequence of having a limited number of representations active in memory. This perspective continues to be well represented in current models of working memory (Cowan, 1988; Cowan, 2001; Oberauer, 2002). This brief historical review shows us how the idea that temporary memory (call it primary, short-term, or working memory) controls attention has a long, rich history that has shaped the development of theories since the dawn of psychology as a science of the mind.

State-of-the-Art Review

The first empirical tests of the proposal that working memory controls attention in the modern era of cognitive psychology were performed by Logan (1978, 1979). He used a dual-task procedure that would become standard in such studies decades later. The procedure involved filling working memory with information and then asking the subjects to perform another task before testing the subjects' memory at the end of the trial. The logic was simple: If working memory plays a critical role in the processing of information for the task performed during the memory retention interval, then performance of that task should be significantly impaired relative to a comparison condition in which the task was performed in isolation. Logan (1978) had subjects discriminate the identity of letters while maintaining digits in memory. He reported that subjects were slower to discriminate the target letters when the task was performed during the memory retention interval (i.e., general dual-task interference that changed the y-intercept), but that the set size effects were unchanged by the working memory load (i.e., the slopes of the reaction time [RT] functions were unchanged; Logan, 1979).

Deploying Attention When Visual Working Memory Is Full

We were among the first group of researchers to adopt a dual-task procedure similar to that of Logan (1978) decades later (Downing, 2000; Downing & Dodds, 2004; Woodman, Vogel, & Luck, 2001). We describe our methods and findings from an example experiment to serve as a concrete illustration. The goal of this empirical work was to test the hypothesis that working memory is critical for the attentional selection of targets that we search for in complex scenes (Woodman et al., 2001; Woodman & Luck, 2010). The paradigm involved three different experimental conditions. In the search-alone condition, subjects performed a search task in isolation. In the memory-alone condition, they performed a working memory task in isolation. In the search-and-memory condition, they performed the search task during the retention interval of the working memory task. Figure 8.1 shows the memory-alone and search-and-memory conditions of Experiment 1 from Woodman and Luck (2010).

Although these experiments were published over 20 years after Logan's studies, the logic remained the same. If you fill visual working memory to capacity with the memory task, then working memory will be unavailable for use in the visual search. If working memory is needed for efficient search, this unavailability of visual working memory should reduce the efficiency of visual search (which is typically assessed by means of the slope of the function relating RT to the set size of the search array). Alternatively, subjects might clear out visual working memory to make it available for the search task. If this were the case, then we should see that they perform extremely poorly on the memory task when performed concurrently with the search task.

Across many experiments of this sort, we found no evidence for either of these predictions (Woodman & Luck, 2010; Woodman et al., 2001). As illustrated in figure 8.2, we found that the slopes of the search functions were not different whether search was performed in isolation (i.e., search alone) or during the retention interval of the memory task (i.e., search and memory). The slope of this function is a standard

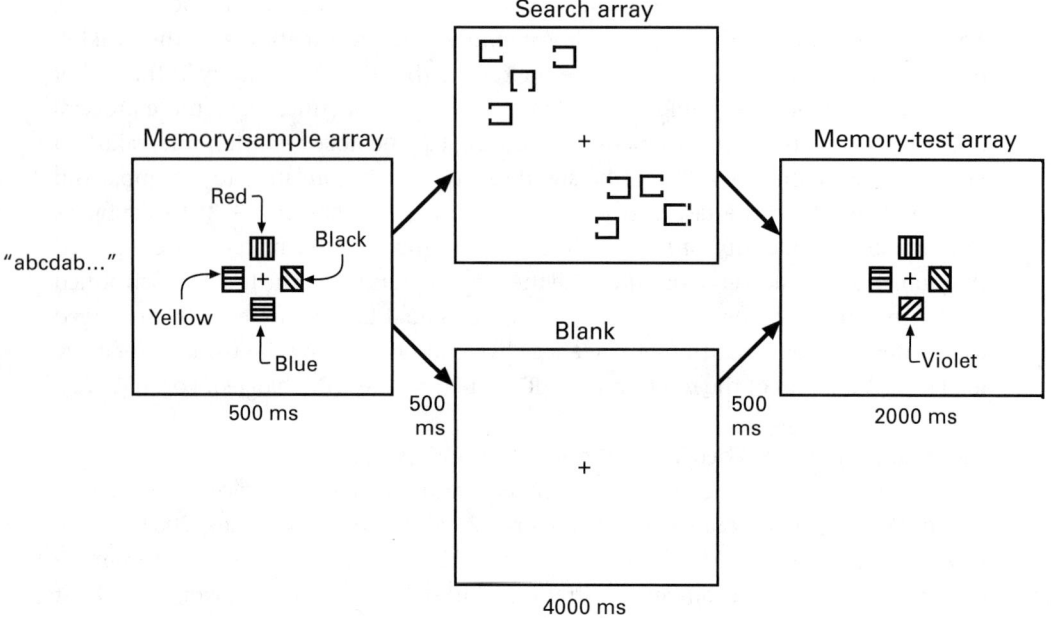

Figure 8.1
An example of the sequence of stimuli presented during the dual-task condition of Experiment 1 of Woodman and Luck (2010). The letters to the left indicate that subjects began each trial by repeating the articulatory-suppression load. There were 500-ms intervals between each stimulus frame in which the screen was blank except for the central fixation point. Reprinted with permission of Psychology Press, Taylor & Francis Group.

measure of the efficiency of the search process because it shows how well the visual system can cope with increased demands. In contrast, the *y*-intercept reflects the processes that occur prior to the search process (e.g., preattentive sensory analysis) and after the search process (e.g., response selection and execution). We consistently observed an increase in the *y*-intercept when search was performed during the memory task, but no change in slope. Thus, the presence of a visual working memory load did not slow the search process but only slowed the processes that precede or follow the search process.

In addition, performance of the memory task showed a consistent but small loss of information from visual working memory when search was performed during the memory task. This loss appears to be a function of presenting any stimulus during the memory retention interval. That is, this drop of between one half and one object's worth of information from visual working memory is observed even when a completely task-irrelevant stimulus is presented (Quinn & McConnell, 1996; Woodman et al., 2001), but not when no search array is presented although expected by the subjects

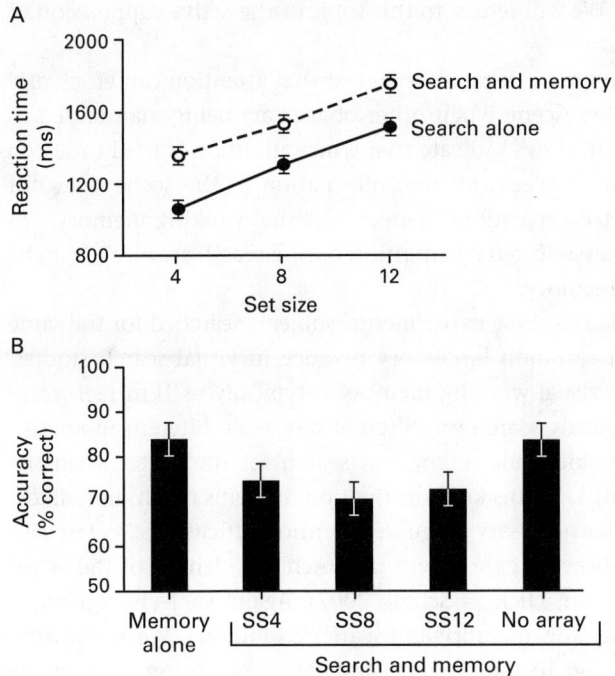

Figure 8.2
The results from Experiment 1 of Woodman and Luck (2010). (A) The visual search RTs from the dual-task condition (dashed lines, empty symbols) and the search-alone condition (solid lines, filled symbols). (B) The change-detection accuracy in the memory-alone condition, left, and the dual-task condition as a function of set size or the presentation of a blank, right. The error bars in this and the subsequent figure show the 95% within-subjects confidence intervals (Loftus & Loftus, 1988). SS, set size. Reprinted with permission of Psychology Press, Taylor & Francis Group.

(see the right bar in figure 8.2B). It may reflect an obligatory storage of the target, or perhaps any object for which a saccade has been planned, in working memory (Hollingworth, Richard, & Luck, 2008).

The findings of these dual-task experiments indicate that storage of information in visual working memory is not necessary for the efficient processing of objects in complex scenes. Recently, interest in this question has increased with further evidence from neuroimaging, event-related potentials (ERPs), and behavior. ERP and behavioral findings have been used to argue that items need to pass through visual working memory when interitem similarity is high within a complex scene (D. E. Anderson, Vogel, & Awh, 2013). In addition, neuroimaging and behavioral findings have suggested that novel nontargets require active suppression during search (Seidl, Peelen, & Kastner, 2012), a process that appears to require access to visual working memory (Rissman,

Gazzaley, & D'Esposito, 2009). We will return to this topic of the active suppression of distractors later in our discussion.

Although this issue continues to be debated, it is clear that attention can efficiently select target objects in a complex scene when other objects are being maintained in visual working memory. These findings indicate that when attention is used to search for and find targets in a complex scene, all the information in the scene does not need to pass through the limited-capacity bottleneck of visual working memory. For example, nontarget items that are selected by attention during search are unlikely to be represented in visual working memory.

In the visual search tasks used in these experiments, subjects searched for the same target on every trial. This is a common laboratory practice in visual search studies, but it does not reflect the way visual working memory is typically used in real-world tasks. In the real world, we typically search for different targets at different moments. For example, when making cookies, one might first search for the butter, then the brown sugar, and then the vanilla. In this kind of situation, it seems much more likely that working memory would be necessary to guide attention efficiently. To test this hypothesis, we conducted a follow-up experiment in which the identity of the target was cued on each trial (Woodman, Luck, & Schall, 2007). Again, subjects performed in memory-alone, search-alone, and memory-and-search conditions. We found that the search slope was much higher in the memory-and-search condition than in the search-alone condition. Thus, there are situations in which working memory plays an absolutely essential role in visual search. However, working memory does not play this role when the target remains constant from trial to trial.

Are Attentional Templates Stored in Visual Working Memory?

The dual-task approach used in the experiments describe so far is a fairly blunt instrument for examining the relationship between attention and working memory. Although it is useful to determine that attention can still be efficiently deployed when visual working memory is full, this does not rule out the possibility that visual working memory enables the deployment of attention in a way that does not require a large amount of storage. Specifically, a number of theories of attention have proposed that the role played by visual working memory typically only requires the storage of one representation.

Many theories of attention propose that visual working memory is used to store a representation of the searched-for target when we process a complex scene (e.g., Bundesen, 1990; Bundesen, Habekost, & Kyllingsbaek, 2005; Desimone & Duncan, 1995; Navalpakkam & Itti, 2005). The simple idea is the top-down control of attention is implemented by holding a representation of the target in visual working memory. For example, when we want to find Sam on the soccer field, we retrieve a representation of Sam from long-term memory, and we maintain that representation in working memory. In doing so, the representation in visual working memory feeds back to the neurons in visual cortex that code for brown hair, brown eyes, the motion of his gait, the red of his shirt and shoes, and so forth. This proposal could explain why the experiments described above found that search displaced a minimal amount of information from visual working memory (i.e., one-half to 1 object's worth of information).

The theoretical proposal that visual working memory is the source of top-down attentional control has garnered considerable empirical attention. Many researchers hypothesized that if visual working memory representations are used to control attention, then maintaining a certain representation in visual working memory should be sufficient to bias attention to similar objects when they appear in our visual field (Carlisle & Woodman, 2011a, 2011b; Downing, 2000; Downing & Dodds, 2004; Hollingworth & Luck, 2009; Houtkamp & Roelfsema, 2006; Olivers, Meijer, & Theeuwes, 2006; Olivers et al., 2011; Soto, Heinke, Humphreys, & Blanco, 2005; Soto et al., 2008; Woodman & Chun, 2006; Woodman & Luck, 2007). In figure 8.3A (plate 1), we show the paradigm that we used in one of our experiments to test this proposal (Woodman & Luck, 2007). The logic is that if we maintain a representation of a red object in visual working memory, then a red object in our visual field should receive the benefit of attention. Most of these experiments involve presenting visual search arrays that either do or do not have a critical item that matches the representation in visual working memory. When the memory-matching item is a distractor (i.e., a nontarget object) in the array, it should attract attention, and this should slow the deployment of attention to the actual target object (compared to when no memory-matching distractor is present).

Figure 8.3B (plate 1) shows the pattern of results obtained from one of these experiments. We found that detection of the visual search target was not significantly slowed by the presence of a memory-matching distractor, provided subjects knew that the target object in the visual search array would never match the item in memory. However, when subjects knew that there was a chance that the target could match the memory item, we found that memory-matching distractors did slow search. We concluded that holding an object in working memory does not cause an obligatory capture of attention by objects that match the features of that object. However, cognitive control processes can allow people to use working memory to guide search when this is adaptive. However, the story did not end with these observations.

Our initial study suggested that people used the contents of working memory to deploy attention in the most adaptive manner. However, other researchers obtained findings suggesting that subjects attend to memory-matching items under all conditions, even when doing so has no benefit for search performance and can only impair search for the target object (e.g., Soto et al., 2005; Soto et al., 2008). Previously, we proposed that such deployments of attention to items matching representations in working memory might be made strategically to improve the memory representation

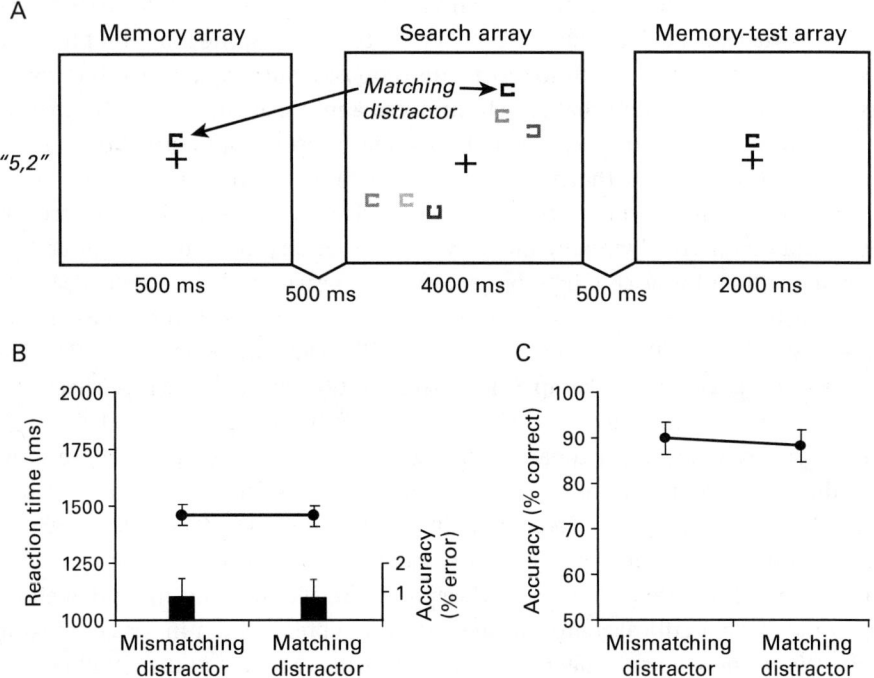

Figure 8.3 (plate 1)
The stimuli and findings of the dual-task experiments examining the relationship between storage of information in visual working memory and the deployment of attention to similar items. The stimuli (A) are from Experiment 2 of Woodman and Luck (2007). Visual search reaction time (B, RT) and memory task accuracy (C, percent correct) are shown for these tasks performed together. Adapted with permission from the American Psychological Association.

being maintained (Woodman & Luck, 2007). Recall that the evidence that memory-matching items are attended comes from dual-task paradigms in which subjects need to remember an item or set of items while performing visual search. The idea is that subjects may attend to the memory-matching items in search arrays not because they cannot help but deploy attention to these items but because they want to deploy attention to these items to help maintain their representations of these items in working memory. We described this as *perceptual resampling* to describe the process of attending to a new perceptual input to help working memory maintenance. That is, by deploying attention to the memory-matching perceptual input, this boosts the working memory representation that needs to be maintained to perform the concurrent memory task accurately. Other researchers have used the term *refreshing* to describe the same idea that directing attention to a representation aids its veridical maintenance in working memory (Raye, Mitchell, Reeder, Greene, & Johnson, 2008; Yi, Turk-Browne, Chun, & Johnson, 2008). This idea provides a logical explanation for why subjects would attend to memory-matching items despite these items' being in opposition to efficient search performance.

The weakness of this perceptual-resampling account is that these strategic attentional deployments do not appear consistent with all deployments of attention to memory-matching items (Balani, Soto, & Humphreys, 2010). For example, in the study of Balani and colleagues (2010), RTs during search were consistent with subjects' attending to memory-matching items; however, subjects were no more accurate for items that reappeared in the visual search arrays than items that did not. As we discuss next, the mixture of results and explanations that we have just reviewed led some researchers to step back and reconsider the theoretical proposals tested with these methods.

The idea that we automatically attend to any input matching a representation in visual working memory is a reasonable extension of the proposal that top-down attentional control comes from representations in visual working memory (Soto et al., 2008). However, alternatives have been proposed where the representations in visual working memory do not directly interact with perceptual selection of inputs without an additional step. In the proposed framework of Olivers and colleagues (Olivers et al., 2011), one visual working memory representation can be in an active state, and only this one active representation controls attention (see also Downing & Dodds, 2004). However, more recent research shows that it is possible to maintain at least two representations in visual working memory that can simultaneously control attention (Beck, Hollingworth, & Luck, 2012).

A slightly different proposal comes from the theory of visual attention framework (Bundesen, 1990; Bundesen & Habekost, 2008; Bundesen et al., 2005). These theorists have proposed that an executive-control process intervenes between storage in visual working memory and perceptual attention that enables a memory representation to influence attentional selection (Bundesen & Habekost, 2008).

Is the Deployment of Attention to Memory-Matching Items Being Actively Suppressed?

Overall, studies have obtained widely divergent results when asking whether attention is automatically attracted to items that match the information currently being stored in working memory (Carlisle & Woodman, 2011a, 2013; Olivers, 2009). This is analogous to the divergent results that have been obtained for decades in studies asking whether salient stimuli automatically capture attention (Egeth & Yantis, 1997; Folk, Remington, & Johnston, 1992; Theeuwes, 1992; Van der Stigchel et al., 2009). We have recently proposed a *signal suppression hypothesis* that can potentially explain many of the divergent results in both of these domains (Sawaki & Luck, 2010, 2014) Specifically, this hypothesis proposes that physically salient objects and objects that match the contents of working memory automatically generate an attentional priority signal (an *attend-to-me* signal). However, this signal can be squashed by a top-down suppression process before attention is actually captured.

Evidence for this hypothesis comes from ERP recordings, in which an *N2pc* ERP component is observed when attention is allocated to a stimulus and a *Pd* ERP component is observed when a stimulus is suppressed. Under conditions that tend to cause behavioral evidence of attentional capture by salient stimuli, previous studies found that a salient but irrelevant stimulus elicits an N2pc component, consistent with capture of attention by the stimulus (e.g., Sawaki & Luck, 2010; Sawaki & Luck, 2014). However, under conditions that minimize behavioral measures of capture, the salient stimulus does not elicit an N2pc component and instead elicits a Pd component (Eimer & Kiss, 2008; Hickey, McDonald, & Theeuwes, 2006) consistent with an active suppression process. Similarly, when a task-irrelevant stimulus array is presented during the delay interval of a visual working memory task, a Pd component is observed when an object in this task-irrelevant array matches the contents of working memory (Sawaki & Luck, 2011). We therefore suggest that at least some of the differences across studies in the involuntary capture of attention by both salient stimuli and memory-matching stimuli reflect differences in the extent to which subjects employ the top-down suppression process reflected by the Pd component to squash the priority signal created by these stimuli. When this suppression process is used consistently, no capture is observed. However, when this suppression process is not used on a proportion of trials, some evidence for capture may be observed.

The active suppression mechanism that we just discussed is likely to be particularly important in several real-world settings. Primates often use eye-movement behavior in socially important ways. For example, among many species of primates, directing gaze to the eyes of an individual that is higher in the social hierarchy is seen as a challenge to that individual's dominance. Clearly, it is adaptive to avoid overt eye movements to such potentially threatening targets to avoid unnecessary social conflict. Because there is such a tight link between covert and overt selection (Hoffman & Subramaniam, 1995), this is likely to involve the suppression of covert processing of such threat stimuli to prevent shifts of gaze, and has the added benefit of suppressing processing of the threatening gaze, reducing the emotional stress that accompanies the processing of such information.

Directly Measuring the Attentional Templates in the Brain

Most of the research described so far used a dual-task approach to look at interactions between attention and working memory. However, in some of our recent work we wanted to take the more basic approach of having subjects perform a visual search task while we eavesdropped on the normal operation of visual working memory with ERPs. Our approach was based on the cued visual search paradigm used by Chelazzi and colleagues in their recordings of single-unit activity from the ventral visual cortex of monkeys (Chelazzi, Duncan, Miller, & Desimone, 1998; Chelazzi, Miller, Duncan, & Desimone, 1993, 2001). On each trial, the monkey was shown a cue stimulus that defined the target for an upcoming visual search task. It was observed that the neurons that coded for the target in the inferotemporal cortex showed an elevated firing rate during the period between the offset of the cue and the onset of the search array. This was interpreted as evidence that the monkeys were actively maintaining a representation of the cue in working memory during this interval, which then biased attention toward the target in the search array. This is like the experiment described earlier (Woodman et al., 2007) in which we cued the target on each trial rather than remaining constant across trials. A memory load interfered with search when the target was cued on each trial, implying that working memory plays a key role in visual search under these conditions.

We have recently used this trial-by-trial cuing approach with ERP recordings in human subjects to provide direct evidence that the cue representation was being held in visual working memory. This research exploited an ERP component known as the contralateral-delay activity (or CDA), a sustained negativity that is observed during the delay interval in working memory paradigms. The CDA is contralateral to the location in the visual field where the to-be-remembered objects were initially presented. The amplitude of the CDA reflects the number of objects maintained in visual working memory and is highly correlated with behavioral measures of visual working memory capacity (Vogel & Machizawa, 2004; Woodman & Vogel, 2008). The CDA has a different time course and voltage distribution over electrodes compared to other lateralized components related to attention and response preparation (McCollough, Machizawa, & Vogel, 2007). We have also observed this ERP component in macaque monkeys and have shown that it is driven by sustained memory-related activity in prefrontal structures (Reinhart et al., 2012). As we discuss next, the CDA can be harnessed to study the involvement of visual working memory in controlling visual attention during visual search.

Figure 8.4 (plate 2) shows the CDA while subjects performed a visual search task in which the identity of the target was cued at the beginning of each trial and changed from trial to trial (Carlisle, Arita, Pardo, & Woodman, 2011; Woodman & Arita, 2011). To provide a balanced sensory input, the cue stimulus included a red item and a green item, and subjects were supposed to search for the shape indicated by the red cue and ignore the green cue (or vice versa). In the interval between the target cue offset and the onset of the search array, we found a sustained negative potential contralateral to the relevant cue item. The CDA amplitude in this interval, before search began, predicted the speed and accuracy of the search response (Carlisle et al., 2011; Reinhart & Woodman, 2014; Woodman & Arita, 2011). Then, we showed that the CDA in this interval was twice as large when two relevant cues were present (indicating two possible target

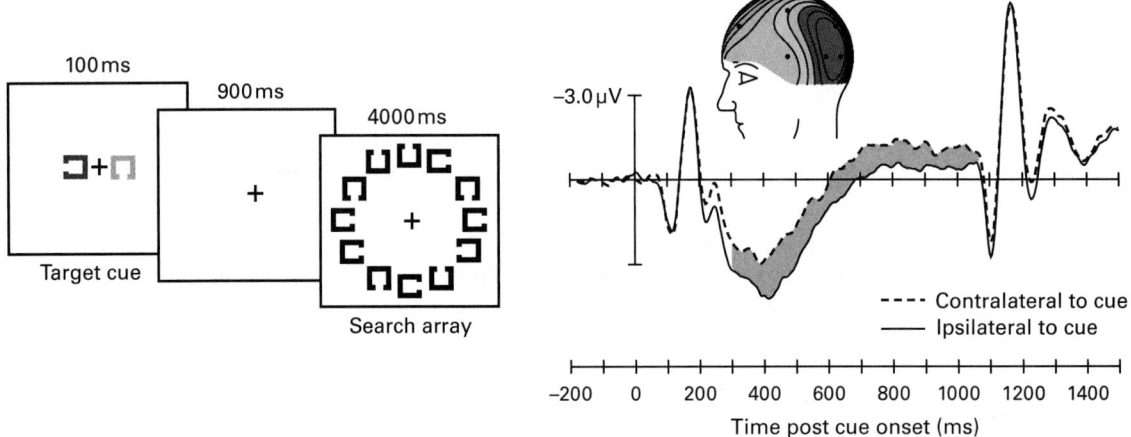

Figure 8.4 (plate 2)
The stimuli and event-related potential findings from Woodman and Arita (2011). Example of
the stimulus sequence (left) and the grand average waveforms from electrodes T5/6, contralateral
(red) and ipsilateral (black) to the location of the cue on each trial (right). The gray region shows
the epoch in which the significant contralateral-delay activity (CDA) was measured, and the inset
shows voltage distribution. The amplitude of the CDA predicted the accuracy of the subsequent
search across subjects ($p < .05$). Adapted with permission from the Association for Psychological
Science and Blackwell Publishing.

items that might occur in the search array). These findings provide strong evidence
that visual working memory representations were used to control the deployment of
visual attention when the task involves looking for a different target on each trial (Carlisle et al., 2011). However, our previous results showing no interference between working memory and visual search in dual-task experiments suggests that working memory
plays little or no role in search when the target identity remains constant on every trial.

The idea is fairly simple: When we are frantically looking for our keys in our kitchen
on our way to work, it is best to hold a representation of those keys in working memory. However, what happens as you look for those keys in room after room? Perhaps
longer term memory systems can take over the job of controlling attention in this situation, freeing up working memory for other tasks.

To test this hypothesis, we conducted experiments in which the same target was
cued across three to seven consecutive trials, and then the cue changed to specify a
different target (Carlisle et al., 2011; Reinhart & Woodman, 2014; Woodman et al.,
2013). We predicted that the amplitude of the CDA would decrease after several presentations of the same target, and this is what we found. As shown in figure 8.5, CDA
amplitude decreased over successive presentations, suggesting the target was no longer

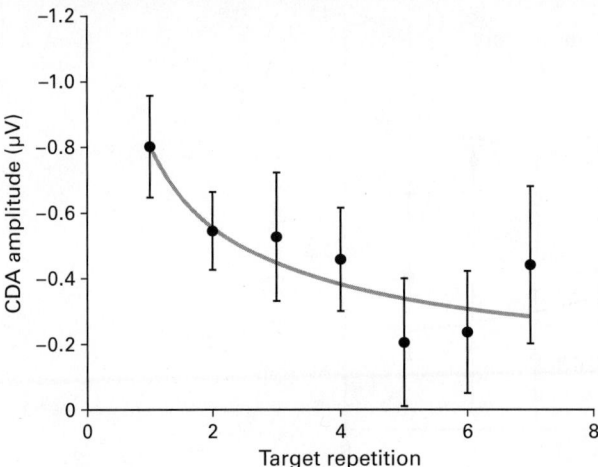

Figure 8.5

The grand average event-related potential results, time-locked to the cue presentation from consecutive groups of trials. The plot shows the contralateral-delay activity (CDA) amplitude across consecutive trials with the same search target. The gray line shows the power-function fit, and the error bars represent ±1 S.E.M. Adapted with permission from the Society for Neuroscience.

represented in visual working memory and that some longer term memory system took over the guidance of attention. Moreover, when subjects searched for the same target shape for the entire experiment, there was no CDA following the target cue (Carlisle et al., 2011). These findings provide direct evidence of a transition from reliance on visual working memory representations to reliance on longer term memory representations in the top-down control of visual attention. As we will discuss further below, these findings also provide strong converging evidence for the idea that the development of automaticity involves a transition from working memory to long-term memory, which has been a cornerstone of many theories of learning and skill acquisition (J. R. Anderson, 1982, 2000; Logan, 1988, 2002; Rickard, 1997) and supported by inferences from behavioral studies (e.g., Woodman et al., 2007). This recent work shows how our measurements of ERPs can be used to more clearly define the roles of various memory representations in controlling attention and shape future theories of attention.

In the study that we just described, it was inferred that long-term memory representations accumulated to enable the takeover of attentional control from visual working memory. Although this is a reasonable inference, we would be far more confident in this conclusion if we could watch this information accumulation in long-term memory. To do this, several recent studies (Reinhart & Woodman, 2014; Woodman et al., 2013) measured a different ERP component, P170, that indexes long-term memory. Voss and colleagues (Voss, Schendan, & Paller, 2010) reported that the amplitude of this frontal

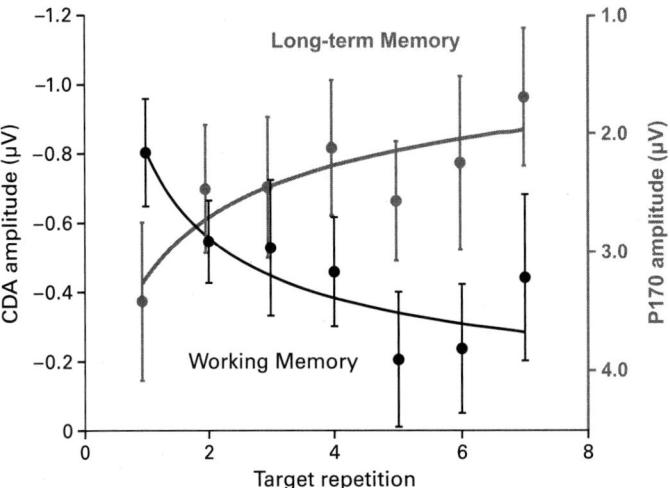

Figure 8.6 (plate 3)
The event-related potential component amplitudes following the target cue onset as a function of target repetition. The P170 amplitude was measured from 150–200 ms postcue and the contralateral-delay activity (CDA) from 300–900 ms postcue. The error bars show ±1 S.E.M. Adapted with permission from the Society for Neuroscience.

positivity was related to repetition priming in long-term memory (Diana, Vilberg, & Reder, 2005; Duarte, Ranganath, Winward, Hayward, & Knight, 2004; Tsivilis, Otten, & Rugg, 2001). As shown in figure 8.6 (plate 3), we found that P170 amplitude increased as the number of trials with a given cue increased, mirroring the decrease in CDA amplitude. This provides positive evidence that long-term memory becomes increasingly recruited as the subject searches for the same target on multiple consecutive trials. Surprisingly, it appears that this transition to the use of long-term memory to control the deployment of attention occurs at the same rate across visual search tasks that vary in difficulty (Gunseli, Olivers, & Meeter, 2014). These observations are important because theories of learning and automaticity consider the automatization of visual search to be a form of repetition priming (Logan, 1990). In addition, these findings appear to conform to the quantitative predictions of at least one theory of learning and automaticity (Logan, 2002) that predicts that working memory representations initially control attention but are made unnecessary as long-term memory representations accumulate.

Coming Full Circle
Recall that we previously mentioned that the idea that working memory representations control attention dates back to James (1890) and Pillsbury (1908). It might seem

that the research we just reviewed demonstrates the inadequacy of these classic statements about the nature of the memory representations that control attention as we search for objects in our environment. Clearly, long-term memory representations are at least equally important in the top-down control of perceptual attention. However, if we return to Pillsbury (1908), we see that he also foreshadowed the idea that the representations in long-term memory are important for the selection of perceptual inputs by attention:

Not only do ideas that are actually present in consciousness at the moment have an influence in determining what impressions shall become conscious, but other experiences, which are much more remote in time and not in consciousness at the moment, also play a part. We can trace these other conditions backward in time, and as they become farther distant they also become more general and harder to trace as individual influence because combined with others in a total complex. (p. 36)

This elegant quote indicates that our recent ERP observations were predicted over a century before they were made. In addition, the above quote foreshadows one of the next obstacles that we will face as we accept the reality that both working memory and long-term memory combine to provide the top-down control of attention. That is, if all of our background knowledge could be contributing to what attention selects at any given moment (e.g., Moores, Laiti, & Chelazzi, 2003), then how do we study the deployment of attention without understanding the intricate dynamics between the representations in long-term memory?

Integration

The laboratory experiments that we discussed in this chapter have important implications for how we operate in our environment outside the laboratory. We began by discussing how representations in memory allow us to find specific objects in our environment. Whether they are our children, food, or threats to our safety, we appear to store representations of what we are looking for in visual working memory when we enter a new environmental context or search for a new target. For example, if we are building a sandwich (e.g., Hayhoe, Shrivastava, Mruczek, & Pelz, 2003), then we need to switch from looking for the bread, to looking for the jelly, the knife, and then the peanut butter. Or if we are working on our car, we need to switch from looking for the bolts to remove to finding the socket needed to remove those bolts. It seems that many real-world tasks like these involve rapidly switching between looking for different specific targets to accomplish the complex task we are performing. The laboratory experiments described in this chapter clearly show that these switches between targets require target representations to be shuttled in and out of visual working memory to perform these real-world tasks. However, the involvement of long-term memory representations in well-practiced skills suggests that this is not the complete story.

As cognitive scientists, we often use driving as an example of a real-world task in which we need to attend to task-relevant information that is vital for our survival and that of others on the roadway (e.g., Kang, Hong, Blake, & Woodman, 2011; Woodman & Vogel, 2005). However, in this task it is likely that long-term memory plays a dominant role in allowing us to search for multiple possible targets at the same time with high efficiency (Wolfe, 2012). In these tasks, visual working memory and long-term memory are likely to work in concert to allow us to attend to possible targets and to be particularly sensitive to certain likely targets (Reinhart & Woodman, 2014). For example, in a school zone we may hold a representation of a child pedestrian in working memory while long-term memory continues to bias attention to trucks, icy patches on the road, distracted drivers, and the flash of red brake lights, leaving the limited capacity of working memory to handle the representation of the child entering the roadway. Consistent with the goal of this book to integrate findings from the laboratory with real-world processing demands, we believe that our ability to use multiple memory representations to simultaneously set attentional weights is critical to understanding the nature of attentional control in any setting (Woodman et al., 2013).

The setting of attentional weights using memory representations brings up an issue that is often neglected in the discussion of attentional templates, but is an issue in the construction of models of attention. Specifically, setting attentional weights is not likely to be a question of just determining the relative importance of a variety of targets. Instead, is seems apparent that learning what is irrelevant is also important. With our limited-capacity processing of visual information, it is important not to waste processing capacity on irrelevant information. In the context of our driving example, if we are going to be able to processes task-relevant information like brake lights or children entering the roadway, then we need to avoid wasting resources on the trees, the clouds drifting by, and the attractive dashboard in our vehicle. This may sound obvious, but the idea that attention needs to be tuned to avoid certain nontargets is not always explicitly dealt with in models of attention (e.g., Bundesen & Habekost, 2008).

A growing body of evidence from the laboratory appears to show how we come to ignore the flashing light on our dashboard or the pop-up banner on our desktop. Vecera, Cosman, Vatterott, and Roper (2014) recently showed that initially task-irrelevant objects that are unique in an array of items capture attention in the laboratory. However, as subjects in these laboratory experiments search for items of a specific color, only items that match that color come to capture attention. Returning to our driving example, this could explain why the onset of a red light on our instrument cluster captures attention because it matches the onset of red brake lights that we are vigilant for while driving. In contrast, the onset of a green light on our dashboard, matching a green traffic light that requires no change in behavior, is not processed to the same degree. At the far extreme, there are the stimuli in our field of view that are completely task irrelevant and would be maladaptive to waste our limited-capacity processing on. For example, if we were to process the vents on our dashboard with an attentional weight above zero, then this means that these stationary, completely irrelevant stimuli in the environment would be robbing task-relevant inputs of our full focus of attention. Instead, the work we have reviewed above indicates that we can set attentional weights to objects such as the vents on the dashboard to levels below zero, suppressing attentional selection of these items completely and setting positive attentional weights for objects that are task relevant. The growing body of work that we have discuss points to long-term memory as the source of the representations that help determine which inputs receive negative versus positive attentional weights (Bundesen & Habekost, 2008; Vecera et al., 2014).

Future Directions

We believe that the next major challenge that we face is how to understand the control mechanisms that allow attention to be guided by working memory representations at certain times and by long-term memory representations at others. Moreover, there are many different types of long-term memories that might compete for the control of attention (e.g., episodic memory, priming, operant conditioning, and classical conditioning). This important topic of future study has been identified by recent theoretical proposals (Olivers et al., 2011) as a great need for our understanding of the top-down control of attention. As Pillsbury (1908) described above, understanding attention could require us to simply solve the problem of understanding memory, a problem of sufficient scale that it could become demoralizing to future attention researchers. Instead of walking away from the entire endeavor, we will review three approaches to understanding the integration of memory and attention that we believe are particularly promising.

The first theoretical perspective that we highlight proposes that attention and memory are not separate cognitive mechanisms at all. As we described briefly in a previous section, this perspective has a long tradition. Norman (1968) proposed the first major model built on the idea that attention and memory are really parts of the same cognitive system. This theoretical perspective has the strength of naturally explaining a tight relationship between attention and different forms of memory. Because working memory representations are just the representations in long-term memory that are within the focus of attention (by some accounts), this perspective can fairly easily accommodate the idea that either these working memory representations or other long-term memory can influence which inputs are selected without proposing different mechanisms for the guidance of attention by working memory and long-term memory. However, with this perspective it is less clear how attention can be used to select certain perceptual inputs while simultaneously maintaining other working memory representations, such as in the dual-task experiments that we reviewed above. This framework could be modified by the addendum that one working memory representation is special in being able to guide attention (Olivers et al., 2011), but it remains to be seen whether this modification will be able to account for how attention is deployed as complex scenes are analyzed (e.g., Beck et al., 2012) and whether overlapping mechanisms of attention and working memory can exhibit sufficient flexibility to account for the dissociations that have been observed (e.g., Woodman & Luck, 2003).

The second theoretical perspective that could account for the use of both working memory and long-term memory representations to control attention is the multiple-component working memory model of Baddeley and colleagues (Baddeley, 1986, 2007; Baddeley & Hitch, 1974; Baddeley & Logie, 1999). The key component in this model that allows it to account for the findings we reviewed here is the central executive. This is an executive-control mechanism, or collection of mechanisms, that supervises the storage of information in working memory, long-term memory, and the selection of new perceptual inputs for further processing. This far-reaching mechanism obviously has the ability to account for how perceptual selection could initially be controlled by a representation in working memory and then later be controlled by representations that accumulate in long-term memory. The only weakness of this theoretical perspective is that it accounts for the dynamics of attentional control across task performance with a homunculus-like mechanism that monitors the contents of multiple memory stores and the demands of the task at hand. Although, as cognitive scientists, we have a natural tendency to favor simpler explanations over those that require such an intelligent agent (Attneave, 1960), it does not mean that this theoretical perspective will ultimately fail.

The final theoretical perspective that we believe may be particularly fruitful in understanding the nature of the memory representations that control attention comes from models of automaticity and skill acquisition. The earliest theories of the learning that underlies our ability to become proficient at new tasks proposed that we initially rely on declarative memory representations that we keep actively in mind. Then, as we become better at the task, we transition to relying on long-term memory representations that guide task performance without our actively maintaining them (Fitts & Posner, 1967). This general idea has been refined and used to account for our ability to become automatic at performing tasks such that our information processing capacity limits are drastically reduced or even eliminated (Schneider & Shiffrin, 1977; Shiffrin & Schneider, 1977). The most sophisticated of these models propose that improved processing is due to both working memory and long-term memory representations racing to guide processing (Logan, 1988, 2002) with the accumulation of long-term memory representations coming to dominate processing and selection with practice. Because this proposal hinges on the simultaneous contributions of both working memory and long-term memory representations, and because of its mathematical specification, we believe that this theoretical framework could potentially account for the body of evidence that we reviewed here, as well as providing novel predictions. However, as we await further modeling work, the jury is still out with regard to whether competing theories can account for the entirety of the body of evidence that we have reviewed here.

How memory representations provide top-down control over the deployment of attention is an active topic of research. Ideally, we are entering a new phase of inquiry where our experiments will be guided by contrasting predictions that allow us to distinguish between the competing theoretical explanations discussed above. We hope that the result will allow us to meet the challenge identified by Pillsbury (1908) to identify and trace the contributions of our memory representations to determining what information is selected by attention.

Box 8.1

Key Points

- The top-down control of attention is governed by representations in both working memory and long-term memory.

- Although the bulk of empirical and theoretical work has focused on how representations in working memory exert control, long-term memory representations appear vital in determining how attention is deployed in the laboratory and the real world.

- These memory representations appear to be used to determine which items are selected for preferential processing, as well as determining which items are actively ignored.

Box 8.2

Outstanding Issues

- Given that some models of working memory propose that attention is just the most activated portion of working memory, how do these models account for the ability of subjects to efficiently select certain inputs while maintaining other information in visual working memory?

- Some models of working memory propose that working memory consists of the focusing of attention on long-term memory representations. If this is true, then how does attention also operate to select sensory inputs for perceptual processing and for storage in working memory?

- What are the key factors that determine when long-term memory representations control attention?

- Given the evidence that long-term memory helps provide top-down control over attention, then can we understand attention without first understanding the encoding, maintenance, and retrieval of information from long-term memory?

- Can models account for the shift from initially using working memory representations to using long-term memory representations to control attention without positing an intelligent central executive that controls this shift?
- Here we focused on experiments that used visual search tasks in which attention needs to select targets in a field of distractors. It remains to be seen if the same dynamics of attentional control (i.e., shifting from working memory to long-term memory representations) occur in other laboratory paradigms, such as spatial cuing, go/no-go tasks, the *n*-back paradigm, and so forth.

Acknowledgments

G.F.W. was supported by grants to from the National Eye Institute (R01-EY019882, P30-EY008126, T32-EY007135) and the National Science Foundation (BCS 09–57072). S.J.L. was supported by a grant from the National Institute of Mental Health (R01-MH076226).

References

Anderson, D. E., Vogel, E. K., & Awh, E. (2013). A common discrete resource for visual working memory and visual search. *Psychological Science, 24*, 929–938. doi:10.1177/0956797612464380.

Anderson, J. R. (1982). Acquisition of a cognitive skill. *Psychological Review, 89*, 369–406.

Anderson, J. R. (2000). *Learning and memory*. New York: Wiley.

Attneave, F. (1960). In defence of homunculi. In W. Rosenblith (Ed.), *Sensory communication* (pp. 777–782). Cambridge, MA: MIT Press.

Baddeley, A. D. (1986). *Working memory*. Oxford, UK: Clarendon.

Baddeley, A. D. (2007). *Working memory, thought, and action*. New York: Oxford University Press.

Baddeley, A. D., & Hitch, G. J. (1974). Working memory. In G. H. Bower (Ed.), *The psychology of learning and motivation* (Vol. 8, pp. 47–90). New York: Academic Press.

Baddeley, A. D., & Logie, R. H. (1999). Working memory: The multiple component model. In A. Miyake & P. Shah (Eds.), *Models of working memory: Mechanisms of active maintenance and executive control* (pp. 28–61). Cambridge, UK: Cambridge University Press.

Balani, A. B., Soto, D., & Humphreys, G. W. (2010). Working memory and target-related distractor effects on visual search. *Memory & Cognition, 38*, 1058–1076.

Beck, V. M., Hollingworth, A., & Luck, S. J. (2012). Simultaneous control of attention by multiple working memory representations. *Psychological Science, 23*, 887–898.

Broadbent, D. E. (1957). A mechanical model for human attention and immediate memory. *Psychological Review, 64*, 205–215.

Brown, J. (1954). The nature of set-to-learn and of intra-material interference in immediate memory. *Quarterly Journal of Experimental Psychology, 6*, 141–148.

Bundesen, C. (1990). A theory of visual attention. *Psychological Review, 97*, 523–547.

Bundesen, C., & Habekost, T. (2008). *Principles of visual attention: Linking mind and brain.* New York: Oxford University Press.

Bundesen, C., Habekost, T., & Kyllingsbaek, S. (2005). A neural theory of visual attention: Bridging cognition and neurophysiology. *Psychological Review, 112*, 291–328.

Carlisle, N. B., Arita, J. T., Pardo, D., & Woodman, G. F. (2011). Attentional templates in visual working memory. *Journal of Neuroscience, 31*, 9315–9322.

Carlisle, N. B., & Woodman, G. F. (2011a). Automatic and strategic effects in the guidance of attention by working memory representations. *Acta Psychologica, 137*, 217–225.

Carlisle, N. B., & Woodman, G. F. (2011b). When memory is not enough: Electrophysiological evidence for goal-dependent use of working memory representations in guiding visual attention. *Journal of Cognitive Neuroscience, 23*, 2650–2664.

Carlisle, N. B., & Woodman, G. F. (2013). Reconciling conflicting electrophysiological findings on the guidance of attention by working memory. *Attention, Perception & Psychophysics, 75*, 1330–1335.

Chelazzi, L., Duncan, J., Miller, E. K., & Desimone, R. (1998). Responses of neurons in inferior temporal cortex during memory-guided visual search. *Journal of Neurophysiology, 80*, 2918–2940.

Chelazzi, L., Miller, E. K., Duncan, J., & Desimone, R. (1993). A neural basis for visual search in inferior temporal cortex. *Nature, 363*, 345–347.

Chelazzi, L., Miller, E. K., Duncan, J., & Desimone, R. (2001). Responses of neurons in macaque area V4 during memory-guided visual search. *Cerebral Cortex, 11*, 761–772.

Cowan, N. (1988). Evolving conceptions of memory storage, selective attention, and their mutual constraints within the human information-processing system. *Psychological Bulletin, 104*, 163–191.

Cowan, N. (2001). The magical number 4 in short-term memory: A reconsideration of mental storage capacity. *Behavioral and Brain Sciences, 24*, 87–185.

Desimone, R., & Duncan, J. (1995). Neural mechanisms of selective visual attention. *Annual Review of Neuroscience, 18*, 193–222.

Diana, R. A., Vilberg, K. L., & Reder, L. M. (2005). Identifying the ERP correlate of a recognition memory search attempt. *Brain Research. Cognitive Brain Research, 24*, 674–684.

Downing, P. E. (2000). Interactions between visual working memory and selective attention. *Psychological Science, 11,* 467–473.

Downing, P. E., & Dodds, C. M. (2004). Competition in visual working memory for control of search. *Visual Cognition, 11,* 689–703.

Duarte, A., Ranganath, C., Winward, L., Hayward, D., & Knight, R. T. (2004). Dissociable neural correlates for familiarity and recollection during the encoding and retrieval of pictures. *Brain Research. Cognitive Brain Research, 18,* 255–272.

Egeth, H. E., & Yantis, S. (1997). Visual attention: Control, representation, and time course. *Annual Review of Psychology, 48,* 269–297.

Eimer, M., & Kiss, M. (2008). Involuntary attentional capture determined by task set: Evidence from event-related brain potentials. *Journal of Cognitive Neuroscience, 208,* 1423–1433.

Fitts, P. M., & Posner, M. I. (1967). *Human performance.* Belmont, CA: Brooks Cole.

Folk, C. L., Remington, R. W., & Johnston, J. C. (1992). Involuntary covert orienting is contingent on attentional control settings. *Journal of Experimental Psychology: Human Perception and Performance, 18,* 1030–1044.

Gunseli, E., Olivers, C. N. L., & Meeter, M. (2014). Effects of search difficulty on the selection, maintenance, and learning of attentional templates. *Journal of Cognitive Neuroscience, 26,* 2042–2054.

Hayhoe, M. M., Shrivastava, A., Mruczek, R., & Pelz, J. B. (2003). Visual memory and motor planning in a natural task. *Journal of Vision, 3,* 49–63.

Hickey, C., McDonald, J. J., & Theeuwes, J. (2006). Electrophysiological evidence of the capture of visual attention. *Journal of Cognitive Neuroscience, 18,* 604–613.

Hoffman, J. E., & Subramaniam, B. (1995). The role of visual attention in saccadic eye movements. *Perception & Psychophysics, 57,* 787–795.

Hollingworth, A., & Luck, S. J. (2009). The role of visual working memory in the control of gaze during visual search. *Attention, Perception & Psychophysics, 71,* 936–949.

Hollingworth, A., Richard, A. M., & Luck, S. J. (2008). Understanding the function of visual short-term memory: Transsaccadic memory, object correspondence, and gaze correction. *Journal of Experimental Psychology. General, 137,* 163–181.

Houtkamp, R., & Roelfsema, P. R. (2006). The effect of items in working memory on the deployment of attention and the eyes during visual search. *Journal of Experimental Psychology: Human Perception and Performance, 32,* 423–442.

James, W. (1890). *The principles of psychology.* New York: Holt.

Kang, M.-K., Hong, S. W., Blake, R., & Woodman, G. F. (2011). Visual working memory contaminates perception. *Psychonomic Bulletin & Review, 18,* 860–869.

Loftus, G. R., & Loftus, E. F. (1988). *Essence of statistics* (2nd ed.). New York: Random House.

Logan, G. D. (1978). Attention in character classification tasks: Evidence for the automaticity of component stages. *Journal of Experimental Psychology. General, 107*, 32–63.

Logan, G. D. (1979). On the use of a concurrent memory load to measure attention and automaticity. *Journal of Experimental Psychology: Human Perception and Performance, 5*, 189–207.

Logan, G. D. (1988). Toward an instance theory of automatization. *Psychological Review, 95*, 492–527.

Logan, G. D. (1990). Repetition priming and automaticity: Common underlying assumptions. *Cognitive Psychology, 22*, 1–35.

Logan, G. D. (2002). An instance theory of attention and memory. *Psychological Review, 109*, 376–400.

McCollough, A. W., Machizawa, M. G., & Vogel, E. K. (2007). Electrophysiological measures of maintaining representations in visual working memory. *Cortex, 43*, 77–94.

Moores, E., Laiti, L., & Chelazzi, L. (2003). Associative knowledge controls deployment of visual selective attention. *Nature Neuroscience, 6*, 182–185.

Navalpakkam, V., & Itti, L. (2005). Modeling the influence of task on attention. *Vision Research, 45*, 205–231.

Norman, D. A. (1968). Toward a theory of memory and attention. *Psychological Review, 75*, 522–536.

Oberauer, K. (2002). Access to information in working memory: Exploring the focus of attention. *Journal of Experimental Psychology: Learning, Memory, and Cognition, 28*, 411–421.

Olivers, C. N. L. (2009). What drives memory-driven attentional capture? The effects of memory type, display type and search type. *Journal of Experimental Psychology: Human Perception and Performance, 35*, 1275–1291.

Olivers, C. N. L., Meijer, F., & Theeuwes, J. (2006). Feature-based memory-driven attentional capture: Visual working memory contents affects visual attention. *Journal of Experimental Psychology: Human Perception and Performance, 32*, 1243–1265.

Olivers, C. N. L., Peters, J. C., Houtkamp, R., & Roelfsema, P. R. (2011). Different states in visual working memory: When it guides attention and when it does not. *Trends in Cognitive Sciences, 15*, 327–334.

Pillsbury, W. B. (1908). *Attention*. New York: Macmillan.

Quinn, J. G., & McConnell, J. (1996). Irrelevant pictures in visual working memory. *Quarterly Journal of Experimental Psychology, A, 49A*, 200–215.

Raye, C. L., Mitchell, K. J., Reeder, J. A., Greene, E. J., & Johnson, M. K. (2008). Refreshing one of several active representations: Behavioral and functional magnetic resonance imaging differences between young and older adults. *Journal of Cognitive Neuroscience, 20*, 852–862.

Reinhart, R. M. G., Heitz, R. P., Purcell, B. A., Weigand, P. K., Schall, J. D., & Woodman, G. F. (2012). Homologous mechanisms of visuospatial working memory maintenance in macaque and human: Properties and sources. *Journal of Neuroscience, 32,* 7711–7722.

Reinhart, R. M. G., & Woodman, G. F. (2014). High stakes trigger the use of multiple memories to enhance the control of attention. *Cerebral Cortex, 24,* 2022–2035.

Rickard, T. C. (1997). Bending the power law: A CMPL theory of strategy shifts and the automatization of cognitive skills. *Journal of Experimental Psychology. General, 126,* 288–311.

Rissman, J., Gazzaley, A., & D'Esposito, M. (2009). The effect of non-visual working memory load on top-down modulation of visual processing. *Neuropsychologia, 47,* 1637–1646.

Sawaki, R., & Luck, S. J. (2010). Capture versus suppression of attention by salient singletons: Electrophysiological evidence for an automatic attend-to-me signal. *Attention, Perception & Psychophysics, 72,* 1455–1470.

Sawaki, R., & Luck, S. J. (2011). Active suppression of distractors that match the contents of visual working memory. *Visual Cognition, 19,* 956–972.

Sawaki, R., & Luck, S. J. (2014). How the brain prevents and terminates shifts of attention. In G. R. Mangun (Ed.), *Cognitive electrophysiology of attention* (pp. 16–29). New York: Elsevier.

Schneider, W., & Shiffrin, R. M. (1977). Controlled and automatic human information processing. I: Detection, search and attention. *Psychological Review, 84,* 1–66.

Seidl, K. N., Peelen, M. V., & Kastner, S. (2012). Neural evidence for distractor suppression during visual search in real-world scenes. *Journal of Neuroscience, 32,* 11812–11819.

Shiffrin, R. M., & Schneider, W. (1977). Controlled and automatic human information processing. II: Perceptual learning, automatic attending, and a general theory. *Psychological Review, 84,* 127–190.

Soto, D., Heinke, D., Humphreys, G. W., & Blanco, M. J. (2005). Early, involuntary top-down guidance of attention from working memory. *Journal of Experimental Psychology: Human Perception and Performance, 31,* 248–261.

Soto, D., Hodsoll, J., Rotshtein, P., & Humphreys, G. W. (2008). Automatic guidance of attention from working memory. *Trends in Cognitive Sciences, 12,* 342–348.

Theeuwes, J. (1992). Perceptual selectivity for color and form. *Perception & Psychophysics, 51,* 599–606.

Tsivilis, D., Otten, L. J., & Rugg, M. D. (2001). Context effects on the neural correlates of recognition memory: An electrophysiological study. *Neuron, 31,* 497–505.

Van der Stigchel, S., Belopolsky, A. V., Peters, J. C., Wijen, J. G., Meeter, M., & Theeuwes, J. (2009). The limits of top-down control of visual attention. *Acta Psychologica, 132,* 201–212.

Vecera, S. P., Cosman, J. D., Vatterott, D. B., & Roper, Z. J. J. (2014). The control of visual attention: Toward a unified account. *Psychology of Learning and Motivation, 60,* 303–347.

Vogel, E. K., & Machizawa, M. G. (2004). Neural activity predicts individual differences in visual working memory capacity. *Nature, 428*, 748–751.

Voss, J. L., Schendan, H. E., & Paller, K. A. (2010). Finding meaning in novel geometric shapes influences electrophysiological correlates of repetition and dissociates perceptual and conceptual priming. *NeuroImage, 49*, 2879–2889.

Wolfe, J. M. (2012). Saved by a log: How do humans perform hybrid visual and memory search? *Psychological Science, 23*, 698–703.

Woodman, G. F., & Arita, J. T. (2011). Direct electrophysiological measurement of attentional templates in visual working memory. *Psychological Science, 22*, 212–215.

Woodman, G. F., Carlisle, N. B., & Reinhart, R. M. G. (2013). Where do we store the memory representations that guide attention? *Journal of Vision, 13*(1), 1–17.

Woodman, G. F., & Chun, M. M. (2006). The role of working memory and long-term memory in visual search. *Visual Cognition, 14*, 808–830.

Woodman, G. F., & Luck, S. J. (2003). Dissociations among attention, perception, and awareness during object-substitution masking. *Psychological Science, 14*, 605–611.

Woodman, G. F., & Luck, S. J. (2007). Do the contents of visual working memory automatically influence attentional selection during visual search? *Journal of Experimental Psychology: Human Perception and Performance, 33*, 363–377.

Woodman, G. F., & Luck, S. J. (2010). Why is information displaced from visual working memory during visual search? *Visual Cognition, 18*, 275–295.

Woodman, G. F., Luck, S. J., & Schall, J. D. (2007). The role of working memory representations in the control of attention. *Cerebral Cortex, 17*, i118–i124.

Woodman, G. F., & Vogel, E. K. (2005). Fractionating working memory: Encoding and maintenance are independent processes. *Psychological Science, 16*, 106–113.

Woodman, G. F., & Vogel, E. K. (2008). Selective storage and maintenance of an object's features in visual working memory. *Psychonomic Bulletin & Review, 15*, 223–229.

Woodman, G. F., Vogel, E. K., & Luck, S. J. (2001). Visual search remains efficient when visual working memory is full. *Psychological Science, 12*, 219–224.

Yi, D.-J., Turk-Browne, N. B., Chun, M. M., & Johnson, M. K. (2008). When a thought equals a look: Refreshing enhances perceptual memory. *Journal of Cognitive Neuroscience, 20*, 1371–1380.

9

Change Blindness and Inattentional Blindness

Daniel Levin and Lewis Baker

Imagine that you are a surgeon in a brightly lit operating room opening your patient's chest cavity in preparation for a delicate lifesaving heart operation. Surrounding you are nurses, an intensely focused anesthesiologist, studious interns, and maybe a neurotic medical student or two, along with an array of instruments that show the patient's vital signs. In such an environment it is crucial to effectively focus attention on relevant information at the precise moment when it is needed. Not only does this require managing information relevant to a complex motor task but it also requires managing a set of socially constrained information-processing priorities. A surgeon identifies anatomy, manipulates internal organs, asks for tools, retrieves information from a machine or a person, and interacts with students when teachable moments arise. Obviously, this environment presents a major challenge to our attentional system. But what if at least some of the relevant information could be made easily available, projected into the surgeon's visual field as a floating display that requires barely a flick of the eyes to read? Wouldn't that be nice?

As we are writing this chapter, advances in display technology are making this possible, and perhaps by the time you are reading it, these displays will be ubiquitous. The trouble is that the potential for easily accessible "head-up" displays to cause disaster has repeatedly been demonstrated. Most recently, surgeons equipped with augmented reality displays have exhibited striking failures to detect unexpected objects (Dixon, Daly, Chan, Vescan, Witterick, & Irish, 2013). In this experiment, surgeons completed an endoscopic navigation task (on a cadaver) using a system that combined a video feed with computer-generated anatomical guidance (see figure 9.1b). Of the 15 surgeons who used the augmented reality system, 14 failed to detect either of two unexpected features (the optic nerve was placed into a cavity where it should not have been, and a screw was put in the image). In contrast, of the 17 surgeons using a standard video system with no superimposed information, 12 saw at least one of the two unexpected features. This is not the first time that attentional failures have been induced by information-enhanced displays. Haines (1991) documented a set of particularly chilling instances in which experienced airline pilots landed their flight simulators

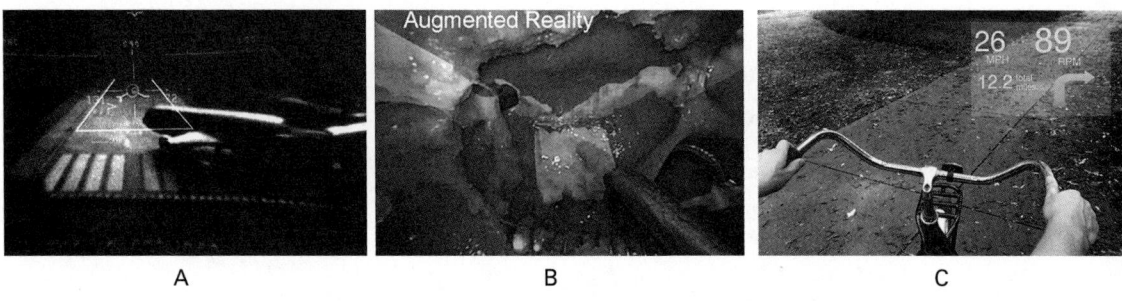

A   B   C

Figure 9.1
Head-up displays. (a) Illustration from Haines (1991) of a pilot who failed to detect a large runway obstruction while using a head-up display. (b) Example of augmented reality display from Dixon et al. (2013). The hole in the upper left quadrant of the display is one of the elements that surgeons missed. (c) Mock-up of navigation display for bicycling.

right on top of airliners taxiing onto the runway before them. The pilots were using a head-up display that projected navigation-relevant information onto the simulator's windshield. They loved it, but apparently they were so focused on the head-up display that they failed to detect the airplane occupying the runway right in front of them (figure 9.1a shows just how enormous the obstruction was). Related findings demonstrate that similar helmet-mounted displays can result in slowed detection of targets, especially when the display is cluttered (Yeh, Merlo, Wickens, & Brandenburg, 2001). Results such as these might reasonably be cause for caution in developing head-up displays for safety-critical applications, but witness recent examples of navigation and information-display applications for Google Glass and other head-up display systems in figure 9.1c (note that we generated this mock-up on the assumption that permission to use such an image would not be forthcoming). In both of these instances, the potential for uncontrolled distraction is enormous.

The problem in all of these instances is twofold. First is the straightforward issue of attention—people fail to detect one stimulus when focusing on another. However, a second issue may be even more important. That is, many of these failures stem from the strategic and metacognitive framework people use to effectively deploy attention in real-world tasks. In many cases, especially those involving novel information displays and novel tasks, viewers' judgments about the likely salience of display elements and about the nature of attention itself strongly affect how they deploy attentional resources. Thus, in the case of head-up displays, judgments about the nature of visual attention are part of the problem: Viewers seem to think that they can attend to one element of a scene and remain aware of other elements in the scene much more effectively when viewing head-up displays than when viewing traditional displays. When viewers are head-down, looking at an information display that is clearly separated from the scene, they have a good sense that each moment they are looking at the display increases the possibility that they will miss something in the outside world. In contrast, this sense may be missing for head-up displays.

In this chapter, we will describe research documenting change blindness (CB) and inattentional blindness (IB) with an eye toward effective application of the knowledge that psychological science has developed about these processes. Our review will begin with a basic description of these phenomena that lays the groundwork for the second part of this review, exploring applications for this work. Thus, our aim is not to replicate previous reviews of the large body of research in these areas (Simons & Levin, 1997; Rensink, 2002; Jensen, Yao, Street, & Simons, 2011). Rather, we focus on specific findings that support the applications we will describe. The initial section of this review will describe some of the prehistory of research documenting IB (a failure to detect unexpected objects) and CB (a failure to detect between-view property changes). We will describe the specific cognitive failures that these phenomena implicate, consider how a range of contextual factors and knowledge can distinguish situations where people detect changes and see unexpected events from those where they fail to do so, and, finally, discuss basic metacognitive work documenting how IB and CB diverge from intuitions that guide attentional deployment. Once we have completed this review, we will describe applications for this work.

Inattentional Blindness and Change Blindness: Historical Context

Research documenting IB was inspired by the dichotic listening paradigm. In the typical dichotic listening experiment, the participant is exposed via headphones to one stream of audio information in each ear and is instructed to attend to one stream while ignoring the other. For example, the participant may be instructed to attend to a series of words in her left ear while ignoring another stream of words presented to her right ear. In a series of well-known experiments, researchers demonstrated that participants could perceive basic sensory properties of the unattended channel but could not explicitly recognize the words in the unattended channel or detect the fact that the unattended words repeated dozens of times (Cherry, 1953; Moray, 1959). However, the possibility that the unattended channel was completely shut out was tempered by research demonstrating that participants did seem able to detect highly salient stimuli such as their own name in the unattended channel (for a review and an interesting follow-up, see Conway, Cowan, & Bunting, 2001), and that stimuli in the unattended channel also appeared capable of causing priming, even to the level of affecting interpretations of ambiguous sentences. The most well-known example of this priming is cited in almost every cognitive psychology text. The attended channel contained a sentence such as "They threw stones at the bank" with an ambiguous word in it (*bank*), which was disambiguated by the contents of the unattended channel. Results indicated that participants were more likely to interpret the attended sentence as referring to a financial institution when the unattended channel contained the word *money* while they were more likely to interpret the attended sentence as referring to a nature scene when the unattended channel contained the word *river* (MacKay, 1973).

Findings that information in the unattended channel could affect subsequent decisions and could sometimes break through to awareness led to the development of attenuation theory (Treisman, 1964), which assumed that unattended information was not completely eliminated from processing, and to late selection theories which assume that attention selects information subsequent to relatively deep preattentive processing (Deutsch & Deutsch, 1963). However, these modifications were themselves subsequently modified. For example, researchers questioned whether unattended channels were, in fact, completely unattended (Treisman, Squire, & Green, 1974), and it can be difficult to replicate the kind of semantic priming results that many of the late selection findings depend upon (for a review, see Bentin, Kutas, & Hillyard, 1995). More recently, Lavie's load theory attempts to explain how the processing of nontarget stimuli is moderated by perceptual and cognitive loads. According to load theory, increased processing of distracting stimuli will occur with increased cognitive load but decreased perceptual load (Lavie, Hirst, de Fockert, & Viding, 2004; Lavie, 2010). This is a large body of research, and so it cannot be reviewed here, but it is important to note that this general pattern of findings has been replicated across many domains, including research on CB.

One of the most compelling follow-ups to the dual-channel experiments was Neisser and Becklen's (1975) experiment testing whether similar principles applied to visual attention. Participants in these experiments viewed a video containing two superimposed streams of action. One stream was a simple basketball game, and the other was a hand-slapping game. Participants could easily attend to one or the other. However, participants rarely noticed "odd" events in the unattended stream such as hand-shaking or the substitution of female basketball players for male players, and Becklen and Cervone (1983) documented even more dramatic failures to detect a woman carrying an umbrella through the basketball game. These experiments arguably established the first instance of IB. IB is clearly similar to the absences of processing observed in the dichotic listening experiments but is perhaps more dramatic because in some cases IB seems to represent a complete failure to detect the *presence* of the unattended channel rather than a failure to process some aspects of the channel.

If IB reflects a failure to be aware of an unattended object, CB reflects a failure to detect visual changes to objects (such as property changes, changes to object location, or the appearance/disappearance of objects), whether those objects are attended or not. As we will review below, some research exploring CB relies upon incidental detection paradigms similar to those used in most of the IB research, but the majority of research in this area uses intentional paradigms where viewers are focused on the task of detecting changes (Rensink, 2002). Therefore, research using the change-detection paradigm really explores two fundamental issues: people's capacity to detect changes when they are engaging a full complement of cognitive resources to the change-detection task, and the degree to which people detect changes by default when they are only trying to understand otherwise meaningful events (Beck, Levin, & Angelone, 2007). In other words some research explores how many changes viewers *can* detect while other research explores the changes they *do* detect. This is important both because these task variations tend to support different conclusions about the representational basis for online vision, and because an effective understanding of task effects may be crucial when applying this work to real-world problems.

In some important ways, research documenting CB was inspired by the early IB work, but more concretely, change-detection tasks had been used for quite some time before CB was a recognized phenomenon. For example, Phillips (1974) asked participants to detect changes between two-dimensional dot arrays as a test of iconic memory, and there was a small controversy when subsequent work appeared to demonstrate that a short-term high-capacity visual buffer might be capable of storing sufficient visual detail to integrate successive views based on a point-by-point visual registration process (Jonides, Irwin, & Yantis, 1982; Breitmeyer, Kropfl, & Julesz, 1982), and the effect was found to be due to CRT phosphor persistence (Jonides, Irwin, & Yantis, 1983). However, a large body of less problematic work—for example, using the partial report technique (Sperling, 1960)—did suggest the existence of some kind of relatively high-capacity short-term visual memory buffer that could plausibly integrate views by matching visual properties.

Contemporary with this early work, other researchers explored visual integration by monitoring fixations and saccades during reading and then testing whether sudden changes to text would affect reading speed, especially if those changes occurred during saccades. This research demonstrated that reading speed was unaffected by eliminating characters outside of a small window around the current fixation point (Rayner & Pollatsek, 1987; see also chapter 12 in the current volume), and one particularly compelling study demonstrated that participants failed to detect case changes to all letters in a sentence (e.g., the word CaSe might change to cAsE; McConkie & Zola, 1979).

The reading work demonstrated that there must be some limit to the comparisons that occur in the visual short-term storage system that supports reading, but it remained possible that such a highly practiced symbol recognition process would be distinctive for its lack of intensive between-view visual comparison. However, it quickly became clear that similar failures to detect changes could occur in natural images. Grimes (1996) monitored eye movements while participants viewed photographic images of scenes and generated large visual changes during saccades. The changes were striking (e.g., in one image the heads of two cowboys depicted in a Western scene suddenly switched places), and viewers generally failed to detect them so long as the changes occurred during saccades. The power of this technique was in transforming a change event from a salient (and comical) visual experience into a nonexperience by simply changing the timing of its occurrence relative to the saccades viewers generated while exploring the scene.

Change Blindness and Inattentional Blindness: State-of-the-Art Review

Some of the follow-ups to the initial IB findings are very well-known and they make clear that the phenomenon can be observed without the ghost-like superimposition characteristic of the Becklen and Neisser videos (Simons & Chabris, 1999) and that IB remains undiminished when the event is associated with an extremely unpleasant noise (e.g., a person who arrives on the scene and screeches her fingernails on a chalkboard; Wayand, Levin, & Varakin, 2005). Other well-known work demonstrated that IB could be studied using simple, more controllable stimuli (Mack & Rock, 1998). In addition, these follow-ups established that some forms of perceptual similarity between the target task and the unattended event could increase detection of the unattended event. For example, the unexpected stimulus is more likely to be detected if it matches the color of the attended stimuli and if it is distinctive relative to other unattended stimuli (Most et al., 2001). Similar to the priming findings from dichotic listening experiments, Mack and Rock (1998) demonstrated that unnoticed stimuli could nonetheless prime subsequent decisions. Other work has demonstrated that IB can occur in the real world, even for objects that have affected navigation behavior. Hyman, Sarb, and Wise-Swanson (2014) demonstrated that some pedestrians failed to notice money hanging from a tree even though it was in their path, requiring them to step aside to avoid it.

The IB work is particularly interesting for how it converges with other recently documented failures to become aware of objects and events. For example, a number of experiments have demonstrated that people sometimes fail to become aware of the fact that attended objects and events are, or are not, relevant to their goals. Solman, Cheyne, and Smilek (2012) demonstrate that participants can fail to recognize that an object they are currently moving (in an interactive visual search display) is the target they are searching for. Conversely, Schooler has studied the phenomenon of "zoning out," whereby people fail to realize that they are not processing central information (see Schooler et al., 2011, for a review), and Levin and Varakin (2004) demonstrated that participants can fail to detect blank gaps of up to 600 ms while viewing videos even though they had good memory for the events surrounding the gaps. Collectively, results such as these might be considered variants of IB on the assumption that IB can include both phenomena in which attention to one location precludes awareness of stimuli at other locations and phenomena in which stimuli at an attended location fail to reach full awareness (Most, 2010).

Quite a bit of research has continued exploring CB. In initial work, researchers sought to induce change-detection failures without the need for eye tracking. A number of researchers repeated the Grimes's (1996) work, but instead of making changes contingent on eye movements, different versions of an image (an original image and a version with a single property change) were alternated, separated by a brief (80-ms) blank-screen interstimulus interval (ISI; Rensink, O'Regan, & Clark, 1997). Alternatively, the postchange image was simply moved (Blackmore, Brelstaff, Nelson, & Troscianko, 1995). Rensink et al.'s technique, referred to as a flicker task, was particularly compelling because it generated repeated failures to detect changes that occurred once per alternation cycle (although it is interesting to note that Hochberg, 1968, briefly mentions using an apparently identical task to demonstrate failures to detect changes to images of faces). Rensink et al. seem to have been the first to describe these failures as "change blindness."

Other studies demonstrated that CB could be observed in a range of circumstances. Levin and Simons (1997) used cuts between shots in simple edited films to mask a wide variety of changes, and they demonstrated that viewers almost never detected changes, even when the sole actor in a scene changed to another person. Simons and Levin (1998) further demonstrated that CB for attended objects could even occur during real-world interactions. In this study, participants were approached on the street and an initial experimenter started a conversation that was interrupted by the brief imposition of an occluder between the two. During the occlusion, an experimenter who was carrying the occluder stayed behind to finish the conversation while the initial experimenter walked off behind the occluder. About half of participants failed to detect changes to their conversation partner, an effect that has been replicated by having the initial experimenter duck behind a counter while the participant signed consent forms (Simons, Chabris, Schnur, & Levin, 2002) or by having the experimenters swap places while the participant photographed them (Levin, Simons, Angelone, & Chabris, 2002). These experiments were the first to demonstrate CB in attended objects, and subsequent research has utilized eye tracking to verify that changes could go undetected even when the viewer clearly fixated on the pre- and postchange objects (O'Regan, Deubel, Clark, & Rensink, 2000; Smith, Lamont, & Henderson, 2012). In addition, recent studies have demonstrated CB even in situations where all of the items in a multi-item display change, so long as the relative proportion of features remains constant (Saiki & Holcombe, 2012).

A number of other more recent studies have explored CB using changes to object arrays, providing a context in which scene complexity and the similarity of pre- and postchange objects can be more easily controlled (Scholl, 2000; Beck & Levin, 2003). Much of this work has continued a tradition of characterizing limits in working memory capacity. For example, Rensink (2000) varied exposure durations for each image to document a leveling of the exposure time–display size function that can be used to infer visual working memory capacity, and a large number of studies now use change detection to ask whether visual working memory storage is limited by the number of objects that must be stored or by the global complexity of visual information that characterizes these objects. In some circumstances, change detection appears to be limited by the number of objects in an array, no matter how many features characterize those objects (Luck & Vogel, 1997). This finding has been taken as support for the view that working memory is best characterized as having some number of "slots" that can each represent one object, no matter how many features the object has. This work is reminiscent of older work on chunking in short-term memory, which shows that this system can overcome informational limits by packing a large number of subelements into a limited number of meaningful units.

As one might imagine, this "features-for-free" view has been controversial. A number of studies have demonstrated that the complexity of the objects does increase perceptual load, especially in situations where viewers must represent multiple values on the same dimension (Alvarez & Cavanagh, 2004). Another factor that seems important in effectively characterizing apparent load is the fidelity with which features must be represented. In situations where the fidelity demands are great, apparent visual working memory capacity is less, and conversely, when only a small number of objects must be represented, fidelity increases (for a review, see Brady, Konkle, & Alvarez, 2011). Finally, several studies have demonstrated that working memory load increases IB (Lavie, 2010; Fougnie & Marois, 2007), and follow-ups to this finding suggest that individuals with high working memory capacity are less susceptible to IB (Seegmiller, Watson, & Strayer, 2011) although this finding is controversial (Bredemeier & Simons, 2012).

One issue with work attempting to define the role of capacity in constraining CB and IB is that it may neglect important questions about the nature of everyday visual experience, which is rarely focused upon effortful attempts to compare features between object arrays. Clearly, this contrast does not bar the successful application of this work. For example, Strayer and Drews (2007) successfully documented increased IB during a driving simulation that was caused by the cognitive load induced by a cell-phone conversation. However, in attempting to understand how vision might work in situations where explicit change detection is not the viewer's primary task, it is important to realize that visual awareness is not an exhaustive measure of visual processing, and that change detection requires a number of cognitive steps, any one of which might or might not meet the viewer's immediate needs. For this reason a range of studies have attempted to characterize what CB and IB mean for online visual experience and to understand which specific representational failures might cause CB.

Most obviously, failures to detect changes to visual properties may occur because the properties have not been represented at all or have been represented in a form that is not detailed enough to differentiate pre- and postchange objects (Levin & Simons, 1997; Blackmore, 2002). These "sparse-representation" views have been supported by the argument that the stability of visual properties in the world and the ease with which those properties can be accessed often make default mental representation of those properties superfluous (O'Regan, 1992). Other support for a sparse-representation hypothesis comes from research demonstrating that infants do not appear to detect visual changes, even to attended objects, until they have verbal labels for those objects (Xu & Carey, 1996). A particularly interesting part of Xu and Carey's argument is that in the absence of labels for objects, infants can represent the objects but often do so in a form that is too general to differentiate objects into anything approaching basic-level categories. If detecting changes is secondary to the generation of verbal labels, then it seems likely that the many objects that typically go unlabeled are also not represented in any particular detail (Simons & Levin, 1998). One important part of this argument is that the typical lab task overestimates the prevalence of visual representation because it forces viewers to create visual representations that might not be necessary in a more typical situation where the viewer's focus is on understanding visual events. It seems obvious but is nonetheless important to emphasize that this sparse-representation view does not claim that visual representations are never created—clearly, viewers can represent objects and visual properties when the task requires it. Instead, the hypothesis is that these representations are not a matter of course when viewing scenes and events. Instead, most durable visual representations are generated in response to specific information-gathering goals that can vary across scenes, kinds of visual task, and viewer intentions.

In several cases, sparse-representation views have been embodied in more explicit models of visual attention derived from the idea that preattentive representations of visual properties are either free-floating and unbound into objects (Treisman & Schmidt, 1982) or bound but otherwise incoherent (Wolfe & Bennett, 1997). According to Rensink's (2000) coherence theory, visual attention is necessary to organize representations of visual properties sufficiently to allow comparisons between views. Because of this, failures to detect changes are likely unless both the pre- and postchange objects are made temporarily coherent by attention. The transitory nature of coherence is a key part of the model, and this has proven controversial, primarily because the model implies that relatively little visual information builds up over views (although the model does include some mechanisms for default representation of information in the form of layout, gist, and scene schemas). Related sparse-representation models focus more directly on factors that generate durable representations while retaining a sparse-representation framework. For example, functionalist approaches such as Hayhoe's and Ballard's argue that visual representations are precisely sculpted to immediate task demands (Ballard, Hayhoe, & Pelz, 1995; Hayhoe, 2000). Particularly compelling support for this hypothesis comes from an experiment by Ballard, Hayhoe, and Peltz (1995), who monitored participants' eye movements during a block-copying task in which participants were asked to match a pattern of colored Lego blocks. Ballard et al.

found that participants would often glance at the model once to select a block of the correct color and again to determine the correct location for the block. This led them to argue that creating visual representations, even when visual capacity is unfilled, is not an effort-free default but rather requires efforts that are typically deployed only when a task demands it. This supports a sparse-representation view especially if one assumes that most everyday tasks do not require the creation of extensive durable representations.

However, any argument in support of a sparse-representation view must contend with the fact that CB is not necessarily evidence for an absence of visual representations and that at least some forms of visual representation do seem to be readily created and retained. Logically, detecting a visual change requires not only representing properties in a form that differentiates the pre- and postchange objects but also comparing those representations across views and remembering the fact that a change has occurred (Simons, 2000). In addition, experiments testing picture memory make clear that viewers can successfully recognize very large numbers of pictures (Nickerson, 1965; Shepard, 1967; Standing, 1973). Although visual memory is far from perfect, even for familiar scenes (Rosielle & Scaggs, 2008) and especially for familiar objects (Nickerson & Adams, 1979), and although it involves an explicit demand to create representations, the sheer number of pictures that can be remembered and the surprising precision of these memories (Konkle, Brady, Alvarez, & Oliva, 2010) at the very least suggest that some kind of basic visual representation can be created without the kind of effortful elaboration necessary to remember other kinds of information such as random digits. In addition, it does appear as though recognition memory for previously viewed objects is more accurate than chance even for a surprise recognition test (Castelhano & Henderson, 2005).

Also, it is possible that experiments requiring explicit reports about changes fail to access representations that might be observed using more sensitive measures. For example, research documenting eye fixations within natural scenes has repeatedly documented better than chance recognition of fixated objects, and these findings have been integrated into theories that describe how visual memories are built up during the course of scene perception (Hollingworth & Henderson, 2002; Henderson & Hollingworth, 2003). In addition, it is possible to demonstrate that viewers who miss changes can nonetheless recognize the changing features, even in an incidental change-detection paradigm (Angelone, Levin, & Simons, 2003; Varakin & Levin, 2006). Finally, a number of researchers have argued that implicit measures of change detection reveal representations where explicit measures have failed to do so. For example, participants who fail to report changes sometimes fixate for a longer period on the changed object (Hollingworth, Williams, & Henderson, 2001), priming has been observed with objects involved in undetected changes (Silverman & Mack, 2006), and Fernandez-Duque and Thornton (2003) found that participants could localize a change that they did not report seeing. However, this work should be interpreted with care given method-ological complexities in interpreting data implying implicit processes (for a review, see Mitroff, Simons, & Franconeri, 2002), and the fact that fixations to changed regions do not always reveal implicit change detection (Ryan & Cohen, 2004).

One means of moving forward from this debate is to leave behind a dichotomy between broad hypotheses about sparse and rich representations. It seems more use-ful to specify which task contexts do and do not produce evidence that visual repre-sentations have been created, have been tracked across views, and can be explicitly reported. For example, it is possible to observe evidence suggesting both the presence and the absence of visual representations in the face of CB in very similar situations. Several studies using incidental change detection have revealed clear evidence of rep-resentational failure associated with CB. In these studies participants were exposed to a real-world conversation-partner change, which was followed by a test of their abil-ity to recognize the changing experimenters in a lineup. Participants who missed the change were at chance in recognizing the experimenters while participants who did see the change were much more accurate (Levin, Simons, Angelone, & Chabris, 2002). A similar experiment demonstrated the same result for a simpler between-view change of the color of a binder that the participants were looking at (Varakin, Levin, & Col-lins, 2007), and in an experiment by Davies and Hine (2007), participants who viewed a video depicting a burglary in which the burglar changed into another person were more accurate in recognizing the burglar if they also detected the change (although it appears as though the participants who missed the change were above chance on the lineup). Finally, Nelson et al. (2011) demonstrated worse lineup performance in participants who missed person substitutions, and they further demonstrated that participants who missed the change were much more likely to falsely recognize the postchange actor as the perpetrator of the depicted crime. This demonstrates a close relationship between CB and failed memory for a visual property.

However, these findings must be contextualized by consideration of the specific incidental encoding circumstances that created this representational failure and by consideration of the fact that the recognition test used to test for visual representa-tions cannot be considered exhaustive. Many of these studies involve real or depicted events in which participants were called upon to manage an online social interaction or to understand the motivations of actors. This likely absorbed many of the process-ing resources that otherwise might have been devoted to generating visual representa-tions. Interestingly, subtle changes to tasks that produce representation failures can produce evidence of representations that are preserved in the face of CB. For example, when participants view person changes on videos depicting simple conversations, they can recognize the changing properties and experimenters reasonably accurately even when they have missed the change (Angelone, Levin, & Simons, 2003). In fact, these experiments revealed no difference in recognition accuracy between participants who did and did not see the change. As mentioned above, similar findings can be obtained for more intentional change-detection experiments using object arrays. Participants can sometimes effectively recognize changing objects even when they had missed the change (Mitroff, Simons, & Levin, 2004; Varakin & Levin, 2006), and change detection can be increased if postchange cues facilitate comparison (Hollingworth, 2003). In addition, a number of studies have demonstrated that CB can be caused when a postchange object overwrites a prechange object (Beck & Levin, 2003; Landman, Spekreijse, & Lamme, 2003).

Collectively, the findings reviewed above make clear that research on CB and IB often underestimates the prevalence of visual representations. However, this does not mean it always does so, and it is important to point out that many studies that support rich representation views involve tasks that explicitly require this kind of encoding. Even studies that explicitly avoid representation-creating requirements (such as the Angelone, Levin, & Simons, 2003, experiments) should be interpreted with caution because the lab environment in itself may induce a specific encoding strategy. After all, anyone who is exposed to a visual stimulus in the context of an experiment is implicitly being told that he or she should encode it in preparation for some kind of test. This may induce some level of elaboration on visual properties that would not otherwise occur. If added elaboration is not associated with added comparison across views, then a wide range of experimental contexts might be expected to induce CB with evidence for preserved representations. However, a number of studies have demonstrated good incidental memory performance for spatial locations (Mandler, Seegmiller, & Day, 1977; Schulman, 1973) and even for objects and scenes in circumstances where the nominal task was clear and unlikely to generate much of an assumption that a memory test would occur (Castelhano & Henderson, 2005).

One thing that may help resolve the apparent contradiction between a sparse-representation view and evidence for preserved or default visual representations is to suggest that many of the strongest sparse-representation findings occur in the context of compelling meaningful events. On this view, the structure of meaningful events creates a strong functional framework that induces the encoding of some features at some points in time while other features are subsumed within the event structure and consequently go unrepresented (or represented and forgotten). Research by Triesch, Ballard, Hayhoe, and Sullivan (2003) illustrates this powerfully. In this experiment, participants completed an object-sorting task in a virtual-reality setting and sometimes failed to detect changes to objects they were looking at and holding in their hand. However, these failures occurred only after participants had decided that the object's specific features were no longer task relevant. Findings such as these make clear the need to consider the role played by specific visual properties in visual events, and it would seem particularly productive to consider CB in the context of theories characterizing the structure of visual events.

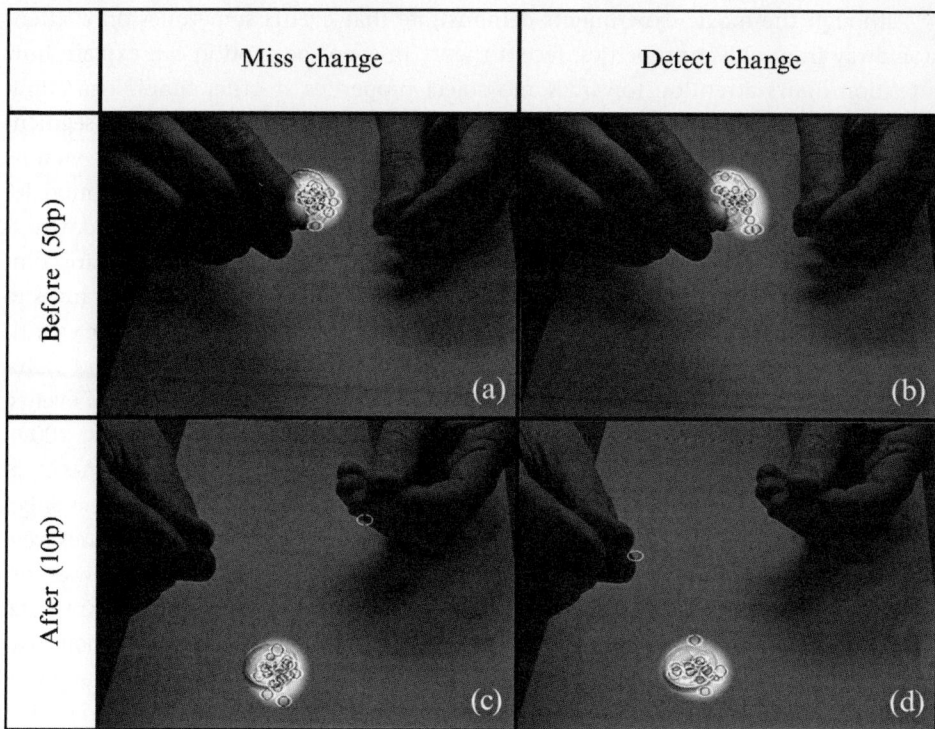

Figure 9.2
Similar eye fixations for successful and unsuccessful change detections in Smith, Lamont, and Henderson (2012). (a, b) Fixations on 50p coin before change. (c, d) Fixations on 10p coin after change.

Several recent studies make a particularly compelling argument that events play a crucial role in guiding attention away from visual properties by demonstrating CB and IB during magic tricks. For example, Kuhn and Tatler (2005) demonstrated that participants think that a cigarette had suddenly disappeared from a magician's hands when it had simply been dropped (in plain view) at a moment when the magician had directed participants' attention away from area where the cigarette was falling. Participants' eye movements were recorded, and it is particularly interesting to note that detection of the cigarette was unrelated to the proximity of the closest fixation to its fall (see also chapter 22 in the current volume). More recently, Smith, Lamont, and Henderson (2012) demonstrated that participants frequently failed to detect the substitution of one coin for another during a magic trick. Again, Smith et al. tracked participants' eye movements and observed that the coin was fixated before and after the change whether or not the change was actually detected (see figure 9.2; see also Memmert, 2006).

Although the magic experiments demonstrate that events sometimes draw attention away from object properties, recent theory in event perception can explain how attention draws attention toward some object properties at other times. The dominant framework for understanding visual event perception is Zacks's event segmentation theory (Zacks et al., 2007). According to event segmentation theory, viewers automatically organize ongoing actions into segmented hierarchical "event models" that include simple low-level descriptions of actions ("stirring") that are nested within higher level abstract descriptions of events (e.g., "making sauce" and "cooking dinner"). A key feature of event segmentation theory is that it specifies circumstances in which visual properties are encoded and forgotten. For example, at the onset of events, "event files" are created for the purpose of tracking actions, inferring goals, and maintaining continuity of spatial relationships. These files are cleared at the end of events, leading to rapid forgetting of irrelevant features (Swallow, Zacks, & Abrams, 2009). Event segmentation theory further proposes that a wide range of visual and conceptual properties are used to determine when events begin and end. These include changes to goals, actors, movements, and spatial locations. For instance, object properties were quickly forgotten after walking through a doorway (a spatial cue signifying a new event) versus walking the same distance within one room (Radvansky, Tamplin, & Krawietz, 2010). In addition, visual attention appears to be particularly focused at event boundaries. For example, detection of event-irrelevant targets is impaired at event boundaries (Huff, Papenmeier, & Zacks, 2012). This complements evidence in spatial cognition and multiple object tracking, which has shown that change detection decreases following disruption to spatial layout (Simons & Wang, 1998; Meyerhoff, Huff, Papenmeier, Jahn, & Schwan, 2011) but improves if there is reliable spatial context (Mou, Xiao, & McNamara, 2008). These findings suggest that the boundaries between events have a special impact on visual property representations, possibly increasing change detection for properties that are potentially central to events while decreasing it for more peripheral objects.

Baker and Levin (2015) recently sought to directly confirm the hypothesis that the structure of everyday events directly affects the representation and comparison of features. They predicted that not only would new events facilitate comparisons of central features but also that spatial disruptions would mimic event boundaries and similarly increase comparisons. Participants viewed movies of short conversations with changes to central features (e.g., the shirt color of an actor) that did or did not co-occur with event boundaries and spatial disruptions (see figure 9.3). Change detection increased both for the new events and spatial disruptions, leading Baker and Levin to propose the *relational trigger* hypothesis, which assumes that spatial changes are basic signals for new events and that both spatial violations and new events increase change detection in preparation for the creation of new event models. A key feature of these experiments is that spatial triggers increased change detection even well after the prechange object

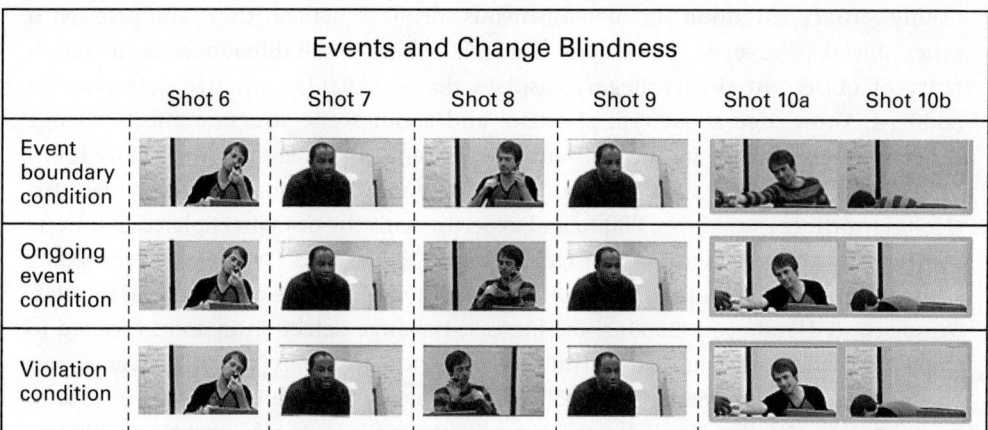

Figure 9.3
Illustration of changes that occurred during the transition to a new event and during a spatial discontinuity in Baker and Levin (2015).

had been visible, suggesting that these triggers operate by inducing comparisons of properties already encoded into working memory. Combined with the above research, this provides evidence in support of a process that attends to features relevant to the comprehension of events.

In addition to constraints imposed by event structure, other forms of broad task-based constraints may affect change detection. Early eye-tracking work by Yarbus (1967) found that different task demands prompted observers to use different eye movements to scan a photograph, and more recent work has demonstrated that visual search and memory tasks are associated with clearly different fixation patterns (Castelhano, Mack, & Henderson, 2009). However, it is not yet clear how, exactly, these differing patterns would affect change detection. It would, for example, be particularly interesting if different tasks could be shown to affect representational processes and comparison processes independently.

The Role of Knowledge, Skill, and between-Participant Differences in Constraining Change Blindness and Inattentional Blindness

Initial work by Pashler (1988) compared change detection for letters and mirror-reversed letters and found that familiarity with the changing objects did not greatly affect change detection (or the pattern of accuracy across variations in ISI). However, more recent work has demonstrated that participants seem to better detect changes in faces than in nonface objects (Ro, Russell, & Lavie, 2001) and in same-race faces compared with cross-race faces (Rhodes, Hayward, & Winkler, 2006). One means of resolving this apparent contradiction might be to hypothesize that faces more readily attract attention in heterogeneous displays where they compete with other objects (Weaver & Lauwereyns, 2011), but that overall differences in the familiarity of objects in single-category displays do not facilitate change detection. In addition, knowledge about typical scenes and about scene structure affects change detection. Objects that are inconsistent with a scene, and therefore carry more informative weight (such as a fire hydrant in a living room), are more frequently detected (Hollingworth & Henderson, 2000), and expertise with the domain depicted in a scene can increase the likelihood of change detection. Sports fans (Werner & Thies, 2000; Laurent et al., 2006) and chess masters (Reingold et al., 2001; Brockmole, Hambrick, Windisch, & Henderson, 2008) show increased change detection in scenes related to their knowledge although this advantage does not generalize either to scenes outside of the experts' domain of expertise or to objects that are causally peripheral to scenes within the experts' domain. Generally, these findings are consistent with previous research testing experts' memory which shows processing advantages that are limited to meaningfully organized scenes (e.g., Chase & Simon, 1973). More recently, Drew, Võ, and Wolfe (2013) demonstrate that expert radiologists are more likely than novices to detect an unexpected stimulus (a small gorilla) while viewing x-rays, although the experts still missed a large number of the intrusions. These findings are interesting because they suggest a global advantage for experts in detecting a novel stimulus in an expertise-related setting, rather than a processing advantage for familiar, central objects in a scene.

In addition to differences in skill, variations in visual awareness have been explored in a range of populations with arguably abnormal attentional processes. A number of recent studies have explored change detection in individuals with autism. The pattern of performance variations is heterogeneous, with some studies demonstrating stronger "automatic change detection" (e.g., oddball detection) in participants with autism (Clery et al., 2013) and some studies demonstrating worse change detection in a more typical change-detection task (Jiang, Capistrano, & Palm, 2014). On one view, autism is associated with enhanced sensory processing and less effective attentional filtering and/or working memory (Miller & McGonigle-Chalmers, 2013; Jiang, Capistrano, & Palm, 2014). This might explain a pattern of enhanced change detection in more perceptual tasks and lessened change detection in more cognitively demanding tasks. In a similar vein, Maccari et al. (2013) find that children with attention deficit/hyperactivity disorder (ADHD) show a larger deficit in change detection than in visual search. Maccari et al. argue that the change-detection task is more effective in isolating the cognitive deficit associated with ADHD because the change-detection task is more strategic than visual search. Other disorders that involve particular preoccupations are associated with increased change detection in specific kinds of object. For example, individuals with Williams syndrome are particularly social, and they show increased change detection for person-related properties during a video depicting a conversation

(Tager-Flusberg, Plesa-Skwerer, Schofield, Verbalis, & Simons, 2007), while individuals with insomnia show increased change detection for sleep-related objects (Marchetti, Biello, Broomfield, MacMahon, & Espie, 2006).

Research has also demonstrated that CB can vary between different broad stimulus domains. For example, Beck, Angelone, and Levin (2004) asked whether CB would vary for properties that viewers know to be stable (e.g., the color of a couch) or unstable (e.g., the location of a cat) over time. On the one hand, participants should detect changes to stable properties more frequently because such "impossible" changes may violate knowledge schemas about a scene (Loftus & Mackworth, 1978). However, one might also expect that changes to unstable objects would be more easily detected if one assumes that attention is naturally drawn to monitoring things that could change (Maule, 1985). In fact, Beck et al. found that changes to unstable objects were more frequently detected, and follow-ups determined that unstable objects benefit from increased comparison of properties across views rather than increased likelihood of fixation or postperceptual decision processes (Beck, Peterson, & Angelone, 2007).

Before moving on to a discussion of CB and IB in everyday contexts, it is important to consider one final issue. That is, very little work has explored the broader strategic context that these failures of visual awareness occur in. This is especially important for understanding complex everyday environments because the visual representations individuals create within these environments often reflects their strategic allocation of attention to different simultaneous tasks. Even in situations such as driving where safety demands a high priority on one's primary task, a multitude of competing demands must be balanced with attention to the road, and this balance becomes especially difficult over long periods characterized by varying levels of vigilance. Not only must one balance attention to different visual tasks but one must also strike an effective balance between attention to the external world and internal thoughts, something that can be quite difficult (Schooler, Reichle, & Halpern, 2004). One thing that can make this balance especially difficult is a failure to appreciate just how much effort it may take to effectively perceive and track specific visual objects.

The challenge people face in making accurate estimates of the need for visual effort is highlighted by a number of reports demonstrating that the vast majority of participants firmly believe that they will not fall prey to CB or to IB (Levin, Momen, Drivdahl, & Simons, 2000; Levin & Angelone, 2008). Overestimates of change detection in particular are large and quite robust, and we have referred to them as "change blindness blindness" (CBB). CBB occurs whether participants are estimating their own or others' performance (Levin, Momen, Drivdahl, & Simons, 2000), and it occurs both for detection of changes that occur rapidly and for changes in which the pre- and postchange views are separated by a long period of time (Levin, Drivdahl, Momen, & Beck,

2002). In addition, participants overestimate people's ability to recognize deletions from photographs of familiar scenes (Rosielle & Scaggs, 2008). CBB can also occur both for incidental and intentional change-detection tasks (Scholl, Simons, & Levin, 2004) although some evidence suggests it is less strong for intentional tasks because participants do not effectively account for the loss in performance characteristic of incidental visual encoding (Beck, Levin, & Angelone, 2007). Participants seem to overestimate the breadth of their attention (Kawahara, 2010), overestimate their ability to remember some kinds of visual information (Levin & Angelone, 2008; Simons & Chabris, 2011), and have difficulty discounting the role of their own knowledge in judging the visibility of blurred images (Harley, Carlsen, & Loftus, 2004).

It is important to point out, however, that these overestimates do not occur for all judgments. In some cases, particularly when participants repeatedly perform a task that varies in difficulty, confidence can be reasonably well calibrated with visual performance (Barthelme & Mamassian, 2010; Levin, Angelone, & Beck 2010). In addition, participants strongly *underestimate* their ability to recognize pictures, even when the task is demonstrated to them (Levin & Angelone, 2008). More recent work has attempted to explain this pattern of overestimate by linking it with participants' understanding of intentional representational systems. Support for this link comes from experiments demonstrating that describing a computerized machine vision system in anthropomorphic terms increases the degree to which participants believe it will be capable of detecting changes and decreases participants' estimates of its ability to remember scenes (Levin, 2012).

Integration: Change Blindness and Inattentional Blindness in Everyday Environments

One of the most compelling things about research documenting CB and IB is that these phenomena were observed right from the start, in tasks using naturalistic stimuli, and even during real-world interactions. Therefore, research exploring everyday applications for these observations is a natural fit for this work. One particularly compelling domain for application is human–computer interaction (HCI). Not only is this an important everyday domain that absorbs a large percentage of people's waking hours, but it presents viewers with an array of meaningful, organized visual scenes in which properties can suddenly change in ways that may defy visual-cognitive assumptions of scene stability. In this section, we review research exploring CB and IB during HCI. Many of these applications are fairly straightforward tests for CB in these environments, but a few studies have gone beyond this and have introduced techniques for lessening CB in environments where a high level of performance is necessary. As we will review below, one of the most important lessons emerging from this work is that it is crucial to study basic constraints on visual attention in the context of task demands and strategic decision making.

Change Blindness in Human–Computer Interaction

One of the most salient features of modern computer interfaces is that they present users with a spatially distributed array of meaningfully structured visual information that simultaneously offers a broad range of interactive possibilities. It is interesting to consider the fact that this was not always the case. Prior to the 1980s most interactions with computers occurred through a text-based command line interface or through simple text-based menus, and it is easy to forget how revolutionary the idea of leveraging real-world visual skills and knowledge about desktops and stacked paper was. During the 1970s many of the concepts underlying the graphical user interface were developed at Xerox's Palo Alto research center. It is particularly telling that the computer that leveraged these ideas was essentially ignored by Xerox, and that the corporation literally allowed Apple and Microsoft to walk into their research lab, view their top-secret machine, and use what they saw to develop interfaces that became the standard for almost *all* of HCI. Thus, it is not surprising that much of the practice and theory in HCI focuses on providing the users a visually rich, well-structured display that can leverage our visual system's apparently inexhaustible capacity to process visual information. In this context, research on CB and IB has the potential to prevent enthusiasm about the graphical user interface from overwhelming users.

Although it is straightforward to predict that CB is common during HCI, there are reasons to believe that this is not a foregone conclusion. The visual information displayed on a computer screen is often task relevant and designed expressly to be noticed by the user, which contrasts with natural visual scenes that were not necessarily designed with a user in mind and that have many irrelevant features. Of course, we wouldn't be discussing CB during HCI if it did not occur, but it is important to consider how broad properties of the HCI task may affect visual attention not only because it can improve the application of the basic research in this context but also because the context may contain important lessons that can guide the basic research. In particular, it may be possible to refine the dichotomy between intentional and incidental tasks with a generalizable description of the relationships between task-relevant and task-irrelevant information during HCI.

A number of studies have demonstrated CB during HCI. For example, Durlach and Chen (2003) demonstrated CB while users interacted with a battle-command interface that represented important vehicles with symbolic icons. In this case, participants failed to detect up to 50% of changes to the sole on-screen icon during an intentional change-detection task in which the icons were momentarily occluded with pop-up windows (see also DiVita et al., 2004). Other researchers have documented CB while users interact with Web pages (Steffner & Schenkman, 2012) and animated geographic displays (Fish, Goldsberry, & Battersby, 2011), and IB has been observed in the form of failures to perceive Web banner ads (e.g., "banner blindness"; Benway & Lane, 1998). Varakin, Levin, and Fidler (2004) documented CB in an e-text reading application. In

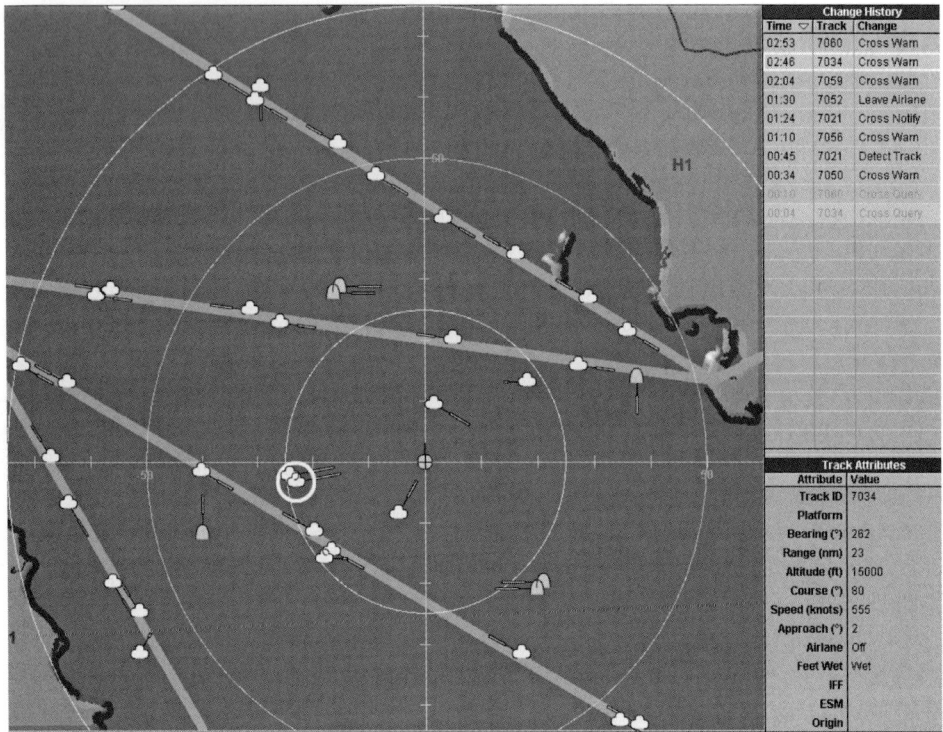

Figure 9.4
The CHEX tool for reducing change blindness during human–computer interaction. The key-change-history window can be seen in the upper right-hand part of the display.

this case, participants were asked to peruse the top layer of a newspaper containing story summaries and then to click on one of the summaries to see a new screen displaying the full story. Simultaneous with the shift to the detail level, a very large change to the screen header was introduced, and about half of the participants failed to detect the change. These experiments are important because they demonstrate that CB can occur not just when viewers are "passively" viewing a stimulus but also when they are interacting with visual information in the form of a graphical user interface (for a review, see Durlach, 2004).

Several HCI researchers have proposed means for reducing the impact of CB. For example, Nowell, Hetzler, and Tanasse (2001) proposed dynamic enhancements to a document visualization system to increase the salience of display changes that mark changes to documents represented in the display, and Nikolic and Sarter (2001) facilitated detection of unexpected mode changes in a simulated cockpit by adding peripheral visual cues. Smallman and St John (2003) developed a tool called CHEX (Change

History made Explicit; see figure 9.4) that monitors a display for potentially important changes and displays them in a sortable list to the side of the main display. Thus, the viewer is reminded about objects that have changed and can use the list to refer back to the location of each changed object. Several experiments have demonstrated that CHEX can improve change detection, especially after a task interruption (Smallman & St. John, 2003; St. John, Smallman, & Manes, 2005). However, a more recent experiment questioned the generality of CHEX's utility. Vallieres, Hodgetts, Vachon, and Tremblay (2012) assessed the utility of CHEX in a naval warfare scenario in which participants were tasked with classifying aircraft as hostile based on a set of potentially changing attributes and preventing hostile aircraft from attacking their ship. Thus, change detection was necessary, but it was subsumed within the threat evaluation task. In this situation, not only did CHEX fail to improve change detection and threat evaluation but the added load from the tool appears to have damaged primary task performance. Vallieres et al. argue that facilitation of change detection is helpful only when it is the primary task and not when it is subsumed within a higher level task. Therefore, a task-by-task analysis of the appropriateness of these cognitive interventions is necessary for them to be effective. The failure of users to detect changes to a subordinate task is similar to failures of typists to be aware of their mistakes because they do not have access to lower level actions subsumed within the higher level goals they are aware of (Logan & Crump, 2010), and the problematic consequences of focusing attention on a subsumed task is consistent with findings demonstrating that focusing attention on simple elements of a complex skilled task can impair performance (Flegal & Anderson, 2008).

In many cases of HCI, specific visual changes are intended to be important signals about action affordances, and these are often missed. For example, the surface/detail level header change tested by Varakin, Levin, and Fidler (2004) was a greatly enhanced variant of the actual change programmed into an e-reader prototype, and one can easily list a large number of other display changes that are intended to signal important interaction opportunities and mode changes. In one recent project, we observed a novice spreadsheet user fail to detect the difference between extending a cell selection by dragging from a cell border and by dragging from the lower right corner. Moving the cursor to these different parts of a cell changes the cursor from an arrow to a hand or to a plus to indicate mode differences, a visual signal that our participant completely missed. This kind of failure is common, and recent research has explored these lapses in novice learners. Kay (2007) extensively documented and classified errors that a range of users produce when learning new spreadsheet functions and found that failures to observe information and consequences of actions were among the most impactful. The latter errors are particularly interesting because they likely represent cases where users produced actions such as key presses but failed to detect the changes those actions induced. Although these failures are clearly consequential, it is important to consider the possibility that users are learning something even when they fail to explicitly detect changes. If so, it may be useful to allow a range of changes to signal interaction opportunities on the assumption that users will gradually acknowledge their presence. Clearly, the usefulness of this strategy depends in part on people's ability to implicitly detect changes and to turn this implicit knowledge into useful actions.

In addition to these basic connections between the visual attention research and HCI, the CHEX example makes clear that higher level forms of cognitive balancing are important. Tasks clearly vary in how visual property monitoring and deeper causal analyses are related. In some cases they are relatively separate and users must be aware of both tasks, but in others one is subsumed within the other. In cases involving some level of subsumption, models of event perception may be particularly relevant as they provide a natural description of the degree to which direct awareness of visual properties is and is not likely to be subsumed within larger meaningful events, and even in the case of subsumed visual tracking, event models can direct researchers toward moments when subsumed properties will be available to awareness. In addition, users are often called upon to make intelligent decisions to effectively balance visual property monitoring with attention to higher level goals and problem solving. If, as the visual metacognition research suggests, users are unaware of how difficult the monitoring tasks can be, they may fail to devote sufficient cognitive resources to them. This issue has been discussed by Varakin, Levin, and Fidler (2004), who argue that a set of specific "illusions of visual bandwidth" derive from metacognitive errors in people's understanding of vision, and that these illusions can explain incorrect assumptions about the usability of visual displays. Similarly, Smallman and Cook (2011) describe a trio of fallacies (that perception is easy, accurate, and complete) that constitute a form of "naïve realism" and demonstrate that computer users overestimate the amount of visual realism that is optimal for performance in a computerized route-drawing task.

The example of head-up navigation aids at the beginning of this chapter is particularly compelling in this context. The NYCycle app (http://nycycle.rga.com/) is currently being developed for Google Glass, with the goal of allowing individuals users of bicycle-rental kiosks to receive directions, information about where they might like to go, and information about the location of drop-off points for their bicycles, all conveniently projected into their visual field. The designers of this system do seem to be aware of the dangers of doing this, and they recommend that users not use the system while they are actually bicycling. As illustrated in their promotional materials, they show a user dutifully using her Google Glass *before* she starts whizzing through a dense urban landscape filled with moving cars, backing-up trucks, aggressive pedestrians, traffic signals, location-marking signs, oblivious photo-taking tourists, cops, wandering children, and enthusiastic dogs.

The thing is, *you don't take Google Glass off*. That's the point. It's always there with all of its interaction affordances, push notifications, and status indicators. So can we really trust users to forgo all of this when it is so easy, and when it likely seems to them as though they can receive this information while they are still looking out into the world? If there is one thing that research on CB and IB make clear, it is that having your eyeballs pointed out into the world doesn't buy you much in terms of being aware of visual properties and unexpected visual events. Even more disturbing is the fact that people do not seem to realize this and may not be particularly good judges of the wisdom of deploying a wide range of technologies that will merge in a range of ways with everyday perception.

These new technologies are likely to be deployed in a range of settings, and so it is important to get beyond simply warning of the visual-cognitive dangers of these deployments. Fortunately, research in visual cognition offers a potentially useful description of the range of perceptual and cognitive constraints that differentiate failure from success in countenancing visual information, and hopefully future research can provide guidelines for safely integrating new forms of seeing into our daily routines.

Concluding Summary and Future Directions

CB and IB demonstrate the striking failure of human observers to be aware of features in their environment. As we have detailed above, the prehistory of inattention research began with investigations into the inability to split attention during a dichotic listening task. Research has since expanded well beyond auditory streams, demonstrating an inability to detect changes at the focus of gaze or in one's own hand.

The phenomena of CB and IB indicate limits to attention that may reveal the representational basis of online vision. On the one hand, failures to detect visual changes may occur because we fail to generate many online visual representations. On the other hand, our ability to recognize hundreds of visual scenes, and the presence of implicit awareness of some visual properties, may reflect a substantially richer set of representation than CB suggests. Ultimately, the prevalence of visual representation may vary across different settings, and it is important to consider the possibility that some task contexts compel the generation of many visual representations while in other more naturalistic settings visual representations are much more tightly constrained by functional needs and are often subsumed within larger tasks. Therefore, just as change detection increases for features pertinent to sports fans and chess players, the absence of change detection may reflect important ways in which visual representations are not pertinent to an understanding of meaningful visual events.

Following this basic research, insights gained from CB and IB studies have been applied in a number of settings. In HCI, CB research can help guide interventions facilitating the effective extraction of information from complex visual computer displays. Perhaps the most dangerous component in this situation is the failure of users to realize their limitations, leading to potentially catastrophic errors. Future exploration into CB can support a wide array of applications ranging in importance from video games to driving.

Box 9.1
Key Points

- Inattentional blindness (IB) is the inability to notice unattended properties, while change blindness (CB) is the failure to detect between-view property changes.

- CB and IB may reveal *sparse representation*, where awareness is limited by attentional selection and encoding, or *rich representation*, where representations are detailed but awareness is limited by memory or comparison.

- Observers show a striking, and possibly dangerous, lack of awareness of their own inattention and change blindness.

- CB and IB have a broad range of applications, including human–computer interaction, communication, law, and medicine.

Box 9.2
Outstanding Issues

- More work is required to understand the tendency to under- or overestimate vulnerability to change blindness in oneself or others.

- Future research may also expand upon the small but growing number of experiments featuring environmental, expertise, or event-driven effects on CB.

References

Alvarez, G. A., & Cavanagh, P. (2004). The capacity of visual short-term memory is set both by visual information load and by number of objects. *Psychological Science, 15*, 106–111.

Angelone, B. L., Levin, D. T., & Simons, D. J. (2003). The roles of representation and comparison failures in change blindness. *Perception, 32*, 947–962.

Baker, L. J., & Levin, D. T. (2015). The role of relational triggers in event perception. *Cognition, 136*, 14-29.

Ballard, D. H., Hayhoe, M. M., & Pelz, J. B. (1995). Memory representations in natural tasks. *Cognitive Neuroscience, 7*, 66–80.

Barthelme, S., & Mamassian, P. (2010). Flexible mechanisms underlie the evaluation of visual confidence. *Proceedings of the National Academy of Sciences, 107*, 20834–20839.

Brady, T. F., Konkle, T., & Alvarez, G. A. (2011). A review of visual memory capacity: Beyond individual items and toward structured representations. *Journal of Vision (Charlottesville, Va.), 11*, 1–34.

Beck, M. R., & Levin, D. T. (2003). The role of representational volatility in recognizing pre-and postchange objects. *Perception & Psychophysics, 65,* 458–468.

Beck, M. R., Angelone, B. L., & Levin, D. T. (2004). Knowledge about the probability of change affects change detection performance. *Journal of Experimental Psychology: Human Perception and Performance, 30,* 778–791.

Beck, M. R., Levin, D. T., & Angelone, B. A. (2007). Change blindness blindness: Beliefs about the roles of intention and scene complexity in change detection. *Consciousness and Cognition, 16,* 31–51.

Beck, M. R., Peterson, M. S., & Angelone, B. L. (2007). The roles of encoding, retrieval, and awareness in change detection. *Memory & Cognition, 35,* 610–620.

Becklen, R., & Cervone, D. (1983). Selective looking and the noticing of unexpected events. *Memory & Cognition, 11,* 601–608.

Bentin, S., Kutas, M., & Hillyard, S. A. (1995). Semantic processing and memory for attended and unattended words in dichotic listening: Behavioral and electrophysiological evidence. *Journal of Experimental Psychology: Human Perception and Performance, 21,* 54–67.

Benway, J. P., & Lane, D. M. (1998). Banner blindness: Web searchers often miss "obvious" links. *Internetworking,* 1.3.

Blackmore, S. (2002). There is no stream of consciousness. *Journal of Consciousness Studies, 9,* 17–28.

Blackmore, S. J., Brelstaff, G., Nelson, K., & Troscianko, T. (1995). Is the richness of our visual world an illusion? Transsaccadic memory for complex scenes. *Perception, 24,* 1075–1081.

Bredemeier, K., & Simons, D. J. (2012). Working memory and inattentional blindness. *Psychonomic Bulletin & Review, 19,* 239–244.

Breitmeyer, B. G., Kropfl, W., & Julesz, B. (1982). The existence and role of retinotopic and spatiotopic forms of visual persistence. *Acta Psychologica, 52,* 175–196.

Brockmole, J. R., Hambrick, D. Z., Windisch, D. J., & Henderson, J. M. (2008). The role of meaning in contextual cueing: Evidence from chess expertise. *Quarterly Journal of Experimental Psychology, 61,* 1886–1896.

Castelhano, M. S., & Henderson, J. M. (2005). Incidental visual memory for objects in scenes. *Visual Cognition, 12,* 1017–1040.

Castelhano, M. S., Mack, M. L., & Henderson, J. M. (2009). Viewing task influences eye movement control during active scene perception. *Journal of Vision, 9,* 1–15.

Chase, W. G., & Simon, H. A. (1973). Perception in chess. *Cognitive Psychology, 4,* 55–81.

Cherry, E. C. (1953). Some experiments on the recognition of speech, with one and two ears. *Journal of the Acoustical Society of America, 25,* 975–979.

Clery, H., Andersson, F., Bonnet-Brilhault, F., Phillippe, A., Wicker, B., & Gomot, M. (2013). fMRI investigation of visual change detection in adults with autism. *NeuroImage: Clinical, 2*, 303–312.

Conway, A. R., Cowan, N., & Bunting, M. F. (2001). The cocktail party phenomenon revisited: The importance of working memory capacity. *Psychonomic Bulletin & Review, 8*, 331–335.

Davies, G., & Hine, S. (2007). Change blindness and eyewitness testimony. *Journal of Psychology, 141*, 423–434.

Deutsch, J. A., & Deutsch, D. (1963). Attention: Some theoretical considerations. *Psychological Review, 70*, 80–90.

DiVita, J., Obermeyer, R., Nygren, T. E., & Linville, J. M. (2004). Verification of the change blindness phenomenon while managing critical events on a combat information display. *Human Factors, 46*, 205–218.

Dixon, B. J., Daly, M. J., Chan, H., Vescan, A. D., Witterick, I. J., & Irish, J. C. (2013). Surgeons blinded by enhanced navigation: The effect of augmented reality on attention. *Surgical Endoscopy, 27*, 454–461.

Drew, T., Võ, M. L. H., & Wolfe, J. M. (2013). The invisible gorilla strikes again: Sustained inattentional blindness in expert observers. *Psychological Science, 24*, 1848–1853.

Durlach, P. J. (2004). Change blindness and implications for complex monitoring and control systems design and operator training. *Human–Computer Interaction, 19*, 423–451.

Durlach, P. J., & Chen, J. Y. C. (2003). Visual change detection in digital military displays. *Proceedings of the Interservice/Industry Training, Simulation, and Education Conference 2003*. Orlando, FL: IITSEC.

Fernandez-Duque, D., & Thornton, I. M. (2003). Explicit mechanisms do not account for implicit localization and identification of change: An empirical reply to Mitroff et al. (2002). *Journal of Experimental Psychology: Human Perception and Performance, 29*, 846–858.

Fish, C., Goldsberry, K. P., & Battersby, S. (2011). Change blindness in animated choropleth maps: An empirical study. *Cartography and Information Science, 38*, 350–362.

Flegal, K. E., & Anderson, M. C. (2008). Overthinking skilled motor performance: Or why those who teach can't do. *Psychonomic Bulletin & Review, 15*, 927–932.

Fougnie, D., & Marois, R. (2007). Executive working memory load induces inattentional blindness. *Psychonomic Bulletin & Review, 14*, 142–147.

Grimes, J. (1996). On the failure to detect changes in scenes across saccades. In K. Atkins (Ed.), *Vancouver studies in cognitive science: Vol. 2: Perception* (pp. 89–110). New York: Oxford University Press.

Haines, R. F. (1991). A breakdown in simultaneous information processing. In G. Obrecht & L. Stark (Eds.), *Presbyopia research: From molecular biology to visual adaptation* (pp. 171–175). New York: Plenum Press.

Harley, E. M., Carlsen, K. A., & Loftus, G. R. (2004). The "saw-it-all-along" effect: Demonstrations of visual hindsight bias. *Journal of Experimental Psychology: Learning, Memory, and Cognition, 30,* 960–968.

Hayhoe, M. M. (2000). Vision using routines: A functional account of vision. *Visual Cognition, 7,* 43–64.

Henderson, J. M., & Hollingworth, A. (2003). Eye-movements, visual memory, and scene representation. In M. A. Peterson & G. Rhodes (Eds.), *Perception of faces, objects, and scenes* (pp. 356–383). Oxford, UK: Oxford University Press.

Hochberg, J. (1968). In the mind's eye. In R. N. Haber (Ed.), *Contemporary theory and research in visual perception* (pp. 309–331). New York: Holt, Rinehart & Winston.

Hollingworth, A. (2003). Failures of retrieval and comparison constrain change detection in natural scenes. *Journal of Experimental Psychology: Human Perception and Performance, 29,* 388–403.

Hollingworth, A., & Henderson, J. M. (2002). Accurate visual memory for previously attended objects in natural scenes. *Journal of Experimental Psychology: Human Perception and Performance, 28,* 113–136.

Hollingworth, A., Williams, C. C., & Henderson, J. M. (2001). To see and remember: Visually specific information is retained in memory from previously attended objects in natural scenes. *Psychonomic Bulletin & Review, 8,* 761–768.

Huff, M., Papenmeier, F., & Zacks, J. M. (2012). Visual target detection is impaired at event boundaries. *Visual Cognition, 20,* 848–864.

Hyman, I. E., Sarb, B. A., & Wise-Swanson, B. M. (2014). Failure to see money on a tree: Inattentional blindness for objects that guided behavior. *Frontiers in Psychology, 5,* 356.

Jensen, M. S., Yao, R., Street, W. N., & Simons, D. J. (2011). Change blindness and inattentional blindness. *Wiley Interdisciplinary Reviews: Cognitive Science, 2,* 529–546.

Jiang, Y. V., Capistrano, C. G., & Palm, B. E. (2014). Spatial working memory in children with high-functioning autism: Intact configural processing but impaired capacity. *Journal of Abnormal Psychology, 123,* 248–257.

Jonides, J., Irwin, D. E., & Yantis, S. (1982). Integrating visual information from successive fixations. *Science, 215,* 192–194.

Jonides, J., Irwin, D. E., & Yantis, S. (1983). Failure to integrate information from successive fixations. *Science, 222,* 188.

Kawahara, J. (2010). Measuring the spatial distribution of the metaattentional spotlight. *Consciousness and Cognition, 19,* 107–124.

Kay, R. H. (2007). The role of errors in learning computer software. *Computers & Education, 49,* 441–459.

Konkle, T., Brady, T. F., Alvarez, G. A., & Oliva, A. (2010). Scene memory is more detailed than you think: The role of categories in visual long-term memory. *Psychological Science, 21,* 1551–1556.

Kuhn, G., & Tatler, B. W. (2005). Magic and fixation: Now you don't see it, now you do. *Perception, 34,* 1153–1161.

Landman, R., Spekreijse, H., & Lamme, V. A. (2003). Large capacity storage of integrated objects before change blindness. *Vision Research, 43,* 149–164.

Laurent, E., Ward, P., Williams, A. M., & Ripoll, H. (2006). Expertise in basketball modifies perceptual discrimination abilities, underlying cognitive processes, and visual behaviours. *Visual Cognition, 13,* 247–271.

Lavie, N. (2010). Attention, distraction, and cognitive control under load. *Current Directions in Psychological Science, 19,* 143–148.

Lavie, N., Hirst, A., de Fockert, J. W., & Viding, E. (2004). Load theory of selective attention and cognitive control. *Journal of Experimental Psychology: General, 133,* 339–354.

Levin, D. T. (2012). Concepts about agency constrain beliefs about visual experience. *Consciousness and Cognition, 21,* 875–888.

Levin, D. T., & Angelone, B. L. (2008). The visual metacognition questionnaire: A measure of intuitions about vision. *American Journal of Psychology, 121,* 1192–1208.

Levin, D. T., Angelone, B. L., & Beck, M. R. (2010). Visual search for rare targets: Distractor tuning as a mechanism for learning from repeated target-absent searches. *British Journal of Psychology, 102,* 313–327.

Levin, D. T., Drivdahl, S. B., Momen, N., & Beck, M. R. (2002). False predictions about the detectability of unexpected visual changes: The role of beliefs about attention, memory, and the continuity of attended objects in causing change blindness blindness. *Consciousness and Cognition, 11,* 507–527.

Levin, D. T., Momen, N. M., Drivdahl, S. B., & Simons, D. J. (2000). Change blindness blindness: The metacognitive error of overestimating change-detection ability. *Visual Cognition, 7,* 397–412.

Levin, D. T., & Simons, D. J. (1997). Failure to detect changes to attended objects in motion pictures. *Psychonomic Bulletin & Review, 4,* 501–506.

Levin, D. T., Simons, D. J., Angelone, B. L., & Chabris, C. F. (2002). Memory for centrally attended changing objects in an incidental real-world change detection paradigm. *British Journal of Psychology, 93,* 289–302.

Levin, D. T., & Varakin, D. A. (2004). No pause for a brief disruption: Failures to detect interruptions to ongoing events. *Consciousness and Cognition, 13,* 363–372.

Loftus, G. R., & Mackworth, N. H. (1978). Cognitive determinants of fixation location during picture viewing. *Journal of Experimental Psychology: Human Perception and Performance, 4,* 565–572.

Logan, G. D., & Crump, M. J. (2010). Cognitive illusions of authorship reveal hierarchical error detection in skilled typists. *Science, 330*, 683–686.

Luck, S. J., & Vogel, E. K. (1997). The capacity of visual working memory for features and conjunctions. *Nature, 390*, 279–281.

Maccari, L., Casagrande, M., Martella, D., Anolfo, M., Rosa, C., Fuentes, L. J., et al. (2013). Change blindness in children with ADHD: A selective impairment in visual search? *Journal of Attention Disorders, 17*, 620–627.

Mack, A., & Rock, I. (1998). *Inattentional blindness.* Cambridge, MA: MIT Press.

MacKay, D. G. (1973). Aspects of the theory of comprehension, memory, and attention. *Quarterly Journal of Experimental Psychology, 25*, 22–40.

Mandler, J. M., Seegmiller, D., & Day, J. (1977). On the coding of spatial information. *Memory & Cognition, 5*, 10–16.

Marchetti, L. M., Biello, S. M., Broomfield, N. M., MacMahon, K. M. A., & Espie, C. A. (2006). Who is pre-occupied with sleep? A comparison of attention bias in people with psychophysiological insomnia, delayed sleep phase syndrome and good sleepers using the induced change blindness paradigm. *Journal of Sleep Research, 15*, 212–221.

Maule, A. J. (1985). The importance of an updating internal representation of the environment in the control of visual sampling. *Quarterly Journal of Experimental Psychology, A, 37*, 533–551.

McConkie, G. W., & Zola, D. (1979). Is visual information integrated across fixations in reading? *Perception & Psychophysics, 25*, 221–224.

Memmert, D. (2006). The effects of eye movements, age, and expertise on inattentional blindness. *Consciousness and Cognition, 15*, 620–627.

Meyerhoff, H. S., Huff, M., Papenmeier, F., Jahn, G., & Schwan, S. (2011). Continuous visual cues trigger automatic spatial target updating in dynamic scenes. *Cognition, 121*, 73–82.

Miller, L., & McGonigle-Chalmers, M. (2013). Exploring perceptual skills in children with autism-spectrum disorders: From target detection to dynamic perceptual discrimination. *Journal of Autism and Developmental Disorders.* doi:10.1007/s10803-013-1977-6.

Mitroff, S. R., Simons, D. J., & Franconeri, S. L. (2002). The siren song of implicit change detection. *Journal of Experimental Psychology: Human Perception and Performance, 28*, 798–815.

Mitroff, S. R., Simons, D. J., & Levin, D. T. (2004). Nothing compares 2 views: Change blindness can result from failures to compare retained information. *Perception & Psychophysics, 66*, 1268–1281.

Moray, N. (1959). Attention in dichotic listening: Affective cues and the influence of instruction. *Quarterly Journal of Experimental Psychology, 11*, 56–60.

Most, S. B. (2010). What's inattentional about inattentional blindness? *Consciousness and Cognition, 19*, 1102–1104.

Most, S. B., Simons, D. J., Scholl, B. J., Jimenez, R., Clifford, E., & Chabris, C. F. (2001). How not to be seen: The contribution of similarity and selective ignoring to sustained inattentional blindness. *Psychological Science, 12,* 9–17.

Mou, W., Xiao, C., & McNamara, T. P. (2008). Reference directions and reference objects in spatial memory of a briefly viewed layout. *Cognition, 108,* 136–154.

Neisser, U., & Becklen, R. (1975). Selective looking: Attending to visually specified events. *Cognitive Psychology, 7,* 480–494.

Nelson, K. J., Laney, C., Fowler, N. B., Knowles, E. D., Davis, D., & Loftus, E. F. (2011). Change blindness can cause mistaken eyewitness identification. *Legal and Criminological Psychology, 16,* 62–74.

Nickerson, R. S. (1965). Short-term memory for complex meaningful visual configurations: A demonstration of capacity. *Canadian Journal of Psychology, 19,* 155–160.

Nikolic, M. I., & Sarter, N. B. (2001). Peripheral visual feedback: A powerful means of supporting effective attention allocation in event-drive, data-rich environments. *Human Factors, 43,* 30–38.

Nickerson, R. S., & Adams, J. J. (1979). Long-term memory for a common object. *Cognitive Psychology, 11,* 287–307.

Nowell, L., Hetzler, E., & Tanasse, T. (2001). Change blindness in information visualization: A case study. *Proceedings of the 7th IEEE Symposium on Information Visualization, San Diego, CA,* 15–22.

O'Regan, J. K. (1992). Solving the "real" mysteries of visual perception: The world as an outside memory. *Canadian Journal of Psychology/Revue Canadienne de Psychologie, 46,* 461–488.

O'Regan, J. K., Deubel, H., Clark, J. J., & Rensink, R. A. (2000). Picture changes during blinks: Looking without seeing, and seeing without looking. *Visual Cognition, 7,* 191–212.

Pashler, H. (1988). Familiarity and visual change detection. *Perception & Psychophysics, 44,* 369–378.

Phillips, W. A. (1974). On the distinction between sensory storage and short-term visual memory. *Perception & Psychophysics, 16,* 283–290.

Radvansky, G. A., Tamplin, A. K., & Krawietz, S. A. (2010). Walking through doorways causes forgetting: Environmental integration. *Psychonomic Bulletin & Review, 17,* 900–904.

Rayner, K., & Pollatsek, A. (1987). Eye movements in reading: A tutorial review. In M. Coltheart (Ed.), *Attention and performance: Vol. 12. The psychology of reading* (pp. 327–362). London: Erlbaum.

Reingold, E. M., Charness, N., Pomplun, M., & Stampe, D. M. (2001). Visual span in expert chess players: Evidence from eye movements. *Psychological Science, 12,* 48–55.

Rensink, R. A. (2000). Visual search for change: A probe into the nature of attentional processing. *Visual Cognition, 7,* 345–376.

Rensink, R. A. (2002). Change detection. *Annual Review of Psychology, 53*, 245–277.

Rensink, R. A., O'Regan, J. K., & Clark, J. J. (1997). To see or not to see: The need for attention to perceive changes in scenes. *Psychological Science, 8*, 368–373.

Rhodes, G., Hayward, W. G., & Winkler, C. (2006). Expert face coding: Configural and component coding of own-race and other-race faces. *Psychonomic Bulletin & Review, 13*, 499–505.

Ro, T., Russell, C., & Lavie, N. (2001). Changing faces: A detection advantage in the flicker paradigm. *Psychological Science, 12*, 94–99.

Rosielle, L. J., & Scaggs, W. J. (2008). What if they knocked down the library and nobody noticed? The failure to detect large changes to familiar scenes. *Memory, 16*, 115–124.

Ryan, J. D., & Cohen, N. J. (2004). The nature of change detection and online representations of scenes. *Journal of Experimental Psychology: Human Perception and Performance, 30*, 988–1015.

Saiki, J., & Holcombe, A. O. (2012). Blindness to a simultaneous change of all elements in a scene, unless there is a change to summary statistics. *Journal of Vision, 12*(3), 1–11.

Scholl, B. J. (2000). Attenuated change blindness for exogenously attended items in a flicker paradigm. *Visual Cognition, 7*, 377–396.

Scholl, B. J., Simons, D. J., & Levin, D. T. (2004). "Change blindness" blindness: An implicit measure of a metacognitive error. In D. T. Levin (Ed.), *Thinking and seeing: Visual metacognition in adults and children* (pp. 145–164). Cambridge, MA: MIT Press.

Schooler, J. W., Reichle, E. D., & Halpern, D. V. (2004). Zoning out while reading: Evidence for dissociations between experience and metaconsciousness. In D. Levin (Ed.), *Thinking and seeing: Visual metacognition in adults and children* (pp. 203–226). Cambridge, MA: MIT Press.

Schooler, J. W., Smallwood, J., Christoff, K., Handy, T. C., Reichle, E. D., & Sayette, M. A. (2011). Meta-awareness, perceptual decoupling and the wandering mind. *Trends in Cognitive Sciences, 15*, 319–326.

Schulman, A. I. (1973). Recognition memory and the recall of spatial location. *Memory & Cognition, 1*, 256–260.

Seegmiller, J. K., Watson, J. M., & Strayer, D. L. (2011). Individual differences in susceptibility to inattentional blindness. *Journal of Experimental Psychology: Learning, Memory, and Cognition, 37*, 785–791.

Shepard, R. N. (1967). Recognition memory for words, sentences, and pictures. *Journal of Verbal Learning and Verbal Behavior, 6*, 156–163.

Silverman, M. E., & Mack, A. (2006). Change blindness and priming: When it does and does not occur. *Consciousness and Cognition, 15*, 409–422.

Simons, D. J. (2000). Current approaches to change blindness. *Visual Cognition, 7*, 1–17.

Simons, D. J., & Chabris, C. F. (1999). Gorillas in our midst: Sustained inattentional blindness for dynamic events. *Perception, 28,* 1059–1074.

Simons, D. J., & Chabris, C. F. (2011). What people believe about how memory works: A representative survey of the US population. *PLoS ONE, 6*(8), e22757.

Simons, D. J., Chabris, C. F., Schnur, T. T., & Levin, D. T. (2002). Evidence for preserved representations in change blindness. *Consciousness and Cognition, 11,* 78–97.

Simons, D. J., & Levin, D. T. (1997). Change blindness. *Trends in Cognitive Sciences, 1,* 261–267.

Simons, D. J., & Levin, D. T. (1998). Failure to detect changes to people during a real-world interaction. *Psychonomic Bulletin & Review, 5,* 644–649.

Simons, D. J., & Wang, R. F. (1998). Perceiving real-world viewpoint changes. *Psychological Science, 9,* 315–320.

Solman, J. F., Cheyne, J. A., & Smilek, D. (2012). Found and missed: Failing to recognize a search target despite moving it. *Cognition, 123,* 100–118.

Smallman, H. S., & Cook, M. B. (2011). Naïve realism: Folk fallacies in the design and use of visual displays. *Topics in Cognitive Science, 3,* 579–608.

Smallman, H. S., & St John, M. (2003). CHEX (Change History Explicit): New HCI concepts for change awareness. *Proceedings of the Human Factors and Ergonomics Society 47th Annual Meeting* (pp. 528–532). Santa Monica, CA: Human Factors and Ergonomics Society.

Smith, T. J., Lamont, P., & Henderson, J. M. (2012). The penny drops: Change blindness at fixation. *Perception, 41,* 489–492.

Sperling, G. (1960). The information available in brief visual presentations. *Psychological Monographs, 74* (Series 498).

Standing, L. (1973). Learning 10,000 pictures. *Journal of Experimental Psychology, 25,* 207–222.

Steffner, D., & Schenkman, B. (2012). Change blindness when viewing Web pages. *Work (Reading, Mass.), 41,* 6098–6102.

St. John, M., Smallman, H. S., & Manes, D. (2005). Recovery from interruptions to a dynamic monitoring task: The beguiling utility of instant replay. *Proceedings of the Human Factors and Ergonomics Society 49th Annual Meeting* (pp. 473–477).

Strayer, D. L., & Drews, F. A. (2007). Cell-phone–induced driver distraction. *Current Directions in Psychological Science, 16,* 128–131.

Swallow, K. M., Zacks, J. M., & Abrams, R. A. (2009). Event boundaries in perception affect memory encoding and updating. *Journal of Experimental Psychology: General, 138,* 236–257.

Tager-Flusberg, H., Plesa-Skwerer, D., Schofield, C., Verbalis, A., & Simons, D. J. (2007). Change detection as a tool for assessing attentional deployment in atypical populations: The case of Williams syndrome. *Cognition, Brain, & Behaviour, 11,* 491–506.

Treisman, A. (1964). Monitoring and storage of irrelevant messages in selective attention. *Journal of Verbal Learning and Verbal Behavior, 3,* 449–459.

Treisman, A., & Schmidt, H. (1982). Illusory conjunctions in the perception of objects. *Cognitive Psychology, 14,* 107–114.

Treisman, A., Squire, R., & Green, J. (1974). Semantic processing in dichotic listening? A replication. *Memory & Cognition, 2,* 641–646.

Triesch, J., Ballard, D. H., Hayhoe, M. M., & Sullivan, B. T. (2003). What you see is what you need. *Journal of Vision, 3,* 86–94.

Vallieres, B. R., Hodgetts, H. M., Vachon, F., & Tremblay, S. (2012). Supporting change detection in complex dynamic situations: Does the CHEX serve its purpose? *Proceedings of the Human Factors and Ergonomics Society 56th Annual Meeting* (pp. 1708–1712).

Varakin, D. A., & Levin, D. T. (2006). How can visual memory be so good if change detection is so bad? Visual representations get rich so they can act poor. *British Journal of Psychology, 91,* 51–77.

Varakin, D. A., Levin, D. T., & Collins, K. (2007). Comparison and representation failures both cause real-world change blindness. *Perception, 36,* 737–749.

Varakin, D. A., Levin, D. T., & Fidler, R. (2004). Unseen and unaware: Applications of recent research on failures of visual awareness for human–computer interface design. *Human–Computer Interaction, 19,* 389–421.

Wayand, J. W., Levin, D. T., & Varakin, D. A. (2005). Inattentional blindness for a noxious multimodal stimulus. *American Journal of Psychology, 118,* 339–352.

Weaver, M. D., & Lauwereyns, J. (2011). Attentional capture and hold: The oculomotor correlates of the change detection advantage for faces. *Psychological Research, 75,* 10–23.

Werner, S., & Thies, B. (2000). Is "change blindness" attenuated by domain-specific expertise? An expert–novices comparison of change detection in football images. *Visual Cognition, 7,* 163–173.

Wolfe, J. M., & Bennett, S. C. (1997). Preattentive object files: Shapeless bundles of basic features. *Vision Research, 37,* 25–43.

Xu, F., & Carey, S. (1996). Infants' metaphysics: The case of numerical identity. *Cognitive Psychology, 30,* 111–153.

Yarbus, A. L. (1967). *Eye movements and vision.* New York: Plenum Press.

Yeh, M., Merlo, J. L., Wickens, C. D., & Brandenburg, D. L. (2003). Head up versus head down: The costs of imprecision, unreliability, and visual clutter on cue effectiveness for display signaling. *Human Factors: The Journal of the Human Factors and Ergonomics Society, 45,* 390–407.

Zacks, J. M., Speer, N. K., Swallow, K. M., Braver, T. S., & Reynolds, J. R. (2007). Event perception: A mind–brain perspective. *Psychological Bulletin, 133,* 273–293.

10

Mind Wandering and Attention

Jonathan Smallwood

Historical Context

Human cognition is fluid, and its flexibility arises in part from the capacity to attend to different mental contents. As is demonstrated by the work covered in this volume, the study of attention has historically favored experimental paradigms in which one source of sensory input (such as vision) was prioritized over another (such as auditory input or information in a different part of space). These studies demonstrated that performance of complex goal-directed actions depends upon our attention system's focusing on task-relevant information from the environment. Often the analogy of a *spotlight* was used to characterize the fact that attention could sharpen the processing of a relevant input (Desimone & Duncan, 1995; Treisman & Gelade, 1980).

In everyday life, however, we often attend to information that is not derived from the environment nor related to the goals of the moment. When our mind wanders to an imagined romantic liaison while driving home from work, or we daydream about a holiday while taking a shower, the spotlight of attention is turned toward mental content that is not derived in a direct fashion from immediate perceptual input.

States of mind wandering and daydreaming have at least two features that research should attempt to understand. First, similar to other conscious states, these experiences are *generated* by the brain and are hence an active state. Thus, an understanding of the daydreaming state requires an explanation of how this experience is represented by the brain, and these may, or may not, be the same processes as those involved in producing experience based on incoming perception. Second, that unlike states of external focus, thinking about a work meeting while commuting depends to a large extent on input that arises from intrinsic, rather than extrinsic changes, and therefore are generated relatively by the *individuals themselves*. The distinct manner by which thoughts are produced without an obvious referent in the external environment may indicate that potentially unique cognitive processes contribute to the initiation of experiences during the daydreaming state.

Box 10.1

Self-Generated Thought

There are many different terms used to describe experiences such as daydreaming or mind wandering. Some focus on the active production of a train of thought that is unrelated to the external task being performed—task-unrelated thought, mind pops, and off-task thoughts. Others focus on their independence from external input, such as stimulus-independent thought. One term that captures both the *active* nature of these experiences and their *independence* from perceptual input is *self-generated thought.*

Self-generated thoughts are distinguished from thoughts and feelings that are produced in a relatively direct way from perceptual input. Perceptually generated experiences include the sensations of warmth, color, or sound that are produced directly from immediate stimulus input. Although these experiences are generated by the individual, they rely in a very direct way on immediate perceptual input. By contrast the process of self-generation allows mental contents to occur which have a less direct dependency on perceptual input. In fact, studies have shown that these experiences can often be completely unrelated to events taking place in the here and now to the extent that they can derail performance on the task in hand.

Self-generation refers to the fact that the mental content is not triggered directly by perception; it does not imply either intention or lack thereof. Self-generated experiences can occur intentionally, such as when we let our minds wander while taking a walk, as well as unintentionally, such as when we mind wander while reading a book.

Self-generated experiences are not synonymous with task-unrelated thought. Engaging in self-generated thought both can be the primary task of the individual (such as when we actively think about a meeting on the commute to work) and can be unrelated to the goal that the individual is pursuing (such as when we mind wander about a failed romance while trying to concentrate at work).

Given the active nature of the daydreaming state, and its capacity to be initiated without a cue in the external environment, it is helpful to consider such states as examples of a broad class of thoughts and actions that are *self-generated*, a term that was first employed in studies of schizophrenia by Chris Frith and others (Frith, 1992; Frith & Done, 1988; Jones & Fernyhough, 2007). Here it is important to note that the term *self* is used in the same sense that a machine (such as a dishwasher) can engage in self-cleaning behavior (e.g., it can initiate an action without being prompted to do so). The term *does not* imply the more complex and philosophical use of the term self that reflects the important question of who we are. Definitional issues surrounding the term self-generated thought are considered in box 10.1.

The capacity to self-generate cognition is one of the features of the daydreaming state that make it both important, and controversial, in studies of attention. An overarching assumption in many fields of cognitive science is that the underlying architecture of the mind can be understood through the examination of how individuals respond to an external stimulus (either explicitly through behavior or implicitly through neural or physiological changes). Although this method provides detailed information on how we engaged in cognition that leads directly to action, it does not provide the ideal method to characterize experiences that emerge through a process of self-generation. Once it is recognized that certain states can emerge without a direct cue in the external environment, it becomes necessary to develop both novel methods and theory to accommodate states like daydreaming into contemporary accounts of attention.

Important theoretical and methodological contributions to our understanding of the daydreaming state arose from the important work in the 1960s by researchers including Jerome Singer, John Antrobus, Leonard Giambra, and Eric Klinger (Antrobus, Antrobus, & Singer, 1964; Giambra, 1989; Klinger, 1978; Singer, 1975a, 1975b). Their pioneering work focused on developing the method of thought sampling that allows self-generated experiences to be characterized even though the process that controls the occurrence of the experience was unknown (and remains so to this day). The technique of thought, or experience, sampling involves capitalizing on the human ability to introspect upon one's experience and provide reports on the contents of conscious thought. This method provides researchers a tool to explore momentary changes in experience that occur over time and is especially valuable for states that arise without an obvious cue in the environment, such as daydreaming or mind wandering. One aspect of the method of thought sampling is that its success depends on the capacity to sample experience in as unbiased a manner as possible. Using this approach, Singer and colleagues determined that the capacity to self-generate thought is inversely related to the rate of external stimulus input in a sustained attention task, such that as task events become faster, the occurrence of self-generated thought declines (Antrobus, 1968; Giambra, 1995). A second important development was the Imaginal Processes Inventory, a self-report instrument that assesses the occurrence of states of daydreaming in daily life (Singer & Antrobus, 1966). In recent years this has been successfully used to explore the links between daydreaming and creativity (Baird et al., 2012) and neural processing (Mason et al., 2007).

Early research into the daydreaming state also fostered the development of important theoretical viewpoints. One particularly valuable avenue of research was the assumption of *current concerns* developed by Eric Klinger and colleagues that suggests that the reason why a given individual tends to focus on a particular aspect of his or her life at any given moment is related to the fact that human goals extend beyond those that are present in a particular moment in time (for a recent review, see Klinger, 2009). Under conditions when the saliency of external events in a task decreases, states of daydreaming increase, as individuals shift focus away from less relevant sensory input to more relevant self-generated mental contents. By emphasizing the attraction of self-generated information when the external environment lacks salient input, the current concerns hypothesis provides an explanation for *why* the mind wanders.

State-of-the-Art Review

Although important groundwork on states of daydreaming took place almost 50 years ago, this topic remained a relatively minor area of psychological research until early in this century. The neglect of self-generated experiences by mainstream psychological research was likely due to the historical legacy of behaviorism that tended to treat introspective reports with skepticism and the development of neuroimaging techniques that allowed covert mental processes to be explored (for reviews, see Callard, Smallwood, Golchert, & Margulies, 2013; Callard, Smallwood, & Margulies, 2012). This section of the review considers recent work on mind wandering and daydreaming with a specific focus on what these experiences can tell us about attention.

Studies from inside and outside of the laboratory suggest that self-generated thoughts occupy as much as half of waking experience (see Schooler et al., 2011; Smallwood & Schooler, 2006), so understanding these experiences has value because it happens frequently in daily life (Killingsworth & Gilbert, 2010). Studies of self-generated thought also suggest that when we attend to internal thought, attention becomes disengaged, or decoupled, from perception (Smallwood, 2013a, 2013c). The mind-wandering state, therefore, demonstrates that it is possible to study attention in a context in which external perception lacks salience. Finally, self-generated thoughts also have both costs and benefits. When self-generated thoughts take the form of absentminded lapses, they compromise ongoing behavior and so inform our understanding of attentional failure (Smallwood, Fishman, & Schooler, 2007). By contrast, they can provide novel solutions to problems, can help us make plans, and can redress negative moods. This complex balance of costs and benefits suggests that optimal cognitive functioning may depend upon balancing both perceptual and self-generated sources of information in an effective manner.

Perceptual and Self-Generated Information Compete for Attention
Self-generated mental states provide a valuable counterpoint to the standard view of attention because they support theoretical positions (e.g., Chun, Golomb, & Turk-Browne, 2011) that we can attend both internally and externally and that these sources of information can often be in competition for limited resources (Smallwood, 2013a; Smallwood & Schooler, 2006). Evidence that self-generated thought can compete with perceptually guided thought comes chiefly from studies that examine changes in processing of external information that occur during states such as daydreaming or mind wandering (known as *perceptual decoupling*), and those that examine how contextual demands impact on the occurrence of self-generated thought.

Box 10.2

Perceptual Decoupling (Franklin, Mrazek, Broadway, et al., 2013; Schooler et al., 2011; Smallwood, 2013a, 2013c)

Studies using a variety of dependent measures have shown that during periods of self-generated thought external input is neglected. This state is known as *perceptual decoupling*.

Evidence. Measures of task performance are compromised during periods of self-generated thought as assessed by errors or response time. Reduced physiological responses to external stimuli are observed in a range of measures. These include eye movements, pupil size, and cortical event-related potentials (see text). This evidence is parsimoniously accounted for by the hypothesis that during self-generated thought attention is functionally disengaged, or decoupled, from external input.

Functions. The function of perceptual decoupling is unclear. It could reflect the resolution of competition between intrinsic and extrinsic information processes in favor of self-generated input. This would help stabilize a train of thought, in this case by reducing external distractions that could derail an internal train of thought. Alternatively, the decoupling of attention from perception may simply be an unrelated consequence of the occurrence of self-generated thought.

Perceptual Decoupling When attention is directed to external information, it functions to facilitate processing of relevant external input, often in the service of an external action. By contrast, when we attend to self-generated information, we become disengaged from sensory input. This attentional shift leads to a phenomenon known as *perceptual decoupling* and is hypothesized to be the reason why self-generated states can lead to error on an external task (Kam & Handy, 2013; Smallwood, 2013a; see box 10.2).

One way to characterize perceptual decoupling is by examining the relationship between self-generated thought and the concurrent processing of external information. A measure of the cortical processing of a stimulus is provided by the amplitude of event-related potentials (ERPs) that are derived from the electroencephalogram. One ERP component, known as the P3, occurs approximately 300 milliseconds after task-relevant events and is thought to index task-related attention (Polich, 1986; Polich & Criado, 2006; Wickens, Kramer, & Donchin, 1984). Studies have found that the P3 is reduced for individuals who experienced high levels of task-unrelated thinking during a task (Barron, Riby, Greer, & Smallwood, 2011). It is also smaller during periods of time when attention is directed to task-unrelated self-generated information than when it is focused on an external task (Macdonald, Mathan, & Yeung, 2011; Smallwood, Beach, Schooler, & Handy, 2008). This reduction in the amplitude of the ERP is not limited to the P3; it also occurs for components that indicate sensory processing of auditory, visual, and tactile domains (Kam et al., 2011; Kam, Xu, & Handy, 2014).

The suppression of the cortical processing of sensory input does not apply equally to all stimuli in the external environment. Task events that deviate strongly from the standard events in a task are not supressed during periods of off-task thought (Kam, Dao, Stanciulescu, Tildesley, & Handy, 2013), nor are they as sensitive to absentminded lapses. For example, task-relevant stimuli that pop out because they have a unique color (color singletons) are sensitive to less absentminded error than those that do not (Smallwood, 2013b). It has been suggested that these boundary conditions to perceptual decoupling reflect an evolutionary advantage that ensures especially salient external events are attended to regardless of an individual's state of attention (Kam, Dao, et al., 2013; Kam & Handy, 2013; Smallwood, 2013b).

Contextual Modulation of Self-Generated Thought   Perceptual decoupling indicates that self-generated and perceptual information compete for the limited resource of attention. Further evidence comes from studies examining the occurrence of task-unrelated thoughts across different task contexts.

Increasing the demand on attention for task-relevant information decreases opportunities for task-unrelated self-generated thought to occur. Tasks with greater perceptual demands tend to increase the focus of information on the task in hand (Forster & Lavie, 2009; Levinson, Smallwood, & Davidson, 2012). Likewise, faster rates of stimulus presentation in a task reduce the occurrence of task-unrelated thought (Antrobus, Coleman, & Singer, 1967; Antrobus, Singer, Goldstein, & Fortgang, 1970; Smallwood et al., 2004; Smallwood, McSpadden, Luus, & Schooler, 2008). Studies have also found that tasks that do not require continuous maintenance of information in working memory increase opportunities for self-generated thought relative to those that do not (Bernhardt et al., 2014; Ruby, Smallwood, Sackur, & Singer, 2013; Smallwood, Brown, et al., 2011; Smallwood, Nind, & O'Connor, 2009; Smallwood, O'Connor, Sudbery, & Obonsawin, 2007; Smallwood, Schooler, et al., 2011; J. M. Smallwood, Baracaia, Lowe, & Obonsawin, 2003). Finally, highly practiced tasks provide greater chance for self-generated thought to occur (Mason et al., 2007; Teasdale et al., 1995). Altogether, these studies show that when a focus on external information is needed to perform a task, the opportunity to engage in self-generated thought is reduced.

Together with evidence for perceptual decoupling, the contextual modulation of self-generated thought demonstrates competition between self-generated and perceptual sources of information for attention. Phenomena such as daydreaming or mind wandering, therefore, illustrate that being attentive is not synonymous with a focus on sensory information—we can actively focus on our thoughts rather than on the world around us. Although this differentiation between attention and perception may seem novel, it would not have seemed so to William James: "*Everyone knows what attention is. It is the taking possession by the mind, in clear and vivid form, of one out of what seem several simultaneously possible objects or trains of thought*" (James, 1890).

Costs and Benefits to Self-Generated Thought

Absentminded Errors: A Consequence of a Wandering Mind   A consequence of the competition between self-generated and perceptual information can be poor performance on an ongoing task. Errors due to task-unrelated thoughts occur across a wide range of task environments, including encoding (Smallwood, Obonsawin, & Heim, 2003; J. M. Smallwood, Baracaia, et al., 2003), sustained attention (McVay & Kane, 2009), and complex span tasks (Mrazek, Smallwood, Franklin, et al., 2012; Mrazek, Smallwood, & Schooler, 2012). In these cases task performance becomes more erratic—error rates increase and responding can become slower and more variable (Allan Cheyne, Solman, Carriere, & Smilek, 2009; Carriere, Seli, & Smilek, 2013; Cheyne, Carriere, & Smilek, 2006; Cheyne, Carriere, Solman, & Smilek, 2011; Seli et al., 2013; Smilek, Carriere, & Cheyne, 2010a). Studies have also shown that this cost to mind wandering occurs in real-world situations, such as while driving automobiles (Galera et al., 2012) and flying planes (Casner & Schooler, 2013). However, perhaps the most obvious cost that occurs when the mind wanders is during tasks in which maintaining a model of ongoing information is critical, and this is a hallmark feature of many aspects of the education system.

For example, the experience of mind wandering during reading is a well-documented illustration of a cost that can occur as a result of self-generated thought (Schooler, Reichle, & Halpern, 2004). Evidence for the deleterious impact of mind wandering during reading comes from the negative correlation between the comprehension of information and the likelihood of engaging in task-unrelated thoughts while reading (Dixon & Bortolussi, 2013; Feng, D'Mello, & Graesser, 2013; Franklin, Smallwood, & Schooler, 2011; Jackson & Balota, 2012; McVay & Kane, 2011; Schooler et al., 2004; Smallwood, Gorgolewski, et al., 2013; Smallwood, McSpadden, & Schooler, 2008; Unsworth & McMillan, 2013). Similar consequences occur if mind wandering occurs during a lecture (Farley, Risko, & Kingstone, 2013; Szpunar, Khan, & Schacter, 2013), suggesting that the two major ways that educational material is delivered could be compromised by the mind-wandering state.

Importantly, studies have shown that the negative consequences of the experience can be linked directly to the material that is presented at the specific moment when task-unrelated thinking occurred. For example, experience sampling studies have found that the material that was presented during an off-task episode is recalled less effectively (Smallwood, McSpadden, & Schooler, 2008), and texts in which task focus is maintained have higher levels of comprehension than those in which task-unrelated thoughts are high (Smallwood, Gorgolewski, et al., 2013). Furthermore, examination of eye movements during reading have also found transient changes in reading behavior that are predictive of both subjective and objective indicators of a lack of attention to the information in the text (Reichle, Reineberg, & Schooler, 2010; Schad, Nuthmann, & Engbert, 2012; Smilek, Carriere, & Cheyne, 2010b).

Although the consequences of task-unrelated thought can be linked to specific periods of time when attention was directed away from the text, these brief lapses can also have a downstream impact on comprehension. In extended narratives readers build a situation model of the text that guides their attention and allows inferences regarding events in the narrative to be made (Zwaan & Radvansky, 1998). Under these conditions the perceptual decoupling associated with mind wandering prevents the information from being encoded and thus prevents readers from updating their model of the narrative (Smallwood, 2011). In one study participants were asked to read a Sherlock Holmes crime novella and the occurrence of task-unrelated thought was assessed throughout the text (Smallwood, McSpadden, & Schooler, 2008). If task-unrelated thinking occurred in the moments when information relevant to the identity of the villain was revealed, readers did not encode this information and as a consequence could not infer the identity of the main protagonist. Presumably, encoding deficits prevented readers from updating their model of the story, which therefore lacked key details necessary to allow them to derive the inferences that would have allowed them to solve the crime. In this way the momentary occurrence of mind wandering can have downstream consequences on tasks such as reading comprehension that depend heavily on updating a situational model of the environment (see box 10.3).

Box 10.3

The Cascade Model of Inattention (Smallwood, 2011; Smallwood, Fishman, et al., 2007)

Certain tasks depend on making a task-relevant action based on immediate perceptual input (such as stopping at a red traffic light). Other tasks depend upon the continual incorporation of information into an ongoing model of the situation (such as reading). The nature of the task demands can determine whether the consequences of task-unrelated thought is limited to the immediate moment or can also damage performance at a later stage.

Under situations in which the task depends on immediate input, the occurrence of task-unrelated thought leads to slower or less accurate behavior. This detriment will manifest if the off-task state coincides with the moment when task-relevant action should occur and will jeopardize performance at that moment.

Under situations in which the task depends on continually updating a situation model, task-unrelated self-generated thought can also lead to downstream consequences. These downstream effects also allow states of off-task thought to influence future performance on a task such as reading.

Freedom from Immediacy: The Utility of Self-Generated Thought Evidence of a link to absentminded error, or mindless reading, indicates that attending to task-unrelated self-generated thought carries a cost. However, the fact that we spend substantial amounts of time engaged in states of mind wandering, or daydreaming (Killingsworth & Gilbert, 2010), suggests that an internal focus to attention could also be important to the human condition. One way that self-generated thought has adaptive value is to allow individuals to give attention to personally relevant goals when the opportunity to directly act is not possible.

Across a wide range of different cultures the content of self-generated thought is often focused on the future, illustrating a prospective bias to the mind-wandering state (Andrews-Hanna et al., 2013; Andrews-Hanna, Reidler, Huang, & Buckner, 2010; Baird, Smallwood, & Schooler, 2011; Smallwood, Brown, et al., 2011; Smallwood & O'Connor, 2011; Smallwood, Schooler, et al., 2011; Song & Wang, 2012; Stawarczyk, Majerus, Maj, Van der Linden, & D'Argembeau, 2011). A content analysis of self-generated thought based on open-ended reports indicated that participants are often considering what to do next (Baird et al., 2011) while other work has suggested that they are focused on near rather than distant episodes (Andrews-Hanna et al., 2013; Andrews-Hanna et al., 2010). Moreover, future related thoughts are often a correlate of a heightened memory for self-relevant information, suggesting a role in a process that helps link an individual's past, present, and future selves together across time (Smallwood, Schooler, et al., 2011). The importance of future events in self-generated thought is an important prediction of the current concerns hypothesis proposed in the 1970s by Eric Klinger and colleagues (Klinger, 1966, 1973, 2013; Klinger & McNelly, 1969).

Self-generated thought may also be related to a capacity to generate novel solutions to problems (Frick, Guilford, Christensen, & Merrifield, 1959; Guilford, 1959). For example, individuals who generate a large number of solutions to hypothetical personal problems tend to experience a lot of self-generated thoughts (Ruby, Smallwood, Sackur, & Singer, 2013). Individual difference in the tendency to daydream also positively predicts a person's proclivity to generate creative solutions (Baird et al., 2012). In addition, creating conditions that favor self-generated thought (such as performing a nondemanding task) tends to increase individuals' capacity to generate creative solutions to problems to which they were already exposed. This evidence suggests periods that favor self-generated thought may facilitate the process of incubation (Mednick, Mednick, & Mednick, 1964) whereby an individual, either consciously or unconsciously, makes progress on a problem on which he or she had reached an impasse (Baird et al., 2012). Altogether, research suggests that the process of self-generated thought may play an important role in individuals' capacity to generate cognitions that help them navigate the complex social world in which they are embedded.

Integration

Studies of self-generated thought indicate that it has both costs and benefits (Mooney-ham & Schooler, 2013), an observation that requires an integrative framework for incorporating this heterogeneous pattern of functional outcomes into general frame-works of attention. In broad terms adaptive cognitive function in the real world is unlikely to depend upon an exclusive focus on either internal or external sources of information. Instead, success in daily life may reflect a capacity to balance attention to both perceptually and self-generated sources of information in an efficient way (J. A. Smallwood & Andrews-Hanna, 2013). Achieving this balance may depend upon moderating the context in which the experience occurs, as well as the content of the experiences themselves. These are known as the *content-regulation* and *context-regulation* hypotheses and are summarized in box 10.4.

The *context-regulation hypothesis* proposes that adaptive cognitive function would entail limiting self-generated material if it would jeopardize ongoing actions and facilitating it when demands are minimal. Individuals with good cognitive control limit their task-unrelated thoughts when external task demands are high, such as during complex span tasks, sustained attention tasks, and reading (McVay & Kane,

---

Box 10.4
Optimizing Cognition (Smallwood & Andrews-Hanna, 2013)
Evidence has shown that self-generated thought has both costs (such as absentminded-ness and unhappiness) and benefits (such as creativity and planning). This indicates that optimal cognition will not depend exclusively on either an internal or an external focus to attention. Instead it will emerge as the efficient management of the *context* in which the episodes emerge and the *content* of the episode (Smallwood & Andrews-Hanna, 2013).

The context-regulation hypothesis. Adaptive cognition entails limiting self-generated experience to situations under which the cost associated with perceptual decoupling is unlikely to unduly jeopardize external ongoing actions. This leads to several predictions for future research. For example, individual differences in the capacity to constrain self-generated thought when external input is important, and facilitate it when it is not, would be a meaningful individual difference that is predictive of measures of success in daily life (such as intelligence or attentional control).

The content-regulation hypothesis. Adaptive cognition depends on regulating the content of self-generated experience to material that is beneficial rather than costly to well-being. This hypothesis leads to a number of novel predictions. For example, rather than exploring the occurrence of self-generated thought as a predictor of affective dysfunction, it follows that states of psychological distress (such as depression) would be associated with differ-ences in the content rather than simply a greater quantity of the experiences.

2009, 2011; Mrazek, Smallwood, Franklin, et al., 2012; Unsworth & McMillan, 2013). Importantly, individuals with good cognitive control also tend to produce more off-task thoughts when the environment is nondemanding (Levinson et al., 2012), and this effect has been observed outside of the laboratory (Kane et al., 2007). Together, these data suggest expertise in attentional control often manifests as variations in the allocation of attention to internal and external sources depending on the demands of the environment.

There is also evidence that the capacity to self-generate thought under nondemanding conditions has a link to an individual's capacity to delay gratification—the ability to disregard smaller immediate rewards in favor of greater rewards in the future (Mischel & Gilligan, 1964). Superior delayed gratification is known to be predictive of positive attributes such as better diet (Weller, Cook, Avsar, & Cox, 2008), greater intelligence (Shamosh & Gray, 2008), fewer problems in addiction (Yoon et al., 2007), and better financial management (Silva & Gross, 2004). Studies have shown that individuals who make patient temporal economic decisions tend to report more off-task thoughts when external demands are low with no increase in the experience under more demanding conditions (Bernhardt et al., 2014; Smallwood, Ruby, & Singer, 2013). In accordance with the context-regulation hypothesis, therefore, patient individuals limit their task-unrelated thinking to conditions in which self-generated thought is less damaging to performance. One important prediction based on the context-regulation hypothesis is that measures of healthy cognitive functioning (such as intelligence or cognitive control) will not be associated with the occurrence of self-generated thoughts across all task environments to the same extent. Instead, certain aspects of beneficial cognitive processing may manifest as individual differences in the capacity to regulate these experiences so that self-generated thoughts are reduced in contexts in which the environment demands continuous attention and facilitated when the task does not require continuous attention.

The *content-regulation hypothesis* suggests that adaptive cognitive function entails limiting the maladaptive content of self-generated thought, such as unpleasant rumination or anxious worries. It has been shown that mind wandering is a correlate of unhappiness, both because it is linked to states of dysphoria (Carriere, Cheyne, & Smilek, 2008; Cheyne et al., 2006; Deng, Li, & Tang, 2012; Smallwood et al., 2004; Smallwood, O'Connor, et al., 2007) and it may be a correlate of negative mood in general (Killingsworth & Gilbert, 2010; Smallwood, Fitzgerald, Miles, & Phillips, 2009). However, studies have shown that the content of self-generated thought can also determine the consequences that the experience has for an individual's well-being. It has been shown that it is past-related self-generated thought that is the main correlate of dysphoria (Poerio, Totterdell, & Miles, 2013; Ruby, Smallwood, Engen, & Singer, 2013; Smallwood & O'Connor, 2011). Moreover, whereas past-related self-generated thought can lead to subsequent unhappiness, a focus on the future tends to lead to a more positive mood some minutes later (Ruby, Smallwood, Engen, & Singer, 2013). Finally, individuals who have good control of attention tend to limit the time they spend engaged in unproductive self-generated thoughts, producing greater future thoughts (Baird et al., 2011). Critically, this is only true for situations when task demands are low (McVay, Unsworth, McMillan, & Kane, 2013). In daily life, self-generated thoughts that are interesting are also linked with higher mood (Franklin, Mrazek, Anderson, et al., 2013). Altogether, this suggests that adaptive function might entail moderating the content of self-generated thoughts. Accordingly one implication of the content-regulation hypothesis is that individual differences in affective style may not predict higher levels of self-generated mental contents per se but rather would be predictive of specific forms of content (such as ruminating about the past).

Future Directions

States of self-generated thought such as daydreaming and mind wandering illustrate basic features of the human mind. Research has revealed our capacity to decouple attention from perceptual input, how common this state is, and the damage that this can do to ongoing action. It has also revealed the advantages that can be gained from attending to self-generated information, especially when the external conditions allow this to occur with minimal risk. Important questions for future studies of daydreaming or mind wandering to address are described in box 10.6.

Despite these advances, questions remain regarding the process through which self-generated thought is regulated and what role attention plays in this process. In particular, the mechanism through which perceptual decoupling occurs remains unclear. One possibility is that it is the result of an active process through which external input is suppressed. This process could be conceived of as analogous to the manner in which the spotlight of attention illuminates other forms of information (Kam, Dao, et al., 2013; Smallwood, 2013a, 2013c). Alternatively, it may be that a failure to attend to perceptual input is an epiphenomenal consequence of the occurrence of task-unrelated thinking (Franklin, Mrazek, Broadway, & Schooler, 2013).

The mechanism through which self-generated thought is regulated is also currently unknown. Studies have shown that individuals' who show a balance between on- and off-task thought are better at noticing lapses in an ongoing task when they occur (Allen et al., 2013). This suggests that meta-cognitive accuracy, and the reactive cognitive control that this allows, may be one way that we regulate our self-generated experiences. It is also possible that processes like proactive cognitive control (Braver, Paxton, Locke, & Barch, 2009) are important in regulating self-generated thought. Understanding the role of meta-cognition, as well as of the contributions made by proactive and reactive cognitive control, in the regulation of self-generated thought is an important avenue for future research.

Although the basic mechanisms that determine states such as mind wandering remain unknown, what we do know can still inform basic attention research. Many important real-world occupations depend on a skill set that includes both perceptual focus (such as sustained attention) and self-generated processes (such as creative problem solving). Under these conditions regulating the context and content of self-generated thought would likely be key in optimizing an individual's success. Attention training studies have examined how mental training (Allen et al., 2012; Slagter et al., 2007; Tang & Posner, 2009) or video-game play (Green & Bavelier, 2003, 2006, 2008; see also chapter 23 in the current volume) can improve attentional or perceptual focus, documenting robust improvements on dependent measures that illustrate external expertise. Yet the studies often lack measures that index less constrained self-generated processes. If both perceptually guided and self-generated thoughts are important elements of an optimal cognitive system, and moreover can often function in opposition, more accurate assessments of the value of the attention training regimes could be gained by including measures of unconstrained self-generated experiences, under both demanding and nondemanding conditions.

Box 10.5

Key Points

- States of daydreaming and mind wandering are active conscious states that depend on the capacity to *self-generate* mental contents. It may also reflect the independence of experience from the ongoing task.

- States of daydreaming and mind wandering have both costs and benefits. The costs and benefits depend on the following:

    1. The content of the experience, which has been shown to modulate the functional outcomes associated with mind wandering, and

    2. The context in which it arises, which determines whether the experience can undermine ongoing action.

Box 10.6

Outstanding Issues

- *What are the mechanisms that govern different aspects of the mind-wandering state such as its continuity, its content, and its occurrence?* This is important in developing a formal model of the component processes that mind wandering entails.

- *What are the challenges in formally integrating states of mind wandering and daydreaming into existing models of attention?* Traditionally, attention researchers have tended to focus on its applications to sensory input whereas states of daydreaming suggest that we can generate and attend to information that is not present in the here and now.

- *How can the costs and benefits associated with mind wandering be formalized?* This is important in understanding the role that this experience plays in our lives in an unbiased manner.

- *How can we maximize the beneficial aspects of our capacity for daydreaming while minimizing its cost?* The advantages to having a mind that can escape from the here and now are substantial, and it is important to bear this capacity in mind when exploring techniques that attempt to improve psychological functioning through attention training.

References

Allan Cheyne, J., Solman, G. J., Carriere, J. S., & Smilek, D. (2009). Anatomy of an error: A bidirectional state model of task engagement/disengagement and attention-related errors. *Cognition*, *111*, 98–113. doi:10.1016/j.cognition.2008.12.009.

Allen, M., Dietz, M., Blair, K. S., van Beek, M., Rees, G., Vestergaard-Poulsen, P., et al. (2012). Cognitive-affective neural plasticity following active-controlled mindfulness intervention [Randomized Controlled Trial Research Support, NIH, Extramural Research Support, Non-U.S. Government]. *Journal of Neuroscience*, *32*, 15601–15610. doi:10.1523/JNEUROSCI.2957-12.2012.

Allen, M., Smallwood, J., Christensen, J., Gramm, D., Rasmussen, B., Jensen, C. G., et al. (2013). The balanced mind: The variability of task-unrelated thoughts predicts error monitoring. *Frontiers in Human Neuroscience*, *7*, 743. doi:10.3389/fnhum.2013.00743.

Andrews-Hanna, J. R., Kaiser, R. H., Turner, A. E., Reineberg, A. E., Godinez, D., Dimidjian, S., & Banich, M. T. (2013). A penny for your thoughts: Dimensions of self-generated thought content and relationships with individual differences in emotional wellbeing. *Frontiers in Psychology*, *4*, 900.

Andrews-Hanna, J. R., Reidler, J. S., Huang, C., & Buckner, R. L. (2010). Evidence for the default network's role in spontaneous cognition [Research Support, NIH, Extramural Research Support, Non-U.S. Government]. *Journal of Neurophysiology*, *104*, 322–335. doi:10.1152/jn.00830.2009.

Antrobus, J. S. (1968). Information theory and stimulus-independent thought. *British Journal of Psychology*, *59*, 423–430.

Antrobus, J. S., Antrobus, J. S., & Singer, J. L. (1964). Eye movements accompanying daydreaming, visual imagery, and thought suppression. *Journal of Abnormal and Social Psychology*, *69*, 244–252.

Antrobus, J. S., Coleman, R., & Singer, J. L. (1967). Signal-detection performance by subjects differing in predisposition to daydreaming. *Journal of Consulting Psychology*, *31*, 487–491.

Antrobus, J. S., Singer, J. L., Goldstein, S., & Fortgang, M. (1970). Mindwandering and cognitive structure. *Transactions of the New York Academy of Sciences, 32*, 242–252.

Baird, B., Smallwood, J., Mrazek, M. D., Kam, J. W., Franklin, M. S., & Schooler, J. W. (2012). Inspired by distraction: Mind wandering facilitates creative incubation. *Psychological Science, 23*, 1117–1122. doi:10.1177/0956797612446024.

Baird, B., Smallwood, J., & Schooler, J. W. (2011). Back to the future: Autobiographical planning and the functionality of mind-wandering. *Consciousness and Cognition*. doi:10.1016/j.concog.2011.08.007.

Barron, E., Riby, L. M., Greer, J., & Smallwood, J. (2011). Absorbed in thought: The effect of mind wandering on the processing of relevant and irrelevant events. *Psychological Science, 22*, 596–601. doi:10.1177/0956797611404083.

Bernhardt, B., Smallwood, J., Tusche, A., Ruby, F. J. M., Engen, II. G., Steinbeis, N., & Singer, T. (2014). Medial prefrontal and anterior cingulate cortical thickness predicts shared individual differences in self-generated thought and temporal discounting. *NeuroImage, 90*, 290–297.

Braver, T. S., Paxton, J. L., Locke, H. S., & Barch, D. M. (2009). Flexible neural mechanisms of cognitive control within human prefrontal cortex [Research Support, NIH, Extramural]. *Proceedings of the National Academy of Sciences of the United States of America, 106*, 7351–7356. doi:10.1073/pnas.0808187106.

Callard, F., Smallwood, J., Golchert, J., & Margulies, D. S. (2013). The era of the wandering mind? Twenty-first century research on self-generated mental activity. *Frontiers in Psychology, 4*, 891.

Callard, F., Smallwood, J., & Margulies, D. S. (2012). Default positions: How neuroscience's historical legacy has hampered investigation of the resting mind. *Frontiers in Psychology, 3*, 321. doi:10.3389/fpsyg.2012.00321.

Carriere, J. S., Cheyne, J. A., & Smilek, D. (2008). Everyday attention lapses and memory failures: The affective consequences of mindlessness. *Consciousness and Cognition, 17*, 835–847. doi:10.1016/j.concog.2007.04.008.

Carriere, J. S., Seli, P., & Smilek, D. (2013). Wandering in both mind and body: Individual differences in mind wandering and inattention predict fidgeting. *Canadian Journal of Experimental Psychology–Revue Canadienne de Psychologie Experimentale, 67*(1), 19–31. doi:10.1037/a0031438.

Casner, S. M., & Schooler, J. W. (2013). Thoughts in flight: Automation use and pilots' task-related and task-unrelated thought. *Human Factors: The Journal of the Human Factors and Ergonomics Society*.

Cheyne, J. A., Carriere, J. S., & Smilek, D. (2006). Absent-mindedness: Lapses of conscious awareness and everyday cognitive failures [Randomized Controlled Trial Research Support, Non-U.S. Government]. *Consciousness and Cognition, 15*, 578–592. doi:10.1016/j.concog.2005.11.009.

Cheyne, J. A., Carriere, J. S., Solman, G. J., & Smilek, D. (2011). Challenge and error: Critical events and attention-related errors. *Cognition*, *121*, 437–446. doi:10.1016/j.cognition.2011 .07.010.

Chun, M. M., Golomb, J. D., & Turk-Browne, N. B. (2011). A taxonomy of external and internal attention. *Annual Review of Psychology*, *62*, 73–101.

Deng, Y.-Q., Li, S., & Tang, Y.-Y. (2012). The relationship between wandering mind, depression and mindfulness. *Mindfulness*, *5*, 124–128.

Desimone, R., & Duncan, J. (1995). Neural mechanisms of selective visual attention [Research Support, Non-U.S. Government Review]. *Annual Review of Neuroscience*, *18*, 193–222. doi:10.1146/ annurev.ne.18.030195.001205.

Dixon, P., & Bortolussi, M. (2013). Construction, integration, and mind wandering in reading. *Canadian Journal of Experimental Psychology–Revue Canadienne de Psychologie Experimentale*, *67*(1), 1–10. doi:10.1037/A0031234.

Farley, J., Risko, E. F., & Kingstone, A. (2013). Everyday attention and lecture retention: The effects of time, fidgeting, and mind wandering. *Frontiers in Psychology*, *4*, 619. doi:10.3389/ fpsyg.2013.00619.

Feng, S., D'Mello, S., & Graesser, A. C. (2013). Mind wandering while reading easy and difficult texts [Research Support, U.S. Government, Non-PHS]. *Psychonomic Bulletin & Review*, *20*, 586–592. doi:10.3758/s13423-012-0367-y.

Forster, S., & Lavie, N. (2009). Harnessing the wandering mind: The role of perceptual load [Research Support, Non-U.S. Government]. *Cognition*, *111*, 345–355. doi:10.1016/j.cognition .2009.02.006.

Franklin, M. S., Mrazek, M. D., Anderson, C. L., Smallwood, J., Kingstone, A., & Schooler, J. W. (2013). The silver lining of a mind in the clouds: Interesting musings are associated with positive mood while mind-wandering. *Frontiers in Psychology*, *4*, 583. doi:10.3389/fpsyg.2013.00583.

Franklin, M. S., Mrazek, M. D., Broadway, J. M., & Schooler, J. W. (2013). Disentangling decoupling: Comment on Smallwood (2013) [Comment Research Support, NIH, Extramural]. *Psychological Bulletin*, *139*, 536–541. doi:10.1037/a0030515.

Franklin, M. S., Smallwood, J., & Schooler, J. W. (2011). Catching the mind in flight: Using behavioral indices to detect mindless reading in real time. *Psychonomic Bulletin & Review*. doi:10.3758/s13423-011-0109-6.

Frick, J. W., Guilford, J. P., Christensen, P. R., & Merrifield, P. R. (1959). A factor-analytic study of flexibility in thinking. *Educational and Psychological Measurement*, *19*, 469–496. doi:10.1177/ 001316445901900401.

Frith, C. D. (1992). *The cognitive neuropsychology of schizophrenia*. Hove, UK: Psychology Press.

Frith, C. D., & Done, D. J. (1988). Towards a neuropsychology of schizophrenia. *British Journal of Psychiatry*, *153*, 437–443.

Galera, C., Orriols, L., M'Bailara, K., Laborey, M., Contrand, B., Ribereau-Gayon, R., et al. (2012). Mind wandering and driving: Responsibility case-control study [Research Support, Non-U.S. Government]. *BMJ (Clinical Research Ed.)*, *345*, e8105. doi:10.1136/bmj.e8105.

Giambra, L. M. (1989). Task-unrelated thought frequency as a function of age: A laboratory study. *Psychology and Aging*, *4*, 136–143.

Giambra, L. M. (1995). A laboratory method for investigating influences on switching attention to task-unrelated imagery and thought. *Consciousness and Cognition*, *4*, 1–21.

Green, C. S., & Bavelier, D. (2003). Action video game modifies visual selective attention. *Nature*, *423*, 534–537. doi:10.1038/Nature01647.

Green, C. S., & Bavelier, D. (2006). Effect of action video games on the spatial distribution of visuospatial attention. *Journal of Experimental Psychology: Human Perception and Performance*, *32*, 1465–1478. doi:10.1037/0096-1523.32.6.1465.

Green, C. S., & Bavelier, D. (2008). Exercising your brain: A review of human brain plasticity and training-induced learning. *Psychology and Aging*, *23*, 692–701. doi:10.1037/A0014345.

Guilford, J. P. (1959). 3 faces of intellect. *American Psychologist*, *14*, 469–479. doi:10.1037/H0046827.

Jackson, J. D., & Balota, D. A. (2012). Mind-wandering in younger and older adults: Converging evidence from the Sustained Attention to Response Task and reading for comprehension [Research Support, NIH, Extramural]. *Psychology and Aging*, *27*, 106–119. doi:10.1037/a0023933.

James, W. (1890). *The principles of psychology*. New York: Holt.

Jones, S. R., & Fernyhough, C. (2007). Thought as action: Inner speech, self-monitoring, and auditory verbal hallucinations. *Consciousness and Cognition*, *16*, 391–399.

Kam, J. W., Dao, E., Farley, J., Fitzpatrick, K., Smallwood, J., Schooler, J. W., et al. (2011). Slow fluctuations in attentional control of sensory cortex. *Journal of Cognitive Neuroscience*, *23*, 460–470. doi:10.1162/jocn.2010.21443.

Kam, J. W., Dao, E., Stanciulescu, M., Tildesley, H., & Handy, T. C. (2013). Mind wandering and the adaptive control of attentional resources. *Journal of Cognitive Neuroscience*, *25*, 952–960. doi:10.1162/jocn_a_00375.

Kam, J. W., & Handy, T. C. (2013). The neurocognitive consequences of the wandering mind: A mechanistic account of sensory–motor decoupling [Review]. *Frontiers in Psychology*, *4*, 725. doi:10.3389/fpsyg.2013.00725.

Kam, J. W., Xu, J., & Handy, T. C. (2014). I don't feel your pain (as much): The desensitizing effect of mind wandering on the perception of others' discomfort. *Cognitive, Affective & Behavioral Neuroscience*, *14*, 286–296. doi:10.3758/s13415-013-0197-z.

Kane, M. J., Brown, L. H., McVay, J. C., Silvia, P. J., Myin-Germeys, I., & Kwapil, T. R. (2007). For whom the mind wanders, and when: An experience-sampling study of working memory and executive control in daily life. *Psychological Science, 18,* 614–621. doi:10.1111/j.1467-9280 .2007.01948.x.

Killingsworth, M. A., & Gilbert, D. T. (2010). A wandering mind is an unhappy mind. *Science, 330,* 932. doi:10.1126/science.1192439.

Klinger, E. (1966). Fantasy need achievement as a motivational construct [Review]. *Psychological Bulletin, 66,* 291–308.

Klinger, E. (1973). Models, context, and achievement fantasy: Parametric studies and theoretical propositions. *Journal of Personality Assessment, 37,* 25–47. doi:10.1080/00223891.1973. 10119826.

Klinger, E. (1978). Modes of normal conscious flow. In K. S. Pope & J. L. Singer (Eds.), *The stream of consciousness: Scientific investigations into the flow of human experience* (pp. 225–258). Boston: Springer.

Klinger, E. (2009). Daydreaming and fantasizing: Thought flow and motivation. In K. D. Markman, W. M. P. Klein, & J. A. Suhr (Eds.), *Handbook of imagination and mental simulation* (pp. 225–239). New York: Psychology Press.

Klinger, E. (2013). Goal commitments and the content of thoughts and dreams: Basic principles. *Frontiers in Psychology, 4,* 415. doi:10.3389/fpsyg.2013.00415.

Klinger, E., & McNelly, F. W., Jr. (1969). Fantasy need achievement and performance: A role analysis. *Psychological Review, 76,* 574–591.

Levinson, D. B., Smallwood, J., & Davidson, R. J. (2012). The persistence of thought: Evidence for a role of working memory in the maintenance of task-unrelated thinking. *Psychological Science, 23,* 375–380. doi:10.1177/0956797611431465.

Macdonald, J. S., Mathan, S., & Yeung, N. (2011). Trial-by-trial variations in subjective attentional state are reflected in ongoing prestimulus EEG alpha oscillations. *Frontiers in Psychology, 2,* 82. doi:10.3389/fpsyg.2011.00082.

Mason, M. F., Norton, M. I., Van Horn, J. D., Wegner, D. M., Grafton, S. T., & Macrae, C. N. (2007). Wandering minds: The default network and stimulus-independent thought [Research Support, NIH, Extramural Research Support, Non-U.S. Government]. *Science, 315,* 393–395. doi:10.1126/science.1131295.

McVay, J. C., & Kane, M. J. (2009). Conducting the train of thought: Working memory capacity, goal neglect, and mind wandering in an executive-control task [Research Support, NIH, Extramural Research Support, Non-U.S. Government]. *Journal of Experimental Psychology: Learning, Memory, and Cognition, 35,* 196–204. doi:10.1037/a0014104.

McVay, J. C., & Kane, M. J. (2011). Why does working memory capacity predict variation in reading comprehension? On the influence of mind wandering and executive attention. *Journal of Experimental Psychology. General.* doi:10.1037/a0025250.

McVay, J. C., Unsworth, N., McMillan, B. D., & Kane, M. J. (2013). Working memory capacity does not always support future-oriented mind-wandering. *Canadian Journal of Experimental Psychology–Revue Canadienne de Psychologie Expérimentale, 67*(1), 41–50.

Mednick, M. T., Mednick, S. A., & Mednick, E. V. (1964). Incubation of creative performance and specific associative priming. *Journal of Abnormal Psychology, 69*, 84–88.

Mischel, W., & Gilligan, C. (1964). Delay of gratification, motivation for the prohibited gratification, and responses to temptation. *Journal of Abnormal Psychology, 69*, 411–417.

Mooneyham, B. W., & Schooler, J. W. (2013). The costs and benefits of mind-wandering: A review. *Canadian Journal of Experimental Psychology–Revue de Psychologie Experimentale, 67*(1), 11–18. doi:10.1037/A0031569.

Mrazek, M. D., Smallwood, J., Franklin, M. S., Chin, J. M., Baird, B., & Schooler, J. W. (2012). The role of mind-wandering in measurements of general aptitude. *Journal of Experimental Psychology. General, 141*, 788–798. doi:10.1037/a0027968.

Mrazek, M. D., Smallwood, J., & Schooler, J. W. (2012). Mindfulness and mind-wandering: Finding convergence through opposing constructs. *Emotion (Washington, D.C.), 12*, 442–448. doi:10.1037/a0026678.

Poerio, G. L., Totterdell, P., & Miles, E. (2013). Mind-wandering and negative mood: Does one thing really lead to another? *Consciousness and Cognition, 22*, 1412–1421.

Polich, J. (1986). Attention, probability, and task demands as determinants of P300 latency from auditory stimuli [Research Support, Non-U.S. Government Research Support, U.S. Government, PHS]. *Electroencephalography and Clinical Neurophysiology, 63*, 251–259.

Polich, J., & Criado, J. R. (2006). Neuropsychology and neuropharmacology of P3a and P3b [Research Support, NIH, Extramural Research Support, Non-U.S. Government Review]. *International Journal of Psychophysiology, 60*, 172–185. doi:10.1016/j.ijpsycho.2005.12.012.

Reichle, E. D., Reineberg, A. E., & Schooler, J. W. (2010). Eye movements during mindless reading [Research Support, Non-U.S. Government]. *Psychological Science, 21*, 1300–1310. doi:10.1177/0956797610378686.

Ruby, F. J. M., Smallwood, J., Engen, H., & Singer, T. (2013). *How self-generated thought shapes mood. PLoS ONE, 8*(10), e77554.

Ruby, F. J. M., Smallwood, J., Sackur, J., & Singer, T. (2013). *Is self-generated thought a means of social problem solving? Frontiers in Psychology, 4*, 962.

Schad, D. J., Nuthmann, A., & Engbert, R. (2012). Your mind wanders weakly, your mind wanders deeply: Objective measures reveal mindless reading at different levels [Research Support, Non-U.S. Government]. *Cognition, 125*, 179–194. doi:10.1016/j.cognition.2012.07.004.

Schooler, J. W., Reichle, E. D., & Halpern, D. V. (2004). Zoning out while reading: Evidence for dissociations between experience and metaconsciousness. In D. T. Levin (Ed.), *Thinking and seeing: Visual metacognition in adults and children* (pp. 203–226). Cambridge, MA: MIT Press.

Schooler, J. W., Smallwood, J., Christoff, K., Handy, T. C., Reichle, E. D., & Sayette, M. A. (2011). Meta-awareness, perceptual decoupling and the wandering mind [Research Support, Non-U.S. Government Review]. *Trends in Cognitive Sciences*, *15*, 319–326. doi:10.1016/j.tics.2011.05.006.

Seli, P., Carriere, J. S., Thomson, D. R., Cheyne, J. A., Martens, K. A., & Smilek, D. (2013). Restless mind, restless body. *Journal of Experimental Psychology: Learning, Memory, and Cognition*. doi:10.1037/a0035260.

Shamosh, N. A., & Gray, J. R. (2008). Delay discounting and intelligence: A meta-analysis. *Intelligence*, *36*, 289–305.

Silva, F. J., & Gross, T. F. (2004). The rich get richer: Students' discounting of hypothetical delayed rewards and real effortful extra credit. *Psychonomic Bulletin & Review*, *11*, 1124–1128.

Singer, J. L. (1975a). *The inner world of daydreaming*. London: Harper & Row.

Singer, J. L. (1975b). Navigating the stream of consciousness: Research in daydreaming and related inner experience. *American Psychologist*, *30*, 727–738.

Singer, J. L., & Antrobus, J. S. (1966). *Imaginal Process Inventory*. New York: Center for Research in Cognition and Affect, Graduate Center, City University of New York.

Slagter, H. A., Lutz, A., Greischar, L. L., Francis, A. D., Nieuwenhuis, S., Davis, J. M., et al. (2007). Mental training affects distribution of limited brain resources [Research Support, NIH, Extramural Research Support, Non-U.S. Government]. *PLoS Biology*, *5*(6), e138. doi:10.1371/journal.pbio.0050138.

Smallwood, J. (2011). Mind-wandering while reading: Attentional decoupling, mindless reading and the cascade model of inattention. *Language and Linguistics Compass*, *5*, 63–77. doi:10.1111/j.1749-818X.2010.00263.x.

Smallwood, J. (2013a). Distinguishing how from why the mind wanders: A process–occurrence framework for self-generated mental activity. *Psychological Bulletin*, *139*, 519–535.

Smallwood, J. (2013b). Penetrating the fog of the decoupled mind: The effects of visual salience in the sustained attention to response task. *Canadian Journal of Experimental Psychology–Revue Canadienne de Psychologie Experimentale*, *67*(1), 32–40. doi:10.1037/A0030760.

Smallwood, J. (2013c). Searching for the elements of thought: Reply to Franklin, Mrazek, Broadway, and Schooler (2013). *Psychological Bulletin*, *139*, 542–547. doi:10.1037/A0031019.

Smallwood, J., & Andrews-Hanna, J. (2013). Not all minds that wander are lost: The importance of a balanced perspective on the mind-wandering state. *Frontiers in Psychology*, *4*, 441. doi:10.3389/fpsyg.2013.00441.

Smallwood, J., Beach, E., Schooler, J. W., & Handy, T. C. (2008). Going AWOL in the brain: Mind wandering reduces cortical analysis of external events [Research Support, NIH, Extramural Research Support, Non-U.S. Government Research Support, U.S. Government, Non-PHS]. *Journal of Cognitive Neuroscience*, *20*, 458–469. doi:10.1162/jocn.2008.20037.

Smallwood, J., Brown, K. S., Tipper, C., Giesbrecht, B., Franklin, M. S., Mrazek, M. D., et al. (2011). Pupillometric evidence for the decoupling of attention from perceptual input during offline thought [Research Support, Non-U.S. Government]. *PLoS ONE, 6*(3), e18298. doi:10.1371/journal.pone.0018298.

Smallwood, J., Davies, J. B., Heim, D., Finnigan, F., Sudberry, M., O'Connor, R., et al. (2004). Subjective experience and the attentional lapse: Task engagement and disengagement during sustained attention. *Consciousness and Cognition, 13,* 657–690. doi:10.1016/j.concog.2004.06.003.

Smallwood, J., Fishman, D. J., & Schooler, J. W. (2007). Counting the cost of an absent mind: Mind wandering as an underrecognized influence on educational performance [Research Support, U.S. Government, Non-PHS]. *Psychonomic Bulletin & Review, 14,* 230–236.

Smallwood, J., Fitzgerald, A., Miles, L. K., & Phillips, L. H. (2009). Shifting moods, wandering minds: Negative moods lead the mind to wander [Controlled Clinical Trial]. *Emotion (Washington, D.C.), 9,* 271–276. doi:10.1037/a0014855.

Smallwood, J., Gorgolewski, K. J., Golchert, J., Ruby, F. J., Engen, H., Baird, B., et al. (2013). The default modes of reading: Modulation of posterior cingulate and medial prefrontal cortex connectivity associated with comprehension and task focus while reading. *Frontiers in Human Neuroscience, 7,* 734. doi:10.3389/fnhum.2013.00734.

Smallwood, J., McSpadden, M., Luus, B., & Schooler, J. (2008). Segmenting the stream of consciousness: The psychological correlates of temporal structures in the time series data of a continuous performance task. *Brain and Cognition, 66,* 50–56. doi:10.1016/j.bandc.2007.05.004.

Smallwood, J., McSpadden, M., & Schooler, J. W. (2008). When attention matters: The curious incident of the wandering mind [Research Support, U.S. Government, Non-PHS]. *Memory & Cognition, 36,* 1144–1150. doi:10.3758/MC.36.6.1144.

Smallwood, J., Obonsawin, M., & Heim, D. (2003). Task unrelated thought: The role of distributed processing. *Consciousness and Cognition, 12,* 169–189.

Smallwood, J., Nind, L., & O'Connor, R. C. (2009). When is your head at? An exploration of the factors associated with the temporal focus of the wandering mind [Research Support, Non-U.S. Government]. *Consciousness and Cognition, 18,* 118–125. doi:10.1016/j.concog.2008.11.004.

Smallwood, J., & O'Connor, R. C. (2011). Imprisoned by the past: Unhappy moods lead to a retrospective bias to mind wandering. *Cognition and Emotion, 25,* 1481–1490. doi:10.1080/02699931.2010.545263.

Smallwood, J., O'Connor, R. C., Sudbery, M. V., & Obonsawin, M. (2007). Mind-wandering and dysphoria. *Cognition and Emotion, 21,* 816–842. doi:10.1080/02699930600911531.

Smallwood, J., Ruby, F. J., & Singer, T. (2013). Letting go of the present: Mind-wandering is associated with reduced delay discounting. *Consciousness and Cognition, 22,* 1–7. doi:10.1016/j.concog.2012.10.007.

Smallwood, J., & Schooler, J. W. (2006). The restless mind [Research Support, U.S. Government, Non-PHS Review]. *Psychological Bulletin, 132*, 946–958. doi:10.1037/0033-2909.132.6.946.

Smallwood, J., Schooler, J. W., Turk, D. J., Cunningham, S. J., Burns, P., & Macrae, C. N. (2011). Self-reflection and the temporal focus of the wandering mind. *Consciousness and Cognition, 20*, 1120–1126. doi:10.1016/j.concog.2010.12.017.

Smallwood, J. M., Baracaia, S. F., Lowe, M., & Obonsawin, M. (2003). Task unrelated thought whilst encoding information. *Consciousness and Cognition, 12*, 452–484.

Smilek, D., Carriere, J. S., & Cheyne, J. A. (2010a). Failures of sustained attention in life, lab, and brain: Ecological validity of the SART [Randomized Controlled Trial Research Support, Non-U.S. Government]. *Neuropsychologia, 48*, 2564–2570. doi:10.1016/j.neuropsychologia.2010.05.002.

Smilek, D., Carriere, J. S., & Cheyne, J. A. (2010b). Out of mind, out of sight: Eye blinking as indicator and embodiment of mind wandering [Research Support, Non-U.S. Government]. *Psychological Science, 21*, 786–789. doi:10.1177/0956797610368063.

Song, X., & Wang, X. (2012). Mind wandering in Chinese daily lives—An experience sampling study [Research Support, Non-U.S. Government]. *PLoS ONE, 7*(9), e44423. doi:10.1371/journal.pone.0044423.

Stawarczyk, D., Majerus, S., Maj, M., Van der Linden, M., & D'Argembeau, A. (2011). Mind-wandering: Phenomenology and function as assessed with a novel experience sampling method. *Acta Psychologica, 136*, 370–381. doi:10.1016/j.actpsy.2011.01.002.

Szpunar, K. K., Khan, N. Y., & Schacter, D. L. (2013). Interpolated memory tests reduce mind wandering and improve learning of online lectures. *Proceedings of the National Academy of Sciences of the United States of America, 110*, 6313–6317. doi:10.1073/pnas.1221764110.

Tang, Y. Y., & Posner, M. I. (2009). Attention training and attention state training [Comparative Study Research Support, Non-U.S. Government Review]. *Trends in Cognitive Sciences, 13*, 222–227. doi:10.1016/j.tics.2009.01.009.

Teasdale, J. D., Dritschel, B. H., Taylor, M. J., Proctor, L., Lloyd, C. A., Nimmo-Smith, I., et al. (1995). Stimulus-independent thought depends on central executive resources [Clinical Trial Randomized Controlled Trial]. *Memory & Cognition, 23*, 551–559.

Treisman, A. M., & Gelade, G. (1980). A feature-integration theory of attention. *Cognitive Psychology, 12*, 97–136.

Unsworth, N., & McMillan, B. D. (2013). Mind wandering and reading comprehension: Examining the roles of working memory capacity, interest, motivation, and topic experience. *Journal of Experimental Psychology: Learning, Memory, and Cognition, 39*, 832–842.

Weller, R. E., Cook, E. W., Avsar, K. B., & Cox, J. E. (2008). Obese women show greater delay discounting than healthy-weight women. *Appetite, 51*, 563–569.

Wickens, C. D., Kramer, A. F., & Donchin, E. (1984). The event-related potential as an index of the processing demands of a complex target acquisition task [Research Support, U.S. Government, Non-PHS]. *Annals of the New York Academy of Sciences, 425*, 295–299.

Yoon, J. H., Higgins, S. T., Heil, S. H., Sugarbaker, R. J., Thomas, C. S., & Badger, G. J. (2007). Delay discounting predicts postpartum relapse to cigarette smoking among pregnant women. *Experimental and Clinical Psychopharmacology, 15*, 176–186.

Zwaan, R. A., & Radvansky, G. A. (1998). Situation models in language comprehension and memory [Review]. *Psychological Bulletin, 123*, 162–185.

11

Scene Perception

Tom Foulsham

For most of the time that our eyes are open, the information occupying the visual field comprises a scene. In this chapter, I will discuss what we know about attention in scene perception. A scene can be defined as a pictorial view of an environment where some-one might act. Scenes consist of multiple items arranged in a regular and meaningful way (e.g., the sky above the ground, a chair next to a table).

There are several reasons why studying scene perception is essential if we are to understand natural human visual attention. First, scenes are ubiquitous. It is rare, out-side of the cognitive science laboratory, for us to look at isolated features or objects.

Second, natural scenes are highly complex. Our environment is often crowded with visual information. How do we act efficiently in such a complex environment? One of the answers to this question is selective attention itself: We attend to particular items and thus focus our limited resources on what is important. Much of this chapter will therefore describe how overt attention is guided toward particular features in a scene.

Third, natural scenes provide a huge amount of structured, semantic information for the observer. Certain objects tend to occur together and in a particular spatial posi-tion. The "gist" of a scene brings with it a range of expectations. Scenes provide a con-text, and visual attention has evolved to make the most of this context.

This brings us to a potential paradox. Attending to the most important objects in a scene is crucial, and yet scene complexity and the limits of peripheral vision mean that we cannot fully identify these objects until we are looking in the right place. Scene con-text seems to aid attention and recognition, and yet one would assume that acquiring this context must in itself entail the perception of defining objects. A partial resolution to this paradox is that scene perception happens more quickly, and with distinct neural processing, than if it were just a case of recognizing a series of objects. Our capacity for perceiving scenes is far greater than would be expected from laboratory studies with simple shapes. As described in the title of Braun (2003), "Natural scenes upset the visual applecart."

I will begin my review by considering research into the rapid perception of gist and some early studies which investigated eye fixations in scenes.

Historical Context

Scene Perception in a Single Glance

Potter and Levy (1969) pioneered the use of natural scenes in cognitive science by using a rapid serial visual presentation (RSVP) paradigm. Surprisingly, participants are able to match target scenes to labels such as "picnic" with above-chance accuracy, even at presentation times as brief as 125 ms. While memory for items in an RSVP is short-lived, observers can recognize scene gist in less than the duration of a single fixation (Potter, Staub, & O'Connor, 2004; Biederman et al., 1974). Scene perception also seems to be rather impervious to manipulations which draw attention elsewhere (Li et al., 2002; but see Cohen, Alvarez, & Nakayama, 2011).

Characterizing the information available in a single glance continues to be of interest for researchers. In particular, Oliva (2005) distinguishes between "conceptual" and "perceptual" gist, with the former referring to the verbal labels used in RSVP experiments. In contrast, perceptual gist permits a general description of the context and structure of the scene. Oliva and colleagues' experiments suggest that this description is based on coarse spatial scale information, and it is feasible that this is available early via the magnocellular pathway. Importantly, this implies that the context and structure of an image can be perceived in a scene-centered fashion, independently of the processes of segmentation, grouping, and object recognition. This context can then be used to support object recognition (Bar, 2004). It is likely that observers also use the context to guide their attention and fixations to the most relevant locations, and I shall now turn to early research using eye tracking which showed just that.

Early Descriptions of Fixation Placement in Scenes

Several classic studies on eye movements in scene perception were carried out before 1970, and these continue to be cited and to inspire more contemporary research.

Following pioneering work on eye movements in reading, Guy Thomas Buswell (1935) published a monograph in which he described the viewing patterns of many different observers looking at pictures of complex scenes and buildings. He found that fixations were partly consistent across participants and concentrated on areas of detail, objects, and people rather than on the background. Buswell also made novel observations regarding the time course of viewing, noting that consistency between viewers decreased over time and fixation durations lengthened.

Buswell (1935) also made the important observation that eye-movement patterns change when a participant views an image for a different purpose (see Wade & Tatler, 2005). When an observer was asked whether there was anyone looking out of the windows in a picture of a building, their fixations covered the regions relevant for this task (i.e., those around the windows). However, it is Al'fred Luk'yanovich Yarbus whose work has become most well-known for revealing the task-specific nature of eye movements. Yarbus (1967) recorded eye movements during scene viewing, and he showed that gaze patterns varied when participants were given different questions to answer about the picture. Thus, Yarbus emphasized that where people look will be different depending on their task. This remains an important constraint for modern researchers. Yarbus's work has had a huge impact on eye tracking and scene perception in general, and his specific findings have been discussed more recently by DeAngelus and Pelz (2009), Greene, Liu, and Wolfe (2012) and Borji and Itti (2014).

Yarbus's suggestion that repeated viewings by the same observer produce similar fixation sequences may have also inspired one of the first computational attempts to link eye movements to perception, described by Noton and Stark (1971). Noton and Stark emphasized the notion of a "scanpath": a sequence of fixations that is particular to a certain viewer and image. Their theory proposed that the eye-movement sequence was an integral part of the encoding and recognition of visual information. Although this claim is not well supported, it can be seen as foreshadowing more modern "embodied" approaches to perception.

Published in the same year as the translation of Yarbus, Mackworth and Morandi (1967) tackled the problem of how to quantify the features in an image. To do so, they compared viewer fixations on each of 64 regions with explicit "informativeness" ratings made by separate participants. This analysis went beyond the qualitative descriptions of Buswell and Yarbus and confirmed that most of the fixations were on regions rated as highly informative. Mackworth and Morandi noted that choosing whether to fixate a region happened before the detailed inspection enabled by the fovea and thus "peripheral vision edited out the redundant stimuli in the pictures" (p. 549).

Loftus and Mackworth (1978), meanwhile, reasoned that an object that is semantically inconsistent with the rest of the scene is more informative and should attract attention. The researchers embedded line drawings of various objects in scenes where they were either consistent (a tractor in a farmyard) or inconsistent (an octopus in the same scene). Participants were indeed more likely to move their eyes to an out-of-place object than to an object that was consistent. The context of the scene was able to influence eye guidance and prompt large saccades straight to the incongruous area. We will return to this finding in the next section.

State-of-the-Art Review: Attention in Scene Perception

When we look at a picture, we make a series of saccades that direct the high resolution of the fovea toward different parts of the environment. Knowledge of the scene is acquired from these regions over a series of discrete fixations (see figure 11.1). Thus, investigations of attentional selection in scenes tend to use eye movements as an index of attention, consistent with the approach known as "active vision" (Findlay & Gilchrist, 2003). Eye movements are a bottleneck, in that they are executed serially,

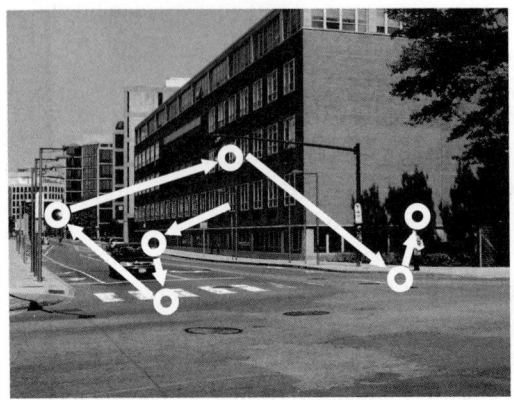

Figure 11.1
Observers make a series of fixations while viewing a natural scene (left panel, circles interspersed with saccades). The fixations of many viewers tend to cluster on interesting or informative locations (right panel).

but in the real world they allow us to maximize the useful signal in an environment that contains much more information than it would be possible to process. The active scanning of natural scenes can therefore tell us a great deal about the guidance of attention.

Fixation and Semantic Inconsistency

Following Loftus and Mackworth (1978), several researchers have investigated attention to incongruous objects. A key question in such studies was whether incongruous objects were indeed fixated *earlier* than congruous objects. Experiments with line drawings by De Graef et al. (1990) and Henderson, Weeks, and Hollingworth (1999) both failed to find an early advantage for incongruous objects. A more robust finding was that oddball objects were fixated more often and for longer.

More recently, researchers have examined these effects in photographs and rendered scenes. Võ and Henderson (2009) found longer gaze durations on both semantically inconsistent objects (e.g., a printer in a kitchen) and "syntactically" inconsistent objects (e.g., a printer floating in the air). However, there was no tendency to fixate these objects earlier. Although Bonitz and Gordon (2008) and Becker et al. (2007) did find such an advantage, this tended to occur later in viewing.

When interpreting these mixed results, it is worth outlining the steps involved in early fixation of an out-of-place object. First, one must understand the context of the scene, and this probably occurs within the first fixation. Second, one must be able to identify an object, or at least its congruency, in extrafoveal vision. This will depend on the size of the object, but even large and conspicuous objects may require foveal inspection. Third, one must decide to move the eyes to this object. This will depend on whether fixating such an object is useful given the current task. Considering these requirements, it is perhaps unsurprising that different stimuli, tasks, and time frames have led to different results. Incongruous objects may also be more visually salient by virtue of having distinctive visual features. When Geoffrey Underwood and I examined the semantic consistency effect, we controlled for the visual saliency of the key objects. When objects were among the most visually salient in the scene, their congruence did not affect how early they were fixated. It was only objects which were not visually conspicuous that were fixated earlier if they were also semantically inconsistent (Underwood & Foulsham, 2006). Moreover, there were no effects of semantic consistency when participants searched for an unrelated target in the same scenes. This effect of task reinforces the finding of Henderson et al. (1999) that when one is searching for target objects, it is *congruent* items that are detected and fixated more quickly. The same effect is reported by Võ and Henderson (2011), who limited extrafoveal processing via a preview of the scene and the use of a gaze-contingent window. This study appears to rule out guidance to semantic inconsistencies on the first glance.

Computational Modeling of Where We Look

There has been an increase in the number of researchers investigating scene viewing, due largely to the development and popularity of computational models which can generate predictions about attention in complex images (see Borji & Itti, 2013, for a recent review of publications spanning cognitive neuroscience and computer vision). I will briefly describe the most influential model, developed by Itti and Koch (2000), before considering how such models have been evaluated in experiments with human observers.

One plausible suggestion for predicting where we attend is that there might be rules for prioritizing regions based only on the features in the image. In particular, it might be best to orient toward regions which stand out from their surroundings. A desirable way of keeping track of this across different features is to have a spatial representation of attention priority: a "master map," an "activation map" or a "saliency map" (Koch & Ullman, 1985). It had proven difficult to implement this idea with complex stimuli, but Itti and Koch (2000) did just that with their saliency map model. The model begins by extracting features (intensity, color, and orientation at different scales) "preattentively" from across the image. These are combined in a center–surround fashion, giving a series of feature maps (representing local contrast at each location), which are normalized and summed into a single saliency map. Thus, this basic saliency map identifies those regions which stand out, and it does so without any attempt to detect objects or meaning. It is a purely bottom-up model. By applying "winner-take-all" processing to the saliency map, the Itti and Koch model predicts that attention will move to the most salient point, and then (following some inhibition) the next most salient and so on. Therefore, it produces explicit predictions about the sequence of locations where people will attend.

There is good evidence that fixated locations have higher luminance contrast and greater edge density than nonfixated locations (e.g., Tatler et al., 2005). As seen in figure 11.1, some parts of an image are more likely to receive inspection than others, and at its simplest the bottom-up approach involves determining what these areas have in common, without recourse to higher cognition. Given limited peripheral resolution, it is plausible that low-level features might resolve the task of directing the eyes, leaving resources free to deal with recognizing and understanding what is at fixation.

Despite this logic, it was clear from the outset that saliency (feature contrast)[1] was unable to predict the variations in attention with task that were observed by Buswell (1935) and Yarbus (1967). While a bottom-up approach might describe the features common to regions that are inspected, it would probably not account for *why* they are fixated. Many have noted that merely correlating features and fixation cannot determine what causes attention to move to this location (Tatler et al., 2005; Henderson et al., 2007). The original saliency model also made the unrealistic simplification that features could be extracted equally well in peripheral vision, which ignores the decreasing resolution present in the visual system. These shortcomings were acknowledged by Itti and Koch (2000).

Testing the Effects of Bottom-Up Saliency

Several authors have confirmed that high visual saliency is *associated with* where people look (e.g., Peters et al., 2005). In Foulsham and Underwood (2008), we compared fixations from a scene memory task with predictions from the Itti and Koch (2000) model in three ways. First, we calculated that around 20% of fixations were on the five most salient regions in the image (greater than the 10% expected according to chance). Second, we calculated the saliency map activation at each fixated point, and again this was greater than chance. Finally, we compared the "scanpath" sequence to the ordered locations predicted by the saliency map model. This comparison showed that the scanpaths were not highly similar to the saliency predictions, with fewer than 1 in 10 fixations occurring on the same point in the order predicted. Thus, the relationship between saliency and fixation was modest.

Compared to nonfixated regions, fixated regions are both more visually salient and more semantically informative (Henderson et al., 2007; Mackworth & Morandi, 1967). Saliency and semantic informativeness are confounded, and correlational designs cannot determine what is causing attention to move to these regions. Thus, one needs to manipulate or control both saliency and relevance separately. This has been accomplished in several different ways. In Foulsham and Underwood (2007), we manipulated the saliency of key objects (pieces of fruit)—for example, by placing them on a contrasting background. During a memorization task, objects that were more salient were attended more often and earlier. In this case, because all the objects were (equally relevant) pieces of fruit, we can be more confident that it is saliency which is attracting attention. An alternative approach is to modify the image, as Nyström and Holmqvist (2008) did by reducing the contrast of key regions of a scene. For meaningful regions (e.g., a face) this contrast reduction had little effect on viewing patterns. Even when participants do look at regions of high luminance contrast, this bias is immediately reversed when the task requires inspection of a different region (Einhäuser, Rutishauser, & Koch, 2008). Moreover, attention is clearly dominated by the goals of the task when we are looking for something that is not salient. We will now turn to those who have investigated search within scenes.

Visual Search in Scenes

How do people find what they are looking for in their complex environment? It is only relatively recently that search has been studied in natural scenes, where there is no clear separation between items and targets, and where distractors are complex and semantically organized (see also chapter 2 in the current volume). The ease with which we can find nonsalient targets means that top-down control must be the major factor in search within scenes. In Underwood and Foulsham (2006) and Foulsham and Underwood (2007), we demonstrated that visual saliency did not significantly affect fixations to objects during search, both when they were targets and when they were not. Henderson et al. (2009) reported a similar finding, arguing that guidance by feature contrast alone is like "searching in the dark" because salient signals are largely irrelevant.

Attention must be guided, therefore, and this guidance can be divided into two sources. First, a searcher has various degrees of information about the *appearance* of the target. If searching for your keys, you probably know their color and shape. If you are asked to find the kettle in someone's kitchen, you might not be sure of its precise color, but you have some knowledge about its likely shape. Consistent with this, participants are faster to locate specific objects ("apple") than objects within a general category ("fruit"; Foulsham & Underwood, 2007), and their fixations are guided more efficiently when shown a picture of the target (compared to a verbal label; Malcolm & Henderson, 2009).

Second, we are guided by our knowledge of object *location* (something which is often underestimated in studies with simpler stimuli). When searching for the kettle in the previous example, you would expect it to be in certain locations (on the counter rather than on the floor or floating in the air). Target objects are found more slowly when they are inconsistent with the setting, presumably because inconsistent items are not strongly associated with a likely position (Henderson et al., 1999). The scene gist can provide information about the layout and likely locations, and as already described this information is available on the very first fixation. Accordingly, the first saccades during search are often toward expected locations (e.g., a roof when searching for a chimney; Eckstein et al., 2006). When the layout is disrupted by dividing it into squares and rearranging them, search is slower and more influenced by the target's saliency (Biederman et al., 1973; Foulsham, Alan, & Kingstone, 2011). Importantly, if guidance were only based on local appearance ("Which part looks like it may contain the target?"), then search within jumbled scenes should be similar to search within normal scenes. It is not, and therefore guidance by expected location must underlie naturalistic search.

We normally have access to the features at an attended location while at the same time we perceive the layout via peripheral vision. A useful technique for controlling these sources of information is the "flash-preview moving window" paradigm introduced by Castelhano and Henderson (2007). In these experiments, participants were presented with a target cue, followed by a brief preview of the scene. The search then proceeded via a gaze-contingent window which masked everything but a small aperture around the currently fixated location. Across several experiments, searchers used conceptual information from the initial glimpse to speed their search. In fact, participants probably don't need *any visual information at all* to begin biasing their search to particular locations. When I asked participants to guess where an object would be located based on only the target identity, they were highly consistent in their guesses, and their consistency predicted search times and attention (Foulsham, 2012; see figure 11.2). A recent investigation of object identities and positions in thousands of scenes shows some surprisingly robust regularities which may underlie our expectations based on gist (Greene, 2013).

Wolfe et al. (2011) describe guidance in scenes in terms of selective and nonselective pathways. The former controls the selection and identification of individual objects, which in scenes probably requires fixation and focal attention. The latter, nonselective pathway uses gist or episodic memory to identify the locations to search. As we shall see in the next section, these sources of information in search have been incorporated into several explicit models of top-down guidance.

Top-Down Models

Rao et al. (2002) were among the first to explicitly model attention to a target in natural scenes. In their model, fixations are based on how well the features at each location correlate with a stored target representation. Possible locations are represented in a map, and gaze moves to the points most similar to the target. Thus, this model is completely reliant on knowledge of target features and does not depend on a priori saliency. Rao et al. show that the model simulates eye movements rather well, and with assumptions such as preferential processing of coarse features, it can account for other phenomena such as "center-of-gravity" saccades where the eyes land between points of interest.

Zelinsky's (2008) target acquisition model (TAM) generalizes and extends some of the principles of the Rao et al. model. Zelinsky shows that a target map based on

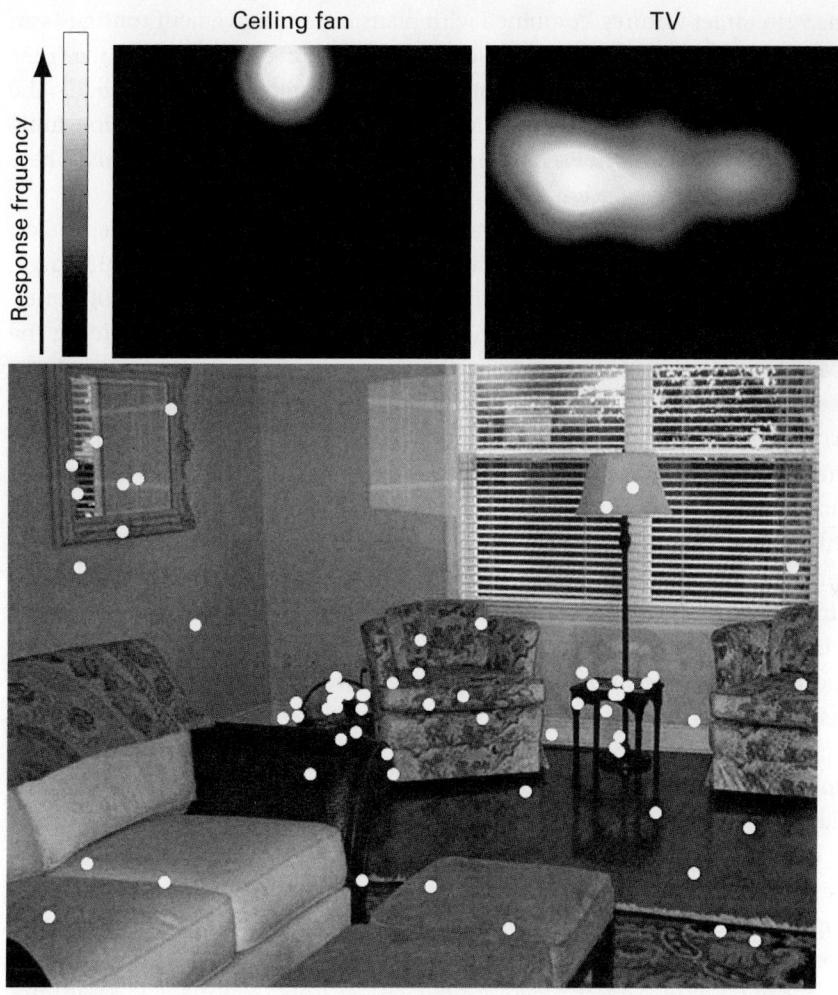

Figure 11.2
When people guess where a named target might be within a picture, they are quite consistent (top panels, with brighter locations guessed more often). Fixations during search for the target "TV" cluster on probable locations, even when the target is absent (bottom panel; see Foulsham, 2012).

correlations with target features, combined with plausible eye-movement routines, can produce a realistic simulation of both simple and naturalistic search. In particular, TAM uses a simulated retina, often missing from saliency map models, which models the drop-off in sensitivity for regions further from the fovea. Thus, focal attention functions to improve the signal-to-noise ratio by moving to target-probable regions, where detailed inspection can take place.

The Rao et al. and Zelinsky models are "top-down" in the sense that they depend on the observer's mental representations (in search this entails knowing what the target is).[2] An alternative way of modeling such guidance is via modulation of a bottom-up saliency map. Navalpakkam and Itti (2005) achieve this by selectively weighting the most relevant features. For example, if searching for something red, the contributions of redness and color to the overall saliency computation can be enhanced.

Torralba et al. (2006) described the first attempt to model guidance based not only on target features but also on where targets are expected to appear. Drawing on data from participants looking for pedestrians, paintings, or mugs within scenes, Torralba et al.'s contextual guidance model combines a bottom-up saliency map with a global pathway providing information on contextually likely locations. The global pathway is a route for recognizing scene gist and activating an associated spatial prior for where a target is likely to appear. When searching for pedestrians, saliency will be modulated by the expectation that pedestrians tend to be found at street level. When looking for paintings, fixations will be constrained to walls.

Modeling the top-down factors that are a hallmark of scene perception requires determining attentional priorities in a particular task, as well as how these priorities should relate to image features and how they are represented in the brain. Although progress is being made within visual search, it remains to be seen how such models can be generalized to all the different sorts of tasks at which humans excel. In such cases it may be more useful to conceptualize eye movements as moving to capture the most information for the current task, or reducing the uncertainty, given the current fixation position (Renninger et al., 2007; Najemnik & Geisler, 2005; Foulsham & Kingstone, 2012).

Systematic Tendencies

One of the difficulties in evaluating the models discussed above is that, even without taking into account the current scene, eye movements are far from random. Instead, the places where people attend are affected by "systematic tendencies" which result in an overall bias toward certain patterns of eye-movement scanning.

One of the most obvious biases is for participants to look in the center of an image. Although this might result from a photographer's tendency to place objects of interest in the center of an image, Tatler (2007) showed that the distribution of image features had only a minor effect on fixations. Neither is the central bias due only to beginning the viewing procedure with a central fixation marker. When participants begin looking at a random location, they normally saccade toward the center (Tatler, 2007; Foulsham et al., 2008). The bias is independent from, but exacerbated by, the practice of centering a scene on a computer monitor (Bindemann, 2010). While all these factors contribute to the central bias, there is also a voluntary or strategic component, perhaps because looking in the center is a convenient place to begin examining a scene.

The saccades made during scene viewing are also biased: Horizontal saccades are most common, and more so than vertical or oblique saccades. In Foulsham et al. (2008), we showed that in a rotated image, the bias toward horizontal saccades follows the orientation of the scene, particularly in outdoor scenes. In interior scenes, vertical saccades were more common, perhaps because interiors tend to contain more vertical edges. There is also a consistent bias toward the left of an image, with around 60% to 70% of first eye movements occurring in this direction (which may be partly due to reading habits; Foulsham et al., 2013; Dickinson & Intraub, 2009).

These patterns, which appear on average across many participants and images, are important because they provide further constraints on where people are likely to look. Surprisingly, these systematic tendencies can be more predictive than complicated saliency models which analyze particular image features. Tatler and Vincent (2009) reported that just knowing where the eye has been and how it tends to move resulted in a more reliable classifier of fixated versus nonfixated locations than image-based models.

Scanpaths and Scene Memory

Yarbus's original characterization of scene viewing emphasized an ordered and repetitive sequence of fixations: a "scanpath." Noton and Stark (1971) suggested that such sequences might be tied to an active process of encoding and retrieval during scene recognition. Specifically, they noticed that scanpaths are sometimes repeated when participants look at an image that they have seen before. They suggested that this was because scanpaths were stored and recapitulated as part of the process of recognition. The recent evidence for this proposal is mixed.

On the one hand, the variability between individuals looking at the same image, and within an individual looking at an image repeatedly, appears to be much higher than suggested by Noton and Stark. It is certainly not the case that individuals look at a scene in the same way every time. Moreover, some images can be recognized very quickly and without executing a sequence of fixations.

On the other hand, given that task and knowledge guide scene viewing in a top-down fashion, it seems likely that an individual's memory might affect this process. In Foulsham and Underwood (2008), using multiple measures, we found that participants' recognition scanpaths were more similar to their own initial eye movements when viewing the same picture than they were to other observers'. Participants do indeed repeat something of their scanpath between viewings, and this has been replicated in subsequent studies (e.g., Harding & Bloj, 2010). However, this effect should not be overstated: Only about 20% to 30% of fixations are the same as those made previously.

Why might observers repeat their scanpaths? Scanpaths are not well captured by saliency (Foulsham & Underwood, 2008) and are particular to an observer, that is, they are idiosyncratic. However, there is no strong evidence that repeating some eye movements has a causal effect on memory. In Foulsham and Kingstone (2013a), we constrained viewing during recognition to a sequence of local regions drawn either from the locations fixated during learning or from a set of control locations. Looking at the same locations in the same order did not cause better recognition. Instead, refixations seem to be a reflection of the retrieval process (Holm & Mäntylä, 2007), and thus they are another example of the top-down nature of active gaze in scenes.

When Objects Are and Are Not Salient

I have argued that much of where people attend in images can be explained by what they are doing and what they know about the scene. However, visual saliency might still play a role by providing a shortcut: a fast and readily available signal as to what is important. A trivial observation is that, in general, participants spend more time looking at objects than at the background. Objects tend to contain variations in color and luminance and to be defined by edges, so we would expect saliency to be higher in objects. Conducting an analysis of thousands of object regions labeled as "interesting" by online observers, Elazary and Itti (2008) found that the saliency map model often tended to select an object as worthy of attention. The authors reasoned that, because their model has no notion of an object, humans may also attend to such regions on the basis of saliency alone. However, there are at least two reasons to doubt this conclusion.

First, object-hood seems to tell us more about where people fixate than visual saliency alone. Einhäuser, Spain, and Perona (2008) showed that observer annotations of objects could predict fixations better than the saliency map, and that when objects had been identified, saliency contributed little. Of course, a different sort of bottom-up model might perform better, and presumably there are some visual features available in peripheral vision which are cues to object presence (Borji, Sihite, & Itti, 2013). Walther and Koch (2006) suggested a model whereby saliency is used to identify "proto-objects," which can then be processed by focal attention and object recognition systems (see also Wischnewski et al., 2010). Feature contrast in objects tends to be highest at the object boundary (i.e., at the edges). However, recent reports show that participants tend to saccade directly to a "preferred viewing location" close to the center of an object (Foulsham & Kingstone, 2013b; Nuthmann & Henderson, 2010). Unlike the pixel-based models which have emerged from computer vision, therefore, human attention seems to act on objects. Attention during scenes may be best understood as a process of parsing into meaningful objects. Understanding how attention supports and is supported by this process is a fruitful area for further research.

Second, there are some objects which receive disproportionate amounts of attention despite being nonsalient. A key example of this is the attention-grabbing nature of the human face (see also chapters 15 and 16 in the current volume). Faces and eyes tend to be rather small and thus should fail to compete with colorful or striking objects such as furniture or foliage. However, as shown by Yarbus (1967), the faces and eyes of other people are looked at early and often whenever they appear in a scene. Birmingham et al. (2009) confirmed that this bias to look at others cannot be explained by visual saliency.

The Timing of Attention in Scenes

So far, I have described a range of research considering *where* we attend when looking at natural scenes. A related set of questions considers *when* fixations are made and how this reflects the time course of scene perception.

Most fixations during scene perception are between 200 and 400 ms in duration. It has often been assumed that the duration of individual fixations reflects processing of the information at that location (and thus objects which are distorted or semantically inconsistent are associated with longer fixations). Trying to memorize an image results in longer average fixation durations than searching for an object, presumably because relatively superficial processing is required to determine whether a given region is the target. Short fixations may also be associated with more exploratory or "ambient" processing, typically during the early part of scene inspection (Unema et al., 2005). Later in viewing, longer fixations may indicate more detailed, "focal" processing. In realistic picture-based tasks, it is likely that attention switches between periods of global and local exploration (Wedel, Pieters, & Liechty, 2008).

It is tempting to conclude that fixation durations reflect only the processing of information at the fovea. However, there are numerous examples of peripheral or global manipulations that affect how long people look. Loschky and McConkie (2002) and van Diepen and d'Ydewalle (2003) used a gaze-contingent display to degrade peripheral information and found that this lengthened average fixation duration. This might be because contextual information is used during object recognition. Importantly, the duration of a single fixation includes the time spent planning the next saccade, so it will also be prolonged when targeting of this saccade is disrupted. Determining the influence of these factors is a challenge, not least because participants may sometimes refixate an object rather than prolong a nonoptimal fixation (Foulsham & Kingstone, 2013b).

The control of fixation duration in scenes has recently been the subject of a formal model, CRISP (Controlled Random-Walk with Inhibition for Saccade Planning; Nuthmann et al., 2010). Fixations in the model are driven by a timer which moves the eyes at a regular rate but is inhibited when required by cognitive or visual difficulties. Experimental support comes from the stimulus-onset-delay paradigm, where the visual scene is masked at the beginning of a critical fixation (Henderson & Pierce, 2008). If fixations were directly controlled by the information being inspected at that time, the duration of these critical fixations should be prolonged in accordance with the duration of the mask. In fact, there is a population of short fixations which occur regardless of the presence of a mask. According to Nuthmann et al., these are predetermined by the timer, and saccade planning has proceeded to a point where they can no longer be delayed. Considering attention during scene perception requires understanding when we move our eyes, which is constrained by the time-limited planning of saccades in parallel with the processing of information at fixation.

Integration: Toward Really Real-World Scene Perception

I have outlined numerous experiments investigating "real-world" scene perception, involving participants looking at drawings or photographs, presented on a screen in controlled conditions. That is, they have investigated depictions of the world, rather than measuring attention in actual real-world conditions. We sometimes spend considerable amounts of time looking at pictures. Laboratory research can tell us about how we perceive artwork (e.g., Nodine et al., 1993) and advertisements (Wedel et al., 2008), and we spend an increasingly large amount of our leisure time looking at computer and television screens.

Nevertheless, the picture-viewing paradigm may not be a very good surrogate for scene perception in the real world (Henderson, 2007; Tatler et al., 2011). Our everyday environment is three-dimensional and multisensory. We are often moving through the world, and it is probably rare for us to look at a static scene. The pictorial views used in experiments are typically smaller than they would appear in the real world, and presenting them on a monitor might induce screen-specific biases. Many experiments in scene perception also use a very general "free-viewing" task, in which participants are asked simply to look at the image. Such purposeless viewing is also probably the exception rather than the rule in natural behavior. For all of these reasons, it is important to consider whether humans attend to scenes in the real world in the same way that they attend to pictures.

Dynamic Scenes
Studying attention in dynamic scenes or video gives an opportunity to study when and where people attend in a situation where stimuli and priorities are changing over time, as they must do in real life.

Bottom-up saliency models appear to do a better job in dynamic settings, with the addition of a channel for detecting motion contrast. For example, Carmi and Itti (2006) recorded eye movements from participants watching short video clips featuring out-door and indoor scenes and television programs. Saliency was a significant predictor of where people looked during video. Moreover, flicker and motion energy were more strongly correlated with fixation locations than was static feature contrast. The stimuli in this study were "MTV-style" clips with frequent cuts. Such rapid onsets tend to syn-chronize people by orienting them to the center of the screen, but they are unusual in everyday life (Tatler et al., 2011). Nevertheless, motion does appear to be a good predic-tor of "attentional synchrony"—the tendency for observers to look in the same place at the same time during video (Mital et al., 2011).

Unlike static depictions of scenes, it normally matters *when* one looks at a certain object, and thus experiments with video can provide insights into tempo-ral attention. Many of us watch movies for pleasure, and further investigation of attention during film promises to reveal much of interest for cinematographers and psychologists alike (see Smith et al., 2012, for a review). However, experiments with dynamic scenes suffer from some of the same problems as those in static scene perception. Videos of the real world, particularly as scripted and edited in Holly-wood, represent an artificial sample of our environment, and studies are limited to small screens and observers who are passively observing the world rather than acting within it.

Virtual Environments

In everyday life, we don't just observe scenes; we inhabit them. However, it is not easy to measure attention in the real environment because we cannot control exactly what appears in the visual field. For this reason, virtual environments can provide a useful tool for those interested in studying immersive scenes. Crucially, these virtual environ-ments allow participants to move around and perform actions. This capacity for action may be crucial for understanding how attention works.

A good example of this approach comes from Rothkopf et al. (2007). Rothkopf et al. set out to examine eye gaze in an extended and realistic task while still permitting experimental control. In their task, participants traversed a virtual walkway (realized via a head-mounted display). Participants were asked to either pick up "litter" blocks, which were randomly distributed around the scene, or to avoid differently colored "obstacle" blocks. Thus, while the virtual environment remained the same, the task priorities were varied. Analysis of gaze locations showed that where people looked was dependent on the task. In the litter task, participants looked mostly at the center of the objects that they were required to pick up. In contrast, during the obstacle task, par-ticipants looked toward the edges of the blocks that they were required to avoid. Thus, both the objects that were selected and the targeting of saccades within these objects were sensitive to the participants' active task. In neither case was the Itti and Koch (2000) model feasible for guiding attention.

Active Vision in a Real Environment

Research studying attention during behavior in a real environment has emphasized the deployment of gaze during active tasks. For example, Land and colleagues have reported detailed descriptions of the sequence of fixations involved in making a cup of tea. In such tasks, visual attention is deployed to the objects which are task relevant, at a time most beneficial for planning subsequent actions (see Tatler et al., 2011; see also chapter 17 in the current volume).

To some extent, such findings mirror those of Yarbus and Buswell, emphasizing the task dependence of attention. However, they also demonstrate the gulf between typical picture-viewing studies and attention in more realistic tasks. Scene viewing studies often address the question of where people look, averaging across an extended period, and with a free or unrealistic task. This may have led to an overemphasis on image-based saliency models. In contrast, studies of attention in real-world tasks have focused on when participants fixate action-relevant items. An important aim for current and future research in scene perception is therefore to reconcile these approaches.

One way to do this is to consider how visual attention might differ in static, dynamic, and real-world contexts. Cristino and Baddeley (2009) recorded naturalistic videos via a camera mounted on a helmet as they walked down the street. New participants then watched these dynamic scenes in the laboratory. It should now come as no surprise that gaze in this naturalistic stimulus was not guided by image salience, and thus where people looked was not affected by spatiotemporal filtering which altered the bottom-up features. Instead, the authors suggested that saccades were planned in a world-centered frame of reference, according to the relevance of objects such as the path and observers' perception of vertical. This points toward one important direction for future research, namely, investigating how attention is coordinated in egocentric and allocentric reference frames within scenes (i.e., relative to oneself vs. other objects, respectively).

Other researchers have used mobile eye tracking to investigate scene-based tasks in the lab. Mack and Eckstein (2011) asked participants to search for real target objects on a tabletop. Targets that were paired with contextually related objects (e.g., headphones and an iPod) were found more quickly, and attention was guided toward these cue objects. This is an example of how contextual guidance in picture viewing (e.g., Torralba et al., 2006) can scale up to more realistic, active situations.

't Hart et al. (2009) and Foulsham, Walker, and Kingstone (2011) tackled the similarity of attention in picture viewing and the real world head on. In both studies, participants' gaze was recorded both in the natural environment and within depictions of this scene in the lab. 't Hart et al. found similarities between real-world gaze and fixations in the lab, especially when the temporal continuity of the stimuli was preserved (i.e., by using video rather than static image frames). In Foulsham, Walker, and Kingstone (2011) participants performed an unconstrained and realistic task—to walk across campus and buy a beverage—while wearing a head-mounted mobile eye tracker. Later, the same participants watched videos filmed by the eye tracker's scene camera while seated in front of a computer screen. In both cases the visual information in the central visual field was the same, but in the walking condition participants were moving around in the real environment. Both walking and watching participants showed a central bias although walkers tended to stay focused on a heading point slightly below the horizon and to shift attention by moving the head rather than making large eye movements. The objects that were fixated also varied. Participants walking in the real environment were more likely to look at the path at their feet and were less likely to look at nearby pedestrians than were observers watching a video. In a follow-up experiment, I have recently returned to the question of whether measuring fixation in randomized, static scenes, can tell us about attention in the real world. Figure 11.3 shows

Figure 11.3
Scenes from the visual field of someone walking in the real world in Foulsham et al. (2011, top panels), along with the individual's point of gaze. When the same scenes are presented as static images to observers in the lab, the distribution of their fixations is quite different (bottom panels; brighter points indicate the most fixated locations).

the fixation distribution of multiple observers looking at a static scene, alongside the gaze of the participant who was observing this scene in the real world. Gazes from static scenes were surprisingly poor at predicting the real-world fixation, particularly when pedestrians or salient objects were present. Thus, it is important to consider both the action-based relevance and the social context of items, which will likely be quite different in real environments.

Future Directions

Natural scenes are more complex than the simple stimuli traditionally used in cognitive psychology. This complexity brings challenges for experimental control and computational modeling. However, it also reveals the rich set of cues provided by our environment that enable us to attend effectively.

When viewing scenes, we bring with us knowledge of objects and where they are likely to be. This knowledge, combined with rapid perception of gist and layout, means that our attention should not be thought of as selecting pixels in a uniform field. Instead, we select relevant objects for our ongoing tasks: searching for the items and information that we need to act appropriately. Thanks to bottom-up models, we can quantify the visual features available for guiding attention. A challenge for the future is to develop equally detailed descriptions of the way that knowledge and task are represented for the control of attention. These descriptions need to be able to explain both where and when we look in complex scenes, as well as how this is constrained by regularities in the oculomotor system.

A second challenge is to supplement the picture-viewing paradigm with a more complete range of naturalistic scenes and tasks. Studying pictorial representations can tell us much about how we combine low- and high-level influences to guide attention, but it cannot tell us everything. Comparing static, dynamic, and real-world environments will allow us to bridge the divide between laboratory and natural behavior. This endeavor is likely to yield considerable insights for researchers studying human visual attention.

Box 11.1
Key Points

- Scene gist is perceived and influences attention from the first glance at an image.
- Fixations within natural scenes select task-relevant objects.
- The sequence and duration of fixations reflects top-down planning and memory, as well as the processing of local stimulus information.

Box 11.2
Outstanding Issues

- How are knowledge and task priorities represented and applied to a scene context when guiding attention?
- A realistic account of scene perception has to look beyond the picture-viewing paradigm.

Notes

1. Confusingly, "saliency" or "salience" can be used to refer to any abstract or neural representation of priority or to describe anything which receives attention. I will use it only in the sense of the bottom-up saliency map model.

2. However, note that the top-down/bottom-up distinction can be problematic (Awh, Belopolsky, & Theeuwes, 2012).

References

Awh, E., Belopolsky, A. V., & Theeuwes, J. (2012). Top-down versus bottom-up attentional control: A failed theoretical dichotomy. *Trends in Cognitive Sciences, 16,* 437–443.

Bar, M. (2004). Visual objects in context. *Nature Reviews. Neuroscience, 5,* 617–629.

Becker, M. W., Pashler, H., & Lubin, J. (2007). Object-intrinsic oddities draw early saccades. *Journal of Experimental Psychology: Human Perception and Performance, 33,* 20–30.

Biederman, I., Glass, A. L., & Stacy, E. W. (1973). Searching for objects in real-world scenes. *Journal of Experimental Psychology, 97,* 22–27.

Biederman, I., Rabinowitz, J. C., Glass, A. L., & Stacy, E. W. (1974). On the information extracted from a glance at a scene. *Journal of Experimental Psychology, 103,* 597–600.

Bindemann, M. (2010). Scene and screen center bias early eye movements in scene viewing. *Vision Research, 50,* 2577–2587.

Birmingham, E., Bischof, W. F., & Kingstone, A. (2009). Saliency does not account for fixations to eyes within social scenes. *Vision Research, 49,* 2992–3000.

Bonitz, V. S., & Gordon, R. D. (2008). Attention to smoking-related and incongruous objects during scene viewing. *Acta Psychologica, 129,* 255–263.

Borji, A., & Itti, L. (2013). State-of-the-art in visual attention modeling. *IEEE Transactions on Pattern Analysis and Machine Intelligence, 35,* 185–207.

Borji, A., & Itti, L. (2014). Defending Yarbus: Eye movements reveal observers' task. *Journal of Vision, 14*(3), 29.

Borji, A., Sihite, D. N., & Itti, L. (2013). Objects do not predict fixations better than early saliency: A re-analysis of Einhäuser et al.'s data. *Journal of Vision, 13*(10), 18.

Braun, J. (2003). Natural scenes upset the visual applecart. *Trends in Cognitive Sciences, 7*, 7–9.

Buswell, G. T. (1935). *How people look at pictures*. Chicago: University of Chicago Press.

Carmi, R., & Itti, L. (2006). Visual causes versus correlates of attentional selection in dynamic scenes. *Vision Research, 46*, 4333–4345.

Castelhano, M. S., & Henderson, J. M. (2007). Initial scene representations facilitate eye movement guidance in visual search. *Journal of Experimental Psychology: Human Perception and Performance, 33*, 753–763.

Cohen, M. A., Alvarez, G. A., & Nakayama, K. (2011). Natural-scene perception requires attention. *Psychological Science, 22*, 1165–1172.

Cristino, F., & Baddeley, R. (2009). The nature of the visual representations involved in eye movements when walking down the street. *Visual Cognition, 17*, 880–903.

DeAngelus, M., & Pelz, J. B. (2009). Top-down control of eye movements: Yarbus revisited. *Visual Cognition, 17*, 790–811.

De Graef, P., Christiaens, D., & d'Ydewalle, G. (1990). Perceptual effects of scene context on object identification. *Psychological Research, 52*, 317–329.

Dickinson, C. A., & Intraub, H. (2009). Spatial asymmetries in viewing and remembering scenes: Consequences of an attentional bias? *Attention, Perception & Psychophysics, 71*, 1251–1262.

Eckstein, M. P., Drescher, B. A., & Shimozaki, S. S. (2006). Attentional cues in real scenes, saccadic targeting, and Bayesian priors. *Psychological Science, 17*, 973–980.

Einhäuser, W., Rutishauser, U., & Koch, C. (2008). Task-demands can immediately reverse the effects of sensory-driven saliency in complex visual stimuli. *Journal of Vision, 8*(2), 2.

Einhäuser, W., Spain, M., & Perona, P. (2008). Objects predict fixations better than early saliency. *Journal of Vision, 8*(14), 18.

Elazary, L., & Itti, L. (2008). Interesting objects are visually salient. *Journal of Vision, 8*(3), 3.

Findlay, J. M., & Gilchrist, I. D. (2003). *Active vision: The psychology of looking and seeing*. New York: Oxford University Press.

Foulsham, T. (2012). "Eyes closed" and "Eyes open" expectations guide fixations in real-world search. In N. Miyake, D. Peebles, & R. P. Cooper (Eds.), *Proceedings of the 34th Annual Conference of the Cognitive Science Society* (pp. 330–335). Austin, TX: Cognitive Science Society.

Foulsham, T., Alan, R., & Kingstone, A. (2011). Scrambled eyes? Disrupting scene structure impedes focal processing and increases bottom-up guidance. *Attention, Perception & Psychophysics, 73*, 2008–2025.

Foulsham, T., Gray, A., Nasiopoulos, E., & Kingstone, A. (2013). Leftward biases in picture scanning and line bisection: A gaze-contingent window study. *Vision Research, 78*, 14–25.

Foulsham, T., & Kingstone, A. (2012). Modelling the influence of central and peripheral information on saccade biases in gaze-contingent scene viewing. *Visual Cognition, 20*, 546–579.

Foulsham, T., & Kingstone, A. (2013a). Fixation-dependent memory for natural scenes: An experimental test of scanpath theory. *Journal of Experimental Psychology. General, 142*, 41–56.

Foulsham, T., & Kingstone, A. (2013b). Optimal and preferred eye landing positions in objects and scenes. *Quarterly Journal of Experimental Psychology, 66*, 1707–1728.

Foulsham, T., Kingstone, A., & Underwood, G. (2008). Turning the world around: Patterns in saccade direction vary with picture orientation. *Vision Research, 48*, 1777–1790.

Foulsham, T., & Underwood, G. (2007). How does the purpose of inspection influence the potency of visual salience in scene perception? *Perception, 36*, 1123–1138.

Foulsham, T., & Underwood, G. (2008). What can saliency models predict about eye movements? Spatial and sequential aspects of fixations during encoding and recognition. *Journal of Vision, 8*(2), 6.

Foulsham, T., Walker, E., & Kingstone, A. (2011). The where, what and when of gaze allocation in the lab and the natural environment. *Vision Research, 51*, 1920–1931.

Greene, M. R. (2013). Statistics of high-level scene context. *Frontiers in Perception Science, 4*, 777.

Greene, M. R., Liu, T., & Wolfe, J. M. (2012). Reconsidering Yarbus: A failure to predict observers' task from eye movement patterns. *Vision Research, 62*, 1–8.

Harding, G., & Bloj, M. (2010). Real and predicted influence of image manipulations on eye movements during scene recognition. *Journal of Vision, 10*(2), 8.

Henderson, J. M. (2007). Regarding scenes. *Current Directions in Psychological Science, 16*, 219–222.

Henderson, J. M., Brockmole, J. R., Castelhano, M. S., & Mack, M. L. (2007). Visual saliency does not account for eye movements during visual search in real-world scenes. In R. van Gompel, M. Fischer, W. Murray, & R. W. Hill (Eds.), *Eye movements: A window on mind and brain* (pp. 537–562). Amsterdam: Elsevier.

Henderson, J. M., Malcolm, G. L., & Schandl, C. (2009). Searching in the dark: Cognitive relevance drives attention in real-world scenes. *Psychonomic Bulletin & Review, 16*, 850–856.

Henderson, J. M., & Pierce, G. L. (2008). Eye movements during scene viewing: Evidence for mixed control of fixation durations. *Psychonomic Bulletin & Review, 15*, 566–573.

Henderson, J. M., Weeks, P. A., & Hollingworth, A. (1999). The effects of semantic consistency on eye movements during complex scene viewing. *Journal of Experimental Psychology: Human Perception and Performance, 25*, 210–228.

Holm, L., & Mäntylä, T. (2007). Memory for scenes: Refixations reflect retrieval. *Memory & Cognition, 35,* 1664–1674.

Itti, L., & Koch, C. (2000). A saliency-based search mechanism for overt and covert shifts of visual attention. *Vision Research, 40,* 1489–1506.

Koch, C., & Ullman, S. (1985). Shifts in selective visual attention: Towards the underlying neural circuitry. *Human Neurobiology, 4,* 219–227.

Li, F. F., VanRullen, R., Koch, C., & Perona, P. (2002). Rapid natural scene categorization in the near absence of attention. *Proceedings of the National Academy of Sciences of the United States of America, 99,* 9596–9601.

Loftus, G. R., & Mackworth, N. H. (1978). Cognitive determinants of fixation location during picture viewing. *Journal of Experimental Psychology: Human Perception and Performance, 4,* 565–572.

Loschky, L. C., & McConkie, G. W. (2002). Investigating spatial vision and dynamic attentional selection using a gaze-contingent multiresolutional display. *Journal of Experimental Psychology. Applied, 8,* 99–117.

Mack, S. C., & Eckstein, M. P. (2011). Object co-occurrence serves as a contextual cue to guide and facilitate visual search in a natural viewing environment. *Journal of Vision, 11*(9), 9.

Mackworth, N. H., & Morandi, A. J. (1967). The gaze selects informative details within pictures. *Perception & Psychophysics, 2,* 547–552.

Malcolm, G. L., & Henderson, J. M. (2009). The effects of target template specificity on visual search in real-world scenes: Evidence from eye movements. *Journal of Vision, 9*(11), 8.

Mital, P. K., Smith, T. J., Hill, R. L., & Henderson, J. M. (2011). Clustering of gaze during dynamic scene viewing is predicted by motion. *Cognitive Computation, 3,* 5–24.

Najemnik, J., & Geisler, W. S. (2005). Optimal eye movement strategies in visual search. *Nature, 434,* 387–391.

Navalpakkam, V., & Itti, L. (2005). Modeling the influence of task on attention. *Vision Research, 45,* 205–231.

Nodine, C. F., Locher, P. J., & Krupinski, E. A. (1993). The role of formal art training on perception and aesthetic judgment of art compositions. *Leonardo, 26,* 219–227.

Noton, D., & Stark, L. (1971). Scanpaths in saccadic eye movements while viewing and recognizing patterns. *Vision Research, 11,* 929–942.

Nuthmann, A., & Henderson, J. M. (2010). Object-based attentional selection in scene viewing. *Journal of Vision, 10*(8), 20.

Nuthmann, A., Smith, T. J., Engbert, R., & Henderson, J. M. (2010). CRISP: A computational model of fixation durations in scene viewing. *Psychological Review, 117,* 382–405.

Nyström, M., & Holmqvist, K. (2008). Semantic override of low-level features in image viewing—Both initially and overall. *Journal of Eye Movement Research, 2*(2), 1–11.

Oliva, A. (2005). Gist of the scene. In L. Itti, G. Rees, & J. K. Tsotsos (Eds.), *Neurobiology of attention* (pp. 251–256). San Diego, CA: Elsevier.

Peters, R. J., Iyer, A., Itti, L., & Koch, C. (2005). Components of bottom-up gaze allocation in natural images. *Vision Research, 45*, 2397–2416.

Potter, M. C., & Levy, E. I. (1969). Recognition memory for a rapid sequence of pictures. *Journal of Experimental Psychology, 81*, 10–15.

Potter, M. C., Staub, A., & O'Connor, D. H. (2004). Pictorial and conceptual representation of glimpsed pictures. *Journal of Experimental Psychology: Human Perception and Performance, 30*, 478–489.

Rao, R. P., Zelinsky, G. J., Hayhoe, M. M., & Ballard, D. H. (2002). Eye movements in iconic visual search. *Vision Research, 42*, 1447–1463.

Renninger, L. W., Verghese, P., & Coughlan, J. (2007). Where to look next? Eye movements reduce local uncertainty. *Journal of Vision, 7*(3), 6.

Rothkopf, C. A., Ballard, D. H., & Hayhoe, M. M. (2007). Task and context determine where you look. *Journal of Vision, 7*(14), 16.

Smith, T. J., Levin, D., & Cutting, J. E. (2012). A window on reality: Perceiving edited moving images. *Current Directions in Psychological Science, 21*, 107–113.

Tatler, B. W. (2007). The central fixation bias in scene viewing: Selecting an optimal viewing position independently of motor biases and image feature distributions. *Journal of Vision, 7*(14), 4.

Tatler, B. W., Baddeley, R. J., & Gilchrist, I. D. (2005). Visual correlates of fixation selection: effects of scale and time. *Vision Research, 45*, 643–659.

Tatler, B. W., Hayhoe, M. M., Land, M. F., & Ballard, D. H. (2011). Eye guidance in natural vision: Reinterpreting salience. *Journal of Vision, 11*(5), 5.

Tatler, B. W., & Vincent, B. T. (2009). The prominence of behavioural biases in eye guidance. *Visual Cognition, 17*, 1029–1054.

't Hart, B. M., Vockeroth, J., Schumann, F., Bartl, K., Schneider, E., Koenig, P., et al. (2009). Gaze allocation in natural stimuli: Comparing free exploration to head-fixed viewing conditions. *Visual Cognition, 17*, 1132–1158.

Torralba, A., Oliva, A., Castelhano, M. S., & Henderson, J. M. (2006). Contextual guidance of eye movements and attention in real-world scenes: The role of global features in object search. *Psychological Review, 113*, 766–786.

Underwood, G., & Foulsham, T. (2006). Visual saliency and semantic incongruency influence eye movements when inspecting pictures. *Quarterly Journal of Experimental Psychology, 59*, 1931–1949.

Unema, P. J., Pannasch, S., Joos, M., & Velichkovsky, B. M. (2005). Time course of information processing during scene perception: The relationship between saccade amplitude and fixation duration. *Visual Cognition, 12*, 473–494.

van Diepen, P., & d'Ydewalle, G. (2003). Early peripheral and foveal processing in fixations during scene perception. *Visual Cognition, 10*, 79–100.

Võ, M. L. H., & Henderson, J. M. (2009). Does gravity matter? Effects of semantic and syntactic inconsistencies on the allocation of attention during scene perception. *Journal of Vision, 9*(3), 24.

Võ, M. L. H., & Henderson, J. M. (2011). Object–scene inconsistencies do not capture gaze: Evidence from the flash-preview moving-window paradigm. *Attention, Perception & Psychophysics, 73*, 1742–1753.

Wade, N., & Tatler, B. W. (2005). *The moving tablet of the eye: The origins of modern eye movement research*. New York: Oxford University Press.

Walther, D., & Koch, C. (2006). Modeling attention to salient proto-objects. *Neural Networks, 19*, 1395–1407.

Wedel, M., Pieters, R., & Liechty, J. (2008). Attention switching during scene perception: How goals influence the time course of eye movements across advertisements. *Journal of Experimental Psychology. Applied, 14*, 129–138.

Wischnewski, M., Belardinelli, A., Schneider, W. X., & Steil, J. J. (2010). Where to look next? Combining static and dynamic proto-objects in a TVA-based model of visual attention. *Cognitive Computation, 2*, 326–343.

Wolfe, J. M., Võ, M. L. H., Evans, K. K., & Greene, M. R. (2011). Visual search in scenes involves selective and nonselective pathways. *Trends in Cognitive Sciences, 15*, 77–84.

Yarbus, A. L. (1967). *Eye movements and vision*. New York: Plenum Press.

Zelinsky, G. J. (2008). A theory of eye movements during target acquisition. *Psychological Review, 115*, 787–835.

12

Eye Movements and Visual Attention during Reading

Mallorie Leinenger and Keith Rayner

Historical Context

The scientific study of skilled reading originated in the mid to late 1800s, and that work is nicely summarized in the classic work *The Psychology and Pedagogy of Reading* by Huey (1908). Well-known researchers of the time, such as Cattell, Dearborn, Dodge, Javal, Woodworth, and Huey himself established many of the basic facts about reading that have stood the test of time. Central to much of that early research was how the eye movements of readers inform us with respect to how they process text. And the relationship between attention and eye movements was very apparent to these early researchers. At the end of chapter 3 (entitled "The Extent of Reading Matter Perceived during a Reading Pause"), Huey wrote the following:

...we are limited, in the amount that can be read during a reading pause, by the inadequacy of the retinal structure, by our inability to attend to more than a few parts of the total picture presented, and by the necessity of our attention's concerning itself with interpretations. (p. 70)

In another classic work, Woodworth (1938) asked, "How much can be read in a single fixation?" and highlighted, like Huey, that information extraction on a given fixation (what Huey termed a "reading pause") is limited. Thus, right from the start there was interest in the relationship between attention, eye movements, and how much can be processed during an individual fixation. In this chapter, we will largely discuss the relationship between attention and eye movements in the context of Huey's chapter on how much information can be processed in a single eye fixation.

The researchers whose work was summarized by Huey were very keen on inferring mental activities from the data that they collected. However, with the onset of the behaviorist movement in American psychology, attempts to make such inferences were rather scarce. Classic research on eye movements in reading was reported by Buswell (1922) and Tinker (1939, 1958), but in contrast to the earlier work, it seems much more descriptive and without as explicit inferences about what is happening in the mind of the reader. Thus, it is not surprising that when Huey's book was reprinted in 1968, we didn't have a much better understanding of the mental processes involved in reading than we had in 1908. Since the 1970s, there has again been considerable interest in trying to infer basic cognitive processes in reading (see Rayner & Pollatsek, 1989; Rayner, Pollatsek, Ashby, & Clifton, 2012). In this chapter, we will review findings related to attention, eye movements, and how much information readers process on each eye fixation.

State-of-the-Art Review

The Coupling of Eye-Movements and Attention

We process incoming visual information when light hits the retina and is transferred to the brain as electrical signals. Importantly, the makeup of the retina is not uniform; rather, different regions of the retina vary in acuity. During normal reading (and other tasks requiring the rapid processing of detail) we execute eye movements in order to bring the region of highest visual acuity, the *fovea*, into alignment with the region we wish to process. The fovea is located in the center of the retina and extends 2° of visual angle in diameter. Visual acuity declines steadily moving away from the center of the fovea through the parafovea (which extends up to 5° of visual angle from fixation) and the periphery (everything beyond the parafovea).

The two basic components of eye movements during reading are *saccades*, the ballistic movements executed to bring new information onto the fovea, and *fixations*, the periods of relative stability between saccades. It is during fixations that information is acquired since visual processing is effectively suppressed while saccades are being executed (Campbell & Wurtz, 1978; Matin, 1974). During the reading of alphabetic languages, fixations tend to be on the order of 225 to 250 ms and average saccades are 7 to 9 letters in length (Rayner, 1998, 2009).[1] The majority of eye movements proceed in the direction of the text (i.e., from left to right for readers of English), with regressions (i.e., saccades that move backward in the text) representing 10% to 15% of eye movements. Additionally, readers only fixate about 70% of words, *skipping* the other 30%.

Eye movements during reading are of interest because they provide fine-grained information about patterns of visual overt attention (i.e., where the eyes are fixating). While it is certainly possible to disengage attention from the point of fixation (Posner, 1980), this covert orienting of attention generally proceeds in a systematic manner, such that a shift in attention to the next saccade target precedes the actual eye movement (Deubel & Schneider, 1996; Kowler, Anderson, Dosher, & Blaser, 1995; Rayner, McConkie, & Ehrlich, 1978). That is, during the time when a saccade is being programmed and executed (generally on the order of 175 ms) covert attention can shift to the next saccade target—in reading, the next word that we want to process—while we wait for the eyes to catch up. Overt attentional shifts can be directly measured using eye tracking to record the eye movements of individuals while they are engaged in the reading process. While the eye-tracking record does not provide a direct measure of covert attention, the use of novel gaze-contingent manipulations (discussed below) allows us to measure the location of covert attention indirectly by assessing parafoveal processing (i.e., the amount of processing that can occur on a parafoveally attended word that has not yet been fixated—see Schotter, Angele, & Rayner, 2012, for a review).

Gaze-Contingent Display-Change Techniques

On any given fixation, how much information is a reader able to process? It certainly is the case that several lines of text may fall on the retina, so how much of that information can a reader make use of on a given fixation? While readers may have the subjective experience that they are able to see the entire page when fixating on it, the fact that they do move their eyes around the page making a series of fixations suggests that they are actually unable to process the entire page in one fixation. However, they also aren't fixating every single letter, so what *is* the area of effective visual processing during reading?

The development of two gaze-contingent display-change techniques has allowed us to better understand how much information can be extracted from a page of text during a single fixation and, specifically, what aspects of information can be extracted from the usable region of text. These are the moving window paradigm and the boundary change paradigm—both of which allow a researcher to dynamically update the display based on the eye movements of the reader (see figure 12.1).

Moving Window Paradigm   One way to measure the range of effective visual processing during normal silent reading, otherwise known as the *perceptual span*, is to manipulate how much of the text can be seen on a given fixation. The moving window paradigm, developed by McConkie and Rayner (1975; see Rayner, 2014, for a recent review of the paradigm), accomplishes this by using a display-change technique where the position of the reader's eyes are monitored using an eye tracker and changes in the text are made contingent on fixation location. Specifically, the text in the fovea and a specific area around it (i.e., the window) is displayed accurately to the reader while text outside of the window is altered or masked (e.g., replaced with strings of *x*s or random letters). In this way, the number of visible letters or words around the point of fixation can be manipulated and updated in real time as the reader moves through the text. If one limits the text that is available within the window, but reading does not differ from reading without a window (i.e., where the entire text is displayed normally), it can be concluded that no information from the region beyond the window was used for further processing. Therefore, the smallest window at which reading precedes normally provides a valid estimate of the perceptual span.

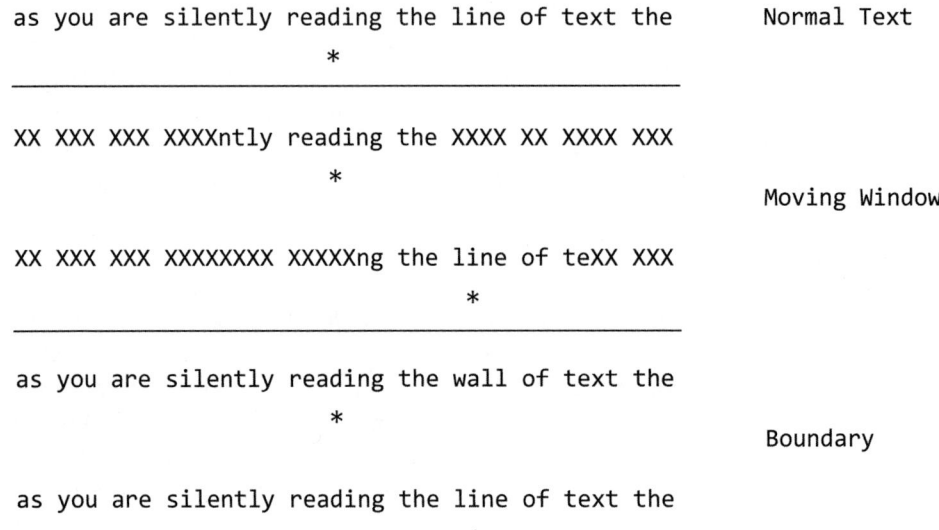

```
as you are silently reading the line of text the          Normal Text
                        *
_____

XX XXX XXX XXXXntly reading the XXXX XX XXXX XXX
                        *
                                                          Moving Window

XX XXX XXX XXXXXXXX XXXXXng the line of teXX XXX
                          *
_____

as you are silently reading the wall of text the
                      *
                                                          Boundary

as you are silently reading the line of text the
                          *
```

Figure 12.1

Examples of the moving window and boundary change paradigms. The first line shows the normal line of text with the fixation location marked with an asterisk. The next two lines show two successive fixations while reading with a 17-character, symmetrical moving window (8 characters available on either side of fixation). The last two lines show an example of the boundary paradigm. The first line shows the text prior to a display change with fixation location marked with an asterisk. Once the eyes cross the invisible boundary (the *e* in *the*), the preview (*wall*) is replaced with the target (*line*).

Using this technique, McConkie and Rayner (1975) tested symmetrical windows and determined that reading proceeded normally with a 31-character window (i.e., a window extending 15 characters on either side of fixation). With 15 characters available on either side of fixation, reading rate did not differ from when the entire line was visible, but smaller windows resulted in reduced reading rates, suggesting that during normal, silent reading, the perceptual span includes the currently fixated word and neighboring words as well.

Interestingly, the perceptual span does not seem to be limited purely because of the drop-off in visual acuity moving away from the fovea. Miellet, O'Donnell, and Sereno (2009) used a novel adaptation of the moving window paradigm in which the letters outside of the center of vision were magnified to counter the drop-off in visual acuity due to retinal eccentricity. They found that parafoveal magnification did not increase the size of the perceptual span, suggesting that attentional constraints rather than acuity limitations are primarily limiting the size of the perceptual span. Additional evidence for attentional constraints comes from the result that readers do not get any useful information from the lines above or below the currently fixated line, even though words in those lines are often closer to fixation (and therefore provide higher visual acuity information) than words that do fall within the perceptual span (Inhoff & Briihl, 1991; Inhoff & Topolski, 1992; Pollatsek, Raney, LaGasse, & Rayner, 1993).

Given that attentional limitations rather than limitations in visual acuity are dictating how much information can be processed on a given fixation, there is no reason that the perceptual span must be symmetrical around the point of fixation. Indeed, McConkie and Rayner (1976) set out to determine whether the perceptual span might in fact be asymmetrical. By independently varying the left and right boundaries of the visible window of text, McConkie and Rayner determined that, when the window extended 14 characters to the right of fixation, but only 4 characters to the left, reading speed was not significantly different than when the window extended 14 characters in either direction. However, when the window extended 14 characters to the left and only 4 characters to the right, reading was considerably slower.

Therefore, the perceptual span is asymmetrical around the point of fixation. Furthermore, it is in fact biased in the direction of attention, extending 3 to 4 characters to the left of fixation (McConkie & Rayner, 1976; Rayner, Well, & Pollatsek, 1980) and 14 to 15 characters to the right of fixation when reading English (McConkie & Rayner, 1975; Rayner & Bertera, 1979), which in general maps onto roughly the currently fixated word and the two upcoming words. In fact, as long as the currently fixated word and the next word are available, and the letters in the rest of the words are replaced with visually similar letters, readers are generally unaware of the manipulation and reading is only about 10% slower than when reading without a window (Rayner, Well, Pollatsek, & Bertera, 1982).

Further evidence that the perceptual span is biased in the direction of upcoming information and is constrained by attentional limitations comes from cross-linguistic research. For example, for readers of Hebrew, which is printed from right to left, the perceptual span asymmetry of English is reversed, extending further to the left than to the right of fixation (Pollatsek, Bolozky, Well, & Rayner, 1981; see Jordan et al., 2014, for reversed asymmetry for Arabic). Additionally, the size of the perceptual span also varies across language as a function of information density, with Chinese readers only utilizing 1 character to the left of fixation and 3 characters to the right (Inhoff & Liu, 1998). Critically, although Chinese readers have smaller physical perceptual spans, their spans are roughly equivalent to English readers' spans when number of words, rather than number of characters, is considered.

Finally, even within a language, and in fact even within a single reader, the perceptual span can change as a function of text difficulty, with larger perceptual spans seen in skilled relative to unskilled or beginning readers (Häikiö, Bertram, Hyönä, & Niemi, 2009; Rayner, 1986; Veldre & Andrews, 2013), faster relative to slower readers (Rayner, Slattery, & Bélanger, 2010), normal relative to dyslexic readers (Rayner, Murphy, Henderson, & Pollatsek, 1989; Rayner, 1983), and when fixating high-frequency relative to low-frequency words (Henderson & Ferreira, 1990). Thus, when more attention is required to process the foveal information because of lower reading skill, slower reading, disabled reading, or more difficult text, readers are less able to make use of upcoming information for parafoveal preprocessing. Additionally, skilled and less-skilled deaf readers have been shown to have larger perceptual spans than would be expected given their reading abilities, suggesting that auditory deprivation can lead to enhanced attentional allocation to the parafovea in deaf readers (Bélanger, Slattery, Mayberry, & Rayner, 2012). Furthermore, it is not only the size of the perceptual span that can change but also the extent and direction of the asymmetry. For example, there is evidence that the perceptual span may extend further than 3 to 4 characters to the left of fixation immediately before a regressive eye movement is executed (Apel, Henderson, & Ferreira, 2012), and the perceptual span of English/Hebrew bilingual readers is asymmetrical toward the right when reading English and toward the left when reading Hebrew (Pollatsek et al., 1981)—two cases where the perceptual span's bias changes in accordance with the deployment of visual attention.

Boundary Change Paradigm   Research using the moving window paradigm has demonstrated that attention is allocated to upcoming words (preceding words in the case of regressions) that are the targets of future fixations. The fact that we can move our attention to an upcoming word and begin to process that word while we wait for our eyes to catch up allows for more efficient processing during reading. However, exactly what information can be gleaned from parafoveal words and integrated across saccades is not clear from the results of the moving window studies. Results from single-word-identification studies suggest that the covert shifting of attention to the location of an upcoming word (or soon to be presented word) may facilitate prelexical processing by modulating either the degree of orthographic feature-level cross talk or the rate of feature uptake (e.g., McCann, Folk, & Johnston, 1992; Risko, Stolz, & Besner, 2010). Beyond attention's facilitation of orthographic feature processing, studies of parafoveal processing that utilize the boundary change paradigm (Rayner, 1975) have revealed important information about what kinds of information may be integrated across saccades. The boundary change paradigm is another gaze-contingent display-change technique in which an invisible boundary is inserted to the left of a predetermined target word (see figure 12.1). While a reader's gaze is to the left of the boundary (prior to fixating the target word), there is an initial preview that is (typically) different from the target word. The initial preview is then replaced with the target word when the eyes cross the boundary during the saccade to the target. Because the change occurs during a saccade when vision is largely suppressed, subjects generally fail to notice it (Slattery, Angele, & Rayner, 2011). Using the boundary change paradigm, then, it is possible to manipulate the amount of overlap between the preview and target word. Faster foveal processing of the target word (measured as shorter fixation times) following a valid preview compared to an invalid preview (e.g., another word, a string of *x*s, a nonword, etc.) is referred to as *preview benefit* (for reviews, see Rayner, 1998, 2009; Schotter et al., 2012). It is also possible to compare the processing of invalid previews to semivalid previews that maintain certain aspects of the target word, such as its orthography, phonology, or semantics, to see if any preview benefit is obtained from integrating the information that was consistent across saccades (i.e., between the preview and the target word).

Research using the boundary change paradigm has revealed that readers generally do not integrate literal representations of the visual information across saccades (McConkie & Zola, 1979; Rayner, McConkie, & Zola, 1980; Slattery et al., 2011). Rather, they process information about a word's length, beginning and ending letters, orthographic codes, phonological codes, and under certain conditions, semantic information (for reviews, see Rayner, 1998, 2009; Schotter, 2013). That is to say that a preview benefit will be observed following a parafoveal preview that overlaps with the target on any of these categories of information.

Direct Control

Eye movements during reading are under direct cognitive control (e.g., Rayner, 1978, 1998—sometimes referred to as the "eye–mind hypothesis" or "eye–mind assumption," e.g., Just & Carpenter, 1976, 1980). In other words, how long a reader spends looking at a word is strongly influenced by linguistic and lexical variables, and therefore, eye movements provide an online index of the cognitive processes underlying language comprehension. Indeed, how long the eyes remain fixated on a given word largely depends on how easy or difficult it is to process. Lexical variables such as how frequent a word is (Inhoff & Rayner, 1986; Rayner & Duffy, 1986), how predictable a word is (Ehrlich & Rayner, 1981; Rayner & Well, 1996), how many meanings a word has (Duffy, Morris, & Rayner, 1988; Sereno, O'Donnell, & Rayner, 2006), and the age at which the word was first acquired (Juhasz & Rayner, 2003, 2006) all influence fixation duration on a given word (see Rayner, 1998, 2009, for reviews). Additionally, higher level discourse information also influences fixation durations. For example, readers tend to make longer fixations at the ends of clauses and sentences, as they are presumably engaged in semantic integration (Rayner, Kambe, & Duffy, 2000) and also fixate longer on information that disambiguates syntactic garden path sentences (Frazier & Rayner, 1982; Levy, Bicknell, Slattery, & Rayner, 2009). Beyond lexical and discourse variables, fixation times are also influenced by low-level visual information like font difficulty (Rayner, Reichle, Stroud, Williams, & Pollatsek, 2006; Slattery & Rayner, 2010) and by individual reader differences like reading skill (Ashby, Rayner, & Clifton, 2005—see also Dambacher, Slattery, Yang, Kliegl, & Rayner, 2013; Rayner & Pollatsek, 1981; Reingold, Reichle, Glaholt, & Sheridan, 2012; and Schad, Risse, Slattery, & Rayner, 2014, for other evidence of direct control).

Perhaps the most convincing evidence in support of the direct control claim is research studying "disappearing text." In these studies, a word disappears or is masked 50 to 60 ms after it is fixated, yet reading proceeds quite normally (Blythe, Liversedge, Joseph, White, & Rayner, 2009; Ishida & Ikeda, 1989; Liversedge, Rayner, White, Vergilino-Perez, Findlay, & Kentridge, 2004; Rayner, Inhoff, Morrison, Slowi-aczek, & Bertera, 1981; Rayner, Liversedge, & White, 2006; Rayner, Liversedge, White, & Vergilino-Perez, 2003). Even though the word is only visually present for a very short amount of time, readers fixate the location for the same amount of time they would have if the word had remained visible—suggesting that the duration of a given fixation is not simply determined by bottom-up, visual processing (in which case we would expect fixations to be shorter when the text disappears), but rather is highly dependent on the ongoing linguistic processing, and therefore subject to direct control. Furthermore, the length of time readers fixate the location is highly dependent on lexical variables like word frequency, indicating quite strongly that the decision of when to move the eyes is dependent on the ongoing cognitive processing of language. Interestingly, while reading proceeds normally when the fixated word is only available for 50 to 60 ms, if the upcoming parafoveal word disappears at the same time, there is a large disruption in reading (Rayner, Liversedge, & White, 2006), again highlighting the important role for attentional shifts and parafoveal preprocessing during reading.

Models of Eye-Movement Control during Reading

Leading models of eye-movement control during reading differ in their assumptions of how many words can be processed at one time. Serial models posit that words are processed and identified one at a time, with processing of one word finishing before processing of the next word can begin. In contrast, parallel models assume that multiple words (as many as three or four) can be processed at the same time. While it is beyond the scope of the current chapter to fully unpack these different models of reading, we will briefly discuss two of the leading models, E-Z Reader, a serial model, and SWIFT, a parallel model.[2] Underlying these models' assumptions of serial or parallel word processing are different implementations of the way and extent to which visual attention is distributed during the reading process. These differences in attentional allocation will be our focus.

The E-Z Reader model (Pollatsek, Reichle, & Rayner, 2006; Reichle, Pollatsek, Fisher, & Rayner, 1998; Reichle et al., 2003) has at its core two basic assumptions: (1) that the signal to move the eyes is determined by lexical processing and (2) that attention is allocated from one word to the next in a strictly serial fashion. In this model, when the eyes fixate on a word (n), there are two stages of lexical processing to complete on that word: early L_1 and late L_2 stages of lexical processing. The L_1 stage is a cursory processing of the word while the L_2 stage involves more thorough processing. Completion of the early stage of lexical processing (L_1) on word n is the signal to begin moving the eyes to the upcoming word (n + 1), though, as previously discussed, the saccade execution may take as long as 175 ms (though this value is actually shorter in current versions of the model). Completion of L_2 on word n is the signal to shift attention to word n + 1. Early lexical processing (L_1) of word n + 1 then can begin prior to an actual fixation on that word (the mechanism for parafoveal preview benefit). Meanwhile, a saccade program to move the eyes from word n to word n + 1 has been generated. Here too there are two stages, M_1, the labile stage, and M_2, the nonlabile stage. While in M_1 saccade programs can be canceled. This comes into play, for example, if L_1 on word n + 1 finishes before M_1 on word n completes; in this case the original saccade program can be canceled and a new saccade to word n + 2 planned (the basis for word skipping). Thus, reading proceeds in this serial fashion, with word identification of the foveal word, followed by a shift in attention to the upcoming word, where parafoveal processing can begin while the system waits for the eyes to "catch up."

In its basic architecture, the SWIFT model (Engbert, Longtin, & Kliegl, 2002; Engbert, Nuthmann, Richter, & Kliegl, 2005; Richter, Engbert, & Kliegl, 2006) shares many key assumptions with the E-Z Reader model; both models have two stages for word identification and two stages for saccadic programming. However, the models differ in their implementations of attentional allocation. In E-Z Reader, attention shifts are strictly serial, while SWIFT assumes that attention is distributed in a gradient over as many as four words, with the implication that processing is spatially distributed and therefore occurs in parallel over multiple words. Furthermore, saccadic programming in SWIFT is not under the direct control of lexical processing to the same extent as in E-Z Reader, and the signals of when and where to move the eyes are decoupled. In SWIFT, the *when* decision is determined by a random timer that triggers a saccade to the next-fixated word (with a foveal inhibition mechanism that can delay saccades when the foveal word is difficult to process). This saccade is not necessarily to the word to the right of fixation however, as the *where* decision is determined by levels of activation across the words in the attentional gradient. These activation levels are continually changing—increasing on a given word during the first stage of lexical processing and decreasing during the second stage—and words with higher levels of activation (those which have received intermediate levels of processing) are more likely to be the targets of saccades.

While both models are able to account for a large amount of the eye-movement data, SWIFT's (and other parallel models') assumption that multiple words can be identified simultaneously is somewhat problematic (Reichle, Liversedge, Pollatsek, & Rayner, 2009). First, it is difficult to account for how parallel models are able to preserve word order and proper syntax when words are not identified in order (Rayner, Angele, Schotter, & Bicknell, 2013). This is not a problem for serial models, as word order falls out naturally from the processing of words in order, one at a time, much like the processing of speech. Even if a mechanism to restore word order is implemented in a parallel model, no current models of word identification allow for more than one word to be identified simultaneously. Reichle et al. (2009) further argued for the serial allocation of attention during reading, citing evidence from visual search that the system needs attention to bind information about an object during visual (conjunctive) search tasks (Treisman & Gelade, 1980). Indeed, Reichle, Vanyukov, Laurent, and Warren (2008) measured eye movements and reaction times as subjects performed visual search tasks that ranged in the level of processing required from very shallow to increasingly complex and approximating the cognitive demands of normal silent reading (asterisk detection, letter detection, rhyme judgment, semantic judgment). Reichle et al. (2008) found an increase in response time, number of fixations, and first fixation duration as the number of words displayed on a given trial increased from 1 to 4 for letter detection, rhyme judgment, and semantic judgment, indicating an increased cost associated with processing multiple words. The same pattern was not observed for asterisk detection, suggesting that this task was a parallel search task, but that the more complex tasks (yet still arguably less complex than normal, silent reading) required serial search to perform. Thus, Reichle et al. (2009) argued that it is unclear how the visual features that comprise a given word can be bound into a representation that is distinct from neighboring words if attention is distributed across them, as is the case according to the SWIFT model.

One criticism that has been raised of serial models is that they have difficulty explaining how fixation durations can be so short (generally on the order of 225–250 ms). Since words take 150 to 300 ms to identify and saccades take 150 to 200 ms to program, the argument is that fixations should then be on the order of 300 to 500 ms, the time it would take to identify the word and then execute an eye movement. Parallel models account for short fixation durations by allowing a word to be identified over multiple fixations so long as it is still located within the attentional gradient. Serial models too can account for the short fixation durations because the time needed to identify a word and the time needed to program a saccade need not be additive. E-Z Reader accounts for short fixation durations by splitting word identification into two stages and allowing lexical identification to complete during the programming of a saccade to the next word. Since saccadic programming and lexical identification are supported by distinct systems (unlike the identification of multiple words), Reichle et al. (2009) argued that it is reasonable to assume that they can co-occur.

Integration: Reading in Everyday Contexts

As can be seen from this short introduction to the study of visual attention during reading, much of what we know has come out of research involving eye tracking during reading. As such, the study of visual attention during reading is well suited to laboratory investigation. Furthermore, laboratory-based eye-tracking studies have a great deal of ecological validity, as subjects are engaged in the task under investigation (i.e., silent reading for comprehension) and are not asked to make decisions or execute additional responses. Indeed, Tinker (1939) reported long ago that overall reading rate and comprehension do not differ between a reader sitting in a chair with a book in hand and reading in an eye-tracking laboratory.

While it is certainly possible to study many aspects of reading outside of the laboratory (e.g., comprehension, memory, words per minute), studying everyday visual attention during reading is not readily adaptable to investigation outside of the lab. The study of visual attention during reading requires knowing where a reader is looking when, and these detailed dynamics of visual attention may not be available for introspection or self-report, nor do they fall out of later tests of comprehension or memory. Instead of moving the investigation of attention during reading out of the laboratory, much research within the laboratory has focused on reading during regular, everyday activities such as browsing the Internet, reading periodicals and advertisements, researching a product for purchase, and reading subtitles (for reviews, see Higgins, Leinenger, & Rayner, 2014; Wedel & Pieters, 2008). Other research has investigated attention during other reading tasks with normal, everyday goals like proofreading (Kaakinen & Hyönä, 2010; Schotter, Bicknell, Howard, Levy, & Rayner, 2014) or researching places to live (Kaakinen & Hyönä, 2014; Kaakinen, Hyönä, & Keenan, 2002), providing insights into the ways specific goals and tasks can affect attentional allocation during reading. While still under investigation in the lab, the ecological validity of such everyday tasks makes generalizing to normal reading outside of the lab even easier. Of course, much of what we know from laboratory investigations can inform our understanding of reading during regular, everyday activities. For example, when reading Web pages, readers are more likely to be distracted by advertisements that are immediately to the right of the body of text being read than ones which run along the top of the text (e.g., Kuisma, Simola, Uusitalo, & Öörni, 2010; Simola, Kuisma, Öörni, Uusitalo, & Hyönä, 2011), perhaps because the ads immediately to the right of the text fall within a reader's perceptual span when they near the end of each line of text—a fact that can be exploited by Web developers and advertisers.

Additionally, as texts are adapted for reading on smaller dynamic displays (e.g., cell phones and smart watches), it will be important for app developers to bear in mind the results of laboratory investigations of eye movements and visual attention when designing reading interfaces, as failing to do so may result in the development of reading aids which in fact disrupt the normal reading process. For example, current attempts to increase reading speed by removing eye movements from the reading process (e.g., using rapid serial visual presentation, where words are presented one at a time, at a fixed location, for a fixed duration—e.g., Spritz) do not take into account that fixations are under direct cognitive control, and therefore such "reading aids" may actually hamper comprehension if readers are unable to control the amount of time spent reading a given word or if they are unable to make a regression to bolster their comprehension (Schotter, Tran, & Rayner, 2014). Furthermore, by presenting words one at a time in an attempt to speed the reading process, such apps do not allow for any parafoveal preprocessing of the upcoming words, thereby removing an aspect of normal reading which has been shown to decrease fixation durations and increase overall reading speed (e.g., Rayner, 2009; Schotter et al., 2012). More effective reading apps will be those that are designed with an understanding of the interaction of visual attention, linguistic processing, and occulomotor control.

Summary and Future Directions

For more than a century, researchers have concerned themselves with the study of attention during reading. Beginning with Huey and contemporaries in the late 1800s and early 1900s, the field of eye movements during reading has sought to determine how much information can be processed on a single fixation, a question that is inherently concerned with attentional constraints. In this chapter we have reviewed the basics of eye movements and visual attention during reading as they inform Huey's and related questions. Research utilizing gaze-contingent display-change techniques has revealed that readers make use of roughly the currently fixated word and the two upcoming words when reading, that this perceptual span is biased in the direction of attention, and that it is limited because of attentional constraints rather than decreasing visual acuity. Furthermore, readers are able to covertly shift their attention to the upcoming parafoveal word to begin processing it prior to actually executing an eye movement, resulting in faster foveal processing when the word is ultimately fixated (parafoveal preview benefit). Evidence primarily from studies of disappearing text demonstrated that eye movements are in fact under the direct control of linguistic processing, as fixation durations on a word are dependent on lexical variables even if the visual information is removed after 50 to 60 ms. Finally, models of eye-movement control during reading differ with respect to how they implement attentional allocation during reading. Serial models argue that attention is allocated to one word at a time with shifts in attention determined by word identification, while parallel models argue that attention is distributed in a gradient over as many as four words at a given time.

Moving forward, future work should continue to test predictions of both the E-Z Reader and SWIFT models of eye-movement control during reading. Since their main point of departure is the extent to which they argue that attention is distributed during normal reading (not at all in the case of E-Z Reader), continued testing of the two models will help us move toward a better understanding of eye-movement control which will necessarily include a better understanding of attentional allocation during reading. Of course, we also believe that there are many good studies that can still be done on various issues that are not directly relevant to a discussion of the models.

Additionally, future work should continue to explore the dynamics of attentional allocation and the perceptual span within an individual as a function of bottom-up factors, like text difficulty, and top down-factors, like task demands and reader goals. For example, previous work has demonstrated that the perceptual span is larger when reading easy text (Henderson & Ferreira, 1990) and when reading text that is irrelevant to a reader's primary goal (Kaakinen & Hyönä, 2014). Taken together, this suggests that readers are processing more parafoveal information when they are reading easy text that is not very relevant or important to them. In such reading situations, what are the implications, for example, for attentional capture by text that is not part of the main text being read (e.g., advertisements) but still falls within the perceptual span? The answers to this and related questions are intrinsically interesting but are also surely relevant, for example, to advertisers hoping to capture the attention of individuals browsing the Internet and periodicals.

Finally, regardless of whether the research is being conducted within or outside of a lab, future research should continue to explore the allocation of visual attention during silent reading for comprehension, as well as during other reading tasks like those under investigation in more recent work on browsing the Internet, researching products to purchase or places to live, and reading subtitles while watching TV. The combination will help us build a more complete understanding of attention during reading across a multitude of reading tasks.

Box 12.1
Key Points

- Research using the moving window paradigm has demonstrated that the perceptual span encompasses roughly the currently fixated word and the two upcoming words, that it is limited because of attentional constraints rather than visual acuity, that it is biased in the direction of attention, and that it varies as a function of text difficulty. Research using the boundary change paradigm has revealed that covert attention can shift to the upcoming word where processing can begin prior to an actual fixation on the word, resulting in preview benefit—faster foveal processing when the word is ultimately fixated.

- Eye movements during reading are under direct cognitive control, and how long the eyes remain fixated on a given word largely depends on how easy or difficult it is to process.

- Models of eye-movement control differ in the extent to which they assume attention (and also lexical processing) is distributed, with serial models (e.g., E-Z Reader) arguing that attention is allocated to one word at a time, and parallel models (e.g., SWIFT) arguing that attention is distributed in a gradient over as many as three to four words simultaneously.

Box 12.2
Outstanding Issues

- How do attention gradient (parallel) models of eye-movement control account for the restoration of word order, and if attention is distributed over multiple words, what is the mechanism for binding the visual features of one word to a distinct linguistic representation separate from that of neighboring words?

- For a given reader, how does the perceptual span change as a function of both bottom-up factors like text difficulty or low-level visual properties of the display and top-down factors like task demands or user goals? How do these different factors interact, and what are the implications for this interactivity on distractibility or memory?

Notes

1. Average saccade length varies across languages as a function of information density. For example, in Chinese, a character-based language, average saccade length is only 2 characters long (Shen, 1927; Wang, 1935; Stern 1978). This corresponds to roughly one word though, as Chinese words range from 1 to 4 characters, with the majority being 1 to 2 characters in length.

2. These are by no means the only models of eye-movement control during reading. For a discussion of these and other models, see Reichle, Rayner, and Pollatsek (2003).

References

Apel, J. K., Henderson, J. M., & Ferreira, F. (2012). Targeting regressions: Do readers pay attention to the left? *Psychonomic Bulletin & Review*, *19*, 1108–1113.

Ashby, J., Rayner, K., & Clifton, C. (2005). Eye movements of highly skilled and average readers: Differential effects of frequency and predictability. *Quarterly Journal of Experimental Psychology*, *58*, 1065–1086.

Bélanger, N. N., Slattery, T. J., Mayberry, R. I., & Rayner, K. (2012). Skilled deaf readers have an enhanced perceptual span in reading. *Psychological Science*, *23*, 816–823.

Blythe, H. I., Liversedge, S. P., Joseph, H. S. S. L., White, S. J., & Rayner, K. (2009). Visual information capture during fixations in reading for children and adults. *Vision Research*, *49*, 1583–1591.

Buswell, G. T. (1922). *Fundamental reading habits: A study of their development*. Chicago: University of Chicago Press.

Campbell, F. W., & Wurtz, R. H. (1978). Saccadic omission: Why we do not see a grey-out during a saccadic eye movement. *Vision Research*, *18*, 1297–1303.

Dambacher, M., Slattery, T. J., Yang, J., Kliegl, R., & Rayner, K. (2013). Evidence for direct control of eye movements during reading. *Journal of Experimental Psychology: Human Perception and Performance*, *39*, 1468–1484.

Deubel, H., & Schneider, W. X. (1996). Saccade target selection and object recognition: Evidence for a common attentional mechanism. *Vision Research*, *36*, 1827–1837.

Duffy, S. A., Morris, R. K., & Rayner, K. (1988). Lexical ambiguity and fixation times in reading. *Journal of Memory and Language*, *27*, 429–446.

Ehrlich, S. F., & Rayner, K. (1981). Contextual effects on word recognition and eye movements during reading. *Journal of Verbal Learning and Verbal Behavior*, *20*, 641–655.

Engbert, R., Longtin, A., & Kliegl, R. (2002). A dynamical model of saccade generation in reading based on spatially distributed lexical processing. *Vision Research*, *42*, 621–636.

Engbert, R., Nuthmann, A., Richter, E. M., & Kliegl, R. (2005). SWIFT: A dynamical model of saccade generation during reading. *Psychological Review*, *112*, 777–813.

Frazier, L., & Rayner, K. (1982). Making and correcting errors during sentence comprehension: Eye movements in the analysis of structurally ambiguous sentences. *Cognitive Psychology*, *14*, 178–210.

Häikiö, T., Bertram, R., Hyönä, J., & Niemi, P. (2009). Development of the letter identity span in reading: Evidence from the eye movement moving window paradigm. *Journal of Experimental Child Psychology*, *102*, 167–181.

Henderson, J. M., & Ferreira, F. (1990). Effects of foveal processing difficulty on the perceptual span in reading: Implications for attention and eye movement control. *Journal of Experimental Psychology: Learning, Memory, and Cognition*, *16*, 417–429.

Higgins, E., Leinenger, M., & Rayner, K. (2014). Eye movements when viewing advertisements. *Frontiers in Psychology*, *5*, 210. doi:10.3389/fpsyg.2014.00210.

Huey, E. B. (1908). *The psychology and pedagogy of reading*. New York: Macmillan. [Republished: Cambridge, MA: MIT Press, 1968.]

Inhoff, A. W., & Briihl, D. (1991). Semantic processing of unattended text during selective reading: How the eyes see it. *Perception & Psychophysics*, *49*, 289–294.

Inhoff, A. W., & Liu, W. (1998). The perceptual span and oculomotor activity during the reading of Chinese sentences. *Journal of Experimental Psychology: Human Perception and Performance*, *24*, 20–34.

Inhoff, A. W., & Rayner, K. (1986). Parafoveal word processing during eye fixations in reading: Effects of word frequency. *Perception & Psychophysics*, *40*, 431–439.

Inhoff, A. W., & Topolski, R. (1992). Lack of semantic activation from unattended text during passage reading. *Bulletin of the Psychonomic Society*, *30*, 365–366.

Ishida, T., & Ikeda, M. (1989). Temporal properties of information extraction in reading studied by a text-mask replacement technique. *Journal of the Optical Society of America*, *6*, 1624–1632.

Jordan, T. R., Almabruk, A. A. A., Gadalla, E. A., McGowan, V. A., White, S. J., Abedipour, L., & Paterson, K. B. (2014). Reading direction and the central perceptual span: Evidence from Arabic and English. *Psychonomic Bulletin & Review*, *21*, 505–511.

Juhasz, B. J., & Rayner, K. (2003). Investigating the effects of a set of intercorrelated variables on eye fixation durations in reading. *Journal of Experimental Psychology: Learning, Memory, and Cognition*, *29*, 1312–1318.

Juhasz, B. J., & Rayner, K. (2006). The role of age of acquisition and word frequency in reading: Evidence from eye fixation durations. *Visual Cognition*, *13*, 846–863.

Just, M. A., & Carpenter, P. A. (1976). Eye fixations and cognitive processes. *Cognitive Psychology*, *8*, 441–480.

Just, M. A., & Carpenter, P. A. (1980). A theory of reading: From eye fixations to comprehension. *Psychological Review*, *87*, 329–354.

Kaakinen, J. K., & Hyönä, J. (2010). Task effects on eye movements during reading. *Journal of Experimental Psychology: Learning, Memory, and Cognition*, *36*, 1561–1566.

Kaakinen, J. K., & Hyönä, J. (2014). Task relevance induces momentary changes in the functional visual field during reading. *Psychological Science*, *25*, 626–632.

Kaakinen, J. K., Hyönä, J., & Keenan, J. M. (2002). Perspective effects on online text processing. *Discourse Processes*, *33*, 159–173.

Kowler, E., Anderson, E., Dosher, B., & Blaser, E. (1995). The role of attention in the programming of saccades. *Vision Research*, *35*, 1897–1916.

Kuisma, J., Simola, J., Uusitalo, L., & Öörni, A. (2010). The effects of animation and format on the perception and memory of online advertising. *Journal of Interactive Marketing*, *24*, 269–282.

Levy, R., Bicknell, K., Slattery, T., & Rayner, K. (2009). Eye movement evidence that readers maintain and act on uncertainty about past linguistic input. *Proceedings of the National Academy of Sciences of the United States of America*, *106*, 21086–21090.

Liversedge, S. P., Rayner, K., White, S. J., Vergilino-Perez, D., Findlay, J. M., & Kentridge, R. W. (2004). Eye movements when reading disappearing text: Is there a gap effect in reading? *Vision Research*, *44*, 1013–1024.

Matin, E. (1974). Saccadic suppression: A review and an analysis. *Psychological Bulletin*, *81*, 899–917.

McCann, R. S., Folk, C. L., & Johnston, J. C. (1992). The role of spatial attention in visual word processing. *Journal of Experimental Psychology: Human Perception and Performance*, *18*, 1015–1029.

McConkie, G. W., & Rayner, K. (1975). The span of the effective stimulus during a fixation in reading. *Perception & Psychophysics*, *17*, 578–586.

McConkie, G. W., & Rayner, K. (1976). Asymmetry of the perceptual span in reading. *Bulletin of the Psychonomic Society, 8*, 365–368.

McConkie, G. W., & Zola, D. (1979). Is visual information integrated across successive fixations in reading? *Perception & Psychophysics, 25*, 221–224.

Miellet, S., O'Donnell, P. J., & Sereno, S. C. (2009). Parafoveal magnification: Visual acuity does not modulate the perceptual span in reading. *Psychological Science, 20*, 721–728.

Pollatsek, A., Bolozky, S., Well, A. D., & Rayner, K. (1981). Asymmetries in the perceptual span for Israeli readers. *Brain and Language, 14*, 174–180.

Pollatsek, A., Raney, G. E., LaGasse, L., & Rayner, K. (1993). The use of information below fixation in reading and in visual search. *Canadian Journal of Experimental Psychology, 47*, 179–200.

Pollatsek, A., Reichle, E. D., & Rayner, K. (2006). Tests of the E-Z Reader model: Exploring the interface between cognition and eye-movement control. *Cognitive Psychology, 52*, 1–56.

Posner, M. I. (1980). Orienting of attention. *Quarterly Journal of Experimental Psychology, 32*, 3–25.

Rayner, K. (1975). The perceptual span and peripheral cues in reading. *Cognitive Psychology, 7*, 65–81.

Rayner, K. (1978). Eye movements in reading and information processing. *Psychological Bulletin, 85*, 618–660.

Rayner, K. (1983). Eye movements, perceptual span, and reading disability. *Annals of Dyslexia, 33*, 163–173.

Rayner, K. (1986). Eye movements and the perceptual span in beginning and skilled readers. *Journal of Experimental Child Psychology, 41*, 211–236.

Rayner, K. (1998). Eye movements in reading and information processing: 20 years of research. *Psychological Bulletin, 124*, 372–422.

Rayner, K. (2009). The thirty fifth Sir Frederick Bartlett Lecture: Eye movements and attention in reading, scene perception, and visual search. *Quarterly Journal of Experimental Psychology, 62*, 1457–1506.

Rayner, K. (2014). The gaze-contingent moving window paradigm: Development and review. *Visual Cognition, 22*, 242–258.

Rayner, K., Angele, B., Schotter, E. R., & Bicknell, K. (2013). On the processing of canonical word order during eye fixations in reading: Do readers process transposed word previews? *Visual Cognition, 21*, 353–381.

Rayner, K., & Bertera, J. H. (1979). Reading without a fovea. *Science, 206*, 468–469.

Rayner, K., & Duffy, S. A. (1986). Lexical complexity and fixation times in reading: Effects of word frequency, verb complexity, and lexical ambiguity. *Memory & Cognition, 14*, 191–201.

Rayner, K., Inhoff, A. W., Morrison, R. E., Slowiaczek, M. L., & Bertera, J. H. (1981). Masking of foveal and parafoveal vision during eye fixations in reading. *Journal of Experimental Psychology: Human Perception and Performance, 7*, 167–179.

Rayner, K., Kambe, G., & Duffy, S. A. (2000). The effect of clause wrap-up on eye movements during reading. *Quarterly Journal of Experimental Psychology, A, 53*, 1061–1080.

Rayner, K., Liversedge, S. P., & White, S. J. (2006). Eye movements when reading disappearing text: The importance of the word to the right of fixation. *Vision Research, 46*, 310–323.

Rayner, K., Liversedge, S. P., White, S. J., & Vergilino-Perez, D. (2003). Reading disappearing text: Cognitive control of eye movements. *Psychological Science, 14*, 385–388.

Rayner, K., McConkie, G. W., & Ehrlich, S. (1978). Eye movements and integrating information across fixations. *Journal of Experimental Psychology: Human Perception and Performance, 4*, 529–544.

Rayner, K., McConkie, G. W., & Zola, D. (1980). Integrating information across eye movements. *Cognitive Psychology, 12*, 206–226.

Rayner, K., Murphy, L. A., Henderson, J. M., & Pollatsek, A. (1989). Selective attentional dyslexia. *Cognitive Neuropsychology, 6*, 357–378.

Rayner, K., & Pollatsek, A. (1981). Eye movement control during reading: Evidence for direct control. *Quarterly Journal of Experimental Psychology, 33*, 351–373.

Rayner, K., & Pollatsek, A. (1989). *The psychology of reading*. Englewood Cliffs, NJ: Prentice-Hall.

Rayner, K., Pollatsek, A., Ashby, J., & Clifton, C. (2012). *Psychology of reading*. New York: Psychology Press.

Rayner, K., Reichle, E. D., Stroud, M. J., Williams, C. C., & Pollatsek, A. (2006). The effect of word frequency, word predictability, and font difficulty on the eye movements of young and older readers. *Psychology and Aging, 21*, 448–465.

Rayner, K., Slattery, T. J., & Bélanger, N. N. (2010). Eye movements, the perceptual span, and reading speed. *Psychonomic Bulletin & Review, 17*, 834–839.

Rayner, K., & Well, A. D. (1996). Effects of contextual constraint on eye movements in reading: A further examination. *Psychonomic Bulletin & Review, 3*, 504–509.

Rayner, K., Well, A. D., & Pollatsek, A. (1980). Asymmetry of the effective visual field in reading. *Perception & Psychophysics, 27*, 537–544.

Rayner, K., Well, A. D., Pollatsek, A., & Bertera, J. H. (1982). The availability of useful information to the right of fixation in reading. *Perception & Psychophysics, 31*, 537–550.

Reichle, E. D., Liversedge, S. P., Pollatsek, A., & Rayner, K. (2009). Encoding multiple words simultaneously in reading is implausible. *Trends in Cognitive Sciences, 13*, 115–119.

Reichle, E. D., Pollatsek, A., Fisher, D. L., & Rayner, K. (1998). Toward a model of eye movement control in reading. *Psychological Review, 105*, 125–157.

Reichle, E. D., Rayner, K., & Pollatsek, A. (2003). The E-Z Reader model of eye-movement control in reading: Comparisons to other models. *Behavioral and Brain Sciences, 26*, 445–476.

Reichle, E. D., Vanyukov, P. M., Laurent, P. A., & Warren, T. (2008). Serial or parallel? Using depth-of-processing to examine attention allocation during reading. *Vision Research, 48*, 1831–1836.

Reingold, E. M., Reichle, E. D., Glaholt, M. G., & Sheridan, H. (2012). Direct lexical control of eye movements in reading: Evidence from a survival analysis of fixation durations. *Cognitive Psychology, 65*, 177–206.

Richter, E. M., Engbert, R., & Kliegl, R. (2006). Current advances in SWIFT. *Cognitive Systems Research, 7*, 23–33.

Risko, E. F., Stolz, J. A., & Besner, D. (2010). Spatial attention modulates feature crosstalk in visual word processing. *Attention, Perception & Psychophysics, 72*, 989–998.

Schad, D. J., Risse, S., Slattery, T., & Rayner, K. (2014). Word frequency in fast priming: Evidence for immediate cognitive control of eye movements during reading. *Visual Cognition, 22*, 390–414.

Schotter, E. R. (2013). Synonyms provide semantic preview benefit in English. *Journal of Memory and Language, 69*, 619–633.

Schotter, E. R., Angele, B., & Rayner, K. (2012). Parafoveal processing in reading. *Attention, Perception & Psychophysics, 74*, 5–35.

Schotter, E. R., Bicknell, K., Howard, I., Levy, R., & Rayner, K. (2014). Task effects reveal cognitive flexibility responding to frequency and predictability: Evidence from eye movements in reading and proofreading. *Cognition, 131*, 1–27.

Schotter, E. R., Tran, R., & Rayner, K. (2014). Don't believe what you read (only once): Comprehension is supported by regressions during reading. *Psychological Science, 25*, 1218–1226.

Sereno, S. C., O'Donnell, P. J., & Rayner, K. (2006). Eye movements and lexical ambiguity resolution: Investigating the subordinate-bias effect. *Journal of Experimental Psychology: Human Perception and Performance, 32*, 335–350.

Shen, E. (1927). An analysis of eye movements in the reading of Chinese. *Journal of Experimental Psychology, 10*, 158–183.

Simola, J., Kuisma, J., Öörni, A., Uusitalo, L., & Hyönä, J. (2011). The impact of salient advertisements on reading and attention on Web pages. *Journal of Experimental Psychology. Applied, 17*, 174–190.

Slattery, T. J., Angele, B., & Rayner, K. (2011). Eye movements and display change detection during reading. *Journal of Experimental Psychology: Human Perception and Performance, 37*, 1924–1938.

Slattery, T. J., & Rayner, K. (2010). The influence of text legibility on eye movements during reading. *Applied Cognitive Psychology, 24*, 1129–1148.

Stern, J. A. (1978). Eye movements, reading and cognition. In J. W. Denders, D. F. Fisher, & R. A. Monty (Eds.), *Eye movements and the higher psychological functions* (pp. 145–155). Hillsdale, NJ: Erlbaum.

Tinker, M. A. (1939). Reliability and validity of eye-movement measures of reading. *Journal of Experimental Psychology, 19,* 732–746.

Tinker, M. A. (1958). Recent studies of eye movements in reading. *Psychological Bulletin, 55,* 215–231.

Treisman, A. M., & Gelade, G. (1980). A feature-integration theory of attention. *Cognitive Psychology, 12,* 97–136.

Veldre, A., & Andrews, S. (2013). Lexical quality and eye movements: Individual differences in the perceptual span of skilled adult readers. *Quarterly Journal of Experimental Psychology, 67,* 703–727.

Wang, F. C. (1935). An experimental study of eye-movements in the silent reading of Chinese. *Experimental School Journal, 35,* 527–539.

Wedel, M., & Pieters, R. (2008). A review of eye-tracking research in marketing. In N. Malhotra (Ed.), *Review of marketing research* (Vol. 4, pp. 123–147). New York: Sharpe.

Woodworth, R. S. (1938). *Experimental psychology.* New York: Holt.

13

Embodied Attention

Richard A. Abrams and Blaire J. Weidler

We are built for action. And so it should come as no surprise that the brain mechanisms that acquire information from the environment, mechanisms of selective attention, depend upon our current action capabilities—even if we don't have an action explicitly planned at the moment. Effectively directed attention is a necessary precursor to action. In this chapter we describe recent research that shows that attention is allocated in a manner that reflects the body's current capabilities—*embodied attention*. By "current capabilities" we limit our treatment mostly to one simple manipulation: the proximity of an observer's hands to the stimuli under consideration. When the hands are near an object, the object can potentially be manipulated, and it is that capability that leads to changes in attention. We discuss these changes next, focusing specifically on the processing that occurs in the absence of any overt or planned action—other chapters in this volume address the interrelationship between action and attention (see, e.g., Pratt et al., chapter 14 in the current volume).

Historical Context

The work that we describe in this chapter fits within the broader perspective known as *embodied cognition* in which one's current physical capabilities or previous experiences are seen as being crucial to cognition more generally (Glenberg, Witt, & Metcalfe, 2013; Wilson, 2002). Evidence for the importance of such embodiment has come from a variety of domains, including not only work on perception (e.g., Bhalla & Proffitt, 1999; Bloesch, Davoli, Roth, Brockmole, & Abrams, 2012; Gibson, 1979; Proffitt, 2006) but also work on conceptual understanding (Lakoff & Johnson, 1980) and language comprehension (Barsalou, 1999). For example, Estes, Verges, and Barsalou (2008) had participants identify letters that were near either the top or the bottom of a video display. Before a letter was presented, however, a word appeared describing an object that might typically be associated with locations high in the visual field (e.g., bird) or low in it (e.g., boot). Estes et al. found that letter identification was affected by the typical location of the cue word—as if the participants had moved their attention to the typical location automatically in response to reading the cue word. Results such as this suggest a close link between our understanding of objects and events in the world and our own sensory experiences with those objects and events.

Much of the thinking about embodiment has its roots in Gibson's (1979) concept of *affordances*. Gibson believed that perception involves the acquisition of information about the behavioral opportunities that are provided by the environment. For example, the handle of a beer mug affords an opportunity for grasping. In a similar manner, the proximity of the hands to nearby objects may afford unique opportunities for behavior—and may also present unique demands to the attention system for information acquisition. Those unique attentional demands are our focus here.

One of the earliest demonstrations of an embodied attention effect was provided by Schendel and Robertson (2004). Their patient, WM, had a brain lesion that caused him to appear blind to stimuli presented to the left of fixation. But when WM extended his left hand toward the video display, much of his vision loss was restored. The extended hand only helped, however, if the display was within reach—showing that the critical factor affecting vision was the potential to interact with the stimuli. Similar results were reported by di Pellegrino and Frassinetti (2000; but see Smith, Lane, & Schenk, 2008), who also studied a patient with a brain lesion. More recent research has extended these observations to individuals without lesions. For example, Dufour and Touzalin (2008) had participants indicate the color or location of briefly presented stimuli that were either near to or far from a hand. They found that participants were more accurate and quicker to detect stimuli near the hand. Reed, Grubb, and Steele (2006) also recorded latencies in response to briefly presented targets. When participants had one hand extended to the display, targets near the hand were detected more quickly than targets that were more distant, even though the target location was not dependent on the location of the outstretched hand. The results suggest that the location of the hand biased the allocation of spatial attention, leading to a detection benefit when the target was nearby. The findings reveal an important role of hand position in the representation of peripersonal space—the space around the body (Brockmole, Davoli, Abrams, & Witt 2013).

Some of the changes in spatial attention just described could be driven by a bias toward one side or the other (the side near the sole extended hand; see Dufour & Touzalin, 2008). More recently, researchers have examined changes in vision when both hands are extended simultaneously toward the stimuli being viewed. For example, Abrams, Davoli, Du, Knapp, and Paull (2008) had participants search for a target

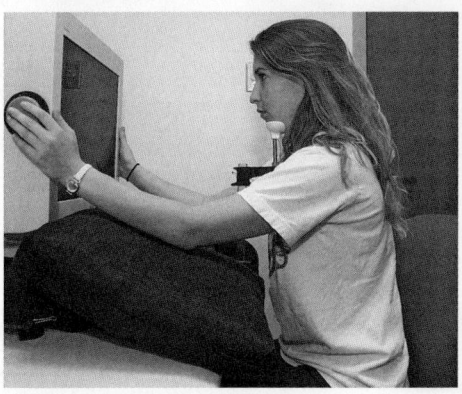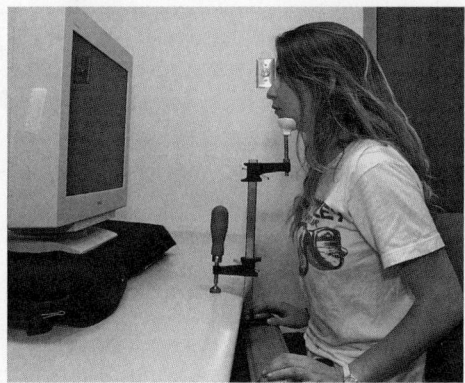

Figure 13.1
Postures used in the hands-near (left) and hands-far (right) condition of many experiments.

letter among distractors while adopting one of the two postures shown in figure 13.1: either with their hands straddling the video display or far away from it. When the hands were near the display, all of the stimuli presented were within reach, and any attentional changes that accrue to reachable objects should affect them. As shown in figure 13.2, Abrams et al. found that participants searched the display at a slower rate when their hands were nearby. (The search rate is determined by the slope of the function relating the search reaction time to the number of elements in the display, with a slower rate evidenced by a steeper slope.) The slower rate, they argued, would permit a more detailed analysis of the objects near the hands, presumably because the potential to interact with the objects would be facilitated by a thorough analysis.

The slowed search near the hands reported by Abrams et al. (2008) could theoretically be caused either by slower *movements* of spatial attention from one item to another or by inhibited *disengagement* of attention prior to movement from one item to the next during the serial search. In a subsequent experiment, Abrams et al. showed that the proximity of the hands to the display did not influence the effectiveness of a peripheral exogenous cue in a cued-detection paradigm, but the hands did reduce the magnitude of inhibition of return that occurred. Inhibition of return is a consequence of the disengagement of attention from an object, and the reduced magnitude suggests that participants were less effective in disengaging attention from objects that were near their hands. Such a finding is consistent with the conclusion that proximity to the hands compels a more detailed analysis.

It may not seem surprising that stimuli are attended differently when there are other stimuli (the hands) nearby. However, researchers have been careful to control for the

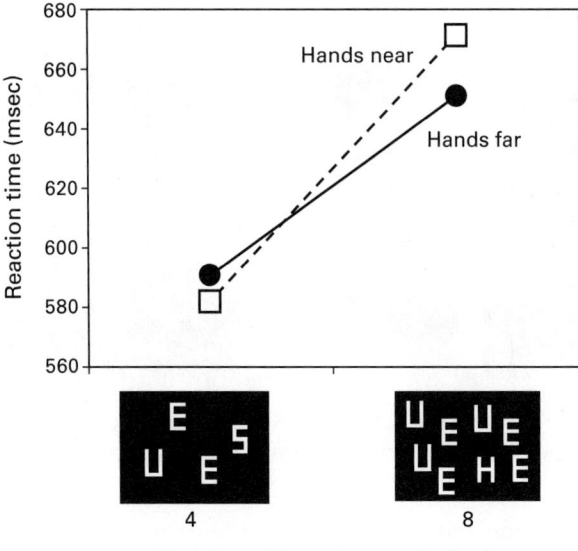

Figure 13.2
Visual search times for displays near to and far from the hands as a function of display size. From Abrams et al. (2008).

potential impact of the hands as mere visual anchors. For example, Reed et al. (2006) compared the effects of an extended hand to that of a wooden board, Abrams et al. (2008; also Reed et al., 2006) studied search when the hands were extended but hidden by a barrier, and Davoli and Abrams (2009) had participants merely *imagine* that their hands were near the stimuli. In each case, performance depended on the proximity (whether real or imagined) of the hands to the display, and not on the presence of visual anchors near the display.

State-of-the-Art Review

The current work on embodied attention falls into three broad categories: (1) research that reveals changes in the speed or efficiency of certain types of processing for objects near the hands—*quantitative* differences in processing, (2) research that demonstrates differences in the nature of the processing that takes place or in the mechanisms presumed to be involved—*qualitative* differences, and (3) work examining spatial and postural factors that might influence the occurrence of hand-proximity effects. We discuss each of these in turn. Finally, we finish the section by briefly considering postural effects other than those involving the hands.

Quantitative Differences in Processing Near the Hands

Delayed Disengagement versus Prioritized Selection   As described previously, some of the early work on hand-proximity effects revealed delayed disengagement of attention for stimuli near the hands (Abrams et al., 2008). Vatterott and Vecera (2013) also examined the possibility of delayed disengagement. They specifically sought to distinguish delayed disengagement near the hand from a hypothesis proposing prioritized attention near the hand. The authors used the *additional singleton paradigm* (e.g., Theeuwes, 1992) to distinguish between the two accounts of how hand proximity affects attention. In this paradigm, participants are asked to search for a shape singleton (e.g., a square among circles); however, on some trials an additional color singleton appears in the search display. The color singleton typically captures attention and slows target detection even though it is task irrelevant. According to Vatterott and Vecera, both the delayed disengagement and prioritized attention accounts would predict that the color singleton distractor should produce more of a cost near an outstretched hand than away from it. This is because either prioritized attention to the distractor or slow disengagement moving away from it will induce a cost. However, the attentional prioritization account additionally predicts that target (the unique shape) detection should be facilitated near the hand whereas the disengagement account does not. As expected, the authors found that a colored singleton's presence did indeed create more of a response time cost when near the hand. Additionally, in support of the delayed disengagement hypothesis, participants performed no faster when the target was near their hand compared to far from their hand.

Attentional Scope   Davoli, Brockmole, Du, and Abrams (2012) sought corroborating evidence for the hypothesis that attentional disengagement is delayed from stimuli near the hands. On each trial of their experiment participants saw compound shapes likes those shown in figure 13.3. The shapes consisted of one of two small elements (the local shape) that were arranged to depict one of two larger objects (the global shape). Participants were asked to first identify the shape at one level of analysis (e.g., the local level) and then a short time later to identify the shape at the other level of analysis (e.g., the global level). Thus, the attentional scope required to accomplish the task changed from one stimulus to the next. Davoli et al. reasoned that if elements near the hands demand particularly thorough evaluation, participants may be delayed at switching between global and local scopes of attention. That is precisely what the authors found: Participants were slower at identifying the second stimulus when stimuli were near the hands compared to when they were far. Importantly, participants were not simply impaired at performing the task near the hands: When the identification task remained the same across the two stimuli (e.g., identify the global shape in both), performance was equivalent in the two hand postures. Instead, when the stimuli were near the hands participants were slower to shift the attentional scope from that

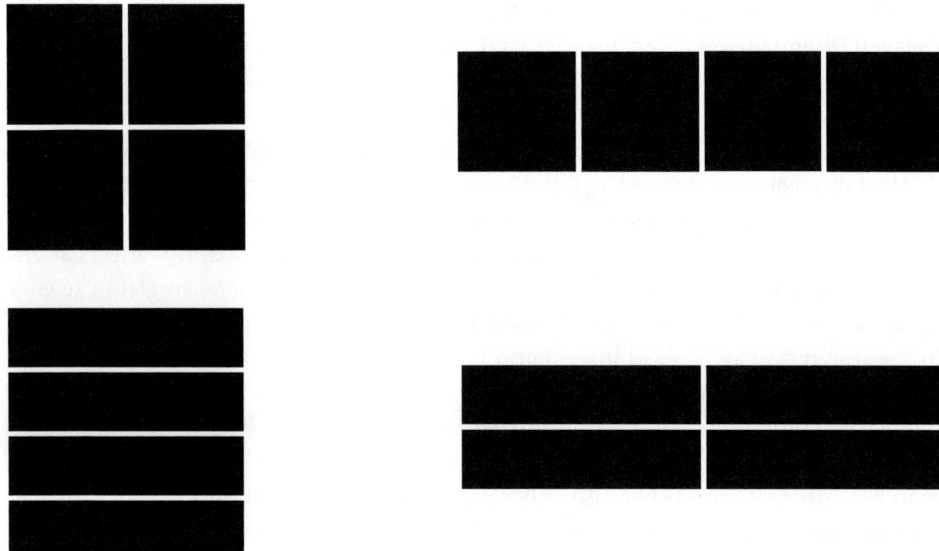

Figure 13.3
Stimuli like those used by Davoli, Brockmole, Du, & Abrams (2012). Participants identified either the local shape or the global shape of each stimulus.

needed for the first identification, and thus they exhibited slowed response times to the second stimulus.

Visual Short-Term Memory   One consequence of prolonged allocation of attention might be an improvement in memory for items near the hands. Tseng and Bridgeman (2011) examined this possibility by studying a form of short-term visual memory. They had participants view two successive displays, each containing multiple objects in different colors, looking for the sole difference between the two displays. Participants were better at identifying the difference when the stimuli were presented near the hands. Such a result is consistent with the more detailed analysis of objects near the hands that might be expected if people are inhibited from disengaging attention from such objects.

Eye Movements   Almost all of the work on hand-proximity effects has examined covert allocation of attention. Only recently have researchers begun to examine the extent to which *overt* movements of attention might be influenced by the proximity of the hands. Suh and Abrams (2013) had subjects produce eye movements either toward stimuli that suddenly appeared (prosaccades) or away from such stimuli (antisaccades).

It has been long known that there is an antisaccade cost: Latencies to look away from an exogenous stimulus are longer, presumably because the participant must disengage attention from the stimulus and program a saccade to a different location (e.g., Fischer & Weber, 1997). If proximity to the hands inhibits the disengagement of attention, then one would expect that the antisaccade cost would be greater for stimuli that are near the hands. That is exactly the pattern that Suh and Abrams (2013) found: Antisaccade latencies were 17 ms slower than prosaccades far from the hands but 38 ms slower than prosaccades near the hands. In a subsequent experiment, the researchers signaled the saccade target by the color of a patch presented suddenly at fixation. In this "endogenous saccade" condition, participants were slower to initiate saccades when their hands were near the display, suggesting that disengagement of attention from the color cue was more difficult near the hands.

Qualitatively Different Processing Near the Hands

The effects of hand proximity described so far could reflect a mechanism that merely prioritizes locations near the hand—either by moving attention to them preferentially or by delaying movements of attention away from them. Such a mechanism may advance (or retard) the processing of some stimuli but would not be expected to *change* such processing. Nevertheless, many researchers have reported qualitative changes in perceptual and cognitive processing for stimuli near the hands. The changes that have been observed are consistent with the suggestion that the space near the hands is important because objects in that space are candidates for action. These changes are reviewed next.

Figure–Ground Segregation   Cosman and Vecera (2010) studied a process thought to occur relatively early in perceptual processing—the segregation of figure from ground. Participants in their experiment viewed stimuli like those shown in figure 13.4 with ambiguous figure and ground regions—either of the two regions might be considered figure. After the presentation, participants were probed with two shapes, one of which had been presented initially, and asked to identify the previously presented shape. Enhanced memory for a shape in such a task serves as an implicit measure of figure–ground segregation. Cosman and Vecera found such an enhancement for shapes that had been presented near an outstretched hand, compared to shapes that were presented near a wooden dowel. Thus, the presence of a nearby hand didn't merely speed processing near the location of the hand—it altered the segregation of figure from ground.

Visual Learning   In the same spirit, Davoli, Brockmole, and Goujon (2012) also considered the possibility that processing of objects near the hands might be unique. They

Figure 13.4
Example of an ambiguous figure–ground stimulus similar to those used by Cosman and Vecera (2010).

had participants search for target letters hidden in complex multicolored shapes, some of which appeared multiple times during the session. Typically, observers find the target more quickly in the repeated images (relative to nonrepeated ones) as the session progresses, revealing that they have learned about the locations of the targets in those images. Indeed, in their first experiment that is what Davoli et al. found—however, the magnitude of the learning was the same for stimuli that were near the hands and those that were far away. In a second experiment, the researchers changed the images in a subtle way: The images that repeated differed in color (but not form) each time they were repeated. In that case, participants were much poorer at learning the images when they were near the hands, suggesting that vision near the hands is biased to item-specific details of the stimuli (i.e., a particular combination of shape and color) at the expense of general attributes (such as the shape in general). These results are consistent with the findings of Tseng and Bridgeman (2011) described earlier. They found improved visual short-term memory for changes in the colors of objects near the hands—an effect that would be expected if the hands produce an attentional bias that favors item-specific processing.

Cognitive Control   Qualitatively different processing may occur in a particular situation if higher order executive functions specify different requirements for the task being performed. Recent evidence suggests that these executive functions may be enhanced

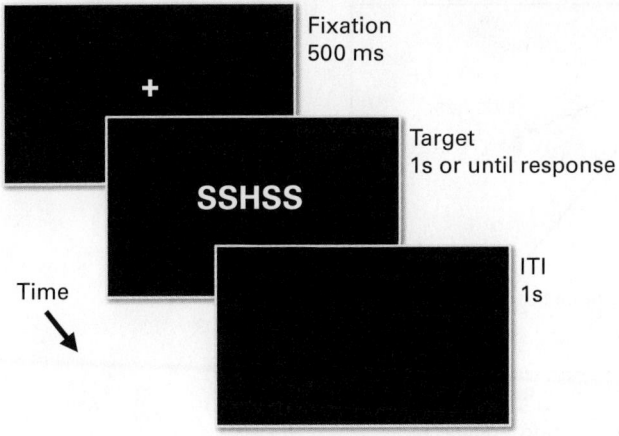

Figure 13.5
Sequence of events on a trial of Weidler and Abrams (2014). Participants identified the central letter while ignoring the flankers. ITI, intertrial interval.

for stimuli that are near the hands. Davoli, Du, Montana, Garverick, and Abrams (2010) provided the first evidence for this possibility. They asked participants to perform a Stroop color-word task (e.g., Stroop, 1935) with stimuli near to and far from the hands. Successful performance of the task requires maintenance of the requirements that the color is to be named and the word is to be suppressed, and a measure of successful performance is the magnitude of the interference produced by incongruent word–color combinations (relative to congruent or neutral ones). Consistent with an enhancement of executive function in near body space, the authors found reduced Stroop interference when participants had their hands near the stimuli.

Weidler and Abrams (2014) also studied executive control mechanisms near the hands. In one experiment, they had participants identify the central letter in a display that contained flanking distractor letters, as shown in the experimental events depicted in figure 13.5. Performance of such a *flanker task* would benefit from executive control mechanisms that maintain the requirement to respond to the target letter and to ignore the distractors. The results are shown in figure 13.6 which displays the accuracy with which participants judged the central letter as a function of the congruency between the central letter and the flanking distractors. Consistent with the enhanced performance in the Stroop paradigm noted earlier (Davoli et al., 2010), participants were less influenced by incongruent flankers when their hands were near the display. While the latter result appears to reflect enhanced cognitive control near the hands, it also remains possible that the benefit in flanker performance was caused by improved

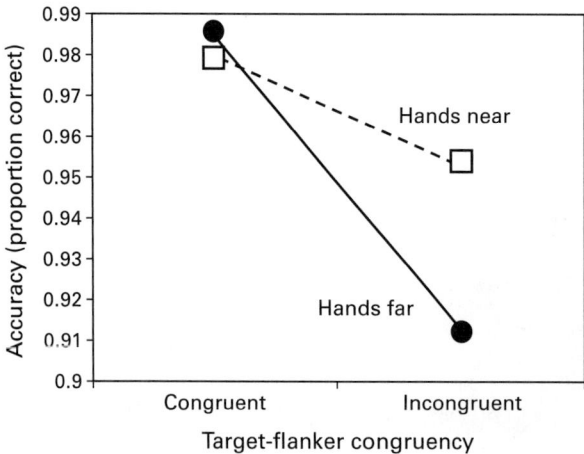

Figure 13.6
Accuracy of central letter identification as a function of the congruency of the distracting flankers separately for hands-near and hands-far postures from Weidler and Abrams (2014).

visual–spatial filtering and not by any change in executive control mechanisms. To rule out that possibility, Weidler and Abrams conducted an additional experiment. In the experiment, participants were asked to indicate either the color or the shape of a centrally presented colored shape. Importantly, on some trials they reported the same attribute that they had reported on the previous trial (e.g., a shape judgment followed by another shape judgment) whereas on other trials they were required to report the other attribute (e.g., a shape judgment followed by a color judgment). In such a *task-switching* paradigm, a common measure of executive control is the reaction time difference between trials that required a task switch and trials that did not. Consistent with an enhancement of cognitive control in near-hand space, the authors found reduced switch costs near the hands. Taken together, the results suggest that at least some of the enhancement of vision near the hands can be attributed to improvements in executive control mechanisms that guide attention.

Electrophysiological Research   If attentional processing is qualitatively different near the hands, then one might expect different brain mechanisms to be involved in such processing. Because attention can influence vision at many stages of processing, the loci of potential hand-proximity effects are also numerous. For example, some of the effects described earlier are consistent with a relatively early influence of hand proximity on attention. These effects include changes in visual search rate (Abrams et al., 2008; Davoli & Abrams, 2009; Weidler & Abrams, 2013) and in figure–ground segregation (Cosman & Vecera, 2010). Other changes induced by hand proximity seem more consistent with an influence later in the processing stream, such as the measures of cognitive control (e.g., Davoli et al., 2010; Weidler & Abrams, 2014) and switches between global and local processing (Davoli, Brockmole, Du, & Abrams, 2012). Some researchers have begun to employ electrophysiological techniques in order to learn more about the locus of hand-proximity effects on attention.

Qian, Al-Aidroos, West, Abrams, and Pratt (2012) measured evoked potentials in response to visual stimuli either near to or far from the hands. Their participants performed a challenging visual discrimination task that required them to direct attention either to an object at fixation or to an object in peripheral vision. During the task, black-and-white checkerboards that were at fixation or in the periphery flickered repeatedly. The checkerboards produced measurable EEG activity, with the question of interest being whether that activity differed for visual events that occurred near to as opposed to far from the hands. Qian et al. found that the P2 component of the response, a voltage change occurring between 160 and 260 ms after stimulus onset, was attenuated in response to stimuli near the hands. However, the attenuation only occurred for stimuli that were in the region that was being attended for the discrimination task—if participants were attending to the periphery, then the P2 was reduced near the hands only for peripheral stimuli, and if they were attending to fixation, then the P2 was reduced near the hands only in response to events at fixation. Earlier components of the EEG response are believed to reflect a narrowing of the spatial focus of attention, whereas the P2 component is thought to reflect selection that occurs at later stages of processing, including stages affected by endogenous, top-down influences such as the demands of the task. Based on these results, hand proximity appears to modulate attention not via early low-level mechanisms but instead by later, higher level mechanisms. The fact that the observed changes only occurred at the attended location is consistent with that conclusion. The results are also consistent with the changes in executive control described earlier (e.g., Weidler & Abrams, 2014)—executive control mechanisms would also be expected to exert an effect later in processing.

Reed, Leland, Brekke, and Hartley (2013) also examined event-related potentials (ERPs). In their task, participants responded to one of two possible shapes and ignored the other. As did Qian et al. (2012), they found that a nearby hand altered late components of the evoked response, and also only for designated target (i.e., attended) stimuli. This suggests that the hands enhance the processing of task-relevant stimuli. In addition, Reed et al. found that the hand altered early components of the ERPs for both target and nontarget stimuli. This early effect may reflect a perceptual or sensory mechanism that amplifies the signal for all stimuli near the hands. Also, as Reed et al. noted, the early ERP effect could reflect a top-down attentional set that stems from the instructions for participants to extend one or the other hand. Qian et al. did not find early effects of hand proximity on the ERPs, but they compared a posture with both hands straddling the display to one in which neither hand was near the display. In contrast, Reed et al. had participants extend only one hand at a time toward one side of the display, and they analyzed responses to stimuli on the two sides separately.

Magnocellular versus Parvocellular Visual Pathways   If the embodiment of attention depends upon our current action capabilities, then one might expect that brain mechanisms associated with action would be more likely to be engaged for stimuli near the hands because those objects can be acted upon. Action is known to be guided primarily by activity along the dorsal visual processing stream, one of two parallel pathways in the visual system. The dorsal stream receives input mainly from the M retinal ganglion cells of the magnocellular visual pathway. Thus, if processing of objects in near-hand space is altered because of the actions that are afforded by the proximity of the hands, enhanced activity of magnocellular mechanisms might be expected for such objects. To test this possibility, Gozli, West, and Pratt (2012) had participants detect either a brief temporal interruption in a visual stimulus or a small (spatial) gap in a visually presented shape either near to or far from the hands. The M cells are known to respond more rapidly than the P retinal ganglion cells of the parvocellular pathway (which projects to the ventral visual processing stream) but to have poorer spatial acuity than the P cells (e.g., Callaway, 1998). As predicted, participants were better at detecting the temporal gap and poorer at detecting the spatial gap for stimuli near the hands, consistent with the known properties of the magnocellular mechanisms. Additional evidence revealing enhanced temporal processing near the hands has been reported by Goodhew, Gozli, Ferber, and Pratt (2013). They showed that the magnitude of object substitution masking—a form of masking that relies upon limitations in temporal discrimination—is attenuated near the hands, as would be expected if the space near the hands benefits from enhanced magnocellular processing. The findings suggest that the availability of a potential target for an action leads to an analysis that is biased toward the brain mechanisms that are more likely to be involved in controlling movement, along with an accompanying shift in the nature of the information that is most effectively processed.

Abrams and Weidler (2014) also examined the magnocellular/parvocellular trade-off in processing for near-hand stimuli. They had participants judge briefly presented sine-wave gratings (Gabor patches), as shown in figure 13.7. The gratings were either vertical or tilted by a small amount in one direction or the other, and they were presented at one of two spatial frequencies. The participant's task was to indicate whether the grating was vertical or tilted. Magnocellular mechanisms are known to be tuned to lower spatial frequencies compared to parvocellular mechanisms (e.g., Callaway, 1998), and hence a bias toward magnocellular processing near the hands would predict a benefit for low spatial frequencies for such stimuli. The observed pattern, shown in figure 13.8,

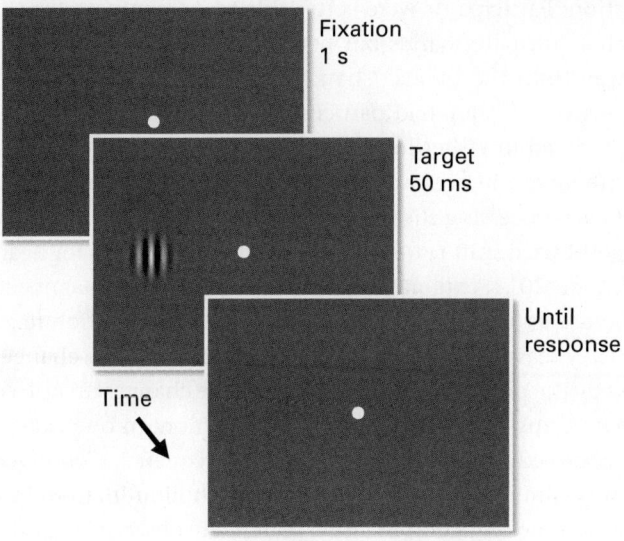

Figure 13.7
Sequence of events on a trial in the experiment by Abrams and Weidler (2014). Participants indicated whether the grating was vertical or tilted slightly.

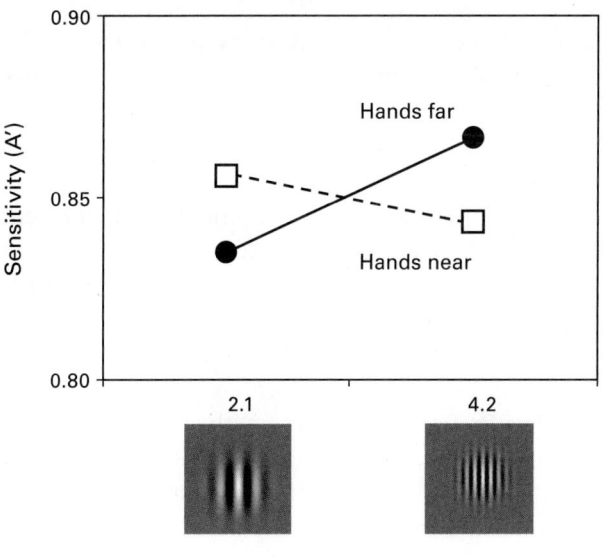

Figure 13.8
Sensitivity to the orientation of the grating as a function of spatial frequency for stimuli presented near to and far from the hands from Abrams and Weidler (2014).

is consistent with that prediction: Participants were better able to discriminate the tilt of low-spatial-frequency gratings near the hands but better at discriminating higher spatial frequency gratings away from the hands. Chan, Peterson, Barense, and Pratt (2013) provided converging evidence. They had participants make judgments about common objects that were rendered in either low- or high-spatial-frequency images. Processing of the low-spatial-frequency images was enhanced in near-hand space, consistent with biased magnocellular processing there.

As a further test of the hypothesized shift toward magnocellular processing for near-hand stimuli, Abrams and Weidler (2014) exploited a known property of the magnocellular pathway—that it is inhibited under long-wavelength (i.e., red) illumination (e.g., Livingstone & Hubel, 1988). They reasoned that red light should attenuate the changes in attention that are produced by the proximity of the hands if those changes do indeed rely upon magnocellular mechanisms. Consistent with that prediction, in one experiment they found that the enhanced sensitivity to low spatial frequencies observed near the hands (figure 13.8) was eliminated under long-wavelength illumination. In a second experiment they measured the reduction in visual search rate observed near the hands (e.g., Abrams et al., 2008). Search rate was slower near the hands compared to far from the hands under green illumination (as in Abrams et al., 2008), but search rate was unaffected by hand proximity under red illumination. Taken together, the results confirm the enhanced activity of magnocellular processing for visual stimuli near the hands.

Spatial and Postural Effects Several lines of research have examined the extent to which changes in the positions or orientation of the hands might influence the embodied attention effects that have been described so far. These are reviewed next.

Distance Effects One approach that has been taken is to examine the spatial extent of the changes that occur. In particular, if the attentional changes serve the purpose of enhancing the analysis of manipulable objects, then the changes would be expected to be limited to objects that are within the reaching space of the hand. Indeed, hand-proximity effects seem to be graded, with the magnitude of the enhancement decreasing with increasing distance from the hand (Adam, Bovend'Eerdt, van Dooren, Fischer, & Pratt, 2012; Reed et al., 2006). That finding is consistent with the suggestion by Schendel and Robertson (2004) and others (e.g., Abrams et al., 2008; Reed et al., 2006) that the effects may be mediated by activity of multimodal neurons identified in several brain areas that are sensitive to tactile stimulation of the hand as well as visual stimulation near the hand (Graziano, Hu, & Gross, 1997). These neurons are known to have visual receptive fields of limited extent, so the effects of hand proximity on attention should also be similarly limited. Tseng and Bridgeman (2011) also explored hand–target distance effects: In a paradigm in which participants held two hands near the display, they found enhancement of the entire display area, without a graded response, in contrast to the effects reported earlier. More work will be needed to establish the spatial extent of the embodied attention effects.

Hand Grasping Space   Other studies have explored the extent to which the orientation or posture of the hand may affect the changes in attention that occur. Reed, Betz, Garza and Roberts (2010) found that attention was biased for objects near to and on the palm side of the hand, but not for objects adjacent to the back of the hand. Objects on the palm side, of course, are in graspable space, and hence the findings are consistent with the conclusion that the attentional changes produced by proximity to the hand stem from the actions that are afforded by a nearby hand. Tseng and Bridgeman (2011) reported a consistent finding from a condition in which participants held one hand above the displayed stimuli and one hand below them. In addition to hand orientation's affecting attention, grasp posture also plays a role. Thomas (2013) found that placing a hand near the display only facilitated target detection when the hand was flat and in the typical palm-in position and hence capable of initiating a power-grasp response, but not when the fingers were pinched together and prepared only for a precision grasp. Additionally, it is worth noting that participants in these experiments were not explicitly planning to produce movements. Instead, the effects arose merely because of the opportunity to grasp into the space containing the stimuli.

Arm Position   Almost all of the experiments investigating the effects of hand proximity on vision have employed a video display that was presented upright in front of the participants, much like the position of a typical computer monitor. As a result, whenever participants held their hands near the stimuli, they were also elevating their hands and extending them away from their bodies. Thus it is possible, in theory, that the observed effects of hand proximity might instead reflect the effects of this unique posture, or be limited to stimuli that were directly in front of the participant, and not be attributable to effects of hand proximity per se. In order to investigate this question, Weidler and Abrams (2013) had participants perform a visual search task with their hands near and far from the display. As can be seen in figure 13.9, the display was flat on a table around waist height, so participants had their arms elevated and their hands extended away from their bodies in the *hands-far* condition whereas their hands were low and near their bodies in the *hands-near* condition. With this postural modification, Weidler and Abrams (2013) found slower search rates when the stimuli were near the hands, replicating the pattern described earlier (Abrams et al., 2008). The results

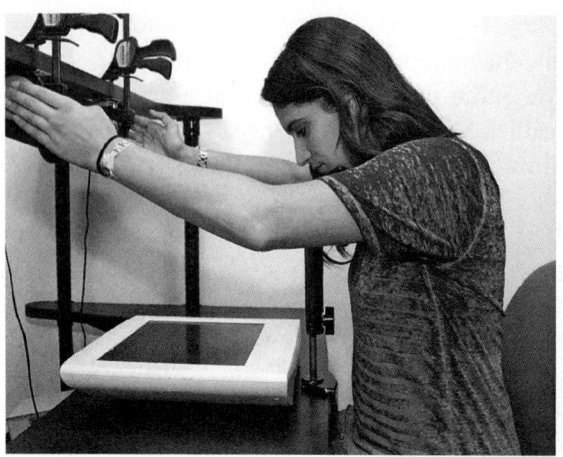

Figure 13.9
Postures studied in the experiment by Weidler and Abrams (2013).

confirm that the effect is driven by the proximity of the hands to the visual stimuli and
not by the effects of a unique posture.

One Hand versus Two An important postural detail that has not been extensively
studied involves the correspondence between the positions of the two hands. In
particular, most of the demonstrations of delayed attentional disengagement have
employed a paradigm in which participants held both of their hands near to or far
from the display simultaneously (e.g., Abrams et al., 2008; Davoli, Brockmole, Du, &
Abrams, 2012) whereas most of the demonstrations of biased attentional selection have
been provided by experiments in which participants held only a single hand near one
part of the display at any one time (e.g., Reed et al., 2006, 2010). If the bases for the
phenomena lie in the fact that stimuli near the hand are also in the receptive field of
multimodal neurons with hand-centered receptive fields, then stimuli would benefit
from either one or two nearby hands. However, it remains possible that some of the
attentional prioritization that has been observed stems from some other factor capable
of inducing a spatial bias. More work will be needed to determine the role played by
this sort of bias, if any.

Other Embodied Influences on Attention
Might there be other changes in the body's capabilities that alter attention in addition
to the effects of hand proximity discussed so far? In particular, being in a standing position may afford people opportunities for action that are not available when seated—
they may be better able to respond to threatening stimuli (by fighting or fleeing), for

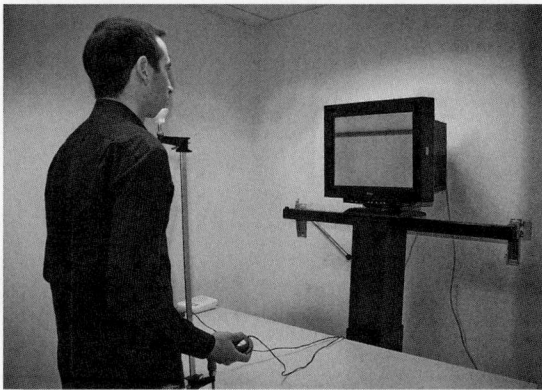

Figure 13.10
The two postures used in the experiments by Davoli et al. (2009).

example. They also may have a better vantage point from which to view interesting or potentially dangerous objects in the environment. In one study examining this, trained baggage screeners performed a visual search through baggage x-ray images while either seated or standing (Drury et al., 2008). There was no overall difference in reaction time between the two postures, but the task may not have had the sensitivity needed to detect subtle effects, leaving open the possibility that standing does lead to changes in attention. To pursue this, in one experiment we (Davoli et al., 2009) had participants perform a Stroop color-word task while in one of the two postures shown in figure 13.10: either seated or standing. The visual display was adjusted between postures in order to ensure that the stimuli did not differ, and subjects responded by pressing handheld buttons in the same manner under both postures. Davoli et al. found that the magnitude of Stroop interference (caused by incongruent color-word stimuli) was dramatically reduced in the standing as opposed to the seated posture: Although participants were 16 ms slower to respond to incongruent (compared to congruent) stimuli when seated, response times were numerically equivalent for congruent and incongruent stimuli when standing.

We noted that standing may lead to embodied changes in attention in part because one of the opportunities better afforded when standing would be to engage in defensive behaviors—such as fighting or fleeing. Similarly, there is also an important defensive advantage associated with enhanced perception near the hands: Sometimes objects near the hands (and other parts of the body) pose a danger or are obstacles that must be avoided. Enhanced attention to such obstacles, or when fighting and fleeing are afforded by one's posture, could yield an important survival advantage. Indeed, the multimodal neurons identified by Graziano and colleagues (e.g., Graziano et al., 1997) that are thought to play a role in the effects of hand nearness on attention (e.g., Schendel

& Robertson, 2004) have also been hypothesized to play a critical role in defense (Graziano & Cooke, 2006). Thus, the embodiment of attention may reflect a sensitivity not only to the body's current capabilities but also to its potential vulnerabilities.

Integration: Embodied Attention in Everyday Contexts

The research discussed in this chapter has illuminated some of the properties of attentional mechanisms that are continually in force—modulating the nature of the processing that takes place to best serve one's current needs. These processes most certainly guide our attention throughout our daily lives. Hence, it may be possible to identify key contexts in which an understanding of embodiment might eventually lead to safer or more efficient and accurate performance of common behaviors that ultimately could lead to a better quality of life.

For example, much of the laboratory research on embodied attention has identified changes in processing that occur for stimuli that are near the hands. In fact, these changes may account in part for the tremendous popularity of handheld electronic devices—they confer a processing advantage due to the enhanced attention devoted to objects near the hands. Given their increasing popularity, laboratory research on embodiment could potentially lead to practical enhancements that could improve the ever-increasing number of tasks that people perform on such devices. For example, handheld devices are frequently used as multitasking platforms—people may switch from a main task (e.g., reading an article) to responding to text messages to checking the weather forecast. Indeed, some operating systems and devices are limited to presenting the display from a single application at any one time. In these cases multitasking is necessary. Given that research has revealed that the costs of switching tasks is less severe for stimuli near the hands (Weidler & Abrams, 2014), participants may actually be better suited at multitasking on handheld devices compared to alternative devices. Of course switching from task to task imposes working memory demands on the user: One must remember the various tasks that are in progress, the current status of each, the manner in which to switch from one to the other, and so on. The working memory advantage enjoyed by stimuli near the hands (e.g., Tseng & Bridgeman, 2011) may help to accomplish these goals. Other operating systems permit multiple-window views of several tasks simultaneously. In these cases it becomes necessary for one to focus attention on a specific portion of the display and to suppress processing of other portions. Many of the changes discussed earlier may facilitate such processes (e.g., Weidler & Abrams, 2014). More generally, the enhanced focusing of attention in near-hand space may confer an overall advantage to handheld devices: It may be easier to ignore extraneous, distracting information when using a handheld device. This might be particularly beneficial given that handheld devices are often used for work outside of the workplace where individuals may be more prone to distraction.

Although we have just indicated many reasons why embodied attention could account for the popularity and ease of use of handheld devices, how can we apply what has been learned in the laboratory to real-world contexts? One domain in which this may be possible is in the classroom. Handheld devices are becoming increasingly utilized there—they are replacing typical desktop computers at a growing rate. Many of the attentional enhancements that accrue to handheld objects may be useful in facilitating learning. For example, it is believed that working memory plays a key role in the learning of a variety of types of information (e.g., Alloway, 2006; Mayer, 2008). Given the enhanced working memory for stimuli near the hands, the use of handheld devices may be an effective aid to learning.

Nevertheless, there may also be some disadvantages to using handheld devices. As noted earlier, people appear to focus on item-specific details for stimuli near the hands at the expense of more general patterns (e.g., Davoli, Brockmole, & Goujon, 2012). Consider the recent study by Brockmole, Davoli, Ehrman, and McNeil (2013). They trained individuals to perform arithmetic problems (such as multiplications) and later tested them by presenting the problems either in the same format that had been studied or in a different but mathematically equivalent one. For example, during training a person might learn that $18 \times 4 = 72$. Later during test the participant might be prompted with $18 \times 4 = ?$, or with $4 \times 18 = ?$ Participants who held their hands near the materials while studying exhibited a performance advantage at test—but only if the test format remained the same as the study format. The hands apparently facilitated the processing of item-specific detail. Nevertheless, those participants who studied with their hands near the display were poorer at answering problems in the new format compared to those who studied with their hands far from the display. Thus, the advantage for the item-specific details came at a cost: Participants who had their hands near the display were less able to generalize beyond the specific example that had been presented. Handheld devices may be advantageous if focusing on specific exemplars is desirable but may be detrimental if generalization of common principles is desired. More work is certainly needed before any definitive conclusions can be reached.

A similar trade-off may exist when one considers the enhanced magnocellular processing that appears to take place near the hands (e.g., Gozli et al., 2012). The advantageous processing of low-spatial-frequency content near the hands comes at the expense of high-spatial-frequency processing (Abrams & Weidler, 2014). The enhancement may facilitate temporal acuity (e.g., Goodhew et al., 2013; Gozli et al., 2012) and extraction of gist information (Chan et al., 2013). However, the impaired processing of high-spatial-frequency content may adversely affect tasks that depend on fine detail, such as reading (Gozli et al., 2012). Taken together, embodied changes in attention have the potential to affect many behaviors that are important in our daily lives. Nevertheless, there is still much to be learned before we will be able to make the best use of the changes.

Summary and Future Directions

Our physical capabilities, and the dangers to which we are exposed, vary from moment to moment. With these changes come varying requirements for the acquisition of information from the environment. Our attentional and cognitive systems appear to be built to flexibly adapt to the changing demands. Enhanced focusing of attention, increased sensitivity to rapid changes, improved visual memory, and enhanced cognitive control are all examples of the adaptations that take place.

While much has been learned about the embodiment of attention, there are many important questions that remain unanswered. For example, several lines of research converge on the conclusion that there is enhanced processing along magnocellular pathways, and reduced processing from parvocellular mechanisms, for near-hand stimuli. Nevertheless, more direct confirmation is needed, and the precise neural mechanisms that implement the trade-off still remain unknown. There has been speculation that multimodal neurons may underlie some of the near-hands effect, but the evidence supporting that is mostly indirect. Some phenomena have been studied by having participants extend a single hand—others have been examined only when two hands straddle the stimuli. More systematic work is needed to learn about the differences that might exist between those two postural manipulations. Also, there appear to be other postural changes (e.g., standing, and changes in hand-grip configuration) that have implications for attentional processing, but very little effort has been devoted to studying them so far. In addition, there may be some applications of embodied attention concepts to domains such as education given the increased utilization of handheld devices. However, given that the spatial (e.g., Adam et al., 2012) and configural (e.g., Thomas, 2013) aspects of hand posture can alter the embodied effects, and that there may be some costs to having stimuli near the hands, application of the laboratory research to real-world situations will require more study.

Box 13.1
Key Points

- Attentional mechanisms are modulated by one's current capabilities so as to acquire information that would best serve current needs.
- Holding one's hands near an object can lead to many changes in the manner in which the object is processed.
- Converging evidence suggests that hand proximity biases processing toward magnocellular visual mechanisms.
- Applications of the knowledge to daily life include possible enhancements to learning, with some caveats.

Box 13.2

Outstanding Issues

- Further research is needed to pinpoint the neural mechanisms underlying processing changes for stimuli near the hands.
- Very little is known about how body posture more generally might influence attentional processing.
- More work is needed to extend research on embodied attention to improvements in our day-to-day lives.

References

Abrams, R. A., Davoli, C. C., Du, F., Knapp, W. H., & Paull, D. (2008). Altered vision near the hands. *Cognition, 107,* 1035–1047. doi:10.1016/j.cognition.2007.09.006.

Abrams, R. A., & Weidler, B. J. (2014). Trade-offs in visual processing for stimuli near the hands. *Attention, Perception & Psychophysics, 76,* 383–390. doi:10.3758/s13414-013-0583-1.

Adam, J. J., Bovend'Eerdt, T. J., van Dooren, F. E., Fischer, M. H., & Pratt, J. (2012). The closer the better: Hand proximity dynamically affects letter recognition accuracy. *Attention, Perception & Psychophysics, 74,* 1533–1538. doi:10.3758/s13414-012-0339-3.

Alloway, T. (2006). How does working memory work in the classroom? *Educational Research Review, 1,* 134–139.

Barsalou, L. W. (1999). Language comprehension: Archival memory or preparation for situated action? *Discourse Processes, 28,* 61–80.

Bhalla, M., & Proffitt, D. R. (1999). Visual–motor recalibration in geographical slant perception. *Journal of Experimental Psychology: Human Perception and Performance, 25,* 1076–1096. doi:10.1037/0096-1523.25.4.1076.

Bloesch, E. K., Davoli, C. C., Roth, N., Brockmole, J., & Abrams, R. A. (2012). Watch this! Observed tool-use affects perceived distance. *Psychonomic Bulletin & Review, 19,* 177–183. doi:10.3758/s13423-011-0200-z.

Brockmole, J. R., Davoli, C. C., Abrams, R. A., & Witt, J. K. (2013). The world within reach: Effects of hand posture and tool-use on visual cognition. *Current Directions in Psychological Science, 22,* 38–44. doi:10.1177/0963721412465065.

Brockmole, J. R., Davoli, C. C., Ehrman, E. K., & McNeil, N. M. (2013). Getting a grip on concepts: Hand position affects access to mathematical knowledge. Paper presented at the meeting of the Psychonomic Society, Toronto, Ontario, Canada.

Callaway, E. M. (1998). Local circuits in primary visual cortex of the macaque monkey. *Annual Review of Neuroscience, 21,* 47–74. doi:10.1146/annurev.neuro.21.1.47.

Chan, D., Peterson, M. A., Barense, M. D., & Pratt, J. (2013). How action influences object perception. *Frontiers in Psychology*, *4*, 443. doi:10.3389/fpsyg.2013.00443.

Cosman, J. D., & Vecera, S. P. (2010). Attention affects visual perceptual processing near the hand. *Psychological Science*, *21*, 1254–1258. doi:10.3758/s13414-012-0339-3.

Davoli, C. C., & Abrams, R. A. (2009). Reaching out with the imagination. *Psychological Science*, *20*, 293–295. doi:10.1111/j.1467-9280.2009.02293.x.

Davoli, C. C., Brockmole, J., Du, F., & Abrams, R. A. (2012). Switching between global and local scopes of attention is resisted near the hands. *Visual Cognition*, *20*, 659–668. doi:10.1080/13506285.2012.683049.

Davoli, C. C., Brockmole, J. R., & Goujon, A. (2012). A bias to detail: How hand position modulates visual learning and visual memory. *Memory & Cognition*, *40*, 352–359. doi:10.3758/s13421-011-0147-3.

Davoli, C. C., Du, F., Bloesch, E. K., Montana, J., Knapp, W. H., & Abrams, R. A. (2009). Posture affects vision and reading. Paper presented at the meeting of the Psychonomic Society, Boston, MA.

Davoli, C. C., Du, F., Montana, J., Garverick, S., & Abrams, R. A. (2010). When meaning matters, look but don't touch: The effects of posture on reading. *Memory & Cognition*, *38*, 555–562. doi:10.3758/MC.38.5.555.

di Pellegrino, G., & Frassinetti, F. (2000). Direct evidence from parietal extinction of enhancement of visual attention near a visible hand. *Current Biology*, *10*, 1475–1477. doi:10.1016/S0960-9822(00)00809-5.

Drury, C., Hsiao, Y., Joseph, C., Joshi, S., Lapp, J., & Pennathur, P. (2008). Posture and performance: Sitting vs. standing for security screening. *Ergonomics*, *51*, 290–307.

Dufour, A., & Touzalin, P. (2008). Improved visual sensitivity in perihand space. *Experimental Brain Research*, *190*, 91–98.

Estes, Z., Verges, M., & Barsalou, L. W. (2008). Head up, foot down: Object words orient attention to objects' typical location. *Psychological Science*, *9*, 93–97. doi:10.1111/j.1467-9280.2008.02051.x.

Fischer, B., & Weber, H. (1997). Effects of stimulus conditions on the performance of antisaccades in man. *Experimental Brain Research*, *116*, 191–200.

Goodhew, S. C., Gozli, D. G., Ferber, S., & Pratt, J. (2013). Reduced temporal fusion in the near-hand space. *Psychological Science*, *24*, 891–900. doi:10.1177/0956797612463402.

Gibson, J. J. (1979). *The ecological approach to visual perception*. Boston, MA: Houghton Mifflin.

Glenberg, A. M., Witt, J. K., & Metcalfe, J. (2013). From the revolution to embodiment: 25 years of cognitive psychology. *Perspectives on Psychological Science*, *8*, 573–585. doi:10.1177/1745691613498098.

Gozli, D. G., West, G. L., & Pratt, J. (2012). Hand position alters vision by biasing through different visual pathways. *Cognition, 124*, 244–250. doi:10.1016/j.cognition.2012.04.008.

Graziano, M. S., & Cooke, D. F. (2006). Parieto-frontal interactions, personal space, and defensive behavior. *Neuropsychologia, 44*, 845–859. doi:10.1016/j.neuropsychologia.2005.09.009.

Graziano, M. S. A., Hu, X. T., & Gross, C. G. (1997). Visual properties of the ventral premotor cortex. *Journal of Neurophysiology, 77*, 2268–2292.

Lakoff, G., & Johnson, M. (1980). *Metaphors we live by*. Chicago: University of Chicago Press.

Livingstone, M., & Hubel, D. (1988). Segregation of form, color, movement, and depth: Anatomy, physiology, and perception. *Science, 240*, 740–749. doi:10.1126/science.3283936.

Mayer, R. E. (2008). Applying the science of learning: Evidence-based principles for the design of multimedia instruction. *American Psychologist, 63*, 760–769. doi:10.1037/0003-066X.63.8.760.

Proffitt, D. R. (2006). Embodied perception and the economy of action. *Perspectives on Psychological Science, 1*, 110–122.

Qian, C., Al-Aidroos, N., West, G., Abrams, R. A., & Pratt, J. (2012). The visual P2 is attenuated for attended objects near the hands. *Cognitive Neuroscience, 3*, 98–104. doi:10.1080/17588928.2012.658363.

Reed, C. L., Betz, R., Garza, J. P., & Roberts, R. J. (2010). Grab it! Biased attention in functional hand and tool space. *Attention, Perception & Psychophysics, 72*, 236–245. doi:10.3758/APP.72.1.236.

Reed, C. L., Grubb, J. D., & Steele, C. (2006). Hands up: Attentional prioritization of space near the hand. *Journal of Experimental Psychology: Human Perception and Performance, 32*, 166–177. doi:10.1037/0096-1523.32.1.166.

Reed, C. L., Leland, D. S., Brekke, B., & Hartley, A. A. (2013). Attention's grasp: Early and late hand proximity effects on visual evoked potentials. *Frontiers in Psychology, 4*, 420. doi:10.3389/fpsyg.2013.00420.

Schendel, K., & Robertson, L. C. (2004). Reaching out to see: Arm position can attenuate human visual loss. *Journal of Cognitive Neuroscience, 16*, 935–943. doi:10.1162/089892904150269.

Smith, D. T., Lane, A. R., & Schenk, T. (2008). Arm position does not attenuate visual loss in patients with homonymous field deficits. *Neuropsychologia, 46*, 2320–2325.

Stroop, J. R. (1935). Studies of interference in serial verbal reactions. *Journal of Experimental Psychology, 18*, 643–662. doi:10.1037/h0054651.

Suh, J., & Abrams, R. A. (2013). Interactions between hand posture and eye movement. Poster presented at the meeting of the Psychonomic Society. Toronto, Ontario, Canada.

Tseng, P., & Bridgeman, B. (2011). Improved change detection with nearby hands. *Experimental Brain Research, 209*, 257–269. doi:10.1007/s00221-011-2544-z.

Theeuwes, J. (1992). Perceptual selectivity for color and form. *Perception & Psychophysics, 51*, 599–606. doi:10.3758/BF03211656.

Thomas, L. E. (2013). Grasp posture modulates attentional prioritization of space near the hands. *Frontiers in Psychology, 4*, 312. doi:10.3389/fpsyg.2013.00312.

Vatterott, D. B., & Vecera, S. P. (2013). Prolonged disengagement from distractors near the hands. *Frontiers in Psychology, 4*, 533. doi:10.3389/fpsyg.2013.00533.

Weidler, B. J., & Abrams, R. A. (2013). Hand proximity—not arm posture—alters vision near the hands. *Attention, Perception & Psychophysics, 4*, 650–653. doi:10.3758/s13414-013-0456-7.

Weidler, B. J., & Abrams, R. A. (2014). Enhanced cognitive control near the hands. *Psychonomic Bulletin & Review, 21*, 462–469. doi:10.3758/s13423-013-0514-0.

Wilson, M. (2002). Six views of embodied cognition. *Psychonomic Bulletin & Review, 9*, 625–636.

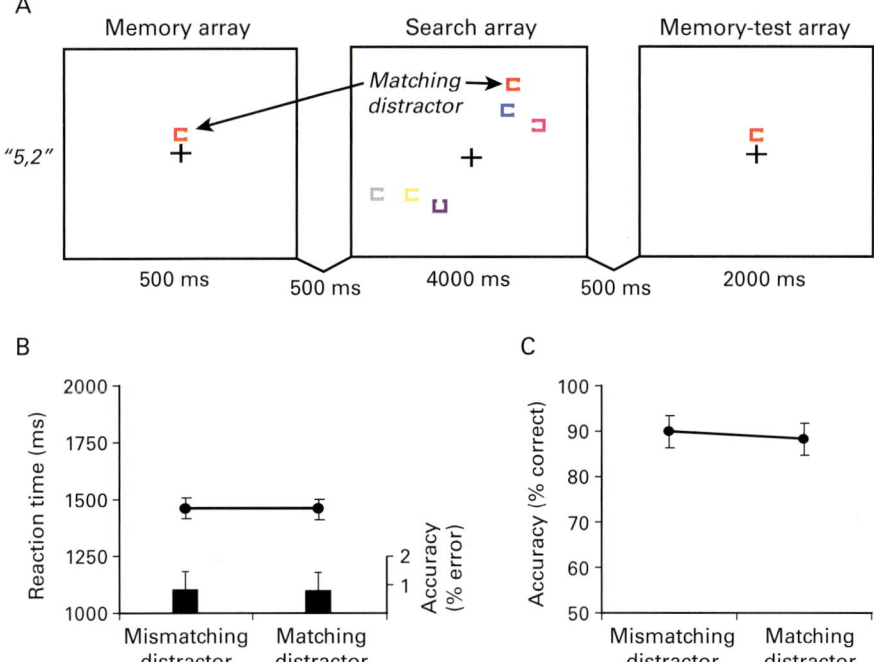

Plate 1 (figure 8.3)

The stimuli and findings of the dual-task experiments examining the relationship between storage of information in visual working memory and the deployment of attention to similar items. The stimuli (A) are from Experiment 2 of Woodman and Luck (2007). Visual search reaction time (B, RT) and memory task accuracy (C, percent correct) are shown for these tasks performed together. Adapted with permission from the American Psychological Association.

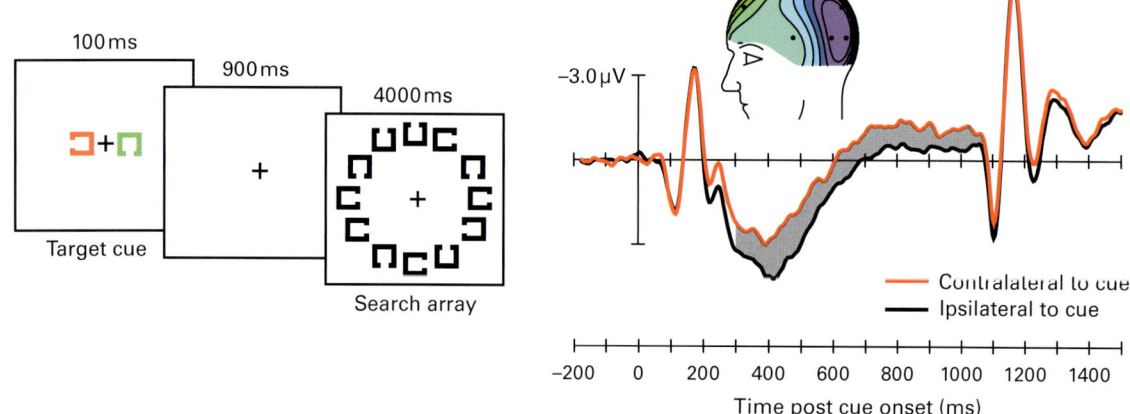

Plate 2 (figure 8.4)

The stimuli and event-related potential findings from Woodman and Arita (2011). Example of the stimulus sequence (left) and the grand average waveforms from electrodes T5/6, contralateral (red) and ipsilateral (black) to the location of the cue on each trial (right). The gray region shows the epoch in which the significant contralateral-delay activity (CDA) was measured, and the inset shows voltage distribution. The amplitude of the CDA predicted the accuracy of the subsequent search across subjects ($p < .05$). Adapted with permission from the Association for Psychological Science and Blackwell Publishing.

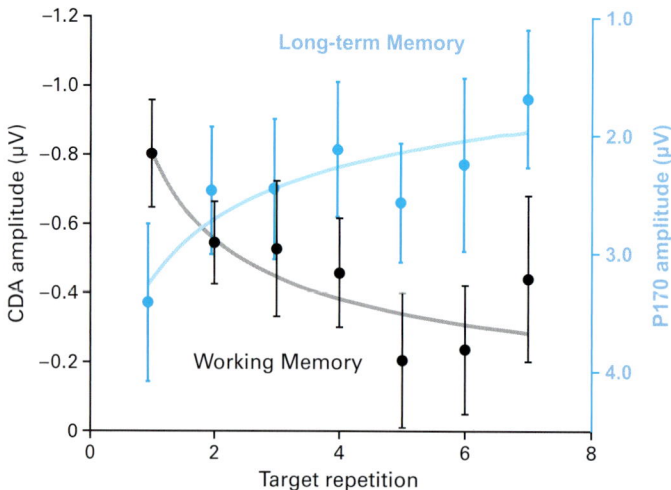

Plate 3 (figure 8.6)

The event-related potential component amplitudes following the target cue onset as a function of target repetition. The P170 amplitude was measured from 150–200 ms postcue and the contralateral-delay activity (CDA) from 300–900 ms postcue. The error bars show ±1 S.E.M. Adapted with permission from the Society for Neuroscience.

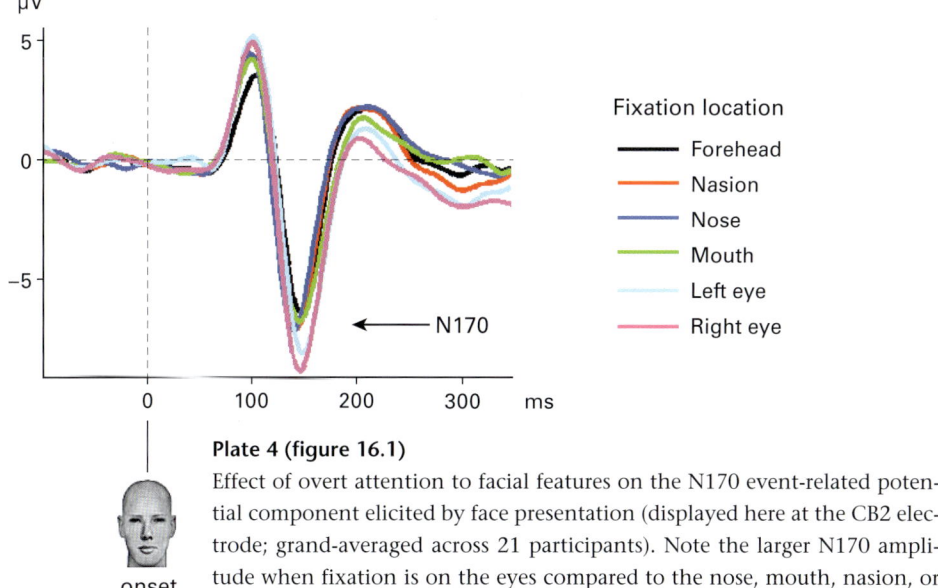

Plate 4 (figure 16.1)

Effect of overt attention to facial features on the N170 event-related potential component elicited by face presentation (displayed here at the CB2 electrode; grand-averaged across 21 participants). Note the larger N170 amplitude when fixation is on the eyes compared to the nose, mouth, nasion, or forehead. Adapted from Nemrodov et al. (2014) with permission.

Plate 5 (figure 20.2)

The location of a fire extinguisher relative to an office door (from Castel, Vendetti, & Holyoak, 2012). The study was partially inspired by the observation, recently illustrated by one of the authors of the article (depicted above), that we are remarkably good at ignoring highly visible objects, and remarkably bad at remembering their location. While taking a building safety training class, the fire safety instructor asked people to note the location of the nearest fire extinguisher, relative to their office. In the safety training class, several people admitted they did not know the location or guessed at the location, and people were told they should learn the location of the nearest fire extinguisher. Upon returning to his office, one of the authors of the eventual field research article (K. H., of Castel, Vendetti, & Holyoak, 2012) made a conscious effort to look for the nearest fire extinguisher and made a startling discovery: The conspicuously placed bright red object in question was right next to his office door, in plain view and literally inches from the doorknob that he had turned for the past 25 years!

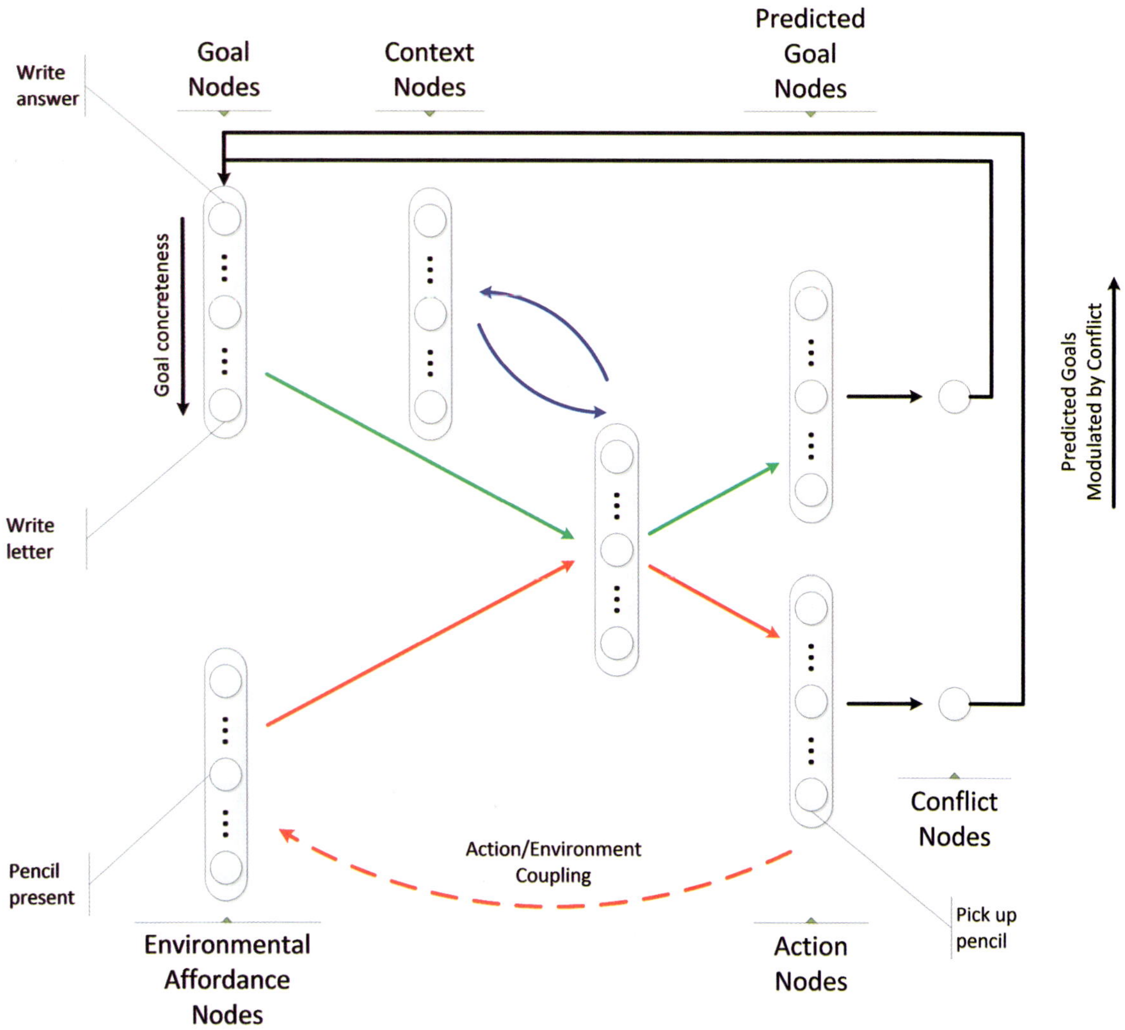

Plate 6 (figure 27.1)

An elaborated version of the goal circuit model with conflict nodes at the outputs. Layers are fully connected, with lines indicating the following mechanisms: the environmental affordance loop (red), the goal circuit loop (green), the task ordering loop (blue), and the modulation of conflict nodes on top-down control (black). Selected nodes are annotated according to the *writing* example given in the text.

14
Action and Attention

Jay Pratt, J. Eric T. Taylor, and Davood G. Gozli

Historical Context

At the very beginnings of the scientific study of psychology in North America, scholars were considering the structure of human actions and the nature of attention as topics of investigation. They were, however, considering them independently of each other. In *The Principles of Psychology*, William James (1890) categorized the varieties of attention into three dichotomies: sensorial or intellectual, immediate or derived, and passive (reflexive) or active (voluntary). These dichotomies continue to resonate in attentional research to this day. Not a full decade later, in 1899, Robert Woodworth, who once studied with James, completed his Ph.D. at Columbia University with a thesis entitled *Accuracy of Voluntary Movement* (Woodworth, 1899). This work set the stage for the modern psychological study of action, and many of Woodworth's initial insights remain valid to this day. Contemporary with James and Woodworth, a possible mechanism for connecting attention and action was proposed by John Dewey (1896) with his "coördination" of sensory and motor processes. In seeking a unifying principle for the fledgling discipline of psychology, Dewey dismissed the notion of a reflex arc— where sensation and action are viewed as the starting point (stimulus) and the end point (response) of a series of disjointed processes of human behavior. Rather, he promoted an uninterrupted cycle of sensorimotor processing as the appropriate framework for understanding human behavior. Accordingly, attention is seen as a key "central process" in guiding the flow of sensorimotor coordination.

By the turn of the 20th century, the trailheads were in place for the scientific exploration of the human attention system, the human action system, and the interactions of the two systems. John Watson's (1913) "Psychology as the Behaviourist Views It," however, set a different path for psychology in North America, and behaviorism would dominate the experimental psychology landscape for the next three decades. The concept of attention, like any other mental construct, simply did not fit within the behaviorist paradigm. The first Hixon Symposium on "Cerebral Mechanisms of Behavior," in 1948, heralded the rise of cognitive science, and with it cognitive psychology.

However, in the framework of information processing, actions were typically studied in the absence of attentional manipulations (e.g., Paul Fitts's, 1954, seminal work on speed–accuracy trade-offs) and attention was typically studied in the absence of actions (e.g., Cherry's, 1953, often cited "cocktail party effect" of attentional selection). Across six decades of research, this separation remains prevalent to this day.

There is, however, an important and ever growing recognition that viewing attention and action in isolation will not offer a realistic representation of either of the two types of processes. In the 1980s, Alan Allport (e.g., Allport, 1980, 1987) argued that cognitive psychology would be best advanced by treating the brain and behavior as its subject matter, instead of an overinterpreted computer metaphor (Driver, 2001). In making his case, Allport pointed out that selection is a key component in both perception and action, and what is selected in one system will ultimately influence what is selected in the other system. While several researchers also made this connection (e.g., Keele & Neill, 1978; Marcel, 1983; Shallice, 1972), Allport (1987) was the first to articulate the notion of "selection-for-action," with the selectivity of the attention system playing a critical role in the planning and control of actions. Coming from a similar neurological starting point, Rizzolatti and his colleagues introduced the premotor theory of attention in 1987. This model completely did away with the separation between attention and action systems and instead hypothesized that the control of attention is drawn from the same neural circuits that are used to control eye movements. Thus, by the end of the 1980s, connections between attention and action had become established enough to essentially form a new subfield of investigation (cf., Allport, 1989; Neumann, 1990).

State-of-the-Art Review

The Premotor Theory of Attention

The premotor theory of attention married action and attention by refuting the prominent idea that attention was functionally and anatomically distinct from the systems that control eye movements. Furthermore, it rejected the assumption that attention was a mental operation upon passively received information. In this account, covert shifts of attention (i.e., shifts of attention without accompanying eye movements) are essentially planned but unexecuted saccades. To support this theory, Rizzolatti and colleagues (1987) demonstrated the meridian effect, which showed that orienting of attention obeys the anatomical constraints of eye movements. The basic finding was that shifting attention between hemifields was more costly than shifting attention equidistantly within a hemifield. This meridian effect is easily explained by a theory that assumes the orienting of attention involves the preparation of eye movements: Moving the eyes to an unexpected location within a hemifield requires reprogramming of the saccade amplitude, but not direction; making a saccade to the contralateral hemifield requires a new motor program altogether. This additional recalculation takes time. Although the eyes did not move in their study, the meridian effect demonstrates that shifts of attention obey the anatomical constraints of eye movements, supporting a premotor theory of attentional shifts.

If the orienting of attention is indeed yoked to the movement of the eyes, then it follows that wherever the endpoint location of a planned saccade is, attention should precede the eye to that same endpoint location. In support of this prediction, preparing a saccade to a region in space increased the efficiency of target detection at that location (Hoffman & Subramaniam, 1995). Likewise, priming from a peripheral object that appears prior to a saccade is only effective when the prime appears at the terminal saccade location (Godijn & Pratt, 2002). These results suggest that preparing a saccade moves attention to the terminal location of the saccade. In a similar study, Deubel and Schneider (1996) showed that it is not possible to direct attention to one location while saccading to another nearby location, indicating a strong, obligatory coupling between saccade preparation and the orienting of attention. The premotor theory of attention predicts that this effect should be reciprocal: Preparing eye movements shifts attention, and shifting attention should assist the preparation of eye movements. Indeed, shifting attention to a peripheral location makes saccades to that location faster and more accurate (Kowler, Anderson, Dosher, & Blaser, 1995). In a similar study, covertly orienting attention altered the trajectories of concurrent saccades, suggesting that the orienting of attention necessarily activates oculomotor circuits (Sheliga, Riggio, & Rizzolatti, 1994; Sheliga, Riggio, & Rizzolatti, 1995). Further corroborating this line of research is evidence that perceptual sensitivity is enhanced to targets likely to appear at attended locations; planning eye movements prepares the observer for perception at that location (Kingstone & Klein, 1991; Hoffmann & Kunde, 1999). Together, these studies demonstrate the reciprocal relationship between orienting attention and eye movements.

If the orienting of attention is necessarily preceded by covert oculomotor preparation, as the premotor theory of attention suggests, then limiting eye movements should also limit the orienting of attention. To test this assertion, Craighero and colleagues had participants perform a classic spatial cuing paradigm (Posner, 1980) with the display positioned 40° to the left or right of the participant's head. This unorthodox positioning of the display meant that participants had to move their eyes to their rotational limit within the orbits in order to fixate on the display's center; they could not move their eyes any further in the temporal direction although they could move their eyes in the nasal direction. A typical cuing effect was found on the nasal side of the display, indicating that attention was free to orient in that direction. By contrast, there was no difference between validly and invalidly cued targets on the temporal side, where eye movements were limited (Craighero, Nascimben, & Fadiga, 2004). A similar study had participants with VI cranial nerve palsy perform the typical spatial cuing paradigm under monocular conditions (Craighero, Carta, & Fadiga, 2001). In their healthy eye, these patients oriented attention normally, showing a delayed detection of invalidly cued targets. In contrast, performing the task with their paretic eye revealed no difference between validly and invalidly cued targets, indicating they did not orient toward the cue. These findings support the idea that orienting of attention depends on the ability to plan and execute eye movements. Interestingly, even congenitally blind individuals demonstrate the link between eye-movement preparation and attention. When attention is directed peripherally by an auditory cue, these individuals showed a robust activation in the oculomotor circuit under fMRI (specifically, the frontal eye fields; Garg, Schwartz, & Stevens, 2007). The same pattern of activation was also observed in sighted controls.

Action-Based Attention

The premotor theory of attention holds that the control of attention is inextricably linked to the preparation of eye movements. However, eye movements are not the only type of actions that might influence selective attention. A critical step to connect selective attention to manual actions was taken by Tipper, Lortie, and Baylis (1992), who argued that attention uses an action-centered representation during reaching. In their task, participants reached for a target on a three-by-three array. Critically, a distractor would appear simultaneously at another location with target onset. Results showed a delayed response time when the distractor was in the reach trajectory, compared to when the distractor was outside the reach trajectory. This result prompted the conclusion that attention is distributed within an action-centered representation during reaching, as interference from the distractors was greatest when they were obstacles to the reaching action. Performing a reaching action changed how stimuli were prioritized, with the reach-relevant distractors receiving priority over the reach-irrelevant distractors. An important extension to these results was reported in a reaching task where a physical obstacle would sometimes impede the reaching action. The obstacle would presumably remove a portion of the display from the action-relevant space. Consistent with this notion, a visual distractor along the reach trajectory interfered less with the action when the physical obstacle was present (Meegan & Tipper, 1999).

The processing cost of a distractor along the path of reaching action is indicative of an inhibited action plan. This idea is supported by research using a similar selective reaching task with older adults and patients with Alzheimer's disease, who suffer from a dramatic loss in the ability to use inhibitory processes (Simone & Baylis, 1997). In this study, older healthy adults showed similar patterns of distractor interference to that found by Tipper et al. (1992). Alzheimer's patients, however, showed greatly enlarged interference effects, despite being able to discriminate the distractors from the targets. It is worth noting that the frame of reference for action-based attention may change with age, as Bloesch, Davoli, and Abrams (2013) found that healthy older adults used a body-centered frame of reference while younger adults used the hand-centered frame reported by Tipper et al. (1992).

To further investigate the nature of action-centered representations of action, Pratt and Abrams (1994) used a similar selective reaching paradigm to pinpoint the sub-movements of reaching that are affected by action-relevant distractors. They had participants move a cursor from a start position to a target appearing in one of three boxes along a single path. On some trials, a distractor could appear at one of the empty locations adjacent to the target. They found that distractors in the path of the cursor delayed response times compared to distractors on the other side of the target, replicating Tipper et al. (1992). They also recorded the time from target onset to movement onset, the time from movement onset to peak velocity (ballistic phase), and the time from peak velocity to the end of the movement (corrective phase). They found that distractors on the path of the reaching action prolonged movement onset and the corrective phase, suggesting that action-centered inhibition affects both the planning and the execution of reaching actions. This result is in agreement with other work showing that distractors that appear in the action-relevant space interfere with reaching responses during both movement preparation and execution (Meegan & Tipper, 1998). These studies show that the planning and execution of action shapes selective attention through action-centered representation of space.

The notion of action-based attention received further support from studies showing that the transport kinematics of a reaching action deflect away from nontargets near the hand in a cluttered environment, supporting the notion that attention uses an action-centered representation during reaching (Tipper, Howard, & Jackson, 1997). However, other studies have shown that reaching trajectories may veer *toward* distractors in a selective reaching task (Welsh, Elliott, & Weeks, 1999). While these studies agree that reaching-relevant distractors are automatically processed, the patterns of data seem in conflict. The response activation model of selective reaching can explain these seemingly contradictory results: When a distracting stimulus is presented sufficiently in advance of the reaching target ($\sim$–750 ms), its representation is inhibited and the movement trajectory deflects away from the distractor's location; when the distracting stimulus is presented closer to target presentation (–250 ms–0 ms), there is insufficient time to inhibit its representation, resulting in a deflection toward the distractor (Welsh & Elliott, 2004; see Howard & Tipper, 1997, for an alternative explanation).

The premotor theory was initially conceived as a link between oculomotor activity and attention. However, paradigms designed to examine the link between eye movements and attentional orienting have been repurposed to show that the allocation of attention is linked to manual motor activity as well. Earlier, we described a study by Deubel and Schneider (1996), who asked participants to prepare a saccade while performing a visual discrimination task at different locations. They found that performance was best when the target appeared at the programmed saccade destination. Deubel, Schneider, and Paprotta (1998) adapted this oculomotor paradigm for guided limb movements. Participants were instructed to prepare a reaching movement to a specific location while also identifying a target embedded among distractors. During the preparation phase of the limb movement, the target was very briefly unmasked. Identification of this target was best when the limb movement was planned to the same location. Further extensions of this paradigm have tested whether attention can be simultaneously allocated to up to two or three action-relevant locations in the scene. In particular, when subjects plan multiple sequential eye movements (Baldauf & Deubel, 2008a), sequential reach movements (Baldauf & Deubel, 2009; Baldauf, Wolf, & Deubel, 2006), or bimanual grasps (Schiegg, Deubel, & Schneider, 2003; Baldauf & Deubel, 2008b), visual selection is enhanced for all of the action-relevant locations (Baldauf & Deubel, 2010). These studies demonstrate that an action–attention link pervades human movement; wherever actions are directed, attention moves as well.

One corollary of the view that attention uses action-centered representations of space is that the information selected by attention during action should depend on the action being performed. To investigate this claim, Welsh and Pratt (2008) had participants detect stimuli using either a point-and-reach or a simple key-press response. On every trial, the participants knew to look for either an onset or an offset. Critically, on some trials, the target would be presented simultaneously with a distractor that would be the opposite stimulus (i.e., if the target was an onset, the distractor would be an offset, or vice versa). The question was whether distractor cost would depend on response type, akin to how a perceptual goal can determine the cost of an irrelevant distractor (Folk, Remington, & Johnston, 1992). Results showed that when making key-press responses, both onset and offset distractors captured attention. This is consistent with the informational demands of the action, as key-press actions are indifferent to the addition or removal of stimulation. In contrast, point-and-reach responses are sensitive to information that remains present throughout the action in order to facilitate the online guidance of action. Consequently, transient onsets captured attention, because they introduce new information, but offsets did not. By demonstrating that the type of action being performed determines the type of events that can capture attention depends on the action, action systems can be viewed as a potential source of attentional control setting. Thus, the line of research beginning with Tipper et al. (1992) revealed how information that is potentially relevant to successful control of action influences selective attention. In the example of reaching, this means a higher selective benefit for new objects that appear on display, compared with those that disappear (Welsh & Pratt, 2008), and higher selective benefit for objects that appear along the reaching trajectory, compared with those that appear outside the trajectory (Festman, Adam, Pratt, & Fischer, 2013). After establishing the sensitivity of attentional processing to action-based spatial representation, the question becomes whether this attentional sensitivity is limited to the mere presence of objects in space or whether it extends to selection across other perceptual dimensions.

Actions and Feature-Based Attention

To investigate the role of actions on attention further, it is useful to compare attentional processes across modes of actions that target different perceptual dimensions. Compared with reaching movements, grasping an object makes a different subset of available visual information pertinent. For example, the object's orientation becomes critical in constraining hand orientation, and the object's size becomes critical in determining grip aperture. Indeed, evidence suggests that attention to both orientation and size is enhanced when observers are in grasping mode compared to when they are in reaching mode. Bekkering and Neggers (2002) employed a visual search task in which a target was defined based on orientation–color feature conjunction. The authors varied the type of action performed toward the target (reaching vs. grasping) and compared the number of error saccades toward distractor features. Critically, they found that in the grasping condition the number of saccades toward the nontarget orientation was lower compared with the reaching condition. The number of error saccades toward the nontarget color, however, did not differ across the two action types. Thus, the intent to perform a reach or grasp altered the efficiency of selecting the action-relevant feature (i.e., orientation).

Grasping has also been shown to enhance attention for size. Wykowska, Schubö, and Hommel (2009) used a dual task, which involved (1) preparing an action (reach vs. grasp), (2) performing a visual search on a separate display, and (3) performing the prepared action. Findings showed that preparing a grasp facilitates visual search for a size singleton, compared to reaching. By contrast, visual search for a luminance singleton benefited from preparing to reach. Both findings of Bekkering and Neggers (2002) and Wykowska et al. (2009) demonstrate how adopting an action mode enhances processing the dimensions that are relevant for successful performance of that action. Interestingly, however, this means that performing grasp movements under divided attention allows relevant dimensions of an irrelevant distractor to affect performance. Castiello and colleagues (Bonfiglioli & Castiello, 1998; Castiello, 1996; Kritikos, Bennett, Dunai, & Castiello, 2000) demonstrated that if subjects grasp a target while visually keeping track of a distractor object (e.g., counting how many times it was illuminated), features of the distractor object (e.g., its size) influence properties of the action (e.g., grasp aperture).

In addition to enhanced selection of action-relevant dimensions, specific action features can facilitate selection of specific visual features congruent with the action. In a change blindness task, Symes et al. (2008) used fruits and vegetables of small (strawberry) or big (apple) size. They found that detection of small targets was facilitated when participants responded with a precision pinch grasp, whereas detection of big targets was facilitated when responses were made with a whole-hand power grasp. Similarly, Ellis and Tucker (2000) found that seeing an irrelevant object that matched the grip size resulted in faster response execution compared to the mismatch condition. Therefore, the type of action one intends (e.g., grasp vs. reach) and the features of that action (e.g., big vs. small grasp) constrain attentional selection in the perceptual domain.

Ideomotor Theory of Action

In addition to the selective advantage for action-relevant features (e.g., size and orientation in grasping), there is yet another way in which actions can influence attention. This class of action-based effects emerge out of the very nature of cognitive representation of actions. In line with Dewey's (1896) insight, there is little difference between action and perception in the terms of their representation, but that the real difference is in the functional role those representations play (Hommel et al., 2001; Prinz, 1997). This assumption is expressed in the ideomotor theory of action, in its claim that actions are represented in terms of their known perceptual effects (James, 1890; for reviews, see Shin, Proctor, & Capaldi, 2010; Stock & Stock, 2004). Consequently, the representation of action effect activates the motor pattern, which in turn brings about the anticipated perceptual action effects. The ideomotor theory has been supported by both behavioral (e.g., Elsner & Hommel, 2001; Kunde, 2001) and imaging studies (e.g., Kühn et al., 2010; Melcher et al., 2008, 2013). Assuming that actions are accompanied by a set of perceptual anticipations, it follows that perceptual events will be treated differently because of the anticipations that accompany actions (cf. O'Regan & Noë, 2001).

In a demonstration of how anticipated action effects can be a source of bias, Hommel (1993), employed a variant of a Simon task (Simon, 1990), in which participants responded to a tone pitch by switching on one of two peripheral lights. The two lights were contralaterally assigned to two keys (e.g., the right key switched on the left light). When participants used lights to respond to auditory stimuli coming from the left or right periphery, a Simon compatibility effect was found between the location of the auditory target and the location of the action effect, inverting the typical Simon effect between key locations and stimuli. In other words, anticipating the action effect (i.e., switching on the light) eliminated the effect of the spatial relationship between manual key press and the target. Similarly, Hommel (2004) eliminated the Stroop effect in a condition where participants' own key press produced a color patch, suggesting that the knowledge of action effect reduced attentional allocation to the interfering stimulus dimension.

Because of the perceptual nature of action representation, actions can also interfere with perception when there is a similarity between the two types of events. Müsseler and colleagues (Müsseler & Hommel, 1997a, 1997b; Müsseler, Wühr, & Prinz, 2000; Wühr & Müsseler, 2001) found that planning a localized response reduces perceptual identification of a compatible spatial stimulus. For instance, while planning to press the "left" key, participants' accuracy of detecting a leftward arrow is reduced when compared to a rightward arrow. Kunde and Wühr (2004) generalized these findings by reporting similar interference effects with other forms of action–percept similarity. For instance, planning a vocal response such as "left" and "blue" could, respectively, interfere with perceptual identification of the word "LEFT" and a blue color patch. This class of findings has been referred to as *action-induced blindness*, highlighting the fact that action planning can have a significant impact on visual cognition.

Recently, Gozli and Pratt (2011) examined a consequence of action-induced blindness in attention by examining the effect of prepared actions on attentional prioritization of abrupt motion signals. Abrupt motion is one of the strongest sources of attentional capture (Abrams & Christ, 2003) and has been shown to resist top-down goal setting (Al-Aidroos, Guo, & Pratt, 2010). Participants in the Gozli and Pratt (2011) study performed a visual search using manual movements along either the horizontal or the vertical axis. The search display on each trial was preceded by uninformative motion signals along the same or different axis as the response axis. Although both types of motions were prioritized compared with static locations, higher attentional priority was given to stimuli that differed from the prepared response. Consistent with action-induced blindness, motion signals along the same axis as the participants' action received reduced attentional priority.

Early explanations of action-induced blindness (Müsseler & Hommel, 1997a, 1997b) were based on inhibition of the action features (MacKay, 1986). According to this account, features associated with an action plan are temporarily inhibited in order to prevent reactivation of action in stimulus-driven manner. Later explanations of action-induced blindness, however, are based on the idea of code occupation (Hommel et al., 2001). The code occupation, introduced in the theory of event coding (TEC; Hommel et al., 2001; Stoet & Hommel, 1999), assumes that the action features are activated during preparation but they are bound to other action features to form a unified action event. Consistent with the code occupation account, Gozli and Pratt (2011) found that action-induced blindness was eliminated by the congruency between the location of the motion signal and action direction. This form of congruency should not influence a suppression-driven blindness to motion stimuli. By contrast, an interference that is driven by code occupation and separability of the events should be sensitive to the congruency among event features.

The code occupation account further assumes that the feature integration phase is enveloped by a larger time window in which features have above-baseline activation but are not yet bound. This assumption was supported by Hommel and Schneider (2002), who presented their participants with two consecutive stimuli, the first of which required a localized response while the second required an identification response. Critically, the identification stimulus was presented early enough that it could reasonably coincide with the preselective action processes, in which action codes are activated but not yet integrated. Hommel and Schneider found that compatibility facilitated identification, which is again inconsistent with an inhibition account of action-induced effects and supports the code occupation assumption of the TEC (cf. Müsseler et al., 2005). In order to induce response activation without response selection, Gozli, Goodhew, Moskowitz, and Pratt (2013) employed uninformative spatial primes (left vs. right). Critically, each response had an acquired action effect (red vs. green). The authors found that the response primes, in addition to biasing response selection, also increased sensitivity to the color associated with the primed response. In short, these results are in line with an account of action planning in which feature activation is followed by feature integration. It seems likely that the feature-activation phase increases attentional bias for action features while the feature-integration phase interferes with selection of action-congruent features (Stoet & Hommel, 1999).

The scope of research on actions and attention is drastically expanded when we consider actions as building blocks of higher cognition (Barsalou, 2008; Gallese & Lakoff, 2005). For instance, concepts referring to manual actions (e.g., "push," "pull") are thought to engage the motor system (Glenberg & Kaschak, 2002), and processing concepts that are associated with a location in space (e.g., "bird," "sky") are thought to engage the oculomotor system (Dudschig et al., 2013; Estes et al., 2008; Gozli, Chow, Chasteen, & Pratt, 2013). In short, concepts do not only refer to objects and states of affair but to patterns of bodily activity (Gallese & Lakoff, 2005). Viewing conceptual understanding as involving sensorimotor tendencies enables us to make specific predictions regarding the way concepts can shape attention. Indeed, tracking the time course of the interaction between concepts and visual attention, Gozli, Chasteen, and Pratt (2013) found that implied spatial meaning of words interferes with visual attention briefly after the word onset whereas after some delay processing is facilitated at the compatible location. The same time course of interaction has been found between language processing and selection of manual actions (Boulenger et al., 2006; Sato et al., 2008). This time course resembles the action-induced modulations of visual processing and could be explained with the same theoretical assumptions (Hommel et al., 2001).

In summary, the empirical evidence clearly points toward the continuity of mental operations and against both the separation of processing *modules* (Anderson, 2010; Barsalou, 2008) and the separation of processing *stages* (Spivey & Dale, 2006). Indeed, in the present framework, a clear distinction between attentional priority and action priority seems implausible (Baldauf & Deubel, 2010; Bisley & Goldberg, 2010; Fecteau & Munoz, 2006). Processes that are referred to as "action selection" seem to parallel those that control allocation of attention. Moreover, anticipated perceptual action effects influence the way concurrent visual events are prioritized. Finally, expanding the view of actions as building blocks of higher cognition expands the scope and implications of the action-based attention research. The way in which conceptual understanding influences perception seems to reflect the same characteristics of their constituent action elements.

Joint Attention

An action-centered representation gives us command over a complex, although often static environment. However, we frequently interact with other people, who are also capable of action. Thus, interacting with others requires that attention represent a nebulous exchange. Attention must select events caused by self and others. To solve this problem, we rely largely on joint representations: mechanisms that coordinate the representation of action across people. The first step of establishing this coordination is the use of joint attention: directing attention to where a partner is attending (Sebanz, Bekkering, & Knoblich, 2006). In its most basic form, joint attention is demonstrated by using eye gaze as a spatial cue for attention. Targets are detected faster when preceded by a picture of a face looking in that direction (Friesen & Kingstone, 1998; Frischen, Bayliss, & Tipper, 2007). Another reliable cue for joint attention is hand gestures. Hands depicted in a grasping posture cue attention to targets that fit the grasp aperture, but inanimate apertures (U-shaped objects) do not (Lindemann, Nuku, Rueschemeyer, & Bekkering, 2011). These studies show how the eyes and hands can direct another person's attention, thereby facilitating a shared representation.

There are several examples in the literature that show the coordination of perception and action can depend on whether a task is performed jointly or alone (Knoblich, Butterfill, & Sebanz, 2011). For example, the Simon effect can be distributed across two people when performed together (Sebanz, Knoblich, & Prinz, 2003). In this task, participants made left or right responses to the color of a stimulus that also pointed left or right. Responses were slow when the response and stimulus were incompatible and fast when they were compatible (Simon, 1990). In some blocks, subjects performed half of the task, responding to only one color while ignoring the other, effectively making it a go/no-go task. Critically, they performed this task either alone or with another person who responded to the other color (a complementary go/no-go task). The compatibility effect (faster compatible/slower incompatible responses) emerged only when performing the go/no-go tasks together, suggesting that the actors represented each other's responses (Sebanz et al., 2003).

Joint attention can also be achieved by attending to a partner's actions. In a joint inhibition of return (IOR) task, participants were slower to process stimuli at locations recently reached to by their partners (Welsh et al., 2005; Welsh et al., 2007). That observing another person's actions can induce IOR is strong evidence that the attentional mechanism can operate on a task representation that is shared between individuals.

A similar demonstration of shared representation in joint actions occurs in demonstrations of Fitts's law. When reaching to targets, movement time (MT) scales to the difficulty of the action (Fitts, 1954). When bimanually tapping between two targets of different difficulties, MTs remain scaled to the harder target, as though compensating to maintain a rhythm (Mottet, Guiard, Ferrand, & Bootsma, 2001). Critically, when two people take aim at targets of different difficulties in an alternating, joint tapping task, their MTs scaled to the harder target, even though each participant's task did not depend on their partner's target (Fine & Amazeen, 2011). This result shows that participants represented their partner's task and achieved an interpersonal rhythm that compensated for the harder of the two tasks.

Finally, like the Simon effect and Fitts's law, action-based effects of reaching trajectory can be distributed across actors in a joint task. Griffiths and Tipper (2009) modified the selective reaching task (Tipper et al., 1992) to include two actors seated across from each other at a narrow table. Participants reached for a target in the presence of a distractor, taking turns with their partner. The researchers concluded that observing another person reach over a distractor evoked a simulation of the same reaching trajectory in later actions. The critical result in support of this conclusion was an upward deflection in the trajectory of the reach when the distractor was absent, after having watched a partner reach for the target when the distractor was present (Griffiths & Tipper, 2009). These effects of joint attention are all explained by assuming a shared task representation: Partners succeed in cooperation because their representations include one another's actions, abilities, and intentions.

Integration

The interaction of attention and action in the prioritization of visual information and the planning of actions is so ubiquitous, so efficient, and typically so successful that it guides our behaviors despite being in the very deep background of awareness. Indeed, the adaptive reciprocal relationship likely only comes to awareness when failures occur, and even then it is only a rough awareness that *something* went wrong rather than exactly *how* it went wrong (e.g., "I didn't see the signpost, even though it was right in front of me, and I drove my car right into it"). Such failures are often ascribed to failures of attention (as in the previous example, selecting the wrong bits of the visual field to send for further processing and inexplicitly discarding information that is both perceptually salient and mission critical) or errors of action planning (reaching for the salt but hitting a glass along the way and spilling the water in it). However, as can be seen by the aforementioned studies, attention and action should be viewed as a continuous cycle of selection and action rather than entirely separate processes whereby one selects and the other acts. Put more eloquently by Dewey (1896), "… both sensation and perception lie inside, not outside the act" (p. 359). Such a view would take us away from looking at human performance as isolated stimulus–response episodes and bring into consideration such concepts as overlearned chains of sensorimotor and ideomotor associations. What James (1890) referred to as "habits" or stabilized tendencies that govern our relationship to our environments can now become a subject of consideration within a framework that assumes a tight coupling between our actions and selective information processing.

While failures in the cycle of attention and action can produce serious, and even spectacular, consequences, these are the exceptions rather than the rule. As noted earlier, the reciprocal relationship of attention's supporting action and action's supporting attention is both very successful and ubiquitous in our everyday lives. Let's take the example of the relatively mundane task of shopping in a typical grocery store. Often, a person enters such a store with a list, written or mental, of items to purchase. Let's suppose that the list includes a favorite type of boxed pasta and a specific type of small mushrooms. An older, modular, and sequential view of cognition might suggest that after allocating attention across the visual scene, the task-salient box of pasta would be selectively attended to, and the relevant feature information for grasping the box (coordinates in space, size, orientation) would then be passed on to the motor system for the planning and production of the necessary limb movement. In other words, selecting the visual information of the desired pasta box would be independent of the action planning for obtaining the box, and vice versa.

A more integrated and continuous view of cognition, where action and attention are viewed as partners in determining behavior, would suggest a very different set of processes are involved in the shopping trip. First, let us examine how attention will affect our actions. When searching for the right-sized mushrooms in a bin of mushrooms, wrong-sized mushrooms along the path of our hands will be more salient for action-based attention and thus will produce more interference than wrong-sized mushrooms beyond the path of our movement (as in Tipper et al.'s, 1992, selective reaching task). If a crowd has gathered to pick over the best remaining mushrooms, and someone has to reach across the passed-over mushrooms at the front of the bin, we might also reach with that avoidant trajectory, even if there are no bad mushrooms in our way (as in the joint attention task of Griffiths & Tipper, 2009). When looking for a specific pasta box, knowing the size of the box would alter our grip aperture, which in turn will aid attention in selecting the proper-sized box from the array of boxes (Symes et al., 2008). If the type of pasta next to our goal pasta box is out of stock, the fact that there is a salient perceptual luminance contrast but no available alternative target (and therefore similar to an offset stimulus) means that lack of product will not interfere with the detection of the goal pasta box (Castiello, 1996). If, however, we are merely scanning the pasta boxes with no specific action goal, the blank space would likely attract attention (Welsh & Pratt, 2008).

Of course, grocery shopping is just one example of how the cycle of attention and action pervades our day-to-day lives. The example is perhaps as mundane as it gets; selecting the wrong-sized mushroom or nonfavorite pasta won't have a significant or long-term effect on our lives. However, pilots selecting and acting on the proper controls in a commercial airliner, drivers navigating through busy city intersections, or surgeons operating on patients, are examples where the same interplay of attention and action is at work but the stakes are much higher. An awareness of how actions shape selection of information can lead to more practical strategies of how attention could be allocated and maintained on crucial information. In other words, when emphasis in training is on acquiring the right kind of attentional tendencies, it may prove useful to reformulate attentional tendencies as action tendencies. An action-oriented view of human performance stands in stark contrast to the passive information-processing view in its potential in offering better strategies for performance success.

Future Directions

The study of action and attention has produced great insights into the cognition of human movement. Much of the great work by scholars in this field is discussed above. However, as in all subjects of inquiry, there remain many fascinating, unanswered questions for the field to address. In this section, we describe what we have identified as some pressing issues of practical and theoretical interest.

An action does not constitute a stable cognitive state but rather is best conceived as a series of continually unfolding processes, which include planning, initiation, and control. Given this, one important question is how do these unfolding action processes parallel the way attentional processes treat sensory input? The TEC (Hommel et al., 2001), by proposing a distinct feature-binding stage that is enveloped by a larger feature-activation stage, makes predictions with regard to the interference and facilitation of concurrent perceptual events. The theoretical work of Thomaschke, Hopkins, and Miall (2012a, 2012b) represents another attempt to distinguish between separate qualitatively distinct stages of interaction between action and vision. The authors highlight how the planning stage is dominated by selection of categorical information (e.g., "left" vs. "right" key press) while the control stage is dominated by continuous information (e.g., distance, size, etc.). The authors further propose an account of facilitation and interference that is grounded in the planning–control distinction (see also Hommel, 2009; Zwickel & Prinz, 2012). Further research is necessary for reconciling the current theories.

A large portion of the changes to our sensory input is caused by our own actions (e.g., movements of the eye and body). Considering the distinction between self-caused and externally caused events would increase the ecological validity of attention research, especially given that the majority of attention research treats observers as passive recipients of information. Considering observers' role in bringing about perceptual information, and the observers' ability to learn and anticipate these changes, can lead to novel and innovative lines of research on human visual attention. Current theories do not exactly predict the attentional consequences of identifying an event as self-caused (e.g., Waszak, Cardoso-Leite, & Hughes, 2012), and the available evidence seems to point toward diverging conclusions that self-caused events are sometimes prioritized (Desantis, Roussel, & Waszak, 2014; Kumar, Manjaly, & Sunny, 2015) and sometimes deprioritized (Cardoso-Leite, Mamassian, Schütz-Bosbach, & Waszak, 2010). Examining how attention operates within a context in which agents distinguish between self-caused and externally caused events represents a new and exciting line of research.

There is an extensive motor control literature on how actions are planned and produced, but not much of this detail has been examined in terms of possible effects on attention. For example, structured perceptual arrays invoke an exception to Fitts's law of speed–accuracy trade-offs (e.g., Adam, Mol, Pratt, & Fischer, 2006; Pratt, Adam, & Fischer, 2007). Does this effect in action planning also influence how attention is allocated to grouped sets of stimuli?

The role of aging on the cycle of action and attention has not been extensively studied to this point in time. It is well-known that various components of the action and attention systems change with advancing age, but few studies have studied the possible interactions. As noted earlier, Bloesch et al. (2013), using a variation of the Tipper et al. (1992) distractor reaching task, found that older adults used egocentric frame of reference rather than the action-based frame of reference used by younger adults. Given this finding, it is likely that older adults will guide their behaviors through different variations of action and attention processes. This insight has practical implications for the way we design and plan environments for an aging population. It also has theoretical implications for the way attention and action systems develop over the life span.

Finally, a criticism that action-minded scientists occasionally level at the conventional paradigms of attention is that most attention studies bear no resemblance to real-world interaction, primarily because all of the traditional research occurs on computerized displays in tiny, darkened rooms. This criticism is legitimate because the real world affords many actions (beyond button presses), which can change how we process information. To some extent, this problem is mitigated by conducting studies in situations that demand real, dynamic, three-dimensional limb movements (e.g., Tipper et al., 1992; Bekkering & Neggers, 2002). Ironically, this criticism is perhaps becoming less externally valid as we spend more and more time acting on computers and other digital devices. Over half of Canadian, American, and European adults own a smartphone (Rooney, 2013), placing a computer display in their pockets at all times. Therefore, a return to digital displays is called for to round out the study on action and attention. This time, however, we should not think of the computer as a passive display of information (i.e., monitor and keyboard), but rather as a flexible action space (i.e., smartphone and tablet). This perspective leads to fruitful questions for future consideration. For example, does attention operate in the space surrounding a cursor with action-related priorities? If so, how should we design the optimal cursor? Human-factors engineering is rich with literature aimed at designing displays, but as of yet the lessons learned from studies on action and attention remain largely unapplied to human–computer interaction. Do the action-related priorities that attention exhibits in reaching and grasping translate to digital movements? Some evidence suggests this is the case, as the perceptual system's selectivity for grasping-relevant object features (e.g., Ganel & Goodale, 2003) extends to actions made with cursors (Janczyk, Pfister, & Kunde, 2013). However, further research is needed to determine how similar real and digital actions really are.

In summary, the study of action and attention has clearly demonstrated the inextricable bond shared by these faculties. Traditionally, psychologists thought of action and attention as separate systems—as though attention existed in a disembodied state that never moved through the world to gain new information or new perspectives. In light of the research and theory discussed in this chapter, we hope the reader will agree that action and attention are two sides of the same coin.

Box 14.1
Key Points

- Attention and eye movements are reciprocally linked; preparing saccades moves attention and moving attention primes saccades. This is evidence for the influential premotor theory of attention, which states that attentional orienting is contingent on the covert preparation of eye movements.

- The link between action and attention goes beyond eye movements. Performing, or intending to perform, a manual action affects selection of available information. Perceptual dimensions that are relevant for successful completion of the action tend to be favored, and perceptual features that match the features of the current action are selected more efficiently. Perceptual anticipation that accompanies an action also influences the way concurrent perceptual events are prioritized.

- Attentional processes take not only one's own actions into account but also the actions and intentions of our co-actors. In other words, selection of information takes place from a representation of environment that can include other people.

Box 14.2
Outstanding Issues

- Psychologists must consider the extent to which attention and action are truly separate, if at all. Attention selects the information necessary for action, and action moves the observer (and manipulates its environment) to new information. Does it make sense to discuss action and attention as though they are separate faculties?

- Compared to passive viewing, how do different aspects of an action, such as planning, initiation, and online control, influence selection of information from the environment? In particular, if an action accompanies perceptual anticipation, how does such anticipation differ from other forms of perceptual expectations that do not involve an action?

References

Abrams, R. A., & Christ, S. E. (2003). Motion onset captures attention. *Psychological Science, 14*, 427–432.

Adam, J. J., Mol, R., Pratt, J., & Fischer, M. H. (2006). Moving farther but faster: An exception to Fitts's law. *Psychological Science, 17*, 794–798.

Al-Aidroos, N., Guo, R. M., & Pratt, J. (2010). You can't stop new motion: Attentional capture despite a control set for color. *Visual Cognition, 18*, 859–880.

Allport, D. A. (1980). Attention and performance. In G. Claxton (Ed.), *Cognitive psychology: New directions* (pp. 112–153). London: Routledge & Kegan Paul.

Allport, D. A. (1987). Selection for action: Some behavioural and neurophysiological considerations of attention and action. In H. Heuer & D. F. Saunders (Eds.), *Perspectives on perception and action* (pp. 395–419). Hillsdale, NJ: Erlbaum.

Allport, D. A. (1989). Visual attention. In M. Posner (Ed.), *Foundations of cognitive science* (pp. 631–682). Cambridge, MA: MIT Press.

Anderson, M. L. (2010). Neural reuse: A fundamental organizational principle of the brain. *Behavioral and Brain Sciences, 33*, 245–266.

Baldauf, D., & Deubel, H. (2008a). Properties of attentional selection during the preparation of sequential saccades. *Experimental Brain Research, 184*, 411–425.

Baldauf, D., & Deubel, H. (2008b). Visual attention during the preparation of bimanual movements. *Vision Research, 48*, 549–563.

Baldauf, D., & Deubel, H. (2009). Attentional selection of multiple goal positions before rapid hand movement sequences: An event-related potential study. *Journal of Cognitive Neuroscience, 21*, 18–29.

Baldauf, D., & Deubel, H. (2010). Attentional landscapes in reaching and grasping. *Vision Research, 50*, 999–1013.

Baldauf, D., Wolf, M., & Deubel, H. (2006). Deployment of visual attention before sequences of goal-directed hand movements. *Vision Research, 46*, 4355–4374.

Barsalou, L. W. (2008). Grounded cognition. *Annual Review of Psychology, 59*, 617–645.

Bekkering, H., & Neggers, S. F. W. (2002). Visual search is modulated by action intentions. *Psychological Science, 13*, 370–374.

Bisley, J. W., & Goldberg, M. E. (2010). Attention, intention, and priority in the parietal lobe. *Annual Review of Neuroscience, 33*, 1–22.

Bloesch, E., Davoli, C., & Abrams, R. A. (2013). Age-related changes in attentional reference frames for peripersonal space. *Psychological Science, 24*, 557–561.

Bonfiglioli, C., & Castiello, U. (1998). Dissociation of covert and overt spatial attention during prehension movements: selective interference effects. *Perception & Psychophysics, 60*, 1426–1440.

Boulenger, V., Roy, A. C., Paulignan, Y., Deprez, V., Jeannerod, M., & Nazir, T. A. (2006). Cross-talk between language processes and overt motor behavior in the first 200 msec of processing. *Journal of Cognitive Neuroscience, 18*, 1607–1615.

Castiello, U. (1996). Grasping a fruit: selection for action. *Journal of Experimental Psychology: Human Perception and Performance, 22*(3), 582.

Cardoso-Leite, P., Mamassian, P., Schütz-Bosbach, S., & Waszak, F. (2010). A new look at sensory attenuation: Action-effect anticipation affects sensitivity, not response bias. *Psychological Science, 21*, 1740–1745.

Cherry, C. (1953). Some experiments on the recognition of speech, with one and two ears. *Journal of the Acoustical Society of America, 25*, 975–979.

Craighero, L., Carta, A., & Fadiga, L. (2001). Peripheral oculomotor palsy affects orienting of visuospatial attention. *Neuroreport, 12*, 3283–3286.

Craighero, L., Nascimben, M., & Fadiga, L. (2004). Eye position affects orienting of visuospatial attention. *Current Biology, 14*, 331–333.

Desantis, A., Roussel, C., & Waszak, F. (2014). The temporal dynamics of the perceptual consequences of action-effect prediction. *Cognition, 132*, 243–250.

Deubel, H., & Schneider, W. X. (1996). Saccade target selection and object recognition: Evidence for a common attentional mechanism. *Vision Research, 36*, 1827–1837.

Deubel, H., Schneider, W. X., & Paprotta, I. (1998). Selective dorsal and ventral processing: Evidence for a common attentional mechanism in reaching and perception. *Visual Cognition, 5*, 81–107.

Dewey, J. (1896). The reflex arc concept in psychology. *Psychological Review, 3*, 357–370.

Driver, J. (2001). A selective review of selective attention research from the past century. *British Journal of Psychology, 92*, 53–78.

Dudschig, C., Souman, J., Lachmair, M., de la Vega, I., & Kaup, B. (2013). Reading "sun" and looking up: The influence of language on saccadic eye movements in the vertical dimension. *PLoS ONE, 8*(2), e56872.

Ellis, R., & Tucker, M. (2000). Micro-affordance: The potentiation of components of action by seen objects. *British Journal of Psychology, 91*, 451–471.

Elsner, B., & Hommel, B. (2001). Effect anticipation and action control. *Journal of Experimental Psychology, 27*, 229–240.

Estes, Z., Verges, M., & Barsalou, L. W. (2008). Head up, foot down object words orient attention to the objects' typical location. *Psychological Science, 19*, 93–97.

Fecteau, J. H., & Munoz, D. P. (2006). Salience, relevance, and firing: a priority map for target selection. *Trends in Cognitive Sciences*, *10*, 382–390.

Festman, Y., Adam, J., Pratt, J., & Fischer, M. (2013). Both hand position and movement direction modulate visual attention. *Frontiers in Perception Science*, *4*, 657.

Fine, J. M., & Amazeen, E. L. (2011). Interpersonal Fitts' law: When two perform as one. *Experimental Brain Research*, *211*, 459–469.

Fitts, P. M. (1954). The information capacity of the human motor system in controlling the amplitude of movement. *Journal of Experimental Psychology*, *47*, 381–391.

Folk, C. L., Remington, R. W., & Johnston, J. C. (1992). Involuntary covert orienting is contingent on attentional control settings. *Journal of Experimental Psychology: Human Perception and Performance*, *18*, 1030.

Friesen, C. K., & Kingstone, A. (1998). The eyes have it! Reflexive orienting is triggered by nonpredictive gaze. *Psychonomic Bulletin & Review*, *5*, 490–495.

Frischen, A., Bayliss, A. P., & Tipper, S. P. (2007). Gaze cueing of attention: Visual attention, social cognition, and individual differences. *Psychological Bulletin*, *133*, 694–724.

Gallese, V., & Lakoff, G. (2005). The brain's concepts: The role of the sensory–motor system in conceptual knowledge. *Cognitive Neuropsychology*, *22*, 455–479.

Ganel, T., & Goodale, M. A. (2003). Visual control of action but not perception requires analytical processing of object shape. *Nature*, *426*, 664–667.

Garg, A., Schwartz, D., & Stevens, A. A. (2007). Orienting auditory spatial attention engages frontal eye fields and medial occipital cortex in congenitally blind humans. *Neuropsychologia*, *45*, 2307–2321.

Glenberg, A. M., & Kaschak, M. P. (2002). Grounding language in action. *Psychonomic Bulletin & Review*, *9*, 558–565.

Godijn, R., & Pratt, J. (2002). Endogenous saccades are preceded by shifts of visual attention: Evidence from cross-saccadic priming effects. *Acta Psychologica*, *110*, 83–102.

Gozli, D. G., Chasteen, A. L., & Pratt, J. (2013). The cost and benefit of implicit spatial cues for visual attention. *Journal of Experimental Psychology. General*, *142*, 1028–1046.

Gozli, D. G., Chow, A., Chasteen, A. L., & Pratt, J. (2013). Valence and vertical space: Saccade trajectory deviations reveal metaphorical spatial activation. *Visual Cognition*, *21*, 628–646.

Gozli, D. G., Goodhew, S. C., Moskowitz, J. B., & Pratt, J. (2013). Ideomotor perception modulates visuospatial cueing. *Psychological Research*, *77*, 528–539.

Gozli, D. G., & Pratt, J. (2011). Seeing while acting: Hand movements can modulate attentional capture by motion onset. *Attention, Perception & Psychophysics*, *73*, 2448–2456.

Griffiths, D., & Tipper, S. P. (2009). Priming of reach trajectory when observing actions: Hand-centred effects. *Quarterly Journal of Experimental Psychology, 62*, 2450–2470.

Hoffmann, J., & Kunde, W. (1999). Location-specific target expectancies in visual search. *Journal of Experimental Psychology: Human Perception and Performance, 25*, 1127–1141.

Hoffman, J. E., & Subramaniam, B. (1995). The role of visual attention in saccadic eye movements. *Perception & Psychophysics, 57*, 787–795.

Hommel, B. (1993). Inverting the Simon effect by intention: Determinants of direction and extent of effects of irrelevant spatial information. *Psychological Research, 55*, 270–279.

Hommel, B. (2004). Coloring an action: Intending to produce color events eliminates the Stroop effect. *Psychological Research, 68*, 74–90.

Hommel, B. (2009). Action control according to TEC (theory of event coding). *Psychological Research, 73*, 512–526.

Hommel, B., Müsseler, J., Aschersleben, G., & Prinz, W. (2001). The theory of event coding (TEC): A framework for perception and action planning. *Behavioral and Brain Sciences, 24*, 849–937.

Hommel, B., & Schneider, W. X. (2002). Visual attention and manual response selection: Distinct mechanisms operating on the same codes. *Visual Cognition, 9*, 392–420.

Howard, L. A., & Tipper, S. P. (1997). Hand deviations away from visual cues: Indirect evidence for inhibition. *Experimental Brain Research, 113*, 144–152.

James, W. (1890). *The principles of psychology* (2 vols.). New York: Holt.

Janczyk, M., Pfister, R., & Kunde, W. (2013). Mice move smoothly: Irrelevant object variation affects perception, but not computer mouse actions. *Experimental Brain Research, 231*, 97–106.

Keele, S. W., & Neill, W. T. (1978). Mechanisms of attention. In E. C. Carterette & M. P. Friedman (Eds.), *Handbook of perception* (Vol. 9, pp. 3–47). New York: Academic Press.

Kingstone, A., & Klein, R. (1991). Combining shape and position expectancies: Hierarchical processing and selective inhibition. *Journal of Experimental Psychology: Human Perception and Performance, 17*, 512–519.

Knoblich, G., Butterfill, S., & Sebanz, N. (2011). Psychological research on joint action: theory and data. In B. Ross (Ed.), *The Psychology of Learning and Motivation, 54* (pp. 59–101). Burlington: Academic Press.

Kowler, E., Anderson, E., Dosher, B., & Blaser, E. (1995). The role of attention in the programming of saccades. *Vision Research, 35*, 1897–1916.

Kritikos, A., Bennett, K. M., Dunai, J., & Castiello, U. (2000). Interference from distractors in reach-to-grasp movements. *Quarterly Journal of Experimental Psychology: Section A, 53*, 131–151.

Kühn, S., Keizer, A., Rombouts, S. A. R. B., & Hommel, B. (2010). The functional and neural mechanisms of action preparation: Roles of EBA and FFA in voluntary action control. *Journal of Cognitive Neuroscience, 23*, 214–220.

Kumar, N., Manjaly, J. A., & Sunny, M. M. (2015). The relationship between action-effect monitoring and attention capture. *Journal of Experimental Psychology. General, 144*(1), 18–23.

Kunde, W. (2001). Response-effect compatibility in manual choice reaction tasks. *Journal of Experimental Psychology: Human Perception and Performance, 27,* 387–394.

Kunde, W., & Wühr, P. (2004). Actions blind to conceptually overlapping stimuli. *Psychological Research, 68,* 199–207.

Lindemann, O., Nuku, P., Rueschemeyer, S. A., & Bekkering, H. (2011). Grasping the other's attention: The role of animacy in action cueing of joint attention. *Vision Research, 51,* 940–944.

MacKay, D. G. (1986). Self-inhibition and the disruptive effects of internal and external feedback in skilled behavior. In H. Heuer & C. Fromm (Eds.), *Generation and modulation of action patterns* (pp. 174–186). Berlin: Springer-Verlag.

Marcel, A. J. (1983). Conscious and unconscious perception: An approach to the relations between phenomenal experience and perceptual processes. *Cognitive Psychology, 15,* 238–300.

Meegan, D. V., & Tipper, S. P. (1998). Reaching into cluttered visual environments: Spatial and temporal influences of distracting objects. *Quarterly Journal of Experimental Psychology, A, 51,* 225–249.

Meegan, D. V., & Tipper, S. P. (1999). Visual search and target-directed action. *Journal of Experimental Psychology: Human Perception and Performance, 25,* 1347–1362.

Melcher, T., Weidema, M., Eenshuistra, R. M., Hommel, B., & Gruber, O. (2008). The neural substrate of the ideomotor principle: An event-related fMRI analysis. *NeuroImage, 39,* 1274–1288.

Melcher, T., Winter, D., Hommel, B., Pfister, R., Dechent, P., & Gruber, O. (2013). The neural substrate of the ideomotor principle revisited: Evidence for asymmetries in action-effect learning. *Neuroscience, 231,* 13–27.

Mottet, D., Guiard, Y., Ferrand, T., & Bootsma, R. J. (2001). Two-handed performance of a rhythmical Fitts task by individuals and dyads. *Journal of Experimental Psychology: Human Perception and Performance, 27,* 1275–1286.

Müsseler, J., & Hommel, B. (1997a). Blindness to response-compatible stimuli. *Journal of Experimental Psychology: Human Perception and Performance, 23,* 861–872.

Müsseler, J., & Hommel, B. (1997b). Detecting and identifying response-compatible stimuli. *Psychonomic Bulletin & Review, 4,* 125–129.

Müsseler, J., Wühr, P., Danielmeier, C., & Zysset, S. (2005). Action-induced blindness with lateralized stimuli and responses. *Experimental Brain Research, 160,* 214–222.

Müsseler, J., Wühr, P., & Prinz, W. (2000). Varying the response code in the blindness to response-compatible stimuli. *Visual Cognition, 7,* 743–767.

Neumann, O. (1990). Visual attention and action. In O. Neumann & W. Prinz (Eds.), *Relationships between perception and action* (pp. 227–267). Berlin: Springer-Verlag.

O'Regan, J. K., & Noë, A. (2001). A sensorimotor account of vision and visual consciousness. *Behavioral and Brain Sciences, 24*, 939–973.

Posner, M. I. (1980). Orienting of attention. *Quarterly Journal of Experimental Psychology, 32*, 3–25.

Pratt, J., & Abrams, R. A. (1994). Action-centered inhibition: Effects of distractors on movement planning and execution. *Human Movement Science, 13*, 245–254.

Pratt, J., Adam, J. J., & Fischer, M. H. (2007). Visual layout modulates Fitts's law: The importance of first and last positions. *Psychonomic Bulletin & Review, 14*, 350–355.

Prinz, W. (1997). Perception and action planning. *European Journal of Cognitive Psychology, 9*, 129–154.

Rizzolatti, G., Riggio, L., Dascola, I., & Umiltá, C. (1987). Reorienting attention across the horizontal and vertical meridians: Evidence in favor of a premotor theory of attention. *Neuropsychologia, 25*, 31–40.

Rooney, B. (2013, May 29). *Europe tops global smartphone penetration.* Retrieved from http://blogs.wsj.com/tech-europe/2013/05/29/europe-tops-global-smartphone-penetration/

Sato, M., Mengarelli, M., Riggio, L., Gallese, V., & Buccino, G. (2008). Task related modulation of the motor system during language processing. *Brain and Language, 105*, 83–90.

Schiegg, A., Deubel, H., & Schneider, W. (2003). Attentional selection during preparation of prehension movements. *Visual Cognition, 10*, 409–431.

Sebanz, N., Bekkering, H., & Knoblich, G. (2006). Joint action: Bodies and minds moving together. *Trends in Cognitive Sciences, 10*, 70–76.

Sebanz, N., Knoblich, G., & Prinz, W. (2003). Representing others' actions: Just like one's own? *Cognition, 88*, B11–B21.

Shallice, T. (1972). Dual functions of consciousness. *Psychological Review, 79*, 383–393.

Sheliga, B. M., Riggio, L., & Rizzolatti, G. (1994). Orienting of attention and eye movements. *Experimental Brain Research, 98*, 507–522.

Sheliga, B. M., Riggio, L., & Rizzolatti, G. (1995). Spatial attention and eye movements. *Experimental Brain Research, 105*, 261–275.

Shin, Y. K., Proctor, R. W., & Capaldi, E. J. (2010). A review of contemporary ideomotor theory. *Psychological Bulletin, 136*, 943–974.

Simon, J. R. (1990). The effects of an irrelevant directional cue on human information processing. In R. W. Proctor & T. G. Reeve (Eds.), *Stimulus–response compatibility: An integrated perspective* (pp. 31–86). Amsterdam: North-Holland.

Simone, P. M., & Baylis, G. C. (1997). Selective attention in a reaching task: Effect of normal aging and Alzheimer's disease. *Journal of Experimental Psychology: Human Perception and Performance, 23*, 595–608.

Spivey, M. J., & Dale, R. (2006). Continuous dynamics in real-time cognition. *Current Directions in Psychological Science, 15,* 207–211.

Stock, A., & Stock, C. (2004). A short history of ideo-motor action. *Psychological Research, 68,* 176–188.

Stoet, G., & Hommel, B. (1999). Action planning and the temporal binding of response codes. *Journal of Experimental Psychology, 25,* 1625–1640.

Symes, E., Tucker, M., Ellis, R., Vainio, L., & Ottoboni, G. (2008). Grasp preparation improves change detection for congruent objects. *Journal of Experimental Psychology: Human Perception and Performance, 34,* 854–871.

Tipper, S. P., Howard, L. A., & Jackson, S. R. (1997). Selective reaching to grasp: Evidence for distractor interference effects. *Visual Cognition, 4,* 1–38.

Tipper, S. P., Lortie, C., & Baylis, G. C. (1992). Selective reaching: Evidence for action-centered attention. *Journal of Experimental Psychology: Human Perception and Performance, 18,* 891–905.

Thomaschke, R., Hopkins, B., & Miall, R. C. (2012a). The planning and control model (PCM) of motorvisual priming: Reconciling motorvisual impairment and facilitation effects. *Psychological Review, 119,* 388–407.

Thomaschke, R., Hopkins, B., & Miall, R. C. (2012b). The role of cue–response mapping in motorvisual impairment and facilitation: Evidence for different roles of action planning and action control in motorvisual dual-task priming. *Journal of Experimental Psychology: Human Perception and Performance, 38,* 336–349.

Waszak, F., Cardoso-Leite, P., & Hughes, G. (2012). Action effect anticipation: Neurophysiological basis and functional consequences. *Neuroscience and Biobehavioral Reviews, 36,* 943–959.

Watson, J. B. (1913). Psychology as the behaviorist views it. *Psychological Review, 20,* 158–177.

Welsh, T. N., & Elliott, D. (2004). Movement trajectories in the presence of a distracting stimulus: Evidence for a response activation model of selective reaching. *Quarterly Journal of Experimental Psychology, A, 57,* 1031–1057.

Welsh, T. N., Elliott, D., Anson, J. G., Dhillon, V., Weeks, D. J., Lyons, J. L., et al. (2005). Does Joe influence Fred's action? Inhibition of return across different nervous systems. *Neuroscience Letters, 385,* 99–104.

Welsh, T. N., Elliott, D., & Weeks, D. J. (1999). Hand deviations toward distractors: Evidence for response competition. *Experimental Brain Research, 127,* 207–212.

Welsh, T. N., Lyons, J., Weeks, D. J., Anson, J. G., Chua, R., Mendoza, J., et al. (2007). Within-and between-nervous-system inhibition of return: Observation is as good as performance. *Psychonomic Bulletin & Review, 14,* 950–956.

Welsh, T. N., & Pratt, J. (2008). Actions modulate attentional capture. *Quarterly Journal of Experimental Psychology, 61,* 968–976.

Woodworth, R. A. (1899). The accuracy of voluntary movement. *Psychological Review, 3,* 1–106.

Wühr, P., & Müsseler, J. (2001). Time course of the blindness to response-compatible stimuli. *Journal of Experimental Psychology: Human Perception and Performance, 27,* 1260–1270.

Wykowska, A., Schubö, A., & Hommel, B. (2009). How you move is what you see: Action planning biases selection in visual search. *Journal of Experimental Psychology: Human Perception and Performance, 35,* 1755–1769.

Zwickel, J., & Prinz, W. (2012). Assimilation and contrast: The two sides of specific interference between action and perception. *Psychological Research, 76,* 171–182.

15
Social Attention

Daniel C. Richardson and Matthias S. Gobel

Imagine the following scenario at a typical conference. An attendee is walking between rows of posters when he catches the eyes of a presenter. This eye contact is a cue that initiates a social interaction, and she asks if he would like to hear about the poster. The presenter looks and points at the first column of the poster, drawing attention to particular sentences and graphs. The listener alternates his gaze between the poster and the presenter's face. At the same time, the presenter is monitoring the listener's gaze to make sure that he is looking at the poster (and hence attending to her message) and that he is looking back at her (to signal his understanding).

While the listener is engaged with the poster, looking away from the presenter's face, she might take the chance to sneak a look at his conference badge. If the presenter doesn't recognize the university on the badge or sees that the listener is lower in academic rank, then she might become more expansive in her gestures and more orientated toward the poster. If she recognizes the university as a prestigious one or sees that the listener is higher in academic rank, then she might orientate more toward the listener and away from the poster. In such instances, the listener now takes over the interaction directing the presenter's attention to specific locations of the poster, and she now follows with his gaze.

Witnessing many such examples, it becomes clear that when visual attention is embedded in the social world, what emerges is a complex interplay between interpersonal communication, a visual context, and the relationship between the people who share it. The goal of this chapter is to examine how these elements—separately and together—can be studied in the laboratory.

The classic understanding of visual attention is that it is determined by both top-down and bottom-up influences (e.g., Henderson, 2003). When visual attention occurs in a social context, there are two additional influences. There are further top-down influences from the beliefs that a viewer has about other people's cognitive states, such as their aspirations, intentions, and desires. And there are further bottom-up influences from the presence of another person as a stimulus, with a particular gaze direction, emotional expression, and social identity. We argue that both of these two influences have received scarce attention in the cognitive literature, and our review tries to change that.

We define social attention as the cognitive process that underlies gazing at or with another person. In its most elementary form, being attentive to where others are looking allows humans to learn where reward and danger are lurking in the environment. In its most sophisticated form, being attentive to where each other is looking allows researchers to discuss new findings at a conference. Indeed, what we term *reciprocal social attention* enables individuals to monitor the success of an interaction, identify problems, and even localize errors. This makes social attention a powerful tool for interpersonal communication, successful cooperation, and human interdependence.

Historical Context

To gauge how people perceive their social environment, cognitive scientists have extensively studied gaze following in carefully controlled laboratory experiments. They have shown that a variety of face-like stimuli can cue attention. Importantly, gaze following is modulated by the social information that is represented in the face, such as whether the person belongs to one's group or not.

Gaze Cueing Paradigms

Following the gaze of others can be used to gather information about one's environment. Indeed, from an evolutionary psychology perspective, gaze following is an important prerequisite for trans-generational learning. A variety of vertebrates, such as ibises, corvids, dogs, goats, dolphins, and primates, follow gaze (Shepherd, 2010). For example, primates have been shown to successfully follow the gaze of both experimenters (e.g., Itakura, 1996) and conspecifics (e.g., Tomasello, Call, & Hare, 1998), and chimpanzees follow the gaze of others in order to appropriate objects in their environment (Itakura & Tanaka, 1998; Tomasello, Hare, & Agnetta, 1999). Human infants follow the gaze of their caregivers from earliest ages (e.g., Farroni, Johnson, Brockbank, & Simion, 2000; Farroni, Massaccesi, Pividori, & Johnson, 2004; Hood, Willen, & Driver, 1998), and social learning in infants is facilitated by social gaze (R. Wu, Gopnik, Richardson, & Kirkham, 2011; R. Wu & Kirkham, 2010). Social attention to conspecifics has been suggested to be an essential precursor for the development of social cognition (Baron-Cohen, 1995; Tomasello, 1995). Thus, shifting attention in the direction of where a conspecific is looking seems to be an important development distinguishing primates and humans from other animals.

In a typical gaze cueing experiment in humans, a face stimulus is presented in the center of the computer screen. The face is usually first portrayed with direct gaze (or closed eyes), followed by averted gaze to the left or to the right, implying eye movements. Subsequently, a target object (e.g., a letter) is presented at one of the two lateral locations, and participants' reaction time for detecting (or identifying) the target is measured. Time and again findings have yielded faster reaction times when targets appeared at locations that were spatially congruent with the averted gaze compared to when targets appeared at locations that were spatially incongruent with the averted gaze (e.g., Bayliss, di Pellegrino, & Tipper, 2004; Driver et al., 1999; Friesen & Kingstone, 1998; Friesen, Ristic, & Kingstone, 2004; Frischen & Tipper, 2004; Hietanen, 1999; Hood et al., 1998; Langton, 2000; Nuku & Bèkkering, 2008; Ristic & Kingstone, 2005; Ristic, Wright, & Kingstone, 2007). These *gaze cueing effects* have been shown to persist up to 3 minutes (Frischen & Tipper, 2006).

Research investigating the effects of social attention in laboratory studies has demonstrated that a variety of stimuli elicit such facilitation effects as long as they resemble faces. Indeed, gaze cueing effects have been replicated with a variety of face stimuli including photographs of faces (e.g., Frischen & Tipper, 2004), computerized faces (e.g., Bayliss et al., 2004; Bayliss, di Pellegrino, & Tipper, 2005), virtual agents (e.g., Nuku & Bekkering, 2008), and even schematic drawings (e.g., Friesen & Kingstone, 1998; Kuhn & Kingstone, 2009). Even when participants were instructed that the shift in gazing behavior is counterpredictive of where the target will appear, they continued to shift their attention in accord with the direction of gaze (e.g., Bayliss & Tipper, 2006; Friesen et al., 2004; Kuhn & Kingstone, 2009). Moreover, task instructions leading participants to perceive an ambiguous stimulus as a face increased cueing effects (Ristic & Kingstone, 2005), and perceiving an ambiguous stimulus as a face was associated with distinct brain activations in neuroimaging studies (Kingstone, Tipper, Ristic, & Ngan, 2004; Tipper, Handy, Giesbrecht, & Kingstone, 2008). Thus, gaze cueing paradigms provide first evidence for people shifting their attention in accord with the visual system of others. Specifically, these studies seem to suggest that social attention depends on perceiving the central stimulus as a social entity (i.e., a face).

Social Identity and Gaze Cueing

Research suggests that the social information conveyed by a face can modulate gaze cueing effects. Social psychology has generated a substantial list of stimulus characteristics that potentially influence gaze cueing effects. For example, more masculine-looking faces lead to greater gaze cueing effects (Jones et al., 2010). Gaze cueing effects have been shown to increase for high- compared to low-status faces (Dalmaso, Pavan, Castelli, & Galfano, 2012). Faces that resemble the onlooker elicit stronger gaze cueing effects than faces that resemble the onlooker less (Hungr & Hunt, 2012). Ingroup membership (Pavan, Dalmaso, Galfano, & Castelli, 2011) and shared political partisanship (Liuzza et al., 2011) have both been shown to increase gaze cueing effects. Taken together, these studies suggest that onlookers' social attention appears most influenced by target faces that are highly relevant (e.g., high-status faces), target faces that might share common goals (e.g., ingroup members) or common opinions (e.g., party members), and target faces that portray important information about the environment. In summary, the social information attributed to a face plays a key role in modulating gaze cueing effects.

Summary

Cognitive scientists have investigated how people perceive their social environment primarily by focusing on gaze-cuing. These early experiments showed that a variety of face-like stimuli can elicit gaze cueing effects, and that specific social dimensions, such as higher social status, amplify gaze cueing effects. Yet, in focusing on gaze cueing paradigms researchers have restricted themselves mostly to examining social attention from an observer's perspective: a subject who passively perceives others. However, allocation of attention in real life is determined by observing others and *being observed* by other people. More recent approaches have therefore focused on the way that people signal information back into the social world through shifting their attention.

State-of-the-Art Review

Being the target of another person's attention has profound effects. Staring can be a threatening gesture (Dovidio & Ellyson, 1982; McNelis & Boatright-Horowitz, 1998), and averting gaze can indicate anxiety and submissiveness (De Waal, 1989; Fox, Mathews, Calder, & Yiend, 2007). Eye contact can also be used in order to deceive others (Mann et al., 2013), to signal social interest (Stass & Willis, 1967), or to signal physical attraction (Mason, Tatkow, & Macrae, 2005). Because observers can capture such information about the actor from being attentive to the actor's gaze, the latter can use gaze shifts to actively signal information back to the observer. Laboratory researchers have recently begun to study how attention is actively employed as a way to signal to and interact with other people.

The Presence of Another Person

Even in the laboratory, experiments do not take place in a social vacuum. To the contrary, measures of attention are susceptible to the influence of the social context in which the experiment takes place. Social psychology has posited for a long time, for example, that the presence of another person influences attention and action (Zajonc, 1965).

In the Stroop color-naming task, participants are usually slower at naming the ink color of a word (e.g., blue) that denotes the name of a different color (e.g., green). Yet, the mere presence of another person reduces Stroop interference effects (Huguet, Galvaing, Monteil, & Dumas, 1999). Muller, Atzeni, and Butera (2004) showed that social comparison with a better performing coactor focuses attention and as a result reduces illusionary conjunctions such as observing that *$* is present when actually only its basic elements are depicted (*S* & /). Beliefs about oneself in comparison to another person, and especially a coactor, seem to consume attentional resources and therefore induce attentional focus on the central cues in the perceptual field (Muller & Butera, 2007; Normand, Bouquet, & Croizet, 2014).

Believing another person is engaged in the same visual task can also increase participants' attention to shared stimuli. Richardson et al. (2012) invited pairs of participants to the laboratory together and instructed them to look at sets of pictures, some with positive valence and some with negative valence. Half of the time, they believed that they were looking at the same images, and half of the time that they were looking at different images. Although they could not see each other or have any interaction, simply knowing that another participant was attending to the same stimuli as they were attending to shifted their attention. When pairs of participants believed that they were looking at the images at the same time, they tended to look toward the more negative images (Richardson et al., 2012). One explanation is that under conditions of joint perception—that is when the stimuli were believed to be shared—participants looked toward the images that they thought their partner would also be looking at (von Zimmermann & Richardson, 2015). Believing their partner was experiencing the same stimuli but did not share the same task (e.g., searching the pictures for an *X* but believing their partner was memorizing the images) did not result in joint perception (Richardson et al., 2012). These findings are consistent with theorizing about social tuning effects, which posit that stimuli that are experienced by other group members become more salient (Shteynberg, 2010). Even the most minimal social context, with no sight of each other or interaction, exerts an influence over gaze. It is through this sensitivity to each other's presence that people can successfully communicate and collaborate, as we will see next.

Social Attention as a Tool for Human Communication

Social attention constitutes an important requirement for successfully acquiring information from others and practicing communication with others. Richardson and Dale (2005) recorded participants' eye movements while they spoke about a TV show and simultaneously looked at an array of pictures showing the cast members. These monologues were then played back to listeners who looked at the same array of pictures. Applying cross-recurrence analysis, Richardson and Dale (2005) found that about 2 seconds after a speaker looked at a picture, the listener was most likely to be looking at it. Speakers and listeners were more likely than chance to look at the same picture within a window of about 6 seconds.

A subsequent study showed that two people coordinate their attention when engaging in an interactive dialogue. In these experiments (Richardson, Dale, & Kirkham, 2007), participants were seated in separate cubicles, looked at the same images on-screen, and talked over the phone. The coordination between their gaze depended upon the common ground between them (Clark, 1996). Both their shared background knowledge and the visual context—as well as the beliefs about what was shared—influenced this attentional coordination (Richardson, Dale, & Tomlinson, 2009).

During conversation, people not only monitor what they can see but also keep track of what their conversation partners can see, what they know, and coordinate their visual attention accordingly. For example, in one condition of Richardson et al.'s (2009) experiment, two participants had an extended phone conversation while looking at an empty grid on-screen. However, for the duration of the conversation they coordinated their gaze around those empty locations because each (falsely) believed that their partner could see something on-screen. In this extreme case of social attention, gaze is being almost entirely driven by the top-down components. The participants can see little that is salient on-screen, they can't see each other, but their eye movements are being determined by what they believe that each other can see. These processes of social attention allow people to negotiate differences in common ground and communicate successfully.

Social Attention as a Tool for Human Cooperation

Social attention also constitutes an important requirement for cooperating with others. For example, inhibition of return is a cognitive mechanism that makes searching more efficient by reducing the chance that locations previously searched are revisited (Klein & MacInnes, 1999). It was demonstrated with the classic spatial cueing paradigm (Posner, Rafal, Choate, & Vaughan, 1985). Attention is cued by a sudden stimulus onset in one location before participants respond to a target. Participants are slower when the target appears in the same location as the cue (Posner et al., 1985; Posner, Snyder, & Davidson, 1980). Interestingly, inhibition-of-return effects do not only exist within individuals but can also occur between people (Tufft, Gobel, & Richardson, 2015; Welsh et al., 2005; Welsh et al., 2007).

Being attentive of another's focus of attention is adaptive in order to successfully coordinate search tasks. Brennan, Chen, Dickinson, Neider, and Zelinsky (2008) asked participants to jointly carry out a visual search task. Depending on the experimental condition, the participant carried out the search either alone, jointly in pairs of participants seeing each other's eye movements recorded by an eye tracker, jointly in pairs of participants talking to each other, or jointly in pairs of participants having access to both each other's vision and voice. Results showed that participants overall performed more efficiently and faster in any of the joint-attention conditions than in the solo-attention condition. In fact, pairs were even more efficient when restricted to using each other's gaze while searching than when using both vision and voice to communicate search strategies.

Neuroimaging studies provide further empirical support for the idea that people change the target of their attention in order to signal information to an interaction partner. Redcay et al. (2010), for example, asked participants to engage in a visual search task alone or jointly with another person whose gaze they could observe in real time via a dual-video-feed setup. The researchers found that attending to a stimulus alone or with another person was represented differently in the human brain. Similar findings were documented when participants were asked to engage in interactive eye-tracking paradigms with an animated virtual character who was controlled by a human agent (Pfeiffer, Timmermans, Bente, Vogeley, & Schilbach, 2011; Wilms, et al., 2010). Such neural correlates are in line with the previously reviewed behavioral findings suggesting that people successfully signal information to their partner through shifts in attention, thereby improving their cooperating in visual search tasks.

Social Attention in Real-Life Situations

When people interact, they bring together their personal values, cultural heritage, and social norms. Socially shared knowledge structures, such as social norms, inform the actor when it is permissible to attend to another person, when it is inappropriate to attend to another person, and when it is actually a social requirement to pay attention to another person in order to acknowledge an ongoing interaction or communication.

Researchers found that social attention changes in situations with potential for social interactions, compared to the isolated experimental conditions during a laboratory study. Foulsham, Walker, and Kingstone (2011) measured the social attention of participants who walked across campus and of participants who watched the video recording of walking across campus from a first-person perspective. Results revealed that participants watching the videos were more likely to attend to people passing by compared to those in the real-life situation. Consistent with these findings, Laidlaw, Foulsham, Kuhn, and Kingstone (2011) found that participants who were sitting in a waiting room attended to prerecorded videos of a confederate for longer periods than they did to a confederate who was actually sitting there in person. Simply making participants believe that they might be watched by others by having them wear an eye tracker changes looking behavior dramatically (Risko & Kingstone, 2011). Presumably, when encountering people in real life, where people can capture each other's social attention, it sometimes is more appropriate to not attend to others.

The potential to interact with others in real life also influences the extent to which people follow each other's attention to objects in the immediate environment. Gallup, Hale, et al. (2012) showed that pedestrians in public environments actively followed the attention of groups of confederates. In fact, following the attention of others increased with the number of confederates looking toward the stimulus before saturating for very large groups. Gallup, Chong, and Couzin (2012) positioned an attractive stimulus in a frequently trafficked corridor and measured whether people would look at it. A hidden camera recorded to what extent a total of 2,882 pedestrians were following other's attention to the attractive stimulus. Interestingly, passersby were more likely to follow the head turns of people walking in front of them, who thus could not see where they were looking, whereas they were less likely to follow the head turns of people walking toward them, who thus could see where they were looking. These studies provide further support for the idea that social norms can influence how people overtly change their social attention when others can see them. In fact, it seems less appropriate to overtly follow the attention of another person when the latter can observe such signals of social interest.

In a different study, Freeth, Foulsham, and Kingstone (2013) found that when being interviewed, interviewees looked more to the face and less to the background in the live condition, where the interviewer was physically present, compared to the video condition, where the interviewer was depicted in a video clip. Another situation where people increase attention to their interaction partner is when they share a meal. D. W.-L. Wu, Bischof, and Kingstone (2013) demonstrated that when eating with another person compared to when eating alone, participants' attention was more drawn away from the surrounding objects to the person in front of them. This effect was amplified among pairs who talked more to each other during the meal. One explanation for increased attention to another person sitting right in front of the person is that people are especially keen to signal their engagement in the interaction and social interest in the other person.

Depending on the situation, social norms can both reduce social attention or increase social attention to other people. Interestingly, reducing and increasing attention both are strong signals that individuals comply with social norms that regulate how much social interest is deemed appropriate in a given situation.

Summary

Research on how social attention is actively signaled in social contexts has yielded important findings. First, even when experimental tasks do not require any interaction between partners, social attention can be influenced by the mere presence of another person in laboratory experiments. Second, since people pay attention to another's presence, they can successfully communicate and collaborate with each other through shifting attention. Third, depending on the situation, people actively disengage attention, engage attention, or follow where other group members are attending to, thereby signaling their compliance with social norms about how much social interest is adequate. Yet, when interacting with others in real life, there is an ongoing interplay of observing and signaling. As we argue next, perhaps social attention is more accurately described as reciprocal.

Integration

Traditional laboratory research has been hampered in its study of social attention. Experimental cubicles place participants in solitary confinement away from other people. There are good methodological reasons for this, of course, but the cost is that the reciprocal nature of social attention that we see in everyday social interactions is excluded from the experimental situation.

As seen so far, studies examining social attention have examined either its perceptual function of gathering information from others or its signaling function of sending information to others. As a result, previous research has focused either on manipulating the stimulus presentation, thereby reducing ecological validity, or on observing social attention in real-life situations, thereby reducing the experimental control over the stimulus presentation. Novel paradigms, however, combine both experimental control over the stimulus presentation and ecological validity. These studies examine how social attention serves to perceive information from others and signal information to others. Because of this continual interaction of perceiving and signaling, we think that social attention is better described as reciprocal in nature, thus representing both of these crucial aspects.

Researchers have started to examine the reciprocal nature of social attention by systematically manipulating how participants construe the experimental situation. Depending on task instructions given by the experimenter, participants will sometimes shift attention in order to perceive information from alleged coactors and other times in order to signal information to alleged coactors, showing that the dual functions of social attention can be experimentally dissociated.

Reciprocal Attention to Nonsocial Stimuli

Our research group recently investigated whether interacting with another person would influence the inhibition-of-return effects within the same individual (Tufft et al., 2015). We employed a classic spatial cueing paradigm (Posner et al., 1980), in which a cue stimulus directs attention to one location of a screen, and participants were then asked to quickly detect a target stimulus that appeared in the same or in a different spatial location. In a novel twist, however, we ran pairs of participants. They sat back-to-back, not interacting, looking at a screen with an eye tracker measuring their gaze. The location of the cue, a red dot, was chosen randomly by the computer, as in the classic paradigm. This was the nonsocial condition. However, in half the trials, we told participants that the cue represented where their partner was looking on the other screen. This was the social condition. Participants' reactions to the subsequent target were compared in the social and nonsocial trials. We found that regardless of the condition, participants looked at the cue for an equal amount of time. However, in the social cue condition, responses to the subsequent target changed: The magnitude of the inhibition-of-return effect was greater. Believing that the cue carries social meaning seems to modulate its later consequences for visual attention.

Increased attention to nonsocial stimuli has also been documented in tasks that require participants to allocate attention selectively while interacting with another person. For example, in the Navon task, participants are presented with one large letter that consists of many smaller letters, so that in order to identify the target letter, attention has to be allocated selectively, inhibiting attending to one of the two features (Navon, 1977). Böckler, Knoblich, and Sebanz (2012) had participants sitting side-by-side perform a joint version of the Navon task. While one participant was instructed to respond to the large letter (i.e., the global stimulus feature), the other participant was instructed to respond to the small letters (i.e., the local stimulus feature). It was found that participants were overall slower when performing the task with a coactor who was instructed to adopt a different focus of attention (Böckler et al., 2012). Similar results were found when pairs of participants engaged in an Eriksen flanker task (Eriksen & Eriksen, 1974), in which a stimulus was flanked by the potential target of the coactor (Atmaca, Sebanz, & Knoblich, 2011). In summary, when interacting with another person in carefully controlled experimental studies, people readily represent where their partner's attention is allocated.

Reciprocal Attention to Social Stimuli

The fact that humans readily attribute attentional states to their interaction partners has been used in gaze cueing paradigms to systematically change participants' allocation of attention. Teufel, Alexis, Clayton, and Davis (2010), for example, used a traditional cueing paradigm in an interactive context, in which the central face cue was a prerecorded video of an experimenter who could either see or could not see the onset of the target stimulus because he was wearing opaque goggles. Attributing the ability to see to the experimenter significantly increased cueing effects (Teufel et al., 2010). In a similar study, Wiese, Wykowska, Zwickel, and Mu (2012) investigated how ascribing mental states to a central cue would affect gaze cueing. Participants saw either a human face or a robot as a central cue on-screen. In a clever twist, however, in some of the trials participants believed that the robot was controlled by a human, providing the robot with intentional states, whereas in other trials the human face was described as a human-like mannequin, precluding any attribution of intentionality. Results revealed that in trials when intentions were ascribed to the central cue (i.e., human face or robot controlled by a human) gaze cueing effects were significantly larger than in trials when no intentional states were ascribed to the central cue (i.e., robot or human-like mannequin; Wiese et al., 2012). Similarly, Pfeiffer et al. (2011) documented that during interactive eye-tracking paradigms, participants' gazing behavior in response to an animated virtual character depended on their beliefs that the virtual character was controlled by a human agent compared to by a computer. In an electroencephalography (EEG) study using a similar paradigm, Wykowska, Wiese, Prosser, and Müller (2014) showed that event-related brain potentials changed more when cues were believed to be controlled by a human but less when believed to be controlled by the machine. Thus, intentionality is distinctly represented in neural correlates, too.

Attributing intentionality—for example, to central cues in gaze cueing paradigms—is a key factor guiding social attention, presumably because it indicates the ability to signal the location of the target. Importantly, it is what one person believes to perceive as another person's signaling that results in dramatic shifts in allocating attention. In everyday life, of course, one person's perceiving and another person's signaling fluidly follow each other and result in reciprocal allocation of attention.

Dissociating the Dual Function of Reciprocal Attention

In most experiments that track gaze to a face, we cannot be certain whether a participant is shifting attention to signal information to others or shifting attention to perceive information from others. In order to understand how people strategically employ social attention in interpersonal interaction, we have dissociated this dual function of reciprocal attention (Gobel, Kim, & Richardson, 2015).

We asked participants to watch a series of video clips of target faces looking directly into the camera. The faces in the video clips could be those of either higher or lower ranked others. We theorized that participants would shift their social attention to targets' eyes to either signal information to targets or perceive information from them. In order to dissociate this dual function of social attention, we manipulated participants' construal of the viewing condition. Participants watched video clips of target faces and were being videotaped at the same time. In some of the trials, we told participants that no one would see their video recording (i.e., one-way viewing). In other trials, however, we told them that the same people from the videos would later return into the laboratory in order to watch participants' video recordings (i.e., two-way viewing). Results showed that beliefs about the viewing condition modulated the allocation of social attention. In the two-way viewing condition, participants increased attention to the eyes of lower ranked target faces. One possible interpretation is that participants shifted attention in order to signal something to lower ranked targets, presumably their superior social standing. In contrast, in the one-way viewing condition, participants increased attention to the eyes of higher ranked target faces. One possible interpretation is that since targets would not observe participants looking at their faces, participants shifted attention in order to encode information from the more relevant targets, that is, targets of higher social rank (Gobel et al., 2015).

Additional evidence for the dual functions of social attention comes from Schilbach et al. (2010), who instructed participants to play an interactive game with another person while measuring their brain activity using functional magnetic resonance imagery (fMRI). In this study, participants either initiated the game, by selecting one object with a shift of their attention, or followed their partner's attentional shift to that object. Results revealed distinct patterns of neural activity, depending on whether participants initiated shifts of attention or followed shifts in attention initiated by their interacting partner (Schilbach et al., 2010). Initiating, which is the act of signaling information, versus following, which is the act of perceiving information, changed dramatically how information about joint attention was represented in the brain. Thus, first evidence has been provided that reciprocal attention can be dissociated into instances of perceiving information from others versus signaling information to others.

Summary

People's beliefs about the experimental situation can modulate social attention. Experimenters have started to use this fact to explore the reciprocal nature of social attention. Making minimal changes to task instructions resulted in dramatic changes in participants' attention allocation. Attention shifted locations when participants were made to believe that they were or were not interacting with another person and when participants were made to believe that the situation fostered perceiving information from others versus signaling information to others. In all of these examples, presenting the same stimuli to all participants, but subtly manipulating participants' construal of the experimental condition, led to pervasive changes in the allocation of attention. This novel approach to studying social attention in laboratory experiments seems a promising gateway to examine real-life phenomena of reciprocal social attention.

Future Directions

We began this chapter by describing a scenario of two researchers interacting during a conference poster session. We described how the presenter, unobserved by the listener, allocated attention to his name badge. The presenter's attention then shifted toward the listener. But why did she shift her attention in the first place? Did she increase her attention to the listener because she wanted to perceive what he was thinking of her poster or because she meant to challenge his critique of her research? Or perhaps she wanted to signal that she thought what he was saying was really interesting? As this example illustrates, understanding the functionality of social attention in real-life situations is extremely difficult.

In this chapter, we have described how social attention can be studied in laboratory experiments. The literature we have reviewed here illustrates how social attention is a cornerstone of successful communication and effective coordination between individuals. Therefore, social attention can be described more accurately, perhaps, as the cognitive process that underlies the exploitation of another's visual system to facilitate human life in social groups.

What remains unanswered, however, is the underlying functionality of social attention in specific situations. For example, why do people with autism spectrum condition reduce attention to the eyes of others (Klin, Jones, Schultz, Volkmar, & Cohen, 2002)? Are they not interested in gathering social information from the eyes, or are they signaling their social disengagement? Why do highly anxious people increase attention toward the eyes of angry faces (Fox et al., 2007)? Are they monitoring for potentially negative feedback, or are they signaling being attentive? While social attention research has made tremendous advances in situating a fundamental cognitive process into social contexts, we now ought to improve our understanding of how this process serves interpersonal interactions. We believe this is the next challenge awaiting social attention researchers.

We propose that one way to better understand the underlying functionality of social attention is the combination of more than one measure. For example, behavioral measures of attention (e.g., reaction times or eye movements) could be combined with physiological measures (e.g., skin conductance or cortisol level) and neuroimaging data (e.g., EEG or fMRI) to further dissociate the dual function of reciprocal social attention.

While we have and always will be inspired by observing how social attention shifts in everyday life situations, we will only be able to fully understand how social attention is functionally allocated if we carry out controlled laboratory experiments. We can then transfer the newly gained knowledge back into the social world and make predictions about when people shift attention and to what end.

Box 15.1

Key Points

- We define social attention as the cognitive process that underlies gazing at or with another person.
- The identities and social characteristics (e.g., social rank) of interacting individuals guide the allocation of social attention.
- Social attention fulfils a dual function. Social attention is employed to perceive information from the world and signal information into the world.
- Interacting individuals fluidly shift attention from perceiving to signaling, and vice versa. Therefore, social attention is best described as reciprocal in nature.

Box 15.2

Outstanding Issues
- Researchers should attempt to dissociate the dual function of reciprocal social attention using a combination of behavioral, physiological, and neuroimaging measures.

Acknowledgments

We thank Jorina von Zimmermann and Miles Tufft for comments on previous versions of this chapter.

References

Atmaca, S., Sebanz, N., & Knoblich, G. (2011). The joint flanker effect: Sharing tasks with real and imagined co-actors. *Experimental Brain Research*, *211*, 371–385.

Baron-Cohen, S. (1995). *Mindblindness: An essay on autism and theory of mind*. Cambridge, MA: MIT Press.

Bayliss, A. P., di Pellegrino, G., & Tipper, S. P. (2004). Orienting of attention via observed eye gaze is head-centred. *Cognition*, *94*, B1–B10.

Bayliss, A. P., di Pellegrino, G., & Tipper, S. P. (2005). Sex differences in eye gaze and symbolic cueing of attention. *Quarterly Journal of Experimental Psychology*, *58*, 631–650.

Bayliss, A. P., & Tipper, S. P. (2006). Predictive gaze cues and personality judgments. *Psychological Science*, *17*, 514–520.

Böckler, A., Knoblich, G., & Sebanz, N. (2012). Effects of a coactor's focus of attention on task performance. *Journal of Experimental Psychology: Human Perception and Performance*, *38*, 1404–1415.

Brennan, S. E., Chen, X., Dickinson, C. A., Neider, M. B., & Zelinsky, G. J. (2008). Coordinating cognition: The costs and benefits of shared gaze during collaborative search. *Cognition*, *106*, 1465–1477.

Clark, H. H. (1996). *Using language*. Cambridge, UK: Cambridge University Press.

Dalmaso, M., Pavan, G., Castelli, L., & Galfano, G. (2012). Social status gates social attention in humans. *Biology Letters*, *8*, 450–452.

De Waal, F. B. M. (1989). *Chimpanzee politics: Power and sex among apes*. Baltimore, MD: John Hopkins University Press.

Dovidio, J. F., & Ellyson, S. L. (1982). Decoding visual dominance: Attributions of power based on relative percentages of looking while speaking and looking while listening. *Social Psychology Quarterly*, *45*, 106–113.

Driver, J., Davis, G., Ricciardelli, P., Kidd, P., Maxwell, E., & Baron-Cohen, S. (1999). Gaze perception triggers reflexive visuospatial orienting. *Visual Cognition*, *6*, 509–540.

Eriksen, B. A., & Eriksen, C. W. (1974). Effects of noise letters upon the identification of a target letter in nonsearch task. *Perception & Psychophysics*, *16*, 143–149.

Farroni, T., Johnson, M. H., Brockbank, M., & Simion, F. (2000). Infants' use of gaze direction to cue attention: The importance of perceived motion. *Visual Cognition*, *7*, 705–718.

Farroni, T., Massaccesi, S., Pividori, D., & Johnson, M. H. (2004). Gaze following in newborns. *Infancy*, *5*, 39–60.

Foulsham, T., Walker, E., & Kingstone, A. (2011). The where, what and when of gaze allocation in the lab and the natural environment. *Vision Research*, *51*, 1920–1931.

Fox, E., Mathews, A., Calder, A. J., & Yiend, J. (2007). Anxiety and sensitivity to gaze direction in emotionally expressive faces. *Emotion, 7*, 478–486.

Freeth, M., Foulsham, T., & Kingstone, A. (2013). What affects social attention? Social presence, eye contact and autistic traits. *PLoS ONE, 8*(1), e53286.

Friesen, C. K., & Kingstone, A. (1998). The eyes have it! Reflexive orienting is triggered by nonpredictive gaze. *Psychonomic Bulletin & Review, 5*, 490–495.

Friesen, C. K., Ristic, J., & Kingstone, A. (2004). Attentional effects of counterpredictive gaze and arrow cues. *Journal of Experimental Psychology: Human Perception and Performance, 30*, 319–329.

Frischen, A., & Tipper, S. P. (2004). Orienting attention via observed gaze shift evokes longer term inhibitory effects: Implications for social interactions, attention, and memory. *Journal of Experimental Psychology. General, 133*, 516–533.

Frischen, A., & Tipper, S. P. (2006). Long-term gaze cueing effects: Evidence for retrieval of prior states of attention from memory. *Visual Cognition, 14*, 351–364.

Gallup, A. C., Chong, A., & Couzin, I. D. (2012). The directional flow of visual information transfer between pedestrians. *Biology Letters, 8*, 520–522.

Gallup, A. C., Hale, J. J., Sumpter, D. J. T., Garnier, S., Kacelnik, A., Krebs, J. R., et al. (2012). Visual attention and the acquisition of information in human crowds. *Proceedings of the National Academy of Sciences of the United States of America, 109*, 7245–7250.

Gobel, M. S., Kim, H. S., & Richardson, D. C. (2015). The dual function of social gaze. *Cognition, 136*, 359–364.

Henderson, J. M. (2003). Human gaze control during real-world scene perception. *Trends in Cognitive Sciences, 7*, 498–504.

Hietanen, J. K. (1999). Does your gaze direction and head orientation shift my visual attention? *Neuroreport, 10*, 3443–3447.

Hood, B. M., Willen, J. D., & Driver, J. (1998). Adult's eyes trigger shifts of visual attention in human infants. *Psychological Science, 9*, 131–134.

Huguet, P., Galvaing, M. P., Monteil, J. M., & Dumas, F. (1999). Social presence effects in the Stroop task: Further evidence for an attentional view of social facilitation. *Journal of Personality and Social Psychology, 77*, 1011–1025.

Hungr, C. J., & Hunt, A. R. (2012). Physical self-similarity enhances the gaze-cueing effect. *Quarterly Journal of Experimental Psychology, 65*, 1250–1259.

Itakura, S. (1996). An exploratory study of gaze-monitoring in nonhuman primates. *Japanese Psychological Research, 38*, 174–180.

Itakura, S., & Tanaka, M. (1998). Use of experimenter-given cues during object-choice tasks by chimpanzees (*Pan troglodytes*), an orangutan (*Pongo pygmaeus*), and human infants (*Homo sapiens*). *Journal of Comparative Psychology, 112*, 119–126.

Jones, B. C., Debruine, L. M., Main, J. C., Little, A. C., Welling, L. L. M., Feinberg, D. R., et al. (2010). Facial cues of dominance modulate the short-term gaze-cuing effect in human observers. *Proceedings of the Royal Society Series B, 277,* 617–624.

Kingstone, A., Tipper, C., Ristic, J., & Ngan, E. (2004). The eyes have it! An fMRI investigation. *Brain and Cognition, 55,* 269–271.

Klein, R. M., & MacInnes, W. J. (1999). Inhibition of return is a foraging facilitator in visual search. *Psychological Science, 10,* 346–352.

Klin, A., Jones, W., Schultz, R., Volkmar, F., & Cohen, D. (2002). Visual fixation patterns during viewing of naturalistic social situations as predictors of social competence in individuals with autism. *Archives of General Psychiatry, 59,* 809–816.

Kuhn, G., & Kingstone, A. (2009). Look away! Eyes and arrows engage oculomotor responses automatically. *Attention, Perception & Psychophysics, 71,* 314–327.

Laidlaw, K. E. W., Foulsham, T., Kuhn, G., & Kingstone, A. (2011). Potential social interactions are important to social attention. *Proceedings of the National Academy of Sciences of the United States of America, 108,* 5548–5553.

Langton, S. R. (2000). The mutual influence of gaze and head orientation in the analysis of social attention direction. *Quarterly Journal of Experimental Psychology, 53,* 825–845.

Liuzza, M. T., Cazzato, V., Vecchione, M., Crostella, F., Caprara, G. V., & Aglioti, S. M. (2011). Follow my eyes: The gaze of politicians reflexively captures the gaze of ingroup voters. *PLoS ONE, 6*(9), e25117.

Mann, S., Ewens, S., Shaw, D., Vrij, A., Leal, S., & Hillman, J. (2013). Lying eyes: Why liars seek deliberate eye contact. *Psychiatry, Psychology and Law, 20,* 452–461.

Mason, M. F., Tatkow, E. P., & Macrae, C. N. (2005). The look of love: Gaze shifts and person perception. *Psychological Science, 16,* 236–239.

McNelis, N. L., & Boatright-Horowitz, S. L. (1998). Social monitoring in a primate group: The relationship between visual attention and hierarchical ranks. *Animal Cognition, 1,* 65–69.

Muller, D., Atzeni, T., & Butera, F. (2004). Coaction and upward social comparison reduce the illusory conjunction effect: Support for distraction–conflict theory. *Journal of Experimental Social Psychology, 40,* 659–665.

Muller, D., & Butera, F. (2007). The focusing effect of self-evaluation threat in coaction and social comparison. *Journal of Personality and Social Psychology, 93,* 194–211.

Navon, D. (1977). Forest before trees: The precedence of global features in visual perception. *Cognitive Psychology, 9,* 353–383.

Normand, A., Bouquet, C. A., & Croizet, J.-C. (2014). Does evaluative pressure make you less or more distractible? Role of top-down attentional control over response selection. *Journal of Experimental Psychology. General, 143,* 1097–1111.

Nuku, P., & Bekkering, H. (2008). Joint attention: Inferring what others perceive (and don't perceive). *Consciousness and Cognition, 17,* 339–349.

Pavan, G., Dalmaso, M., Galfano, G., & Castelli, L. (2011). Racial group membership is associated to gaze-mediated orienting in Italy. *PLoS ONE, 6*(10), e25608.

Pfeiffer, U., Timmermans, B., Bente, G., Vogeley, K., & Schilbach, L. (2011). A nonverbal Turing test: Differentiating mind from machine in gaze-based social interaction. *PLoS ONE, 6*(11), e27591.

Posner, M. I., Rafal, R. D., Choate, L. S., & Vaughan, J. (1985). Inhibition of return: Neural basis and function. *Cognitive Neuropsychology, 2,* 211–228.

Posner, M. I., Snyder, C. R., & Davidson, B. J. (1980). Attention and the detection of signals. *Journal of Experimental Psychology. General, 109,* 160–174.

Redcay, E., Dodell-Feder, D., Pearrow, M. J., Mavros, P. L., Kleiner, M., Gabrieli, J. D. E., et al. (2010). Live face-to-face interaction during fMRI: A new tool for social cognitive neuroscience. *NeuroImage, 50,* 1639–1647.

Richardson, D. C., & Dale, R. (2005). Looking to understand: The coupling between speakers' and listeners' eye movements and its relationship to discourse comprehension. *Cognitive Science, 29,* 1045–1060.

Richardson, D. C., Dale, R., & Kirkham, N. Z. (2007). The art of conversation is coordination: Common ground and the coupling of eye movements during dialogue. *Psychological Science, 18,* 407–413.

Richardson, D. C., Dale, R., & Tomlinson, J. M. (2009). Conversation, gaze coordination, and beliefs about visual context. *Cognitive Science, 33,* 1468–1482.

Richardson, D. C., Street, C. N. H., Tan, J. Y. M., Kirkham, N. Z., Hoover, M. A., & Ghane Cavanaugh, A. (2012). Joint perception: Gaze and social context. *Frontiers in Human Neuroscience, 6,* 194.

Risko, E. F., & Kingstone, A. (2011). Eyes wide shut: Implied social presence, eye tracking and attention. *Attention, Perception & Psychophysics, 73,* 291–296.

Ristic, J., & Kingstone, A. (2005). Taking control of reflexive social attention. *Cognition, 94,* B55–B65.

Ristic, J., Wright, A., & Kingstone, A. (2007). Attentional control and reflexive orienting to gaze and arrow cues. *Psychonomic Bulletin & Review, 14,* 964–969.

Schilbach, L., Wilms, M., Eickhoff, S. B., Romanzetti, S., Tepest, R., Bente, G., et al. (2010). Minds made for sharing: Initiating joint attention recruits reward-related neurocircuitry. *Journal of Cognitive Neuroscience, 22,* 2702–2715.

Shepherd, S. V. (2010). Following gaze: Gaze-following behavior as a window into social cognition. *Frontiers in Integrative Neuroscience, 4,* 5.

Shteynberg, G. (2010). A silent emergence of culture: The social tuning effect. *Journal of Personality and Social Psychology, 99*, 683–689.

Stass, J. W., & Willis, F. N. (1967). Eye contact, pupil dilation, and personal preference. *Psychonomic Science, 7*, 375–376.

Teufel, C., Alexis, D. M., Clayton, N. S., & Davis, G. (2010). Mental-state attribution drives rapid, reflexive gaze following. *Attention, Perception & Psychophysics, 72*, 695–705.

Tipper, C. M., Handy, T. C., Giesbrecht, B., & Kingstone, A. (2008). Brain responses to biological relevance. *Journal of Cognitive Neuroscience, 20*, 879–891.

Tomasello, M. (1995). Origins of human communication. In C. Moore & P. Dunham (Eds.), *Joint attention: Its origins and role in development* (pp. 103–130). Hillsdale, NJ: Erlbaum.

Tomasello, M., Call, J., & Hare, B. (1998). Five primate species follow the visual gaze of conspecifics. *Animal Behaviour, 55*, 1063–1069.

Tomasello, M., Hare, B., & Agnetta, B. (1999). Chimpanzees, *Pan troglodytes*, follow gaze direction geometrically. *Animal Behaviour, 58*, 769–777.

Tufft, M. R. A., Gobel, M. S., & Richardson, D. C. (2015). Social eye cue: how knowledge of another person's attention changes your own. In P. Bello, M. Guarini, M. McShane, & B. Scassellati (Eds.), *Proceedings of the 37th Annual Conference of the Cognitive Science Society*. Austin, TX: Cognitive Science Society.

von Zimmermann, J., & Richardson, D. C. (2015). Joint perception. In S. Obhi & E. Cross (Eds.), *Shared representations: Sensorimotor foundations of social life*. Cambridge, UK: Cambridge University Press.

Welsh, T. N., Elliott, D., Anson, J. G., Dhillon, V., Weeks, D. J., Lyons, J. L., et al. (2005). Does Joe influence Fred's action? Inhibition of return across different nervous systems. *Neuroscience Letters, 385*, 99–104.

Welsh, T. N., Lyons, J., Weeks, D. J., Anson, J. G., Chua, R., Mendoza, J., et al. (2007). Within- and between-nervous-system inhibition of return: Observation is as good as performance. *Psychonomic Bulletin & Review, 14*, 950–956.

Wiese, E., Wykowska, A., Zwickel, J., & Mu, H. J. (2012). I see what you mean: How attentional selection is shaped by ascribing intentions to others. *PLoS ONE, 7*(9), e45391.

Wilms, M., Schilbach, L., Pfeiffer, U., Bente, G., Fink, G., & Vogeley, K. (2010). It's in your eyes: Using gaze feedback to create truly interactive paradigms for social cognitive and affective neuroscience. *Social Cognitive and Affective Neuroscience, 5*, 98–107.

Wu, D. W.-L., Bischof, W. F., & Kingstone, A. (2013). Looking while eating: The importance of social context to social attention. *Scientific Reports, 3*, 2356.

Wu, R., Gopnik, A., Richardson, D. C., & Kirkham, N. Z. (2011). Infants learn about objects from statistics and people. *Developmental Psychology, 47*, 1220–1229.

Wu, R., & Kirkham, N. Z. (2010). No two cues are alike: Depth of learning during infancy is dependent on what orients attention. *Journal of Experimental Child Psychology, 107,* 118–136.

Wykowska, A., Wiese, E., Prosser, A., & Müller, H. J. (2014). Beliefs about the minds of others influence how we process sensory information. *PLoS ONE, 9*(4), e94339.

Zajonc, R. B. (1965). Social facilitation. *Science, 149,* 269–274.

16

Attention to Eyes in Face Perception

Roxane J. Itier

Historical Context

Humans are a fundamentally social species, and social interactions are mediated by accurate extraction of social cues from the face. The human face provides a plethora of cues such as identity, gender, age, emotion, attractiveness, and even trustworthiness and is thus the most important social stimulus we process daily. In the past few decades, tremendous progress has been made toward unraveling the cognitive neuroarchitecture underlying face processing although many unanswered questions remain. One question concerns the role that attention to the eyes plays in face perception.

Through the course of evolution, changes to the morphology of the human face have made the position of the eyes more salient (Emery, 2000). The human eye itself has evolved tremendously, with the largest ratio of exposed sclera size of all primate species and an elongated shape, which are thought to have allowed a better eye mobility and a better detection of gaze direction. Humans are indeed the only primate species with white sclera that contrasts with the dark iris (Kobayashi & Kohshima 1997), making gaze direction accurately visible up to 5 meters away (Gamer & Hecht, 2007). These physical changes have accompanied important changes in the human brain which are believed to be linked to the emergence of social cognition (Emery, 2000), with a complex network of brain areas involved in the processing of faces and their attributes, especially gaze (see Itier & Batty, 2009, for a review).

In this chapter I will first review the literature showing that attention is attracted by the eyes of the face and will expand on the role of this attention to eyes in face perception as measured behaviorally and by eye-movement studies. I will then take a different perspective on attention to eyes by reviewing the event-related potential (ERP) literature on face and eye perception. Finally, I will try to integrate these two (unfortunately) separate literatures and discuss how these laboratory studies help us understand attention to eyes in the real world. This review is far from exhaustive and focuses mainly on face perception and recognition tasks, leaving aside the literature on attention to gaze and attention orienting by gaze.

State-of-the-Art Review

Attention to the Eyes and Its Role in Face Perception as Revealed by Eye-Movement Monitoring and Behavioral Studies

The monitoring of eye movements is a well-known technique in the study of attention. During visual exploration eye location is used as a measure of overt attention and new information is only acquired during fixation as vision is suppressed during saccades (Henderson et al., 2003; Rayner, 2009). Where people look is thus often used as a measure of what they attend to. Pioneering work showed that viewers' attention is drawn to informative areas of the scene and that object saliency (typically of a low-level nature such as with respect to contrast, luminance, spatial frequencies, etc.) influences what part of the scene is fixated (Rayner, 2009; Yarbus, 1967). However, eye movements are also goal directed and depend heavily on the type of stimulus being viewed and on the task constraints (Henderson et al., 2003; Rayner, 2009).

Laboratory studies using eye-movement monitoring during the exploration of complex scenes containing people have shown that in free-viewing conditions, participants look more at the faces than the objects of the scene, that they look more at the eyes within the face, and that the eyes are also revisited many times during the course of exploration (Birmingham et al., 2008a, 2008b; Klin et al., 2002). This attraction to the eye region is not driven by the eye saliency (Birmingham et al., 2009) but most likely by the need to decipher the social content of the scene (Birmingham et al., 2008a, 2008b; Smilek et al., 2006). In research exploring centrally presented faces devoid of background context, it has also often been reported that the eyes are the most attended facial feature (e.g., Althoff & Cohen, 1999; Arizpe et al., 2012; Barton et al., 2006; Bindemann et al., 2009; Heisz & Shore, 2008; Henderson et al., 2001; Henderson et al., 2005; Janik et al., 1978; Laidlaw et al., 2012) although many inconsistencies remain.

In general, the internal features comprising the eyes, the mouth, and the nose are viewed more than the external features such as face contour, ears, hair, and so forth (Althoff & Cohen, 1999; Henderson et al., 2005; Walker-Smith et al., 1977), and this external/internal feature looking difference is linked to familiarity and task. People looked more at the internal features of familiar faces but more at the external features of unfamiliar faces in a matching task while in a familiarity judgment task the internal features were looked at more regardless of face familiarity (Stacey et al., 2005). During free viewing, Althoff and Cohen (1999) showed that the greater viewing of eyes than other features was more pronounced for famous than nonfamous faces regardless of the task (fame judgment or emotional judgment). The eyes were also looked at more than the other facial features during the learning of unfamiliar faces (Henderson et al., 2001, 2005). Importantly, the accuracy on a following memory task was lower when eye movements were prevented during learning than when they were allowed, suggesting that these eye movements (and thus overt attention to eyes) facilitated learning of face identity (Henderson et al., 2005). More recently Heisz and Shore (2008) showed that, as an unfamiliar face became familiar over the course of several exposures, fewer fixations were made overall but the eyes were looked at more than other features regardless of familiarity. In addition, in the recall task (but not in the old/new judgment), eyes were also looked at more as familiarity increased, at the expense of other facial features which were looked at less (Heisz & Shore, 2008). These data are in contrast to the greater scanning of the eye region for novel compared to famous faces reported by Barton and colleagues (2006). This difference might be due to the use of a morphing procedure in the latter study where one identity was progressively morphed into a new one and participants were asked to respond as to which identity they were seeing. Barton et al. (2006) however reproduced the finding of a general decreased exploration of famous compared to novel faces reported previously (Althoff & Cohen, 1999) and argued that the decreased sampling of the eyes in famous faces reflected the strength of the preexisting representation of the face identity in memory, with better representation of the upper face part containing the eyes than the lower face part.

The reason why the eyes are attended more than other features in these identity judgments is likely because the eye region provides many cues that are specific to each individual (eye color and shape, specific distances between elements like pupils, eyebrows, eyelids, eyelashes, etc.). The importance of the eye region in face recognition has been shown by studies in which masking the eyes had a larger detrimental effect on recognition accuracy than masking the nose or mouth (e.g., McKelvie, 1976) although a review of subjective reports, verbal description, and feature alteration and reconstruction suggested the importance of eyes varied greatly with stimuli, tasks, and designs (Shepherd et al., 1981). Detection of eye changes is enhanced in familiar compared to unfamiliar faces, an effect not seen for other facial features (O'Donnell & Bruce 2001). Eyes seem important even for face detection, which is slowed down when eyes are masked but not when other features are masked (Lewis & Edmonds, 2003), and for face categorization, which is also slowed down when one eye is slightly shifted above the other (Cooper & Wojan, 2000). Detecting a face in noise also mainly requires seeing a symmetrical pair of eyes (Paras & Webster, 2013).

Response classification techniques have also shown that the eye region is the most diagnostic facial area for identity recognition (Haig, 1985; Schyns et al., 2002; Sekuler et al., 2004). Some studies suggest that eyes may not be utilized properly in some cases of prosopagnosia (Bukach et al., 2008; Caldara et al., 2005; Orban de Xivry et al., 2008), the inability to recognize faces. Patient PS is considered a rare case of "pure" prosopagnosia (Rossion et al., 2003), and the response classification technique *Bubbles* revealed that in a recognition task, PS was using mostly the mouth region and not at all the eyes to perform the task, in contrast to control participants who used mostly the eyes (Caldara et al., 2005). Eye tracking showed that, although PS looked mostly at the mouth and much less at the eyes of familiar faces, she did look at eyes but did not seem to use them for face recognition and her scanning behavior was unrelated to her recognition accuracy (Orban de Xivry et al., 2008). These results suggest that what is looked at is not necessarily used for the task at hand, which is problematic for the interpretation of eye movements and might be the reason for the inconsistent findings reported in eye-tracking studies.

Other studies confirm that the link between eye movements and information processing during face perception is not straightforward. One study manipulated the type of information processing employed by participants (holistic, i.e., taking all the elements of the face as a whole, vs. analytical, i.e., using one element at a time) and found that whereas "holistic processors" fixated the eyes and nose the most, "analytical processors" mainly focused on the particular facial feature they used for subsequent processing (Schwarzer et al., 2005). These results supported the view that holistic and featural processing modes elicit different scanning behaviors and that the functional use of facial information revealed by eye movements varies across individuals. However a study using the face composite task, a classic hallmark of holistic processing (Young et al., 1987), while recording eye movements, reached a different conclusion. In this task increased mistakes and response times are seen when participants are asked to match two identical top parts of a face aligned with distinct bottom parts, and these effects are thought to be due to holistic processing. That is, the bottom part of the aligned face influences the perception of the fixated top part, yielding the illusion of a perceived new identity even when both top parts are identical. This effect disappears when the parts are misaligned. de Heering et al. (2008) found that, despite a clear behavioral composite effect, the distribution of the fixations on the top part was similar in the aligned and misaligned conditions, suggesting that holistic face perception and its disruption can be dissociated from gaze behavior.

Similar inconsistencies have been reported regarding the scanning of inverted faces. Putting faces upside down impacts their perception and recognition, and this so-called inversion effect is much larger for faces than for other object categories (Yin, 1969). Forty years of research has shown that inversion impairs mainly holistic processing (Rossion, 2009) and impacts faces more than objects because faces are processed holistically while objects are processed mainly on the basis of their features (see Tanaka & Gordon, 2011, for a review). In a recognition study, Henderson et al. (2001) reported that face inversion had little influence on the distribution of fixations on the face. Similarly, eyes were looked at more than mouths regardless of face orientation in free-viewing conditions (Laidlaw et al., 2012). In contrast, in identity judgments using a morphing procedure, participants viewed the eyes and nose more than the mouth in upright faces but viewed the mouth more than the eyes in inverted faces, regardless of face familiarity (Barton et al., 2006). Again, these different results likely stem from differences in task and design between studies.

The role of attention to the eyes is further complicated by recent studies suggesting eye movements are culture dependent. In one study comparing learning, recognition, and categorization of faces by race, Western Caucasians showed a triangular pattern of fixations across the eyes and the mouth while East Asian observers focused on the central region of the face, and this eye-movement pattern was similar across the three tasks (Blais et al., 2008). This cultural difference in fixation pattern was confirmed in another study for face learning and recognition of human and animal faces (Kelly et al., 2010). Similar patterns of eye fixations were also seen in children 7 to 12 years of age compared to adults of the same cultural groups (Chinese and Caucasians from the United Kingdom), and these patterns strengthened with age, supporting the view that culture shapes eye movements from early childhood (Kelly et al., 2011). Attention to the eyes thus seems culture dependent. In Caucasians, a recent study reported a shift from more fixations to the mouth to more fixations to the eyes with age in children 4 to 10 years old during free viewing (Meaux et al., 2014), also suggesting that attention to the eyes reflects complex underlying processes that change with age.

An important question is whether eyes grab attention reflexively. During the perception of emotional faces presented for only 150 ms, the first saccade was made predominantly to the eye region (Gamer et al., 2013). The authors argued that these first saccades were reflexive eye movements toward the eyes because they were triggered by the stimulus and were executed irrespective of whether the face was still present on the screen or not. In a study in which participants performed a gaze-direction task (is the face looking at you or away) and a head-orientation task (is the face in front view or in 3/4 view) on the same face stimuli, the first saccade was directed to the eye region in 90% of the trials in the gaze judgment task but in 50% of the trials in the head-orientation task (Itier, Villate, & Ryan, 2007), showing that initial attention to the eyes can be goal oriented and modulated by task demands. However, the fact that even when they were completely task irrelevant (as in the head judgment) eyes were fixated first in 50% of the trials suggested that task demands could not completely suppress spontaneous attention to the eyes. Similarly, when participants' task was to not look at the eyes or mouth, viewing of the eyes was diminished but still above chance level whereas viewing of the mouth was successfully suppressed (Laidlaw et al., 2012). Complete avoidance of the eyes was, however, possible when faces were inverted, suggesting that attention to eyes is both reflexive and volitional and that holistic processing plays a central role in this automatic attraction to eyes.

Other studies have in contrast reported that the first eye movements are not directed at the eyes per se. In one study in which a saccade was needed to refocus gaze on the centrally presented face before exploring it, the first fixation landed around the eyes but mostly on the nose when faces were in front view; in mid-profile views it landed on the innermost eye, but in profile views it landed in between the eye and the ear (Bindemann et al., 2009). These findings suggested a center-of-gravity effect in saccade programming where initial eye saccades are drawn to the central region of a target configuration. Thus, fixations would land around the eyes only when they coincide with the center of gravity and would not reflect true attraction to the eyes. After this center-of-gravity effect, however, the next eye movements recorded were mostly to the eyes (Bindemann et al., 2009).

In a study in which participants first moved their eyes toward the face before recognizing it, only a certain number of fixations were allowed before a mask was presented (Hsiao & Cottrell, 2008). Recognition performance was above chance level with only one fixation (the fixation that was necessary to first view the face) although maximal performance was reached with a second fixation. Subsequent fixations did not improve performance, suggesting that two fixations suffice for accurate face recognition and this advantage was not due to the available viewing time. Importantly, both first fixations were always made around the center of the nose, below the eyes (Hsiao & Cottrell 2008). Across three face tasks (gender discrimination, emotion recognition, and identification), the first fixation was also found to be just below the eyes, and this position was functionally important as forcing participants to maintain gaze away from their preferred point of fixation substantially degraded performance (Peterson & Eckstein, 2012). Importantly, optimal position of this first fixation was not the center of the face or monitor but was best predicted by a "foveated" ideal observer that integrates information optimally across the face while being constrained by the decreased quality of the information from fovea to periphery. These results are consistent with the view that gaze is first positioned strategically to allow as much information gathering as possible through holistic processing. Furthermore, for face recognition, these strategic gaze positions vary across individuals and are stable over time, with some people being consistently closer to the eyes while other are consistently closer to the nose (Peterson & Eckstein, 2013).

To summarize, the functional role of attention to the eyes in various face-perception tasks remains unclear. Attention to the eyes is influenced by participants' culture and age, study design, and task constraints. Furthermore, this attention to eyes is not always found as identity recognition can be achieved with only one fixation situated *below* the eyes (Hsiao & Cottrell, 2008; Peterson & Eckstein, 2013), in a little over 200 ms and presumably using holistic processing, a finding in line with studies of the neural activity linked to face perception as reviewed below.

The Role of Attention to the Eyes in Face Perception as Revealed by Event-Related Potentials

The event-related potential (ERP) technique is well suited to the study of the neural activity underlying cognitive events, and its remarkable temporal resolution makes it ideal to study the timing of these events. The face-perception literature using the ERP technique has focused on the N170 component, an early negative ERP component recorded on the scalp between 130 and 200 ms after face onset (Bentin et al., 1996; George et al., 1996; Jeffreys, 1996). The N170 amplitude is larger for faces than for a variety of other visual objects (Bentin et al., 1996; Eimer, 2000; Itier & Taylor, 2004; Itier et al., 2006) and has thus been termed "face sensitive." A clear N170 response is seen to human faces presented as schematic drawings, sketches, paintings, or photographs (Sagiv & Bentin, 2001) but also to animal faces (Itier et al., 2006, 2011).

As mentioned earlier, inversion impairs holistic processing and impacts face perception and recognition much more than object perception. Studies have shown that the N170 is greatly impacted by face inversion which delays the component and increases its amplitude (Bentin et al., 1996; Itier & Taylor, 2002; Rossion et al., 1999), while the N170 to objects is unaffected (Rossion et al., 2000) or just delayed by inversion (Itier et al., 2006; Kloth et al., 2013). This N170 face-inversion effect suggests that the N170 component reflects face holistic processing. Further support for this claim comes from the composite face illusion described earlier. The N170 is also increased in amplitude for misaligned faces compared to whole (aligned) faces (Letourneau & Mitchell 2008), reflecting the disruption of holistic processing. Finally, Mooney figures (made solely of black and white blobs) and Arcimboldo paintings (faces made up of fruits and vegetables by the sixteenth-century artist Arcimboldo) elicit a larger N170 when the stimulus is perceived as a face than when it is perceived as a nonsense figure (George et al., 2005) or as fruits/vegetables (Caharel et al., 2013). These results support the holistic view as one needs to extract the global face shape information from the collection of black and white blobs or fruits and vegetables in order to perceive a face. They also demonstrate that the N170 is linked to the actual perception of a face. The N170 is thus viewed as the earliest reliable marker of face perception which is holistic (Rossion & Caharel 2011).

However, the N170 is as large and sometimes larger to isolated eye regions than to whole faces (Bentin et al., 1996; Itier, Alain, Sedore, & McIntosh, 2007; Itier et al., 2006, 2011; Taylor, Edmonds, McCarthy, & Allison, 2001; Taylor, Itier, Allison, & Edmonds, 2001) from 4 years of age (Taylor, Edmonds, et al., 2001) whereas isolated mouths or noses elicit a smaller and much delayed response (Bentin et al., 1996; Nemrodov & Itier, 2011; Taylor, Itier, et al., 2001). This eye sensitivity is also more pronounced for human than animal eyes (Itier et al., 2011), and the inversion of eye regions does not yield the classic N170 inversion effect (Itier, Alain et al., 2007; Itier et al., 2006, 2011; Kloth et al., 2013), confirming that whole faces are processed holistically while eye regions are processed as a feature. Bentin et al. (1996) initially suggested that the N170 reflected the activity of an eye detector, but the fact that taking the eyes out of the face did not impact the N170 amplitude (Eimer, 1998; Itier, Alain, et al., 2007; Itier et al., 2011; Kloth et al., 2013) argued against it. The eye detector idea was proposed again

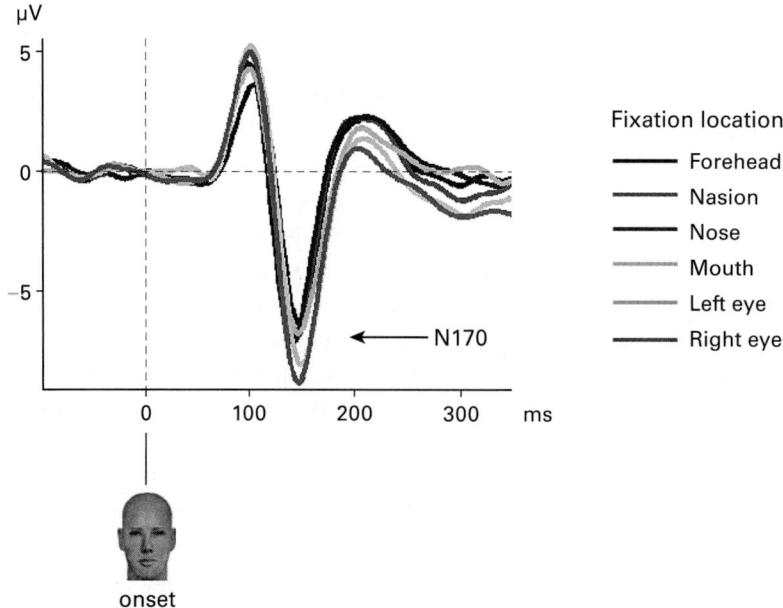

Figure 16.1 (plate 4)
Effect of overt attention to facial features on the N170 event-related potential component elicited by face presentation (displayed here at the CB2 electrode; grand-averaged across 21 participants). Note the larger N170 amplitude when fixation is on the eyes compared to the nose, mouth, nasion, or forehead. Adapted from Nemrodov et al. (2014) with permission.

recently in light of new findings showing that the N170 is in fact sensitive to overt attention to the eyes of faces.

Nemrodov et al. (2014) used an eye tracker to enforce fixation on different face locations while concurrently recording ERPs. They found that, regardless of face orientation, the N170 was larger when fixation was on an eye (see figure 16.1, plate 4) than on other fixation locations (nose, mouth, nasion—between the two eyes—or forehead). This larger response to eyes disappeared in eyeless faces, demonstrating it was due to the presence of an eye in fovea. Finally, the classic N170 inversion effect was seen at all fixation locations in intact faces but was maximally reduced for fixation on the (empty) eye locations in eyeless faces, while fixation on the mouth in eyeless faces yielded a normal face-inversion effect. These results show that overt attention to facial features modulates the neural activity linked to holistic face perception. The results also advocate in favor of an eye detector whose activity is recorded during the N170 window. If there is an eye detector in play, how is the face processed holistically?

The Lateral Inhibition, Face Template and Eye Detector Based (LIFTED) Model: A New Account of Holistic and Featural Processing   The complex pattern of results obtained by Nemrodov et al. (2014) led to a new theoretical account of face perception. The model proposes that eyes are first detected by an eye detector mechanism and act as an anchor point from which the position of facial features is coded according to an upright human face template (a front view face with a typical face configuration defined by two eyes above a nose itself above a mouth). The information in fovea is overrepresented in the primary visual cortex compared to the perifoveal and peripheral information (Daniel & Whitteridge, 1961). This cortical magnification should result in a stronger input from fixated features (in fovea) than nonfixated features (outside of fovea) into the higher face-sensitive visual areas whose activity is recorded on the scalp as the N170 component. However, as reviewed earlier, the processing of the upright face is holistic. To achieve this holistic processing, an inhibition mechanism was proposed where neurons coding for information in fovea are inhibited by neurons coding perifoveal information through lateral inhibitions (horizontal connections between neurons). This mechanism cancels out the overrepresentation of foveated information, ensuring that information from the rest of the face participates in the recorded N170 (see figure 16.2). These lateral inhibitions would act as the "glue" between features, ensuring holistic processing of the upright face. The strength of these lateral inhibitions would depend on the retinotopic distances between the foveated feature and the perifoveal features, making holistic processing dynamic and dependent on the size of the face and thus on the distance at which it is seen. When fixated, eyes (and thus the activity of the eye detector) would be inhibited in an upright face, just like the other fixated features. However, depending on the face size, the angular distances between the eyes and the other features could be too great for the remaining facial information to completely inhibit the eye detector activity. We believe this is the case in Nemrodov et al. (2014): The face was large enough to prevent a full inhibition of the eyes, thereby revealing the eye detector activity.

The lateral inhibition, face template, and eye detector (LIFTED) model suggests that both foveated and peripheral information contribute to the scalp recorded N170 component, but differently depending on face orientation. While upright faces are processed holistically, inverted faces are processed mostly featurally (Tanaka & Gordon, 2011). This featural processing would simply arise from a lack of lateral inhibitions. The fixated feature would thus be overrepresented (cortical magnification). At the neuronal level this would mean a larger contribution of the fixated feature to the overall neural activation (compared to when the face is upright), in addition to the contribution of peripheral visual information, yielding an increase in N170 amplitude (the face-inversion effect). At the behavioral level, this overrepresentation of the fixated feature would explain the narrowing of the perceptual field in inverted faces for which features have to be processed sequentially and independently because of the loss of holistic

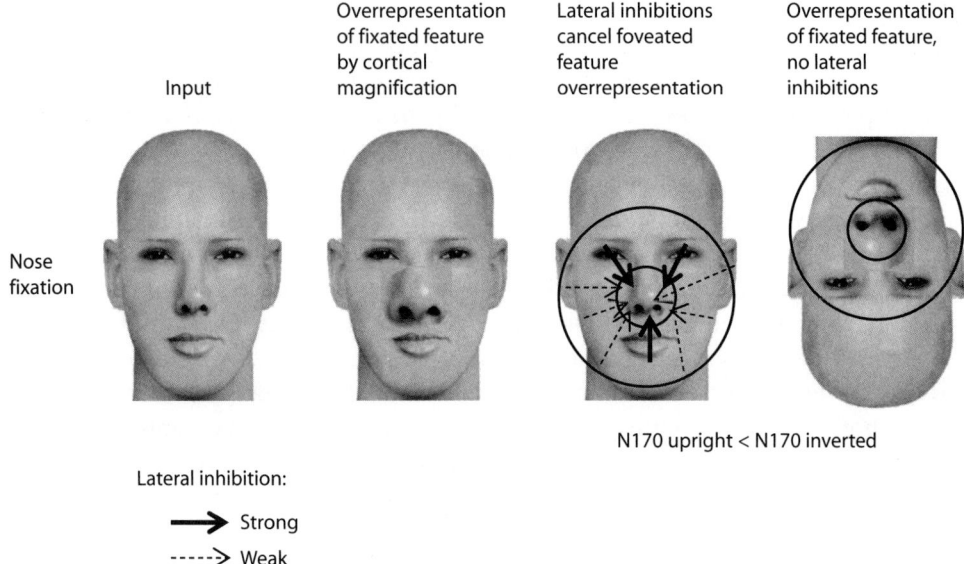

Figure 16.2

Schematic representation of how lateral inhibitions work according to the lateral inhibition, face template, and eye detector model with fixation on the nose as an example. The smaller circles represent the fovea; the larger circles represent the parafovea, that is, the approximate distance from fixation under which peripheral features contribute to the inhibition and to the overall neuronal response (N170). The overrepresentation of the fixated feature due to cortical magnification is represented as a disproportionately large feature. In upright faces, this overrepresentation is canceled by lateral inhibitions (dashed arrows) from neurons coding for parafoveal features onto neurons coding for the foveated feature, enabling holistic processing of the face. The strength of the lateral inhibitions depends on the distance between the fixated and peripheral features (stronger when distances are shorter, represented by thicker arrows). In inverted faces, lateral inhibitions are no longer seen, and the N170 represents the added contribution of the neurons coding for the foveated feature and the neurons coding for the parafoveal features, hence yielding a larger N170 for inverted than upright faces. Adapted from Nemrodov et al. (2014) with permission.

processing (Rossion, 2009), requiring viewers to change their face scanning pattern toward more fixations on specific features (e.g., Van Belle et al., 2010) in order to collect all relevant facial information to do the task at hand (e.g., recognize the identity). This simple inhibition mechanism explains holistic processing of an upright face, and the cancellation of these inhibitions explains the face-inversion effect. According to this theory, objects are processed just like inverted faces (i.e., without lateral inhibition).

This theory describes a novel neuronal account of face perception as indexed by the N170 ERP component where overt and covert attention to eyes plays a central role in holistic processing. Let's note that the sensitivity of the N170 to eyes within the face was also suggested by studies combining the Bubbles technique and ERP recordings in gender and emotion categorization tasks (e.g., Schyns et al., 2003; Schyns et al., 2007). Although the N170 seems sensitive to eyes within the face, the hypothesis of an eye detector anchoring the face percept at the eyes implies that eyes are detected before holistic processing occurs, that is, before the N170 component. Schyns et al. (2007) have proposed an integration of facial features across time, starting with the eyes about 50 ms before the N170, that is, during the occurrence of the P1, the ERP component preceding the N170 which typically occurs between 100 and 130 ms after face onset. Together, these findings suggest that eyes attract attention very early on during the course of visual processing, and that the integration of the facial information occurs in a second step. The N170 would in fact reflect the end of that integration process. That is, in less than 200 ms after face onset, the visual system has detected the eyes and integrated all the facial information (foveal and perifoveal) into a holistic face percept.

Integration—Merging Eye-Movement and ERP Findings: What Is the Role of Attention to the Eyes in Face Perception?

The recent ERP findings reviewed above suggest eyes play a central role in holistic face processing which occurs in less than 200 ms and can be monitored by the N170 component. Attention to the eyes seems to occur even before that, possibly between 100 and 150 ms after face onset, and both attention to eyes and holistic processing are largely independent from eye movements. Some eye-movement studies suggest that fixating around the nose and eye region rather than on the eyes per se is the most efficient strategy to maximize information gathering (e.g., Bindemann et al., 2009; de Heering et al., 2008; Hsiao & Cottrell, 2008; Orban de Xivry et al., 2008; Peterson & Eckstein, 2012, 2013), presumably through holistic processing.

Note that holistic processing implies the use of covert attention to include perifoveal information in the face percept. The size at which the face is viewed is thus critical. A face seen at a distance is perceived as a small whole, and eye movements are not needed if the entire face falls within fovea. Visual acuity falls very quickly outside of fovea (which is only 1–2° of visual angle; e.g., Kniestedt & Stamper, 2003), and the details of the face can only be extracted by moving the eyes and putting each element of the face in fovea. For faces that are larger than the fovea (e.g., a real life size face seen at a normal social interaction distance) eye movements are thus necessary to put the various features in fovea for fine-grain analysis. In most eye-movement studies the face is large enough to promote eye movements, and on some occasions the face is too large for a complete holistic processing through covert attention and eye movements would then be necessary to encode the identity of the person. This size factor is simply ignored in most studies, and face size varies quite a lot between studies. If the LIFTED model is correct, this size effect would also matter a lot in the earliest stages of vision (see also McKone, 2009, for variations of holistic processing—measured behaviorally— with viewing distance).

Holistic processing is especially important when faces are presented briefly, as is often done in ERP studies. In fact most ERP studies present faces for 500 ms or less to precisely avoid the contamination of the neural recordings by eye movements. In eye-movement studies, however, faces are presented for a few seconds. Assuming the face size is small enough to promote identity learning in two fixations through holistic processing (e.g., Hsiao & Cottrell, 2008), what are participants doing during the several remaining seconds in which the face is presented on the screen? And why are they going to the eyes more often than to other features? Eye-movement studies suggest that people look at the eyes to decipher the social content of the scene (Birmingham et al., 2008a, 2008b; Smilek et al., 2006). Indeed, eyes allow inferences about mental states, emotions, and intentions (Itier & Batty, 2009). This natural tendency to infer mental states might be the reason why, even during simple exploration of face photographs and even though the identity of the face has been extracted in less than half a second, participants often look more at the eyes than the rest of the face. This makes sense given that during natural face interaction the identity is extracted immediately but once we know who we are talking to we continue to explore people's faces. In real life faces are constantly moving as people talk, express emotions, and look around, which incites us to go back to the eyes to ensure eye contact and continually decipher possible states of mind and others' attention (see Peterson & Eckstein, 2013, for a similar view).

Future Directions

This comparison between typical lab studies and real-life situations brings us to the issue of the ecological validity of the work described in this chapter. Most ERP research uses face photographs that have been manipulated to eliminate low-level features known to influence the early stages of visual processing and their associated early neural responses, such as contrast, luminance, or spatial frequencies. Often, face pictures are presented in gray scale rather than in color and consist of internal features contained in oval shapes with no hair, ears, neck, or other cues that might influence recognition or attract attention (e.g., mustache, mole, etc.). While these controlled conditions are necessary to unravel the specific mechanisms associated with face processing, they create stimuli that are unnatural and very different from the faces of people we encounter in everyday life. They also lose completely the potential for social interaction, which has been shown to alter significantly social attention and eye movements (e.g., Laidlaw et al., 2011) as well as neural activity. For instance, the N170 is similar between live faces and face photographs (Pönkänen et al., 2008) but is altered by gaze manipulations mostly in live conditions (Pönkänen et al., 2011). Similarly, while it is clearly acknowledged that people suffering from autism spectrum condition (ASC) present with impairments in face recognition, facial emotion discrimination, and the use of eye gaze in joint attention tasks (Itier & Batty, 2009), many studies have failed to report clear deficits in lab settings (e.g., Nation & Penny, 2008), and one possible reason for that is the lack of social meaning that face photographs evoke compared to real-life faces. This is all the more probable since in natural viewing conditions people with ASC clearly avoid the eyes (e.g., Klin et al., 2002) and impairment in joint attention is part of the ASC diagnosis. Bridging the gap between lab settings and real-life situations is the current challenge we all face but that will need to be overcome to really understand face perception and how its breaks down in specific disorders.

Conclusions

The eyes capture attention, but the role of this attention to eyes in face perception remains unclear. Recent findings in the ERP literature support the idea of an eye detector mechanism and suggest that eyes play a central role even in the earliest stages of vision. Eyes might act as an anchor point from which the percept of a face is derived. The LIFTED model provides a new neuronal account of early face processing where the perception of the whole face is achieved by inhibiting visual information in fovea and especially the activity of the eye detector. In this framework, the distance at which the face is viewed is critical. Once the face is perceived (through holistic processing) and its identity recognized, attention still seems to go back to the eyes. The functional role of this attention capture by eyes in later stages of visual processing is still unclear although very likely linked to our natural tendency to infer mental states from the eye region and use eye gaze to infer others' attention. While face stimuli used in laboratory studies still provide a useful approximation of the cognitive events involved in face perception, the social dimension of real faces mainly conveyed by eyes might influence even the earliest stages of vision. The face processing field will have to move toward more ecological settings in order to completely unravel the neurocognitive events underlying face perception and how it breaks down in specific disorders such as autism spectrum disorder.

Acknowledgments

Roxane J. Itier was supported by the Canada Research Chair (CRC) Program, the Natural Sciences and Engineering Research Council of Canada (NSERC), and an Early Researcher Award from the Ontario Government.

Box 16.1

Key Points

- Attention to eyes as measured by eye movements varies with many factors including task demands, size and type of face stimuli, and age and culture of the participants.
- Attention to eyes modulates the early neural activity linked to holistic face processing (N170 ERP component).

Box 16.2

Outstanding Issues

- If attention to eyes is driven by an eye detector, how does this detector operate?
- What drives individual optimal gaze position (between eyes and nose) in face perception?
- What does the attraction to the eye region seen in natural viewing conditions reflect?

References

Althoff, R. R., & Cohen, N. J. (1999). Eye-movement-based memory effect: A reprocessing effect in face perception. *Journal of Experimental Psychology: Learning, Memory, and Cognition*, *25*, 997–1010.

Arizpe, J., Kravitz, D. J., Yovel, G., & Baker, C. I. (2012). Start position strongly influences fixation patterns during face processing: Difficulties with eye movements as a measure of information use. *PLoS ONE*, *7*(2), e31106.

Barton, J. J., Radcliffe, N., Cherkasova, M. V., Edelman, J., & Intriligator, J. M. (2006). Information processing during face recognition: The effects of familiarity, inversion, and morphing on scanning fixations. *Perception*, *35*, 1089–1105.

Bentin, S., Allison, T., Puce, A., Perez, E., & McCarthy, G. (1996). Electrophysiological studies of face perception in humans. *Journal of Cognitive Neuroscience*, *8*, 551–565.

Bindemann, M., Scheepers, C., & Burton, A. M. (2009). Viewpoint and center of gravity affect eye movements to human faces. *Journal of Vision (Charlottesville, Va.)*, *9*(2), 1–16.

Birmingham, E., Bischof, W. F., & Kingstone, A. (2008a). Social attention and real-world scenes: The roles of action, competition and social content. *Quarterly Journal of Experimental Psychology*, *61*, 986–998.

Birmingham, E., Bischof, W. F., & Kingstone, A. (2008b). Gaze selection in complex social scenes. *Visual Cognition*, *16*, 341–355.

Birmingham, E., Bischof, W. F., & Kingstone, A. (2009). Saliency does not account for fixations to eyes within social scenes. *Vision Research, 49*, 2992–3000.

Blais, C., Jack, R. E., Scheepers, C., Fiset, D., & Caldara, R. (2008). Culture shapes how we look at faces. *PLoS ONE, 3*(8), e3022.

Bukach, C. M., Le Grand, R., Kaiser, M. D., Bub, D. N., & Tanaka, J. W. (2008). Preservation of mouth region processing in two cases of prosopagnosia. *Journal of Neuropsychology, 2*, 227–244.

Caharel, S., Leleu, A., Bernard, C., Viggiano, M. P., Lalonde, R., & Rebai, M. (2013). Early holistic face-like processing of Arcimboldo paintings in the right occipito-temporal cortex: Evidence from the N170 ERP component. *International Journal of Psychophysiology, 90*, 157–164.

Caldara, R., Schyns, P., Mayer, E., Smith, M. L., Gosselin, F., & Rossion, B. (2005). Does prosopagnosia take the eyes out of face representations? Evidence for a defect in representing diagnostic facial information following brain damage. *Journal of Cognitive Neuroscience, 17*, 1652–1666.

Cooper, E. E., & Wojan, T. J. (2000). Differences in the coding of spatial relations in face identification and basic-level object recognition. *Journal of Experimental Psychology: Learning, Memory, and Cognition, 26*, 470–488.

Daniel, P. M., & Whitteridge, D. (1961). The representation of the visual field on the cerebral cortex in monkeys. *Journal of Physiology, 159*, 203–221.

de Heering, A., Rossion, B., Turati, C., & Simion, F. (2008). Holistic face processing can be independent of gaze behaviour: Evidence from the composite face illusion. *Journal of Neuropsychology, 2*(Pt 1), 183–195.

Eimer, M. (1998). Does the face-specific N170 component reflect the activity of a specialized eye processor? *Neuroreport, 9*, 2945–2948.

Eimer, M. (2000). The face-specific N170 component reflects late stages in the structural encoding of faces. *Neuroreport, 11*, 2319–2324.

Emery, N. J. (2000). The eyes have it: The neuroethology, function and evolution of social gaze. *Neuroscience and Biobehavioral Reviews, 24*, 581–604.

Gamer, M., & Hecht, H. (2007). Are you looking at me? Measuring the cone of gaze. *Journal of Experimental Psychology: Human Perception and Performance, 33*, 705–715.

Gamer, M., Schmitz, A. K., Tittgemeyer, M., & Schilbach, L. (2013). The human amygdala drives reflexive orienting towards facial features. *Current Biology, 23*, R917–R918.

George, N., Evans, J., Fiori, N., Davidoff, J., & Renault, B. (1996). Brain events related to normal and moderately scrambled faces. *Brain Research. Cognitive Brain Research, 4*, 65–76.

George, N., Jemel, B., Fiori, N., Chaby, L., & Renault, B. (2005). Electrophysiological correlates of facial decision: Insights from upright and upside-down Mooney-face perception. *Brain Research. Cognitive Brain Research, 24*, 663–673.

Haig, N. D. (1985). How faces differ—A new comparative technique. *Perception, 14*, 601–615.

Heisz, J. J., & Shore, D. I. (2008). More efficient scanning for familiar faces. *Journal of Vision (Charlottesville, Va.)*, *8*(1), 9–10.

Henderson, J. M., Falk, R., Minur, S., Dyer, F. C., & Mahadevan, S. (2001). Gaze control for face learning and recognition by humans and machines. In T. F. Shipley & P. J. Kellman (Eds.), *From fragments to objects—Segmentation and grouping in vision* (pp. 463–481). Amsterdam: Elsevier Science.

Henderson, J. M., Williams, C. C., Castelhano, M. S., & Falk, R. J. (2003). Eye movements and picture processing during recognition. *Perception & Psychophysics*, *65*, 725–734.

Henderson, J. M., Williams, C. C., & Falk, R. J. (2005). Eye movements are functional during face learning. *Memory & Cognition*, *33*, 98–106.

Hsiao, J. H., & Cottrell, G. (2008). Two fixations suffice in face recognition. *Psychological Science*, *19*, 998–1006.

Itier, R. J., Alain, C., Sedore, K., & McIntosh, A. R. (2007). Early face processing specificity: It's in the eyes! *Journal of Cognitive Neuroscience*, *19*, 1815–1826.

Itier, R. J., & Batty, M. (2009). Neural bases of eye and gaze processing: The core of social cognition. *Neuroscience and Biobehavioral Reviews*, *33*, 843–863.

Itier, R. J., Latinus, M., & Taylor, M. J. (2006). Face, eye and object early processing: What is the face specificity? *NeuroImage*, *29*, 667–676.

Itier, R. J., & Taylor, M. J. (2002). Inversion and contrast polarity reversal affect both encoding and recognition processes of unfamiliar faces: A repetition study using ERPs. *NeuroImage*, *15*, 353–372.

Itier, R. J., & Taylor, M. J. (2004). N170 or N1? Spatiotemporal differences between object and face processing using ERPs. *Cerebral Cortex*, *14*, 132–142.

Itier, R. J., Van Roon, P., & Alain, C. (2011). Species sensitivity of early face and eye processing. *NeuroImage*, *54*, 705–713.

Itier, R. J., Villate, C., & Ryan, J. D. (2007). Eyes always attract attention but gaze orienting is task-dependent: Evidence from eye movement monitoring. *Neuropsychologia*, *45*, 1019–1028.

Janik, S. W., Wellens, A. R., Goldberg, M. L., & Dell'osso, L. F. (1978). Eyes as the center of focus in the visual examination of human faces. *Perceptual and Motor Skills*, *47*, 857–858.

Jeffreys, D. A. (1996). Evoked potential studies of face and object processing. *Visual Cognition*, *3*, 1–38.

Kelly, D. J., Liu, S., Rodger, H., Miellet, S., Ge, L., & Caldara, R. (2011). Developing cultural differences in face processing. *Developmental Science*, *14*, 1176–1184.

Kelly, D. J., Miellet, S., & Caldara, R. (2010). Culture shapes eye movements for visually homogeneous objects. *Frontiers in Psychology*, *1*, 6.

Klin, A., Jones, W., Schultz, R., Volkmar, F., & Cohen, D. (2002). Visual fixation patterns during viewing of naturalistic social situations as predictors of social competence in individuals with autism. *Archives of General Psychiatry, 59,* 809–816.

Kloth, N., Itier, R. J., & Schweinberger, S. R. (2013). Combined effects of inversion and feature removal on N170 responses elicited by faces and car fronts. *Brain and Cognition, 81,* 321–328.

Kniestedt, C., & Stamper, R. L. (2003). Visual acuity and its measurement. *Ophthalmology Clinics of North America, 16,* 155–170.

Kobayashi, H., & Kohshima, S. (1997). Unique morphology of the human eye. *Nature, 387,* 767–768.

Laidlaw, K. E. W., Foulsham, T., Kuhn, G., & Kingstone, A. (2011). Potential interactions are important to social attention. *Proceedings of the National Academy of Sciences of the United States of America, 108,* 5548–5553.

Laidlaw, K. E. W., Risko, E. F., & Kingstone, A. (2012). A new look at social attention: Orienting to the eyes is not (entirely) under volitional control. *Journal of Experimental Psychology: Human Perception and Performance, 38,* 1132–1143.

Letourneau, S. M., & Mitchell, T. V. (2008). Behavioral and ERP measures of holistic face processing in a composite task. *Brain and Cognition, 67,* 234–245.

Lewis, M. B., & Edmonds, A. J. (2003). Face detection: Mapping human performance. *Perception, 32,* 903–920.

McKelvie, S. J. (1976). The role of eyes and mouth in the memory of a face. *American Journal of Psychology, 89,* 311–323.

McKone, E. (2009). Holistic processing for faces operates over a wide range of sizes but is strongest at identification rather than conversational distances. *Vision Research, 49,* 268–283.

Meaux, E., Hernandez, N., Carteau-Martin, I., Martineau, J., Barthelemy, C., Bonnet-Brilhault, F., et al. (2014). Event-related potential and eye tracking evidence of the developmental dynamics of face processing. *European Journal of Neuroscience, 39,* 1349–1362.

Nation, K., & Penny, S. (2008). Sensitivity to eye gaze in autism: Is it normal? Is it automatic? Is it social? *Development and Psychopathology, 20,* 79–97.

Nemrodov, D., Anderson, T., Preston, F., & Itier, R. J. (2014). Early sensitivity for eyes within faces: A new neuronal account of holistic and featural processing. *NeuroImage, 97,* 81–94.

Nemrodov, D., & Itier, R. J. (2011). The role of eyes in early face processing: A rapid adaptation study of the inversion effect. *British Journal of Psychology, 102,* 783–798.

O'Donnell, C., & Bruce, V. (2001). Familiarisation with faces selectively enhances sensitivity to changes made to the eyes. *Perception, 30,* 755–764.

Orban de Xivry, J. J., Ramon, M., Lefevre, P., & Rossion, B. (2008). Reduced fixation on the upper area of personally familiar faces following acquired prosopagnosia. *Journal of Neuropsychology*, 2(Pt 1), 245–268.

Paras, C. L., & Webster, M. A. (2013). Stimulus requirements for face perception: An analysis based on "totem poles." *Frontiers in Psychology*, 4, 18.

Peterson, M. F., & Eckstein, M. P. (2012). Looking just below the eyes is optimal across face recognition tasks. *Proceedings of the National Academy of Sciences of the United States of America*, 109, E3314–E3323.

Peterson, M. F., & Eckstein, M. P. (2013). Individual differences in eye movements during face identification reflect observer-specific optimal points of fixation. *Psychological Science*, 24, 1216–1225.

Pönkänen, L. M., Alhoniemi, A., Leppänen, J. M., & Hietanen, J. K. (2011). Does it make a difference if I have an eye contact with you or with your picture? An ERP study. *Social Cognitive and Affective Neuroscience*, 6, 486–494.

Pönkänen, L. M., Hietanen, J. K., Peltola, M. J., Kauppinen, P. K., Haapalainen, A., & Leppänen, J. M. (2008). Facing a real person: An event-related potential study. *Neuroreport*, 19, 497–501.

Rayner, K. (2009). The 35th Sir Frederick Bartlett lecture: Eye movements and attention in reading, scene perception, and visual search. *Quarterly Journal of Experimental Psychology*, 62, 1457–1506.

Rossion, B. (2009). Distinguishing the cause and consequence of face inversion: The perceptual field hypothesis. *Acta Psychologica*, 132, 300–312.

Rossion, B., & Caharel, S. (2011). ERP evidence for the speed of face categorization in the human brain: Disentangling the contribution of low-level visual cues from face perception. *Vision Research*, 51, 1297–1311.

Rossion, B., Caldara, R., Seghier, M., Schuller, A. M., Lazeyras, F., & Mayer, E. (2003). A network of occipito-temporal face-sensitive areas besides the right middle fusiform gyrus is necessary for normal face processing. *Brain*, 126, 2381–2395.

Rossion, B., Delvenne, J. F., Debatisse, D., Goffaux, V., Bruyer, R., Crommelinck, M., et al. (1999). Spatio-temporal localization of the face inversion effect: An event-related potentials study. *Biological Psychology*, 50, 173–189.

Rossion, B., Gauthier, I., Tarr, M. J., Despland, P., Bruyer, R., Linotte, S., et al. (2000). The N170 occipito-temporal component is delayed and enhanced to inverted faces but not to inverted objects: An electrophysiological account of face-specific processes in the human brain. *Neuroreport*, 11, 69–74.

Sagiv, N., & Bentin, S. (2001). Structural encoding of human and schematic faces: Holistic and part-based processes. *Journal of Cognitive Neuroscience*, 13, 937–951.

Schwarzer, G., Huber, S., & Dummler, T. (2005). Gaze behavior in analytical and holistic face processing. *Memory & Cognition, 33*, 344–354.

Schyns, P. G., Bonnar, L., & Gosselin, F. (2002). Show me the features! Understanding recognition from the use of visual information. *Psychological Science, 13*, 402–409.

Schyns, P. G., Jentzsch, I., Johnson, M., Schweinberger, S. R., & Gosselin, F. (2003). A principled method for determining the functionality of brain responses. *Neuroreport, 14*, 1665–1669.

Schyns, P. G., Petro, L. S., & Smith, M. L. (2007). Dynamics of visual information integration in the brain for categorizing facial expressions. *Current Biology, 17*, 1580–1585.

Sekuler, A. B., Gaspar, C. M., Gold, J. M., & Bennett, P. J. (2004). Inversion leads to quantitative, not qualitative, changes in face processing. *Current Biology, 14*, 391–396.

Shepherd, J., Davies, G., & Ellis, H. (1981). Studies of cue saliency. In G. Davies, H. D. Ellis, & J. Shepherd (Eds.), *Perceiving and remembering faces* (pp. 105–131). New York: Academic Press.

Smilek, D., Birmingham, E., Cameron, D., Bischof, W., & Kingstone, A. (2006). Cognitive ethology and exploring attention in real-world scenes. *Brain Research, 1080*, 101–119.

Stacey, P. C., Walker, S., & Underwood, J. D. (2005). Face processing and familiarity: Evidence from eye-movement data. *British Journal of Psychology, 96*, 407–422.

Tanaka, J. W., & Gordon, I. (2011). Features, configuration, and holistic face processing. In A. J. Calder, G. Rhodes, M. J. Johnson, & J. V. Haxby (Eds.), *The Oxford handbook of face perception* (pp. 177–194). New York: Oxford University Press.

Taylor, M. J., Edmonds, G. E., McCarthy, G., & Allison, T. (2001). Eyes first! Eye processing develops before face processing in children. *Neuroreport, 12*, 1671–1676.

Taylor, M. J., Itier, R. J., Allison, T., & Edmonds, G. E. (2001). Direction of gaze effects on early face processing: Eyes-only versus full faces. *Brain Research. Cognitive Brain Research, 10*, 333–340.

Van Belle, G., De Graef, P., Verfaillie, K., Rossion, B., & Lefevre, P. (2010). Face inversion impairs holistic perception: Evidence from gaze-contingent stimulation. *Journal of Vision (Charlottesville, Va.), 10*(5), 10.

Walker-Smith, G. J., Gale, A. G., & Findlay, J. M. (1977). Eye movement strategies involved in face perception. *Perception, 6*, 313–326.

Yarbus, A. (1967). *Eye movements and vision*. New York: Plenum Press.

Yin, R. K. (1969). Looking at upside-down faces. *Journal of Experimental Psychology, 81*, 141–145.

Young, A. W., Hellawell, D., & Hay, D. C. (1987). Configurational information in face perception. *Perception, 16*, 747–759.

II
Attention in the Real World

17

Everyday Visual Attention

Benjamin W. Tatler and Michael F. Land

A fundamental function of visual attention is to manage the information that allows the organism to carry out actions necessary for survival effectively and appropriately. Decisions about when and where to shift attention in natural behavior are therefore likely to be intimately linked to the information demands of the current actions and behavioral goals. It is appropriate to consider attention not as an isolated system—as is often the case in laboratory-based paradigms—but as part of a broader network of vision, action, and planning during interactions with the environment. If our goal is to understand how we orient attention in our day-to-day activities, then it is important to study attentional orientation in the context of natural, everyday behaviors.

It is worth considering what we mean by attention here. Previously we have argued that attention is not just the system that directs the eyes but a broader network that involves planning, motor action, and vision (Land & Tatler, 2009). The close coordination between gaze control to locate objects, visual control to supply information and monitor actions, and actions themselves suggests that all of these functions are likely to be under common control (see figure 17.1). Attention might provide the common control that stitches the individual systems together.

If we are to produce a valid account of everyday attention, it is important to ensure that our measurements are made in an ecologically valid context: That is, we must record data in realistic environments and within behaviors that are as close to natural as possible (Kingstone, Smilek, & Eastwood, 2008). This is especially the case if we accept that attention is intricately enmeshed in the task structure and actions that we engage in (Land & Tatler, 2009). Any disruption of natural behavior has the potential to alter the task structure, resulting in attentional allocation unlike that which would support the natural, undisrupted behavior (Kingstone et al., 2008).

Of course, conducting experiments with everyday behaviors in real-world settings presents considerable challenges for traditional experimental approaches to measuring attention: If we wish to study truly natural settings, it is very unlikely that we can control extraneous variables or replicate stimulus conditions across individuals precisely. Furthermore, variations in the way behaviors are completed can make data difficult to

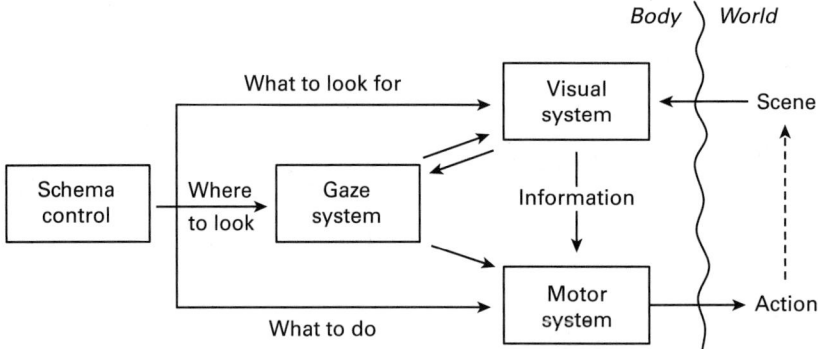

Figure 17.1
Interactions between gaze, vision, and motor control in behavioral coordination. Adapted from Land & Tatler (2009).

interpret (e.g., there are many ways to make a cup of tea). This large natural variation from both the environment and the individual means that it is tempting to suggest that simplifications must be made to either the environment, the behavior, or both in order to collect meaningful data. However, either of these simplifying steps introduces the risk of disrupting natural behavior and potentially the manner in which attention is deployed in unexpected ways. The preponderance of reductionist approaches in the study of attention has resulted partly from this lack of control over environmental variables and individual variations in behavior when studying natural tasks and partly from the limits imposed by the suitability and availability of instruments for measuring attention allocation in natural settings and behaviors. In the sections that follow, we will consider the development of mobile eye-tracking devices for measuring overt attentional allocation in natural contexts, the manner in which the attentional deployment underlying natural behavior is influenced by simplifications to the environment or task, and what we have learned about attentional allocation in everyday behaviors.

Historical Context

Measuring Attention in the Real World
If we are to measure attention in everyday situations, we must first consider what aspects of attention it is possible to measure. In laboratory settings, paradigms have been developed to measure covert allocation of attention, that is, changes in the locus of attention without an overt eye movement (e.g., Posner, 1980; Wolfe, 1998). However, such paradigms are almost impossible to implement in natural settings because they tend to require onset events in peripheral vision, measurable responses in the

absence of saccades toward peripheral locations, brief presentation times, and often the forced holding of fixation.

Covert and overt behavior are usually linked (Deubel & Schneider, 1996; Rizzolatti, Riggio, Dascola, & Umiltá, 1987; Shepherd, Findlay, & Hockey, 1986), and the predominant behavior that we display to a peripheral event is to move our eyes to it prior to making a response (Findlay & Gilchrist, 2003). Thus, although many of the approaches for studying covert attention in the lab are difficult to achieve during natural behavior and would require its disruption, studying overt attention is both feasible and minimally disruptive.

In natural situations, measuring overt allocation of attention via eye movements is a tractable option. Here the limitations are technological rather than paradigmatic. Such measurements require eye trackers that can record with adequate spatial and temporal accuracy and yet are portable. If they are to allow us to study natural behavior, mobile eye trackers should interfere with such behavior as little as possible. These devices should be minimally obtrusive for the wearer.

Historical Development of Eye Tracking: From Eyecups to Teacups

If we are to understand overt allocation of visual attention, we must monitor when and where the eyes are directed. Perhaps the earliest effective technique that was developed to monitor eye movements was to form an afterimage by viewing a bright light source and then asking the subject to observe the movements of the afterimage (see Wade & Tatler, 2005). This technique was instrumental in revealing the saccade-and-fixate nature of eye-movement behavior both during postrotational nystagmus (Wells, 1792) and during reading (Hering, 1879). Subsequent devices were developed that allowed objective records of eye movements to be made. The first eye-tracking devices recorded movements of the eyes mechanically, via physical attachments to the eyes. For example, Delabarre (1898) used a plaster-of-paris eyecup to attach a lever to the eye and record its deflections on a smoked rotating drum. These devices were problematic, however, requiring anesthesia of the tracked eye and imparting artifacts into the recorded traces due to the inertia from the eyecups and attachments. An important step toward a less disruptive means of measuring eye movements was provided by Dodge and Cline (1901) when they used photographic techniques to record light reflected from the surface of the eye. This first device paved the way for many of the eye trackers that are used in contemporary research. While eye trackers that used reflected light offered a less obtrusive and behaviorally disruptive way of measuring eye movements, most such systems require that the subject's head be stabilized for precise measurement of eye movements and use recording equipment that is nonportable and housed within laboratories.

The need for mobile eye trackers has been recognized for some time as an important way of bringing eye-tracking research into more ecologically valid settings. Hartridge and Thompson (1948) developed the first head-mounted eye tracker, and this was refined by Shackel (1960) and Mackworth and Thomas (1962) to allow greater freedom for the participant. Thomas (1968) used a head-mounted eye tracker that comprised a single camera on a helmet that recorded both the scene in front of the wearer's head and (via a series of prisms and lenses) the corneal reflections from a light source positioned just below the eye. While portable, this tracker was rather bulky because of the size and weight of the ciné camera mounted on the helmet and did not allow the wearer unrestricted natural movement in all settings. Furthermore, the periscope optics used to obtain the corneal reflection occluded the view of the tracked eye. As a result this tracker did not fully satisfy the requirement of being unobtrusive for the wearer during natural behavior.

Further developments in mobile eye tracking had to wait until the early 1990s, by which time video cameras were smaller and computers were available to assist in analysis of the recorded video data. Land (e.g., Land, 1993) built a mobile eye tracker that used a single head-mounted video camera, together with mirrors to record simultaneously an image of the scene and the eye (see figure 17.2). In this system, gaze direction

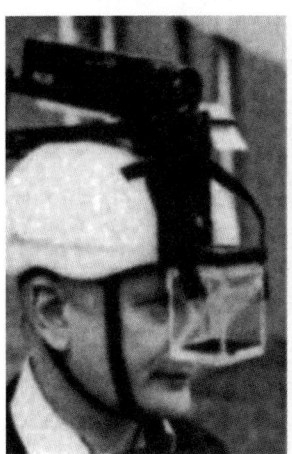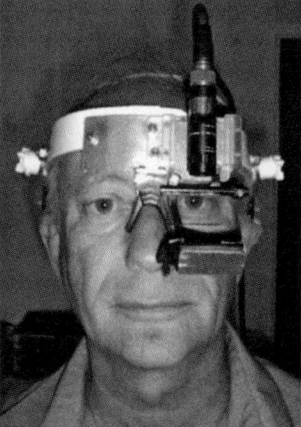

Figure 17.2
(Left) An early helmet-mounted mobile system developed by Land. (Middle) A sample frame from the processed video data collected using the eye tracker shown in the left panel. Here gaze direction has been estimated offline using a computer-generated model of the eye. The bright spot (highlighted by the black arrow in this example frame) indicates the fitted estimate of gaze direction. The images of the eye and scene are reversed and the eye image is inverted because of mirrors in the mobile tracker. (Right) A descendant of the eye tracker shown on the left that was used in a variety of studies of natural behavior in the late 1990s (including Land et al., 1999) and throughout the 2000s. Left and middle images from Land (1993).

was calculated by fitting a model of the eyeball around the iris in each frame of the recorded data. This eye tracker and its immediate descendants were used in studies of a wide range of real-world activities including driving (Land & Lee, 1994), table tennis (Land & Furneaux, 1997), tea making (Land, Mennie, & Rusted, 1999), musical sight-reading (Furneaux & Land, 1999), and cricket (Land & McLeod, 2000).

Since these studies, a number of commercial systems have been developed, and these are now increasingly used in research across a wide range of fields of study. These systems are becoming increasingly lightweight as cameras continue to become smaller, and gaze estimation is now automated in all commercially available systems. At present, however, there remain several limitations with existing mobile eye-tracking devices that must be considered when planning studies of attention allocation in natural situations. First, the spatial and temporal resolutions of mobile devices are well below those of desk-mounted eye trackers: Many mobile systems have sampling rates of 30 Hz, with some systems reaching 60 or 120 Hz, whereas desk-mounted systems reach 2000 Hz; typically, with mobile systems the experimenter can hope to achieve a spatial accuracy of about 1° of visual angle or perhaps as good as 0.5° with careful calibration and optimal conditions, yet accuracies better than 0.5° are expected and achieved readily in desk-mounted systems. These limitations in spatial and temporal resolution necessarily limit the questions that can be asked about attention allocation in natural situations: They are not suitable for measuring precise timings of saccades nor for examining the precise landing points of saccades. However, these limitations should not be overemphasized: Saccade landing points in natural behavior are not particularly precise, as we will see later, and the gap of 100 ms or more between saccades means that even at 30 Hz saccade landing points can be unambiguously attributed to individual components of complex actions. Second, most systems (with only a few exceptions at present) record monocularly rather than binocularly: This introduces imprecision in calibration accuracy over changes in depth due to the distance between the (scene) camera and the eyeball. Such errors can be reduced by minimizing the distance between the scene camera and eyeball but must be considered in studies where data are required over large changes in depth at short distances. Third, while software has been developed to automate the frame-by-frame fitting of models to estimate gaze direction, further analyses remain largely reliant on manual frame-by-frame coding of video data. Detecting saccades remains a difficult problem for mobile systems due to the co-occurrence of head and eye movements in natural behavior (e.g., Land, 2004), and assigning fixations to objects often requires manual coding. These analytical limitations impose severe costs on conducting real-world studies and limit the amount of data that can feasibly be collected for any study. It is likely that software solutions to issues of saccade detection and possibly even assignment of fixations to objects will be developed over the next few years, but for now analysis time is a considerable issue in any study of everyday attention.

Research Questions from Lab to Life

At the heart of understanding overt attention allocation in natural behavior is the need to understand what factors underlie decisions about when and where to move the eyes. These questions are not unique to studies of attention in everyday settings. Indeed, since eye-movement research was in its infancy, researchers have been interested in the selection priorities that underlie fixation selection. A key theme in this field that defined early research and continues to define contemporary research is the extent to which eye movements are targeted on the basis of low-level stimulus properties or higher level goals (see Tatler, Hayhoe, Land, & Ballard, 2011). Certainly, it has always been clear that stimulus properties can influence fixation behavior (McAllister, 1905), but even from the earliest studies of viewing behavior, it was recognized that the relationship between the form of the patterns viewed and the eye movements of the observer was not as close as early researchers had expected (Stratton, 1906). Moreover, the great variation in fixation patterns between individuals (McAllister, 1905) made it clear that factors other than stimulus properties were likely to be involved in allocating foveal vision. In light of evidence gathered from observers viewing the Müller-Lyer illusion (Judd, 1905), Poggendorff illusion (Cameron & Steele, 1905), and Zöllner illusion (Judd & Courten, 1905), Judd (1905) came to the conclusion that

the actual movements executed are in no small sense responses to the verbal stimuli which the subject receives in the form of general directions. The subject reacts to the demands imposed upon him by the general situation. ... The whole motive for movement is therefore not to be sought in the figures themselves. (pp. 216–217)

Classic work by Buswell (1935) and Yarbus (1967) later extended these ideas to complex scenes, demonstrating that instructions given to participants prior to viewing profoundly influenced inspection behavior. In contemporary research, studies of natural behavior are often used to illustrate the intimate link between task and fixation behavior and thus the importance of internal goals rather than stimulus properties for fixation selection.

State-of-the-Art Review

Why Study Everyday Attention in the Real World?

By establishing where the eyes land during an action sequence, we can work out the strategy the brain uses to acquire the information needed to control the action. This allows us to see the events that occur immediately before they become manifest as actions, and so takes us that much nearer to the internal machinery that generates behavior. In this context, we will use a wider definition of attention than that generally assumed in vision research, where its main role is to direct first an internal pointer and then gaze. However, a gaze shift is only part of an action. The motor system needs to be programmed to perform the action which the gaze change heralds, and the visual system itself needs to know what it has to look for (the orientation of a handle on a mug, the level of liquid in a jug, etc.). A system with outputs to many parts of the brain has to be involved in coordinating these somewhat disparate components of an action. We assume here that these are all aspects of a descending attention system, which has connections to much more than the eye-movement system. Here we make the case that the study of this wider attention system is only possible in the context of real actions that are not based on visual paradigms alone.

The benefits of observational breadth seen in real-world studies come at a cost: The rigor and repeatability of lab-based studies are simply not attainable. The technical and analytical limits of mobile eye trackers, and the limited range of paradigms that can be carried out in real-world settings, do constrain what can be measured. In particular, extraneous variables make it hard to replicate stimulus conditions, and such variations in environmental conditions can influence findings considerably. For example, if we wish to study driving in an urban setting, variations such as traffic density and flow will present problems for interpreting findings if they are not accounted for in analyses. Indeed, spatial and temporal allocation of overt attention changes considerably depending upon whether the driver is stopped in traffic or driving in (slowly) moving traffic (Land & Tatler, 2009). In terms of spatial allocation of attention, gaze was directed at either the car in front or oncoming cars 77% of the time when driving at 7 mph, with only 6% of the time spent looking at off-road targets. Conversely, when the driver was stopped in traffic, gaze was directed to off-road locations for 75% of the time with little time spent looking at the car in front (22%) or oncoming cars (6%). In terms of temporal allocation of attention, gaze events when the driver was stopped in traffic were more than twice as long on average (1.14 seconds) than when the driver was moving (0.54 seconds). Thus, data collected during an urban drive will be strongly influenced by the traffic conditions and flow during the drive, over which the experimenter has little control (see figure 17.3).

One option, if we wish to exact control on environmental variables and to collect large data sets over a variety of paradigms, is to simplify the environment by using virtual reality or laboratory paradigms. However, in doing so we run the risk of disrupting the natural behavior and collecting data that do not represent attentional allocation in the real-world setting that we are interested in. A compelling demonstration of the dangers of simplifying the setting or task is provided by Dicks, Button, and Davids (2010) in the context of goal tending in football (soccer for North American readers). Previous research on goal tending has—for obvious reasons—tended to simplify the stimulus (using videos of people kicking a ball rather than a live kicker) and has used simplified verbal or manual responses by the goal keeper to indicate how he should have moved to intercept the ball and save the goal (Abernethy, 1990; Savelsbergh, Williams, van der Kamp, & Ward, 2002). Dicks et al. (2010) showed that gaze behavior

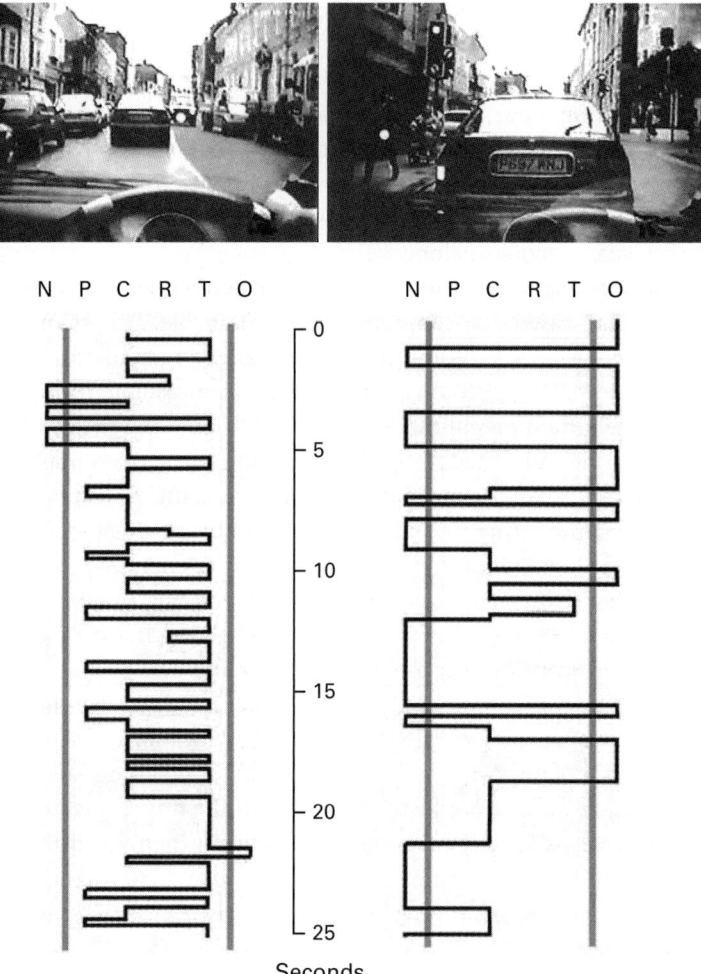

Figure 17.3

Allocation of gaze when driving slowly in traffic (left) and when stationary (right) to nearside, off-road locations (N); parked vehicles on near-side (P); car in front (C); on or above open roadway (R); oncoming traffic (T); and off-side, off-road locations (O). From Land & Tatler (2009).

when diving to save a ball following a kick taken by a real player was different to that when either (1) the response was simplified to a simple manual or verbal indication of how the goal keeper should move to intercept the ball or (2) a life-sized video of a player kicking the ball was shown rather than the live kick. Thus, simplifying either the stimulus or the manner of response leads to allocation of overt attention that is unlike that observed in the natural behavior.

Given these concerns, it is clear that any account of attention allocation in everyday situations must be derived from studies of intact natural behaviors. In the remainder of this chapter how attention is allocated in everyday situations is considered by drawing primarily upon findings from two everyday situations: carrying out domestic tasks, such as making a cup of tea or a sandwich, and locomotion in real-world environments.

Attention in Everyday Tasks

Domestic tasks such as making a cup of tea or a sandwich provide ideal situations for studying everyday attention because they are very familiar tasks to most of us and are representative of the kind of activities we naturally engage in. Moreover, these tasks are relatively complex in their structure; typically they are assumed to have hierarchical organization into goals and subgoals (Brewer & Dupree, 1983; Schwartz, Reed, Montgomery, Palmer, & Mayer, 1991; Schwartz et al., 1995) with a requirement to attentionally supervise the sequencing of these subgoals (Norman & Shallice, 1986) in order to successfully complete the behavior. The relatively complex organizational structure of these tasks, together with the fact that they require the use of a range of different objects and tasks, makes them a rich domain for understanding natural behavior (Land & Hayhoe, 2001). Almost all the activities we engage in require movement either during their performance or around it, and this movement must itself be supervised in order to ensure our safe passage through the environment. Thus, in any task attention is not solely supervising progress through goals and subgoals but also through the environment itself. Studies of domestic tasks such as making tea (Land et al., 1999) and making a sandwich (Hayhoe, 2000; Hayhoe, Shrivastava, Mruczek, & Pelz, 2003), together with studies of walking (e.g., Patla & Vickers, 2003; Rothkopf, Ballard, & Hayhoe, 2007), have revealed key insights into the spatial and temporal allocation of overt attention in everyday tasks.

Spatial Allocation of Attention   A striking and immediate observation in any study of attention in real-world settings is the intimate link between gaze and the current action goals, and this is found not only in domestic tasks but in a wide range of studies of real-world activities (Land & Tatler, 2009). A link between attention and cognitive goals is unsurprising as it has long been recognized that where we attend is influenced by our higher level goals (Buswell, 1935; Kowler, 1990; Yarbus, 1967). However, it is the strength of this coupling in natural tasks that is striking: When participants engaged in the task of making a sandwich in a cluttered environment containing many task-irrelevant distractors, Hayhoe et al. (2003) found that 84% of fixations fell on task-relevant objects, compared to only 52% on task-relevant items prior to commencing the task. In tea making it was similarly rare that participants would look at objects other than those immediately relevant to the task (Land et al., 1999). This ubiquitous finding in real-world research has implications for how we approach models and descriptions of attentional allocation in natural tasks: Attention allocation in space will be driven by the distribution of task-relevant objects in the environment and will be targeted almost exclusively toward those objects that are needed for completing the task. This in itself is a challenge for models of attention allocation based upon statistical properties of the environment (e.g., Itti & Koch, 2000). Moreover, not only do we look preferentially at objects required for completing our behavioral goals, but the manner in which vision is distributed over these objects depends upon the nature of our intended interaction with the objects. For two classes of visually similar objects, Rothkopf et al. (2007) showed that fixations were directed to the margins of objects that the observer intended to avoid, but to the center of objects that they intended to intercept when walking through a virtual environment. Descriptions based upon the statistics of environments or visual features are likely to be severely limited in their ability to describe attention allocation in natural tasks (Tatler et al., 2011).

While fixation allocation is generally tightly linked to ongoing actions, it is not clear whether all fixations that are made during natural tasks are strictly necessary for coordinating vision and action, nor is it straightforward to discern the function of some fixations. Indeed in tea making about a third of fixations could be confidently attributed to serving actions—these were the ones that initially targeted an object that was soon to be manipulated (Land et al., 1999). Of this third of fixations, Land et al. suggested that most could be classified as serving one of four purposes: (1) *locating* an object that will later be used in the task; (2) *directing* action toward an object by maintaining fixation on the goal of the action, for example, fixating an object while reaching toward it; (3) *guiding* two objects or object parts relative to each other by making alternating fixations between the two objects; and (4) *checking* whether some condition required for the task has been met, such as watching the level of water in the kettle. The remaining two thirds of fixations in tea making were less easily categorized as serving a particular purpose. Typically these "secondary" fixations were within a few degrees of the initial fixation. For example, during the waiting period for the kettle to boil, saccades continued to be executed and fixations were made to various locations on the kettle yet were not clearly serving any particular information-gathering purpose (see figure 17.4). It may be that the differing timescales over which manual actions and saccades operate introduce a certain amount of free time or potential for redundancy

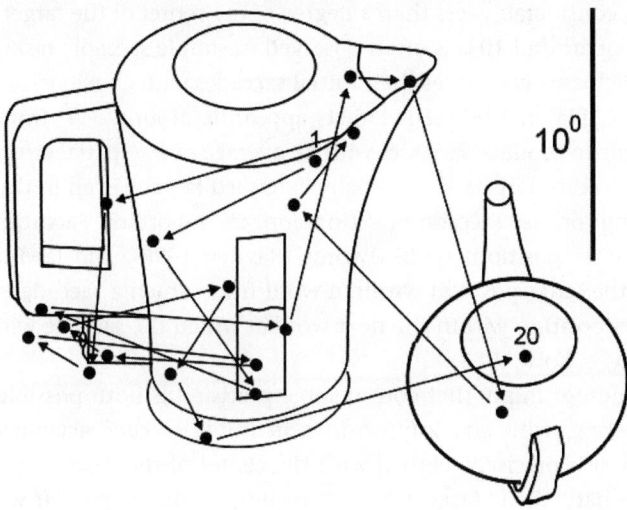

Figure 17.4
Fixations while waiting for the kettle to boil. From Land & Tatler (2009).

in fixation allocation. Object manipulations in everyday tasks tend to last around 3 seconds on average (Land et al., 1999), yet we launch on average around three saccades per second. It may be that these extra fixations, roughly two-thirds of the total while making tea, are not strictly necessary for information gathering and action organization to achieve our behavioral goals. The possibility that only a subset of the fixations we make might be necessary for task completion is supported by evidence from a patient who was unable to make eye saccades and instead relocated gaze with energetically costly head movements: This patient was able to make tea entirely normally but only made around one-third as many gaze relocations compared to individuals with intact saccadic systems (Land, Furneaux, & Gilchrist, 2001).

Given evidence from laboratory-based studies of saccade targeting precision, it would seem logical to assume that a key objective of the visual system is to bring the fovea precisely and directly to bear upon the exact detail in the environment that requires inspection. When responding to a sudden onset target in peripheral vision, initial saccades often land slightly short of the target (Becker, 1991), and when this happens, the initial hypometric saccade is followed by a small corrective saccade. Corrective saccades imply a desire for precision in saccade targeting, and the range of distances over which they occur reveals the desired level of targeting precision. That these corrective saccades can be found after hypometric saccades to targets less than 10° away (Abrams, Meyer, & Kornblum, 1989; Becker, 1972)—when the primary saccade brings the eye to within a degree of the target—suggests a goal of bringing the center of the fovea to within considerably less than a degree of the center of the target. Moreover, while hypometry of around 10% is often observed in simple saccadic tasks, when required to do so, participants can target their initial saccades with considerably improved precision (Kowler & Blaser, 1995). For targets appearing around 3–4° from fixation, participants were able to produce saccades with an average of 1% spatial error, and this was the case for both point targets and spatially extended targets. Even in the more complex task of reading, precise saccade targeting appears important: Saccades appear to target particular letter positions within words (Rayner, 1998), and lawful relationships exist between the letter position within a word from which a saccade is launched and the likely letter position within the next word at which the saccade will land (McConkie, Kerr, Reddix, & Zola, 1988).

These laboratory-based findings imply that considerable precision is both possible and central to saccade targeting, with any imprecisions in initial saccade accuracy being corrected until the fovea is precisely aligned with the center of the desired target. However, evidence from naturalistic tasks is less compelling in this respect. If we can use the existence and extent of corrective saccades as an indicator of the desired precision with which fixations are targeted, then we can use the distance between fixations and the targets of our actions as indicators of desired targeting accuracy in natural tasks. Johansson, Westling, Bäckström, and Flanagan (2001) found that when participants were engaged in a task that required a small object to be lifted to a goal location while avoiding collision with an obstacle en route, vision was directed to the key locations for successful completion of the task (contact points between digits and the object, between the object and goal, and the protruding corner of the obstacle) as would be expected. However, precise targeting of these key locations did not seem to be essential: Once the eyes were within 1.5–2° of the target location, no further corrections were made. In domestic tasks estimates of saccadic precision seem even coarser: When acting on objects during tea making, saccades were found to have amplitudes peaking between 5° and 10° (Land & Hayhoe, 2001; Land et al., 1999). This implies that if particular objects or parts of objects were being targeted, albeit rather imprecisely, gaze would be likely to land within about 5° of the center of the target. It might be tempting to use such estimates of precision derived from these more natural settings as indicators that spatial allocation of attention is somewhat sloppy in real-world tasks, but it should be remembered that objects in the real world occupy much larger visual angles than targets in laboratory-based paradigms. For example, an electric kettle held at arms length can occupy around 15° × 20°. With such large target objects, extrafoveal vision will be sufficient for gathering much of the information required for manipulating the objects, and so there is a lesser requirement to precisely bring the fovea to bear on a particular part of the object.

Taken together, these findings from everyday tasks suggest that we allocate spatial attention to objects that we are using to complete our behavioral goals by getting the eyes close enough to sample the required information. The varying precision with which the fovea is aligned with target objects, the potential profligacy of eye movements, and the tight coupling between vision and current actions mean that the approaches often taken in laboratory-based paradigms for inferring the principles that underlie overt attentional allocation may not be appropriate for *everyday* natural tasks. A common approach for trying to understand laboratory-based scene viewing is to use the locations that are fixated to reverse engineer an understanding of factors guiding inspection. Typically, local information within 0.5–1° of the center of fixation is extracted from scenes and used to construct and evaluate models of fixation selection (e.g., Borji & Itti, 2013; Judd, Durand, & Torralba, 2012). This approach makes three assumptions that may not hold for natural tasks. First, it is assumed that visual information at the fovea played a key role in the decision that brought the eyes to their current location. While this is likely to be the case to at least some extent, there arise situations in natural tasks where the eyes are directed to locations that are visually unremarkable at the time that they are targeted but are about to become interesting and informative. For example, fixations are directed to locations where objects will be placed in domestic tasks (Hayhoe et al., 2003; Land et al., 1999) or through which balls will pass in cricket and squash, but where they have not yet arrived (Hayhoe, McKinney, Chajka, & Pelz, 2012; Land & McLeod, 2000). These locations are essentially "empty" when targeted, and local information at these locations is unlikely to have contributed to the saccade targeting decision. Second, these reverse-engineering approaches assume that the information that contributed to the saccade targeting decision is localized to within 0.5–1° of the center of the proceeding foveal location: In most computational approaches, visual features are extracted and analyzed only over a small area around the center of gaze (see Tatler et al., 2011). Given the varying precision of saccade targeting in natural tasks, this assumption is unlikely to hold. Third, these approaches assume that all fixations are targeted in the same way whereas we have seen in natural tasks that some fixations may be less central to goal completion than others and may be less strictly targeted and coordinated. These three issues mean that using eye-tracking data collected from everyday task settings may not reveal the factors that underlie the fine-grained moment-to-moment decisions about where to allocate overt attention. However, an understanding of task goals and behavioral organization are more likely to reveal the principles and priorities that underlie attention allocation in these situations.

Temporal Allocation of Attention   The tight coupling between attention and ongoing actions highlights the importance of not only allocating attention to the relevant spatial locations but doing so at the right time. Ballard et al. (1992) described this spatiotemporal coupling between vision and action as a "do-it-where-I'm-looking" strategy for coordinating vision and action. When participants arranged colored squares on

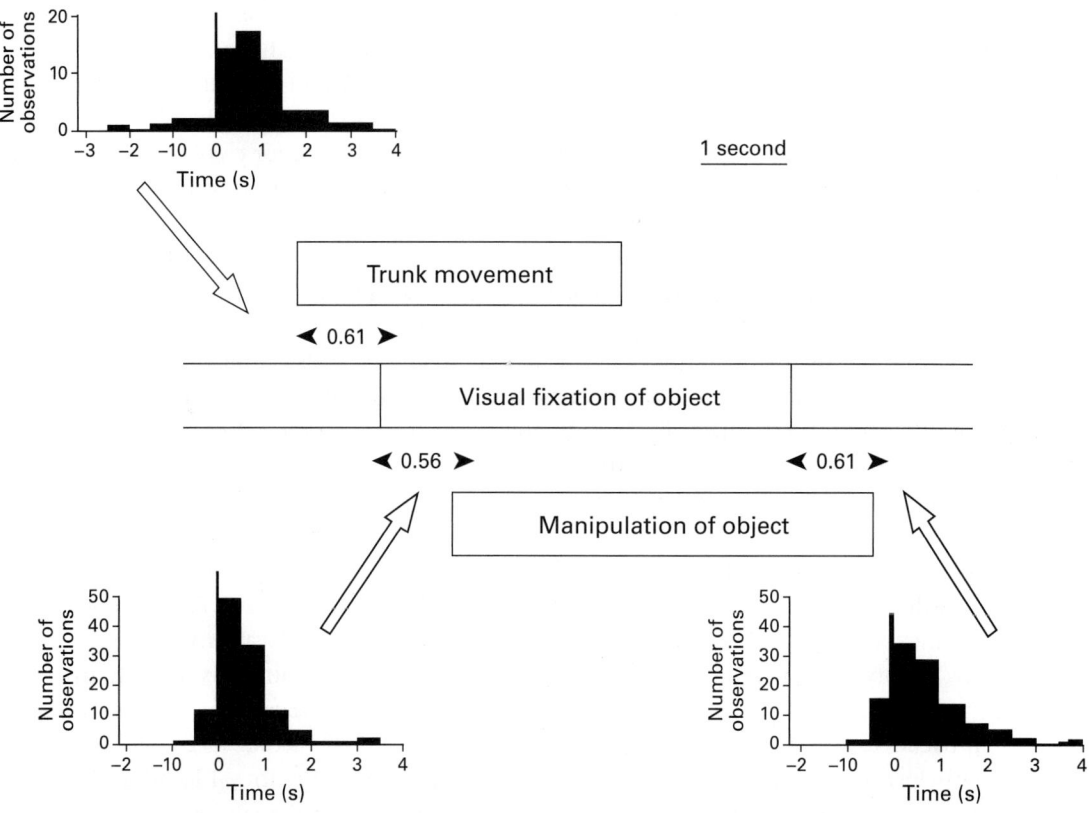

Figure 17.5

Temporal coordination of body rotations toward, visual fixation of, and manipulation of an object while making a cup of tea (adapted from Land et al., 1999). This sequence was described as an object-related act by Land et al.

a display to replicate a viewed pattern, vision and action were tightly coupled, with vision rarely allocated to any location other than the currently manipulated square and actions rarely performed on squares that were not being fixated. However, the onset of overt attention to and interaction with an object are rarely synchronous in natural tasks, with vision (i.e., an appropriate saccade) typically leading action by around 0.5–1 second (Land & Tatler, 2009). Indeed, when participants were making a cup of tea, each manipulation of an object in the task followed a very similar pattern in the manner in which vision and action were coordinated (Land et al., 1999): Manual contact with an object was preceded by a saccade to that object, which was itself preceded by a body turn toward the object if one was necessary (see figure 17.5). Thus, body and gaze orientation were used to bring overt attention to bear upon the target object some 0.5–1 second before manual contact with it. Given the regularity with which this pattern of coordination was observed and the consistency of timing between vision and action, Land et al. (1999) proposed that this spatiotemporal relationship between vision and manipulation offered an irreducible behavioral unit in natural tasks, which they referred to as the object-related act (ORA). They proposed that the ORA refined previous ideas about the lowest levels of behavioral organization in hierarchal models of task performance (Norman & Shallice, 1986; Schwartz et al., 1991, 1995), placing visual monitoring and vision–action coordination at the heart of this level of behavioral organization. Work across a wide variety of natural behaviors has found spatio-temporal coordination between vision and action that is remarkably consistent with that described by Land et al. (1999) within ORAs: Vision precedes action by about 0.5–1 second in reading aloud (Buswell, 1920), copy typing (Inhoff, Briihl, Bohemier, & Wang, 1992), walking (Patla & Vickers, 2003), and driving (Land & Lee, 1994; Land & Tatler, 2001).

The prevalence of this 0.5–1 second delay between vision and action in self-paced tasks suggests that the temporal allocation of gaze is under relatively strict control across very different natural tasks. However, this delay is not universal, and in some situations there is simply not the opportunity to get the eyes this far ahead of action. For example, in ball sports such as cricket (Land & McLeod, 2000) and squash (Hayhoe et al., 2012) the speed with which the ball moves precludes such a lead time for the eyes. In such situations, however, successful, skilled performance of the task still relies on getting the eyes ahead of the ball at key points in the ball's flight: typically at or around the locations at which the ball bounces and changes direction. Successful completion of tasks such as these seems to depend upon the correct temporal allocation of attention: In cricket good and bad batsmen alike will look at the location on the pitch where the ball bounces. The difference is that good batsmen will direct their eyes to this location about 100 ms before the ball arrives at the bounce point, whereas poor batsmen will direct their eyes to the same location as much as 100 ms later, by which time the ball may have already bounced (see figure 17.6; Land & McLeod, 2000).

The link between correct temporal allocation of attention and skill level in ball sports raises questions about the possible link between skill and the typically observed 0.5–1 second delay between vision and action in everyday tasks. The everyday tasks that we engage in and that have been studied are necessarily tasks with which we have considerable experience, and even if we are unfamiliar with a particular real-world task, the behavior is likely to be built upon the foundations of other well-learned actions and subtasks. It may therefore be that the 0.5–1 second delay is in some way the end state of a process of learning visuomotor coordination to perform skilled actions. Sailer, Flanagan, and Johansson (2005) investigated how learning interacts with the spatio-temporal allocation of gaze in a visuomotor task. Their task required participants to

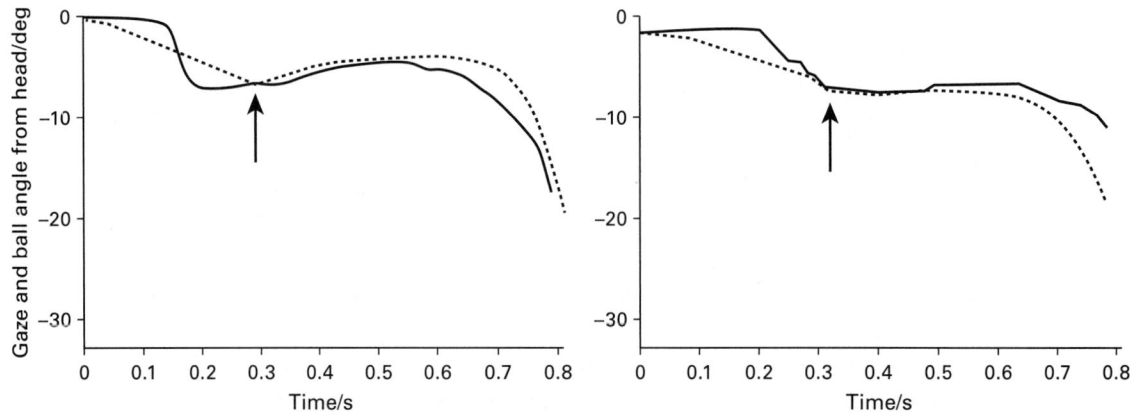

Figure 17.6
Gaze behavior of a good batsman (left) and weak batsman (right) while facing a very short delivery. Gaze direction is shown in each plot as the solid line. The ball's trajectory is shown as the dotted line. The good batsman brought his eye to the location of the bounce around 100 ms before the ball arrived. In contrast, the arrival of the poor batsman's gaze and the ball at the bounce point are almost simultaneous. Redrawn from Land & McLeod (2000).

guide a cursor to a series of targets on a monitor, controlled by a device with initially unknown mappings between actions and movements of the cursor. The task was initially very difficult, but over a period of about 20 minutes participants became quite skilled at controlling the cursor. Of particular interest here is that the temporal relationship between gaze and the cursor changed dramatically over the learning period. Initially gaze lagged the cursor movements in time. However, by the time the participants were skilled at the task, gaze was allocated in an anticipatory manner. Moreover the timing was such that gaze led the cursor movements by around 0.4 seconds, which is in line with the typically observed lead by gaze over action. A similar progression toward a greater lead time by the eyes over action can be found when comparing learner drivers to more experienced drivers (Land, 2006).

It is clear from examining gaze allocation in natural tasks that if we are to understand how attention is allocated in everyday situations, then we must account for the manner in which attention is allocated both in space and in time. Timing is rarely emphasized in accounts (or indeed models) of fixation behavior when viewing static scenes but is clearly central to natural tasks. This presents considerable challenges for existing models of attention allocation (Tatler et al., 2011) and may mean that alternative theoretical and modeling frameworks are more appropriate as a starting point for understanding attention allocation in everyday tasks (Hayhoe & Ballard, 2005; Tatler et al., 2011).

Flexibility in Attention Allocation   In attempting to understand attention allocation we search for regularities—in particular, the factors that guide the eyes. Indeed in this chapter we have sought such "rules" and described everyday attention in this manner, notably the almost universal 0.5–1 second delay between vision and action. However, this search for regularity and common underlying principles neglects one of the most striking aspects of how we allocate attention in natural tasks: the flexibility with which gaze is used to gather information. When reaching for previously visible targets, we typically use a combination of peripheral vision and memory to plan saccades and reaching movements, but the relative reliance on each of these sources of information is flexible rather than fixed, with greater reliance on memory when visibility of peripheral targets is reduced (Brouwer & Knill, 2007). The involvement of memory in shifts of attention and gaze is also emphasized by the fact that we can target objects that are out of sight (Land, 2014). Thus, attention shifts and reaches are planned on the basis of the most reliable evidence rather than set rules governing the relative contributions of different sources of evidence.

Similar flexibility in the manner in which attention is allocated can be seen in walking. Jovancevic-Misic and Hayhoe (2009) showed that when we are walking toward other people, what we learn about how someone is likely to behave when we encounter them is used to adapt the manner in which we attend to that person when we encounter them again. Oncoming pedestrians were assigned roles as potential colliders (who were asked to walk on collision courses toward the participant on each encounter) or avoiders (who were asked to avoid collision courses). Participants rapidly learned who the potential colliders were and adapted their gaze behavior such that they looked sooner and for longer at the potential colliders than at the avoiders. When the oncoming pedestrians switched roles, participants were able to adapt their responses after only a few encounters. Thus, not only can we learn about when and where to attend "on the fly," but also this learning can be rapidly adapted to changes in the environment.

While general rules can be drawn up for the temporal coupling between vision and action—that vision and action co-occur but that vision precedes action by around 0.5–1 second—departures from these rules reveal online flexibility of visuomotor coordination. In tasks such as tea making and sandwich making, fixations are sometimes made to objects that are not the current target of action but will be used in the next few seconds (Hayhoe et al., 2003; Land et al., 1999). These "look-ahead" fixations (Pelz & Canosa, 2001) seem to aid subsequent attentional allocation to objects (Mennie, Hayhoe, & Sullivan, 2007). There is a longer lead of vision over action at the start of manipulations of objects that have been the target of a "look-ahead" fixation than for objects that have not been fixated during a previous action (Mennie et al., 2007). These look-ahead fixations therefore appear to be purposeful components that serve the planning of subsequent attentional allocation for action. Presumably, the opportunity to look ahead to the target of a future action is afforded by the fact that there is some

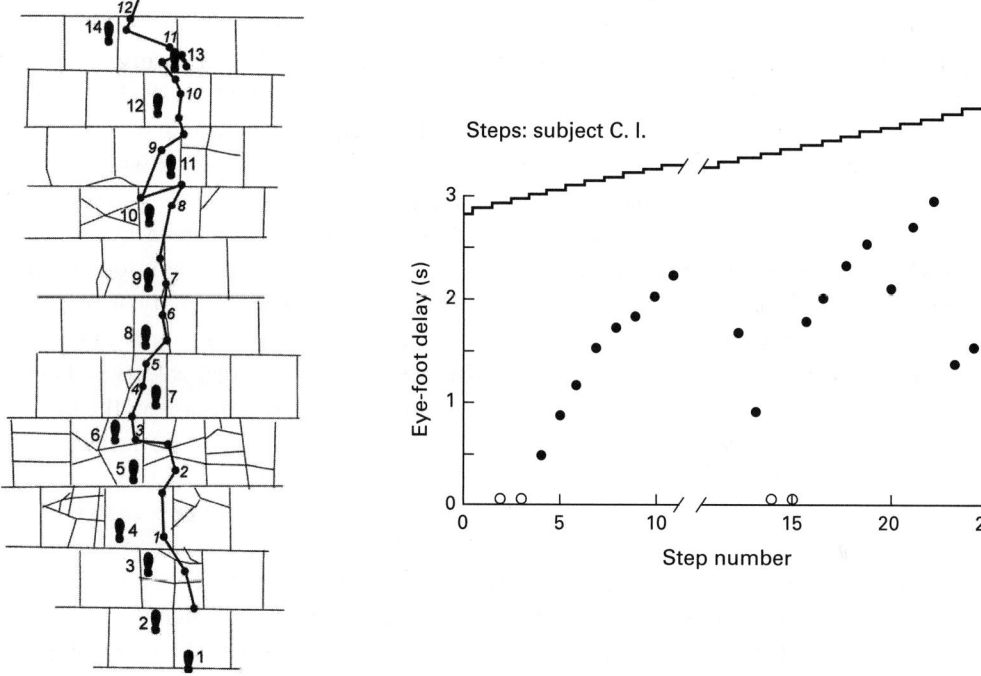

Figure 17.7

(Left) Fixations and footfalls as one participant walked on paving stones with the instruction to avoid stepping on the cracks. Numbered fixations show those that occurred simultaneously with each correspondingly numbered footfall. Fixations are around two steps ahead of each footfall, corresponding to an eye–footfall delay of 1.11 seconds on average. (Right) Eye–footfall delay for one participant walking up stairs. At the start of each ascent, gaze is directed to locations not far ahead of each footfall, but as the stairs are climbed, gaze targets locations further ahead of each footfall. Data in both plots are from unpublished work in Land's lab (previously discussed in Land & Tatler, 2009).

redundancy in fixations during ORAs, as we have discussed earlier in this chapter. At times when actions do not require strict monitoring, the attentional system can take advantage of these opportunities to look ahead and acquire information useful to the next part of the ongoing behavior.

Similar departures from typical temporal coordination of vision and action can be found when walking. While in some situations—such as walking on rough terrain or being required to place footfalls precisely (see figure 17.7)—gaze is allocated to terrain about 0.5–1 second ahead of each footfall (Patla & Vickers, 2003), in situations where terrain is predictable this coupling is less strict. For example, when walking up stairs, vision initially precedes footfalls by less than a second, but this delay increases as the steps are climbed and reaches about 2 to 3 seconds by the top of a flight of stairs (see figure 17.7; Land & Tatler, 2009). Presumably this increase in eye–foot latency reflects decreasing need to monitor each footfall on the predictable staircase and thus a possibility to allocate gaze further ahead. Ultimately, on level terrain, the ground ahead is rarely looked at. Similarly, walking in natural settings involves not only the need to correctly place one foot in front of the other but also the need to monitor, and if necessary adjust, behavior in response to obstacles and other pedestrians, and the density and threat of these will vary considerably depending upon the place and time at which we are walking. In such situations, attending to the ground around a second ahead of footfalls is not the dominant visual strategy: Foulsham, Walker, and Kingstone (2011) found that while participants made some fixations to locations on the path just ahead, more time was spent looking at objects (e.g., lampposts, trees, buildings).

These examples of departures from typical spatiotemporal coupling between vision and action suggest that the systems that underlie attentional allocation in everyday tasks are able to take advantage of opportunities to allocate gaze proactively to other aspects of the environment from which information will aid future behavior.

Head Movements in Everyday Attention

Usually both eyes and head contribute to gaze shifts, but not in a rigidly predictable manner. Goossens and van Opsal (1997) found that for gaze shifts up to about 40° the head contributes about a third of the total gaze movement, but this can vary between 20% and 70% for different individuals. Even for the same individual the standard deviation of the head contribution remains high. For shifts greater than 40° the head inevitably has a greater role, as the eyes approach their limit of travel in the orbit. Thus, for small gaze changes, head movements do not seem to be a very good predictor of attention shifts, compared with the gaze changes themselves. However, there is good evidence that there are differences in the timing and dynamics of head movements made to different kinds of attention shift. In a driving simulator study, Doshi and Trivedi (2012) found that while gaze shifts induced by the appearance of external stimuli involved delayed and weak head movements, gaze shifts associated with planned task changes led to earlier and stronger head movements. Thus, whether the cause of the gaze shift is "bottom-up" or "top-down" does affect the balance between eye and head contributions. The relative contribution of eyes and head to gaze shifts also seems to depend upon whether the gaze shift is made in the presence of manual action or not. Epelboim et al. (1995) showed that gaze shifts when looking at colored pegs on a tabletop were accomplished with minimal involvement of the head, but when participants were required to touch the same pegs, head movements contributed strongly to the accompanying gaze shifts. It is also quite possible to affect this balance consciously. If one does not wish to draw attention to where one is looking, the simplest precaution is simply not to move one's head, or even to move it in the opposite direction to the gaze shift. This clearly implies that there is some element of predictability to head movements that can be picked up by others.

Social Factors in Everyday Visual Attention

It is interesting to note that in the majority of studies of natural behavior, participants have been monitored as they work in isolation toward a behavioral goal. Where others have been present in the environment, any contribution that their presence may have made to attention allocation has largely been ignored. This is an interesting and somewhat surprising omission given the rich and growing understanding we have of the importance of social context upon attention allocation (see also chapter 15 in the current volume).

Where we attend—both covertly and overtly—is profoundly influenced by where another individual is looking. When shown images containing people, observers preferentially fixate the eyes of others (Birmingham, Bischof, & Kingstone, 2009). Not only are eyes sought out but the direction in which they point is followed. This has been shown in laboratory-based paradigms using modified Posner paradigms (Friesen & Kingstone, 1998; Ricciardelli, Bricolo, Aglioti, & Chelazzi, 2002) and in images of more complex scenes (Castelhano, Wieth, & Henderson, 2007). In more realistic situations, however, the propensity to seek and follow gaze cues from others seems to be tempered by other social factors. In particular, we seem to prefer to avoid looking at and following gaze cues of others in many social settings, perhaps in order to avoid unwanted social interaction with strangers. When waiting in a room, observers were far less likely to look at another individual who was present in the room than when the individual was shown on a monitor screen within the room (Laidlaw, Foulsham, Kuhn, & Kingstone, 2011). Similarly, passersby were found to be more likely to look in the same direction as others when approaching from behind them than when approaching from in front of them (Gallup, Chong, & Couzin, 2012). This result is again consistent with the possibility that we do not engage in gaze seeking and following behaviors when there is a risk that these might be noticed by the person providing the gaze cues and thus might trigger an unwanted social interaction with that individual. Real social settings not only involve the presence of others (and their gaze cues) but also interaction with others. Gaze cues have long been recognized as important communicative tools (e.g., Argyle, 1988)—and this likely underlies their importance in influencing attention allocation in others—but are rarely the sole means of communication. More commonly, gaze and language are used together in communication. During interactions, gaze cues result in modifications of spoken language (Clark & Krych, 2004) and language influences where people look (Altmann & Kamide, 1999). Moreover, it appears that the manner in which we seek and follow gaze cues from another individual may be modulated by the relative informativeness of those gaze cues and concurrent language, with gaze cues sought out and followed only when concurrent spoken language does not provide sufficiently precise information to complete a task (Macdonald & Tatler, 2013).

Taken together, these findings suggest that any study of everyday attention must be conducted in the social context in which it naturally occurs, and researchers must consider the importance of others in the environment in determining how a participant allocates his or her attention. In social settings, attention allocation is influenced not only by where others are attending but also by the mere presence of others with whom we might potentially interact. Avoiding social interaction with others appears to result in some degree of internal monitoring of our own attention allocation to result in avoidance of eye contact and shared loci of attention.

Attention allocation can be influenced not only by the actual social presence of another but even by the implied social presence and monitoring by another that an eye tracker imparts. Indeed, Risko and Kingstone (2011) showed that when participants believed their eye movements were being monitored, they avoided looking at a swimsuit calendar in the laboratory, whereas participants who believed their eye movements were not being monitored looked at the calendar. This result poses interesting challenges for eye-tracking research in general, but specifically for any attempts to study attention allocation in natural tasks, where we wish to observe the natural, uncensored behavior rather than behavior that is shaped by the implied social monitoring of the eye tracker worn during the study.

Integration

Our understanding of attention allocation in everyday settings has important implications for the manner in which we study and understand attention allocation in laboratory-based research. From the evidence reviewed above we can argue that everyday attention research not only provides insights into long-running and ongoing debates in the literature but also redefines the questions that should be at the heart of laboratory-based study.

How we allocate attention in space and the factors that underlie spatial selection decisions have been at the heart of studies of eye movements and attention since the beginning of eye-tracking research and continue to dominate much of the ongoing work on visual attention. In the laboratory, research has focused on attempting to describe the extent to which attention is allocated on the basis of low-level stimulus properties and/or higher level internal goals. In particular, research over the last two or three decades has been defined by attempts to quantify the involvement of low-level stimulus properties on (overt) attention allocation. This focus is in large part due to the development of computational models of image salience that allow quantification of stimulus properties, much of which can be attributed to the pioneering work of Itti and colleagues (e.g., Itti & Koch, 2000; Itti, Koch, & Niebur, 1998) for computing image salience in complex scenes. Any computational model provides the opportunity to assess the association between the computed property in the image and fixation selection statistically. It is this approach of quantifying the association between low-level image properties and fixation selection that has dominated recent research on overt attentional allocation in complex scene viewing (see Tatler et al., 2011, for a review). It is becoming clear that laboratory-based findings may struggle to extend to natural settings. For example, a large fraction of fixation behavior when one is viewing images on a computer monitor appears to arise from a bias to fixate at and around the center of the monitor irrespective of the content of the displayed image (Tatler, 2007; Tseng, Carmi, Cameron, Munoz, & Itti, 2009).

While laboratory-based study of the spatial allocation of attention in scenes has focused on quantifying low-level factors in eye guidance, study of real-world behavior has focused on understanding the role of task and action in attention allocation. These two approaches are somewhat complementary, but given the strength of the relationship between task goals and fixation allocation in natural tasks, it is important to question whether the laboratory-based focus on image properties is appropriate for understanding everyday allocation of attention. It should be noted, however, that the methods for quantifying image properties in laboratory-based studies cannot readily be applied in real environments. This is because any calculation based on the video record from the eye tracker will necessarily be subject to translation through the camera optics and any digitization. Similarly, many low-level features will vary depending upon lighting conditions, which vary considerably over time in natural environments. Thus, accurate measurement of image properties at fixated locations is impractical in natural settings.

As we have seen in the previous section, findings from real-world tasks could be used to suggest that the laboratory-based focus on spatial allocation of attention may be somewhat misplaced. We have argued that given the intimate link between vision and action, the lack of precise foveal targeting often seen in real behaviors, and the possible profligacy of fixations, understanding what is at the fovea on a fixation-by-fixation basis may not be as informative as is often assumed in laboratory-based studies of eye guidance. In contrast, the temporal relationship between vision and action is more tightly controlled and more consistent across a broad range of real-world activities. We argue that understanding this temporal relationship may be more informative for understanding everyday visual attention than attempting to understand spatial allocation alone. Laboratory-based approaches typically ignore time as a relevant dimension in scene viewing; however, there are, of course, some notable exceptions: Noton and Stark (1971) argued that the order of fixations was crucial for perception and encoding of scenes, and these ideas are echoed in more recent work by Foulsham, Underwood, and colleagues (e.g., Foulsham & Kingstone, 2013; Underwood, Humphrey, &

Foulsham, 2009). The temporal dynamics of a real environment during natural behavior are hard to capture suitably in a laboratory-based paradigm. Not only does the environment change over time in complex ways but the task goals, actions we perform, and objects we require vary considerably as we progress through a task. Thus, while real-world study might suggest that laboratory-based study of attention allocation in complex scenes would benefit from a shift in emphasis toward the temporal dimension, effectively capturing the temporal dynamics during real behavior poses a considerable challenge for laboratory-based study.

Future Directions

There is growing appreciation of the need to study attention allocation in the context of real-world behaviors, as is evident from the contributions to this volume and the increasing use of real behaviors in contemporary research. A key component of this research agenda is the study of everyday behaviors such as those we have touched upon in this chapter. While studies of attention allocation during natural behavior began in the mid-20th century, we are still very much in the early phases of such research. Only recently has technology permitted effective and widely accessible devices for recording eye movements in unconstrained settings and similarly recently has recognition of the need to study real-world behavior become widespread. There is clearly much to do in this domain, and the coming years will likely see significant advances in our understanding of and approaches to the study of everyday attention.

As has always been the case, at the heart of developments in real-world study of attention allocation will be the advancement of technology. Smaller cameras will allow increasingly unobtrusive eye trackers. This is important not just for the wearers' comfort but also because less obtrusive and obvious trackers will provide the opportunity for gathering data in situations where subjects are less aware (or at least less frequently reminded) that their eye movements are being monitored. The importance of the potential for covert monitoring of eye movements is evident from recent work by Risko and Kingstone (2011). Not only will lighter and subtler optics benefit research but better software solutions for detecting saccades and fixations and processing video data will greatly improve the research potential for real-world study. If we can streamline these analytical processes, which are still often manual, it will be possible to collect and work on data sets much larger than are typically used in real-world research at present.

We have argued that there is a need to understand the factors that underlie decisions about how the eyes are allocated in space and time in natural tasks. Ultimately, the challenge is to be able to identify the underlying principles for these decisions with sufficient precision to allow us to construct models of attention allocation in everyday behaviors. This is a considerable challenge given that so much of our visual behavior seems to be driven by high-level factors such as task goals and these factors are very hard to quantify and generalize across tasks. One promising approach in this respect that is likely to play a prominent role in the future development of this field is to describe attention allocation in terms of behavioral reward. Ballard and colleagues have developed models based upon reward that have been successful in generating human-like behavior and (overt) attention allocation in walking tasks (Ballard & Hayhoe, 2009; Rothkopf & Ballard, 2010; Sprague, Ballard, & Robinson, 2007). In these models attention is used to reduce uncertainty about the state of an attended subgoal, with uncertainty about unattended subgoals increasing over time. Decisions about which subgoal to attend to are then based on the principle of reducing uncertainty by selecting the subgoal with maximum uncertainty at the time of the decision. These models are in their relative infancy but have the potential to provide a theoretical framework that is appropriate and tractable across tasks and that can explain allocation in space and time in a single account (for a discussion of these ideas, see Tatler et al., 2011).

It is surprising that social factors have so frequently been neglected or omitted from studies of everyday behaviors. The presence of and interaction with other individuals profoundly influences how we allocate attention. This is especially evident given recent work on social attention in increasingly real-world settings (see also chapter 15 in the current volume). It is therefore important that any account of everyday visual attention must embrace the influence of social context, and it seems very likely that this will form an important component of the future development of our understanding of everyday attention.

Conclusion

We have seen in this chapter that spatial allocation of attention may be underpinned by different guiding principles than we would expect from laboratory-based investigations of attention allocation. We further argued that time is of central importance to our understanding of attention allocation in natural tasks and that any account of everyday attention must account for how attention is allocated in both space and time. However, while general rules may be drawn up for attentional engagement in the real world, it is clear that there is remarkable flexibility with which we utilize and acquire information, and this is evident in the departures from normal spatiotemporal coupling between vision and action that are frequently observed. At the outset of this chapter we argued that if we are to understand attention in everyday situations, then we can only really do so by studying what occurs in those settings, but we recognize that the inherent variability of natural environments is antithetical to typical psychological study. However, it is this environmental variability and how we respond to it that reveals the flexibility with which we guide our attention and coordinate our behaviors, and which provides some of the most important insights into everyday attention.

Box 17.1
Key Points

- An ecologically valid account of visual attention requires the study of gaze during undisrupted natural behavior. To this end, overt shifts of attention should be measured with minimal disruption to everyday activities.

- Gaze allocation in space is tightly coupled to task goals, with very few fixations to objects not relevant to the current task. However, the precision with which gaze is targeted depends on the requirements of the action, and often it is sufficient for the fovea to be within a few degrees of the relevant location.

- Gaze leads action by between 0.5 and 1 second in many real-world situations, with body movement leading gaze movements by a similar interval when required. Even in tasks when such a long lead by the eye is not possible, skilled behavior still requires that the eyes are ahead of action at key points.

Box 17.2
Outstanding Issues

- Given tight temporal control of vision with respect to action, it may be that models and frameworks for understanding everyday attention allocation require a shift of emphasis from spatial (as has been the emphasis in the lab) to temporal analysis.

- Technological advances are still needed to minimize intrusion by eye trackers in order to allow collection of data under more natural conditions, and to improve data coding and analysis.

- Social factors profoundly influence attention allocations, and they do so differently in the real world than in the lab. Future accounts of everyday attention should encompass the social nature of human activity.

References

Abernethy, B. (1990). Expertise, visual search, and information pick-up in squash. *Perception, 19,* 63–77.

Abrams, R. A., Meyer, D. E., & Kornblum, S. (1989). Speed and accuracy of saccadic eye movements: Characteristics of impulse variability in the oculomotor system. *Journal of Experimental Psychology: Human Perception and Performance, 15,* 529–543.

Altmann, G. T. M., & Kamide, Y. (1999). Incremental interpretation at verbs: Restricting the domain of subsequent reference. *Cognition, 73,* 247–264.

Argyle, M. (1988). *Bodily communication* (2nd ed.). London: Methuen.

Ballard, D. H., & Hayhoe, M. M. (2009). Modeling the role of task in the control of gaze. *Visual Cognition, 17*, 1185–1204.

Ballard, D. H., Hayhoe, M. M., Li, F., Whitehead, S. D., Frisby, J. P., Taylor, J. G., & Fisher, R. B. (1992). Hand–eye coordination during sequential tasks. *Philosophical Transactions of the Royal Society of London. Series B: Biological Sciences, 337*, 331–339.

Becker, W. (1972). The control of eye movements in the saccadic system. *Bibliotheca Ophthalmologica, 82*, 233–243.

Becker, W. (1991). Saccades. In R. H. S. Carpenter (Ed.), *Vision and visual dysfunction* (Vol. 8, pp. 95–137). Basingstoke, UK: Macmillan.

Birmingham, E., Bischof, W. F., & Kingstone, A. (2009). Get real! Resolving the debate about equivalent social stimuli. *Visual Cognition, 17*, 904–924.

Borji, A., & Itti, L. (2013). State-of-the-art in visual attention modeling. *IEEE Transactions on Pattern Analysis and Machine Intelligence, 35*, 185–207.

Brewer, W. F., & Dupree, D. A. (1983). Use of plan schemata in the recall and recognition of goal-directed sections. *Journal of Experimental Psychology: Learning, Memory, and Cognition, 9*, 117–129.

Brouwer, A., & Knill, D. (2007). The role of memory in visually guided reaching. *Journal of Vision, 7*(5), 1–12.

Buswell, G. T. (1920). *An experimental study of the eye–voice span in reading. Supplementary Educational Monographs No. 17*. Chicago: University of Chicago Press.

Buswell, G. T. (1935). *How people look at pictures: A study of the psychology of perception in art*. Chicago: University of Chicago Press.

Cameron, E. H., & Steele, W. M. (1905). The Poggendorff illusion. *Psychological Monographs, 7*(1), 83–111.

Castelhano, M. S., Wieth, M., & Henderson, J. M. (2007). I see what you see: Eye movements in real-world scenes are affected by perceived direction of gaze. In L. Paletta & E. Rome (Eds.), *Attention in cognitive systems: Theories and systems from an interdisciplinary viewpoint* (pp. 251–262). Berlin: Springer.

Clark, H. H., & Krych, M. A. (2004). Speaking while monitoring addressees for understanding. *Journal of Memory and Language, 50*, 62–81.

Delabarre, E. B. (1898). A method of recording eye movements. *American Journal of Psychology, 9*, 572–574.

Deubel, H., & Schneider, W. X. (1996). Saccade target selection and object recognition: Evidence for a common attentional mechanism. *Vision Research, 36*, 1827–1837.

Dicks, M., Button, C., & Davids, K. (2010). Examination of gaze behaviors under in situ and video simulation task constraints reveals differences in information pickup for perception and action. *Attention, Perception & Psychophysics, 72*, 706–720.

Dodge, R., & Cline, T. S. (1901). The angle velocity of eye movements. *Psychological Review, 8*, 145–157.

Doshi, A., & Trivedi, M. M. (2012). Head and eye dynamics during visual attention shifts in complex environments. *Journal of Vision, 12*(2), 9, 1–16.

Epelboim, J. L., Steinman, R. M., Kowler, E., Edwards, M., Pizlo, Z., Erkelens, C., et al. (1995). The function of visual search and memory in sequential looking tasks. *Vision Research, 35*, 3401–3422.

Findlay, J. M., & Gilchrist, I. D. (2003). *Active vision: The psychology of looking and seeing.* Oxford, UK: Oxford University Press.

Foulsham, T., & Kingstone, A. (2013). Fixation-dependent memory for natural scenes: An experimental test of scanpath theory. *Journal of Experimental Psychology. General, 142*, 41–56.

Foulsham, T., Walker, E., & Kingstone, A. (2011). The where, what and when of gaze allocation in the lab and the natural environment. *Vision Research, 51*, 1920–1931.

Friesen, C. K., & Kingstone, A. (1998). The eyes have it! Reflexive orienting is triggered by nonpredictive gaze. *Psychonomic Bulletin & Review, 5*, 490–495.

Furneaux, S., & Land, M. F. (1999). The effects of skill on the eye–hand span during musical sight-reading. *Proceedings. Biological Sciences, 266*, 2435–2440.

Gallup, A. C., Chong, A., & Couzin, I. D. (2012). The directional flow of visual information transfer between pedestrians. *Biology Letters, 8*, 520–522.

Goossens, H. H. L. M., & van Opsal, A. J. (1997). Human eye–head coordination in two dimensions under different sensorimotor conditions. *Experimental Brain Research, 114*, 542–560.

Hartridge, H., & Thompson, L. C. (1948). Methods of investigating eye movements. *British Journal of Ophthalmology, 32*, 581–591.

Hayhoe, M. M. (2000). Vision using routines: A functional account of vision. *Visual Cognition, 7*, 43–64.

Hayhoe, M. M., & Ballard, D. H. (2005). Eye movements in natural behavior. *Trends in Cognitive Sciences, 9*, 188–194.

Hayhoe, M. M., McKinney, T., Chajka, K., & Pelz, J. B. (2012). Predictive eye movements in natural vision. *Experimental Brain Research, 217*, 125–136.

Hayhoe, M. M., Shrivastava, A., Mruczek, R., & Pelz, J. B. (2003). Visual memory and motor planning in a natural task. *Journal of Vision, 3*, 49–63.

Hering, E. (1879). Über Muskelgeräusche des Auges. *Mathematischnaturwissenschaftliche Klasse, 79*, 137–154.

Inhoff, A. W., Briihl, D., Bohemier, G., & Wang, J. (1992). Eye–hand span and coding of text during copytyping. *Journal of Experimental Psychology: Learning, Memory, and Cognition, 18*, 298–306.

Itti, L., & Koch, C. (2000). A saliency-based search mechanism for overt and covert shifts of visual attention. *Vision Research, 40*, 1489–1506.

Itti, L., Koch, C., & Niebur, E. (1998). A model of saliency-based visual attention for rapid scene analysis. *IEEE Transactions on Pattern Analysis and Machine Intelligence, 20*, 1254–1259.

Johansson, R. S., Westling, G., Bäckström, A., & Flanagan, J. R. (2001). Eye–hand coordination in object manipulation. *Journal of Neuroscience, 21*, 6917–6932.

Jovancevic-Misic, J., & Hayhoe, M. M. (2009). Adaptive gaze control in natural environments. *Journal of Neuroscience, 29*, 6234–6238.

Judd, C. H. (1905). The Müller–Lyer illusion. *Psychological Monographs, 7*(1), 55–81.

Judd, C. H., & Courten, H. C. (1905). The Zöllner illusion. *Psychological Monographs, 7*(1), 112–139.

Judd, T., Durand, F., & Torralba, A. (2012). *A benchmark of computational models of saliency to predict human fixations* (Tech. Rep. No. MIT-CSAIL-TR-2012–001). Cambridge, MA: MIT Computer Science and Artificial Intelligence Laboratory.

Kingstone, A., Smilek, D., & Eastwood, J. D. (2008). Cognitive ethology: A new approach for studying human cognition. *British Journal of Psychology, 99*, 317–340.

Kowler, E. (1990). The role of visual and cognitive processes in the control of eye movement. In E. Kowler (Ed.), *Eye movements and their role in visual and cognitive processes* (pp. 1–70). Amsterdam: Elsevier.

Kowler, E., & Blaser, E. (1995). The accuracy and precision of saccades to small and large targets. *Vision Research, 35*, 1741–1754.

Laidlaw, K. E., Foulsham, T., Kuhn, G., & Kingstone, A. (2011). Potential social interactions are important to social attention. *Proceedings of the National Academy of Sciences of the United States of America, 108*, 5548–5553.

Land, M. F. (1993). *Eye–head coordination during driving*. Le Touquet, France: IEEE Systems, Man and Cybernetics.

Land, M. F. (2004). The coordination of rotations of eyes, head and trunk in saccadic turns produced in natural situations. *Experimental Brain Research, 159*, 151–160.

Land, M. F. (2006). Eye movements and the control of actions in everyday life. *Progress in Retinal and Eye Research, 25*, 296–324.

Land, M. F. (2014). Do we have an internal model of the outside world? *Philosophical Transactions of the Royal Society of London. Series B, Biological Sciences, 369*, 20130045.

Land, M. F., & Furneaux, S. (1997). The knowledge base of the oculomotor system. *Philosophical Transactions of the Royal Society of London. Series B, Biological Sciences, 352*, 1231–1239.

Land, M. F., Furneaux, S. M., & Gilchrist, I. D. (2001). The organization of visually mediated actions in a subject without eye movements. *Neurocase, 8*, 80–87.

Land, M. F., & Hayhoe, M. M. (2001). In what ways do eye movements contribute to everyday activities? *Vision Research, 41*, 3559–3565.

Land, M. F., & Lee, D. N. (1994). Where we look when we steer. *Nature, 369*, 742–744.

Land, M. F., & McLeod, P. (2000). From eye movements to actions: How batsmen hit the ball. *Nature Neuroscience, 3*, 1340–1345.

Land, M. F., Mennie, N., & Rusted, J. (1999). The roles of vision and eye movements in the control of activities of daily living. *Perception, 28*, 1311–1328.

Land, M. F., & Tatler, B. W. (2001). Steering with the head: The visual strategy of a racing driver. *Current Biology, 11*, 1215–1220.

Land, M. F., & Tatler, B. W. (2009). *Looking and acting: Vision and eye movements in natural behaviour.* Oxford, UK: Oxford University Press.

Macdonald, R. G., & Tatler, B. W. (2013). Do as eye say: Gaze cueing and language in a real-world social interaction. *Journal of Vision, 13*(4), 6, 1–12.

Mackworth, N. H., & Thomas, E. L. (1962). Head-mounted eye-movement camera. *Journal of the Optical Society of America, 52*, 713–716.

McAllister, C. N. (1905). The fixation of points in the visual field. *Psychological Monographs, 7*(1), 17–53.

McConkie, G. W., Kerr, P. W., Reddix, M. D., & Zola, D. (1988). Eye movement control during reading: I. The location of initial eye fixations in words. *Vision Research, 28*, 1107–1118.

Mennie, N., Hayhoe, M. M., & Sullivan, B. (2007). Look-ahead fixations: Anticipatory eye movements in natural tasks. *Experimental Brain Research, 179*, 427–442.

Norman, D. A., & Shallice, T. (1986). Attention to action: Willed and automatic control of behaviour. In R. J. Davidson, G. E. Schwartz, & D. Shapiro (Eds.), *Consciousness and self-regulation: Advances in research and theory* (Vol. 4, pp. 1–18). New York: Plenum Press.

Noton, D., & Stark, L. (1971). Eye movements and visual perception. *Scientific American, 224*, 34–43.

Patla, A. E., & Vickers, J. N. (2003). How far ahead do we look when required to step on specific locations in the travel path during locomotion. *Experimental Brain Research, 148*, 133–138.

Pelz, J. B., & Canosa, R. (2001). Oculomotor behavior and perceptual strategies in complex tasks. *Vision Research, 41*, 3587–3596.

Posner, M. I. (1980). Orienting of attention. *Quarterly Journal of Experimental Psychology, 32*, 3–25.

Rayner, K. (1998). Eye movements in reading and information processing: 20 years of research. *Psychological Bulletin, 124*, 372–422.

Ricciardelli, P., Bricolo, E., Aglioti, S. M., & Chelazzi, L. (2002). My eyes want to look where your eyes are looking: Exploring the tendency to imitate another individual's gaze. *Neuroreport, 13*, 2259–2264.

Risko, E. F., & Kingstone, A. (2011). Eyes wide shut: Implied social presence, eye tracking and attention. *Attention, Perception & Psychophysics, 73*, 291–296.

Rizzolatti, G., Riggio, L., Dascola, I., & Umiltá, C. (1987). Reorienting attention across the horizontal and vertical meridians: Evidence in favor of a premotor theory of attention. *Neuropsychologia, 25*(1A), 31–40.

Rothkopf, C. A., & Ballard, D. H. (2010). Credit assignment in multiple goal embodied visuomotor behavior. *Frontiers in Psychology, 1*, 173.

Rothkopf, C. A., Ballard, D. H., & Hayhoe, M. M. (2007). Task and context determine where you look. *Journal of Vision, 7*(14), 1–20.

Sailer, U., Flanagan, J. R., & Johansson, R. S. (2005). Eye–hand coordination during learning of a novel visuomotor task. *Journal of Neuroscience, 25*, 8833–8842.

Savelsbergh, G. J., Williams, A. M., van der Kamp, J., & Ward, P. (2002). Visual search, anticipation and expertise in soccer goalkeepers. *Journal of Sports Sciences, 20*, 279–287.

Schwartz, M. F., Montgomery, M. W., Fitzpatrick-DeSalme, E. J., Ochipa, C., Coslett, H. B., & Mayer, N. H. (1995). Analysis of a disorder of everyday action. *Cognitive Neuropsychology, 12*, 863–892.

Schwartz, M. F., Reed, E. S., Montgomery, M. W., Palmer, C., & Mayer, N. H. (1991). The quantitative description of action disorganisation after brain damage: A case study. *Cognitive Neuropsychology, 8*, 381–414.

Shackel, B. (1960). Review of the past and present in oculography (p. 57). Medical electronics proceedings of the second international conference. London: Hiffe.

Shepherd, M., Findlay, J. M., & Hockey, R. J. (1986). The relationship between eye-movements and spatial attention. *Quarterly Journal of Experimental Psychology, A, 38*, 475–491.

Sprague, N., Ballard, D. H., & Robinson, A. (2007). Modeling embodied visual behaviors. *ACM Transactions on Applied Perception, 4*, 11.

Stratton, G. M. (1906). Symmetry, linear illusions and the movements of the eye. *Psychological Review, 13*, 82–96.

Tatler, B. W. (2007). The central fixation bias in scene viewing: Selecting an optimal viewing position independently of motor biases and image feature distributions. *Journal of Vision* , *7*(14), 4, 1–17.

Tatler, B. W., Hayhoe, M. M., Land, M. F., & Ballard, D. H. (2011). Eye guidance in natural vision: Reinterpreting salience. *Journal of Vision, 11*(5), 1–23.

Thomas, E. L. (1968). Movements of the eye. *Scientific American, 219*, 88–95.

Tseng, P.-H., Carmi, R., Cameron, I. G., Munoz, D. P., & Itti, L. (2009). Quantifying center bias of observers in free viewing of dynamic natural scenes. *Journal of Vision* , *9*(7), 4, 1–16.

Underwood, G., Humphrey, K., & Foulsham, T. (2009). Saliency and scan patterns in the inspection of real-world scenes. *Visual Cognition, 17*, 812–834.

Wade, N. J., & Tatler, B. W. (2005). *The moving tablet of the eye: The origins of modern eye movement research*. Oxford, UK: Oxford University Press.

Wells, W. C. (1792). *An essay upon single vision with two eyes: Together with experiments and observations on several other subjects in optics*. London: Cadell.

Wolfe, J. M. (1998). Visual search. In H. Pashler (Ed.), *Attention* (pp. 13–73). London: University College London Press.

Yarbus, A. L. (1967). *Eye movements and vision*. New York: Plenum Press.

18
Attention and Driving

David L. Strayer

Historical Context and Overview

Most of us take driving for granted. But operating an automobile is the single riskiest activity that most readers of this chapter engage in on a regular basis. In fact, motor vehicle crashes are the leading cause of accidental injury deaths in the United States and are the leading cause of all deaths for people between the ages 1 to 33 and 56 to 71 (National Safety Council [NSC], 2010). Driving is a complex skill that takes years to master. Indeed, plots of fatal crash rates normalized by million miles driven steadily decline from novice/teen drivers until crash rates asymptote around 25 years of age (National Highway Traffic Safety Administration [NHTSA], 2009). However, even when crash rates are at asymptote, driving still represents a relatively risky activity. Driving also provides a unique opportunity to examine attention in action in an everyday context. Some aspects of driving can become relatively automatic, requiring little attention to be performed well. For example, skilled drivers can maintain lane position on predictable sections of the highway with little thought, and, intriguingly, paying more attention to these aspects of driving can sometimes degrade performance (Medeiros-Ward, Cooper, & Strayer, 2014). However, other aspects of driving, particularly those associated with reacting to unexpected or unpredictable events, require focused attention for successful performance.

Inattentive and distracted driving occurs when a driver fails to allocate sufficient attention to activities critical to safe driving (Regan, Hallett, & Gordon, 2011). In many circumstances, this involves diverting attention from driving to a concurrent nondriving activity. The degree to which driving is altered by a secondary task provides a road map for understanding the relationship between attention and driving (Strayer et al., 2013). Some of the "old standards" commonly paired with driving include talking to passengers, eating, drinking, lighting a cigarette, shaving, applying makeup, and listening to the radio (Stutts et al., 2003). However, the last decade has seen an explosion of new wireless technologies that have made their way into the automobile, enabling a host of new sources of driver distraction (e.g., sending and receiving e-mail or text messages, communicating via a cellular device, watching movies, using the Internet, etc.). It is likely that these newer sources are more distracting because they are more cognitively engaging and because they are often performed over more sustained periods of time (Strayer & Drews, 2007a).

In fact, driver distraction is increasingly recognized as a significant source of injuries and fatalities on the roadway. The NHTSA estimated that inattention accounted for 25% of all police-reported crashes (Ranney, Mazzae, Garrott, & Goodman, 2000; Wang, Knipling, & Goodman, 1996). Other estimates have suggested that inattention was a factor in as many as 35% to 50% of all crashes (Sussman, Bishop, Madnick, & Walker, 1985). More recently, data from the 100-car naturalistic driving study (Dingus et al., 2006) found that inattention was a factor in 78% of all crashes and near crashes, making it the single largest crash causation factor in their analysis.

Public concern over driver distraction has mirrored the advances in electronic technology in the automobile. For example, distracted driving became a cause for concern shortly after radios became a standard feature in many vehicles. In fact, in the early 1930s several states considered legislation banning listening to the radio while driving (none were enacted into law). More recently, the use of cellular phones by motorists became routine by the mid-1990s, and New York State enacted the first law prohibiting handheld cell-phone use in 2001. Currently, 14 states have enacted laws prohibiting a driver's use of a handheld cell phone, but all states allow the use of a hands-free device (Governors Highway Safety Association [GHSA], 2014). Similarly, the use of a cellular device to send and receive text messages became commonplace by the mid-2000s, and in 2007 Washington State became the first to pass a law banning the practice while driving. Currently, 44 states prohibit texting while operating a motor vehicle (GHSA, 2014).

The objective of this chapter is to explore the relationship between attention and driving, focusing primarily on situations where the performance of a concurrent nondriving activity adversely affects driving. We will begin by considering how divided attention impacts several mental activities that are important for establishing good awareness of the driving environment. The focus here will be on cognitive sources of distraction associated with the diversion of attention from driving rather than structural sources associated with taking the eyes off the road or the hands off the wheel (e.g., to send or receive a text message), which also adversely affect traffic safety. Next, we consider why performing one or more secondary activities while driving is disruptive, thereby helping to shed light on the architecture underlying attention and dual-task processing (both on the roadway and in the laboratory). Finally, we consider important new lines of inquiry that focus on individual differences in multitasking and also on the interrelationship between attention and automaticity in the context of driving.

State-of-the-Art Review

A driver's situation awareness reflects the dynamic mental model of the driving environment (e.g., Durso, Rawson, & Girotto, 2007; Endsley, 1995; Horrey, Wickens, & Consalus, 2006). Situation awareness in driving is dependent upon several psychological processes including *scanning* specific areas for indications of threats, *predicting* where threats might materialize if they are not visible, *identifying* threats and objects in the scenario when they occur, *deciding* whether an action is necessary and what action is necessary, and *executing* appropriate *responses—SPIDER* for short (Fisher & Strayer, 2014). SPIDER comprises an active set of psychological processes of which the quality of processing is often dependent upon limited capacity of attention (Kahneman, 1973). When drivers engage in secondary activities (e.g., talking on a cell phone), attention is often diverted from driving, thereby impairing performance on these SPIDER-related processes (Regan, Hallett, & Gordon, 2011; Regan & Strayer, 2014). Figure 18.1 presents the relationship between the allocation of limited-capacity attention, the SPIDER-related processes, and their bidirectional links to a driver's situation awareness. SPIDER is important for establishing and maintaining good situation awareness, which is important for coordinating and scheduling the SPIDER-relevant processes. In the following paragraphs, the effects of divided attention on driving performance are reviewed with a particular focus on these SPIDER-related processes.

Visual Scanning

A number of studies have demonstrated that as secondary-task workload increases, drivers tend to fixate more on objects immediately in front of their vehicle and less on dashboard instrumentation, side or rearview mirrors, and objects in the periphery (Cooper, Medeiros-Ward, & Strayer, 2013; Engström, Johansson, & Östlund, 2005; Harbluk, Noy, Trbovich, & Eizenman, 2007; He, Becic, Lee, & McCarley, 2011; Recarte & Nunes, 2000; Reimer, 2009; Reimer, Mehler, Wang, & Coughlin, 2012; Victor, Harbluk, & Engström, 2005; Tsai et al., 2007). This tendency for drivers engaged in a secondary, nonvisual task to concentrate their gaze on the center of the roadway may also influence lane position variation incurred by glances to peripheral objects. Indeed, the general finding is that drivers tend to steer in the direction of visual gaze and gaze in the direction they intend to steer (Readinger et al., 2002; Rogers, Kadar, & Costall, 2005; Wilson, Chattington, & Marple-Horvat, 2008; but see Cooper, Medeiros-Ward, & Strayer, 2013, for a different interpretation). In any event, diverting attention from driving leads to a pattern of gaze concentration directed toward the forward roadway that can have adverse effects on drivers' situation awareness because they fail to scan the periphery for potential threats (e.g., a pedestrian crossing the roadway).

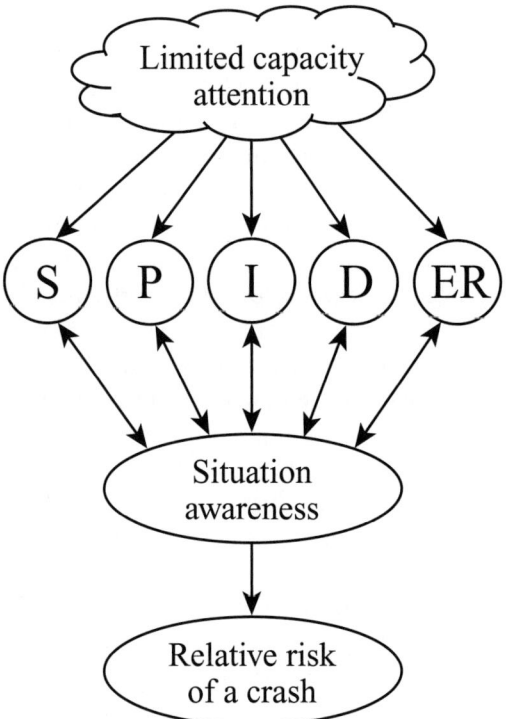

Figure 18.1
The SPIDER model of attention in driving. Spider is an acronym standing for scanning, predicting, identifying, deciding on, and executing a response. These processes are dependent on limited-capacity attention and inform a driver's situation awareness. Under distraction, the driver's situation awareness is impaired, and this increases the relative risk of a crash.

Predicting Hazards

Diverting attention from driving to perform secondary tasks that are not related to the safe operation of a motor vehicle can also degrade the anticipation and prediction of hazards in the driving scene. For example, Taylor et al. (2013) monitored participant's eye movements as they drove in a driving simulator in undivided and divided attention (i.e., concurrent hands-free cell-phone use) conditions. These authors were particularly interested in looking at anticipatory glances in locations where a potential hazard might present itself, for instance, a truck parked on the side of the street that obscured the drivers' view of a crosswalk where a pedestrian could appear. Experienced drivers who were not distracted by secondary tasks were more likely to glance toward the location of a potential hazard whereas drivers who were talking on a cell phone exhibited a 50% decline in anticipatory glances to safety-critical locations. Note that this sort of impairment is an example where expectancy-driven search of the driving environment is impaired. Interestingly, these anticipatory glances are often absent in novice drivers, who have not yet learned where potential hazards might occur (Underwood, Chapman, Bowden, & Crundall, 2002). In this respect, Taylor et al. (2013) suggest that dividing attention makes an experienced driver perform more like a novice driver. This pattern of impaired scanning for potential hazards has also been observed with drivers operating a motor vehicle on residential streets with pedestrian crosswalks and four-way intersections using a wide array of cognitive secondary tasks (Strayer et al., 2013).

The impaired visual scanning (the S in SPIDER) and the degraded prediction of hazards at safety-critical locations (the P in SPIDER) are examples of bottom-up and top-down effects of attention on driving, respectively. In the former case, drivers under secondary-task load tend to concentrate their gaze on the forward roadway and neglect salient objects in the periphery when they appear—a restriction of bottom-up visual attention. In the latter case, the diversion of attention from driving degrades expectancy-driven top-down processing, and this occurs even when there is no object in the driving environment. These findings document the critical role that visual attention plays in driving and how performing attention-demanding secondary tasks affects both bottom-up and top-down processing of the driving environment (e.g., Strayer et al., 2013). Other researchers have also distinguished between bottom-up salience and top-down pertinence as separate factors in visual attention (Bundesen, 1990; Logan & Gordon, 2001; Posner & Snyder, 1975; Schneider & Shiffrin, 1977; Shiffrin & Schneider, 1977), and the driving data indicate that both factors are impaired in dual-task conditions.

Identification

When drivers perform a secondary task that diverts attention from driving, identification of objects in the line of sight is often impaired, resulting in a form of *inattentional blindness* (Mack & Rock, 1998; Simons & Chabris, 1999; Strayer & Drews, 2007b). For example, the recognition memory for objects in the driving environment was reduced by 50% when the driver was talking on a hands-free cell phone (Strayer, Drews, & Johnston, 2003; Strayer & Johnston, 2001; see also McCarley et al., 2004). Notably, this study used an eye tracker to determine what drivers were looking at and only tested the memory for objects on which drivers fixated; nevertheless, the use of a cell phone halved what the driver noticed in the driving scene. Subsequent studies varied the safety relevance of objects in the driving scene (e.g., pedestrians, other moving vehicles, parked cars, billboards, etc.) to see if more safety-relevant objects were protected from inattentional blindness; however, there was no relationship between safety relevance and susceptibility to inattentional blindness (Strayer, Cooper, & Drews, 2004). While multitasking, drivers do not appear to strategically prioritize the processing of safety-critical information in the driving scene over the cell-phone conversation.

Further evidence of cell-phone-induced inattentional blindness comes from studies measuring the real-time brain activity associated with processing a lead vehicle's brake lights when participants were following that vehicle in a driving simulator. Compared to single-task conditions, Strayer and Drews (2007b; see also Strayer et al., 2013) found that the amplitude of the P300 component of the event-related brain potential, a measure sensitive to the allocation of attention to the task (e.g., Sirevaag, Kramer, Coles, & Donchin, 1989; Sirevaag et al., 1993), was reduced by 50% when the drivers were talking on a hands-free cell phone. Thus, drivers using a cell phone fail to see information in the driving scene because they do not encode it as well as they do when they are not distracted by a cell-phone conversation. It should be obvious that inattentional blindness to safety-critical objects in the driving environment can have an adverse effect on a driver's situation awareness.

Decision Making

Good decision making is important for safe driving. Drivers must evaluate several sources of information when deciding to change lanes in traffic or make a left turn in the presence of oncoming traffic. When drivers divert attention from driving, they often fail to fully evaluate the alternative sources of information. Cooper et al. (2009) examined the decision of drivers to changes lanes in low-, medium-, and high-density traffic. Compared to conditions of undivided attention, when drivers were talking on a hands-free cell phone, they were more likely to make unsafe lane changes. Similarly, Cooper and Zheng (2002) used a gap acceptance task to study drivers' decision making when making left-turn maneuvers in the presence of oncoming traffic. When drivers were listening and responding to verbal messages, their ability to process all the information necessary for making a safe turn was reduced. Drivers misjudged the gap size and the speed of oncoming vehicles, and this was most apparent on wet roadways. In both studies, divided attention led to unsafe decision making that increased the risk of a crash.

The Lane Change Task (ISO 26022, 2010) is an International Organization for Standardization (ISO) method for examining simple decision making in the context of driving under secondary-task load. In the Lane Change Task, participants control a rudimentary simulator on a three-lane section of road without traffic and they are instructed to change lanes when they recognize a sign pointing to the left lane (e.g., "↑XX"), the center lane (e.g., "X↑X"), or the right lane (e.g., "XX↑"). The Lane Change Task is exquisitely sensitive to secondary-task load (Mattes, 2003; Mattes & Hallén, 2009). For example, Mattes (2003) compared a variety of visual and cognitive secondary tasks that diverted attention from driving and found that lane change performance, as measured by deviations from a normative model, systematically decreased as the workload of the secondary task increased.

Execution of a Response

One of the hallmark signatures of divided attention is a slowing of reaction time to imperative events—in this context, when something in the driving environment requires a response (e.g., initiating a braking response to a child running across the street), those reactions are often delayed (for meta-analyses of brake reaction time under secondary-task load, see Caird, Willness, Steel, & Scialfa, 2008; Horrey & Wickens, 2006; Ishigami & Klein, 2009). The effect of secondary-task load on brake reaction time is magnified as the perceptual load in the driving environment increases, for example, as when the number of vehicles on the roadway increases (Strayer, Drews, & Johnston, 2003). Moreover, a secondary task, such as a cell-phone conversation, tends to positively skew the reaction time distributions, so that the late responses are particularly delayed (Ratcliff & Strayer, 2014). Brown, Lee, and McGehee (2001) found that sluggish brake reactions such as these can increase both the likelihood and the severity of a motor vehicle collision.

Because of the consistent sensitivity of reaction time measures to secondary-task load, a new effort is being undertaken to standardize a protocol for reaction time measurement while driving (ISO, 2012). In a go/no-go variant of the detection reaction time (DRT) task used by Strayer et al. (2013) to measure cognitive workload while driving an automobile in a residential neighborhood, a red/green LED light was positioned using a headband so that it had a fixed position relative to the head and did not interfere with the participant's central field of view. Both the color of the light and the interval between trials were randomized, and participants were instructed to respond to the green light as quickly as they could by depressing a microswitch that was attached to the participants' left thumb, but to not respond to the red light. The resulting reaction time and accuracy data provided a much more finely calibrated metric than the more traditional measures of brake reaction time and following distance (which often covary, making unambiguous interpretation difficult).

Situation Awareness

A driver's situation awareness reflects his or her mental model of the driving environment. It is informed and updated by the SPIDER-related processes and facilitates expectancy-based processing of the driving scene. Fisher and Strayer (2014) used an order-of-processing model to show how a small decrease in the likelihood that a driver successfully completes any one of the SPIDER-related activities cascades throughout the system to compromise a driver's situation awareness. In essence, a distracted driver is less situationally aware of the driving environment than a nondistracted driver. In this context, even small lapses in situation awareness can lead to poor performance

(Endsley, 1995). For example, a 5% decrease in the likelihood of any particular SPIDER activity's being completed successfully doubles the relative risk of a crash (Fisher & Strayer, 2014). This modeling effort is informative because it is often counterintuitive how small changes in the component processes can translate into large impairments in driving performance. Further refinement of the modeling is still needed to fully understand the temporal dynamics and contingencies of the SPIDER-related processes under secondary-task load.

Two additional factors are important to consider in discussions concerning driver distraction and traffic safety. The first is the duration of a secondary task that is concurrently performed while driving. Drivers may attempt to strategically perform secondary tasks when they perceive the demands of driving to be low (e.g., while stopped at a traffic light). However, as the duration of an interaction increases, the ability of a driver to accurately predict the driving demands decreases. For example, changing a radio station places demands on visual attention, but the duration of the activity is relatively short (e.g., a few seconds). By contrast, a cell-phone conversation may extend for several minutes, and the conditions that were in effect at the initiation of a call may change considerably over this interval. In general, secondary tasks that compete for attention for longer intervals will have greater impact on traffic safety than activities with shorter durations.

The second factor is the exposure rate of an activity. The more that drivers engage in a distracting activity, the greater the impact to public safety. For example, the risk of being in a motor vehicle accident increases by a factor of four when drivers are talking on a cell phone (e.g., McEvoy et al., 2005; Redelmeier & Tibshirani, 1997). What compounds the risk to public safety is that at any daylight hour it is estimated that 10% of drivers on U.S. roadways are talking on their cell phone (Glassbrenner, 2005). The effect on traffic safety of a performing a secondary task while driving is determined by the product of the distraction caused by that activity and the exposure rate of the activity (NSC, 2010).

Integration

Driver distraction has received considerable attention for the role it plays in injuries and crashes on the roadway. However, from a theoretical perspective, it is also important to consider *why* diverting attention from driving degrades performance. Situations where drivers keep their eyes on the road and their hands on the wheel but divert attention from the road to perform a secondary task that is not associated with operating the vehicle provide a relatively pure case of cognitive distraction (Strayer et al., 2013). As reviewed above, the SPIDER-related processes associated with the safe operation of a motor vehicle are impaired when attention is diverted from driving to perform a secondary task. Bergen et al. (2013) describes two broad classes of interpretation that have been proposed to account for this type of dual-task interference.

Domain-general models of attention argue that people have limited computational or attentional resources to distribute across the various tasks that they are engaged in, and when two tasks require more of these resources at the same time than are available, performance on one or both tasks is impaired (Kahneman, 1973; Navon & Gopher, 1979). That is, domain-general accounts implicate a common resource from which the various activities compete. Furthermore, it is assumed that there is a monotonic relationship between the allocation of attention and performance on resource-limited activities. Figure 18.1 provides an example of this sort of model of attention as it relates to driving performance.

By contrast, *domain-specific* models of attention argue that interference is caused by an overlap in specific mental operations of one sort or another. One example from the domain-specific class of models is multiple resource theory (Wickens, 1980, 1984), which posits that there are separate pools of processing resources for visual and auditory inputs, separate resources for verbal and spatial codes of processing, and separate pools for verbal and manual responses. Another domain-specific account is couched in terms of code conflict (Navon & Miller, 1987); if two tasks use similar representational codes that need to be used at the same time, then they can produce "cross talk" and thereby impair performance on one or both tasks (Pashler, 1994).

It would appear that domain-general models provide a better account of the pattern of interference caused by an auditory–verbal–vocal cell-phone conversation on a visual–spatial–manual driving task. On the surface, these two tasks would seem to have very little overlap in processing; hence there should be little cross talk between them. However, it is noteworthy that many of the cognitively demanding secondary tasks that have been studied involve speech-based interactions, and there is a general tendency for interference to increase as speech production increases. For example, listening to the radio or an audiobook produces much less interference than talking on either a handheld or hands-free cell phone (Strayer et al., 2013). This pattern is troubling given the recent push for new vehicles to use voice-based commands to control the climate and infotainment systems in the vehicle. While language can be a powerful interface for these applications, it comes at a cost of increased cognitive load that can impair driving.

Bergen et al. (2013) suggested that language-mediated interactions performed in the course of driving can cause domain-specific interference because of an overlap in the mental hardware required to process language and to drive. Drivers were asked to verify sentences that involved visual imagery (e.g., "The letters on a stop sign are white"), motor imagery (e.g., "To open a jar, you turn the lid counterclockwise"), or abstract sentences (e.g., "The capital of North Dakota is Bismarck"). Brake reaction time was equally delayed by all sentences, in support of domain-general interpretations of dual-task interference. However, following distance increased for motor imagery and visual imagery sentences, but not for abstract sentences, providing evidence for domain-specific dual-task interference. Bergen et al. (2013) argued that language use interferes with driving when processing the linguistic content overlaps with the cognitive operations used to control the vehicle.

The preceding documents the fact that cell-phone conversations impair driving. But what about other conversations engaged in while driving? In particular, do in-vehicle conversations with a passenger impair driving to the same extent as cell-phone conversations? One way to examine this issue is to compare the crash risk while conversing on a cell phone (established as a four-fold increase, Redelmeier & Tibshirani, 1997; McEvoy et al., 2005) with the crash risk when there is another adult in the vehicle. Epidemiological evidence (Rueda-Domingo et al., 2004; Vollrath, Meilinger, & Kruger, 2002) indicates that the crash rate drops below 1.0 when there is an adult passenger in the vehicle (i.e., there is a slight safety advantage for having another adult passenger in the vehicle). Given that in many instances the passenger and the driver are conversing, these findings would seem to be at odds with the suggestion that any conversation task diverts attention from driving.

Drews, Pasupathi, and Strayer (2008; see also Consiglio, Driscoll, Witte, & Berg, 2003; Crundall, Bains, Chapman, & Underwood, 2005; and Hunton & Rose, 2005) provided a direct comparison of passenger and cell-phone conversations by having pairs of participants serve as the driver or as an interlocutor (1) on a cell phone or (2) as a passenger seated next to the driver in a driving simulator. The driver's task was to exit the highway at the rest stop and park the vehicle. Passenger conversations did not significantly degrade navigation performance whereas half of the drivers engaged in a cell-phone conversation failed to take their assigned exit. Video analysis revealed that the passenger often actively engaged in supporting the driver by pointing out hazards, helping to navigate, and reminding the driver of the task (i.e., exiting at the rest stop). In other cases, the conversation was temporally halted during a difficult section of driving and then resumed when driving became easier. These real-time adjustments to the conversation based on the demands of driving were not evident in cell-phone conversations. In effect, the passenger acted as another set of eyes that helped the driver control the vehicle, and this sort of activity is not afforded by cell-phone conversations.

Future Directions

Individual Differences

There is a growing literature on individual differences in attention and cognitive control that is important in the context of driving. For example, Watson and Strayer (2010) asked 200 participants to perform the Operation Span task (Kane & Engle, 2002, 2003) and a driving simulator task separately and then perform the two tasks concurrently.

Consistent with the SPIDER-related impairments described above, about 98% of the participants showed substantial declines in performance on both tasks in dual-task conditions compared to single-task baseline levels. However, 2% of the participants showed no measurable decline in dual-task performance. Subsequent neuroimaging of these "supertaskers" found a pattern of neural efficiency in regions of prefrontal cortex that have been associated with multitasking performance in other contexts (Medeiros-Ward, Watson, & Strayer, 2015). Interestingly, there is a negative relationship between individual differences in working memory capacity, as measured by the Operation Span task, and the propensity to multitask while driving in the real world (Sanbonmatsu, Strayer, Medeiros-Ward, & Watson, 2013). This relationship is disquieting because there is a positive correlation between measures of working memory capacity and driving performance (Watson et al., 2013). It will be important for future research to better explore the relationship between individual differences in attention and cognitive control in complex real-world environments such as driving.

Mind Wandering

A common experience often reported by drivers is that they suddenly "come to" and realize that they have no recollection of their recent driving experience. This sort of mind-wandering episode represents an interesting situation where attention is directed at internal thought rather than at an external source of distraction or directed toward the task of driving. Mind-wandering episodes tend to be associated with boredom or with drivers' being familiar with a route (Yanko & Spalek, 2013). In fact, epidemiological studies have found an association between mind-wandering episodes and an increased crash risk (Galera et al., 2012). Driving simulator studies of mind wandering using either a self-caught procedure (He et al., 2011) or a probe-caught procedure (Yanko & Spalek, 2014) have found that brake reaction to imperative events is slowed when drivers are mind wandering. Interestingly, unlike distraction from external sources (e.g., a cell-phone conversation) which are associated with an *increase* in headway, when participants were mind wandering, their headway *decreased*. Yanko and Spalek (2014) speculated that this might reflect distinct attentional networks: the former associated with the frontal-attentional network and the latter associated with the default network.

So why don't drivers immediately run off the road when they are mind wandering? Medeiros-Ward, Cooper, and Strayer (2014) found that some aspects of driving become sufficiently automatic with practice that they can be performed without attention. In fact, when participants were driving on a predictable/familiar section of roadway, diverting attention from steering actually improved lane maintenance whereas attending to steering degraded lane maintenance. These authors suggested that complex skilled behaviors as diverse as playing a musical instrument (Shaffer, 1976), typing (Logan & Crump, 2011), and driving an automobile are supported by a hierarchical control network that coordinates the interaction between automatic encapsulated routines

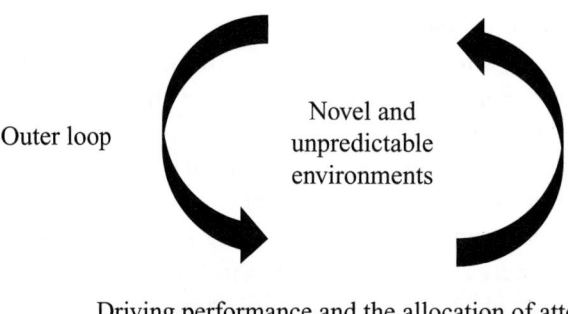

Driving performance and the allocation of attention

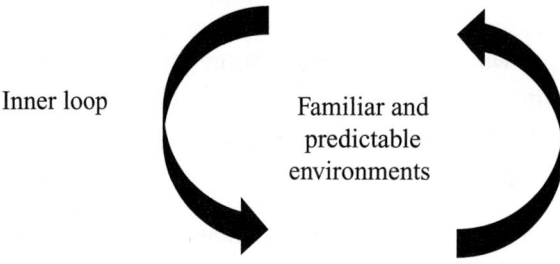

Figure 18.2
Hierarchical control and driving. Performance based on the outer attention-demanding loop gets better with *more* attention allocated to the task and gets worse with *less* attention allocated to the task. By contrast, performance based on the encapsulated inner loop gets better with *less* attention allocated to the task and gets worse with *more* attention allocated to the task (Medeiros-Ward, Cooper, & Strayer, 2014; see also Logan & Crump, 2011).

and limited-capacity attention. Figure 18.2 provides an illustration of the hierarchical relationship between outer and inner loops of control in driving. For example, unpredictable gusts of wind can make the task of maintaining lane position dependent on outer loop processing. With unpredictable wind gusts, the diversion of attention from outer loop processing disrupts driving performance. However, in predictable driving environments (e.g., a well-marked straight road without wind gusts), the diversion of attention from an encapsulated inner loop process actually improves lane maintenance (Medeiros-Ward, Cooper, & Strayer, 2014; Salvucci, 2006; Salvucci & Beltowska, 2008). Importantly, even though diversion of attention may improve lane maintenance, drivers may very well miss their exit if they divert attention from driving (e.g., Drews, Pasupathi, & Strayer, 2008).

Self-Regulation
Finally, with regard to the use of secondary in-vehicle tasks (e.g., talking on a cell phone), experimental studies using driving simulators or instrumented vehicles (Caird, et al., 2008; Horrey & Wickens, 2006; Strayer et al., 2013; see also the epidemiological

studies by Redelmeier & Tibshirani, 1997; McEvoy et al., 2005) have produced strikingly different estimates of driving impairment and crash risk than the correlation-based naturalistic studies of driving (Dingus et al., 2006; Klauer et al., 2014). Although there are several potential reasons for the discrepant results, one untested hypothesis is that it stems from a driver's self-regulation of the secondary-task activities based on driving demand.

Following Braver, Gray, and Burgess (2007), it is suggested that there are two forms of self-regulation in driving: *proactive* and *reactive*. An example of proactive self-regulation is when a driver decides in advance not to use a cell phone when he or she is operating a motor vehicle. An example of reactive self-regulation is when a driver moderates his or her usage in *real time* based upon driving difficulty or perception of driving errors. Reactive self-regulation may also involve trading off different aspects of driving performance when multitasking. For example, a driver may slow down when he or she is talking on a cell phone, and this change in behavior may be a manifestation of self-regulation. However, at this point the reactive self-regulation interpretation can be criticized for circular logic in that there is often no direct link establishing when a behavior is a form of adaptive self-regulation or, instead, a by-product of the diversion of attention from driving. Future research is needed to address this important issue.

Conclusions

Driving is a complex skill that is dependent upon limited-capacity attention. The SPIDER framework identifies several cognitive processes that are impaired when drivers divert attention from driving. There are obvious traffic-safety concerns associated with driver distraction. The patterns of interference in this real-world context may also help illuminate the mechanisms of dual-task performance. Future research is needed to more fully explicate the role that individual differences, personality factors, and the ability to self-regulate play in this context.

Box 18.1
Key Points

- Driving is a complex skill. With experience, some aspects of driving can become relatively automatic, requiring little attention for successful performance. Other aspects of driving depend on attention to be performed well.
- Good situation awareness of the driving environment depends on scanning, predicting, identifying, deciding on, and executing responses (SPIDER) that rely on limited-capacity attention. Diverting attention from driving results in an impoverished mental model of the driving environment, increasing the relative risk of a crash.

Box 18.2

Outstanding Issues

- Future research is needed to better understand the role that individual differences play in self-regulatory behavior while driving.
- The relationship between cognitive distraction and crash risk needs to be quantified.
- The utility of hierarchical control theory in understanding other complex real-world behaviors should be explored.

References

Bergen, B., Medeiros-Ward, N., Wheeler, K., Drews, F., & Strayer, D. L. (2013). The crosstalk hypothesis: Language interferes with driving because of modality-specific mental simulation. *Journal of Experimental Psychology. General, 142*, 119–130.

Braver, T., Gray, J., & Burgess, G. (2007). Explaining the many varieties of working memory variation: Dual mechanisms of cognitive control. In A. R. A. Conway, C. Jarrold, M. J. Kane, A. Miyake, & J. N. Towse (Eds.), *Variation in working memory* (pp. 76–106). New York: Oxford University Press.

Brown, T. L., Lee, J. D., & McGehee, D. V. (2001). Human performance models and rear-end collision avoidance algorithms. *Human Factors, 43*, 462–482.

Bundesen, C. (1990). A theory of visual attention. *Psychological Review, 97*, 523–547.

Caird, J. K., Willness, C. R., Steel, P., & Scialfa, C. (2008). A meta-analysis of the effects of cell phones on driver performance. *Accident Analysis & Prevention, 40*, 1282–1293.

Consiglio, W., Driscoll, P., Witte, M., & Berg, W. P. (2003). Effect of cellular telephone conversations and other potential interference on reaction time in a braking response. *Accident Analysis & Prevention, 35*, 494–500.

Cooper, J. M., Medeiros-Ward, N., & Strayer, D. L. (2013). The impact of eye movements and cognitive workload on lateral position variability in driving. *Human Factors, 55*, 1001–1014.

Cooper, J. M., Vladisavljevic, I., Medeiros-Ward, N., Martin, P. T., & Strayer, D. L. (2009). Near the tipping point of traffic stability: An investigation of driving while conversing on a cell phone in simulated highway traffic of varying densities. *Human Factors, 51*, 261–268.

Cooper, P. J., & Zheng, Y. (2002). Turning gap acceptance decision-making: The impact of driver distraction. *Journal of Safety Research, 33*, 321–335.

Crundall, D., Bains, M., Chapman, P., & Underwood, G. (2005). Regulating conversation during driving: A problem for mobile telephones? *Transportation Research Part F: Traffic Psychology and Behaviour, 8*, 197–211.

Dingus, T. A., Klauer, S. G., Neale, V. L., Petersen, A., Lee, S. E., Sudweeks, J., et al. (2006). *The 100-car naturalistic driving study: Phase II—Results of the 100-car field experiment* (DOT Report No. HS 810 593). Washington, DC: U.S. Department of Transportation.

Drews, F. A., Pasupathi, M., & Strayer, D. L. (2008). Passenger and cell-phone conversation during simulated driving. *Journal of Experimental Psychology. Applied, 14*, 392–400.

Durso, F., Rawson, K., & Girotto, S. (2007). Comprehension and situation awareness. In F. T. Durso, R. Nickerson, S. T. Dumais, S. Lewandowsky, & T. Perfect (Eds.), *The handbook of applied cognition* (2nd ed., pp 163–164). Chichester, UK: Wiley.

Endsley, M. R. (1995). Towards a theory of situation awareness in dynamic systems. *Human Factors, 37*, 32–64.

Engström, J., Johansson, E., & Östlund, J. (2005). Effects of visual and cognitive load in real and simulated motorway driving. *Transportation Research Part F: Traffic Psychology and Behaviour, 8*, 97–120.

Fisher, D. L., & Strayer, D. L. (2014). Modeling situation awareness and crash risk. *Annals of Advances in Automotive Medicine, 5*, 33–39.

Galera, C., Orriols, L., M'Bailera, K., Laborey, M., Contrand, B., Ribereau-Gayon, R., et al. (2012). Mind wandering and driving: Responsibility case–control study. *British Medical Journal, 345*, e8105.

Governors Highway Safety Association. (2014). Distracted driving laws. Downloaded July 28, 2014, from http://www.ghsa.org/html/stateinfo/laws/cellphone_laws.html.

Glassbrenner, D. (2005). *Traffic safety facts research note: Driver cell phone use in 2005—Overall results* (Report No. DOT HS 809 967). Washington, DC: National Center for Statistics and Analysis, National Highway Traffic Safety Administration.

Harbluk, J. L., Noy, Y., Trbovich, P. L., & Eizenman, M. (2007). An on-road assessment of cognitive distraction: Impacts on drivers' visual behavior and braking performance. *Accident Analysis & Prevention, 39*, 372–379.

He, J., Becic, E., Lee, Y., & McCarley, J. S. (2011). Mind wandering behind the wheel: Performance and oculomotor correlates. *Human Factors, 53*, 13–21.

Horrey, W. J., & Wickens, C. D. (2006). Examining the impact of cell phone conversations on driving using meta-analytic techniques. *Human Factors, 48*, 196–205.

Horrey, W. J., Wickens, C. D., & Consalus, K. P. (2006). Modeling drivers' visual attention allocation while interacting with in-vehicle technologies. *Journal of Experimental Psychology. Applied, 12*, 67–78.

Hunton, J., & Rose, J. M. (2005). Cellular telephones and driving performance: The effects of attentional demands on motor vehicle crash risk. *Risk Analysis, 25*, 855–866.

Ishigami, Y., & Klein, R. M. (2009). Is a hands-free phone safer than a handheld phone? *Journal of Safety Research, 40*, 157–164.

International Organization for Standardization. (2010). Road vehicles—Ergonomic aspects of transport information and control systems—Simulated Lane Change Test to assess in-vehicle secondary task demand [ISO 26022:2010].

International Organization for Standardization. (2012). Road vehicles—Transport information and control systems—Detection-Response Task (DRT) for assessing selective attention in driving [ISO TC 22 SC 13 N17488, Working draft; under development by Working Group 8 of ISO TC22, SC 13].

Kahneman, D. (1973). *Attention and effort*. Englewood Cliffs, NJ: Prentice-Hall.

Kane, M., & Engle, R. (2002). The role of prefrontal cortex in working-memory capacity, executive attention, and general fluid intelligence: An individual-differences perspective. *Psychonomic Bulletin & Review, 9*, 637–671.

Kane, M., & Engle, R. (2003). Working memory capacity and the control of attention: The contributions of goal neglect, response competition, and task set to Stroop interference. *Journal of Experimental Psychology. General, 132*, 47–70.

Klauer, S. G., Guo, F., Simons-Morton, B. G., Ouiment, M. C., Lee, S. E., & Dingus, T. A. (2014). Distracted driving and risk of road crashes among novice and experienced drivers. *New England Journal of Medicine, 370*, 54–59.

Logan, G. D., & Crump, M. J. C. (2011). Hierarchical control of cognitive processes. In B. Ross (Ed.), *The psychology of learning and motivation* (Vol. 54, pp. 1–27). Burlington, MA: Academic Press.

Logan, G. D., & Gordon, R. D. (2001). Executive control of visual attention in dual-task situations. *Psychological Review, 108*, 393–434.

Mack, A., & Rock, I. (1998). *Inattentional blindness*. Cambridge, MA: MIT Press.

Mattes, S. (2003). The Lane Change Task as a tool for driver distraction evaluation. In H. Strasser, H. Rascher, & H. Bubb (Eds.), *Quality of work and products in enterprises of the future* (pp. 57–60). Stuttgart, Germany: Ergonomia Verlag.

Mattes, S., & Hallén, A. (2009). Surrogate distraction measurement techniques: The Lane Change Test. In M. Regan, J. Lee, & K. Young (Eds.), *Driver distraction* (pp. 107–122). Boca Raton, FL: CRC Press.

McCarley, J. S., Vais, M., Pringle, H., Kramer, A. F., Irwin, D. E., & Strayer, D. L. (2004). Conversation disrupts scanning and change detection in complex visual scenes. *Human Factors, 46*, 424–436.

McEvoy, S. P., Stevenson, M. R., McCartt, A. T., Woodward, M., Haworth, C., Palamara, P., et al. (2005). Role of mobile phones in motor vehicle crashes resulting in hospital attendance: A case-crossover study. *British Medical Journal, 331*, 428–433.

Medeiros-Ward, N., Cooper, J. M., & Strayer, D. L. (2014). Hierarchical control and driving. *Journal of Experimental Psychology. General, 143*, 953–958.

Medeiros-Ward, N., Watson, J. M., & Strayer, D. L. (2015). On supertaskers and the neural basis of extraordinary multitasking. *Psychonomic Bulletin & Review, 22*, 876–883.

National Highway Traffic Safety Administration. (2009). *Traffic safety facts research note: Fatal crashes involving young drivers.* Retrieved from http://www-nrd.nhtsa.dot.gov/Pubs/811218.PDF

National Safety Council. (2010). Understanding the distracted brain: Why driving while using hands-free cell phones is risky behavior. Retrieved from http://www.nsc.org/DistractedDrivingDocuments/Cognitive-Distraction-White-Paper.pdf.

Navon, D., & Gopher, D. (1979). On the economy of the human-processing system. *Psychological Review, 86*, 214–255.

Navon, D., & Miller, J. (1987). Role of outcome conflict in dual-task interference. *Journal of Experimental Psychology: Human Perception and Performance, 13*, 435–448.

Pashler, H. (1994). Dual-task interference in simple tasks: Data and theory. *Psychological Bulletin, 116*, 220–244.

Posner, M. I., & Snyder, C. R. R. (1975). Attention and cognitive control. In R. L. Solso (Ed.), *Information processing and cognition: The Loyola symposium* (pp. 55–85). Hillsdale, NJ: Erlbaum.

Ranney, T., Mazzae, E., Garrott, R., & Goodman, M. (2000). NHTSA driver distraction research: Past, present and future. Retrieved from http://www-nrd.nhtsa.dot.gov/departments/Human%20Factors/driver-distraction/PDF/233.pdf.

Ratcliff, R., & Strayer, D. L. (2014). Modeling simulated driving with a one-boundary diffusion model. *Psychonomic Bulletin & Review, 21*, 577–589.

Readinger, W. O., Chatziastros, A., Cunningham, D. W., Bülthoff, H. H., & Cutting, J. E. (2002). Gaze-eccentricity effects on road position and steering. *Journal of Experimental Psychology. Applied, 8*, 247–258.

Recarte, M. A., & Nunes, L. M. (2000). Effects of verbal and spatial-imagery tasks on eye fixations while driving. *Journal of Experimental Psychology. Applied, 6*, 31–43.

Redelmeier, D. A., & Tibshirani, R. J. (1997). Association between cellular-telephone calls and motor vehicle collisions. *New England Journal of Medicine, 336*, 453–458.

Regan, M. A., Hallett, C., & Gordon, C. P. (2011). Driver distraction and driver inattention: Definition, relationship and taxonomy. *Accident Analysis & Prevention, 43*, 1771–1781.

Regan, M. A., & Strayer, D. L. (2014). Towards an understanding of driver inattention: Taxonomy and theory. *Annals of Advances in Automotive Medicine, 58*, 5–13.

Reimer, B. (2009). Impact of cognitive task complexity on drivers' visual tunneling. *Transportation Research Record: Journal of the Transportation Research Board, 2138*, 13–19.

Reimer, B., Mehler, B., Wang, Y., & Coughlin, J. F. (2012). A field study on the impact of variations in short-term memory demands on drivers' visual attention and driving performance across three age groups. *Human Factors, 54*, 454–468.

Rueda-Domingo, T., Lardelli-Claret, P., Luna-del-Castillo, J. de D., Jimenez-Moleon, J. J., García-Martın, M., & Bueno-Cavanillas, A. (2004). The influence of passengers on the risk of the driver causing a car collision in Spain: Analysis of collisions from 1990 to 1999. *Accident Analysis & Prevention, 36,* 481–489.

Rogers, S. D., Kadar, E. E., & Costall, A. (2005). Gaze patterns in the visual control of straight-road driving and braking as a function of speed and expertise. *Ecological Psychology, 17,* 19–38.

Salvucci, D. D. (2006). Modeling driver behavior in a cognitive architecture. *Human Factors, 48,* 362–380.

Salvucci, D. D., & Beltowska, J. (2008). Effects of memory rehearsal on driver performance: Experiment and theoretical account. *Human Factors, 50,* 834–844.

Sanbonmatsu, D. M., Strayer, D. L, Medeiros-Ward, H., & Watson, J. M. (2013). Who multi-tasks and why? Multi-tasking ability, perceived multi-tasking ability, impulsivity, and sensation seeking, *PLoS ONE, 8*(1), e54402.

Schneider, W., & Shiffrin, R. M. (1977). Controlled and automatic human information processing: I. Detection, search, and attention. *Psychological Review, 84,* 1–66.

Shaffer, L. H. (1976). Intention and performance. *Psychological Review, 83,* 375–393.

Shiffrin, R. M., & Schneider, W. (1977). Controlled and automatic human information processing: II. Perceptual learning, automatic attending, and a general theory. *Psychological Review, 84,* 127–190.

Simons, D. J., & Chabris, C. F. (1999). Gorillas in our midst: Sustained inattentional blindness for dynamic events. *Perception, 28,* 1059–1074.

Sirevaag, E. J., Kramer, A. F., Coles, M. G., & Donchin, E. (1989). Resource reciprocity: An event-related brain potential analysis. *Acta Psychologica, 70,* 77–97.

Sirevaag, E. J., Kramer, A. F., Wickens, C. D., Reisweber, M., Strayer, D. L., & Grenell, J. H. (1993). Assessment of pilot performance and mental workload in rotary wing aircraft. *Ergonomics, 9,* 1121–1140.

Strayer, D. L., Cooper, J. M., & Drews, F. A. (2004). What do drivers fail to see when conversing on a cell phone? In *Proceedings of the 48th Annual Meeting of the Human Factors and Ergonomics Society* (pp. 2213–2217).

Strayer, D. L., Cooper, J. M., Turrill, J., Coleman, J., Medeiros-Ward, N., & Biondi, F. (2013). *Measuring cognitive distraction in the automobile.* Washington, DC: AAA Foundation for Traffic Safety.

Strayer, D. L., & Drews, F. A. (2007a). Multi-tasking in the automobile. In A. Kramer, D. Wiegmann, & A. Kirlik (Eds.), *Attention: From theory to practice* (pp. 121–133). Oxford, UK: Oxford University Press.

Strayer, D. L., & Drews, F. A. (2007b). Cell-phone induced inattention blindness. *Current Directions in Psychological Science, 16,* 128–131.

Strayer, D. L., Drews, F. A., & Johnston, W. A. (2003). Cell phone induced failures of visual attention during simulated driving. *Journal of Experimental Psychology. Applied, 9,* 23–52.

Strayer, D. L., & Johnston, W. A. (2001). Driven to distraction: Dual-task studies of simulated driving and conversing on a cellular phone. *Psychological Science, 12,* 462–466.

Stutts, J., Feaganes, J., Rodman, E., Hamlet, C., Meadows, T., Rinfurt, D., et al. (2003). *Distractions in everyday driving.* Washington, DC: AAA Foundation for Traffic Safety.

Sussman, E. D., Bishop, H., Madnick, B., & Walker, R. (1985). Driver inattention and highway safety. *Transportation Research Record, 1047,* 40–48.

Taylor, T., Pradhan, A. K., Divekar, G., Romoser, M., Muttart, J., Gomes, R., et al. (2013). The view from the road: The contribution of on-road glance-monitoring technologies to understanding driver behavior. *Accident Analysis & Prevention, 58,* 175–186.

Tsai, Y., Viirre, E., Strychacz, C., Chase, B., & Jung, T. (2007). Task performance and eye activity: Predicting behavior relating to cognitive workload. *Aviation, Space, and Environmental Medicine, 5,* B176–B185.

Underwood, G., Chapman, P., Bowden, K., & Crundall, D. (2002). Visual search while driving: Skill and awareness during inspection of the scene. *Transportation Research Part F: Traffic Psychology and Behaviour, 5,* 87–97.

Victor, T. W., Harbluk, J. L., & Engström, J. A. (2005). Sensitivity of eye-movement measures to in-vehicle task difficulty. *Transportation Research Part F: Traffic Psychology and Behaviour, 8,* 167–190.

Vollrath, M., Meilinger, T., & Kruger, H. P. (2002). How the presence of passengers influences the risk of a collision with another vehicle. *Accident Analysis & Prevention, 34,* 649–654.

Wang, J. S., Knipling, R. R., & Goodman, M. J. (1996). The role of driver inattention in crashes: New statistics from the 1995 Crashworthiness Data System. *40th Annual Proceedings of the Association for the Advancement of Automotive Medicine* (pp. 377–392). Vancouver, Canada.

Watson, J. M., Lambert, A. E., Cooper, J. M., Boyle, I. V., & Strayer, D. L. (2013). On attentional control and the aging driver. In M. Garner, R. Zheng, & R. Hill (Eds.), *Engaging older adults with modern technology: Internet use and information access needs* (pp. 20–32). Hershey, PA: IGI Global.

Watson, J. M., & Strayer, D. L. (2010). Supertaskers: Profiles in extraordinary multi-tasking ability. *Psychonomic Bulletin & Review, 17,* 479–485.

Wickens, C. D. (1980). The structure of attentional resources. In R. S. Nickerson (Ed.), *Attention and performance VIII* (pp. 239–257). Hillsdale, NJ: Erlbaum.

Wickens, C. D. (1984). Processing resources in attention. In R. Parasuraman & R. Davies (Eds.), *Varieties of attention* (pp. 63–101). New York: Academic Press.

Wilson, M., Chattington, M., & Marple-Horvat, D. E. (2008). Eye movements drive steering: Reduced eye movement distribution impairs steering and driving performance. *Journal of Motor Behavior, 40,* 190–202.

Yanko, M. R., & Spalek, T. M. (2013). Rote familiarity breeds inattention: A driving simulator study. *Accident Analysis & Prevention, 57,* 80–86.

Yanko, M. R., & Spalek, T. M. (2014). Driving with the wandering mind: The effect that mind-wandering has on driving performance. *Human Factors, 56,* 260–269.

19

Serial versus Concurrent Multitasking: From Lab to Life

Paul W. Burgess

"Multitasking" is a very recent term. Its first use is widely attributed to Witt and Ward (1965) in an IBM technical manual describing a computer operating system. However, its application in describing human behavior is much more recent, and a key issue is emerging. This is that "multitasking" is being used to describe (at least) two situations that might be distinct from each other in psychological terms.

The first situation is where people are actively trying to do two or more tasks at the same time. Examples might be driving while making a phone call or reading while simultaneously watching a film. Many recent examples are also given in the literature in relation to "media multitasking," that is, attempting to attend simultaneously to stimuli from multiple media sources (e.g., Ophir et al., 2009; see also chapter 23 in the current volume).

The second is where people have a "to-do" list of jobs or tasks in their minds and are trying to optimize their behavior in order to achieve the best outcome in relation to them. An example might be cooking a complex meal while also preparing the house for guests and dealing with the demands of children.

When people in everyday life describe themselves as "multitasking," they may be using the term to refer to either of these situations almost interchangeably. And indeed, organized everyday life when seen over substantial time periods can often include examples of both, sometimes with the first form embedded within the second in complex ways. In both situations, there is an intention to perform more than one task, typically within a certain time frame. However, in the second situation there is no attempt to perform more than one task *at the same time*. Following Salvucci and Taatgen (2008), the term *"concurrent multitasking"* will be used to describe the first situation. For the second situation, this chapter will introduce the term *"serial multitasking."* This paper will argue that there are significant differences between these two forms of multitasking, and that recent evidence from cognitive neuroscience may give some clues as to fruitful ways of pursuing these differences.

Historical Context

Although the term "multitasking" is a recent invention, it would be wrong to conclude that research relevant to human multitasking was not carried out before the invention of this term. In other words, it is the term "multitasking" that is new, rather than research into the corresponding mental phenomena.

Until recently, dealing with the demands of multiple tasks was known as "multiple-task performance." This was, in practice, most often the study of dual-task type paradigms where more than two tasks were used. In other words, it was the study of *concurrent multitasking*. The first book (to my knowledge) on that topic was Diane Damos's *Multiple-Task Performance*, published less than 25 years ago (Damos, 1991). In that book, Brookings and Damos (1991) point out that research into the nascent field of individual differences in multiple-task performance dates back more than 100 years, to, for example, the dual-tasking studies of Binet (1890) and Sharp (1899). They made the point that previously the most obviously relevant constructs or behavioral descriptions under examination were neither "multitasking" nor "multiple-task performance" but were theoretical abilities described variously as "power of distribution of attention," "division of attention," and "timesharing." This last term is especially important since it is prima facie a behavioral description rather than a construct one and so may well refer either to concurrent or serial multitasking. Ackerman, Schneider, and Wickens (1984) provided a definition: "Time-sharing ability refers to the ability to perform multiple tasks in combination" (p. 71). Thus, in practice, "time-sharing" was synonymous with the term "concurrent multitasking" as it is defined today. It is the study of this behavior (concurrent multitasking) which has been the overwhelming focus for studies and theorizing in the cognate sciences, in particular experimental psychology, and more recently, computational modeling (see, e.g., Kieras et al., 2000). There have been various interesting demonstrations from these experiments that may have relevance for understanding everyday life performance. For instance, Schumacher et al. (2001) have argued that after only moderate practice, some people can manage almost perfect time-sharing between two basic choice reaction tasks, and Ophir et al. (2009) demonstrated that heavy "media multitaskers" in everyday life performed more poorly on a test of task-switching ability, arguing that this is likely due to a reduced ability to filter out interference from irrelevant material.

Historically, however, a recurring theme for the study of multitasking has concerned less the determinants of performance than whether there is a specific ability that is tapped by multitasking situations beyond that required for performance of the individual tasks themselves. From the earliest days this was bound up with notions of "intelligence." One of the earliest views was expounded by McQueen (1917), which is a publication of his D.Sc. thesis from 1915 at the University of London, United Kingdom. The topic of the research was attributed to Charles Spearman, who was the laboratory director, and provided guidance for McQueen's research until rejoining the Army on the outbreak of war. This link with Spearman may have been influential in the initial hypothesis for the research. McQueen wrote as follows:

We certainly believe … that the assumption is without justification that a single test … affords a reliable measure of an individual's "power of distribution" of attention…. We shall show … evidence indicating that there is no such general power as a power to distribute the attention successfully. (p. 4)

Spearman was of course the principal proponent of the notion of the existence in the mind of a general intelligence factor (or "g"). The main evidence for Spearman's hypothesis was the repeated finding of "positive manifold" (i.e., more correlations that were positive than would be expected by chance) in correlation matrices of large numbers of people performing various disparate tasks. Perhaps it would have been awkward—or at least a complication—for the g hypothesis if there was an ability (multitasking) that existed outside the specialist skills required for the individual subtasks. To have "specific factors" playing such a superordinate role in behavior might challenge the purity of the notion of g. Thus, from the earliest days there may have been skepticism about the idea that multitasking involved an ability that was distinct from general intelligence. We will return to the issue of multitasking and intelligence below, because cognitive neuroscience has recently offered some startling new evidence that speaks to it, suggesting that this skepticism was not warranted.

Despite this early skepticism, the search for a specific "multitasking ability" continued. It was hampered however by the lack of information-processing models that specified what operations such an ability would support, and the relation of this ability to others within the cognitive system (e.g., Ackerman et al., 1984). Ackerman et al. (1984) presented the view quite clearly. They contend that when considering multitasking data from any one individual, there are specific sets of variance attributable to individual differences in ability to perform the subtasks, but also another quite separate source of variance that represents the person's time-sharing (also known as multitasking) ability. They proposed, as an example, that where two subtasks are being considered, approximately 25% of the score might be attributable to multitasking ability (with 37.5% attributable to each of the subtasks, assuming zero error variance). So, for Ackerman et al., multitasking definitely represents a special ability above and beyond those required to perform the subtasks themselves. As mentioned above, this chapter will outline modern cognitive neuroscience evidence that supports such a view, and it even suggests basic information about the neural substrates of multitasking ability.

The confidence of Ackerman's view contrasted with the main bulk of thinking from the 1970s and 1980s. Brookings and Damos (1991) reviewed the evidence for a separate multitasking ability emerging from both factor analytic and non–factor analytic studies during that period, usually involving dual-task-type paradigms. With notable exceptions (e.g., Sverko, Jerneic, & Kulenovic, 1983), these studies failed to find evidence of such an ability. However, Brookings and Damos pointed out that most, if not all, of the principal studies had major limitations in the methodological design and/or analysis that would have limited the chances of discovery of one (i.e., an ability that would be used to deal with many different types of multitasking situation, as opposed to abilities that might be involved in dealing with only a limited subset of them) should one exist.

There were nevertheless some interesting investigations. For instance, one way in which a general multitasking ability might be demonstrated would be if training in multitasking could be shown to generalize to other situations. Damos and Wickens (1980) examined performance on two task combinations to try to discover evidence that multitasking (or in their terminology, "timesharing") skills are learned with practice and can transfer between task combinations. The first task combination consisted of a short-term memory task and a classification task; the second consisted of two identical one-dimensional compensatory tracking tasks, requiring participants to keep a moving circle centered within a horizontal track by using a control stick. One task was controlled by each hand. The experiment used three groups of participants. The first received dual-task training on both task combinations. The second received single-task training on the first task combination and received dual-task training on the second combination (the tracking). The third was given dual-task training on the second (tracking) combination only (i.e., no single-task training). By contrasting these groups, Damos and Wickens (1980) were able to examine the degree to which performance on the dual-task tracking task benefited from prior dual-task exposure. They found that transfer of learning was positive: Earlier dual-task training did seem to develop a generalizable dual-task skill. However, as pointed out by Lintern and Wickens (1991), the amount of transfer of time-sharing adaptation from one situation to another was modest relative to the skill learning demonstrated on the individual subtasks. Moreover, they contend, "Most timesharing skills may be specific to a given task combination and may not be of the generic type" (p. 136). This begs the question of the definition of a multitasking ability.

In these ways, the issue of whether there are specific multitasking skills has historically been contentious. One suggestion presented here is that the source of this contention is that there may be different abilities that underpin behavior in different multitasking situations. Therefore, it might have been more fruitful to have sought "multitasking abilities" (e.g., independent of g) rather than look for evidence of a single common one. Thus, a wider consideration of the range of multitasking situations than has been considered historically, and especially in relation to multitasking demands that present themselves in everyday life, may yield more definitive experimental results.

State-of-the-Art Review

It should be clear from the foregoing historical context that the emerging field of "multitasking" has several issues at the heart of it. The first of these is that there is little formal distinction concerning whether the term refers to a behavior, a construct, or a situation. In other words, does it refer to the act of doing several things at once—in which case are examples like walking while talking examples of multitasking? Or does it refer to a theoretical ability, in the same way as does, for example, the term "working memory"? Or does it refer to a situation that is presented to a participant (e.g., three novel tasks are presented to a person with the instruction that the person should attempt to do them all "at the same time")? At various times in the past, terms have been used that seem to relate very strongly to the more recent term "multitasking" that might have been thought of as constituting any one of these.

In this treatise, "multitasking" will be used in reference to the third kind of definition. In other words, it will be used as defining a set of features of a situation that are presented to an individual. Ideally we then should be able to hone these features into a cardinal set that would distinguish multitasking situations or paradigms from any other (e.g., a dual-task paradigm, or a working memory paradigm, etc.). If this definition of the term is adopted, it leaves open the matter of whether people who are multitasking are actually "doing several tasks at once." This is useful since it may sidestep the traditional concerns of dual- or multiple-tasking (e.g., the nature of the psychological refractory period (PRP), or the switch cost in task-switching paradigms; Monsell & Driver, 2000). It is not that these issues are not important—far from it—but more that there is an opportunity with the new term to reorient the concerns of the area toward the everyday, at least for those who see that as desirable—in other words, to create experimental paradigms as starting points for investigations that deliberately model the demands of "real-world," or "everyday," situations. Such a starting point may help offset the danger of becoming "phenomenon-driven" (Newell, 1973; Monsell & Driver, 2000) and thus encourage the paradigms used for research to be more closely aligned to the everyday situations for which experimental paradigms are usually intended to be a model. As Kieras et al. (2000) put it,

Future research must use a wide variety of empirical procedures to investigate multitask performance. This investigation should extend beyond basic laboratory paradigms like the task-switching and PRP procedures, which … come nowhere near to engaging the whole host of executive mental processes that people presumably have. Rather, to explore these processes … overlapping-task procedures with complex realistic tasks and unpredictable stimulus–response event sequences will be needed. (p. 710)

As we will see below, it is exactly these sorts of procedures that have proved a starting point for one particular line of investigation of multitasking within cognitive neuroscience.

Serial versus Concurrent Multitasking

Interestingly, an "everyday situation" is taken as the reference situation in the theoretical paper of Salvucci and Taatgen (2008). This paper aims to form a model of the demands made by concurrent multitasking. However, these researchers do not take as a starting point a typical multiple-tasks type of experimental paradigm. Instead, the cardinal example of a situation requiring multitasking that they describe is cooking in a busy kitchen. They imagine a cook trying to fulfill multiple food orders simultaneously. While some tasks can be carried out in parallel (e.g., baking fish in an oven while pasta boils on a stove), in many situations this cannot happen because either the cook needs to use the same resource (e.g., oven) for two separate tasks or the cook is physically unable to perform two distinct complex actions at the same time. Salvucci and Taatgen (2008) point out that in this situation the cook needs to delay performing one action while another is carried out (p. 103). This everyday example illustrates well how multitasking in *everyday life* is often not a matter of a person's attempting to do two tasks at the same time as is asked of a participant in typical experimental concurrent multitasking paradigms (e.g., those historically called dual- or multiple-task paradigms). Instead, much of the behavior is more closely aligned to the definition of serial multitasking presented here. Thus, we may have the beginning of a discontinuity between the demands of everyday life and those of experimental practice.

Of course, I provide absolutely no evidence whatsoever for this contention (i.e., that much of dealing with several tasks in everyday life takes the form of serial rather than concurrent multitasking). But the motivation for it is to try to capture the essence of what people in "everyday life" (i.e., outside academia) mean by the term, in the hope that such a endeavor might guard against "phenomena-driven" experimental work—or at least that the phenomena under examination are those observed in the "real world" rather than only the lab.

So let us consider what an "everyday" definition of a multitasking situation might look like. For this argument, let us leave aside the issue for the moment of concurrent versus serial multitasking and seek the broadest level of definition that encompasses all forms. We cannot start and end with the characteristic of "doing several tasks at once" since people routinely perform two tasks at the same time: It depends what you call a "task." So for instance, someone who is walking while holding a conversation could be argued to be performing two tasks simultaneously (i.e., transport to a destination and having a discussion) and therefore "dual tasking." If one then adds the behavior where that person is also opening an umbrella, then prima facie this might be "multitasking." However, if one stopped a thousand people on a busy London street and asked them to give an example of a typical situation that requires "multitasking," few would probably give an example this mundane. Instead, running through the examples that one would get would probably be the following characteristics:

Characteristic 1: There are several tasks that are desired to be performed, and for some reason they cannot be done in pure sequence (i.e., task 1, task 2, task 3, etc.). Instead, adequate performance will either (1) require interleaving of responses, thoughts, or actions in relation to the different tasks (e.g., task 1, task 2, task 1, task 3, task 2, etc.) and/or (2) require the person to be *trying* to do more than one task "at the same time." It is the "trying" that is the important feature here since it is a moot point as to whether many complex mental tasks can actually be performed "at the same time" when examined at the millisecond level. In this way (2) may actually be (1) when observed scientifically.

Characteristic 2: There is *novelty* in either the *combination* or *order* of the subtasks being attempted. Highly overlearned sequences of coordinated behaviors, such as occurs, for instance, when an experienced driver is driving (e.g., checking the rearview mirror while switching on an indicator) is unlikely to be described as "multitasking." However, introduce a novel requirement (or even better, several) into the sequence (e.g., ask the driver to count how many times he touches the brake pedal in a 5-minute period), and most people would probably consider that as an example of some form of multitasking.

Characteristic 3: This might refer to the distinctness of the tasks and therefore the level of interference between them. Thus, talking to two people in a three-way conversation is unlikely to be seen as a strong example of "multitasking" behavior. However, using two telephones to conduct two different conversations with different people at the same time probably would be, despite the fact that here there are also three people talking, and only one may be talking at a time, just as before. So captured in this "everyday knowledge" is perhaps a notion of holding some distinct stream of thought in mind (in this case the currently unattended conversation) while indulging in another stream of thought in relation to another goal or activity (the other conversation).

Of course, there are many observations one might make about these characteristics. For instance, different multitasking situations will differ in the degree to which they show these characteristics. And the characteristics themselves are not easy to measure. Novelty, in particular, presents all manner of measurement and methodological problems (Burgess, 1997).

Another aspect is the intent of the participants (see Characteristic 1) as much as the situation that is presented to them. Indeed, one of the most common ways in which multitasking situations differ from each other is in the intent of the participants and the way they conceive of the parameters of the task. The priority they assign is particularly important. In dual-task paradigms or multiple-task ones, the participant is typically instructed to try to perform both tasks at once. This does not mean however that the two tasks necessarily are given equal priority. A good example is Gopher (1993), who assigned equal priorities to two subtasks (visuomanual tracking and a choice reaction time [CRT] task) in one group of participants who were then given a series of dual-task-type practice sessions. In another group, the priorities for the subtasks were varied: Sometimes it was the tracking task that was given priority, sometimes the CRT, and sometimes they were assigned equal priority. Gopher found that variable-priority training led to better performance on the two subtasks than fixed-priority training.

This characteristic—that the participant is trying to perform two or more tasks at the same time (whether this is psychologically possible or not) is prima facie a feature of *concurrent* multitasking paradigms.

However, as mentioned above, many everyday situations are not like the situation presented by Gopher. Instead, they often require *serial* multitasking. This is where, for practical or other reasons, only one task can be attempted at a time. The participants are not trying to do multiple tasks at the same time, but they are trying to get several tasks done within a certain time period, and for one reason or another they are not able to do each task from start to finish in strict sequence.

While experimental psychology has concerned itself principally with studying concurrent multitasking situations, human neuropsychology (and to a lesser extent, cognitive neuroscience more broadly) has largely concerned itself with the study of serial multitasking situations (with a few notable exceptions, e.g., Baddeley et al., 1997). One major difference between the two fields is that neuropsychology is driven to a greater degree by observation of people (in this case, patients) in everyday life. And it is serial multitasking situations where many patients with neurological damage to a certain part of the frontal lobes show a particular difficulty. These neuropsychological studies have clear implications for creating information-processing models of the differing demands of these two multitasking situations, and so the data are highly relevant to experimental psychology and cognitive science, even though this is not often recognized.

Everyday Multitasking Deficits in Neurological Patients: Lessons for Building Information-Processing Models

What have we learned from neuropsychological studies about the nature of the organization of the processes involved in serial multitasking? Well, the most obvious lesson from neurological patients is that deficits in serial multitasking ability can be seen in the context of intact intellect, as well as intact episodic memory, language, visuomotor skills, and visuoperceptual skills, as well as intact ability on most commonly used problem-solving tasks. This was first demonstrated empirically by Shallice and Burgess (1991). Thus, neuropsychology provides strong evidence for a special (i.e., independent of *g*) multitasking facility of the kind that was sought by experimental psychologists for many decades, with rather equivocal results.

Shallice and Burgess (1991) described three people who had all suffered frontal lobe damage following traumatic brain injury. All three had no significant impairment on formal tests of perception, language, and intelligence, and two of them performed well on a variety of traditional tests of novel problem solving. However, they all demonstrated severe problems in everyday life, especially with situations that required serial multitasking. Shallice and Burgess (1991) invented two new tests of multitasking to quantify these problems. The first is called the Multiple Errands

Test (MET) and is a real-life multitasking test typically carried out in a shopping precinct. Participants are pretrained on a set of rules (e.g., no shop should be entered other than to buy something; spend as little money as possible) and then are asked to undertake a number of tasks while following these rules. The tasks vary in terms of complexity (e.g., buy a small brown loaf vs. discover the exchange rate of a particular foreign currency yesterday), and there are a number of "hidden" problems in the tasks that have to be appreciated and the possible course of action evaluated. For instance, the participants are required to write and send a postcard, but (1) they are given no pen, (2) they are instructed not to use anything not bought on the street to help them, and (3) they are also told that they need to spend as little money as possible. In this way, the task is very "open-ended" or "ill-structured," meaning that there are many possible courses of action, and it is up to the individual to determine for themselves which one they will choose, and to a certain extent, what constitutes acceptable performance.

Clearly, a task as complex as the MET taps many different abilities that relate to novel problem solving, prioritization, and so forth on top of pure multitasking ability. Thus, in order to measure the multitasking component in a more controlled way, Shallice and Burgess invented a test called the Six Element Test (SET). This required subjects to swap between three simple subtasks, each divided into two sections, within 15 minutes, while following some arbitrary rules (e.g., "You cannot do part A of a subtask followed immediately by part B of the same subtask"). They were told that they should attempt at least some items of all subtasks. It was up to the participant to decide when to switch tasks since no signal was given, and although a clock was present, it was covered, so that checking it has to be a deliberate action of lifting the cover to check it. Fifteen minutes was not long enough to complete all the subtasks (or even all of one subtask), so the participants needed to remember to switch from one task to another before reaching completion, so that they would have at least attempted all subtasks. The SET therefore has a strong component of voluntary time-based task switching (see Burgess et al., 2009, 2011; Gonen-Yaacovi & Burgess, 2012, for review).

All three cases reported by Shallice and Burgess all performed both the MET and SET at a level below the 5th percentile compared with age- and IQ-matched controls. On the MET the subjects made several different types of error. For instance, they found themselves having to go into the same shop several times when they could have bought all they needed at one visit. They also forgot to carry out tasks that they intended to do, and they often broke the tasks rules. There were also many examples of curious social behavior and interactions (e.g., offering sexual favors in lieu of payment). These kinds of social issues are a regular occurrence in this type of patient, and it may potentially be instructive as to the basis of, for example, social attention (see chapter 15 in the current volume for an outline of the mechanics of social attention). Shallice and Burgess (1991) termed this kind of behavioral disorganization in the context of preserved intellect and other cognitive functions the "strategy application disorder." This term was chosen as to be reasonably neutral about a range of difficulties that these patients might show upon further investigation. But one defining feature seemed to be impairment in situations requiring serial multitasking.

On the Specificity of the Multitasking Impairment in Neurological Patients

The demonstration that serial multitasking impairments can be seen in the context of, for example, intact IQ is a strong indication for a specific ability that underpins at least some forms of multitasking. However, one might argue that the types of tasks used above tap more than just multitasking abilities. For instance, they may require planning, "monitoring" (i.e., keeping a record of where you are in a goal list), and remembering of the task rules. Thus, Burgess et al. (2000) used the group lesion study methodology to investigate whether impairments in these sorts of stages of performance (e.g., planning) were a root cause of multitasking impairments. Sixty acute neurological patients and 60 age- and IQ-matched healthy controls were administered a multitasking test similar to the SET. Before participants began the test, their ability to learn the task rules (by both spontaneous and cued recall) was measured. They were then asked how they intended to do the test, and a measure of the complexity and appropriateness of their plans was gained. This enabled us to look at whether their failures could be due to poor planning. The participants then performed the task itself, and by comparing what they did with what they had planned to do, a measure of "plan following" was made. Multitasking performance itself was measured as the number of task switches minus the number of rule breaks. After these stages were finished, subjects were asked to recollect their own actions by describing in detail what they had done. Finally, delayed memory for the task rules was examined.

This procedure enabled us to demonstrate the specificity of the deficit in patients who fail the test in the context of good planning, comprehension, and (retrospective) memory abilities. Critically, we found that patients with left-hemisphere rostral prefrontal cortex (PFC) lesions (i.e., the most anterior regions of PFC, variously also referred to as frontopolar cortex or Brodmann's Area [BA] 10), when compared with patients with lesions elsewhere, showed a significant multitasking impairment despite no significant impairment on any variable except the one reflecting multitasking performance. In other words, despite being able to learn the task rules, form a plan, remember their actions, and say what they should have done, they nevertheless did not do what they said that they intended to do.

Notably, perhaps, we have recently found that patients with rostral PFC lesions show problems with certain tasks which require time perception (Volle et al., 2011). How this relates to multitasking problems we do not currently know. However, it may turn out to be instructive (e.g., see chapter 3 in the current volume for discussion of temporal attention).

Whatever the precise cause, the link between serial multitasking deficits and damage to rostral PFC has been replicated using the group lesion method, first by Dreher et al. (2008) and then by John Duncan's group in Cambridge, United Kingdom (Roca et al., 2010, 2011). Roca et al.'s (2010) study is particularly noteworthy in the current context in that it showed that multitasking deficits were still seen even after the scores were adjusted for change in a specific measure of "fluid intelligence" (Cattell's Culture Fair test). So after 25 years of research into serial multitasking deficits in humans, we seem to have reached as good an agreement as scientists usually can that serial multitasking is a behavior that makes demands upon a resource that is independent from "IQ," and that there is an area within rostral PFC that seems at least in part to support it.

Integration

The Cognitive Demands of Serial Multitasking

If serial multitasking performance is not strongly related to IQ, or a range of other cognitive/problem-solving abilities, then what might be the processes that support it? This section will consider some possibilities.

Clearly, the situations presented to the participants when they are asked to do the MET and SET are very different from, for example, a typical dual-task or multiple-task paradigm used in the experimental psychology literature. Specifically, the characteristics of the MET and SET have been described by Burgess and Wu (2013) and are detailed in table 19.1. Characteristic 6 is not a necessary feature and may not occur in most

Table 19.1
Characteristics of Typical Serial Multitasking Situations in Everyday Life according to Burgess (2000)

---

1. A number of discrete and different tasks have to be completed.

2. Performance on these tasks has to be dovetailed in order to be time-effective.

3. Because of either cognitive or physical constraints, only one task can be performed at any one time.

4. The appropriate stages for returns to task are not signaled directly by the situation.

5. There is no moment-by-moment performance feedback of the sort that participants in many laboratory experiments will receive. Typically, failures are not signaled at the time they occur.

6. Unforeseen interruptions, sometimes of high priority will occasionally occur, and things will not always go as planned.

7. Tasks usually differ in terms of priority, difficulty, and the length of time they will occupy.

8. People decide for themselves what constitutes adequate performance.

---

experimental situations (like the SET when given clinically) but is an added and important determinant of performance in typical everyday life.

The finding of a specific link between serial multitasking deficits and rostral PFC damage invites (although it does not provide) a first-pass constraining of the possible candidate mental processes that support the behavior to those that have been associated with rostral PFC function. So what are they?

Rostral PFC structures have been found to support, at least in part, a wide range of mental abilities. These include analogical reasoning, maintenance and execution of delayed intentions (also known as prospective memory), source and context memory, mentalizing, mind wandering and other stimulus-independent mental phenomena, time perception, humor judgment, and task initiation speed (Burgess, Gonen-Yaacovi, & Volle, 2012; Burgess & Wu, 2013; Gilbert et al., 2007, 2012). One theory of the role of rostral PFC in cognition is that it underpins an attentional "gateway" that allows one to control the degree to which one is engaging in stimulus-independent thought (e.g., an inner mental dialogue) versus attending to external stimuli. This theory is known as the "gateway hypothesis," and the theoretical construct underpinning this mental ability is known as "XN attending control" (Burgess et al., 2005; Burgess, Gilbert, Okuda & Simons, 2006; Burgess, Dumontheil, & Gilbert, 2007; Burgess & Wu, 2013).

Why should the information processing associated with these types of behaviors be particularly associated with performance in serial multitasking situations, especially everyday-type ones? One possibility is that while it is the case that the kinds of mental phenomena associated with rostral PFC function may have little relation to performance of the kinds of highly constrained and very structured tasks that are typically administered to participants in a brain scanner, or in an experimental psychology study, this may not be true of very open-ended serial multitasking tasks like those that have the characteristics outlined in table 19.1. In typical functional magnetic resonance imaging (fMRI) experimental tasks, for instance, it is ideal to have a large number of repeated trials or responses (often presented in pseudorandom order to avoid artifact), and it is nearly always very clear to the participants what exactly is required of them, and whether they are succeeding or not. However this is *not* typical of many or maybe even most tasks in everyday life, at least outside the classroom or a very repetitive job. In more "realistic" experimental tasks, or open-ended or "ill-structured" everyday life situations, metacognitive introspections about, for example, how well one is doing, or what is expected of one, may well be highly helpful to task performance. In this way there may be consistent patterns of *perceptual decoupling* occurring (for more information about these kinds of phenomena, please see chapter 10 in the current volume). Thus, recent research into the functions of rostral PFC is suggesting that there are many cognitive abilities that are specifically designed to help individuals to be competent in everyday life, but are not at all well-captured in many experimental situations.

In my lab, we have been examining the plausibility of the hypothesis that XN attending control (see above) is a central feature to the performance of tasks which involve activation of a delayed intention, which is a central feature of many serial multitasking situations (see Characteristic 4 in table 19.1). Benoit et al. (2012) designed an fMRI paradigm which parametrically crossed XN control with a task involving maintaining and executing a delayed intention. We found that there was considerable overlap in the brain regions that showed blood-oxygen-level-dependent (BOLD) signal change during the XN control manipulation and those involved in conditions involving a delayed intention (see also Burgess et al., 2011; Volle et al., 2011). These results are consistent with the hypothesis that being able to effect good control over the direction of one's attending is an important determinant of being able to schedule and interleave a series of tasks within a certain time frame (i.e., serial multitasking).

By contrast, perhaps concurrent multitasking makes fewer demands upon processes involved in dealing with delayed intentions. This might be because the "goal list" or "goal instructions" for each task are being consciously rehearsed, or are, to use the framework of John Duncan, integrated into a single "task model" (Duncan et al., 2008). If this was the case, then one might expect in neuroimaging experiments to see a different pattern of activation during concurrent multitasking compared with serial multitasking.

Etienne Koechlin and his colleagues have examined this situation directly. Instead of "multitasking," Koechlin has used the term "cognitive branching" to describe "a domain-general core function that enables the brain to contingently put on hold one behavioural option/task in order to perform another one" (Koechlin 2013, p. 285). This provides a rather good description of the situation being investigated by researchers interested in multitasking (especially serial multitasking), and indeed, in later work (e.g., Dreher et al., 2008) the construct "cognitive branching" was redescribed as "multitasking." The task that Koechlin introduced was ingenious. There were four conditions, and in each, individual letters (appearing in either upper- or lowercase) from the word "tablet" were presented one at a time to participants. In the first, control condition, participants were asked to judge whether two successively presented letters were in immediate succession in the word "tablet" (only uppercase letters were presented). In the second condition ("delay condition"), participants were also asked to decide if two successively presented uppercase letters occurred in immediate succession in the word "tablet." But this time they were asked to ignore lowercase letters. This meant that there was a delay period for uppercase letters. The third condition was a dual-task condition. Here, participants also had to decide if successively presented letters (paying attention to both upper- and lowercase letters) were in immediate succession in the word "tablet." They also had to decide if each first letter indicating a case change was the letter *T* (or *t*). The fourth condition was the multitasking condition. Here, they essentially had to do both of the previous tasks. So for successive uppercase letters, or for successive lowercase letters, participants decided whether the current letter followed immediately the previously presented letter in the word "TABLET" (or "tablet"). And they also had to decide whether each first letter signifying a case change was the letter T (or t).

Koechlin et al. (1999) found bilateral BOLD signal increases during the dual-task condition in the posterior dorsolateral PFC BA 9, the middle frontal gyrus, close to the precentral gyrus (BA 8), and the lateral inferior parietal lobule (BA 40). There were no specific brain regions identified in the delayed condition. However, in the multitasking condition there were BOLD signal increases found bilaterally in dorsal aspects of the rostral PFC (i.e., the frontal pole, or BA 10). BOLD signal increase in these regions only occurred in the multitasking condition, and was similar in the other conditions, including the control condition. This meant that the rostral PFC activations could not be explained by differences in "task difficulty" as indexed by task performance since that would have been reflected in a gradual increase from the control to the delayed task dual-task and branching conditions. One of the key aspects of Koechlin's study was that it gives evidence that dual-task performance and serial multitasking are not identical in terms of the neural architecture that supports them. It is tempting to conclude that perhaps this is the result of concurrent performance of more than one task, and serial performance of them, being psychologically different.

Future Directions

Research into the cognitive and brain basis of multitasking behavior has, in scientific terms, only just started. There are early signs that the criticisms that were made of dual-tasking research 30 years or so ago (e.g., lack of theoretical underpinning; paradigm-bound research) may not also be made of this emerging field. There are useful theoretical models emerging (e.g., Salvucci & Taatgen, 2008), partly because we now have methods for theory testing that were only incipient back then (e.g., computational modeling). And we now have demonstrations of neurological patients with circumscribed lesions that have caused circumscribed multitasking problems, in effect answering a question that is nearly 100 years old: whether there is a specific multitasking ability beyond that which is required for performance of the subtasks themselves. If patients with rostral PFC (BA 10) damage can be absolutely unimpaired at the subtasks themselves but show catastrophic impairment in organizing their performance when many tasks are put together, then prima facie this is strong evidence that putting several subtasks together makes demands beyond the sum of the individual subtasks. Moreover, there is confirmatory evidence in terms of localization from neuroimaging—again, a new method that did not start to exert itself until the mid-1990s.

One main current focus in the cognitive neuroscience of multitasking is trying to understand how the brain supports the ability to remember to carry out an intended act at a future time (i.e., prospective memory) since this seems highly relevant to understanding serial multitasking. Dual-task or concurrent multitasking–type paradigms seem to make demands upon different parts of the brain than does serial multitasking. Since those regions are often also closely associated with "working-memory"–type paradigms, one possible explanation consistent with these results is that concurrent multitasking ability may be relatively more related to active rehearsal or maintenance of a goal set whereas serial multitasking may stress relatively more those processes relating to delayed intentions. Accordingly, for serial multitasking, a key concern for behavioral study becomes that of "intention cost," first demonstrated behaviorally (as far as I am aware) by Burgess et al. (2003, figure 1, p. 910). This is the cost of performance of the current task you are performing caused by maintaining a delayed intention, in this case, to do the next scheduled task. There are echoes here of Witt and Ward (1965), who said, "Half-completed tasks, in a multitasking operation, tie up the system facilities and prevent other tasks from proceeding" (p. 58). How this relates to, for example, the aspect of novelty of the subtasks and the combinations of them is however not known, and this may be a promising avenue of enquiry for experimental psychologists.

An attempt at a simple behavioral analysis of the characteristics of serial multitasking has been attempted here. However, it is preliminary, and the defining characteristics of concurrent multitasking situations are as yet not well established either. Clearly, we would all agree that "multitasking situations" occur when someone is dealing with multiple tasks or demands. But what might we say beyond that, in terms of instructions to participants, or their aims, or the nature of the subtasks? How much does one have to alter a subtask (and in what way) before it can be considered a "different task"? Are well-rehearsed combination of actions and behaviors (like walking while talking) examples of multitasking, and if not, why not? It seems that the issue of the novelty of the combination of actions or tasks is likely key to our definition, but there is relatively little work in that area as yet.

A fascinating possibility for understanding the basis of multitasking is presented by Messinger et al. (2009), who claim to have discovered "multitasking neurons" in the PFC of primates. In their experiment, monkeys were required to attend to a visual stimulus at one location while remembering a second place. Most neurons specialized in either attentional or mnemonic processing. However, about one-third of the cells responded to features of both, and they termed these "multitasking neurons." They argue that these neurons afford an advantage for situations where two or more demands are made simultaneously. They do this by having stronger tuning than more specialized cells and expressing preferred directions for attention and for working memory (they may even oppose one another). In this way, Messinger et al. argue that pairs of multitasking neurons can represent cognitive parameters more efficiently than pairs that include even a single specialized cell. This view may have received further momentum by the recent finding from Takeuchi et al. (2013), who gave healthy young adults 4 weeks of concurrent multitasking training and examined the brain changes that occurred over that period (in regional gray matter volume and also in resting functional connectivity). They found that concurrent multitasking training was accompanied by increased regional gray matter volume in three subregions of the PFC (left lateral rostral PFC, dorsolateral PFC, and the left inferior frontal junction) as well as a set of regions outside the frontal lobes (left posterior parietal cortex, left temporal and lateral occipital regions). They also found decreased resting-state connectivity between the right dorsolateral PFC and the ventral anterior cingulate cortex. Takeuchi et al. (2013) argue that these results may suggest that there are some neural changes that are not tightly restricted to specific task conditions. There are resonances here perhaps with John Duncan's (2001) adaptive coding theory of the properties of PFC. The notion that frontal neurons may have a less tightly specialized functional capacity than neurons elsewhere in the cortex is an attractive one. However, at least for rostral PFC, there is considerable evidence now for a large degree of functional specialization within this large brain region (Burgess et al., 2007; Burgess, Gonen-Yaacovi, & Volle, 2012; Burgess & Wu, 2013), and it does not seem implausible that there may be neurons that are highly involved with, for example, maintaining an intention over a delay period. Whether this sort of function can be linked very specifically to a particular behavior once we understand the operations they perform at the information-processing level remains to be discovered.

However, at this very early stage of investigation into multitasking, many of the basic questions have not yet even been asked, never mind answered. For instance, there is surprisingly little integration between the literature on task switching (see chapter 7 in the current volume for a review) and that of multitasking, despite its obvious relevance. A good place to start seeking these questions will be to look at the behaviors in everyday life that we label as multitasking, ask what demands they make, and then create experimental analogues of them. This is already occurring in some fields (e.g., Hancock et al., 2003). This chapter argues that this approach in the neuropsychology of complex behaviors has been fruitful. It contrasts with a "construct" (rather than behavioral) led approach that, for example, *first* posits a multitasking construct (independent of a particular behavior) and then builds a paradigm to test that theory, with little regard to whether the resulting task is anything like situations that are encountered in everyday life either now or in human history. One danger of the latter approach is that we may end up studying situations that are so divorced from those that the cognitive system will have developed to deal with (in an evolutionary sense) that we may be examining curiosities of the system rather than mechanisms fundamental to mainstream behavior (Burgess, Alderman, et al., 2006).

Box 19.1

Key Points

- Concurrent multitasking occurs when a person is attempting to do several tasks at the same time (e.g., talking on a mobile phone while shopping).

- Serial multitasking occurs when people are interleaving several tasks but can only do one at a time. This interleaving often occurs because one task cannot be finished before another must start (e.g., when cooking a meal).

- Serial and concurrent multitasking may make different cognitive demands.

- Serial and concurrent multitasking may be supported, at least in part, by different brain structures, or at least by operation of the same brain regions to different degrees.

- Serial multitasking ability seems to a surprising degree to be unrelated to general intellectual ability, as measured by many current IQ tests. It is not yet well established whether concurrent multitasking is similarly unrelated.

Box 19.2

Outstanding Issues

- In what ways do open-ended multitasking tests (i.e., where the participants decide for themselves what to do and when) make different cognitive demands from ones where the optimal course is decided by the experimenter?

- In what ways are well-practiced multitasking situations different from novel ones at information-processing and neural levels?

- What are the respective roles of dorsolateral and rostral prefrontal cortex in multitasking?

- Do serial and concurrent multitasking abilities change over the life span in the same way?

- Theoretical models of all forms of multitasking are urgently needed.

References

Ackerman, P. L., Schneider, W., & Wickens, C. D. (1984). Deciding the existence of time-sharing ability: A combined methodological and theoretical approach. *Human Factors, 26,* 71–82.

Baddeley, A., Della Sala, S., Papagno, C., & Spinnler, H. (1997). Dual-task performance in dysexecutive and nondysexecutive patients with a frontal lesion. *Neuropsychology, 11,* 187–194.

Benoit, R. G., Gilbert, S. J., Frith, C. D., & Burgess, P. W. (2012). Rostral prefrontal cortex and the focus of attention in prospective memory. *Cerebral Cortex, 22,* 1876–1886.

Binet, A. (1890). La concurrence des états psychologiques. *Revue Philosophique de la France et de l'étranger, 29*, 138–155.

Brookings, J. B., & Damos, D. L. (1991). Individual differences in multiple-task performance. In D. L. Damos (Ed.), *Multiple-task performance* (pp. 363–386). London: Taylor & Francis.

Burgess, P. W. (2000). Real-world multitasking from a cognitive neuroscience perspective. In S. Monsell & J. Driver (Eds.), *Control of cognitive processes: Attention and performance XVIII* (pp. 465–472). Cambridge, MA: MIT Press.

Burgess, P. W. (1997). Theory and methodology in executive function research. In P. Rabbitt (Ed.), *Theory and methodology of frontal and executive function* (pp. 81–116). Hove, UK: Psychology Press.

Burgess, P. W., Alderman, N., Forbes, C., Costello, A., Coates, L. M.-A., Dawson, D. R., et al. (2006). The case for the development and use of "ecologically valid" measures of executive function in experimental and clinical neuropsychology. *Journal of the International Neuropsychological Society, 12*, 1–16.

Burgess, P. W., Alderman, N., Volle, E., Benoit, R. G., & Gilbert, S. J. (2009). Mesulam's frontal lobe mystery re-examined. *Restorative Neurology and Neuroscience, 27*, 493–506.

Burgess, P. W., Dumontheil, I., & Gilbert, S. J. (2007). The gateway hypothesis of rostral prefrontal cortex (Area 10) function. *Trends in Cognitive Sciences, 11*, 290–298.

Burgess, P. W., Gilbert, S. J., Okuda, J., & Simons, J. S. (2006). Rostral prefrontal brain regions (Area 10): A gateway between inner thought and the external world? In W. Prinz & N. Sebanz (Eds.), *Disorders of volition* (pp. 373–396). Cambridge, MA: MIT Press.

Burgess, P. W., Gonen-Yaacovi, G., & Volle, E. (2011). Functional neuroimaging studies of prospective memory: What have we learnt so far? *Neuropsychologia, 49*, 2246–2257.

Burgess, P. W., Gonen-Yaacovi, G., & Volle, E. (2012). Rostral prefrontal cortex: What neuroimaging can learn from human neuropsychology. In B. Levine & F. I. M. Craik (Eds.), *Mind and the frontal lobes* (pp. 47–92). New York: Oxford University Press.

Burgess, P. W., Scott, S. K., & Frith, C. D. (2003). The role of the rostral frontal cortex (area 10) in prospective memory: a lateral versus medial dissociation. *Neuropsychologia, 41*, 906–918.

Burgess, P. W., Simons, J. S., Dumontheil, I., & Gilbert, S. J. (2005). The gateway hypothesis of rostral PFC function. In J. Duncan, L. Phillips, & P. McLeod (Eds.), *Measuring the mind: Speed, control and age* (pp. 215–246). Oxford, UK: Oxford University Press.

Burgess, P. W., Veitch, E., Costello, A., & Shallice, T. (2000). The cognitive and neuroanatomical correlates of multitasking. *Neuropsychologia, 38*, 848–863.

Burgess, P. W., & Wu, H.-C. (2013). Rostral prefrontal cortex (Brodmann Area 10): Metacognition in the brain. In D. T. Stuss & R. T. Knight (Eds.), *Principles of frontal lobe function* (2nd ed., pp. 524–534). New York: Oxford University Press.

Damos, D. L. (1991). *Multiple-task performance*. London: Taylor & Francis.

Damos, D. L., & Wickens, C. D. (1980). The identification and transfer of timesharing skills. *Acta Psychologica, 46*, 15–39.

Dreher, J.-C., Koechlin, E., Tierney, M., & Grafman, J. (2008). Damage to the fronto-polar cortex is associated with impaired multitasking. *PLoS ONE, 3*(9), e3227. doi:10.1371/journal. pone.0003227.

Duncan, J. (2001). An adaptive coding model of neural function in prefrontal cortex. *Nature Reviews. Neuroscience, 2*, 820–829. doi:10.1038/35097575.

Duncan, J., Parr, A., Woolgar, A., Thompson, R., Bright, P., Cox, S., et al. (2008). Goal neglect and Spearman's g: Competing parts of a complex task. *Journal of Experimental Psychology. General, 137*, 131–148.

Gilbert, S. J., Bird, G., Frith, C. D., & Burgess, P. W. (2012). Does "task difficulty" explain "task-induced deactivation"? *Frontiers in Cognition, 3*, 125.

Gilbert, S. J., Dumontheil, I., Simons, J. S., Frith, C. D., & Burgess, P. W. (2007). Comment on "Wandering minds: The default network and stimulus-independent thought." *Science, 317*, 43.

Gonen-Yaacovi, G., & Burgess, P. W. (2012). Prospective memory: The future for future intentions. *Psychologica Belgica, 173*(52/2–3), 173–204.

Gopher, D. (1993). The skill of attention control: Acquisition and execution of attention strategies. In D. Meyer & S. Kornblum (Eds.), *Attention and performance XIV: Synergies in experimental psychology, artificial intelligence, and cognitive neuroscience—A silver jubilee volume* (pp. 299–322). Cambridge, MA: MIT Press.

Hancock, P. A., Lesch, M., & Simmons, L. (2003). The distraction effects of phone use during a crucial driving maneuver. *Accident Analysis & Prevention, 35*, 501–514.

Kieras, D. E., Meyer, D. E., Ballas, J. A., & Lauber, E. J. (2000). Modern computational perspectives on executive mental processes and cognitive control: Where to from here? In S. Monsell & J. Driver (Eds.), *Control of cognitive processes: Attention and performance XVIII* (pp. 681–712). Cambridge, MA: MIT Press.

Koechlin, E. (2013). Motivation, control, and human prefrontal executive function. In D. T. Stuss & R. T. Knight (Eds.), *Principles of frontal lobe function* (2nd ed.). New York: Oxford University Press.

Koechlin, E., Basso, G., Pietrini, P., Panzer, S., & Grafman, J. (1999). The role of the anterior prefrontal cortex in human cognition. *Nature, 399*, 148–151.

Lintern, G., & Wickens, C. D. (1991). Issues for acquisition and transfer of timesharing and dual-task skills. In D. L. Damos (Ed.), *Multiple-task performance*. London: Taylor & Francis.

McQueen, E. N. (1917). The distribution of attention. *British Journal of Psychology, 11* (Monograph Supplements, 5).

Messinger, A., Lebedev, M. A., Kralik, J. D., & Wise, S. P. (2009). Multitasking of attention and memory functions in the primate prefrontal cortex. *Journal of Neuroscience, 29*, 5640–5653.

Monsell, S., & Driver, J. (2000). Banishing the control homunculus. In S. Monsell & J. Driver (Eds.), *Control of cognitive processes: Attention and performance XVIII* (pp. 3–32). Cambridge, MA: MIT Press.

Newell, A. (1973). You can't play twenty questions with nature and win. In W. A. Chase (Ed.), *Visual information processing* (pp. 283–308). New York: Academic Press.

Ophir, E., Nass, C., & Wagner, A. D. (2009). Cognitive control in media multitaskers. *Proceedings of the National Academy of Sciences of the United States of America, 106*, 15583–15587. doi:10.1073/pnas.0903620106.

Roca, M., Parr, A., Thompson, R., Woolgar, A., Torralva, T., Antoun, N., et al. (2010). Executive function and fluid intelligence after frontal lobe lesions. *Brain, 133*, 234–247.

Roca, M., Torralva, T., Gleichgerrcht, E., Woolgard, A., Thompson, R., Duncan, J., et al. (2011). The role of Area 10 (BA10) in human multitasking and in social cognition: A lesion study. *Neuropsychologia, 49*, 3525–3531.

Salvucci, D. D., & Taatgen, N. A. (2008). Threaded cognition: An integrated theory of concurrent multitasking. *Psychological Review, 115*, 101–130.

Schumacher, E. H., Seymour, T. L., Lauber, E. J., Kieras, D. E., Meyer, D. E., Glass, J. M., et al. (2001). Virtually perfect time sharing in dual-task performance: Uncorking the central cognitive bottleneck. *Psychological Science, 12*, 101–108.

Shallice, T., & Burgess, P. W. (1991). Deficits in strategy application following frontal lobe damage in man. *Brain, 114*, 727–741.

Sharp, S. (1899). Individual psychology: A study in psychological method. *American Journal of Psychology, 10*, 329–391.

Sverko, B., Jerneic, Z., & Kulenovic, A. (1983). A contribution to the investigation of time-sharing ability. *Ergonomics, 26*, 151–160.

Takeuchi, H., Taki, Y., Nouchi, R., Hashizume, H., Sekiguchi, A., Kotozaki, Y., et al. (2013). Effects of multitasking-training on gray matter structure and resting state neural mechanisms [Epub ahead of print]. *Human Brain Mapping, 12*, 17. doi:10.1002/hbm.22427.

Volle, E., Gonen-Yaacovi, G., de Lacy Costello, A., Gilbert, S. J., & Burgess, P. W. (2011). The role of rostral prefrontal cortex in prospective memory: A voxel-based lesion study. *Neuropsychologia, 49*, 2185–2198.

Witt, B. I., & Ward, L. (1965). IBM operating system/360 concepts and facilities (IBM Systems Reference Library File No. S360-36, Form C28-6535-0).

20

Attention and Incidental Memory in Everyday Settings

Alan D. Castel, Meenely Nazarian, and Adam B. Blake

Historical Context

Much of the information that we acquire from the world is done incidentally. That is, we are not actively trying to memorize certain events, faces, or objects in our environment or songs we hear. This raises an important question regarding the link between attention and memory and how this functions in real-world settings. Do we need to attend to something in order to remember it, and without consciously attending to something, will we somehow remember it? In this chapter, we explore and review how attention and memory are linked in real-world settings, and how and when incidental encoding can lead to strong memories, as well as fleeting and reconstructive representations.

Memory is often a product of attention, and of how often one encounters, uses, and retrieves the information in question. A recent study has shown the ability to preserve details of over 2,000 images, demonstrating the capacity of long-term memory to store details of objects we attend to (Brady, Konkle, Alvarez, & Oliva, 2008). People are also remarkably good at recognizing scenes they were briefly shown, even when tested among hundreds of other scenes (Nickerson, 1965), and names and faces of high school classmates from 50 years ago (Bahrick, Bahrick, & Wittlinger, 1975). However, the simple presentation (and multiple presentations) of information does not always lead to good memory (e.g., Berkerian & Baddeley, 1980). Myers (1916), in one of the first studies of what was called incidental perception, showed that people underestimated the size of a one-dollar bill but overestimated the size of a five-dollar bill although both of these tendencies appear to decrease as age and experience increases. Another classic study regarding everyday attention and memory for common objects examined people's memory for the features of common coins. Specifically, Nickerson and Adams (1979) asked participants to draw the features and layout of an American penny from memory. Although most people have seen the penny many times, participants had difficulty recalling all of the features in the correct location (see figure 20.1). While some renderings are more accurate than others, people often place features in

(a)   (b)

Figure 20.1

The classic "memory for a penny" study by Nickerson and Adams (1979), in which participants were asked to draw the features and layout of an American penny from memory. While most people have seen the penny many times, the recall task is an especially difficult and frustrating one. (a: left panel) The actual American penny. (b: right panel): Examples of drawings by participants who were asked to draw the penny from memory. While some renderings have certain degrees of accuracy, oftentimes people place features in incorrect locations or misremember features from the other side of the penny (e.g., "United States of America").

incorrect locations or misremember features from the other side of the penny. More recent work has extended this notion to other more relevant objects and information, such as the keypads of calculators and telephones (Rinck, 1999), letters on keyboards (Liu, Crump, & Logan, 2010; Snyder, Ashitaka, Shimada, Ulrich, & Logan, 2014), the layout of frequently used elevator buttons (Vendetti, Castel, & Holyoak, 2013), ubiquitous and popular logos (Blake, Nazarian, & Castel, 2015), and aspects of road signs, such as the stop sign (Martin & Jones, 1998). These findings suggest that constant exposure, interaction, and use do not necessarily lead to accurate spatial recall but may allow for more general, gist-based memory (Wolfe, 1998).

Memory is a reconstructive process (Bartlett, 1932), and the findings from the historical and influential penny study support that this process is often at play. Specifically, the memory representations for the coins often involve blending features from other coins or from the other side of the penny (see also Rubin & Kontis, 1983; Jones & Martin, 1997). This pattern holds even with things that we think we see frequently, likely pay attention to more often, are designed for recognition and memorability, and feel we should remember better, like the Apple logo (Blake, Nazarian, & Castel, 2015). In the Apple logo study that we did in our laboratory at UCLA, we asked participants to first draw, and then later recognize the Apple logo. Participants thought they would do much better at drawing and recognizing the very frequently seen Apple, relative to their actual performance, especially when they made their predictions before completing the memory task, offering some novel metacognitive insight regarding potential overconfidence in these types of situations. Moreover, their drawings often exhibited features present on real apples or other minimalistic or similar logos, but were not in fact part of the actual Apple logo, suggesting that their memory for the logo was a reconstruction that began with the gist, or essence, of an apple-like logo. However, with explicit practice with verbal materials and songs, memory can be exceptionally good, such as when rehearsing and recalling the Star-Spangled Banner (Rubin, 1977). In addition, even without explicit awareness, a frequently heard song can get "stuck in your head," despite very little attention or effort used to intentionally encode the song (Hyman et al., 2013). More recent research involving memory for coins has shown that when asked to deliberately remember their features, after a brief presentation, people in fact show accurate memory (Marmie & Healy, 2004), suggesting important differences in the efficiency of incidental (or passive) versus intentional encoding of the features.

State-of-the-Art Review of Selected Area

Does frequent exposure to certain information ensure a strong memory representation, and if we pay attention to something, will we later remember it? These are common and important questions that illustrate the sometimes elusive link between attention and memory. Understanding how attention can influence memory is critical for long-term retention and learning, both in intentional and incidental learning situations. The present chapter selectively reviews how attention and memory are linked in ecologically valid contexts and real-world settings, with an emphasis on real-world tasks, materials, and goals. While there is a large literature on divided attention and explicit memory (see Craik, Govoni, Naveh-Benjamin, & Anderson, 1996), the present review focuses on the role of attention in incidental, and more everyday, settings. Understanding the manner in which attention can influence memory has implications for many different real-world problems, including classroom learning, multitasking, training, and advertising to name a few.

Researchers interested in memory and attention often study these processes in the laboratory, allowing for control over a number of important variables. However, as Neisser (1982) states, this may make us study the wrong questions in the wrong context if we want to be able to translate any of the findings in the laboratory to real-world settings (but see Banaji & Crowder, 1989). Similarly, Jenkins's (1979) tetrahedral model of memory experiments emphasizes the sensitivity of memory (and presumably attention) to context, such that performance in a given situation is determined by interactions between four categories of variables: participant characteristics and goals, the cognitive strategy that is necessary for good performance, the nature of the to-be-remembered materials, and the manner in which one assesses performance. In the present review, several recent studies will be outlined that capture how attention and memory interact in everyday settings, as well as implications for how to improve memory.

Inattentional Blindness and Amnesia in Everyday Settings

To better examine how attention and incidental memory encoding operate in real-world settings, several recent studies have examined how well people remember the spatial location of objects in the environment. Take, for example, the location of the nearest fire extinguisher, relative to your office, or in your home. Do you know where it is? In a workplace environment, fire extinguishers are placed in locations that make them easily accessible and in plain view, such that they can be quickly located. However, despite having viewed these bright red objects many times, people may be unaware of their precise locations or even the fact that they have seen them so often (see figure 20.2, plate 5). To examine this issue, we (Castel, Vendetti, & Holyoak, 2012) were interested in whether people could accurately remember and locate the nearest fire extinguisher in their workplace setting and, critically, whether people display impairments in being able to remember and locate these highly visible and potentially lifesaving devices. We tested the ability of occupants of an office building (in our Psychology Department at UCLA) to recall the location of the nearest fire extinguisher (there were six on each floor), as well as other objects (e.g., clock, drinking fountain). Despite years of exposure to these potentially very important objects, a majority failed to remember the location of the nearest fire extinguisher. However, they were able to locate it relatively quickly when asked to search for it, suggesting that seeing is not the same as noticing. Several people also remarked that it was probably near the elevator (even if one was much closer), suggesting that sometimes people infer the locations, have a general or "gist-based" notion, or misremember certain aspects, without relying on actual memory for a specific (nearest) location (cf. Bartlett, 1932; Intraub & Richardson, 1989; Loftus, 1992; Wolfe, 1998). The results support an important distinction between "seeing" and "noticing" objects and reveal a novel form of inattentional amnesia for salient objects (see also Vo & Wolfe, 2012; Wolfe, 1999; Wolfe, Horowitz, & Kenner, 2005; Wolfe, Alvarez, Rosenholtz, Kuzmova, & Sherman, 2011).

The ability to locate the fire extinguisher might reflect mechanisms of attentional priority and goal-directed attention (e.g., Anderson, Laurent, & Yantis, 2011; Castel, McGillivray, & Friedman, 2012; Yantis & Johnson, 1990). Although people may not remember the location of the nearest fire extinguisher, which could be considered beneficial since unnecessary information would not be retained in memory, they can, however, locate a bright red object when this becomes goal relevant (hopefully, in the case of a real fire). It may be that when goals become activated, people can execute action-specific programs (e.g., Cañal-Bruland & van der Kamp, 2009) that enable them to locate a previously unnoticed fire extinguisher. The fire extinguisher study

Figure 20.2 (plate 5)
The location of a fire extinguisher relative to an office door (from Castel, Vendetti, & Holyoak, 2012). The study was partially inspired by the observation, recently illustrated by one of the authors of the article (depicted above), that we are remarkably good at ignoring highly visible objects, and remarkably bad at remembering their location. While taking a building safety training class, the fire safety instructor asked people to note the location of the nearest fire extinguisher, relative to their office. In the safety training class, several people admitted they did not know the location or guessed at the location, and people were told they should learn the location of the nearest fire extinguisher. Upon returning to his office, one of the authors of the eventual field research article (K. H., of Castel, Vendetti, & Holyoak, 2012) made a conscious effort to look for the nearest fire extinguisher and made a startling discovery: The conspicuously placed bright red object in question was right next to his office door, in plain view and literally inches from the doorknob that he had turned for the past 25 years!

also created an important learning event via the initial failed retrieval, as participants were asked to locate the nearest fire extinguisher after they indicated verbally that they did not know the location. People typically found the fire extinguisher in less than 5 seconds, although sometimes they overlooked the closest option. In a follow-up study, conducted with the same participants after a 2-month "retention interval," all of the earlier tested participants remembered the location of the nearest fire extinguisher, suggesting that the earlier failed retrieval (and/or self-revelation that they did or did not know the location of the nearest fire extinguisher) served as an effective intervention that enhanced later memory for these lifesaving devices. This follow-up study also offers important insight on the ability to recall information that is considered salient and necessary; when participants were initially asked for the location of these devices, the location became a salient issue, further explaining why the follow-up study showed improved results. Perhaps even being required to *use* a fire extinguisher would also lead to improved recall of its spatial location.

Although people had poor memory for the location of fire extinguishers, one reason may be that they rarely had to interact with these objects, and as mentioned, people may be wasting cognitive resources to retain unnecessary information. Much like coins, people likely become habituated, and attention is not actively directed toward the objects in question. In another related study, we (Vendetti, Castel, & Holyoak, 2013) examined how people remember the spatial layout of the buttons on a frequently used elevator panel to determine if physical interaction (rather than simple exposure) would ensure effective incidental encoding of spatial information. Participants who worked in an eight-story office building (again, our convenient real-world setting, the UCLA Psychology Department) displayed very poor recall for the elevator panel (see examples of drawings in figure 20.3) but above-chance performance when asked to recognize the panel among several options. Interestingly, performance was related to how often and how recently the person had used the elevator. In contrast to their poor memory for the spatial layout of the elevator buttons, most people readily recalled small distinctive graffiti on the elevator wall. In a more implicit test, in which participants entered the elevator but the labels on the buttons were covered, the majority of participants were able to locate their office floor and eighth floor buttons when asked to point toward these buttons when in the actual elevator. However, identification was very poor for other floors (including the first floor), suggesting that even frequent interaction with information does not always lead to accurate spatial memory.

Similar findings exist in other domains, in that people have relatively poor recall for frequently encountered information, including stop signs (Martin & Jones, 1998), and even frequently encountered logos (e.g., Apple, Google) that are thought and designed to be highly memorable (see Blake, Nazarian, & Castel, 2015). However, in advertising, when certain information is more central to the focus of attention, or used when communicating, this information may be well encoded and retained (see chapter 25 in the

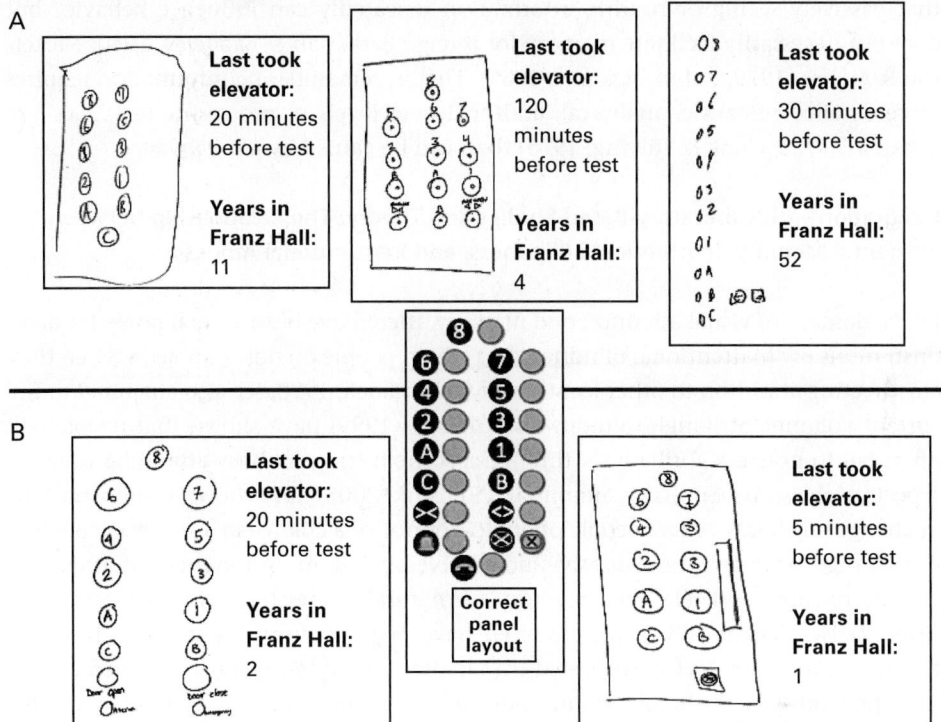

Figure 20.3
Examples of elevator panels drawn by participants, from Vendetti, Castel, and Holyoak (2013). The correct button layout representation is shown in the center. (A: top panel) Examples of incorrect drawings from participants, which varied from those that were somewhat close (i.e., had 2 columns and all 8 floors) to those whose drawing was not closely representative of the actual elevator button panel. (B: bottom panel) Examples of correct drawings from participants. Note: Participants were scored as having a correct button panel layout if they correctly drew the buttons representing the A–8 floors. Thus, participants' drawings could be scored as correct even if they did not correctly depict service buttons.

current volume). For example, professional soccer teams' sponsors advertising directly on the jerseys of the players, or the embedding of a company name in a team name (e.g., New York Red Bulls and Bayer 04 Leverkusen), can enhance memory because of attentional focus as well as repeated retrieval and use of the information during conversation. In addition, background music has been shown to influence consumer choices, such that people who were exposed to either French or German music were more likely to buy French or German wine at a supermarket (North, Hargreaves, & McKendrick, 1999). Taken together, these findings are broadly consistent with other work indicating that passively seeing or hearing information repeatedly can influence behavior but does not necessarily facilitate memory for it (e.g., Berkerian & Baddeley, 1980; Nickerson & Adams, 1979; Rubin & Kontis, 1983). That is, remembering information requires more detailed semantic, analytical, and/or deeper levels of processing (e.g., Craik & Lockhart, 1972; Craik & Tulving, 1975) than can be gained by a brief glance.

Integration with Laboratory-Based Studies and Theory: The Relationship between Incidental Memory, Inattentional Blindness, and Inattentional Amnesia

In the domain of visual attention and memory, there have been several powerful demonstrations of "inattentional blindness," in which people do not see objects when they are directing attention to other locations (Mack & Rock, 1998; see also chapter 9 in the current volume). Strikingly, Simons and Chabris (1999) have shown that people will often fail to notice a gorilla walk through a scene if they are busy attending to other aspects of the scene. Similarly, Simons and Rensink (2005) have shown demonstrations of change blindness, where people often do not notice a change in a scene, even when this change occurs with previously studied objects that are in the focus of attention. Despite having good relational memory for natural scenes (e.g., Hollingworth, Williams, & Henderson, 2001), people often have poor memory for specific objects in question. Wolfe (1999) has suggested that inattentional amnesia occurs when people have previously seen the objects in question but do not have a specific memory for having seen these objects, possibly because of a failure of attentional control and interruption in encoding the objects in memory. This has been shown in real-world settings, in which people will show inattentional blindness and amnesia for very prominent and important buildings (e.g., the library on a university campus; see figure 20.4, from Rosielle & Scaggs, 2008). The result of not noticing large-scale changes to highly familiar environments (such as the library not present in a familiar picture of campus) suggests that visual long-term memory for familiar scenes lacks the precision to be able to effectively identify even large-scale changes. In addition, other participants were asked to estimate the difficulty of change detection for another individual who would be asked to perform the same task. Those individuals felt that others should easily be able to notice such a large change, suggesting a metacognitive illusion regarding the ability to detect these changes (see also Levin, 2002). It is likely that the fire extinguisher falls prey to inattentional blindness and subsequent object-based amnesia and could also be subject to habituation effects, leading to poorer memory for it. As shown in other laboratory-based visual search tasks, inattentional amnesia and inefficiencies in visual search can influence how we locate and remember objects in real-world settings (e.g., Kingstone et al., 2003; Vo & Wolfe, 2012; Wolfe, Horowitz, & Kenner, 2005; Wolfe, Alvarez, Rosenholtz, Kuzmova, & Sherman, 2011) although we might not always be aware of the effect of inattentional blindness (Levin, 2002).

Original scene   Altered scene

Figure 20.4
Stimuli from the change detection study entitled "What if they knocked down the library and no-body noticed? The failure to detect large changes to familiar scenes" by Rosielle and Scaggs (2008). College students were asked to identify what (if anything) was wrong with the pictures of a highly familiar scene from their college campus and were surprisingly poor at identifying the seemingly large change to the familiar environment (i.e., the missing library building in the altered scene). In addition, other participants were asked to estimate the difficulty of change detection for *another* individual who would be asked to perform the same task, and these participants felt that these other participants *should* be able to detect this seemingly large change, suggesting a metacognitive illusion regarding the ability to detect these changes.

Another mechanism at play when recalling frequently encountered information is a visualization strategy, and subsequent reconstruction of this visual image during retrieval. This reconstruction can lead to predictable and theoretically informative memory errors. For example, attempting to visualize the spatial layout of the elevator panel may trigger more erroneous generic information that can interfere with the actual memory for the correct layout, possibly due to schema-based intrusions (see also Rubin & Kontis, 1983) and the reliance on a sometimes accurate gist-based approach (Wolfe, 1998). Reliance on reconstructive visualization can also lead to vivid and convincing false memories. People will sometimes falsely remember objects that were not actually presented in a scene if their schema for that type of scene dictates that certain objects should indeed be present. For example, people will falsely remember that a graduate student's office had books on the bookshelf, even if no books were actually present (Brewer & Treyens, 1981). In addition, Roediger and McDermott (1995) showed that people would falsely recall words that were not presented in lists. In this study, participants were asked to study a list of words (awake, dream, blanket, etc.), and when tested for recall, they would frequently falsely recall the critical lure, or the word "sleep." A similar mechanism may be the culprit when people have vivid, but not always accurate, "flashbulb" memories of events that are actually more reconstructive in nature (see Schmolck, Buffalo, & Squire, 2000).

Object or item distinctiveness can also play an important role in how well we remember the presence or absence of certain objects in the real world, in terms of capturing attention and leading to good memory for the object information in question. In the elevator study, a particularly striking finding was that a remarkable number of participants incorrectly recalled the location of the floor indicator (the illuminating and dynamic sign that indicates what floor the elevator is at), something that people will often attend to when traveling in an elevator, sometimes to avoid making eye contact with others in the elevator. In striking contrast, participants' recall was remarkably good for the unusual animal-like graffiti (shaped like either a dog or a cat, depending on the elevator) that was present in both elevators. This disparity in recall may be related to the degree of match each object had in terms of the relation to an elevator schema. The actual position of the elevator floor indicator (top right, above the button panel) may have conflicted with a prior elevator layout schema (indicator in top middle, centered above the doors), creating interference, whereas the peculiar graffiti presumably did not match an elevator schema at all, therefore triggering allocation of attention, noticeable distinctiveness, and hence better encoding into memory (Hunt & McDaniel, 1993; Parker, Wilding, & Akerman, 1998; von Restorff, 1933). It may be that the human memory and attention systems may become tuned to ignore information that is constantly present, as there is no functional reason one needs to encode the precise details of the Apple logo, except perhaps to detect or spot counterfeit logos (which may be more prevalent in the growing market of pseudo-Apple products), or the location of the fire extinguisher unless there is an actual fire. We may rely on an archival system of information such that only necessary information is stored and highly active in memory while less relevant details are not accessible (Nickerson, 1980). In addition, memory for specific episodes may become more semantic in nature, such that we have a more generic memory for certain events, as well as a more personal memory for other events. Roediger and Crowder (1976) showed that people will tend to remember the first few presidents, as well as the most recent presidents, suggesting that primacy and recency effects can occur for semantic memory. This is likely due to recency and frequency of use when recalling information that is essentially encoded incidentally, and these factors could also influence how people recall historic events.

Ways to Enhance Memory and Intentional Learning

One way to enhance memory is through some form of effortful and semantic processing of information, and often this will provide the necessary first step of capturing attention. In real-world settings, this can have important implications for learning in classrooms, as well as marketing. Many people, including students of all ages, have the intuition that items that are easier to process are then easier to remember, even when this is not true (Koriat, 2008). Interestingly, for those who view intelligence as a changeable construct, creating challenges or difficulties in education and during study are viewed as more engaging and helpful. The people who tend to fall prey to the idea that easy learning means easy remembering tend to also have the notion that intelligence is a fixed trait (Miele, Finn, & Molden, 2011). In addition, Werth and Strack (2003) showed that questions and answers that were easy to read, as opposed to difficult, produced higher ratings of judgment that the participant would answer the question correctly. Additionally, another study showed that the fluency of encoding is associated with predictions of improved recall (Hertzog, Dunlosky, Robinson, & Kidder, 2003). However, research has shown that if the to-be-learned material is processed in a way that challenges the learner to a certain degree, learning is enhanced for this material, a concept known as "desirable difficulties" (Bjork, 1994; McDaniel & Butler, 2010). Desirable difficulties have been shown to enhance learning in classroom settings. For example, Diemand-Yauman et al. (2011) demonstrated that text in a disfluent typeface (e.g., Monotype Corsiva) was remembered better than text in a clear typeface (e.g., Arial). Similarly, Sungkhasettee, Friedman, and Castel (2011) found that people recalled inverted words better than upright words, even though people's predictions regarding memorability did not differentiate between the upright and inverted words. In a real-world setting, this desirable difficulty can be exploited by marketing and advertising, by making certain information more distinctive, as well as more difficult to process (see figure 20.5). Accordingly, this could have the (perhaps intentional) effect of leading to attentional capture of this information, as well as better memory for the inverted and somewhat otherwise-bland/not memorable name of the "The Standard" hotel.

Retrieval practice can enhance attention and memory for information that may be the subject of inattentional blindness. The "field study" setting offers a relatively simple but powerful form of intervention training that may enhance memory for the nearest fire extinguisher through subsequent retrieval practice. For example, in the fire extinguisher study, participants in the initial session engaged in a form of errorful learning (a desirable difficulty), often failing to know the location of the nearest fire extinguisher, but then finding it, which likely enhanced memory when tested in the 2-month follow-up. This form of initial failed retrieval and retrieval practice (e.g., Bjork, 1994; Karpicke & Roediger, 2008; Kornell, Hays, & Bjork, 2009; McGillivray & Castel, 2010) could allow for the potent formation of an episodic memory (cf. Tulving, 1983) for the location of the fire extinguisher, thus acquiring information that could potentially prove to be lifesaving in the event of a fire. Recent work suggests that survival processing can enhance memory (e.g., Nairne & Pandeirada, 2010), and that memory can and should be tuned to adaptive survival mechanisms, such as remembering the location of a lifesaving object. Asking people to search for a survival-relevant object during training may enhance the likelihood of finding this object

Figure 20.5
This iconic yet simplistic Hollywood hotel sign can capture attention due to the distinctive inverted word, even when the word itself not very unique (or standard). The processing of inverted words can also confer memory benefits (e.g., Sungkhasettee et al., 2011), as people engage in an additional effortful process for these words, resulting in what may be a desirable difficulty (Bjork, 1994) in terms of how this additional processing can lead to greater retrieval strength at a later time.

during an actual fire, or when searching visual displays for dangerous and rare items, and when training people to use surveillance systems (Fleck & Mitroff, 2007; Gelernter, 2013; Wolfe, Horowitz, & Kenner, 2005).

In terms of training and transfer, it would be informative to know if the awareness of the location of the fire extinguisher in one context (e.g., at one's office) may enhance attention and memory for fire extinguishers in other settings, possibly as a result of failing to initially find it in the first context. Anecdotally, the authors and many participants in the fire extinguisher study later noted that fire extinguishers in other settings and locations now appear to "pop out" at them, suggesting that retrieval failure can lead to enhanced noticing. This may also be related to the observation, for example, that when people buy a new car, they then begin to notice the prevalence of that same car that they did not notice before, perhaps as a result of attentional focus and personal relevance of the objects in question.

While the current chapter focused on incidental memory as a by-product of attention, there is a large literature on how attention (or lack of attention) impacts intentional learning. In terms of episodic memory and attention, there are a number of studies demonstrating that impairments in attention lead to poorer memory, especially in terms of remembering associations (Naveh-Benjamin, 2000; Castel & Craik, 2003). Divided attention also leads to what is known as the cocktail party effect, in which hearing one's name in a seemingly unattended channel captures attention even when people were not attempting to monitor this source of input (Moray, 1959), although this effect may depend on working memory capacity (Conway, Cowan, & Bunting, 2001). Distraction can lead to significant brain-related activity that impairs learning (Foerde, Knowlton, & Poldrack, 2006). Distraction, and divided attention, also plays an important role in real-world settings when trying to remember names and faces, as well as in how talking on a cell phone can lead to poor memory and inattentional blindness (Hyman et al., 2010; Strayer, Drews, & Johnston, 2003), and when the mind wanders while reading (see chapter 10 in the current volume).

Summary and Future Directions

In summary, the present review highlights how attention and memory are linked in real-world settings, and how and when incidental encoding can lead to strong memories, as well as fleeting and reconstructive representations. Several mechanisms are involved when incidental memory is compromised, including habituation, inattentional blindness and amnesia, as well as lack of semantic processing. Methods to improve the products of incidental memory include retrieval practice, goal-based attention, distinctness, and the use of desirable difficulties that result in enhanced processing.

It is critical that future studies build off of the important insight gathered from laboratory-based attention and memory studies. The use of more ecologically relevant materials, in combination with more real-world settings and in vivo field studies, can provide a complementary theoretical approach for studying cognition and have more translational impact (e.g., Dodd et al., 2012; Kingstone et al., 2003; Kingstone, Smilek, Birmingham, Cameron, & Bischof, 2005; Smilek et al., 2007). This translational approach emphasizes the observation and description of human behavior, as well as the personal and subjective reports that often accompany people's behavior (including their successes and failures), as people engage in tasks in real-world situations.

The use of new technology plays a major role in how we modify attention and what we attempt to remember. For example, we now heavily rely on Google or Wikipedia as a source of knowledge, which can lead to good incidental memory for where to find information (the virtual location, Web site, or electronic file folder on our computer), even when we are unable to recall the actual information in question (Sparrow, Liu, & Wegner, 2011). There may be good reasons not to burden one's memory for certain information, as long as it can be made accessible when needed. For example, one may not require a precise representation of the location of a fire extinguisher because one understands that fire extinguishers are placed in a systematic manner, and *should* be relatively easy to find when needed. To reduce information "overload," relying on one's external access to information may in fact be advantageous.

In terms of student learning in classroom settings, as students often use laptops to take notes during lectures, this practice of typing (rather than writing or scribbling) can also lead to people's "transcribing lectures" in a verbatim manner rather than focusing on important conceptual information (e.g., Bui, Myerson, & Hale, 2013; Mueller & Oppenheimer, 2014). Other important directions for future study include how attention and memory are modified during new learning practices and in stressful settings. For example, the role of attention during intentional learning has important implications when learning takes place in potentially distracting or stressful settings (e.g., military training), and in eyewitness memory situations that involve stress (e.g., Morgan, Hazlett, Baranoski, Doran, Southwick, & Loftus, 2007), as well as when learning is self-guided or involves dynamic learning environments (e.g., in large virtual classrooms; see Farley, Risko, & Kingstone, 2013; Risko et al., 2012; Szpunar, Moulton, & Schacter, 2013). The role of attention during incidental and intentional learning, and the associated subjective confidence, has direct implications for eyewitness testimony (see also chapter 21 in the current volume), as well as for how the face of an older criminal can disappear when situated in a tourist and retirement friendly beach city (as was the case with Whitey Bulger, the notorious Boston mobster who went unnoticed in Santa Monica, California, for over a decade). Furthermore, future studies should investigate the effect expertise has on incidental memory. Participants showed poor spatial recall for the locations of fire extinguishers but performed better when asked again during a 2-month follow-up study. It may be that object relevance plays a key role, and that perhaps volunteer firefighters will demonstrate better recall (and motivated memory) as it is part of their training to locate and use fire extinguishers or other fire-relevant equipment, often in stressful situations.

Most of the studies described in the present review test college undergraduates as the sample. It is important to know how these findings generalize to people of varying ages, backgrounds, and levels of motivation. Thus, a fruitful avenue for further study is understanding how cognitive aging can lead to important changes in how one allocates attention, and also what one tries to remember. For instance, some research in this area suggests that aging can lead to changes in goal-based attention and not simply declines in global attention resources (see Castel, McGillivray, & Friedman, 2012; Castel, 2008). In addition, the potential inability to control attention (or greater distraction) in older adults may lead to the surprising by-product of incidental memory for potential information that can be used for creative problem solving (e.g., Kim, Hasher, & Zacks, 2007). Finally, a parallel future direction is investigating the degree to which people are aware of how attention can modify incidental learning, and whether meta-cognitive awareness of attention and memory shortcomings (e.g., Levin, 2002) can improve performance, such as when people become aware that they have been mind wandering, or, contrary to their own intuitions, are not aware of the location of the nearby fire extinguisher, despite having seen it every day.

Box 20.1

Key Points

- Seeing or hearing something many times does not mean it will be well remembered. Many studies have shown that simple repetition and passive encoding do not lead to effective learning and memory, and this may be counter to what many learners might think.

- Inattentional blindness and inattentional amnesia can result for frequently encountered objects and information, such that people may stop noticing (and/or not be able to recall) the location of the nearest fire extinguisher or the button panels on an elevator.

- Retrieval failure, and awareness of inattentional amnesia, can result in a potent learning event, if people seek to test, and restudy, the information in question, and this mechanism has important theoretical and practical value.

Box 20.2

Outstanding Issues

- New technology allows us to be able to off-load information, as opposed to using our own attention and memory, but can also lead to distraction, and future research will assess how this distraction can impair learning and memory in the (virtual) classroom.

- Metacognitive awareness of attention and memory illusions can be informative, and future research will be needed to better understand why we do/do not appreciate the principles of attention and memory in everyday settings, and how this can inform training and education.

Acknowledgments

We appreciate the helpful comments and suggestions from Matthew Rhodes, Michael Dodd, and several anonymous reviewers.

References

Anderson, B. A., Laurent, P. A., & Yantis, S. (2011). Value-driven attentional capture. *Proceedings of the National Academy of Sciences of the United States of America, 108*, 10367–10371.

Bahrick, H. P., Bahrick, P. O., & Wittlinger, R. P. (1975). Fifty years of memory for names and faces: A cross-sectional approach. *Journal of Experimental Psychology. General, 104*, 54–75.

Banaji, M. R., & Crowder, R. G. (1989). The bankruptcy of everyday memory. *American Psychologist, 44*, 1185–1193.

Bartlett, F. C. (1932). *Remembering: A study in experimental and social psychology.* Cambridge, UK: Cambridge University Press.

Berkerian, D. A., & Baddeley, A. D. (1980). Saturation advertising and the repetition effect. *Journal of Verbal Learning and Verbal Behavior, 19*, 17–25.

Bjork, R. A. (1994). Memory and metamemory considerations in the training of human beings. In J. Metcalfe & A. Shimamura (Eds.), *Metacognition: Knowing about knowing* (pp. 185–205). Cambridge, MA: MIT Press.

Blake, A. B., Nazarian, M., & Castel, A. D. (2015). The Apple of the mind's eye: Everyday attention, metamemory and reconstructive memory for the Apple logo. *The Quarterly Journal of Experimental Psychology, 68*, 858–865. doi: 10.1080/17470218.2014.1002798

Brady, T. F., Konkle, T., Alvarez, G. A., & Oliva, A. (2008). Visual long-term memory has a massive storage capacity for object details. *Proceedings of the National Academy of Sciences of the United States of America, 105*, 14325–14329.

Brewer, W. F., & Treyens, J. C. (1981). Role of schemata in memory for places. *Cognitive Psychology, 13*, 207–230.

Bui, D. C., Myerson, J., & Hale, S. (2013). Notetaking with computers: Exploring alternative strategies for improved recall. *Journal of Educational Psychology, 105*, 299–309.

Cañal-Bruland, R., & van der Kamp, J. (2009). Action goals influence action-specific perception. *Psychonomic Bulletin & Review, 16*, 1100–1105.

Castel, A. D. (2008). The adaptive and strategic use of memory by older adults: Evaluative processing and value-directed remembering. In A. S. Benjamin & B. H. Ross (Eds.), *The psychology of learning and motivation* (Vol. 48, pp. 225–270). London: Academic Press.

Castel, A. D., & Craik, F. I. M. (2003). The effects of aging and divided attention on memory for item and associative information. *Psychology and Aging, 18*, 873–885.

Castel, A. D., McGillivray, S., & Friedman, M. C. (2012). Metamemory and memory efficiency in older adults: Learning about the benefits of priority processing and value-directed remembering. In M. Naveh-Benjamin & N. Ohta (Eds.), *Memory and aging: Current issues and future directions* (pp. 245–270). New York: Psychology Press.

Castel, A. D., Vendetti, M., & Holyoak, K. J. (2012). Fire drill: Inattentional blindness and amnesia for the location of fire extinguishers. *Attention, Perception & Psychophysics*, *74*, 1391–1396.

Conway, A. R., Cowan, N., & Bunting, M. F. (2001). The cocktail party phenomenon revisited: The importance of working memory capacity. *Psychonomic Bulletin & Review*, *8*, 331–335.

Craik, F. I. M., Govoni, R., Naveh-Benjamin, M., & Anderson, N. D. (1996). The effects of divided attention on encoding and retrieval processes in human memory. *Journal of Experimental Psychology. General*, *125*, 159–180.

Craik, F. I. M., & Lockhart, R. S. (1972). Levels of processing: A framework for memory research. *Journal of Verbal Learning and Verbal Behavior*, *11*, 671–684.

Craik, F. I. M., & Tulving, E. (1975). Depth of processing and the retention of words in episodic memory. *Journal of Experimental Psychology. General*, *104*, 268–294.

Diemand-Yauman, C., Oppenheimer, D. M., & Vaughan, E. B. (2011). Fortune favors the bold (and the italicized): Effects of disfluency on educational outcomes. *Cognition*, *118*, 111–115.

Dodd, M. D., Weiss, N., McDonnell, G. P., Sarwal, A., & Kingstone, A. (2012). Gaze cues influence memory … but not for long. *Acta Psychologica*, *141*, 270–275.

Farley, J., Risko, E. F., & Kingstone, A. (2013). Everyday attention and lecture retention: The effects of time, fidgeting, and mind wandering. *Frontiers in Psychology*, *4*, 619.

Fleck, M. S., & Mitroff, S. R. (2007). Rare targets are rarely missed in correctable search. *Psychological Science*, *18*, 943–947.

Foerde, K., Knowlton, B. J., & Poldrack, R. A. (2006). Modulation of competing memory systems by distraction. *Proceedings of the National Academy of Sciences of the United States of America*, *103*, 11778–11783.

Gelernter, J. (2013, November). Effective threat detection for surveillance. In *2013 IEEE International Conference on Technologies for Homeland Security (HST)* (pp. 290–296). IEEE.

Hertzog, C., Dunlosky, J., Robinson, A. E., & Kidder, D. P. (2003). Encoding fluency is a cue used for judgments about learning. *Journal of Experimental Psychology: Learning, Memory, and Cognition*, *29*, 22–34.

Hollingworth, A., Williams, C. C., & Henderson, J. M. (2001). To see and remember: Visually specific information is retained in memory from previously attended objects in natural scenes. *Psychonomic Bulletin & Review*, *8*, 761–768.

Hunt, R. R., & McDaniel, M. A. (1993). The enigma of organization and distinctiveness. *Journal of Memory and Language*, *32*, 421–445.

Hyman, I. E., Jr., Boss, S. M., Wise, B. M., McKenzie, K. E., & Caggiano, J. M. (2010). Did you see the unicycling clown? Inattentional blindness while walking and talking on a cell phone. *Applied Cognitive Psychology*, *24*, 597–607.

Hyman, I. E., Jr., Burland, N. K., Duskin, H. M., Cook, M. C., Roy, C. M., McGrath, J. C., et al. (2013). Going Gaga: Investigating, creating, and manipulating the song stuck in my head. *Applied Cognitive Psychology, 27,* 204–215.

Intraub, H., & Richardson, M. (1989). Wide-angle memories of close-up scenes. *Journal of Experimental Psychology: Learning, Memory, and Cognition, 15,* 179–187.

Jenkins, J. J. (1979). Four points to remember: A tetrahedral model of memory experiments. In L. S. Cermak & F. I. M. Craik (Eds.), *Levels of processing in human memory* (pp. 429–446). Hillsdale, NJ: Erlbaum.

Jones, G. V., & Martin, M. (1997). Handedness dependency in recall from everyday memory. *British Journal of Psychology, 88,* 609–619.

Karpicke, J. D., & Roediger, H. L. (2008). The critical importance of retrieval for learning. *Science, 319,* 966–968.

Kim, S., Hasher, L., & Zacks, R. T. (2007). Aging and a benefit of distractibility. *Psychonomic Bulletin & Review, 14,* 301–305.

Kingstone, A., Smilek, D., Birmingham, E., Cameron, D., & Bischof, W. F. (2005). Cognitive ethology: Giving real life to attention research. In J. Duncan, L. Phillips, & P. McLeod (Eds.), *Measuring the mind: Speed, control and age.* (pp. 341–358). Oxford, UK: Oxford University Press.

Kingstone, A., Smilek, D., Ristic, J., & Eastwood, J. D. (2003). Attention, researchers! It is time to take a look at the real world. *Current Directions in Psychological Science, 12,* 176–180.

Koriat, A. (2008). Easy comes, easy goes? The link between learning and remembering and its exploitation in metacognition. *Memory & Cognition, 36,* 416–428.

Kornell, N., Hays, M. J., & Bjork, R. A. (2009). Unsuccessful retrieval attempts enhance subsequent learning. *Journal of Experimental Psychology: Learning, Memory, and Cognition, 35,* 989–998.

Levin, D. (2002). Change blindness blindness: As visual metacognition. *Journal of Consciousness Studies, 9,* 111–130.

Liu, X., Crump, M. J. C., & Logan, G. D. (2010). Do you know where your fingers have been? Explicit knowledge of the spatial layout of the keyboard in skilled typists. *Memory & Cognition, 38,* 474–484.

Loftus, E. F. (1992). When a lie becomes memory's truth: Memory distortion after exposure to misinformation. *Current Directions in Psychological Science, 1,* 121–123.

Mack, A., & Rock, I. (1998). *Inattentional blindness.* Cambridge, MA: MIT Press.

Marmie, W. R., & Healy, A. F. (2004). Memory for common objects: Brief intentional study is sufficient to overcome poor recall of US coin features. *Applied Cognitive Psychology, 18,* 445–453.

Martin, M., & Jones, G. V. (1998). Generalizing everyday memory: Signs and handedness. *Memory & Cognition, 26,* 193–200.

McDaniel, M. A., & Butler, A. C. (2010). A contextual framework for understanding when difficulties are desirable. In A. S. Benjamin (Ed.), *Successful remembering and successful forgetting: Essays in honor of Robert A. Bjork* (pp. 175–199). New York: Psychology Press.

McGillivray, S., & Castel, A. D. (2010). Memory for age–face associations: The role of generation and schematic support. *Psychology and Aging, 25,* 822–832.

Miele, D. B., Finn, B., & Molden, D. C. (2011). Does easily learned mean easily remembered? It depends on your beliefs about intelligence. *Psychological Science, 22,* 320–324.

Moray, N. (1959). Attention in dichotic listening: Affective cues and the influence of instructions. *Quarterly Journal of Experimental Psychology, 11,* 56–60.

Morgan, C. A., III, Hazlett, G., Baranoski, M., Doran, A., Southwick, S., & Loftus, E. (2007). Accuracy of eyewitness identification is significantly associated with performance on a standardized test of face recognition. *International Journal of Law and Psychiatry, 30,* 213–223.

Mueller, P. A., & Oppenheimer, D. M. (2014). The pen is mightier than the keyboard: Advantages of longhand over laptop note-taking. *Psychological Science, 25,* 1159–1168.

Myers, G. C. (1916). Incidental perception. *Journal of Experimental Psychology, 1,* 339–350.

Nairne, J. S., & Pandeirada, J. N. S. (2010). Adaptive memory: Ancestral priorities and the mnemonic value of survival processing. *Cognitive Psychology, 61,* 1–22.

Naveh-Benjamin, M. (2000). Adult age differences in memory performance: Tests of an associative deficit hypothesis. *Journal of Experimental Psychology: Learning, Memory, and Cognition, 26,* 1170–1187.

Neisser, U. (1982). Memory: What are the important questions? In U. Neisser & I. E. Hyman, Jr., (Eds.), *Memory observed* (pp. 3–18). New York: Worth.

Nickerson, R. S. (1965). Short-term memory for complex meaningful visual configurations: A demonstration of capacity. *Canadian Journal of Psychology, 19,* 155–160.

Nickerson, R. S. (1980). Motivated retrieval from archival memory. In J. H. Flowers (Ed.), *Nebraska symposium on motivation, 1980: Cognitive processes* (pp. 73–119). Lincoln: University of Nebraska Press.

Nickerson, R. S., & Adams, M. J. (1979). Long-term memory for a common object. *Cognitive Psychology, 11,* 287–307.

North, A. C., Hargreaves, D. J., & McKendrick, J. (1999). The influence of in-store music on wine selections. *Journal of Applied Psychology, 84,* 271–276.

Parker, A., Wilding, E., & Akerman, C. (1998). The von Restorff effect in visual object recognition memory in humans and monkeys: The role of frontal/perirhinal interaction. *Journal of Cognitive Neuroscience, 10,* 691–703.

Rinck, M. (1999). Memory for everyday objects: Where are the digits on numerical keypads? *Applied Cognitive Psychology, 13,* 329–350.

Risko, E. F., Anderson, N., Sarwal, A., Engelhardt, M., & Kingstone, A. (2012). Everyday attention: Variation in mind wandering and memory in a lecture. *Applied Cognitive Psychology*, *26*, 234–242.

Roediger, H. L., III, & Crowder, R. G. (1976). A serial position effect in recall of United States presidents. *Bulletin of the Psychonomic Society*, *8*, 275–278.

Roediger, H. L., III, & McDermott, K. B. (1995). Creating false memories: Remembering words not presented in lists. *Journal of Experimental Psychology: Learning, Memory, and Cognition*, *21*, 803–814.

Rosielle, L. J., & Scaggs, W. J. (2008). What if they knocked down the library and nobody noticed? The failure to detect large changes to familiar scenes. *Memory*, *16*, 115–124.

Rubin, D. C. (1977). Very long-term memory for prose and verse. *Journal of Verbal Learning and Verbal Behavior*, *16*, 611–621.

Rubin, D. C., & Kontis, T. C. (1983). A schema for common cents. *Memory & Cognition*, *11*, 335–341.

Schmolck, H., Buffalo, E. A., & Squire, L. R. (2000). Memory distortions develop over time: Recollections of the OJ Simpson trial verdict after 15 and 32 months. *Psychological Science*, *11*, 39–45.

Simons, D. J., & Chabris, C. F. (1999). Gorillas in our midst: Sustained inattentional blindness for dynamic events. *Perception*, *28*, 1059–1074.

Simons, D. J., & Rensink, R. A. (2005). Change blindness: Past, present, and future. *Trends in Cognitive Sciences*, *9*, 16–20.

Smilek, D., Eastwood, J. D., Reynolds, M. G., & Kingstone, A. (2007). Metacognitive errors in change detection: Missing the gap between lab and life. *Consciousness and Cognition*, *16*, 52–57.

Snyder, K. M., Ashitaka, Y., Shimada, H., Ulrich, J. E., & Logan, G. D. (2014). What skilled typists don't know about the QWERTY keyboard. *Attention, Perception & Psychophysics*, *76*, 162–171.

Sparrow, B., Liu, J., & Wegner, D. M. (2011). Google effects on memory: Cognitive consequences of having information at our fingertips. *Science*, *333*, 776–778.

Strayer, D. L., Drews, F. A., & Johnston, W. A. (2003). Cell phone–induced failures of visual attention during simulated driving. *Journal of Experimental Psychology. Applied*, *9*, 23–32.

Sungkhasettee, V. W., Friedman, M. C., & Castel, A. D. (2011). Memory and metamemory for inverted words: Illusions of competency and desirable difficulties. *Psychonomic Bulletin & Review*, *18*, 973–978.

Szpunar, K. K., Moulton, S. T., & Schacter, D. L. (2013). Mind wandering and education: from the classroom to online learning. *Frontiers in Psychology*, *4*, 495.

Tulving, E. (1983). *Elements of episodic memory*. New York: Oxford University Press.

Vendetti, M., Castel, A. D., & Holyoak, K. J. (2013). The floor effect: Impoverished spatial memory for elevator buttons. *Attention, Perception & Psychophysics*, *75*, 636–643.

Vo, M. L.-H., & Wolfe, J. M. (2012). When does repeated search in scenes involve memory? Looking at versus looking for objects in scenes. *Journal of Experimental Psychology: Human Perception and Performance, 38*, 23–41.

von Restorff, H. (1933). Uber die wirkung von bereichsbildun-gen im spurenfeld [On the effect of spheres formations in the trace field]. *Psychologische Forschung, 18*, 299–342.

Werth, L., & Strack, F. (2003). An inferential approach to the knew-it-all-along phenomenon. *Memory, 11*, 411–419.

Wolfe, J. M. (1998). Visual memory: What do you know about what you saw? *Current Biology, 8*, R303–R304.

Wolfe, J. M. (1999). Inattentional amnesia. In V. Coltheart (Ed.), *Fleeting memories* (pp. 71–94). Cambridge, MA: MIT Press.

Wolfe, J. M., Alvarez, G. A., Rosenholtz, R., Kuzmova, Y. I., & Sherman, A. M. (2011). Visual search for arbitrary objects in real scenes. *Attention, Perception & Psychophysics, 73*, 1650–1671.

Wolfe, J. M., Horowitz, T. S., & Kenner, N. (2005). Rare items often missed in visual searches. *Nature, 435*, 439–440.

Yantis, S., & Johnson, D. N. (1990). Mechanisms of attentional priority. *Journal of Experimental Psychology: Human Perception and Performance, 16*, 812–825.

21
Eyewitness Memory

Kerri L. Pickel

Historical Context

Since the mid-1970s, eyewitness memory has emerged as a topic of strong interest within the field of psychology. Inspired by Loftus's classic demonstrations of the misinformation effect, researchers collected data showing that witnesses' reports are far from perfect. Their accuracy is affected by numerous factors, including "system" variables that can be controlled by police investigators (e.g., the wording of questions during interviews, lineup composition and instructions) and "estimator" variables that cannot be controlled (e.g., the witness's age and level of anxiety). The implications for the legal system are immense. Since 1989, more than 300 individuals convicted of crimes in the United States have been exonerated through DNA testing, and in approximately 75% of these cases eyewitness error was a factor, making this the single greatest contributor to wrongful convictions (Innocence Project, n.d.). As a result, researchers have endeavored to understand the cognitive mechanisms that progressively engage from the moment a legally relevant event begins to unfold to the witness's testimony before a jury. At the beginning of this progression, attentional processes will determine what, if anything, the witness later remembers about the event and the culprit.

State-of-the-Art Review

This chapter focuses on one of the estimator variables that influences the accuracy of witnesses' descriptions of a perpetrator and their ability to identify the perpetrator in a lineup: allocation of attention during the crime event. For the present purposes, *attention* is defined as the concentration of mental effort. Witnesses commonly attempt to divide their attention among multiple stimuli, but because attentional resources are finite, such attempts will result in fewer resources allocated to any one stimulus. For example, a convenience store customer who stumbles upon a robbery while approaching the checkout counter might not only study the robber's face in an attempt to remember it but also allocate some attention to the parking lot outside to see if help is on the way or to her thoughts and fears regarding her safety (Lane, 2006). This division of resources will reduce the level of attention paid to the robber. Given this competition among stimuli, can the witness effectively encode forensically relevant details about the robber so that she will remember and report them later?

Effects of Divided Attention during Encoding

Insufficient attention allocated to a target individual can have various consequences, three of which will be discussed here. First, the witness might fail to remember the individual at all when questioned later on. Second, the witness might recall seeing the individual but not remember his or her face very well and therefore might not perform well when asked to view a lineup. Third, insufficient attention can increase suggestibility.

Inattentional Blindness   Sometimes individuals devote little or no attention to a stimulus, with the result that they do not subsequently remember it at all. In other words, the possibility exists that someone could completely miss seeing a perpetrator commit a crime. Chabris, Weinberger, Fontaine, and Simons (2011; see also Spiegel, 2011, or chapter 9 in the current volume) described a 1995 case in which a Boston police officer, Kenneth Conley, was convicted of perjury and obstruction of justice after he testified that he did not notice two other officers savagely beating a suspect even though he ran right past the attack while chasing a different suspect. The beating victim eventually turned out to be an African American undercover officer, and Conley, who is White, was accused of lying to protect his colleagues.

Chabris et al. thought Conley could have experienced *inattentional blindness*, defined as an observer's failure to notice fully visible stimuli when he or she is engaged in an attentionally demanding task (Most, Scholl, Clifford, & Simons, 2005). Previous research had demonstrated that this phenomenon can occur while observers view complex scenes. In one well-publicized study, Simons and Chabris (1999) asked participants to watch a video of two teams of players, one wearing white shirts and the other black shirts, passing basketballs to other members of their team while moving around a small open area. The participants were randomly assigned to watch either the white or the black team and to perform either a relatively easy task (counting the passes made by their assigned team) or a difficult one (keeping separate counts of the aerial and bounce passes made by their assigned team). In one version of the video, a confederate in a gorilla suit walked directly among the group of players, passing through the participants' region of attentional focus and remaining visible for 5 seconds. Remarkably, a substantial proportion of participants said they did not see the gorilla. Inattentional blindness was more likely when the task was harder; 54% of the participants performing the difficult task noticed the gorilla versus 62.5% performing the easy task. Additionally, a greater proportion of participants counting the black team's passes reported seeing the gorilla (70.5%) compared to those watching the white team (46%). Given that both the gorilla and the black team's shirts were dark in color, this last result indicates that unexpected stimuli that share visual features with attended stimuli are more often detected.

Following up on this experiment, Simons and Chabris (1999) tested the limits of the inattentional blindness effect by creating an additional video in which the gorilla was more salient. This time, she was visible for 9 seconds instead of 5 seconds, and she paused in the center of the room to face the camera and thump her chest. A new group of participants watched this video while performing the easy counting task and watching the white team. Even under these conditions, which should have led to relatively high rates of detection, only 50% of the participants noticed the gorilla.

In a naturalistic study of inattentional blindness, Hyman, Boss, Wise, McKenzie, and Caggiano (2010) observed that individuals walking through a plaza on a university campus were less likely to notice a clown wearing brightly colored clothing and riding a unicycle if they were conversing on a cell phone rather than listening to a music player or simply walking. The authors proposed that having a cell-phone conversation (but not listening to music or walking) is sufficiently demanding so as to divert attention away from other stimuli, even ones that should be quite salient (research involving attention and cell-phone use will be discussed in more detail in a following section).

Drawing upon findings like these, Chabris et al. (2011) tested their hypothesis that Officer Conley might have experienced inattentional blindness by attempting to simulate the events that occurred that night in 1995. They asked participants to pursue a confederate who jogged along a 400-meter route through a university campus while counting the number of times he touched his head (apparently to mimic the mental effort Conley might have expended while considering how to catch and overpower his suspect). The participants ran past three other confederates who staged a loud fight 8 meters from the route. After completing the route, the participants were asked first whether they had seen anything unusual and then whether they had seen a fight. When tested at night and with the fight visible for approximately 15 seconds, only 35% of the participants noticed. Chabris et al. repeated the procedure during the daytime and with the fight visible for a longer duration (approximately 30 seconds), but even then only 56% said they saw the fight. An additional experiment showed that increasing the task difficulty by requiring participants to count head touches the runner made with his left and right hands separately reduced the proportion who noticed the fight (42%) whereas eliminating the counting task increased it (72%). The authors concluded that, although only Officer Conley knows whether he saw the assault on the undercover officer, it is at least possible that he did not notice it. Conley was initially sentenced to 34 months of incarceration but successfully appealed.

In sum, the existing research shows that witnesses can completely fail to notice a person committing a crime if their attention is directed elsewhere, especially if they are performing a mentally demanding task or if the visual appearance of the perpetrator is inconsistent with the characteristics of the stimuli to which they are already attending. In such circumstances, the witnesses would not be helpful to detectives who need a physical description of the perpetrator or a lineup identification.

Divided Attention and Memory for Faces   Although police investigators may be interested in various forensically relevant details, such as a perpetrator's height, body type, and clothing color, it is obviously advantageous if witnesses can recognize his or her face. Divided attention during encoding can impair this ability, which could make witnesses unable to identify the perpetrator in a lineup, or, even worse, they may identify an innocent suspect.

Reinitz, Morrissey, and Demb (1994) asked participants to view faces either while counting stimuli that appeared concurrently or while performing no additional task. Compared to controls, participants whose attention was divided between studying the faces and the counting task were less able to recognize the faces they had seen. They were also less likely than controls to say they "remembered" these faces (i.e., that they explicitly recollected seeing them previously as opposed to merely feeling that they were familiar).

Interestingly, dividing attention while encoding faces seems to have different effects depending on the gender of the witness and the target. Palmer, Brewer, and Horry (2013) pointed out that previous research has shown that female but not male viewers attend more to own-gender faces and subsequently recognize those faces better. One reason for these findings could be developmental. As adults, both men and women may preferentially attend to own-gender faces, but women have a longer personal history of doing so, given that all infants tend to interact more with female adults and therefore spend more time attending to female faces. Thus, men's attentional preference for male faces is "undermined by an early history of selective attention for female faces" (p. 363). Another possible explanation is that women are more interested than men in initiating and maintaining social interactions with others of their own gender.

Based on this prior research, Palmer et al. (2013) predicted that dividing attention during encoding would impair female witnesses' recognition of female faces more than it would impair their recognition of male faces. Moreover, the authors expected the attention manipulation to influence male witnesses' recognition of male and female faces equally. The witnesses watched a video depicting target individuals engaged in various everyday activities. In the divided-attention condition, the witnesses pressed keys to respond to auditory tones varying in pitch. In the full-attention condition, they could hear the tones but were not required to make any responses to them. After watching the video, the witnesses viewed two lineups (each of which was either target-present or target-absent in different conditions), one for a male target from the video and one for a female target.

As predicted, divided attention reduced female witnesses' lineup performance with regard to the female target but had no significant effect on their performance regarding the male target. Unexpectedly, male witnesses performed equally well when trying to identify the female target, regardless of attention condition, but divided attention decreased their performance with respect to the male target. This pattern of results with male witnesses conflicts with some other findings (Palmer et al., 2013), so more work in this area is clearly needed.

Divided Attention and Suggestibility *Eyewitness suggestibility* refers to witnesses' tendency to report details consistent with misleading suggestions made to them after they observed a target event. For example, Schooler, Gerhard, and Loftus (1986) showed witnesses a slide sequence of a traffic accident. The critical slide showed a car at an intersection. Half of the witnesses saw a yield sign in this slide, and half saw the same slide with the sign removed. Next, witnesses completed a written questionnaire that included 17 items about the event. Witnesses who did not view the sign read a question that suggested its presence ("Did another car pass the red Datsun while it was stopped at the yield sign?") whereas those who did view the sign answered a different, nonmisleading question. Finally, all witnesses completed a final memory test asking whether they remembered several objects from the event depicted in the slides. Of the witnesses who did not see the sign, 25% reported that they did. Further research indicated that several variables (e.g., the amount of elapsed time between the event and exposure to the misleading information) influence the likelihood that witnesses will claim to remember suggested details or objects.

As Lane (2006) noted, after years of study researchers eventually concluded that errors due to misleading suggestions are probably source misattributions such that witnesses mistakenly ascribe information derived from a postevent source to the target event. Source monitoring is a demanding process in which individuals examine a memory to see whether it contains characteristics typical of perceived events, such as perceptual and contextual details, or characteristics more associated with imagined events, such as information related to the cognitive processes involved. Top-down processes may also be engaged—for example, to evaluate plausibility. As an illustration, consider a witness in Schooler et al.'s experiment who saw no yield sign. During the final memory test, when asked whether there was a yield sign, this witness might remember something about a yield sign and try to determine where he encountered this information. Inspecting his memory, he might realize that perceptual details such as the exact height of the sign are sparse, as are contextual details like its precise location on the street corner. Moreover, he remembers some cognitive operations that arose in response to encountering information about the sign; for example, he thought to himself that it made sense for a yield sign to be placed at the intersection to slow down traffic moving in the same direction as the Datsun. Although it would not be implausible or inconsistent with other known information for a yield sign to be present, the witness might conclude that, given his overall review of his memory, the source of the sign must have been the questionnaire, not the target event.

Because source monitoring is effortful, Lane (2006) hypothesized that dividing witnesses' attention during encoding of the target event would reduce the ability to perform this task effectively. His participants watched a slide sequence of a maintenance man stealing from an office. The divided-attention witnesses completed a secondary task simultaneously, whereas the full-attention witnesses completed that task later, so that they had nothing else to do while watching the slides. Next, the witnesses answered a questionnaire that suggested the presence of six nonexistent objects. As the final step, they listened to a list of 25 objects and indicated for each whether they saw it in the slides and whether they read about it in the postevent questionnaire. Note that witnesses could report encountering an object in both sources or neither. Compared to full-attention witnesses, those in the divided-attention condition were less likely to make correct source attributions for suggested objects as well as for objects that actually appeared in the slides.

Some Reasons for Division of Attention during Encoding

Divided attention can occur because, while in the presence of a target individual, witnesses simultaneously try to attend to one or more other tasks or stimuli. As a result, details associated with the perpetrator may not be encoded well, and the witnesses may be unable to provide very accurate or complete descriptions later on. This section will review three tasks that can elicit attention from witnesses: encoding information about a weapon held by a perpetrator, trying to comprehend accented speech, and judging the veracity of an individual.

Weapon Focus If witnesses allocate a large amount of attention to a weapon brandished by a perpetrator, then they are probably not paying a high level of attention to forensically relevant details associated with the perpetrator's identity. The *weapon-focus effect* is that witnesses tend to direct their attention toward a weapon held by a perpetrator, which causes them to remember the perpetrator's physical features and clothing less accurately than they would have without the weapon's presence. Numerous studies have demonstrated the effect with many types of weapons and using not only slide sequences and videos but also live, staged events (for a review, see Fawcett, Russell, Peace, & Christie, 2013). Moreover, meta-analyses show reliable effects for both identification accuracy and the accuracy of witnesses' descriptions of the perpetrator although the size of the former is smaller (Fawcett et al., 2013).

In one of the first experimental investigations of weapon focus (Loftus, Loftus, & Messo, 1987), participants viewed a slide sequence depicting customers moving through the order line at a fast-food restaurant. On four critical slides a target individual held either a handgun or a personal check. He extended the object toward the cashier, who gave him some cash. Compared to the witnesses who saw the check, those in the weapon condition were less likely to identify the target in a photo lineup (in two experiments), and they provided poorer descriptions of him (in one experiment). An important aspect of this study is that the authors collected evidence using an eye-tracking device that the witnesses directed considerable visual attention toward the gun. Eye fixations on the gun were more frequent and lasted longer than fixations on the check, suggesting that there was a trade-off between the gun and the target in the weapon condition such that paying more attention to one meant paying less attention to the other (see also Biggs, Brockmole, & Witt, 2013).

Researchers have not yet settled the question of why weapons at crime scenes attract attention. One possibility, which was originally founded on Easterbrook's (1959) cue-utilization hypothesis, is that the weapon elevates witnesses' anxiety as they recognize it as a threatening object that could be used to injure or kill them or someone else. In turn, this increased anxiety enhances the effects of competition among stimuli, as Mather and Sutherland (2011) have shown; specifically, high-priority stimuli that are already receiving a relatively large amount of resources will attract even more attention while attention to low-priority stimuli will be further diminished. Although priority is determined partly by bottom-up cues, such as perceptual contrast, it is also influenced by top-down variables, such as the viewer's goals. Assuming that witnesses would define a weapon as informative and important to the goal of self-preservation, anxiety should amplify their tendency to attend to the weapon at the expense of low-priority stimuli, such as the perpetrator's articles of clothing, that are less relevant to the goal. Some data from weapon-focus investigations (e.g., Davies, Smith, & Blincoe, 2008) support this anxiety/threat explanation. It is important to note, however, that not all witnesses to crimes, including ones involving weapons, feel intense fear while observing the event (Pickel, 2007).

An alternative explanation for the weapon-focus effect is that weapons might attract attention because they seem unusual in many environments, and observers tend to look longer and more often at unusual rather than expected objects as they attempt to reconcile the presence of the incongruous items with the scene context (Gordon, 2004). Thus, for example, our convenience store customer would know from past experience that such establishments usually contain a cash register, packages of candy on shelves, and soda machines, and therefore she would expect to see these objects. She would not, however, expect to see a 9-mm pistol in someone's hands. Upon entering through the front door, she would acquire a gist for the visual scene that greeted her and begin identifying objects within 100 ms (Gordon, 2004). If a gun were present, she would recognize its inconsistency with the convenience store schema and begin attending preferentially to it. She might feel fear, but that would not be necessary to elicit the weapon-focus effect. Several studies (e.g., Hope & Wright, 2007; Mitchell, Livosky, & Mather, 1998; Pickel, 1999, 2009) provide evidence in favor of the unusualness hypothesis.

Anxiety/threat and unusualness may both contribute to the weapon-focus effect, as suggested by Fawcett et al.'s (2013) meta-analysis. The two explanations are not incompatible, and either may be sufficient to produce the effect. In any case, it is clear that a visible weapon attracts witnesses' attention so that they subsequently remember the perpetrator less accurately.

Comprehending Accented Speech Another situation in which divided attention during encoding may impair memory for the perpetrator is when witnesses must try to understand a message spoken with an accent. Perpetrators sometimes talk to victims or bystander witnesses, giving them instructions or asking questions, or witnesses might overhear a perpetrator conversing with an accomplice. In such cases, comprehending this communication might be important for survival. Moreover, through travel or immigration, witnesses might interact with a perpetrator who speaks with what they perceive as a foreign accent. Previous research (e.g., Munro & Derwing, 1995) indicates that listeners have to work harder to understand accented versus unaccented messages. Accented speech includes deviations from native speaker pronunciation norms, so listeners may have trouble identifying some of the speaker's phonetic segments or words. Noting these results, Josh Staller and I (Pickel & Staller, 2012) predicted that trying to comprehend accented rather than unaccented messages would interfere with witnesses' ability to encode information about a perpetrator simultaneously. Specifically, we thought that witnesses' attempts to split their attention between two demanding tasks (semantically processing the message and encoding the perpetrator's appearance) would lead to worse performance on the forensically relevant, latter task.

We first used a secondary visual task to verify that processing an accented message is in fact more effortful than processing unaccented speech. Next, we conducted two experiments in which participants watched a video of a target individual delivering a brief message and subsequently tried to remember information about him. For example, in the first of these experiments, witnesses were asked to imagine that they were driving a car and had stopped for a red light when a carjacker jumped in and started giving orders. Like actual victims would, they were asked to attend to his message in order to avoid getting hurt. In one condition, the carjacker spoke with the same Midwestern U.S. accent that our witnesses did (which they would perceive as no accent), and in another condition he spoke with an Irish accent. Witnesses who heard the Irish accent described the carjacker less accurately than those who heard the Midwestern accent, and they also were less likely to identify his voice in a voice lineup. Furthermore, listening to a more complex message containing a greater number of details, which should require more effort to understand than a simpler message, caused witnesses to report less correct information about the perpetrator.

Judging Veracity   Another task that witnesses could try to perform while observing a suspect or perpetrator is judging his or her veracity. For instance, a store manager might evaluate the truthfulness of a customer who denies being a shoplifter's accomplice. The manager might let the customer go, only to realize her involvement later, at which point he would want to be able to describe her for police.

Determining whether someone is lying requires substantial attentional resources (e.g., Reinhard & Sporer, 2008). To detect lies, people monitor and evaluate cues they believe indicate deception, including nonverbal behaviors (e.g., gaze aversion, fidgeting), and cues within the content of the target's message (e.g., logical inconsistency). Notice that these deception cues are not forensically relevant details, like height and hair color, that police investigators usually hope to obtain from witnesses. Teresa Kulig, Heather Bauer, and I (Pickel, Kulig, & Bauer, 2013) hypothesized that, because judging veracity is demanding, performing this task should interfere with concurrent attempts to encode details associated with a suspect, making witnesses' subsequent reports less accurate than they would otherwise be.

Our participants watched a video of a suspect delivering a brief message. Beforehand, we asked the witnesses either to decide whether the suspect was lying or simply to observe. After watching, the witnesses completed a questionnaire asking them to provide details about the suspect's appearance and to recall his or her message as completely as possible. As expected, witnesses who judged veracity remembered the suspect's appearance and message less accurately than those who simply observed him or her. Additionally, inducing witnesses to be suspicious about the suspect's truthfulness amplified the memory effect, apparently by prompting witnesses to allocate even more resources to the judgment task as they scrutinized the suspect closely.

In follow-up research (Pickel, Klauser, & Bauer, 2014), Brittney Klauser, Heather Bauer, and I found that, like inducing suspicion, motivating witnesses to try hard to judge veracity accurately intensified the memory impairment effect. Motivation encourages witnesses to work harder to monitor cues associated with deception (Reinhard & Sporer, 2008), with the consequence that the judgment task consumes a greater amount of attentional resources.

We also discovered another intriguing result. As explained above, judging veracity leads witnesses to scrutinize the suspect. In actuality, they inspect cues they believe signal deception (e.g., gaze aversion), but we conjectured that the experience of carefully studying the suspect would cause witnesses to develop the false sense that they had elaboratively encoded forensically relevant details about him or her (e.g., hair color, height). In turn, we thought this belief would increase their self-reported levels of certainty about the accuracy of their memory and would also inflate their ratings of other "testimony-relevant judgments" (Wells & Bradfield, 1998) about the quality of their witnessing experience and their performance. The results supported our hypothesis. Although judging veracity impaired their memory of the suspect, the witnesses who performed this task ironically expressed greater certainty than controls in their physical descriptions (in one of two experiments) and in their ability to remember the suspect's message, and they reported having a better view of the suspect and paying more attention to his or her face. Also, in one experiment witnesses who judged the suspect's veracity said they had a clearer image of him in their memory compared to controls. These inflated ratings are important because jurors' verdicts are greatly affected by witnesses' expressions of certainty.

Within the context of this chapter, the major finding from this line of research is that judging a suspect's veracity and attempting to encode information about him or her are both tasks that require a high level of attentional resources. As a result, successfully performing them concurrently may be problematic for witnesses.

Integration

In order to encode details about a crime event and its perpetrator that will be valuable to police investigators, witnesses must complete two steps. First, they must notice the perpetrator and begin allocating attention to him or her. Second, they must encode sufficient information related to the perpetrator even though there may be pressure to perform another task at the same time. Two theoretical frameworks, each addressing one of these encoding steps, are briefly outlined below. The frameworks may be useful in illuminating the mechanisms underlying each and in guiding future research. Other theoretical accounts may also be applicable, but these two will be helpful illustrations of the general attentional processes that may unfold during encoding.

Allocating Attention

The first step for witnesses is to notice the perpetrator in the act of committing a crime. As described above in the section on inattentional blindness, it is possible for witnesses to fail to notice even an unexpected and dramatic event, particularly if they are already engaged in an attentionally demanding task. How do different factors interact to determine whether witnesses will notice the perpetrator and begin attending to him or her?

Most et al. (2005) attempted to answer this question by building on Neisser's (1976) concept of the "perceptual cycle," in which individuals perceive, interpret, and reinterpret information using both bottom-up and top-down processes. The basic idea is that certain stimulus properties can elicit an orienting response, at which point expectations and schemas can guide attentional exploration. As additional information is detected, it adjusts observers' interpretations of the stimuli that may be present and helps direct further exploration. As Most et al. explained, "This cycle of attentional guidance continuously enriches the emerging representations and modifies the observer's expectations, eventually leading to a conscious percept" (p. 224). The entire process happens rapidly.

Using data from experiments in which participants viewed shapes or human faces on computer displays, Most et al. (2005) refined Neisser's conceptualization. They proposed that individuals engaged in an effortful task might automatically shift their attention to an object because of its salience. This capture of attention will probably be transient and will not lead to conscious perception unless supplementary sustained processing occurs. According to the authors, sustained attentional processing depends upon *attentional set*, or the observers' predisposition to receive specific types of information. Specifically, if the properties of a newly encountered stimulus match those of the target stimuli to which observers are already attending, sustained attentional processing is more likely to follow the transient attentional shift to the new stimulus. Thus, a greater proportion of Simons and Chabris's (1999) participants reported seeing the dark-colored gorilla if they had been counting the passes made by the black rather than the white team.

Because observers cannot constantly and perfectly maintain an attentional set, Most et al. (2005) allowed that there could be points in time when a salient stimulus would be especially likely to be noticed, even if its properties are dissimilar to those of the target stimuli. Moreover, attentional sets can be voluntarily changed. For example, a convenience store customer waiting her turn in the checkout line might have been calculating the amount of money needed to pay for her purchases before noticing a man shouting at the clerk to empty the register, but then she could adopt a different goal that involved attending to the robber.

Multiple Resource Theory

After witnesses have become aware of a perpetrator, they nevertheless might not devote their full attention to him or her because of some concurrent task that also demands attention. What happens as witnesses try to multitask? Research on this issue reveals that, when participants attempt to complete two tasks at once, performance on at least one often suffers, although several variables can moderate this effect. Many recent studies of multitasking focus on cell-phone use while driving (see chapter 18 in the current volume). In some, driving performance is evaluated by behavioral responses such as reaction time to a traffic signal. More relevant to the present discussion are measures involving participants' ability to encode and remember objects encountered while driving.

Strayer and Drews (2007) asked participants to drive in a simulator that realistically recreates real-world driving environments. Their simulator was constructed using the dashboard instrumentation, steering wheel, and foot pedals from a Ford Crown Victoria sedan with an automatic transmission. It has three high-resolution displays that afford a 180° field of view, and it includes software that allows participants to navigate through various lifelike scenarios under assorted traffic conditions. An eye tracker monitors participants' eye fixations.

In a series of experiments, the authors compared participants' performance in a dual-task condition (driving while conversing with a confederate using a hands-free phone) with their performance in a single-task condition (just driving). In the first experiment, participants completed a surprise recognition test for objects in the environment after they finished driving a simulated route. Dual-task participants were much less likely than single-task participants to recognize road signs upon which they had fixated, even when controlling for the duration of the fixation. Experiment 2 showed that participants' ratings of the relevance of various objects (e.g., pedestrians, billboards) to safe driving were not correlated with recognition memory, implying that drivers do not strategically reallocate attention from less relevant objects to the phone conversation while maintaining high attention to very relevant objects. Finally, in the third experiment, participants followed a pace car that braked randomly on a simulated freeway. The authors measured the amplitude of the P300 component of the event-related potential corresponding to the onset of the pace car's brake lights. The P300 indicates the amount of attention allocated to a stimulus. The P300 amplitude was 50% smaller in the dual-task versus the single-task condition, suggesting that "drivers using a cell phone fail to see information in the driving scene because they do not encode it as well as they do when they are not distracted by the cell-phone conversation" (Strayer & Drews, 2007, p. 130).

Studies like these serve as the foundation for theoretical accounts of dual-task performance. One successful model is multiple resource theory (MRT; Wickens, 2002, 2008). According to MRT, mental resources are limited and allocatable. Every task requires some amount of resources to support it, and difficult tasks require more than easier ones. Further, resources are organized into distinct dimensions with different levels, and some dimensions are nested inside others. For example, the "stages of processing" dimension contains the perceptual and the cognitive levels. Inside the perceptual level is the "modalities" dimension containing the visual and auditory levels. Each level represents a sort of compartment of resources that are reserved for certain kinds of tasks. Thus, a visual task would draw resources from the visual level within the modalities dimension but would not draw from the auditory level. However, a very difficult visual task might deplete its modality-specific resources and require general perceptual resources that exist within the perceptual compartment but that are not specifically assigned to either modalities level.

In sum, two tasks can be performed simultaneously if there are enough available resources to support both. Success is obviously more likely if the tasks are dissimilar (i.e., they pull resources from different rather than the same levels along a particular dimension). However, task difficulty matters as well. As explained above, two dissimilar tasks can still compete for common resources, as when very challenging tasks deplete their respective level-specific resources and require supplementation from a larger common pool. Thus, whether the tasks involve different levels or not, it may be impossible to perform them together if they are sufficiently taxing. In this case, the task prioritized by the individual may be completed satisfactorily while performance on the secondary task falters. Of course, it is also possible that the individual will make errors on both tasks.

The experiences of eyewitnesses can be conceptualized using MRT. For instance, consider a driver who is stopped at a red light when a carjacker jumps into the front seat of her car. If someone had asked her on the previous day whether witnesses should try to remember a carjacker's appearance so that they can later help police find and arrest him, she would have agreed. Thus, studying the robber is one of her goals. However, it is not her only goal. The carjacker begins giving instructions, demanding that she drive him to a certain location using a particular route. To avoid injury, she must listen to his directions carefully so she can follow them without error. In this way, comprehending the carjacker's message becomes the driver's primary task, and encoding the robber's physical appearance is relegated to secondary status. Although these two tasks involve different modalities (i.e., the primary task is auditory and the secondary is visual), they may each be effortful enough to compete for common perceptual resources. In addition, the primary task's difficulty will increase if the carjacker speaks with an accent or if his message includes a high level of detail (Pickel & Staller, 2012). Performance on at least one of the tasks must suffer, and because processing the message is prioritized, encoding the carjacker's appearance is more likely to be impaired. Specifically, prioritizing the message may render the witness less able to encode the carjacker's features and clothing elaborately; she may not notice as many physical details, engage in semantic-level processing, or connect newly acquired information to other details in the environment and to knowledge in her long-term memory.

MRT predicts that, like interpreting an accented message, executing any primary task (e.g., judging the perpetrator's veracity, attending to a weapon) while trying to encode a perpetrator's appearance can impair the encoding task. A pair of tasks may be similar or heterogeneous, but they will nevertheless end up competing for the same general resources if they are both challenging enough.

The discussion thus far assumes that encoding the perpetrator's appearance is always the secondary task, with some other task taking priority. Actually, however, MRT allows for individuals to decide where to direct their attention and for attentional allocation to change over time. In fact, focus on a weapon is not inevitable, and witnesses who have been educated about the weapon-focus effect can remember the appearance of an armed perpetrator as accurately as the appearance of an unarmed one (Pickel, Ross,

& Truelove, 2006). Although allocation policy issues are explicitly incorporated into MRT, Wickens (2008) has commented that the model could be improved by a better understanding of the variables that govern it.

Although not currently addressed within MRT, it seems reasonable to suppose that an individual's overall amount of available resources fluctuates over time—for example, in response to stress or anxiety (Deffenbacher, Bornstein, Penrod, & McGorty, 2004; Easterbrook, 1959). Therefore, witnesses experiencing high levels of fear may display performance decrements on both the primary and the secondary tasks as their total resource capacity shrinks. Consistent with this proposal, Josh Staller and I (Pickel & Staller, 2012) discovered that witnesses remembered both the perpetrator's message and appearance less accurately if the message was highly threatening rather than low in threat. We argued that hearing the threats caused a defensive stress response in witnesses that diminished overall attentional capacity.

Future Directions

In theory, any concurrent task that witnesses perform while observing a perpetrator or suspect could impair subsequent memory for that individual if the task is effortful enough. Moreover, the more demanding the task, the greater the impact it should have on memory. This presents a problem for police investigators trying to obtain useful information from witnesses. How can researchers help?

One step forward could be to extend current findings by trying to identify additional tasks, besides those already known to researchers, that witnesses are likely to perform, to learn more about how they prioritize tasks, and to understand how increases in anxiety moderate prioritization. Witnesses' behaviors during encoding, however, are not under the control of police investigators, so there is no simple way to eliminate their negative effects on memory even after these behaviors have been recognized. On the other hand, witnesses could be asked during police interviews what mental activities they engaged in while in the presence of the perpetrator as a way of taking into account the possibility that those activities decreased the accuracy of the memory report (Pickel et al., 2014). This precaution has limited value, though; knowing that a witness was multitasking enables one to predict the relative validity of his or her statement but not to establish its absolute accuracy (Wickens, 2008).

Alternatively, researchers could examine whether witnesses could be taught to control their attentional allocation more effectively so that, when it is safe to do so, they will prioritize the task of encoding descriptive information about the perpetrator. Some previous data suggest that this tactic could be feasible (Pickel et al., 2006). Convenience store clerks, bank tellers, and other individuals who are relatively likely to become witnesses to crimes could be targeted for this type of training. It is unknown, however, how beneficial such training would actually be or how long its effects would last.

Another approach that some have suggested is to supplement laboratory studies with archival analyses involving the descriptive reports and lineup identifications of actual eyewitnesses. Lab simulations are sometimes criticized for being too artificial. For example, unlike real witnesses, lab participants know their responses will not affect the course of another person's life, they may feel less anxiety or fear during the critical event, and they are typically questioned minutes after the event rather than hours, days, or months later. Analyzing real-world cases may appear to offer a way to test the actual effects of different factors, such as divided attention, that have been examined in lab studies. However, archival studies bring their own set of problems (Horry, Halford, Brewer, Milne, & Bull, 2014; Pickel, 2007). One of the most important is the absence of ground truth, meaning that researchers usually cannot determine with certainty who the perpetrator was or what he or she looked like at the time of the crime, which makes it impossible to evaluate the accuracy of witnesses' identifications and descriptive reports. Additionally, predictor variables may be confounded; for example, the amount of attention the witness allocates to the culprit's face may be positively correlated with exposure time. Other variables that could affect accuracy may go unmeasured and uncontrolled, such as the sobriety of the witness at the time of the crime. Another potential problem is that the behavior of police investigators (e.g., nonverbal cues) could alter the witness's lineup decision. As a final example, the witnesses included in an archival study may represent a biased sample because police detectives may choose not to interview or present a lineup to witnesses who seem reluctant, uncooperative, unreliable, or deceptive.

Horry et al. (2014) observed that "archival studies are often considered to be tests of whether effects found in the laboratory generalize to the field" (p. 106). Because of crucial methodological issues, however, the authors argued that it is not meaningful to compare the results of these two types of investigations. Furthermore, they concluded that the usefulness of archival research is fairly limited:

Well-designed and appropriately analyzed archival studies can possibly tell us something about the factors that are related to a specific population of cases: those in which the police choose to test an eyewitness and in which the identity of the perpetrator is unknown. Whether this knowledge is particularly valuable is unclear. (p. 106)

Nevertheless, researchers could try to continue to identify and correct the problems that exist within archival studies in general and then find ways to explore attentional allocation during actual crime events.

In summary, attentional processes engaged during encoding clearly delineate the accuracy and completeness of witnesses' subsequent memory of a suspect or perpetrator. Police investigators, prosecutors, and jurors may wish to discover what else a witness was doing while observing a crime event and to take this information into account when evaluating the witness's report.

Box 21.1

Key Points

- Witnesses' allocation of attention during a crime event determines the accuracy of their memory of the perpetrator.

- Witnesses can experience inattentional blindness, failing to notice a fully visible perpetrator because they are performing an attentionally demanding task and therefore do not engage in sustained attentional processing of the perpetrator.

- Even when witnesses notice a perpetrator, divided attention during encoding may impair their subsequent memory for details associated with him or her.

Box 21.2

Outstanding Issues

- It is important to identify additional cognitive tasks that witnesses are likely to perform, to learn more about how they prioritize tasks, and to understand how increases in anxiety moderate prioritization.

- More work is required to understand whether witnesses could be taught to control their attentional allocation more effectively so as to prioritize the task of encoding descriptive information about the perpetrator.

- More naturalistic research approaches are also important—although common approaches, such as archival research studies, present unique challenges.

References

Biggs, A. T., Brockmole, J. R., & Witt, J. K. (2013). Armed and attentive: Holding a weapon can bias attentional priorities in scene viewing. *Attention, Perception & Psychophysics, 75*, 1715–1724. doi:10.3758/s13414-013-0538-6.

Chabris, C. F., Weinberger, A., Fontaine, M., & Simons, D. J. (2011). You do not talk about Fight Club if you do not notice Fight Club: Inattentional blindness for a simulated real-world assault. *Perception, 2*, 150–153. doi:10.1068/i0436.

Davies, G. M., Smith, S., & Blincoe, C. (2008). A "weapon focus" effect in children. *Psychology, Crime & Law, 14*, 19–28. doi:10.1080/10683160701340593.

Deffenbacher, K. A., Bornstein, B. H., Penrod, S. D., & McGorty, E. K. (2004). A meta-analytic review of the effects of high stress on eyewitness memory. *Law and Human Behavior, 28*, 687–706. doi:10.1007/s10979-004-0565-x.

Easterbrook, J. A. (1959). The effect of emotion on cue utilization and the organization of behavior. *Psychological Review, 66*, 183–201.

Fawcett, J. M., Russell, E. J., Peace, K. A., & Christie, J. (2013). Of guns and geese: A meta-analytic review of the "weapon focus" literature. *Psychology, Crime & Law, 19*, 35–66. doi:10.1080 /1068316X.2011.599325.

Gordon, R. D. (2004). Attentional allocation during the perception of scenes. *Journal of Experimental Psychology: Human Perception and Performance, 30*, 760–777. doi:10.1037/0096-1523.30.4.760.

Hope, L., & Wright, D. (2007). Beyond unusual? Examining the role of attention in the weapon focus effect. *Applied Cognitive Psychology, 21*, 951–961. doi:10.1002/acp.1307.

Horry, R., Halford, P., Brewer, N., Milne, R., & Bull, R. (2014). Archival analyses of eyewitness identification test outcomes: What can they tell us about eyewitness memory? *Law and Human Behavior, 38*, 94–108. doi:10.1037/lhb0000060.

Hyman, I. E., Boss, M., Wise, B. M., McKenzie, K. E., & Caggiano, J. M. (2010). Did you see the unicycling clown? Inattentional blindness while walking and talking on a cell phone. *Applied Cognitive Psychology, 24*, 597–607. doi:10.1002/acp.1638.

Innocence Project. (n.d.). Eyewitness misidentification. Retrieved from http://www .innocenceproject.org/understand/Eyewitness-Misidentification.php

Lane, S. M. (2006). Dividing attention during a witnessed event increases eyewitness suggestibility. *Applied Cognitive Psychology, 20*, 199–212. doi:10.1002/acp.1177.

Loftus, E. F., Loftus, G. R., & Messo, J. (1987). Some facts about "weapon focus." *Law and Human Behavior, 1*, 55–62.

Mather, M., & Sutherland, M. R. (2011). Arousal-biased competition in perception and memory. *Perspectives on Psychological Science, 6*, 114–133. doi:10.1177/1745691611400234.

Mitchell, K. J., Livosky, M., & Mather, M. (1998). The weapon focus effect revisited: The role of novelty. *Legal and Criminological Psychology, 3*, 287–303.

Most, S. B., Scholl, B. J., Clifford, E. R., & Simons, D. J. (2005). What you see is what you set: Sustained inattentional blindness and the capture of awareness. *Psychological Review, 112*, 217–242. doi:10.1037/0033-295X.112.1.217.

Munro, M. J., & Derwing, T. M. (1995). Processing time, accent, and comprehensibility in the perception of native and foreign-accented speech. *Language and Speech, 38*, 289–306.

Neisser, U. (1976). *Cognition and reality: Principles and implications of cognitive psychology*. San Francisco: Freeman.

Palmer, M. A., Brewer, N., & Horry, R. (2013). Understanding gender bias in face recognition: Effects of divided attention at encoding. *Acta Psychologica, 142*, 362–369. doi:10.1016/j. actpsy.2013.01.009.

Pickel, K. L. (1999). The influence of context on the "weapon focus" effect. *Law and Human Behavior, 23*, 299–311.

Pickel, K. L. (2007). Remembering and identifying menacing perpetrators: Exposure to violence and the weapon focus effect. In R. C. L. Lindsay, D. Read, D. Ross, & M. P. Toglia (Eds.), *The handbook of eyewitness psychology: Vol. 2. Memory for people* (pp. 339–360). Mahwah, NJ: Erlbaum.

Pickel, K. L. (2009). The weapon focus effect on memory for female versus male perpetrators. *Memory, 17*, 664–678. doi:10.1080/09658210903029412.

Pickel, K. L., Klauser, B. M., & Bauer, H. M. (2014). The cost of detecting deception: Judging veracity makes eyewitnesses remember a suspect less accurately but with more certainty. *Applied Cognitive Psychology, 28*, 314–326. doi:10.1002/acp.2991.

Pickel, K. L., Kulig, T. C., & Bauer, H. M. (2013). Judging veracity impairs eyewitnesses' memory of a perpetrator. *Memory, 21*, 843–856. doi:10.1080/09658211.2013.765486.

Pickel, K. L., Ross, S. J., & Truelove, R. S. (2006). Do weapons automatically capture attention? *Applied Cognitive Psychology, 20*, 871–893. doi:10.1002/acp.1235.

Pickel, K. L., & Staller, J. B. (2012). A perpetrator's accent impairs witnesses' memory for physical appearance. *Law and Human Behavior, 36*, 140–150. doi:10.1037/h0093968.

Reinhard, M.-A., & Sporer, S. L. (2008). Verbal and nonverbal behavior as a basis for credibility attribution: The impact of task involvement and cognitive capacity. *Journal of Experimental Social Psychology, 44*, 477–488. doi:10.1016/j.jesp.2007.07.012.

Reinitz, M. T., Morrissey, J., & Demb, J. (1994). Role of attention in face encoding. *Journal of Experimental Psychology: Learning, Memory, and Cognition, 20*, 161–168.

Schooler, J. W., Gerhard, D., & Loftus, E. F. (1986). Qualities of the unreal. *Journal of Experimental Psychology: Learning, Memory, and Cognition, 12*, 171–181.

Simons, D. J., & Chabris, C. F. (1999). Gorillas in our midst: Sustained inattentional blindness for dynamic events. *Perception, 28*, 1059–1074.

Spiegel, A. (2011). Why seeing (the unexpected) is often not believing. Retrieved from http://www.npr.org/2011/06/20/137086464/why-seeing-the-unexpected-is-often-not-believing

Strayer, D. L., & Drews, F. A. (2007). Cell-phone-induced driver distraction. *Current Directions in Psychological Science, 16*, 128–131. doi:10.1111/j.1467-8721.2007.00489.x.

Wells, G. L., & Bradfield, A. L. (1998). "Good, you identified the suspect": Feedback to eyewitnesses distorts their reports of the witnessing experience. *Journal of Applied Psychology, 83*, 360–376.

Wickens, C. D. (2002). Multiple resources and performance prediction. *Theoretical Issues in Ergonomics Science, 3*, 159–177. doi:10.1080/14639220210123806.

Wickens, C. D. (2008). Multiple resources and mental workload. *Human Factors, 50*, 449–455. doi:10.1518/001872008X288394.

22

Attention and Misdirection: How to Use Conjuring Experience to Study Attentional Processes

Gustav Kuhn and Robert Teszka

Magic is one of the oldest art forms, dating back to the ancient Egyptians—some of the ancient scrolls report a magician baffling his audience by decapitating the head of a live goose only to restore the animal and bring it back to life (Christopher, 2006). Since written records began, we have been fascinated with the way conjurors allow us to experience the impossible. Although magic tricks have evolved over time, many of the basic principles remain the same. Misdirection is a fundamental principle underlying most magic tricks, and it has been suggested that "misdirection plays such an important role in magic that one might say that magic is misdirection and misdirection is magic" (Hugard, 1960, p. 115). Many magic books define misdirection as a form of deception in which the attention of an audience is manipulated (Bruno, 1978; Fitzkee, 1945; Hay, 1972). Although attention plays an important role in misdirection, there is more to misdirection than meets the eye.

There is a clear link between misdirection and attention, and magicians' applied knowledge can potentially provide new insights into attention and awareness (Kuhn, Amlani, & Rensink, 2008; Macknik et al., 2008; Resnink & Kuhn, 2014). In this chapter we provide a road map explaining how this applied knowledge can be used to study visual attention. Magicians' theories of misdirection provide a valuable starting point since much of their applied knowledge has been formalized in theoretical accounts of misdirection. However, understanding the attentional mechanisms requires experimental research. We outline how misdirection can be studied experimentally and discuss some of the potential advantages of this approach (e.g., high ecological validity, discovering novel principles of attention). Although misdirection is central to magic, defining misdirection in a meaningful way is far from trivial, which is why we start by discussing the concept of misdirection.

Historical Context

Magic tricks are generally described in terms of methods and effects. The effect is the magical event experienced by the spectator (e.g., a rabbit appears), and the method refers to the way in which this seemingly impossible event was created (e.g., misdirecting attention and thus preventing the audience from noticing the magician secretly sneaking the rabbit into the hat). Loosely defined, misdirection is used to prevent the spectator from noticing the method used to create the effect while still experiencing the effect (Lamont & Wiseman, 1999). This definition has two parts to it. On one level, misdirection is used to prevent people from seeing an event. This is an important aspect of misdirection, but it is clearly not sufficient. Imagine a situation in which the magician asks you to look at the person behind you, after which he reaches into his secret pocket, grabs the rabbit, and places it into his top hat. Once you orient your attention back to the stage, the magician waves his magic wand and produces the rabbit. Misdirection only works if observers don't attribute misdirection as a potential method for the effect, which is why the spectator should not be aware of the misdirection (Lamont & Wiseman, 1999).

Although attention plays an important role, misdirection is not limited to manipulating attention (Kuhn, Caffaratti, Teszka, & Rensink, 2014), nor is it confined to magic. For example, politicians are often accused of using misdirection to deflect the media's attention away from less desirable issues. For the purpose of this review we will focus on attentional misdirection and the way it can be used to study attentional processes. Some of our early pioneers in psychological research showed a keen interest in studying magic scientifically (Binet, 1894; Triplett, 1900). More recently, it has been suggested that scientists could learn much from studying conjuring scientifically (Demacheva, Ladouceur, Steinberg, Pogossova, & Raz, 2012; Kuhn, Amlani, & Rensink, 2008; Macknik et al., 2008; Rensink & Kuhn, 2014). Intuitively, the connection between attention and misdirection is clear, and it is easy to see how some of the magician's real-world experience in attentional manipulation could be exploited. That said, studying misdirection in a meaningful way is challenging. How do you measure whether the misdirection was effective? What cognitive mechanisms are influenced by the misdirection? Which principle of misdirection are you studying?

Below we discuss different approaches that have been used to study misdirection in relation to attention. We will also discuss the advantages that this approach has over more traditional experimental methodologies.

How Can We Study Misdirection?

Magicians' Real-World Experience
Good magic requires a comprehensive understanding of how to misdirect someone's attention. Becoming a professional magician typically requires around 10,000 hours of training, and much of this training takes the form of performing in front of live audiences (Rissanen, Palonen, Pitkänen, Kuhn, & Hakkarainen, 2013). Professional magicians use the audience's reaction as feedback on their performance and technique.

Years of experience enables the practicing magician to acquire real-world knowledge about how to misdirect attention. This knowledge about attentional techniques has been thoroughly tested in front of live audiences, a process that is not that dissimilar to the way in which scientists test their theories of attention in the lab. If the magician's concept of misdirection is incorrect, the secret will be spotted and the magic trick will fail. For misdirection to be effective, the principles must be both powerful (i.e., everybody's attention must be misdirected) and reliable (i.e., it must always work). This real-world experience in misdirection provides a useful starting point for studying attentional processes.

Most magicians are reluctant to explain how individual tricks are done, and accessing this knowledge as an outsider can be challenging. However, it is important to note that since the first magic book in 1584 (Scot, 1584) thousands of magic books and magic tricks have been published, and many magicians are happy to talk to lay people about general principles in magic (e.g., see the explosion of instructional DVDs and online videos). The main challenge lies in the fact that most knowledge about misdirection is described as part of individual magic tricks and is embedded within other less relevant technical details. Most nonmagicians will struggle to fully understand the descriptions simply because they lack the relevant background information, similar to the way nonscientists may struggle to understand a psychology journal article. In the following sections we will summarize this specialist knowledge with the aim of providing the reader with enough detail to apply it in experimental research on attention.

The Tarbell mail-order course in magic was an early work that devoted a section specifically to psychology and to misdirection (Tarbell, 1927/1971) and became enormously influential among magicians after being collected as a series of books in the 1950s. One of these lessons specifically focuses on misdirection and discusses several principles that had been written about in separate works. It states outright that "wherever you direct their attention, the audience will look there" and that this can be done because the audience follows your eyes, showing that magicians have intuitively understood gaze following. The lesson also discusses the principle of naturalness and the way in which awkward and unusual hand positions attract attention. Several other early works discuss misdirection, but we will turn to books that focus substantially or entirely on defining and understanding misdirection.

Dariel Fitzkee's (1945) *Magic by Misdirection* covers several principles in depth, with specific usage of psychological terminology. He argued that a core principle of misdirection involves the control of attention, and he devoted separate chapters to different methods of manipulating attention. The efficacy of all the methods is based on the theory that all misdirection acts by manipulating the spectator's cognitive faculties. It is important to note that Fitzkee uses a much broader definition of attention than we would use today. Fitzkee and many other magicians use the term "attention" to describe our awareness, and there is a strong implicit assumption that attention and awareness are directly linked. Spectators are unaware of the method behind the effect because misdirection influences not only their awareness of events but also their reasoning and interpretation of those events. Core techniques for doing this include creating interest in some events by heightening the audience's anticipation that something is about to happen or by making the event unusual. Conversely, interest in an area is reduced by having the event seem natural in context or by creating a moment of relaxation (after an object appears or when a trick appears to be over). Fitzkee's work is comprehensive, but it does not attempt to classify or organize the principles.

Joseph Bruno (1978) proposed a more systematic approach in *Anatomy of Misdirection*, where he argues that misdirection can either distract, divert, or relax attention. Distraction is a key principle that magicians use to divide attention between multiple events, and it works because the spectators cannot keep track of everything that is happening. Bruno argues that distraction works as long as it makes sense in context, but that it is essentially a rather crude form of misdirection. Magicians are therefore advised to divert attention rather than distract it. According to Bruno, diversion means holding attention. The performance appears to be uninterrupted, and the spectators feel that they were watching everything the whole time. Magicians hold attention by creating interest. They can highlight a special object or important action, or take advantage of the climax of an effect that just occurred. While the spectators attend to the diversion, they are unaware of any secret methods that are carried out elsewhere. Finally there is the principle of relaxation, where spectators assume that nothing is happening. During a pause, or after the conclusion to an effect, spectators do not focus their attention, and so the magician can utilize his or her secret method to set up an effect. The magician can also actively relax attention (or reduce interest) by repeating an action normally until it is no longer worth attending to, at which point a "fake" version of the action can be performed so that a secret method can be applied.

In *Conjurors' Psychological Secrets* S. H. Sharpe (1988) constructed a more academic classification of misdirection, with reference to early psychological research available at the time (Triplett, 1900). Crucially, his framework distinguishes between attention and inattention. In this theory, misdirection has two dimensions. Methods in which the magician deliberately acts to change aspects of the environment, such as the physical appearance or an object's potential importance (active misdirection) are contrasted to methods that depend on the stimuli's intrinsic properties rather than the magician's involvement, such as using common objects instead of suspicious plastic props (passive misdirection). The other factor is the purpose of the misdirection, which could be to disguise (in order to avoid attention) or to distract by deliberately shifting attention. These two factors combine to create four types of misdirection.

More recently, *Magic in Theory* proposed a taxonomy of misdirection that was intended to be used by academic psychology researchers (Lamont & Wiseman, 1999). Misdirection is divided into physical and psychological misdirection. The former deals with manipulating attention, while the latter concerns manipulating people's suspicions. This captures the distinction between attentional and nonattentional processes that was only implicit in the previous works. Physical misdirection involves not only directing where the spectator is looking but also when they look, by reducing attention at the time of the method and increasing it at the time of the magical effect. Psychological misdirection reduces suspicion about the true method by influencing nonattentional processes such as spectators' ability to reason about the method and how they interpret the actions they see.

State-of-the-Art Review

How can we use the applied knowledge above to further our scientific understanding of attention? One of the main challenges is that most theories of misdirection use rather loose definitions of attention, and some of these definitions do not necessarily correspond to the concepts used by researchers. Science relies on precise definitions and clearly specified theories about specific mechanisms. Understanding the psychological mechanisms underlying these misdirection theories therefore requires further scientific investigation. Although misdirection theories are based on real-world experience, they are ultimately based on anecdotal evidence, and in some cases even the magician's intuition may be wrong. For example, it is assumed that visual social cues (e.g., eye gaze) are among the most effective of cues for misdirecting attention (Fitzkee, 1945; Lamont & Wiseman, 1999; Sharpe, 1988). While there is much empirical evidence to support the use of visual social cues (Kuhn & Land, 2006; Kuhn, Tatler, & Cole, 2009; Tatler & Kuhn, 2007), their use may be limited to certain situations (Cui, Otero-Millan, Macknik, King, & Martinez-Conde, 2011). Theories of misdirection provide a valuable starting point for attention research. However, these theories cannot be taken at face value and must be evaluated with experimental evidence.

Controlled Experimentation

Studying misdirection scientifically has the potential of informing us about new attentional phenomena and allows us to develop novel paradigms to study attention and awareness. However, the translation from the magician's stage to the lab also poses challenges. Most of the principles discussed above have so far not been studied experimentally, and we hope that the methodological issues discussed here will help future researchers to translate between magic theory and scientific experimentation.

For controlled experimentation, it is appropriate to isolate some of these principles and focus on those that have applications outside the context of a particular magic trick. Next, we need to decide on how to measure whether the misdirection was effective or not. Misdirection is intended to prevent people from noticing the method,

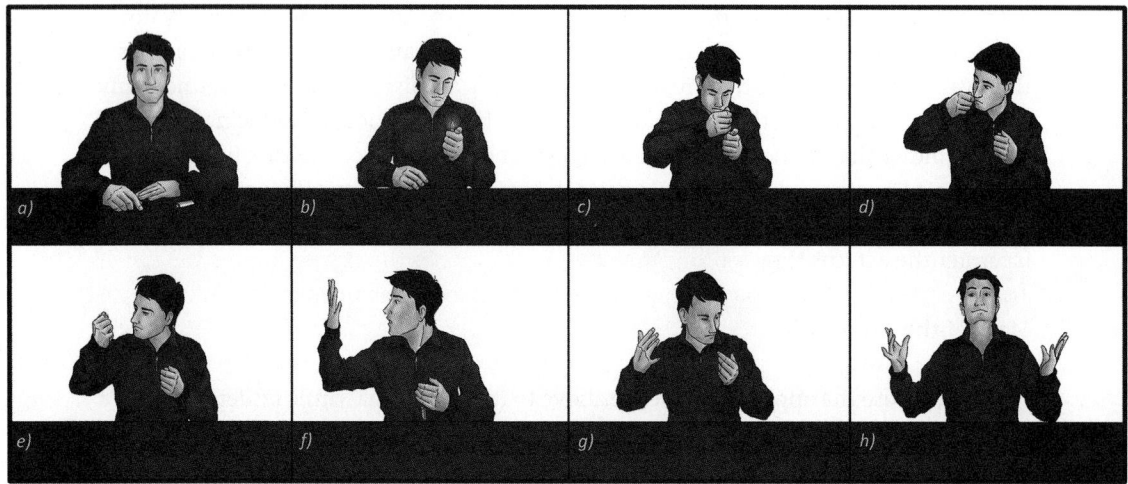

Figure 22.1
(a) The magician is seated at a table across from the viewer. A lighter is on the table. (b) He picks up the lighter and flicks it on. (c–f) He pretends to take the flame away and make it vanish, providing a gaze cue as misdirection away from his other hand. At (f), the lighter is visibly dropped into his lap. (g–h) The lighter appears to have vanished. Figure from Friebertshauser, Teszka, and Kuhn (2014).

and it therefore makes intuitive sense to evaluate whether this was achieved or not. The most direct way of doing this is to ask people whether they saw how the trick was done. However, not all methods are visible, which is why some researchers have chosen to develop paradigms in which misdirection is used to prevent people from seeing a potentially visible event. Kuhn and Tatler (2005) developed a misdirection paradigm where the observer experiences a magic trick in which the magician apparently makes a lighter and a cigarette disappear. The method for this effect is very simple in that the magician just drops the items into his lap (see figure 22.1 for illustration). Crucially, this takes place in full view so that the dropping items are fully visible. The magician employs misdirection that prevents most of the participants from noticing the dropping items although it happens in full view. As the method is potentially visible, people's detection of the event (seeing the items being dropped) provides a direct index of whether the misdirection was effective or not. In subsequent experiments the magician made only a lighter disappear, but the rationale remained the same (Kuhn et al., 2009). Importantly, if the same trick is repeated, most participants notice the dropping lighter, thus illustrating that the method was indeed potentially visible (Kuhn & Tatler, 2005; Kuhn et al., 2009). Anthony Barnhart has developed a paradigm in which misdirection is used to prevent participants from noticing a coin that visibly moves from one cup to the others (Barnhart & Goldinger, 2014). The rationale behind this paradigm is the same in that the effectiveness of the misdirection is measured by establishing whether the event was perceived or not.

Others have used misdirection to prevent participants from noticing the change in an object and thus induced a perceptual failure similar to change blindness. For example, T. J. Smith, Lamont, and Henderson (2013) designed a study in which the back of a deck of cards changed from blue to red. Participants were asked to count the number of cards that were being dealt on the table. The backs of the first five cards were blue, but the remaining cards had red backs. Although the change in color was fully visible, most participants failed to notice the change. In our own work we have used sleight of hand to visibly change the back of a packet of playing cards (Kuhn, Teszka, Tenaw, & Kingstone, submitted). If misdirection is employed at the point at which the color change occurs, most participants fail to notice it. People's ability to detect the change provides a useful index of the success of misdirection.

Misdirection has also been used to make people perceive events that have never taken place. For example, in the vanishing ball-illusion, the magician pretends to throw a ball up in the air while secretly palming it in his hand (Kuhn & Land, 2006; see figure 22.2). The magician misdirected people's attention and their expectations, which resulted in an illusory percept of seeing the ball leaving the hand (Kuhn & Land, 2006; Triplett, 1900). Here people's perception, or memory, of the event provided an index of whether the misdirection was effective or not. Using a similar approach, Cui et al. (2011) used a video clip in which a magician pretended to toss a coin from one hand to the other. Participants were then asked to indicate which hand the coin was in. If participants claimed the coin was in the target hand, the misdirection was considered to be effective.

One of the downsides of using visual experience as a measure of misdirection is that we can only probe awareness at discrete points in time, and paradigms need to be developed that allow for a visual probe. Magicians generally assume that misdirection involves manipulating where we look, and the terms "spatial attention" and "eye movements" are often treated as the same (e.g., Lamont & Wiseman, 1999). Eye movements provide us with an online measure of visual attention, and where we look influences the information that is being selected (Findlay & Gilchrist, 2003). Most experimental studies on misdirection have included eye-movement measures. Several of these studies have shown that misdirection is very effective at influencing where people look (Cui et al., 2011; Kuhn & Findlay, 2010; Kuhn & Land, 2006; Kuhn & Tatler, 2005; Kuhn, Tatler, Findlay, & Cole, 2008; T. J. Smith et al., 2013), and social cues, for example, are particularly effective (Kuhn et al., 2009). Rather than focusing on where people look, others have focused on the types of eye movements people make in response to different types of misdirection cues (Otero-Millan, Macknik, Robbins, & Martinez-Conde, 2011).

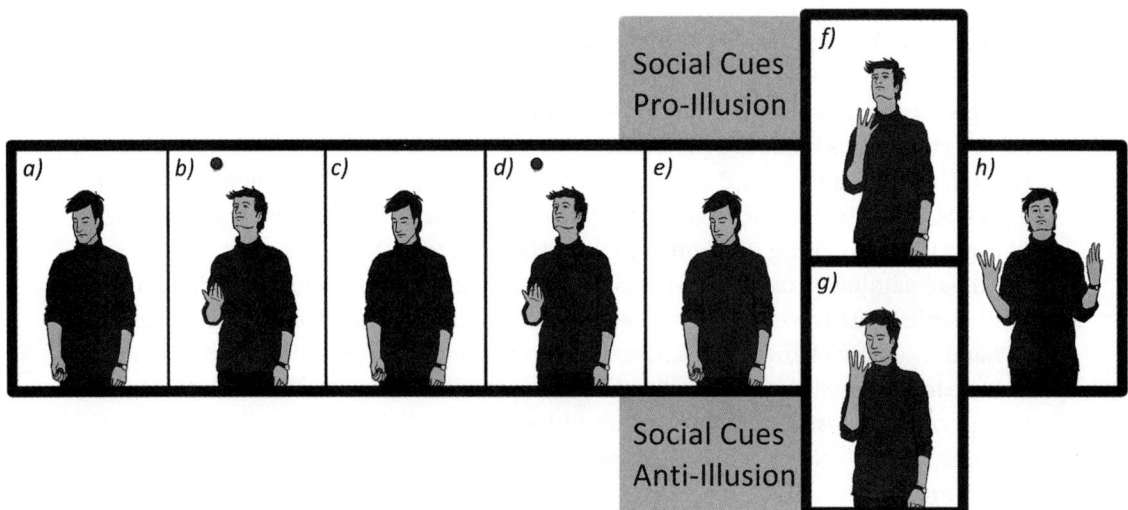

Figure 22.2
The magician is standing across from the viewer, holding a small ball. (a–e) He throws the ball in the air twice, following its trajectory with his gaze. At (f), he pretends to throw the ball, but actually retains it in his hand. In the pro-illusion condition he looks up, following the implied trajectory of the ball, while in the anti-illusion condition he continues looking at his hand (g). (h) The ball appears to have vanished. Figure from Friebertshauser et al. (2014).

Eye movements provide us with a valuable window into the mind. They allow us to measure people's focus of attention during the course of a magic routine in a relatively unobtrusive manner. However, it is important to note that where people look does not necessarily tell us what they see. Many of the studies that have combined eye-movement measures with perceptual probes have shown that there is very little relationship between where people look and whether they detect a sustained event such as the dropping of the lighter (Kuhn & Findlay, 2010; Kuhn & Tatler, 2005) or notice a visual change (T. J. Smith, Lamont, & Henderson, 2012). This dissociation between where people look and their visual experience can provide us with valuable insights into the relationship between eye movements and visual awareness (Mack & Rock, 1998; Memmert, 2006).

Studying misdirection allows us to investigate attentional processes within a more naturalistic context. However, experiments in misdirection require a certain level of reduction and experimental control. Our first experimental studies investigating attentional misdirection were conducted in highly naturalistic settings. In these studies we were concerned that the experimental setting (i.e., doing magic in a lab) would have adverse effects on how people would perceive the magic trick, and so we chose to conduct the eye-tracking studies in a university bar (Kuhn & Tatler, 2005). Participants watched the magician perform the misdirection trick in which a lighter and a cigarette disappeared while we monitored their eye movements using a real-world eye tracker. The misdirection was extremely effective at preventing participants from noticing the dropping lighter, and the magician tried his best to ensure that each subsequent performance was as similar as possible (Tatler & Kuhn, 2007). However, there was, of course, no way of ensuring that all of the performances were identical, and measuring eye movements within this real-world context presents a whole range of technical challenges. In subsequent experiments the same misdirection trick was prerecorded and presented on a computer monitor. Although the misdirection was somewhat less effective, it still worked (Kuhn, Tatler, et al., 2008). The fact that magic tricks work when presented on a computer monitor should probably come as little surprise, as television is a popular medium for this type of performance. More recently, we conducted a study investigating the importance of real face-to-face interactions in misdirection by verbal and visual social cues (Kuhn, Teszka, & Kingstone, submitted). To our surprise, the cues were just as effective when the magic trick was presented on a monitor as when it was performed in a face-to-face situation. In the video context participants spent more time looking at the face than in the face-to-face context (see also Laidlaw, Foulsham, Kuhn, & Kingstone, 2011), suggesting that there may be some subtle attentional differences. However, on the whole these results are good news for attention researchers as they demonstrate that misdirection can work effectively when presented on a computer monitor.

In most attention paradigms participants are exposed to a large number of experimental trials and we average performances across these trials. One of the key rules in magic states that magic tricks should never be performed more than once using the same method. Indeed previous studies have shown how repeating the same trick dramatically increases people's ability to detect the event (Kuhn & Findlay, 2010; Kuhn & Tatler, 2005). Simply telling someone what they are about to see increases detection and significantly changes where they look (Kuhn & Tatler, 2005; Kuhn et al., 2009; Kuhn, Tatler, et al., 2008). Limiting your experimental design to a single trial certainly creates challenges. However, this is not that different from other attention paradigms, such as inattentional blindness, where participants are typically exposed to one critical trial (Mack & Rock, 1998). Others, however, have studied misdirection by presenting participants with several repetitions of the same trick (Cui et al., 2011; Otero-Millan et al., 2011), and there may be situations in which it is possible to design multiple-trial experiments.

Advantages of Studying Misdirection

Now that we have addressed the methodologies on offer to study misdirection, we will focus on some of the potential advantages the scientific study of misdirection offers for research on attention.

Studying Attention with Ecological Validity without Compromising Experimental Control

There has been much debate on the merits and pitfalls of studying attentional processes within the confined surroundings of the experimental laboratory (Kingstone, Smilek, & Eastwood, 2008; Kingstone, Smilek, Ristic, Friesen, & Eastwood, 2003). Studying attentional processes in the context of misdirection allows researchers to investigate attentional processes in more naturalistic environments. The importance of ecological validity has become particularly apparent in the field of social attention (Risko, Laidlaw, Freeth, Foulsham, & Kingstone, 2012; Skarratt, Cole, & Kuhn, 2012) but may also apply to other issues.

Tatler and Kuhn (2007) used a correlational design to investigate the relationship between the magician's gaze and the observers' locus of fixation. In their experiment participants were in a face-to-face interaction and the eye movements were monitored using a real-world eye tracker. There was a significant correlation between where the magician looked and the areas fixated by the observer. However, there were clearly many instances where the gaze was not followed. A follow-up study used a similar misdirection paradigm where participants watched different videos in which the magician made a lighter disappear by dropping it into his lap (Kuhn et al., 2009). Crucially, in one of the conditions, social cues (i.e., where the magician looked) were used to misdirect attention away from the drop, while in the other condition, the magician looked straight at the dropping lighter. Participants were much more likely to notice the drop when the magician looked at the dropping lighter. These results demonstrated that gaze cues influence both where people look and attentional processes that are independent of eye movements (i.e., covert attention). It should be noted that although this experiment used a single trial per participant, the difference was highly reliable.

Social attention has also been studied using the vanishing-ball illusion (Kuhn & Land, 2006). Here the magician pretended to throw a ball up in the air when in fact it remained concealed in his hand. Crucially, in one of the conditions, the magician's gaze followed the imaginary trajectory of the ball, while in the other condition, he looked at the hand concealing the ball (see figure 22.2). The misdirection that was supported by the social cues was far more effective than when the magician looked at the hand, supporting the view that gaze cues are important in driving people's expectations. When questioned, most participants claimed that they merely looked at the ball, yet the eye-movement data revealed that they spent a large proportion of the time looking at the face and made strategic use of the magician's social cues. These results suggest that we make more use of social cues than we would intuitively think. In a very recent study using segments of the cups and balls routine (Tachibana, 2014) it was shown that individuals who spend more time looking at the face were more misdirected. Others, however, have shown that if there is a rapid action that is confined to a small spatial location, gaze may not add much (Cui et al., 2011).

Misdirection paradigms have also been used to investigate individual differences in response to social cues. Impairments in social interactions are one of the hallmarks of autism spectrum disorder (ASD; American Psychiatric Association, 1994), but it has been difficult to find this difference using traditional gaze cuing paradigms (Nation & Penny, 2008). Some have argued that these differences are not found because typical gaze cuing paradigms do not tap into social processes (Kingstone, 2009; Nation & Penny, 2008). The vanishing-ball illusion relies heavily on social cues, which makes it an ecologically valid tool to investigate social attention. Kuhn et al. (2010) compared the effectiveness of this illusion between a group of high-functioning adults with ASD compared to a group of typically developing individuals. If individuals with ASD are less influenced by social cues, we would expect them to be less susceptible to this illusion. Rather surprisingly, the opposite pattern of results was found, in that the individuals with ASD were significantly more susceptible and the ASD group spent the same amount of time looking at the magician's face. However, they were significantly slower to fixate the face than the control participants. The control participants were extremely efficient at using the magician's social cues to help them fixate the ball once it was thrown in the air, something the ASD group struggled to do. These results suggest that by adulthood individuals with ASD do look at faces, but they are not as efficient at using the cues provided by the face to execute eye movements to the right place at the right time. Social cues form a pivotal aspect of misdirection (Lamont & Wiseman, 1999; Sharpe, 1988), which is one way misdirection offers a valuable technique for investigating social attention within a more naturalistic context.

Misdirection deals with influencing people's awareness and, as such, offers an intuitive tool to investigate the relationship between attention and awareness. Most magicians presume that the key to misdirecting the audience's attention involves misdirecting where they look (e.g., Lamont & Wiseman, 1999). One of the most consistent findings from the research on misdirection has been the lack of relationship between gaze fixation and awareness of the event. For example, in the cigarette and lighter trick there was no relationship between where people looked and whether they detected the drop or not (Kuhn & Findlay, 2010; Kuhn & Tatler, 2005). Similarly, in the study in which the back of the cards changed color, there was no relationship between eye position and participants' noticing the color change (T. J. Smith et al., 2013), or when one coin was changed into another one (T. J. Smith et al., 2012). Our own work (Kuhn, Teszka, Tenaw, & Kingstone, submitted) in which sleight of hand is used to change the back of a deck of playing cards dovetails with these results. Most research investigating the relationship between attention and awareness (Koch & Tsuchiya, 2007) or overt and covert attention (D. T. Smith & Schenk, 2012) has used rather artificial experimental paradigms. Misdirection allows researchers to investigate the relationship between attention and awareness and different forms of attention (e.g., overt vs. covert) within a more naturalistic context.

Clearly, there are many other paradigms that can be used to investigate attentional processes in a more naturalistic context, but these often require elaborate setups (Simons & Levin, 1998). Magic has a clear practical advantage in that it is portable, it is quick, and it does not require setups that go beyond the magician's skills.

Discovering Novel Attentional Phenomena

Many of the misdirection principles used by magicians correspond to attentional processes that are studied by researchers. For example, the use of social cues in misdirecting attention clearly corresponds to work on social attention. One of the most exciting advantages of using misdirection to study attentional processes lies in potentially discovering novel attentional phenomena that have so far been ignored by more traditional attention researchers. For example, pickpocket Apollo Robbins has claimed that curved movements produce stronger misdirection than do straight movements. Otero-Millan and colleagues (2011) investigated this claim and revealed that people who watch curved movements generate more smooth pursuit eye movements than people who observe straight movements. More research is required to evaluate the true significance of these findings, but the authors of this paper suggest that curved motion trajectories, as used by some magicians, may result in different attentional processes than do straight trajectories.

In our own lab we have investigated several magic principles that may highlight novel attentional processes. For example, we are investigating the way in which auditory social cues, such as addressing a person using a question, influence attentional processes. Traditionally, social attention has been studied in the visual domain. Magicians often use their voice and speech to misdirect attention (Lamont & Wiseman, 1999; Tamariz, 2007), and we have been exploring the effectiveness of these verbal cues on visual attention. Rather intriguingly, these auditory social cues influence where people look but have no effect on information encoding.

Empirical work on misdirection is clearly still in its early stages, but given the breadth of applied knowledge gained by magicians, we are likely to discover some interesting phenomena. While some of these principles are primarily relevant to the context of magic, others may be generalized and could shed light onto some interesting attentional principles.

Integration

How Does Work on Misdirection Link to More Traditional Laboratory Research?

Since Yarbus's (1967) seminal book on eye movements, there has been much interest in understanding where people look and the mechanisms that drive this overt attentional selection process (e.g., see chapter 11 in the current volume). In scene viewing experiments, participants are asked to view images while their eye movements are recorded.

The central goal of these studies is to model and predict the areas that are being fixated. These types of studies have revealed that people are more likely to fixate areas that are of interest (Henderson, Brockmole, Castelhano, & Mack, 2007). It has also become apparent that certain low-level perceptual properties are fixated more frequently than others. For example, participants are more likely to fixate areas that have high spatial frequencies as well as edges and high contrasts (Mannan, Ruddock, & Wooding, 1997). These and many other findings have led some to suggest that our eye movements are controlled by low-level stimulus features that are extracted early on in the visual process (Itti & Koch, 2001).

An important objective in misdirection is to manipulate where the audience looks. This form of misdirection is directly related to research in the field of scene viewing. Most misdirection theories emphasize the importance of low-level perceptual characteristics (e.g., passive misdirection). For example, magicians often use bright colors or high luminance contrasts (e.g., fire) to capture the audience's attention. Similarly, different types of movements and visual transients are used to orient attention (Lamont & Wiseman, 1999). Within the magic literature it is well established that the appearance of an object (such as when, e.g., a ball appears under a cup) forms a powerful cue to attract attention, and magicians will exploit this to carry out a secret method (e.g., picking up an extra ball) while the audience's attention has been distracted.

Research on scene viewing has also revealed that our attention is influenced by the presence of others (e.g., see chapter 15 in the current volume). For instance, people's eye movements are particularly attracted to faces (Birmingham, Bischof, & Kingstone, 2008; Yarbus, 1967) and areas that are looked at by others (Fletcher-Watson, Leekam, Benson, Frank, & Findlay, 2008). Clearly, theories of misdirection have much to say about how attention is influenced by social cues.

Our eye movements are also driven by our goals and intentions. It is in this area where misdirection could offer the most insight. Although low-level features play an important role in misdirection, magicians frequently manipulate observers' expectations and interest through top-down processes; this involves capitalizing on numerous processes that interact with one another. Magic tricks seldom happen in isolation—a magician rarely just makes an object disappear. Magic tricks are framed within intricate routines that are carefully choreographed, and these routines are used to manipulate aspects of the environment that are important for a successful performance. For example, the magician may ask you to shuffle a deck of cards. In your mind shuffling the cards offers you the chance of mixing them and ensuring that they have not been tampered with. The true purpose of this action may be to distract your attention from the magician, who will use this moment to carry out a secret method. Magicians may also choreograph the routine in ways that will give certain objects a particular importance. Once primed, you are more likely to attend to that object at the expense of attending to the location in which the secret method is being conducted. These top-down forms of misdirection are just as important as the more low-level ones. Similar conclusions have been reached in research on scene viewing.

Misdirection involves distorting people's awareness though manipulating attentional processes, and similarities have been drawn between misdirection and inattentional blindness (Kuhn & Findlay, 2010). In inattentional blindness, participants fail to notice a visually salient event if they are engaged in an attentionally demanding distractor task (Mack & Rock, 1998; Most, Scholl, Clifford, & Simons, 2005; see also chapter 9 in the current volume). For example, in the gorilla illusion, participants fail to notice the gorilla walking across the scene if they are engaged in counting the number of times the ball has been passed from one basketball player to the next (Simons & Chabris, 1999). It has been suggested that the processes responsible for people's failure in noticing the lighter being dropped are analogous to inattentional blindness (Kuhn & Findlay, 2010). Misdirection clearly has similarities with inattentional blindness, but there are also differences (Memmert, 2010; Memmert & Furley, 2010). Most (2010) has argued that the main difference between inattentional blindness and misdirection is that the former results from general attentional demands while misdirection is due to spatial attentional manipulations. In the misdirection paradigm, this is clearly the case. However, if we look at misdirection more generally, much of misdirection involves manipulating nonspatial attentional processes (e.g., time misdirection, temporal characteristics).

We study selective attention because our limited processing capacity requires us to prioritize certain aspects of the information. Traditionally, selective attention has been studied using a spatial metaphor whereby people attend to discrete locations (Posner, 1980) or objects (Egly, Driver, & Rafal, 1994). However, we frequently attend to particular points in time (e.g., see chapter 3 in the current volume). For example, traffic lights turn amber right before turning green. This is intended to allow one to anticipate the go signal and therefore respond more rapidly. Indeed there are lots of situations in which one deploys one's attention to points in time rather than merely spatial locations (Coull & Nobre, 1998). In misdirection, these temporal fluctuations in attention are just as important as, if not more important than, spatial aspects. There are very few instances where the method and the effect take place at the same time. Instead, the method is typically carried out before the effect is revealed. For example, in the false transfer, the magician pretends to transfer a coin from one hand to the next. He then typically waits a few seconds with the hand pretending to conceal the coin before the coin is revealed to have disappeared. This temporal separation between method and the effect is typically known as time misdirection (Fraps, 2014). The result of separating the method from the effect is that the audience will not know which aspect of the magic trick will be important and they will be less likely to attend to it. This is the reason why magicians rarely tell their audience what they are about to do before the method has been carried out. Keeping the audience in suspense prevents them from suspecting where, but also when, the method will take place.

Magicians carefully choreograph their routines to allow them to manipulate temporal aspects of attention. Magicians talk about creating times of high and low interest, and the misdirection works by carrying out the method during the low points. Low points of attention are points immediately after an effect or after a joke. People are generally more relaxed after a joke and are less likely to suspect that the magician will carry out a secret method. However, in reality, these offbeat moments are the moments in which the methods typically take place, and magicians often strategically make a joke just before carrying out their method. Body language is also often used to misdirect temporal aspects of attention. For example, Slydini, who was a true master of misdirection, describes ways in which body posture is used to misdirect attention (Ganson, 1980). During times at which a method took place he would relax into his chair; then he would move forward to focus attention before revealing the effect.

Future Directions

Years of real-world experience have allowed magicians to accumulate vast amounts of practical experience in misdirecting various aspects of our attentional system. This practical experience potentially offers valuable insights into attentional processes. Although there has been a long tradition of drawing parallels between magic and psychology, advances had been relatively slow (Lamont & Henderson, 2009). However, the last decade has seen a vast surge in interest in the field as well as a large number of scientific papers investigating different aspects of misdirection and magic more generally (Kuhn, Amlani, & Rensink, 2008; Macknik et al., 2008; Rensink & Kuhn, 2014).

The scientific study of misdirection is clearly still in its infancy, and much progress remains to be made to advance the knowledge transfer. For one, we need a framework to study misdirection. Each paper that is published in the field will help advance our scientific understanding of the processes, as well as help develop new methodologies. For example, there has been much debate about the relationship between misdirection and inattentional blindness, and this constructive dialogue will help advance the field (Kuhn & Tatler, 2011; Memmert, 2010; Memmert & Furley, 2010; Moran & Brady, 2010; Most, 2010).

More structured approaches to advancing the field may also be of use. Magicians are clear masters of misdirection, and their main concern lies in finding methods that allow them to manipulate observers' awareness effectively, regardless of what psychological or neurological mechanisms are involved. As researchers, we are interested in the mechanisms and in understanding *why* certain principles work. One of the first steps in advancing this field of research would be to develop a psychological taxonomy of misdirection (Kuhn et al., 2014). This taxonomy would help organize the vast amounts of applied knowledge that have been developed by the magic community. In addition to helping to organize the knowledge, it would also allow researchers with no experience in magic to study misdirection without revealing the secrets of individual tricks.

Although the general concept of using misdirection to study attention is intuitive, most research in the field has been conducted by individuals who have some experience in magic. Firsthand conjuring experience is advantageous, but it is not a requirement. The secretive nature of magic might alienate some researchers from using misdirection in their own research. However, it is important to note that this secretive nature of magic is superficial, and most magicians are happy to collaborate with scientists and are typically more than happy to share their techniques. A direct knowledge transfer between magicians and scientists is clearly beneficial, but the scientific study of misdirection does not require personal contact with magicians. Investigating misdirection does not require a full understanding of magic. The key is for scientists to focus on certain aspects of this process. We hope that this review has helped describe some of the processes involved using misdirection to study attention and that the road map provided here will help researchers from different fields to benefit from this applied knowledge.

Box 22.1
Key Points

- Magicians have acquired much applied knowledge about misdirecting attention which attention researchers are now starting to exploit.
- Magicians' knowledge about misdirection has been formalized in theories of misdirection which provides a starting point for scientific investigations.
- Misdirection research has relevance to existing laboratory traditions such as scene viewing and social attention research.
- Using misdirection to study attention allows for high ecological validity and offers the potential to discover new attentional mechanisms.

Box 22.2
Outstanding Issues

- What are the psychological mechanisms involved in misdirection? Can we develop a meaningful taxonomy of misdirection?
- What is the best way of incorporating this real-world knowledge into the study of cognition and attention?

- Misdirection also manipulates top-down knowledge by influencing reasoning and memory to reduce interest and suspicion. How do these nonattentional processes interact with attention?
- A magic trick is a special case of social interaction where one of the people is untrustworthy. How much of the findings from magic research generalize outside of the context of magic? How do social factors such as status or cultural background interact with misdirection principles?
- Does expertise in magic or training in misdirection principles have an effect on attentional processing or other cognitive abilities?

References

American Psychiatric Association. (1994). *Diagnostic and statistical manual of mental disorders* (4th ed.). Washington, DC: APA.

Barnhart, A. S., & Goldinger, S. D. (2014). Blinded by magic: Eye-movements reveal the misdirection of attention. *Frontiers in Psychology, 5.* doi: 10.3389/fpsyg.2014.01461.

Binet, A. (1894). Psychology of prestidigitation. *Annual Report of the Board of Regents of the Smithsonian Institution* (pp. 555–571). Washington, DC.

Birmingham, E., Bischof, W. F., & Kingstone, A. (2008). Gaze selection in complex social scenes. *Visual Cognition, 16,* 341–356.

Bruno, J. (1978). *Anatomy of misdirection.* Baltimore, MD: Stoney Brook Press.

Christopher, M. (2006). *The illustrated history of magic.* New York: Carroll & Graf.

Coull, J. T., & Nobre, A. C. (1998). Where and when to pay attention: The neural systems for directing attention to spatial locations and to time intervals as revealed by both PET and fMRI. *Journal of Neuroscience, 18,* 7426–7435.

Cui, J., Otero-Millan, J., Macknik, S. L., King, M., & Martinez-Conde, S. (2011). Social misdirection fails to enhance a magic illusion. *Frontiers in Human Neuroscience, 5,* 103. doi:10.3389/fnhum.2011.00103.

Demacheva, I., Ladouceur, M., Steinberg, E., Pogossova, G., & Raz, A. (2012). The applied cognitive psychology of attention: A step closer to understanding magic tricks. *Applied Cognitive Psychology, 26,* 541–549. doi:10.1002/acp.2825.

Egly, R., Driver, J., & Rafal, R. D. (1994). Shifting visual attention between objects and locations—Evidence from normal and parietal lesion subjects. *Journal of Experimental Psychology. General, 123,* 161–177.

Findlay, J. M., & Gilchrist, I. D. (2003). *Active vision.* Oxford, UK: Oxford University Press.

Fitzkee, D. (1945). *Magic by misdirection.* Pomeroy, OH: Lee Jacobs Productions.

Fletcher-Watson, S., Leekam, S. R., Benson, V., Frank, M. C., & Findlay, J. M. (2008). Eye-movements reveal attention to social information in autism spectrum disorder. *Neuropsychologia*, *47*, 248–257.

Fraps, T. (2014). Time and magic—Manipulating subjective temporality In D. L. V. Arstila (Ed.), *Subjective time: The philosophy, psychology, and neuroscience of temporality.* Cambridge, MA: MIT Press.

Friebertshauser, A., Teszka, R., & Kuhn, G. (2014). Eyetracking magic video stimuli summaries. doi:http://dx.doi.org/10.6084/m9.figshare.1183468

Ganson, L. (1980). *The magic of Slydini.* Bideford, Devon, UK: Supreme Magic.

Hay, H. (1972). *The amateur magician's handbook* (3rd ed.). New York: New American Library.

Henderson, J. M., Brockmole, J. R., Castelhano, M. S., & Mack, M. L. (2007). Visual saliency does not account for eye movements during search in real-world scenes. In R. P. G. van Gompel, M. H. Fischer, W. S. Murray, & R. L. Hill (Eds.), *Eye movements: A window on mind and brain* (pp. 537–562). Oxford, UK: Elsevier.

Hugard, J. (1960, March). Misdirection. *Hugard's Magic Monthly*, p. 115.

Itti, L., & Koch, C. (2001). Computational modelling of visual attention. *Nature Reviews. Neuroscience*, *2*, 194–203.

Kingstone, A. (2009). Taking a real look at social attention. *Current Opinion in Neurobiology*, *19*, 52–56.

Kingstone, A., Smilek, D., & Eastwood, J. D. (2008). Cognitive ethology: A new approach for studying human cognition. *British Journal of Psychology*, *99*, 317–340.

Kingstone, A., Smilek, D., Ristic, J., Friesen, C. K., & Eastwood, J. D. (2003). Attention, researchers! It is time to take a look at the real world. *Current Directions in Psychological Science*, *12*, 176–180.

Koch, C., & Tsuchiya, N. (2007). Attention and consciousness: Two distinct brain processes. *Trends in Cognitive Sciences*, *11*, 16–22.

Kuhn, G., Amlani, A. A., & Rensink, R. A. (2008). Towards a science of magic. *Trends in Cognitive Sciences*, *12*, 349–354. doi:10.1016/j.tics.2008.05.008.

Kuhn, G., Caffaratti, H. A., Teszka, R., & Rensink, R. A. (2014). A psychologically-based taxonomy of misdirection. *Frontiers*, *5*, 1392.

Kuhn, G., & Findlay, J. M. (2010). Misdirection, attention and awareness: Inattentional blindness reveals temporal relationship between eye movements and visual awareness. *Quarterly Journal of Experimental Psychology*, *63*, 136–146. doi:10.1080/17470210902846757.

Kuhn, G., Kourkoulou, A., & Leekam, S. R. (2010). How magic changes our expectations about autism. *Psychological Science*, *21*, 1487–1493. doi:10.1177/0956797610383435.

Kuhn, G., & Land, M. F. (2006). There's more to magic than meets the eye. *Current Biology, 16,* R950–R951.

Kuhn, G., & Tatler, B. W. (2005). Magic and fixation: Now you don't see it, now you do. *Perception, 34,* 1155–1161. doi:10.1068/p3409bn1.

Kuhn, G., & Tatler, B. W. (2011). Misdirected by the gap: The relationship between inattentional blindness and attentional misdirection. *Consciousness and Cognition, 20,* 432–436. doi:10.1016/j. concog.2010.09.013.

Kuhn, G., Tatler, B. W., & Cole, G. G. (2009). You look where I look! Effect of gaze cues on overt and covert attention in misdirection. *Visual Cognition, 17,* 925–944.

Kuhn, G., Tatler, B. W., Findlay, J. M., & Cole, G. G. (2008). Misdirection in magic: Implications for the relationship between eye gaze and attention. *Visual Cognition, 16,* 391–405.

Kuhn, G., Teszka, R., Tenaw, N., & Kingstone, A. (submitted). Don't be fooled! Top-down inhibition of attentional orienting in response to social cues in face-to-face and video context.

Laidlaw, K. E. W., Foulsham, T., Kuhn, G., & Kingstone, A. (2011). Potential social interactions are important to social attention. *Proceedings of the National Academy of Sciences of the United States of America, 108,* 5548–5553. doi:10.1073/pnas.1017022108.

Lamont, P., & Henderson, J. M. (2009). More attention and greater awareness in the scientific study of magic. *Nature Reviews. Neuroscience, 10,* 241.

Lamont, P., & Wiseman, R. (1999). *Magic in theory.* Hatfield, UK: University of Hertfordshire Press.

Mack, A., & Rock, I. (1998). *Inattentional blindness.* Cambridge, MA: MIT Press.

Macknik, S. L., King, M., Randi, J., Robbins, A., Teller, Thompson, J., & Martinez-Conde, S. (2008). Attention and awareness in stage magic: Turning tricks into research. *Nature Reviews. Neuroscience, 9,* 871–879.

Mannan, S. K., Ruddock, K. H., & Wooding, D. S. (1997). Fixation sequences made during visual examination of briefly presented 2D images. *Spatial Vision, 11,* 157–178.

Memmert, D. (2006). The effects of eye movements, age, and expertise on inattentional blindness. *Consciousness and Cognition, 15,* 620–627.

Memmert, D. (2010). The gap between inattentional blindness and attentional misdirection. *Consciousness and Cognition, 19,* 1097–1101. doi:10.1016/J.Concog.2010.01.001.

Memmert, D., & Furley, P. (2010). Beyond inattentional blindness and attentional misdirection: From attentional paradigms to attentional mechanisms. *Consciousness and Cognition, 19,* 1107–1109. doi:10.1016/J.Concog.2010.06.005.

Moran, A., & Brady, N. (2010). Mind the gap: Misdirection, inattentional blindness and the relationship between overt and covert attention. *Consciousness and Cognition, 19,* 1105–1106.

Most, S. B. (2010). What's "inattentional" about inattentional blindness? *Consciousness and Cognition, 19*, 1102–1104.

Most, S. B., Scholl, B. J., Clifford, E. R., & Simons, D. J. (2005). What you see is what you set: Sustained inattentional blindness and the capture of awareness. *Psychological Review, 112*, 217–242.

Nation, K., & Penny, S. (2008). Sensitivity to eye gaze in autism: Is it normal? Is it automatic? Is it social? *Development and Psychopathology, 20*, 79–97.

Otero-Millan, J., Macknik, S. L., Robbins, A., & Martinez-Conde, S. (2011). Stronger misdirection in curved than in straight motion. *Frontiers in Human Neuroscience, 5*, 133. doi:10.3389/fnhum.2011.00133.

Posner, M. I. (1980). Orienting of attention. *Quarterly Journal of Experimental Psychology, 32*, 3–25.

Rensink, R. A., & Kuhn, G. (2014). A framework for using magic to study the mind. *Frontiers in Psychology, 5*. doi: 10.3389/fpsyg.2014.01508.

Risko, E. F., Laidlaw, K. E., Freeth, M., Foulsham, T., & Kingstone, A. (2012). Social attention with real versus reel stimuli: Toward an empirical approach to concerns about ecological validity. *Frontiers in Human Neuroscience, 6*. doi:10.3389/fnhum.2012.00143.

Rissanen, O., Palonen, T., Pitkänen, P., Kuhn, G., & Hakkarainen, K. (2013). Personal social networks and the cultivation of expertise in magic: An interview study. *Vocations and Learning, 6*, 347–365.

Scot, R. (1584). *The discovery of witchcraft*. London: Elliot Stock.

Sharpe, S. H. (1988). *Conjurers' psychological secrets*. Calgary, Alberta, Canada: Hades.

Simons, D. J., & Chabris, C. F. (1999). Gorillas in our midst: Sustained inattentional blindness for dynamic events. *Perception, 28*, 1059–1074.

Simons, D. J., & Levin, D. T. (1998). Failure to detect changes to people during a real-world interaction. *Psychonomic Bulletin & Review, 5*, 644–649. doi:10.3758/bf03208840.

Skarratt, P. A., Cole, G. G., & Kuhn, G. (2012). Visual cognition during real social interaction. *Frontiers in Human Neuroscience, 6*, 9. doi:10.3389/fnhum.2012.00196.

Smith, D. T., & Schenk, T. (2012). The premotor theory of attention: Time to move on? *Neuropsychologia, 50*, 1104–1114. doi:10.1016/j.neuropsychologia.2012.01.025.

Smith, T. J., Lamont, P., & Henderson, J. M. (2012). The penny drops: Change blindness at fixation. *Perception, 41*, 489–492. doi:10.1068/p7092.

Smith, T. J., Lamont, P., & Henderson, J. M. (2013). Change blindness in a dynamic scene due to endogenous override of exogenous attentional cues. *Perception, 42*, 884–886. doi:10.1068/p7377.

Tachibana, R. (2014). The effects of social misdirection on magic tricks: How deceived and undeceived groups differ. *i-Perception, 5*(3), 143–146.

Tamariz, J. (2007). *The five points to magic*. Seattle, WA: Hermetic Press.

Tarbell, H. (1971). Magic as a science. In R. W. Read (Ed.), *The Tarbell course in magic* (Vol. 1). Brooklyn, NY: D. Robbins. (Original work published 1927)

Tatler, B. W., & Kuhn, G. (2007). Don't look now: The magic of misdirection. In R. P. G. van Gompel, M. H. Fischer, W. S. Murray, & R. L. Hill (Eds.), *Eye movements: A window on mind and brain* (pp. 697–714). Oxford, UK: Elsevier.

Triplett, N. (1900). The psychology of conjuring deceptions. *American Journal of Psychology, 11,* 439–510.

Yarbus, A. L. (1967). *Eye movements and vision.* New York: Plenum Press.

23

Action Video Games in the Spotlight: The Case of Attentional Control

Daphne Bavelier and Julia Föcker

Historical Context

Early video games, such as *PONG* in the 1970s or *Pac-Man* in the 1980s, have become icons of another era. With bulky geometric features and only a handful of actions, they are a far cry from the sophistication of today's video games. Yet, already, in those early days, researchers were asking about the impact of such game play on brain and behavior. One of the first quantitative reports of a behavioral change in video-game players concerned eye–hand coordination (Griffith, Voloschin, Gibb, & Bailey, 1983). Visuomotor coordination, as measured within a rotary pursuit task, was found to be superior in video-game players as compared to non–video game players, a result similar to the recent report that new computer users gain in eye–hand coordination (Wei et al., 2014). In these early studies, video-game players were selected as those individuals playing between 2 to 59 hours of video-game play per week, in the past 2 to 99 months; these were contrasted to individuals who played less than 10 hours total during the previous 6 months. These selection criteria tellingly exemplify what it meant to be or not to be a gamer at the time.

Performance gain was not restricted to visuomotor coordination; it was also noted in tasks as varied as mental rotation or divided attention (Greenfield, Brannon, & Lohr, 1994; Greenfield, deWinstanley, Kilpatrick, & Kaye, 1994). For example, habitual video-game players outperformed non–video game players in a mental paper-folding task, which required participants to imagine folding a two-dimensional pattern into a three-dimensional box (Greenfield, Brannon, & Lohr, 1994). In that study, habitual versus novice players were selected based on whether participants scored above or below 100,000 points on the game *The Empire Strikes Back* (Greenfield, Brannon, & Lohr, 1994).

A critical question quickly arose. Did these changes in behavior reflect participants' self-selection, with those attracted toward video-game play having better skills to start with, or rather did they represent a causal impact of video-game play on performance? To answer this question, training studies were carried out asking participants to play games and evaluating the impact of game play on tasks ranging from very simple stimulus–response mapping to higher cognitive processes. Guiding a marble on a three-dimensional grid and preventing it from falling, as is required in *Marble Madness*, was used to measure transfer effects from game play to spatial skills in children and was found to enhance spatial skills equally in boys and girls (Subrahmanyam & Greenfield, 1994). A few other studies evaluated the possibility of enhancing spatial skills by training individuals within a specific game environment. One such study required players to rotate falling geometric shapes and fitting them into horizontal lines in *Tetris* (Okagaki & Frensch, 1994); others required players to guide an object through a maze while trying to shoot enemies in *Targ* (Gagnon, 1985), to drive a tank through a battle zone in *Battlezone* (Gagnon, 1985), or to control a spaceship in a three-dimensional space in *Zaxxon* (Dorval & Pepin, 1986). These studies highlighted the potential benefits of these games on spatial cognition.

The exact source of these changes in behavior became an important question. Improvements could be due to mere test–retest improvements by which repeated test taking facilitates performance. To control for this possible confound, a few investigators extended the training study design to include both an experimental group and a test–retest control group (Greenfield, Brannon, & Lohr, 1994; Greenfield, deWinstanley, et al., 1994; Okagaki & Frensch, 1994). Visuospatial attentional skills were measured before and after randomly assigning participants either to an experimental group that played the game *Robot Battle* for 5 hours or to a control, test–retest group that only performed pre- and posttest at about the same time intervals as the experimental group (Greenfield, deWinstanley, et al., 1994). The video-game-trained group showed reduced reaction times (RTs) in a task that required them to shift attention across the visual field, as compared to the test–retest control group, indicating benefits beyond test–retest advantages.

In these early studies, the video-game features—for example, controlling a spaceship in a 3-D space, guiding an object through a maze, dividing attention across the battlefield—intuitively made sense with the skill enhanced, and thus it was concluded that video-game play characteristics led to beneficial effects. Alternative accounts exist, however. Motivational or social factors, whereby participants assigned to an active condition tend to do better than those asked to carry on with their business-as-usual activities, remained a concern. By design, test–retest control groups tend to have less contact with experimenters and feel less engaged in the study; as a result, they may be less motivated to perform well, a recognized effect in social psychology termed the Hawthorne effect (Parsons, 1974). To address this concern, an active control group is needed. Subrahmanyam and Greenfield (1994) were among the first to have included such an active control group. Their training study contrasted an experimental group required to play the video game *Marble Madness* to a control group required to play a computerized word game *Conjecture*. Greater improvement after *Marble Madness* than after *Conjecture* unambiguously established the role of *Marble Madness* in inducing faster performance in a visuospatial task. While relatively rare at the time, designs with active control groups have now become standard in the field of cognitive training and behavioral interventions (Green, Strobach, & Schubert, 2014; Klingberg, 2010).

In sum, the early days of the research on how video-game play affects behavior already suggested improved visuospatial skills and possibly attention in video-game players of the time. They also made clear that for all intended practical purpose more complex and differentiated experimental designs were needed to understand the source of the behavioral improvements noted in video-game players.

State-of-the-Art Review—The Impact of Action Video Games on Brain and Behavior

Over the past 10 years, research on the impact of video-game play on human behavior has gained momentum. The most studied genre so far is fast-paced, action-packed, first- or third-person-shooter games such as *Call of Duty* or *Unreal Tournament* (Green & Bavelier, 2012). Recently, other game genres have received scrutiny as well. This is all the more important in that different video-game genres have distinct effects on brain and behavior, calling for more fine-grained studies.

Variety of Skills Altered by Action Video-Game Play

Action video-game play, an often decried activity in society, has been documented to have unsuspected positive impacts (for reviews, see Green & Bavelier, 2012; Spence & Feng, 2010). It results in enhanced perceptual skills (Appelbaum, Cain, Darling, & Mitroff, 2013; Li, Polat, Makous, & Bavelier, 2009), lesser visual masking (Li, Polat, Scalzo, & Bavelier, 2010; Pohl et al., 2014), and a more veridical perception of the time of events (Donohue, Woldorff, & Mitroff, 2010). It has been associated with enhanced selective attention in space, in time, and to objects (Green & Bavelier, 2003), better divided attention (Dye & Bavelier, 2010; Green & Bavelier, 2003, 2006a), and reduced attentional capture whether orienting attention with or without eye movements (Chisholm & Kingstone, 2012; Chisholm, Hickey, Theeuwes, & Kingstone, 2010). Finally, it also benefits perceptual decision making (Green, Pouget, & Bavelier, 2010), mental rotation (Feng, Spence, & Pratt, 2007) and task switching (Cain, Landau, & Shimamura, 2012; Colzato, van Leeuwen, van den Wildenberg, & Hommel, 2010; Green, Sugarman, Medford, Klobusicky, & Bavelier, 2012; Karle, Watter, & Shedden, 2010; Strobach, Frensch, & Schubert, 2012). So far, action video games have been shown to improve performance mostly in the visual domain (Green & Bavelier, 2007; Green, Li, & Bavelier, 2010; Li et al., 2009; Li et al., 2010) with some extensions to audition (Green, Pouget, et al., 2010) and multisensory perception (Donohue et al., 2010).

While more work is certainly needed to further delineate the impact of action game play, the variety of behaviors it alters is already rather surprising. Indeed, a vast literature documents how training makes perfect, but also how training typically results in improvement on the trained task with only limited transfer to other, even related tasks. For example, expert *Tetris* players excel at manipulating *Tetris*-like shapes, but it is not clear they benefit from a general improvement in visuospatial cognition (Destefano, Lindstedt, & Gray, 2011). Interestingly, playing *Tetris* for only 10 to 20 hours has been shown to enhance mental rotation and visuospatial skills (De Lisi & Wolford, 2002; Sims & Mayer, 2002; Terlecki, Newcombe, & Little, 2008; Uttal et al., 2013). This pattern of results suggests an inverted U-shaped curve when it comes to transfer, whereby transfer is maximal at an intermediate stage of training when the trained task has become familiar but is still effortful. Once automatization occurs and expertise develops, transfer becomes more limited. While such a U-shaped curve remains to be tested, it has long been known that transfer and expertise are indeed not optimized by the same training regimen (Schmidt & Bjork, 1992).

Establishing the Impact of Action Video Games and Further Methodological Considerations

The impact of action game play on behavior has been established mostly in young adults, with some studies suggesting the same trends in children 7 to 17 years of age, especially when it comes to augmenting attentional skills (Dye, Green, & Bavelier, 2009; Dye & Bavelier, 2010). Typically, the performance on the skill under study is compared between habitual action video-game players (5 hours of action game play on average per week over at least the previous 6 months) and individuals who seldom play (playing 1 hour or less per week of any game genre and having done so for the past year). However, as reviewed above, training studies are key to rule out possible confounds. Several have been carried out contrasting an experimental group required to play action video games to an active control group also required to play popular entertainment video games, but from different genres, such as strategy or social games (Boot, Kramer, Simons, Fabiani, & Gratton, 2008; Feng et al., 2007; Green & Bavelier, 2003, 2006a, 2007; Green, Li, et al., 2010; Green et al., 2012; Li et al., 2009; Li et al., 2010; Oei & Patterson, 2013; Strobach et al., 2012; Wu & Spence, 2013; Wu et al., 2012). Other studies have been carried where the control group is required to play the same game as the experimental group but a nonadaptive version of it, confined to the easiest levels (Anguera et al., 2013; Olesen, Westerberg, & Klingberg, 2004).

Training studies are of great relevance for practical purposes. First, they help in identifying those game features that may best foster brain plasticity and performance enhancements. By contrasting two equally engaging games that have different features, they allow researchers to infer those game features important for impact above and beyond fun and engagement. Second, they provide a baseline for games for impact, as those should be held to the strict standard of providing more benefits than either conventional interventions or video games built for entertainment. Third, they allow a better understanding of interindividual variations in cognitive training and are essential for providing a richer understanding of how to intervene successfully across the population at large.

In the past few years, several concerns have been raised about the interpretation of the beneficial effects documented in action video-game studies (Boot, Blakely, & Simons, 2011; Green et al., 2014; Kristjansson, 2013). First, studies comparing overtly selected action game players and nonplayers may suffer from expectation biases. In short participants, if they know they are expected to do better, may indeed do better, be it because they try harder, have more self-confidence, or experience greater arousal. While a valid criticism, in the case of action video games several studies using covert participant recruitment also report positive effects on attention, establishing that effects are still observed in the absence of expectation biases (Appelbaum et al., 2013; Buckley, Codina, Bhardwaj, & Pascalis, 2010; Colzato, van den Wildenberg, Zmigrod, & Hommel, 2013; Donohue et al., 2010; Donohue, James, Eslick, & Mitroff, 2012; Dye et al., 2009; Dye & Bavelier, 2010; Hutchinson & Stocks, 2013; for work with some covert and some overt groups, see Clark, Fleck, & Mitroff, 2011; Trick, Jaspers-Fayer, & Sethi, 2005). One may ask why we should continue to carry out studies with overt recruitment knowing that they suffer from clear shortcomings. The answer is rather pragmatic. Only 5% of the undergraduate population matches the inclusion criteria for action video players, and the percentage becomes even smaller for non–video game players. A study aiming at covertly recruiting 10 participants in each group therefore requires testing about 200 participants. The time and cost for such a large study is difficult to argue for, especially given that such group comparison studies cannot rule out self-selection biases. As already discussed in early studies, people who play action video games may be good at the skill tested to begin with, and maybe those superior skills give them an edge in game play and are the reason why they play. Self-selection and not game play would then account for the effect. Studies comparing expert populations allow one to establish the presence of group differences, but training studies are necessary to establish a role for the training regimen per se.

A second concern is that even training studies may suffer from expectation biases. Participants cannot be blind to the video game they are requested to play. The notion of a placebo control is therefore impossible to implement, unlike what is the standard in drug testing. One cannot prevent participants from developing beliefs about the video game used for training, and it is obvious that the training itself cannot look physically identical to the user across the two treatments. This state of affair has been heavily criticized as a methodological flaw by some authors (Boot, Simons, Stothart, & Stutts, 2013) when it is a mere fact that applies to any behavioral interventions, be it music training, cognitive behavioral therapy, meditation, or aerobic exercise to cite a few. There is no question that advancing the field calls for new methodologies. We and others have proposed that the use of an active control that triggers similar engagement, fun, and flow as the experimental intervention is key in controlling for biases, along with the use of tasks with many experimental conditions such that behavioral changes are expressed in term of not just main effects but also interactions (Green et al., 2014). Advancing this new methodological framework through more and more refined training design will be of tremendous value as one considers how to translate intervention studies carried in the laboratory to everyday applications.

Below, we review more systematically why video-game experience may be seen to have such broad impact and provide some of the main lessons that can be drawn from the growing body of literature.

Greater Attentional Control as a Mediating Mechanism

A recurrent finding in the literature is that action video-game training has the potential to enhance a surprisingly broad range of skills. A major question, therefore, concerns the mechanisms at play that could explain such wide impact. Here we review the hypothesis that enhanced attentional control acts as a common underlying mechanism to explain the variety of effects documented as a result of action game play. Efficient distribution of attention to task-relevant stimuli and efficient suppression of sources of noise or distraction are hypothesized to facilitate the processing of task-relevant stimulus dimensions. In turn, a better model of the task at hand can be developed and more accurate representations inferred, allowing players to make more informed decisions about their environment (Bavelier, Green, Pouget, & Schrater, 2012; Green, Pouget, et al., 2010).

Behavioral Evidence for Enhanced Attentional Control Consistent with this view, many behavioral studies point to enhanced attentional control in action video gamers as compared to nongamers (Green & Bavelier, 2012). Attentional control has been characterized as the set of mechanisms by which processing resources are flexibly allocated as the context of a task or the goal of an individual changes. These processes include enhancement of neural responses for task-relevant dimensions and suppression of responses to potentially distracting information, as well as the constant monitoring and reevaluation of task set as environmental conditions change (Corbetta, Patel, & Shulman, 2008).

Action video-game players appear to benefit from such enhanced flexibility in attentional resource allocation through two main changes: (1) greater attentional resources and (2) better ability at suppressing distractors and sources of noise. Several studies using the flanker compatibility effect point to greater attentional resources in action video-game players. Based on the attentional load theory, under low-load conditions, spared attentional resources automatically spill over to the remaining information in the environment, leading to distractor processing and large compatibility effects in flanker tasks (Lavie, Hirst, de Fockert, & Viding, 2004). As a result all populations show large flanker effects at low load that reduce as load increases. Interestingly, flanker effects remain relatively high in action video-game players as compared to non–action video-game players as load is initially increased, in accordance with greater resources (Dye et al., 2009; Green & Bavelier, 2003, 2006a). While greater compatibility effects are often interpreted as a lack of control, in the case of video-game players, the large differences in RTs between compatible and incompatible distractor trials is observed concurrently with their incompatible RTs being faster than the compatible RTs of non–video-game players. Clearly, action video-game players do not suffer from inefficient processing, even in the face of greater compatibility effects. Instead, action video-game players appear to possess more resources relative to non–video-game players, which allow them to keep processing distracting items at higher load task than non–action video-game players. Crucially, while action video-game players show greater compatibility effects in typical Erikson flanker tasks, in tasks designed to have very high load and thus exhaust their resources, they appear to more efficiently suppress distractors or sources of noise than nongamers, in accordance with greater attentional control (Krishnan, Kang, Sperling, & Srinivasan, 2013; Mishra, Zinni, Bavelier, & Hillyard, 2011).

By contrast, few changes have been observed in bottom-up attention, despite the fact that such exogenous pull of attention is quite common in action video games (Castel, Pratt, & Drummond, 2005; Hubert-Wallander, Green, & Bavelier, 2011; West, Stevens, Pun, & Pratt, 2008). The findings of reduced attentional and oculomotoric capture effects in action video-game players, in the context of a similar initial exogenous or automatic pull of attention, are also consistent with enhanced control over the allocation of spatial attention in the context of little to no changes in exogenous attention (Cain, Prinzmetal, Shimamura, & Landau, 2014; Chisholm & Kingstone, 2012; Chisholm et al., 2010; but see Heimler, Pavani, Donk, & van Zoest, 2014).

Brain Imaging and Electrophysiological Evidence for Greater Attentional Control Recent brain imaging and electrophysiological studies comport with the view that attentional control is changed for the better by action video-game play. For instance, electrophysiological data complement the findings of enhanced attention skills in action video-game players by documenting greater suppression of task-irrelevant information and thus more efficient filtering of potential distractors (Krishnan et al., 2013; Mishra et al., 2011; Wu et al., 2012). A reduced recruitment of frontoparietal areas in action video-game players as attentional demand increases from low to high load in a visual search task also points to a more efficient attentional system (Bavelier, Achtman, Mani, & Föcker, 2012).

Anatomical differences in gray matter complement this view. One study investigating cortical thickness in 152 14-year-old adolescents has provided evidence for a positive association between cortical thickness and the duration of video-game play in areas related to attention and executive control, such as the left frontal eye fields and the left dorsolateral prefrontal cortex (Kühn, Lorenz, et al., 2014). Along the same line, differences in gray matter have been observed in expert video-game players, whereby increased gray matter in the posterior parietal cortex, one of the main nodes in the attentional control network, was seen to correlate with better working memory performance (Tanaka et al., 2013). While Kühn, Lorenz, et al. (2014) did not differentiate between different video-game genres, the finding that different games have different impacts on behavior suggests that it will be important to do so in future anatomical studies.

To summarize, a variety of studies documents enhanced attentional control after action game play. Greater attentional control may mediate the broad range of performance enhancements noted in action gamers by enabling them to focus on task-relevant features while efficiently suppressing sources of noise or distractions.

Integration and Future Directions—From the Bench to Everyday Life

The broad range of transfer effects seen after action game play has increased the interest in using games as rehabilitative and/or teaching tools. To this end, it becomes critical to understand training design characteristics that contribute to attentional changes and other behavioral benefits. We review those below and then discuss examples of the everyday-life impact of video-game play.

Training Design Characteristics for Broad Impact
Inducing Sustained Effects   To be of use for health or education purposes, the effects of a video-game regimen must be sustained over time. The few training studies that have followed trainees for several months after the end of their game play regimen (ensuring no game play between the end of training and follow-up tests several months later) document long-lasting effects, with improvements in mental rotation still noted 5 months after the end of a 10-hour training study (Feng et al., 2007) and improved vision noted up to 1 year after the end of a 50-hour training study (Li et al., 2009; Li et al., 2010).

Although there are no studies documenting which design characteristics ensure long-lasting effects, the duration and schedule of training are likely critical. The most successful training studies exploited a key feature of learning-distributed practice—whereby small regular doses of behavioral intervention are required from the trainee (Anguera et al., 2013; Feng et al., 2007; Green & Bavelier, 2003). It is likely that the same amount of practice massed in a short period of time will be less successful (Stafford & Dewar, 2014; see for different outcomes based on massed vs. distributed schedule; Green, Pouget, et al., 2010; van Ravenzwaaij, Boekel, Forstmann, Ratcliff, & Wagenmakers, 2014). Thus, whether a video game is used for rehabilitation of function in a patient or for educational purposes, small (20–40 minute), regular, daily doses of video-game play distributed over several weeks are likely to become the norm.

Game Features That May Foster Brain Plasticity   The research on action video-game play, by contrasting action video games with other game genres, provides some pointers as to the game features that may foster enhanced attentional control and broad transfer effects. Those features important to guarantee engagement and time on task, such as reward and intrinsic motivation, are certainly of interest. For instance, several studies point to the recruitment of reward-related areas, such as the mesolimbic system, during video-game play (Cole, Yoo, & Knutson, 2012; Hoeft, Watson, Kesler, Bettinger, & Reiss, 2008; Katsyri, Hari, Ravaja, & Nummenmaa, 2013; Koepp et al., 1998; Mathiak et al., 2011). While necessary, these features may not be sufficient as exemplified by the fact that training studies in young adults typically contrast action video games to other commercially available titles also reported to be highly engaging and fun (e.g., *The Sims* or *Tetris*). Yet, greater improvements have been noted in action trainees as compared to control trainees (Feng et al., 2007; Green & Bavelier, 2006a, 2006b, 2007; Green, Pouget, et al., 2010; Green et al., 2012; Li et al., 2009; Li et al., 2010; Oei & Patterson, 2013; Strobach et al., 2012; Wu & Spence, 2013; Wu et al., 2012).

Game-specific features, such as control and pacing, appear worth considering carefully. While it remains unclear how much the sense of control and agency matters to foster brain plasticity, motor functions are known to be quite plastic (Sanes & Donoghue, 2000). The importance of letting the players be in control of their actions is illustrated by a few failed training studies in our laboratory. These studies used video games where the players did not control either the movement of their character or the precision of their shooting. Although only suggestive, these failures point to an important role for motor control in a training regimen. In accordance with this view, games that highlight spatial navigation, such as, for example, playing *Super Mario* on a portable Nintendo dual-screen console, lead to increases in gray matter, especially in brain structures responsible for spatial navigation such as the right hippocampus (Kühn, Gleich, Lorenz, Lindenberger, & Gallinat, 2014), but also in the dorsolateral prefrontal cortex, an important structure for cognitive control. A systematic assessment of the role of control in the effects seen so far should be a promising avenue of research.

Another feature of interest is pacing or the fact that to be efficient a game has to require timed decision. The importance of matching the pacing in the game to the capability of the player is well illustrated in the older adult literature. Older adults asked to train on an off-the-shelf action video game failed to show any effect of training but also failed to comply with their training regimen (Boot, Champion, et al., 2013). That video games designed for young adults are too fast for elderly populations, and thus discouraging, is not entirely surprising. When the game is appropriately tuned to the pace of older adults, reduced task switching, and multitasking costs, as well as enhanced working memory and visual short-term memory, have been reported (Anguera et al., 2013; Basak, Boot, Voss, & Kramer, 2008; Belchior et al., 2013; Mayas, Parmentier, Andres, & Ballesteros, 2014; Toril, Reales, & Ballesteros, 2014). The most successful interventions, however, are the ones that individually adapt the pace of the training to each player. This technique was initially pioneered in the Advanced Cognitive Training for Independent and Vital Elderly (ACTIVE) study, a large clinical trial using a gamified visual search task, the Useful Field of View, in a large population of older adults (Jobe et al., 2001). Participants were trained with an adaptive version of the Useful Field of View that tracked their 75% correct performance by shortening display duration as they improved. This study established not only improved performance on the Useful Field of View but also transfer to unrelated constructs such as self-rated health up to 5 years post-training (Wolinsky et al., 2010). More recently, *NeuroRacer,* a driving video game in which players must try to keep their car on the road while monitoring whether to react or not to various signs that appear along the way, exploited the same concept. For maximal effect, the game was developed to adapt to the players' skill level, maintaining their performance at about 75% correct (Anguera et al., 2013). Training revealed performance benefits that extended to untrained cognitive control abilities such as sustained attention and working memory.

Other critical features in fostering learning and transfer have been hypothesized, such as the need for divided attention, the need for prediction and error monitoring, and a balance between exploratory and exploitative behaviors. Even violence has been discussed as a "motivator" in the context of video-game play, and possibly a factor that enhances the enjoyment of the player. Yet, a recent study comparing the same game, but with one version showing avatars dying in gruesome fashion after being shot and the other beaming adversaries into the air before they dissolve, reveals no effect on the enjoyment of the players (Przybylski, Ryan, & Rigby, 2009). While this study calls into question the contribution of violence and gore to the effects seen after action video-game play, threat and the regulation of approach and avoidance behaviors it entails may be critical factors in fostering brain plasticity. It will be important in future studies to systematically evaluate the role of these various game features in eliciting greater attentional control and broad behavioral benefits.

Interindividual Differences in Training Outcome   A key practical issue concerns the possibility of fostering improvements in all trainees, even in the face of wide variations among the population. Variations may have their roots in factors as varied as gender, socioeconomic status, or genetics.

Gender is a well-established source of interindividual difference for skills such as mental rotation. Interestingly, action video-game play, as well as other game play which more directly taps mental rotation skills, such as *Tetris*, enhances mental rotation (Feng et al., 2007; Subrahmanyam & Greenfield, 1994; Uttal et al., 2013). This raises the question of whether video games could be used as a tool to reduce gender differences in visuospatial skills. In a first study, Feng et al. (2007) showed improved mental rotation skills after playing 10 hours of the first-person-shooter game *Medal of Honor* and confirmed that females improved more than males. However, in a follow-up study, this greater increase was attributed to the fact that females initially exhibit lower skill level on visuospatial tests, giving them more room to improve (Spence, Yu, Feng, & Marshman, 2009). Thus, while video-game play in some of its forms may help equalize visuospatial abilities across males and females, a major societal concern remains, as males are more likely to play those fast-paced, action-packed games shown to foster visuospatial cognition. It would seem urgent to develop games that enhance spatial cognition but within a game world that is attractive to females.

Another important interindividual factor concerns socioeconomic status and degree of education. Research on action video games has so far focused on young, educated adults at the height of their perceptual, attentional, and cognitive skills. While showing that enhancements are still possible in such fit populations is certainly of interest, it leaves unanswered the question of whether such training would also be beneficial to all young healthy individuals in the population at large. Video games have been shown to be excellent tools as preparation for future learning (Arena & Schwartz, 2014), as well as a means to provide a safety net for those children that have dropped out of school (Goldin et al., 2014). More research on which populations benefit from which game features is, however, greatly needed.

Recent work looking at interindividual variations in cognitive control further underscores the importance of taking into consideration trainees' characteristics. For instance, Colzato, van den Wildenberg, and Hommel (2014) established greater transfer effects from an action video-game training regimen to a task-switch paradigm (a standard measure of cognitive flexibility and control) depending on the Valin (Val)/ Methionin (Met) genotype of the trainees. For instance, Val/Val individuals showed greater enhancement in cognitive flexibility as compared to Met carriers (Val/Met, Met/ Met) after an action game play training. Even though these results should be interpreted with caution because of several methodological issues (small sample size, lack of a control group), they highlight the importance of further understanding interindividual variations when it comes to devising training regimens.

Everyday-Life Impact of Video-Game Play

Video games are quickly shifting from mere entertainment products to key interventions tools. One of the first studies to document transfer from video-game training to real life was that of Gopher, Weil, and Bareket (1994), who showed that training on a specially designed computer game enhanced the performance of Israeli Air Force pilots. Since then anecdotic reports of video games being used for athletes, workforce training, car racing, or drone control abound, yet few controlled studies are available (Chiappe, Conger, Liao, Caldwell, & Vu, 2013; Hayes & Silberman, 2007). We review the handful of studies that document the impact of video-game play well beyond the laboratory.

Improved Reading Skills   A recent study established enhanced reading in dyslexic children after the use of a subset of mini-games selected to contain action features from *Rayman Raving Rabbids* (Franceschini et al., 2013). Ten Italian dyslexic children trained on action mini-games showed greater improvement in attention and speed of reading than the 10 children trained on other *Rayman* mini-games selected to contain mostly reactive play. The authors argued that playing the action mini-games improved visuospatial attention, a key determinant of reading skills, at least in Italian readers in which the mapping of orthography to phonology is quite transparent. Further studies are needed to confirm the link between attention and dyslexia in other language systems and replicate those results with a large sample of children (see Bavelier et al., 2013, for further discussion).

Improved Vision after Video-Game Training in Amblyopic Patients   The fact that video-game play can improve visual functions even in young healthy adults raises the question of whether those games may be used as rehabilitation tools in low-vision patients such as those suffering from amblyopia. Amblyopia, or lazy eye, is a consequence of abnormal binocular vision early in life, which results in physiological alterations in the visual cortex. In addition to reduced visual acuity and contrast sensitivity, amblyopia has been also associated with deficiencies in several aspects of attention (Sharma, Levi, & Klein, 2000).

Despite the deep-rooted belief that amblyopia cannot be treated once the patient has reached adulthood, recent studies propose that training adult amblyopic patients with games such as *Medal of Honor* or *Tetris* leads to improved visual acuity and reduced suppression as well as improved attentional functions (Hess et al., 2012; Li et al., 2013; Li, Ngo, Nguyen, & Levi, 2011). In the case of patients, improved vision was not restricted to the action game training group. Patients who were trained with *Tetris* or *SimCity* also showed improved vision (Hess et al., 2012; Li et al., 2011). The fact that these latter games do not elicit improvement in vision in younger adults, but do so in amblyopic patients, is in line with the importance of matching game requirements to patients' skills, as discussed above with respect to older adults. Commercially available action video games designed for young adults in their prime are typically too fast and demanding for clinical populations, especially at the outset of training.

Endoscopic Surgeons Another practical application of game play is the training of endoscopic surgeons. It was first noted by a group at Beth Israel that surgeons playing video games outperformed surgeons who were not playing any kind of video games (Rosser et al., 2007). In a follow-up study, undergraduate medical students who either played video games or who had not played any kinds of video games were compared (Kennedy, Boyle, Traynor, Walsh, & Hill, 2011). Psychomotor abilities measured on a laparoscopic simulator were enhanced in video-game players. However, other skills such as perceptual ability and visuospatial skills did not differ across the two groups. While such changes could have been expected, this study did not differentiate between action game players and other genres, preventing any firm conclusions, as different video games may nurture different effects. More relevant is a training study in which surgical novices trained on the action video game *Half Life* improved their performance in virtual-reality endoscopic surgical simulators compared to a nonactive control group who did not receive any kind of training. A third group trained on the video game *Chessmaster* also showed more improvements than the control group although the improvements were not as broad as in the action-trained group (Schlickum, Hedman, Enochsson, Kjellin, & Fellander-Tsai, 2009).

Thus, while action video games may lead to slightly broader transfer, the need for visuomotor control that is typical of any video games may, in and of itself, be helpful for laparoscopic surgery skills.

Conclusion

The accumulated body of work over the past 10 years makes it clear that video games are not just mere entertainment products. The burgeoning video-game research documents important changes in brain organization and behavior after video-game play. We argue that enhanced attentional control might be one underlying mechanism, which explains the surprisingly large range of transfer effects documented so far after action game play. Yet, if we are to leverage the power of this medium for learning, much more research is required to better identify the game components that drive brain plasticity, as well as those aspects of behavior that may be affected.

Box 23.1
Key Points

- Action video-game play affects a broad range of behavioral functions as well as their neural bases. We propose that enhanced attentional control is one of the driving factors which acts as a common underlying mechanism to explain the variety of these effects.

- There is a common agreement that intervention studies should include an active control group and use a distributed schedule of practice. These same rules apply to the field of action video-game training. Whether we consider action video-game training or other forms of behavioral interventions, further methodological advances are needed to allow the identification of truly efficient game features, above and beyond already known determinants of success such as motivation, mind-set, or arousal.
- Anecdotal evidence of the impact of action video-game play on everyday life abounds. Well-controlled studies documenting such transfer are much less numerous, however. A handful of studies do point to some everyday life transfer, such as improving vision in amblyopic patients or surgery skills in laparoscopic surgeons. This should be an important direction for future work.

Box 23.2
Outstanding Issues

- Which game characteristics enhance brain plasticity and learning? Which ones allow transfer across domains and skills?
- How do attention and learning relate? What is the contribution of attentional factors to learning, and vice versa, which learning mechanisms may alter attentional control?
- Given the large interindividual variability in attention and learning, can we capitalize on the adaptability of video games to ensure beneficial impact in the population at large? And if so, how can the medium be best individualized to provide the greatest clinical or educational impact?

Acknowledgments

This work was partially support by grants from the Office of Naval Research and the Swiss National Foundation (100014_140676) to D.B.

References

Anguera, J. A., Boccanfuso, J., Rintoul, J. L., Al-Hashimi, O., Faraji, F., Janowich, J., et al. (2013). Video game training enhances cognitive control in older adults. *Nature, 501,* 97–101.

Appelbaum, L. G., Cain, M. S., Darling, E. F., & Mitroff, S. R. (2013). Action video game playing is associated with improved visual sensitivity, but not alterations in visual sensory memory. *Attention, Perception & Psychophysics, 75,* 1161–1167.

Arena, D. A., & Schwartz, D. L. (2014). Experience and explanation: Using videogames to prepare students for formal instruction in statistics. *Journal of Science Education and Technology*, *23*, 538–548.

Basak, C., Boot, W. R., Voss, M. W., & Kramer, A. F. (2008). Can training in a real-time strategy video game attenuate cognitive decline in older adults? *Psychology and Aging*, *23*, 765–777.

Bavelier, D., Achtman, R. L., Mani, M., & Föcker, J. (2012). Neural bases of selective attention in action video game players. *Vision Research*, *61*, 132–143.

Bavelier, D., Green, C. S., Pouget, A., & Schrater, P. (2012). Brain plasticity through the life span: Learning to learn and action video games. *Annual Review of Neuroscience*, *35*, 391–416.

Bavelier, D., Green, C. S., & Seidenberg, M. S. (2013). Cognitive development: Gaming your way out of dyslexia? *Current Biology*, *23*, R282-3. doi: 10.1016/j.cub.2013.02.051.

Belchior, P., Marsiske, M., Sisco, S. M., Yam, A., Bavelier, D., Ball, K., et al. (2013). Video game training to improve selective visual attention in older adults. *Computers in Human Behavior*, *29*, 1318–1324.

Boot, W. R., Blakely, D. P., & Simons, D. J. (2011). Do action video games improve perception and cognition? *Frontiers in Psychology*, *2*(226), 1–6.

Boot, W. R., Champion, M., Blakely, D. P., Wright, T., Souders, D. J., & Charness, N. (2013). Video games as a means to reduce age-related cognitive decline: Attitudes, compliance, and effectiveness. *Frontiers in Psychology*, *4*, 31.

Boot, W. R., Kramer, A. F., Simons, D. J., Fabiani, M., & Gratton, G. (2008). The effects of video game playing on attention, memory, and executive control. *Acta Psychologica*, *129*, 387–398.

Boot, W. R., Simons, D. J., Stothart, C., & Stutts, C. (2013). The pervasive problem with placebos in psychology: Why active control groups are not sufficient to rule out placebo effects. *Perspectives on Psychological Science*, *8*, 445–454.

Buckley, D., Codina, C., Bhardwaj, P., & Pascalis, O. (2010). Action video game players and deaf observers have larger Goldmann visual fields. *Vision Research*, *50*, 548–556.

Cain, M. S., Landau, A. N., & Shimamura, A. P. (2012). Action video game experience reduces the cost of switching tasks. *Attention, Perception & Psychophysics*, *74*, 641–647.

Cain, M. S., Prinzmetal, W., Shimamura, A. P., & Landau, A. N. (2014). Improved control of exogenous attention in action video game players. *Frontiers in Psychology*, *5*, 69.

Castel, A. D., Pratt, J., & Drummond, E. (2005). The effects of action video game experience on the time course of inhibition of return and the efficiency of visual search. *Acta Psychologica*, *119*, 217–230.

Chiappe, D., Conger, M., Liao, J., Caldwell, J. L., & Vu, K. P. L. (2013). Improving multi-tasking ability through action videogames. *Applied Ergonomics*, *44*, 278–284.

Chisholm, J. D., Hickey, C., Theeuwes, J., & Kingstone, A. (2010). Reduced attentional capture in action video game players. *Attention, Perception & Psychophysics, 72*, 667–671.

Chisholm, J. D., & Kingstone, A. (2012). Improved top-down control reduces oculomotor capture: The case of action video game players. *Attention, Perception & Psychophysics, 74*, 257–262.

Clark, K., Fleck, M. S., & Mitroff, S. R. (2011). Enhanced change detection performance reveals improved strategy use in avid action video game players. *Acta Psychologica, 136*, 67–72.

Cole, S. W., Yoo, D. J., & Knutson, B. (2012). Interactivity and reward-related neural activation during a serious videogame. *PLoS ONE, 7*(3).

Colzato, L. S., van den Wildenberg, W. P., & Hommel, B. (2014). Cognitive control and the COMT Val(1)(5)(8)Met polymorphism: Genetic modulation of videogame training and transfer to task-switching efficiency. *Psychological Research, 78*, 670–678.

Colzato, L. S., van den Wildenberg, W. P., Zmigrod, S., & Hommel, B. (2013). Action video gaming and cognitive control: Playing first person shooter games is associated with improvement in working memory but not action inhibition. *Psychological Research, 77*, 234–239.

Colzato, L. S., van Leeuwen, P. J., van den Wildenberg, W. P., & Hommel, B. (2010). DOOM'd to switch: Superior cognitive flexibility in players of first person shooter games. *Frontiers in Psychology, 1*, 8.

Corbetta, M., Patel, G., & Shulman, G. L. (2008). The reorienting system of the human brain: From environment to theory of mind. *Neuron, 58*, 306–324.

De Lisi, R., & Wolford, J. L. (2002). Improving children's mental rotation accuracy with computer game playing. *Journal of Genetic Psychology, 163*, 272–282.

Destefano, M., Lindstedt, J. K., & Gray, W. D. (2011). Use of complementary actions decreases with expertise. In L. Carlson, C. Hölscher, & T. Shipley (Eds.), *Proceedings of the 33rd Annual Conference of the Cognitive Science Society* (pp. 2709–2014). Austin, TX: Cognitive Science Society.

Donohue, S. E., James, B., Eslick, A. N., & Mitroff, S. R. (2012). Cognitive pitfall! Videogame players are not immune to dual-task costs. *Attention, Perception & Psychophysics, 74*, 803–809.

Donohue, S. E., Woldorff, M. G., & Mitroff, S. R. (2010). Video game players show more precise multisensory temporal processing abilities. *Attention, Perception & Psychophysics, 72*, 1120–1129.

Dorval, M., & Pepin, M. (1986). Effect of playing a video game on a measure of spatial visualization. *Perceptual and Motor Skills, 62*, 159–162.

Dye, M. W., & Bavelier, D. (2010). Differential development of visual attention skills in school-age children. *Vision Research, 50*, 452–459.

Dye, M. W., Green, C. S., & Bavelier, D. (2009). The development of attention skills in action video game players. *Neuropsychologia, 47*, 1780–1789.

Feng, J., Spence, I., & Pratt, J. (2007). Playing an action video game reduces gender differences in spatial cognition. *Psychological Science, 18*, 850–855.

Franceschini, S., Gori, S., Ruffino, M., Viola, S., Molteni, M., & Facoetti, A. (2013). Action video games make dyslexic children read better. *Current Biology, 23*, 462–466.

Gagnon, D. (1985). Videogames and spatial skills—An exploratory study. *ECTI-Educational Communication and Technology Journal, 33*, 263–275.

Goldin, A. P., Hermida, M. J., Shalom, D. E., Elias Costa, M., Lopez-Rosenfeld, M., Segretin, M. S., et al. (2014). Far transfer to language and math of a short software-based gaming intervention. *Proceedings of the National Academy of Sciences of the United States of America, 111*, 6443–6448.

Gopher, D., Weil, M., & Bareket, T. (1994). Transfer of skill from a computer game trainer to flight. *Human Factors, 36*, 387–405.

Green, C. S., & Bavelier, D. (2003). Action video game modifies visual selective attention. *Nature, 423*, 534–537.

Green, C. S., & Bavelier, D. (2006a). Effect of action video games on the spatial distribution of visuospatial attention. *Journal of Experimental Psychology: Human Perception and Performance, 32*, 1465–1478.

Green, C. S., & Bavelier, D. (2006b). Enumeration versus multiple object tracking: The case of action video game players. *Cognition, 101*, 217–245.

Green, C. S., & Bavelier, D. (2007). Action-video-game experience alters the spatial resolution of vision. *Psychological Science, 18*, 88–94.

Green, C. S., & Bavelier, D. (2012). Learning, attentional control, and action video games. *Current Biology, 22*, R197–R206.

Green, C. S., Li, R. J., & Bavelier, D. (2010). Perceptual learning during action video game playing. *Topics in Cognitive Science, 2*, 202–216.

Green, C. S., Pouget, A., & Bavelier, D. (2010). Improved probabilistic inference as a general learning mechanism with action video games. *Current Biology, 20*, 1573–1579.

Green, C. S., Strobach, T., & Schubert, T. (2014). On methodological standards in training and transfer experiments. *Psychological Research, 78*, 756–772.

Green, C. S., Sugarman, M. A., Medford, K., Klobusicky, E., & Bavelier, D. (2012). The effect of action video game experience on task-switching. *Computers in Human Behavior, 28*, 984–994.

Greenfield, P. M., Brannon, C., & Lohr, D. (1994). Two-dimensional representation of movement through three-dimensional space: The role of video game expertise. *Journal of Applied Developmental Psychology, 15*, 87–103.

Greenfield, P. M., deWinstanley, P., Kilpatrick, H., & Kaye, D. (1994). Action video games and informal education: Effects on strategies for dividing visual attention. *Journal of Applied Developmental Psychology, 15*, 105–123.

Griffith, J. L., Voloschin, P., Gibb, G. D., & Bailey, J. R. (1983). Differences in eye–hand motor coordination of video-game users and non-users. *Perceptual and Motor Skills, 57,* 155–158.

Hayes, E., & Silberman, L. (2007). Incorporating video games into physical education. *Journal of Physical Education, Recreation & Dance, 78*(3), 7.

Heimler, B., Pavani, F., Donk, M., & van Zoest, W. (2014). Stimulus- and goal-driven control of eye movements: Action videogame players are faster but not better. *Attention, Perception & Psychophysics, 76,* 2398–2412.

Hess, R. F., Thompson, B., Black, J. M., Machara, G., Zhang, P., Bobier, W. R., et al. (2012). An iPod treatment of amblyopia: An updated binocular approach. *Optometry (St. Louis, Mo.), 83*(2), 87–94.

Hoeft, F., Watson, C. L., Kesler, S. R., Bettinger, K. E., & Reiss, A. L. (2008). Gender differences in the mesocorticolimbic system during computer game-play. *Journal of Psychiatric Research, 42,* 253–258.

Hubert-Wallander, B., Green, C. S., & Bavelier, D. (2011). Stretching the limits of visual attention: The case of action video games. *Wiley Interdisciplinary Reviews: Cognitive Science, 2,* 222–230.

Hutchinson, C. V., & Stocks, R. (2013). Selectively enhanced motion perception in core video gamers. *Perception, 42,* 675–677.

Jobe, J. B., Smith, D. M., Ball, K., Tennstedt, S. L., Marsiske, M., Willis, S. L., et al. (2001). ACTIVE: A cognitive intervention trial to promote independence in older adults. *Controlled Clinical Trials, 22,* 453–479.

Karle, J. W., Watter, S., & Shedden, J. M. (2010). Task switching in video game players: Benefits of selective attention but not resistance to proactive interference. *Acta Psychologica, 134,* 70–78.

Katsyri, J., Hari, R., Ravaja, N., & Nummenmaa, L. (2013). The opponent matters: Elevated fMRI reward responses to winning against a human versus a computer opponent during interactive video game playing. *Cerebral Cortex, 23,* 2829–2839.

Kennedy, A. M., Boyle, E. M., Traynor, O., Walsh, T., & Hill, A. D. K. (2011). Video gaming enhances psychomotor skills but not visuospatial and perceptual abilities in surgical trainees. *Journal of Surgical Education, 68,* 414–420.

Klingberg, T. (2010). Training and plasticity of working memory. *Trends in Cognitive Sciences, 14,* 317–324.

Koepp, M. J., Gunn, R. N., Lawrence, A. D., Cunningham, V. J., Dagher, A., Jones, T., et al. (1998). Evidence for striatal dopamine release during a video game. *Nature, 393,* 266–268.

Krishnan, L., Kang, A., Sperling, G., & Srinivasan, R. (2013). Neural strategies for selective attention distinguish fast-action video game players. *Brain Topography, 26,* 83–97.

Kristjansson, A. (2013). The case for causal influences of action videogame play upon vision and attention. *Attention, Perception & Psychophysics, 75,* 667–672.

Kühn, S., Gleich, T., Lorenz, R. C., Lindenberger, U., & Gallinat, J. (2014). Playing Super Mario induces structural brain plasticity: Gray matter changes resulting from training with a commercial video game. *Molecular Psychiatry, 19*, 265–271.

Kühn, S., Lorenz, R., Banaschewski, T., Barker, G. J., Buchel, C., Conrod, P. J., et al. (2014). Positive association of video game playing with left frontal cortical thickness in adolescents. *PLoS ONE, 9*(3).

Lavie, N., Hirst, A., de Fockert, J. W., & Viding, E. (2004). Load theory of selective attention and cognitive control. *Journal of Experimental Psychology. General, 133*, 339–354.

Li, J., Thompson, B., Deng, D., Chan, L. Y., Yu, M., & Hess, R. F. (2013). Dichoptic training enables the adult amblyopic brain to learn. *Current Biology, 23*, R308–R309.

Li, R. J., Polat, U., Makous, W., & Bavelier, D. (2009). Enhancing the contrast sensitivity function through action video game training. *Nature Neuroscience, 12*, 549–551.

Li, R. J., Polat, U., Scalzo, F., & Bavelier, D. (2010). Reducing backward masking through action game training. *Journal of Vision, 10*(14), 33, 1–13.

Li, R. W., Ngo, C., Nguyen, J., & Levi, D. M. (2011). Video-game play induces plasticity in the visual system of adults with amblyopia. *PLoS Biology, 9*(8).

Mathiak, K. A., Klasen, M., Weber, R., Ackermann, H., Shergill, S. S., & Mathiak, K. (2011). Reward system and temporal pole contributions to affective evaluation during a first person shooter video game. *BMC Neuroscience, 12*, 66.

Mayas, J., Parmentier, F. B., Andres, P., & Ballesteros, S. (2014). Plasticity of attentional functions in older adults after non-action video game training: A randomized controlled trial. *PLoS ONE, 9*(3).

Mishra, J., Zinni, M., Bavelier, D., & Hillyard, S. A. (2011). Neural basis of superior performance of action videogame players in an attention-demanding task. *Journal of Neuroscience, 31*, 992–998.

Oei, A. C., & Patterson, M. D. (2013). Enhancing cognition with video games: A multiple game training study. *PLoS ONE, 8*(3).

Okagaki, L., & Frensch, P. A. (1994). Effects of video game playing on measures of spatial performance: Gender effects in late adolescence. *Journal of Applied Developmental Psychology, 15*, 33–58.

Olesen, P. J., Westerberg, H., & Klingberg, T. (2004). Increased prefrontal and parietal activity after training of working memory. *Nature Neuroscience, 7*, 75–79.

Parsons, H. M. (1974). What happened at Hawthorne? New evidence suggests the Hawthorne effect resulted from operant reinforcement contingencies. *Science, 183*, 922–932.

Pohl, C., Kunde, W., Ganz, T., Conzelmann, A., Pauli, P., & Kiesel, A. (2014). Gaming to see: Action video gaming is associated with enhanced processing of masked stimuli. *Frontiers in Psychology, 5*, 70.

Przybylski, A. K., Ryan, R. M., & Rigby, C. S. (2009). The motivating role of violence in video games. *Personality and Social Psychology Bulletin, 35*, 243–259.

Rosser, J. C., Jr., Lynch, P. J., Cuddihy, L., Gentile, D. A., Klonsky, J., & Merrell, R. (2007). The impact of video games on training surgeons in the 21st century. *Archives of Surgery, 142*(2), 181–186; discusssion 186.

Sanes, J. N., & Donoghue, J. P. (2000). Plasticity and primary motor cortex. *Annual Review of Neuroscience, 23*, 393–415.

Schlickum, M. K., Hedman, L., Enochsson, L., Kjellin, A., & Fellander-Tsai, L. (2009). Systematic video game training in surgical novices improves performance in virtual reality endoscopic surgical simulators: A prospective randomized study. *World Journal of Surgery, 33*, 2360–2367.

Schmidt, R. A., & Bjork, R. A. (1992). New conceptualizations of practice—Common principles in three paradigms suggest new concepts for training. *Psychological Science, 3*, 207–217.

Sharma, V., Levi, D. M., & Klein, S. A. (2000). Undercounting features and missing features: Evidence for a high-level deficit in strabismic amblyopia. *Nature Neuroscience, 3*, 496–501.

Sims, V. K., & Mayer, R. E. (2002). Domain specificity of spatial expertise: The case of video game players. *Applied Cognitive Psychology, 16*, 97–115.

Spence, I., & Feng, J. (2010). Video games and spatial cognition. *Review of General Psychology, 14*, 92–104.

Spence, I., Yu, J. J. J., Feng, J., & Marshman, J. (2009). Women match men when learning a spatial skill. *Journal of Experimental Psychology: Learning, Memory, and Cognition, 35*, 1097–1103.

Stafford, T., & Dewar, M. (2014). Tracing the trajectory of skill learning with a very large sample of online game players. *Psychological Science, 25*, 511–518.

Strobach, T., Frensch, P. A., & Schubert, T. (2012). Video game practice optimizes executive control skills in dual-task and task switching situations. *Acta Psychologica, 140*, 13–24.

Subrahmanyam, K., & Greenfield, P. M. (1994). Effect of video game practice on spatial skills in girls and boys. *Journal of Applied Developmental Psychology, 15*, 13–32.

Tanaka, S., Ikeda, H., Kasahara, K., Kato, R., Tsubomi, H., Sugawara, S. K., et al. (2013). Larger right posterior parietal volume in action video game experts: A behavioral and voxel-based morphometry (VBM) study. *PLoS ONE, 8*(6).

Terlecki, M. S., Newcombe, N. S., & Little, M. (2008). Durable and generalized effects of spatial experience on mental rotation: Gender differences in growth patterns. *Applied Cognitive Psychology, 22*, 996–1013.

Toril, P., Reales, J. M., & Ballesteros, S. (2014). Video game training enhances cognition of older adults: A meta-analytic study. *Psychology and Aging, 29*, 706–716.

Trick, L. M., Jaspers-Fayer, F., & Sethi, N. (2005). Multiple-object tracking in children: The "Catch the Spies" task. *Cognitive Development, 20*, 373–387.

Uttal, D. H., Meadow, N. G., Tipton, E., Hand, L. L., Alden, A. R., Warren, C., et al. (2013). The malleability of spatial skills: A meta-analysis of training studies. *Psychological Bulletin, 139,* 352–402.

van Ravenzwaaij, D., Boekel, W., Forstmann, B. U., Ratcliff, R., & Wagenmakers, E. J. (2014). Action video games do not improve the speed of information processing in simple perceptual tasks. *Journal of Experimental Psychology. General, 143,* 1794–1805.

Wei, K., Yan, X., Kong, G., Yin, C., Zhang, F., Wang, Q., & Kording, K. P. (2014). Computer use changes generalization of movement learning. *Current Biology, 24,* 82–85.

West, G. L., Stevens, S. A., Pun, C., & Pratt, J. (2008). Visuospatial experience modulates attentional capture: Evidence from action video game players. *Journal of Vision, 8*(16), 13, 1–9.

Wolinsky, F. D., Mahncke, H., Vander Weg, M. W., Martin, R., Unverzagt, F. W., Ball, K. K., et al. (2010). Speed of processing training protects self-rated health in older adults: Enduring effects observed in the multi-site ACTIVE randomized controlled trial. *International Psychogeriatrics, 22,* 470–478.

Wu, S., & Spence, I. (2013). Playing shooter and driving videogames improves top-down guidance in visual search. *Attention, Perception & Psychophysics, 75,* 673–686.

Wu, S., Cheng, C. K., Feng, J., D'Angelo, L., Alain, C., & Spence, I. (2012). Playing a first-person shooter video game induces neuroplastic change. *Journal of Cognitive Neuroscience, 24,* 1286–1293.

24
Stress and Attention

Peter A. Hancock and Gerald Matthews

Historical Context and State-of-the-Art Review

Introduction to Stress

In any chapter on stress and attention that appears in a *Handbook on Attention*, the authors should focus mostly on stress since the reader will already have considerable understanding of the concept of attention from the other contributions. Thus, we begin here with the definitions of stress with the aim to later understand how such stress affects attentional capacities and, as we shall find, vice-versa. Stress is a ubiquitous influence on human existence, as indeed it is on all living things (Selye, 1958). It forms a major evolutionary shaping force that affects behavior in a generally consistent and predictable manner. In terms of human psychological research, stress has been largely regarded as an appraisal-based influence. This follows the work of Richard Lazarus (e.g., Lazarus & Folkman, 1984), who pointed out that for most individuals there are certain forms of activity that might generally be considered extremely stressful. For example, jumping out of an aircraft and free-fall parachuting would be, for most people, a highly stressful and truly a fight-or-flight experience. However, there are certain individuals who actually enjoy and embrace such activities. Inevitably then, from the appraisal perspective, stress is a relational concept where we must consider the task at hand and the environmental context in which such performance occurs as well as the character of the individual involved. Thus, what for one person proves highly stressful for another, according to Lazarus's definition, need not necessarily be stressful at all. This makes eminent sense from a psychological perspective because this is how individuals do differentially appraise all psychological aspects of the challenges that face them. However, appraisal is a necessary but not sufficient description of stress since it omits certain critical forms of challenge such as, for example, its physiological dimensions.

In terms of physiological challenge, we as human beings all share certain common characteristics. Thus, different forms of physiological demand prove stressful to all humans regardless of how they appraise them. Indeed, this argument can be extended to all living systems. The high variability, in the form of individual differences, that we see exhibited in response to the variously appraised psychological challenges become much more constrained when we consider reactions to physiological forms of under-load or overload. For example, the immediate deprivation of oxygen, the long-term deprivation of food, and the shorter-term deprivation of water each proves to be stress-ful regardless of how each person "appraises" the situation. It is true there will still be individual differences as to the onset rate that such reactions occur and that some will panic while others will be stoic or even frozen. However, in the end, we find that all individuals are put in extremis by such deprivations. Thus, at the outer boundaries of stress, in this case of physiological demand, we find that all humans experience stress and an unavoidable or obligatory decrement in their capacity to respond (Plessow, Fischer, Kirschbaum, & Goschke, 2011). This fact is true physiologically, but it is also true physically. This is because humans are not just living systems but are also physical entities. When any physical entity meets with problematic levels and types of energy in the environment, it experiences direct physical damage. This is what Haddon (1970) called meeting *"ecological tigers"* (and see Hancock, 2005) by which he meant ranges of kinetic, chemical, and thermal energy (etc.) that were incompatible either with contin-ued life or physical existence or both. Our news media are sadly replete with environ-mental and technical disasters where such "tigers" are unleashed and people not only fail as living systems but are obliterated as physical objects (Cooke, 2013).

As we now begin to appreciate stress from a multilevel perspective, we can see that it can be a simple physical force imposed upon the physical body, it can be of a physio-logical nature imposed upon the basic living processes, or it can be, as is most appropri-ate to the present chapter, an appraised form of psychological demand. From a formal cognitive science perspective (Matthews, 2001), the effects of stress on attention could be variously attributed to changes in neural functioning (Plessow, Schade, Kirschbaum, & Fischer, 2012), to changes in the virtual constructs of standard cognitive models such as attentional resource supply and/or working memory capacity (Schwabe, Wolf, & Oitzl, 2010), or to individuals' high-level performance goals (Kofman, Meiran, Green-berg, Balas, & Cohen, 2006) and beliefs about how to attain their goals. The psycho-logical dimension is almost ubiquitously a relational dimension, and so Lazarus is not wrong in his definition. However, the relation between person and task environment is represented first within the physical body and then within parameters of the func-tional cognitive architecture as well as within those high-level self-regulative processes emphasized by Lazarus. Under the moderate levels of stress that typify laboratory experiments, we may see empirical dissociations between these levels; for example, psychophysiological indices are only modestly predictive of information-processing measures, at best (Matthews, 2001). However, at the extremes of stress, under the pres-sures of evolutionary imperatives, such reactions to stress coalesce. Indeed, if they did not, neither human beings nor any living system would, on average, survive.

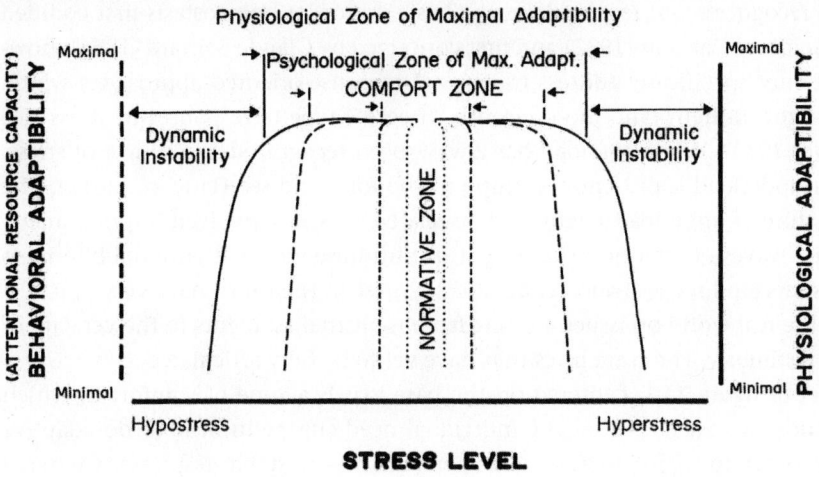

Figure 24.1

The extended-U theory of stress and response capacity. After Hancock, P. A., & Warm, J. S., 1989. A dynamic model of stress and sustained attention. *Human Factors, 31,* 519–537. Courtesy of the Human Factors and Ergonomics Society.

Stress and Response Capacity

From the foregoing, it is evident that stress responses vary most widely across the central part of the possible spectrum of environmental demands and that there are outer boundaries at which response variability begins to decrease and failure begins to occur for everyone. In the extended-U model of stress and performance (see figure 24.1) these outer boundaries are represented as threshold limits of progressive dynamic instability. These zones of incipient failure are in contrast with the more normative functioning in the stable central levels between extremes of hypostress and hyperstress. This descriptive conception, as shown in figure 24.1, can serve to guide us as to how attention is focused under stress and how attention is a tool for survival which belongs to the cognitive arsenal of weapons by which humans are able to navigate successfully through a world that potentially threatens them and consequently has certain forms of stress intrinsic to it.

The Hancock and Warm (1989) model (see figure 24.1) recognizes these basic aspects of stress effects and looks to encapsulate them first in a descriptive and subsequently in a causal model. Figure 24.1 recognizes that the stress or demands of the environment can range from an insufficiency or underload of any energetic value (hypostress) to an excess or overload of that environmental form of demand (hyperstress). The curves shown in the body of the illustration indicate that for the vast range of these external, "input" forms of underload and overload, humans can adjust or adapt such that their performance, or more generally their behavior, is sustained at relatively stable levels. This recognizes and recapitulates the basic notion of homeostasis first codified by Walter Bradford Cannon (1932) and first conceived by Claude Bernard (1865); however, it does not specifically address traditional response-oriented approaches which emphasize fight-or-flight (and more recently "freeze") type of responses to stress (see also Cannon, 1932). While overload has always been recognized as a source of stress, the issue of underload is only now coming to the fore (and see Hancock, 2013). The insidious nature of underload might well be related to issues involved in phenomena like temporal waves of attention that might in turn influence inattentional blindness or mindlessness (and see Eastwood et al., 2012). These in their turn have very practical impacts in the real world on issues ranging from industrial accidents to the veracity of eyewitness testimony. These are links that have yet to be fully articulated.

In respect of figure 24.1, centered on the base axis is a zone of comfort in which ambient conditions are such that the individual need engage in little or no adaptive response in order to maintain successful performance or stable response. Centered within the envelope of comfort is a normative zone which can be associated with the notion of "flow" (Csikszentmihalyi, 1990). In this state, the individual appears to "anticipate" environmental and task demands and so expresses no classical stress response of any measurable degree. Highly allied to the concept of the "zone of optimal functioning" (see, e.g., Kamata, Tenenbaum, & Hanin, 2002), flow is an idealized state in which all expectations about immediate future events seem to be fulfilled such that the individual reports "being out in front of the game." Typically, such reports emanate from formal sports in which the rules are well-known, well-practiced, and predictable to a greater degree than our less structured parts of normal existence, although flow states are known there also. Here, attention seems diffuse since the normally expected "focus" on reducing the uncertainty of information ambiguity has now been obviated by the fact that the world appears, at least for a short period of time, perfectly predictable. Since the world always remains uncertain and potentially threatening, flow states are presumably rare occurrences "in the wild."

Outside of these respective regions of dynamic stability, the level of ambient demand eventually becomes sufficient to induce progressive response failure. In figure 24.1, two such "levels" of failure are identified as being psychological adaptability and physiological adaptability, respectively. However, in reality these are useful only as discipline-based descriptive identifiers since the main seat of psychological functioning, that is, the brain, always remains one of the body's organs anyway, although admittedly physiologically a very specialized one. At some level of external demand the individual's capacity to respond becomes impaired. This impairment is not represented by a smooth and progressive degradation, as described in the traditional and classic inverted-U approach to stress (Yerkes & Dodson, 1908). Rather, it is more of a sudden brisance of failure (Hancock, 2009). In real-world conditions, people persist in the face of extremes of stress up to the very limits of their capacity. It is only in the arid conditions of the sterile laboratory that undergraduate participants gradually give up, on average, to provide the graceful degradation the inverted-U presents (and see Hancock & Ganey, 2003).

The form of failure, as shown in figure 24.1, is a descending exponential. Significant levels of change in psychological capacity are coincident with the first beginnings of change in physiological stability (and see Hancock, 1982). This assertion denies the postulation of one of the earliest stress concepts termed "physiological adequacy." This medical precept stated that as long as physiological stability could be sustained, there would be no change in "psychological" (i.e., task) performance. Empirically, this simply proves to be incorrect. However, sadly, many real-world legislative thresholds for different forms of occupational stress were initially founded upon this fallacy and some still remain so to this day. Figure 24.1 thus also shows that psychological dimensions of stress and attention are not simply of theoretical or conceptual interest but are issues that impact working people every day of their lives (for one specific example, see Weltman, & Egstrom, 1966; for an example of effects on international standards, see Hancock & Vasmatzidis, 2003). In the same way that unconsciousness is the end point of overstressed psychological functioning, so death is the end point of overstressed physiological functioning.

The causal mediator in psychological functioning in the Hancock and Warm model is attentional capacity. Stress acts to "drain" these attentional resources and so represents a form of dual-task demand in which adapting to stress can be regarded as a task in itself. This led Hancock and Warm to propose that the immediate or proximal task in hand, that is, what the person is doing or trying to do, represents the primary form of stress in the most stable condition. One way of alleviating this proximal stress is simply to give up (and see Hancock & Caird, 1993), but in many real-world circumstances this is not a viable option. The degree to which certain physiological "fuels" subserve the notion of attentional resources and the specific way such resources are drained by stress (e.g., Is this an overall reduction, or are the degrading effects experienced in only "local" brain areas?) are still matters of ongoing neuroscience investigations (see Fairclough & Houston, 2004; Kurzban, 2010). However, a unified vision of psychological performance deterioration and neurochemical insufficiency now looks at least to be a feasible future proposition.

A Brief View of Attention

Having surveyed the issue of stress, we now have to look, albeit briefly, at the issue of attention, which has been the subject of extensive study in psychology as well as neuroscience (e.g., James, 1890; Wachtel, 1967). Attention represents the critical linking process which mediates between individuals and their perceived environment. Consequently, attention is of very practical concern to many constituencies and not merely to psychological researchers and theorists alone (see, e.g., Dewar & Olson, 2002; Moskowitz & Sharma, 1974). Here, we approach attention from an evolutionary perspective. That is, we wish primarily to see it as a capacity of humans that promotes their successful survival.

So what then is attention? James (1890), of course, advanced the notion that attention was a taking hold of the mind in order to process some stimuli to the exclusion of others. However, in important ways this is mostly a descriptive and rather tautological statement. Thus, we need to identify the many mediating constructs which fracture the threat of definitional circularity. Attention derives from the fact that humans, like all living beings, are not able to assimilate everything in their environment at any one time. Now this might appear obvious because each neonate is essentially bootstrapped into this situation of necessary inadequacy. However, this assertion of insufficiency need not be quite so obvious as it first might seem. An organism may conceivably be perfectly adjusted to its environment such that it only processes what is necessary for survival. That organism would then basically ignore or even perhaps be unaware of anything unnecessary to its survival. In reality, humans experience this insufficiency quandary partly because the nature of their obligatory information itself changes and evolves. Attention is most certainly not uniquely human. However, the contemporary character of human attention is driven by the fact that there is now an overwhelming excess of information in our fabricated, technical environment. Thus, attention itself almost certainly coevolves with the nature of the natural and built environments that organisms experience (and see Smith & Hancock, 1995). As our technology grows, so the gap between our assimilation capacities and the wealth of burgeoning information also expands. Our insufficiency of attention, and our meta-cognitive recognition of this shortfall, grows accordingly. Recognition of this vast gap between sufficient attention for survival and the proliferation of modern, almost obligatory, sources of information results in stress. However, such an information-processing gap may also drive our curiosity and indeed our interest in the topic of attention itself.

Attention then is a selective process by which some things are admitted to further processing and others are not. Modern environments prove to be replete with "*thieves of attention*" that seek to hijack this active and dynamic process for their own purposes, not the least of which is the modern pop-up advertisement. Notwithstanding the nominal "failure" as characterizing modern human attention, it is true that in general we have, over the millennia, proved to be very successful survivors. Work in experimental laboratories illustrating what humans cannot do is often derived from artificial situations, such as dichotomous listening tasks, in which multiple sources of information immediately and intentionally bombard individuals in order specifically to expose their limits. When such failures do occur, the enthusiastic researcher is tempted to draw general principles, but such conclusions may be iatrogenically misdirected (Hancock, 2013). For example, Gibson (1979) cogently disputes the validity of generating any general theory founded on observations of exceptional percepts. In actuality, real-world attention is very effective, and it is rare that individuals fail in a spectacular manner even in our modern world. Of course, on a statistical basis across large numbers, there will be a sufficient number of these failures to attract social recognition (Perrow, 1984). These exceptional failures must be carefully examined before we can safely extrapolate from them.

Attention is a relational property that cannot be defined without understanding the ambient environment. This is why neuroscientific searches for attention are a necessary but not sufficient step in understanding the whole concept. One thing neuroscience does allow us to explore is the multimodal nature of attention. In the wider world, attention is overwhelmingly thought of in terms of vision. If we look at the amount of cortex accorded to each sensory system, we do find visual overdominance. However, the idea of attention as vision underestimates the contribution by the other sensory modalities (e.g., Spence & Driver, 2004) which often provide sources of orientation which help direct vision in a complex "holarchy" of interactions (Koestler, 1978).

In the "natural" environment, then, we find that humans are able to pay attention to considerable amounts of information and are magnificently successful as adaptive animals but these "vestigial" processing capacities may be poorly suited to the "modern" information world. The problem here is that our environment has changed so radically and rapidly that we have not had sufficient evolutionary time to adjust (and see Linnell, Caparos, de Fockert, & Davidoff, 2013). Although our technology has helped to mediate this particular evolutionary saltation, tools have been only partially successful in this endeavor. Indeed, poor impoverished and inefficient technologies bring their own added attentional demands. Today, we have even started to invent names of syndromes for such attentional shortfalls. However, these nominal syndromes are largely symptomatic of the aforementioned dissociations between resident attentional capacities versus the rate of modern changes in society (Gleick, 2000). Thus, both chronic and acute attention changes as a function of the environment which it encounters. The rate of evolution of attention itself promises to covary with the radical change in our informational landscape. Thus, as well as being selective, attention is a nonstationary phenomenon, and this makes a simple definition even more difficult.

In contrast, the primary function(s) that attention supports are relatively straightforward; these are the basic needs for food, water, shelter, procreation, and so forth. With the culturally advantageous capacity to transmit salient solutions, as generated and passed on through tools such as language, mathematics, and logic as well as physical artifacts, humans now prove dominant in the current environment. Such technologies serve to short-circuit the necessity for the ab initio learning of these capacities in each generation. Today, traditional "natural" sources of threat represent statistically minor levels of danger to modern humans (unless individuals specifically desire them, e.g., thrill seekers). The vestigial processes of attention then still function sufficiently well in the modern world. Attention still captures things that are novel, which is itself a reflection of the resident capacity for information foraging (Pirolli, 2007; Pirolli, & Card, 1995). Novelty is sought since it provides high levels of informational return for the search effort invested. This information search gradient is why, for example, we feel so pressured to answer our phone. Since telephones are now potentially connected to about 30% of all humans on the planet, the formal reduction of uncertainty for a ringing telephone is about 1 in 3 billion (assuming no identifiable ring tone). The ringing cue provides such a gradient and promises such a degree of uncertainty reduction that humans basically have an almost impossible time ignoring it. Intensity is another attentional vector which evidently relates to survival. Intensity implies stimuli that are close by, and such stimuli remain potential threats. Thus, for threat mitigation and for potential benefit, something that is novel, intense, and/or unusual will attract attention. Sensory systems generate an orientating response in order to understand and resolve ambiguities that are generated. However, in our modern world, such orientations are often maladaptive and, in and of themselves, induce stress reactions.

This survey of attention, albeit problematically brief, allows us to begin to generate a synthesis with stress theory. What happens to attention, the selective and evolving capacity to assimilate information, as stress increases? Let us first examine this in relation to the boundaries of the psychological or appraised nature of stress. As one begins to appraise an environment as stressful, the inevitable tendency is that one focuses on the nature and source of that threat as well as its potential avenues of resolution. After all, organisms that do not resolve such threats cease to exist. Stress then focuses attention on the immediate proximal problem, that is, the here and now (and see Arnsten, 2009). This means that in the prefrontal cortex, in which human beings extend their phenomenological being by considering future planning, stress curtails activity (and see also Hancock, 2010; Killingsworth & Gilbert, 2010). Thus, under extremes of stress we do not generally see mind wandering or mindlessness, we do not see revelry or planning for the distant future because those organisms that plan for the future when the immediate present constitutes an imminent threat are organisms that don't tend to exist in that future. Phenomenologically, what we see is a collapse of space in time over which the individual exerts his or her attention in order to resolve immediate conflict. In short, stress reverses the phylogenetic development of attention.

As noted earlier, appraisal is a highly idiographic process, and so some people tend to appraise a stress where the vast majority of individuals see no stress. And there are certain pathologies in which individuals act as though they are under constant and extreme levels of stress when to the vast majority of others there is essentially no stress at all. These can range from stress-induced immunological vulnerabilities and their variously associated problems to post-traumatic stress disorder (PTSD), where the individual relives the stress of life events that are now no longer present (Beidel et al., 2011). This gives us two ends of a spectrum, one of which is the pathology that comes along with such hypersensitivity and the other end being a hyposensitivity or exhibiting a "devil-may-care," almost insensate attitude to sources of threat. The latter individual, although well adapted to certain exceptional conditions (e.g., combat), is in danger from both the normal environment and the local social community. Thus, under stress, we see the incipient collapse of the phenomenological experience of extended reality to the here and now.

Stress and the Narrowing of Attention

Phenomenologically, then, under extremes of stress, individuals "narrow" their range of attention (for original works, see Bacon, 1974; Calloway, & Dembo, 1960; Cornsweet, 1969; Easterbrook, 1959; Leibowitz, & Appelle, 1969; Reeves & Bergum, 1972). This narrowing was originally postulated as a spatial phenomenon (e.g., Bursill, 1958) although more recent investigations have shown this to be only part of the story. Such narrowing is directed to the cues of greatest perceived salience, these being the ones that promise to resolve the proximal threat and thus reduce the appraised stress. However, an area that has not yet been evaluated sufficiently is the equivalent collapse in time; these stress-induced effects are often referred to as temporal distortion (see Eagleman, 2008; Hancock & Weaver, 2005). Although it is harder both empirically and methodologically to substantiate these latter effects, the present notion is that stress narrows attention to the here and now, that is, in both space and time.

This whole process has also been termed "cognitive tunneling" since the narrowing is not always sensory in nature (Dirkin & Hancock, 1985). The experiences of pilots in high-performance aircraft, who under g forces experienced visual tunneling, then grayout, and then black-out, had a strong impact in perpetuating the notion that "tunneling" was a sensory rather than a cognitive effect. The notion of a cognitive component was first postulated by Bursill (1958) in the presence of heat stress (see also Leibowitz et al., 1972) and then demonstrated by Hockey (1970a, 1970b) under acoustic stress. Subsequently, Hancock and Dirkin (1982) showed that individuals could "narrow" to spatially peripheral locations, demonstrating that narrowing took place to the information sources of greatest salience, which need not necessarily be in the center of the visual field. Such observations on cognition are by no means new since, as Dr. Samuel Johnson (1709–1784) is reported as noting, "*Depend upon it sir when a man knows he is to be hanged in a fortnight* [two weeks], *it concentrates his mind wonderfully.*"

We might also ask whether there is such a phenomenon as "cognitive expansiveness." That is, might the opposite of attentional narrowing also be feasible? The idea that the range of attention in space and time increases when we perceive less stress is a reasonable one. Indeed, it might provide some degree of potential conceptual support for many of the claims of activities such as meditation (MacLean, et al., 2010) and cognitive recovery in nominally "natural" environments. However, following Selye's

(1958) original observations, we must remember that happy, euphoric, or so-called "eustress" states are not necessarily "stress-free" but rather bring their own unique challenges. Thus, stress does not always derive from noxious or unpleasant stimuli. Rather, stress can derive, in many cases, from pleasant circumstances such as a promotion or marriage which may be considered hedonic changes. Holmes and Rahe (1967) noted many situations where we have positive life changes, and those changes are major events which can represent very pleasant circumstances yet remain stressful. Thus, stress may be appraised as a direct result of changing the environment around us. So change is the particular axis upon which we can colocate both stress and attention. This leads to an interesting paradox in that novel situations are stressful because of changes; intense situations are stressful because of the rate of change, and so in highly complex fluctuating environments attention is already taxed by the changes that are occurring but people may very well appraise such changes as stressful in and of themselves. Thus, stress and attention necessarily interact, especially if the task at hand is viewed as the proximal stress to be confronted. Having looked at the narrowing phenomenon, let us now examine the thorny issue of individual differences before we look to extract general principles and final conclusions.

Individual Differences in Stress and Adaptation

Stress and adaptation help us to understand pervasive individual differences in attention. Individual difference factors are conventionally divided into trait and state factors. Traits are stable characteristics, including ability and personality, whereas states represent transient psychological qualities, such as emotions. Personality research identifies traits for stress vulnerability, such as neuroticism, but such dimensions are often rather weakly associated with task performance under stress. The most extensively researched dimension is trait anxiety, which is associated both with weak endogenous control of attention and with prioritization of the processing of threat-related stimuli (Eysenck & Derakshan, 2011). Here, we focus mostly on state factors, which are more proximal influences on attention, and which may correspond more closely to the dynamic adaptive processes of interest than do the temporally more stable traits.

As a singular term, the label "stress" does not do full justice to the various phenomenological experiences of individuals confronted with attentional challenges. Different individuals may claim subjectively to feel overwhelmed, alienated, distracted, or challenged since many descriptive terms are used under the umbrella of stress. Different experiences of stress, which may be accompanied by different physiological responses, may be understood as different modes of adaptation to task challenges. When task demands are excessive, one mode of response is to increase voluntary effort and engage in subgoal management to maintain performance. This comes at the cost of personal discomfort and strain, accompanied by sympathetic arousal and changes in the EEG indicative of effort, including frontal theta (see Hockey, 1997; Hockey, Nickel, Roberts, & Roberts, 2009). An alternate response mode is to lower one's task standards or shed lower priority subtasks, a pattern of response which is characteristic of fatigue (Hockey, 1997; Matthews, Desmond, Neubauer, & Hancock, 2012).

It is important to work with multidimensional models of stress states to systematize individual differences in the experience of stressful task environments. An important advance has come with the psychometric identifications of three distinct dimensions of the stress state, that is, distress, worry, and task engagement (Matthews et al., 2002). Each represents a distinct "transactional theme" (Lazarus, 1999) that specifies the core adaptive relationship between the person and the task environment. Distress is defined by feelings of tension and negative mood accompanied by a sense of loss of control and confidence. Changes in distress are driven by perceptions of high workload (Matthews et al., 2002), including being overwhelmed, as well as by the cognitive stress processes specified in the Lazarus (1999; Lazarus & Folkman, 1984) stress model, that is, appraisals of threat and lack of control coupled with coping by focusing attention on personal concerns and emotions (Matthews et al., 2013). The adaptive relationship here is one of damage control when overloaded with task demands. In terms of the Hancock and Warm (1989) extended-U model, the person senses impending hyper-stress and oncoming dynamic instability and looks to respond accordingly. Consistent with Hockey's (1997) compensatory control model, performance may be maintained through increased effort, but deficits are seen when the task requires substantial executive control, as when multitasking is required (Matthews & Campbell, 2010). Studies of anxiety suggest that inhibition of task-irrelevant stimuli is vulnerable to impairment (Eysenck & Derakshan, 2011).

State anxiety is experienced as worry as well as emotional distress, that is, intrusive thoughts that divert resources and working memory from task-related processing. Often, worry is distracting and more detrimental to attention than anxious emotion (Zeidner, 1998). In this case, the adaptive mode is one of pulling back from the immediate task to process the personal relevance of the task context, for example, when a test-anxious student contemplates the price of failure on an examination (Wells & Matthews, 1994; Zeidner, 1998). Worry is future oriented. In relation to the extended-U, the person chooses to peer over the precipice of the hyperstress zone of dynamic instability by anticipating performance failure, in effect shrinking his or her comfort zone. Worry is a consequence of combining emotion-focused coping with avoidance coping in the form of mental disengagement (Matthews et al., 2013). In clinical anxiety, excessive worry and "worry about worry" may lead to the major attentional abnormalities that accompany generalized anxiety disorder and other conditions such as PTSD (Wells & Matthews, 2006).

The third fundamental dimension is task engagement, which represents a eustress state of attentional focus, energetic arousal, and task-directed motivation. It is driven by high levels of challenge appraisal and task-focused coping. It represents a transactional theme of willing commitment of effort to the task (Matthews et al., 2002, 2012). It can be conceived as a form of "eustress" in Selye's (1958) original terminology. Task engagement is the state dimension most directly associated with performance on attentionally demanding tasks including controlled visual search and sustained focused attention (Matthews, Warm, Reinerman, Langheim, et al., 2010). It may be seen as a marker for availability of attentional resources mobilized by task challenge. Engagement extends the width of the stable range of the extended-U (Hancock & Warm, 1989) by enhancing the person's ability to compensate for dynamic instabilities induced by both hypo- and hyperstress. Task engagement may be especially important at the hypostress pole, as evidenced by the relationship between low engagement and substantial performance decrement on vigilance tasks (Matthews, Warm, Reinerman, Langheim, et al., 2010). Low engagement is a core feature of fatigue states, as the person becomes simultaneously tired and alienated from the task (Saxby et al., 2013).

Studies of vigilance illustrate how individual differences in performance change under stress and how they can be understood in relation to cognitive theory. Broadly, when attention must be sustained during a simple signal-detection task, environmental stressors often impair performance, but effects are inconsistent, with stress sometimes facilitating attention (e.g., Hancock, 1984; Hancock & Vasmatzidis, 2003; Szalma & Hancock, 2011). Inconsistency in the stress–performance association might reflect moderator effects of both the type of stress induced and task demands. To further complicate the picture, vigilance tasks themselves are stressful even in the absence of any environmental factors (Hancock, 1984; Warm, Parasuraman, & Matthews, 2008). Mapping the multiple stress state responses observed helps to clarify and differentiate stress effects. For example, Matthews, Warm, Shaw, and Finomore (2014) had participants perform two visual vigilance tasks. The first task was a short, high-workload task requiring detection of rapidly presented masked character stimuli. It was used to evaluate individual differences in engagement, distress, and worry responses. The second, longer task simulated monitoring a tactical display for symbols representing moving enemy vehicles. Both vigilance tasks showed the perceptual sensitivity decrement typical of high-workload detection tasks (Warm, Dember, & Hancock, 1996; Warm et al., 2008), and both elicited a patterned stress response combining high distress and low task engagement. As in other studies (see Matthews, Warm, Reinerman, Langheim, & Saxby, 2010, for a review), the task engagement response to the first task predicted both overall sensitivity and sensitivity decrement on the second task. However, distress was only weakly associated with vigilance. Thus, task engagement appears to be the critical element of stress response for sustaining attention to simple stimuli, and distress may be epiphenomenal. Indeed, Helton, Matthews, and Warm (2009) showed that increased engagement fully mediated an otherwise perplexing facilitation of vigilance by loud jet engine noise. Moderator effects of task workload imply that engagement may be a marker for availability of attentional resources (Matthews, Warm, Reinerman, Langheim, & Saxby, 2010). Other attentional processes may be sensitive to other elements of stress; for example, distress has been linked to working memory and dual tasking (Matthews & Campbell, 2010) consistent with evidence that hypothalamic–pituitary–adrenal axis (HPA) activation may be associated with impaired cognitive control of multitasking (Plessow, Kiesel, & Kirschbaum, 2012).

Vigilance studies also highlight the strengths and weaknesses of neurophysiological explanations of individual differences in attention. Early studies suggested that the vigilance decrement was accompanied by loss of cortical arousal, but it proved difficult to show that de-arousal was a direct cause of attentional impairment (Davies & Parasuraman, 1982). Furthermore, traditional arousal theory has severe limitations as an explanation for individual differences in performance (Matthews & Gilliland, 1999). However, use of psychophysiological measures that are more closely tied to processing than standard arousal indices may be more informative. The vigilance decrement is reliably accompanied by decreased cerebral blood flow velocity (CBFV) in the right hemisphere (Warm et al., 2008). Furthermore, similar workload factors control both performance and hemodynamic responses, implying that CBFV may index attentional resource allocation to task processing (Warm et al., 2008). Matthews, Warm, Reinerman, Langheim, et al. (2010) found that individual differences in CBFV response to a high-workload short task battery predicted vigilance on a subsequent task, consistent with such a resource interpretation. However, structural equation modeling showed that, although correlated, subjective task engagement and CBFV predicted vigilance independently. Pragmatically, prediction may be optimized by combining subjective and objective stress response, and it may be difficult to capture the functionally relevant part of the subjective response with psychophysiological measures. Similarly, Matthews, Reinerman-Jones, Barber, and Abich (2015) indicated that while both psychophysiological and subjective workload measures were diagnostic of task demands in a multitasking environment, the various measures were largely independent of one another. Brain imaging studies of vigilance suggest that sustenance of attention is controlled by multiple regions, including prefrontal cortex, anterior insula, and parietal areas, as well as various subcortical structures (Langner & Eickhoff, 2013). Stress factors might impinge in various parts of this widely distributed brain network. Hence, it may be difficult to index holistic network functioning via psychophysiology, and subjective measures may pick up otherwise inaccessible variance in functioning, as well as strategic control processes such as effort investment (Matthews, Warm, Reinerman, Langheim, & Saxby, 2010; Matthews et al., 2014).

A focus on state factors also highlights the dynamic nature of individual differences in stress response. In the standard literature on anxiety (see, e.g., Zeidner, 1998), high state worry is often seen as a static causal influence on attentional impairment. However, appraisals of oneself as distracted may feed back to amplify worry (or reduce it if worry can be reinterpreted as a motivator). Just as attention itself can be seen as effect rather than cause (Johnston & Dark, 1986), causal paths between stress states and attentional processes may be bidirectional and variable over time as the person follows one of several adaptational trajectories shaped by appraisal and coping processes (Matthews, 2001). Over longer time periods, learned attentional strategies may also shape personality. Various techniques are available for training attention to enhance emotion regulation and well-being (Wadlinger & Isaacowitz, 2011). Thus, relationships between stress and attention must be seen in a wider context of self-regulation over varying time spans (Wells & Matthews, 1994).

The dynamic interplay between stress and attention in the individual is mediated by executive control processes. As elaborated below, a pivotal attribute of humans is their ability to base action on the mental modeling of future events, a facility that requires executive control in the form of planning, inhibition of currently prepotent but future-irrelevant stimuli, and switching attention between current and future contingencies. Using Posner's Attention Network Test (Fan et al., 2002), Matthews and Zeidner (2012) showed that engagement and distress states were correlated (in opposite directions) with executive processing speed. Stress states are dynamically linked to executive processes. For example, worry is associated with overly negative future projections and excessive attention to them, engagement with effortful pursuit of systematic plans and distress and anxiety with disruption of control as the person crisis-manages immediate overload (Eysenck & Derakshan, 2011; Matthews et al., 2013). Having looked at these various aspects of psychological stress appraisal, we can now offer a brief synthetic overview.

Integration and Future Directions

It is possible then, given the foregoing observations, to provide a somewhat newer perspective on stress and attention at this juncture. This requires that we conceive that not all of the effects of stress are actually resident in any one individual human being. The typical focus of both physiology and psychology is to emphasize Protagoras's notion that "man is the measure of all things," thus fixing the "unit of analysis" very firmly on the individual human being. However, it is possible that neither stress nor attention is necessarily exclusively resident in the individual per se but rather, to some degree, is part of the greater environment. If this postulate is a step too far to take conceptually in one stride, let us think of an intermediary case, as, for example, in group or team performance. We can imagine that any teams such as an intensive care unit or a pediatric surgery group are collectives that inevitably encounter stress because lives are at stake. Here we can see that information need not be resident in the head of any one single individual; it could in fact be an emergent property of the overall group's experience and actions. Hence, we can advance the notion that stress is a systemic property and that there is also a collective attention on behalf of the group. What we might expect, but do not yet have sufficient evidence for, is the way such teams then "narrow" their focus and/or collectively adapt to novel, intense, or intrusive information. Anecdotally, we believe such phenomena occur, as for example in the 1972 crash of Eastern Airlines Flight 401 in the Everglades (National Transportation Safety Board, 1973). In practical worlds where people often work in groups, such conceptual and applied understanding would be of great practical value, especially where faults and failures occur in such group processes (see, e.g., Cannon-Bowers & Salas, 1998). The opportunity to integrate a social theory of stress and attention with a person-based one is an intriguing vista for future research and provides an excellent demonstration of how work on stress and attention can open new doors for understanding basic mechanisms of attention. Do individual models apply to groups and vice-versa? Such work is of importance and progressing.

Summary and Conclusions

What conclusions can be drawn concerning the influence of stress on human attention? The contemporary position is rather clear. Stress acts to recapitulate and reverse evolution. That is, human beings are primarily characterized by their strong and atypical growth of the prefrontal cortex, sometimes termed degree of "encephalization." Human brain development itself is not so particularly unusual, except to the degree that the region of growth of the prefrontal cortex is so quantitatively different from the growth of all other brain regions. This growth has created an organism whose adaptive advantage lies in its ability to project into the future (Hancock, 2010). Such planning capacity improves adaptive response to such a degree that human beings are currently the dominant species. The price for this advantage has been to create a veritable "theater of the absurd." That is, human beings expend much effort, in terms of cognitive resources, envisioning many possible events (both during normal consciousness and during sleep/dreaming), many of which inevitably never occur (Hancock, 2010). To the degree that they do, the human is able to anticipate uncertainties in the environment whereas purely reactive organisms have to wait for developments. The price here is paid in terms of unique humancentric iatrogenic conceptualizations such as time and rationality, among many others. Stress foreshortens this horizon of imagination to the here and now. That is, human beings become much more like other reactive animals as stress increases. "Narrowing" phenomena provide empirical confirmation of this collapse. In this reactive mode, humans experience much greater levels of systemic stress as a result of the increases in uncertainty and thus the elevation in perceived level of threat. How this overall orchestration is managed within the active brain is an endeavor for neuroscience to unravel and is not broached in this necessarily brief chapter. However, this is not to say that significant progress in this direction has not been attained (see, e.g., Liston, McEwen, & Casey, 2009; Robbins & Arnsten, 2009; Pabst, Schoofs, Pawlikowski, Brand, & Wolf, 2013; Pabst, Brand, & Wolf, 2013;

McEwen, 1998). We should also note that neither have we surveyed the vast literature on the biochemical bases of the stress response, again because of space limitations, although we certainly recognize that this is a whole other dimension to the stress story. Insufficient levels of stress and challenge would condemn us to a life of ennui and boredom; excessive levels of stress mean we live in frantic desperation. Nominally, we have evolved in an information-replete environment to pursue and sustain a happy medium between these extremes, mediated by our developed and developing attentional capacities. However, now we are faced with complex, artificial (technical) worlds which raise new challenges which human response capacities are often poorly attuned to address. Stress is now a social epidemic, and as the rate of technical change increases, it promises neither to abate, nor to disappear—watch this space!

Box 24.1

Key Points

- Stress is the shaping force of evolution.
- Stress and attention are mutually embracive. Stress demands attention; attention mediates stress.
- Humans deal well with the stresses of vestigial environments; modern environments generate novel and sometimes overwhelming demands.
- Appraised stress differs for each individual; predicting response to each person's idiographic appraisal presents great challenges.
- In both chronic and acute forms, attention evolves with the acute and chronic forms of stress which impinge upon it.

Box 24.2

Outstanding Issues

- More work evaluating the influence of stress on time (i.e., temporal distortion) is needed.
- A better understanding of how stress operates in complex group tasks (e.g., in an intensive care unit) is needed.

Acknowledgments

We would like to acknowledge the help of the editors and two unknown reviewers, whose insightful comments have much improved our present offering.

References

Arnsten, A. F. (2009). Stress signalling pathways that impair prefrontal cortex structure and function. *Nature Reviews. Neuroscience, 10,* 410–422.

Bacon, S. (1974). Arousal and the range of cue utilization. *Journal of Experimental Psychology, 102,* 81–87.

Beidel, D. C., Frueh, B. C., Uhde, T. W., Wong, N., & Mentrikoski, J. (2011). Multicomponent behavioral treatment for chronic combat-related posttraumatic stress disorder: A randomized controlled trial. *Journal of Anxiety Disorders, 25,* 224–231.

Bernard, C. (1865). *The cahier rouge. Claude Bernard and experimental medicine.* Cambridge, MA: Schenkmen.

Bursill, A. E. (1958). The restriction of peripheral vision during exposure to hot and humid conditions. *Quarterly Journal of Experimental Psychology, 10,* 113–129.

Calloway, E., & Dembo, D. (1960). Narrowed attention: A psychological phenomenon that accompanies a certain physiological change. *Archives of Neurology and Psychiatry, 79,* 74–90.

Cannon, W. B. (1932). *The wisdom of the body.* New York: Norton.

Cannon-Bowers, J. A., & Salas, E. (1998). (Eds.). *Making decisions under stress: Implications for individual and team training.* Washington, DC: American Psychological Association.

Cooke, T. (2013). *Atlas of history's greatest disasters and mistakes.* New York: Metro Books.

Cornsweet, D. M. (1969). Use of cues in the visual periphery under conditions of arousal. *Journal of Experimental Psychology, 80,* 14–18.

Csikszentmihalyi, M. (1990). *Flow: The psychology of optimal experience.* New York: HarperCollins.

Davies, D. R., & Parasuraman, R. (1982). *The psychology of vigilance.* London: Academic Press.

Dewar, R. E., & Olson, P. (2002). (Eds.). *Human factors in traffic safety.* Tucson, AZ: Lawyers & Judges.

Dirkin, G. R., & Hancock, P. A. (1985). An attentional view of narrowing: The effect of noise and signal bias on discrimination in the peripheral visual field. In I. D. Brown, R. Goldsmith, K. Coombes, & M. A. Sinclair (Eds.). *Ergonomics International 85: Proceedings of the Ninth Congress of the International Ergonomics Association* (pp. 751–753). Bournemouth, UK, September.

Eagleman, D. M. (2008). Human time perception and its illusions. *Current Opinion in Neurobiology, 18,* 131–136.

Easterbrook, J. A. (1959). The effect of emotion on cue utilization and the organization of behaviour. *Psychological Review, 66,* 183–201.

Eastwood, J. D., Fischen, A., Fenske, M. J., & Smilek, D. (2012). The unengaged mind: Defining boredom in terms of attention. *Perspectives on Psychological Science, 7,* 482–495.

Eysenck, M. W., & Derakshan, N. (2011). New perspectives in attentional control theory. *Personality and Individual Differences, 50,* 955–960.

Fairclough, S. H., & Houston, K. (2004). A metabolic measure of mental effort. *Biological Psychology, 66,* 177–190.

Fan, J., McCandliss, B. D., Sommer, T., Raz, A., & Posner, M. I. (2002). Testing the efficiency and independence of attentional networks. *Journal of Cognitive Neuroscience, 14,* 340–347.

Gibson, J. J. (1979). *The ecological approach to visual perception.* Boston: Houghton Mifflin.

Gleick, J. (2000). *Faster.* New York: Hachette Digital.

Haddon, W., Jr. (1970). On the escape of tigers: An ecologic note. *Technology Review, 72,* 44–47.

Hancock, P. A. (1982). Task categorization and the limits of human performance in extreme heat. *Aviation, Space, and Environmental Medicine, 53,* 778–784.

Hancock, P. A. (1984). Environmental stressors. In J. S. Warm (Ed.), *Sustained attention in human performance* (pp. 103–144). London: Wiley.

Hancock, P. A. (2005). The tale of a two-faced tiger. *Ergonomics in Design, 13,* 23–29.

Hancock, P. A. (2009). Performance on the very edge. *Military Psychology, 21,* S1– S68.

Hancock, P. A. (2010). The battle for time in the brain. In J. A. Parker, P. A. Harris, & C. Steineck (Eds.), *Time, limits and constraints: The study of time XIII* (pp. 65–87). Leiden, The Netherlands: Brill.

Hancock, P. A. (2013). In search of vigilance: The problem of iatrogenically created psychological phenomena. *American Psychologist, 68,* 97–109.

Hancock, P. A., & Caird, J. K. (1993). Experimental evaluation of a model of mental workload. *Human Factors, 35,* 413–429.

Hancock, P. A., & Dirkin, G. R. (1982). Central and peripheral visual choice-reaction time under conditions of induced cortical hyperthermia. *Perceptual and Motor Skills, 54,* 395–402.

Hancock, P. A., & Ganey, H. C. N. (2003). From the inverted-U to the extended-U: The evolution of a law of psychology. *Journal of Human Performance in Extreme Environments, 7*(1), 5–14.

Hancock, P. A., & Vasmatzidis, I. (2003). Effects of heat stress on cognitive performance: The current state of knowledge. *International Journal of Hyperthermia, 19,* 355–372.

Hancock, P. A., & Warm, J. S. (1989). A dynamic model of stress and sustained attention. *Human Factors, 31,* 519–537.

Hancock, P. A., & Weaver, J. L. (2005). On time distortions under stress. *Theoretical Issues in Ergonomics Science, 6,* 193–211.

Helton, W. S., Matthews, G., & Warm, J. S. (2009). Stress state mediation between environmental variables and performance: The case of noise and vigilance. *Acta Psychologica, 130,* 204–213.

Hockey, G. R. J. (1970a). Effect of loud noise on attentional selectivity. *Quarterly Journal of Experimental Psychology, 22*, 28–36.

Hockey, G. R. J. (1970b). Signal probability and spatial location as possible bases for increased selectivity in noise. *Quarterly Journal of Experimental Psychology, 22*, 37–42.

Hockey, G. R. J. (1997). Compensatory control in the regulation of human performance under stress and high workload: A cognitive energetical framework. *Biological Psychology, 45*, 73–93.

Hockey, G. R. J., Nickel, P., Roberts, A. C., & Roberts, M. H. (2009). Sensitivity of candidate markers of psychophysiological strain to cyclical changes in manual control load during simulated process control. *Applied Ergonomics, 40,* 1011–1018.

Holmes, T. H., & Rahe, R. H. (1967). The social readjustment rating scale. *Journal of Psychosomatic Research, 11*, 213–218.

James, W. (1890). *The principles of psychology*. New York: Holt.

Johnston, W. A., & Dark, V. J. (1986). Selective attention. *Annual Review of Psychology, 37*, 43–75.

Kamata, A., Tenenbaum, G., & Hanin, Y. L. (2002). Individual zone of optimal functioning (IZOF): A probabilistic estimation. *Journal of Sport & Exercise Psychology, 24*, 189–208.

Killingsworth, M. A., & Gilbert, D. T. (2010). A wandering mind is an unhappy mind. *Science, 330*, 932.

Koestler, A. (1978). *Janus: A summing up*. New York: Random House.

Kofman, O., Meiran, N., Greenberg, E., Balas, M., & Cohen, H. (2006). Enhanced performance on executive functions associated with examination stress: Evidence from task-switching and Stroop paradigms. *Cognition and Emotion, 20*, 577–595.

Kurzban, R. (2010). Does the brain consume additional glucose during self-control tasks? *Evolutionary Psychology, 8*, 244–259.

Langner, R., & Eickhoff, S. B. (2013). Sustaining attention to simple tasks: A meta-analytic review of the neural mechanisms of vigilant attention. *Psychological Bulletin, 139*, 870–900.

Lazarus, R. S. (1999). *Stress and emotion: A new synthesis*. New York: Springer.

Lazarus, R. S., & Folkman, S. (1984). *Stress, appraisal and coping*. New York: Springer.

Leibowitz, H., Abernethy, C., Buskirk, E., Bar-Or, O., & Hennessy, R. (1972). The effect of heat stress on reaction time to centrally and peripherally presented stimuli. *Human Factors, 14*, 155–160.

Leibowitz, H., & Appelle, S. (1969). The effect of a central task on luminance thresholds for peripherally presented stimuli. *Human Factors, 11*, 387–392.

Linnell, K. J., Caparos, S., de Fockert, J. W., & Davidoff, J. (2013). Urbanization decreases attentional engagement. *Journal of Experimental Psychology: Human Perception and Performance, 39*, 1232–1247.

Liston, C., McEwen, B. S., & Casey, B. J. (2009). Psychosocial stress reversibly disrupts prefrontal processing and attentional control. *Proceedings of the National Academy of Sciences of the United States of America, 106*, 912–917.

MacLean, K. A., Ferrer, E., Aichele, S. R., Bridwell, D. A., Zanesco, A. P., Jacobs, T. L., et al. (2010). Intensive meditation training improves perceptual discrimination and sustained attention. *Psychological Science, 21*, 829–839.

Matthews, G. (2001). Levels of transaction: A cognitive science framework for operator stress. In P. A. Hancock & P. A. Desmond (Eds.), *Stress, workload and fatigue* (pp. 5–33). Mahwah, NJ: Erlbaum.

Matthews, G., & Campbell, S. E. (2010). Dynamic relationships between stress states and working memory. *Cognition and Emotion, 24*, 357–373.

Matthews, G., Campbell, S. E., Falconer, S., Joyner, L., Huggins, J., Gilliland, K., et al. (2002). Fundamental dimensions of subjective state in performance settings: Task engagement, distress and worry. *Emotion (Washington, D.C.), 2*, 315–340.

Matthews, G., Desmond, P. A., Neubauer, C., & Hancock, P. A. (Eds.). (2012). *The handbook of operator fatigue*. Farnham, Surrey, UK: Ashgate.

Matthews, G., & Gilliland, K. (1999). The personality theories of H.J. Eysenck and J.A. Gray: A comparative review. *Personality and Individual Differences, 26*, 583–626.

Matthews, G., Reinerman-Jones, L. E., Barber, D. J., & Abich, J. (2015). The psychometrics of mental workload: Multiple measures are sensitive but divergent. *Human Factors, 57*, 125–143.

Matthews, G., Szalma, J., Panganiban, A. R., Neubauer, C., & Warm, J. S. (2013). Profiling task stress with the Dundee Stress State Questionnaire. In L. Cavalcanti & S. Azevedo (Eds.), *Psychology of stress: New research* (pp. 49–90). Hauppauge, NY: Nova Science.

Matthews, G., Warm, J. S., Reinerman, L. E., Langheim, L. K., & Saxby, D. J. (2010). Task engagement, attention and executive control. In A. Gruszka, G. Matthews, & B. Szymura (Eds.), *Handbook of individual differences in cognition: Attention, memory and executive control* (pp. 205–230). New York: Springer.

Matthews, G., Warm, J. S., Reinerman, L. E., Langheim, L., Washburn, D. A., & Tripp, L. (2010). Task engagement, cerebral blood flow velocity, and diagnostic monitoring for sustained attention. *Journal of Experimental Psychology. Applied, 16*, 187–203.

Matthews, G., Warm, J. S., Shaw, T. H., & Finomore, V. S. (2014). Predicting battlefield vigilance: A multivariate approach to assessment of attentional resources. *Ergonomics*. Online publication. doi:10.1080/00140139.2014.899630.

Matthews, G., & Zeidner, M. (2012). Individual differences in attentional networks: Trait and state correlates of the ANT. *Personality and Individual Differences, 53*, 574–579.

McEwen, B. S. (1998). Stress, adaptation, and disease: Allostasis and allostatic load. *Annals of the New York Academy of Sciences, 840*, 33–44.

Moskowitz, H., & Sharma, S. (1974). Effect of alcohol on peripheral vision as a function of attention. *Human Factors, 16*, 174–180.

National Transportation Safety Board (1973). *Eastern Airlines L-1011, Miami, Florida, 20 December 1972 (Report NTSB-AAR-94/07)*. Washington, DC: NTSB.

Pabst, S., Brand, M., & Wolf, O. T. (2013). Stress and decision making: A few minutes make all the difference. *Behavioural Brain Research, 250*, 39–45.

Pabst, S., Schoofs, D., Pawlikowski, M., Brand, M., & Wolf, O. T. (2013). Paradoxical effects of stress and an executive task on decisions under risk. *Behavioral Neuroscience, 127*, 369–379.

Perrow, C. (1984). *Normal accidents: Living with high-risk technologies*. New York: Basic Books.

Pirolli, P. L. (2007). *Information foraging theory: Adaptive interaction with information*. Oxford, UK: Oxford University Press.

Pirolli, P., & Card, S. (1995). Information foraging in information access environments. In *Proceedings of the SIGCHI Conference on Human Factors in Computing Systems* (pp. 51–58). New York: ACM Press/Addison-Wesley.

Plessow, F., Fischer, R., Kirschbaum, C., & Goschke, T. (2011). Inflexibly focused under stress: Acute psychosocial stress increases shielding of action goals at the expense of reduced cognitive flexibility with increasing time lag to the stressor. *Journal of Cognitive Neuroscience, 23*, 3218–3227.

Plessow, F., Kiesel, A., & Kirschbaum, C. (2012). The stressed prefrontal cortex and goal-directed behaviour: Acute psychosocial stress impairs the flexible implementation of task goals. *Experimental Brain Research, 216*, 397–408.

Plessow, F., Schade, S., Kirschbaum, C., & Fischer, R. (2012). Better not to deal with two tasks at the same time when stressed? Acute psychosocial stress reduces task shielding in dual-task performance. *Cognitive, Affective & Behavioral Neuroscience, 12*, 557–570.

Reeves, F., & Bergum, B. (1972). Perceptual narrowing as a function of peripheral cue relevance. *Perceptual and Motor Skills, 35*, 719–724.

Robbins, T. W., & Arnsten, A. F. T. (2009). The neuropsychopharmacology of fronto-executive function: Monoaminergic modulation. *Annual Review of Neuroscience, 32*, 267–287.

Saxby, D. J., Matthews, G., Warm, J. S., Hitchcock, E. M., & Neubauer, C. (2013). Active and passive fatigue in simulated driving: Discriminating styles of workload regulation and their safety impacts. *Journal of Experimental Psychology: Applied, 19*, 287–300.

Schwabe, L., Wolf, O. T., & Oitzl, M. S. (2010). Memory formation under stress: Quantity and quality. *Neuroscience and Biobehavioral Reviews, 34*, 584–591.

Selye, H. (1958). *The stress of life*. London: Elsevier.

Smith, K., & Hancock, P. A. (1995). Situation awareness is adaptive, externally-directed consciousness. *Human Factors, 37*, 137–148.

Spence, C., & Driver, J. (2004). (Eds.). *Crossmodal space and crossmodal attention*. Oxford, UK: Oxford University Press.

Szalma, J. L., & Hancock, P. A. (2011). Noise effects on human performance: A meta-analytic synthesis. *Psychological Bulletin, 137*, 682–707.

Wachtel, P. (1967). Conception of broad and narrow attention. *Psychological Bulletin, 68*, 417–429.

Wadlinger, H. A., & Isaacowitz, D. M. (2011). Fixing our focus: Training attention to regulate emotion. *Personality and Social Psychology Review, 15*, 75–102.

Warm, J. S., Dember, W. N., & Hancock, P. A. (1996). Vigilance and workload in automated systems. In R. Parasuraman & M. Mouloua (Eds.), *Automation and human performance: Theory and applications* (pp. 183–200). Mahwah, NJ: Erlbaum.

Warm, J. S., Parasuraman, R., & Matthews, G. (2008). Vigilance requires hard mental work and is stressful. *Human Factors, 50*, 433–441.

Wells, A., & Matthews, G. (1994). *Attention and emotion: A clinical perspective*. Hove, UK: Erlbaum.

Wells, A., & Matthews, G. (2006). Cognitive vulnerability to anxiety disorders: An integration. In L. B. Alloy & J. H. Riskind (Eds.), *Cognitive vulnerability to emotional disorders* (pp. 303–325). Hillsdale, NJ: Erlbaum.

Weltman, G., & Egstrom, G. (1966). Perceptual narrowing in novice drivers. *Human Factors, 8*, 499–506.

Yerkes, R. M., & Dodson, J. D. (1908). The relation of strength of stimulus to rapidity of habit-formation. *Journal of Comparative Neurology and Psychology, 18*, 459–482.

Zeidner, M. (1998). *Test anxiety: The state of the art*. New York: Plenum Press.

25

Attention Research in Marketing: A Review of Eye-Tracking Studies

Michel Wedel

Academic marketing research studies exchanges between individuals and/or institutions; the practice of marketing facilitates these exchanges through research on consumer needs and the development, distribution, pricing, promotion, and advertising of products and services to cater to those needs. The rapid growth of the use of eye tracking in commercial marketing research in the last decade is driven by the recognition that marketing often involves visual stimuli, as well as by technological innovations and sharp declines in the costs of eye-tracking devices. Firms are using eye-tracking research to support the design and evaluation of product design, advertising, Web sites, packaging, and shelf layout. Research companies are conducting many thousands of eye-tracking studies each year.

As miniature cameras have become ubiquitous, they are enabling eye-movement recording in desktop, laptop, and tablet computers, as well as in interactive billboards, kiosks, smart-TVs, and smartphones. Research agencies capitalize on this in the collection of eye-movement data, and it is revolutionizing human interaction with these electronic devices. Gaze-based rendering is beginning to improve the effectiveness of digital billboards, games, and point-of-sale displays, and enhance consumers' shopping experiences in brick-and-mortar, Internet, and virtual stores. Advertisers will soon be charged based on how long consumers look at their ads. As interactive digital devices are recording more and more of consumers' gaze, vast amounts of information about humans' visual behavior in their day-to-day lives is becoming available. This will enable unprecedented insights into consumer information acquisition and decision making in real-life situations.

Unfortunately, however, marketing practice as yet fails to tap the full potential of eye-tracking data because applied researchers stop short of reporting simple graphical summaries such as heat maps. Once their novelty fades, users will realize that heat maps presented in isolation do not provide fundamental insights and lack actionability. Advances in academic marketing research, reviewed in this chapter, address this problem and are finding their way into marketing practice (Wedel & Pieters, 2008).

Historical Context

Emerging Recognition of the Central Role of Attention
In the early 1900s, Nixon (1924) and Poffenberger (1925) first applied eye-movement research to evaluate advertisements. While Nixon (1924) manually recorded eye movements of consumers, the use of eye cameras improved efficiency and accuracy (Karslake, 1940). Yet, there was a period of relative paucity in research, possibly caused by the commonly held but erroneous view that attention is only a first stage toward higher cognitive processes, which was instilled by hierarchical processing models such as the model of Attention, Interest, Desire, and Action (AIDA) (Starch 1923; Strong 1925). Once it became established that eye movements are tightly coupled with visual attention and that information acquisition and other cognitive processes can be inferred from them, there was a revival of research in the 1970s (Kroeber-Riel, 1979; Russo & Rosen, 1975; Treistman & Gregg, 1979; van Raaij, 1977). The 1990s saw a surge of interest in eye-tracking research, in part driven by advances in infrared recording technology. This initially involved descriptive studies that documented attention to ads, nutrition labels, and alcohol and cigarette warnings (Krugman et al., 1994; Lohse, 1997; Fox et al., 1998; Fletcher et al., 1995). From then on, further improvements in eye-tracking technology and the recognition of the central role of attention in consumer behaviors led to the increased application and testing of theories and findings from psychology. In many cases these theories were supported, and in some cases they were extended. Importantly, this enabled the evaluation of long-standing beliefs in marketing practice on the effectiveness of its visual stimuli.

Researchers in psychology (Kingstone et al., 2003; Zelinsky, 2008) have argued for the study of human behavior in natural contexts. Marketing provides rich contexts in which consumers engage with complex stimuli (Web sites, catalogs, shelves, ads) through a variety of tasks (exploration, search, choice). To illustrate, figure 25.1 shows plots of fixations collected with Tobii eye trackers during visual exploration of two print ads. However, multiple unobservable cognitive processes simultaneously affect eye movements in natural contexts. While experiments look at a few causes in isolation, statistical models need to be used to disentangle the influences of multiple mechanisms that operate simultaneously in real-life settings. Since around 2000, Bayesian statistics has played a central role in the modeling of visual attention in marketing.

Bayesian Models of Attention and Eye Movements
The foundation of the Bayesian statistical framework is Bayes' rule, a normative formulation of how subjective beliefs should be updated in the light of new evidence. While it has a long history, it became highly popular after the development of Markov Chain Monte Carlo (MCMC) sampling methods in the 1990s, which greatly facilitated

Figure 25.1
Eye movements of five participants on an advertisement for Ford (top panel reprinted with permission of Ford Motor Company) and on an advertisement for Travelers (bottom panel reprinted with permission of Travelers; photo credit: John Offenbach).

the estimation of models. In marketing, Bayesian models have been applied to eye-movement data in three different ways.

First, a Bayesian formulation of analysis of variance and generalized linear models addresses the challenges of analyzing eye-movement data for the purpose of hypothesis testing (see Pieters & Wedel, 2007, for an example). This improves on traditional analyses because it accommodates the small samples common in experimental research, allows for a wide range of distributional assumptions that reflect the measurement properties of eye-movement metrics, and deals with missing values, latent variables, unobserved individual differences, and hierarchical data structures. This is a statistical approach, which leads to more sound inference and testing of hypotheses (Kruschke, 2013).

Second, the Bayesian approach enables models to be formulated that represent cognitive concepts from first principles. They reflect predictions on how multiple unobserved attention processes may have affected the recorded eye movements. Estimating these often complex integrated models is made possible by MCMC technology and enables one to disentangle the underlying attention processes and recover the key parameters characterizing them (see, e.g., Stüttgen, Boatwright, & Monroe, 2012). This is a formal approach, which enables simultaneous inferences on several underlying cognitive processes from eye-movement data (Lee, 2011).

Third, Bayes' rule can be used as a theory of the mechanisms of human visual perception, such as the integration of input from multiple sources, as well as top-down influences on attention (Kersten, 1999. Applications in marketing are yet rare. This is a theoretical approach, which lends itself to estimation with MCMC methodology (see, e.g., van der Lans, Pieters, & Wedel, 2008a).

Most of the models developed for eye-movement analysis combine some of the features of each of the three above approaches—statistical, formal, theoretical—depending on whether or not theory is sufficiently detailed and on whether or not Bayesian updating is a plausible mechanism. They have provided many novel insights, reviewed next.

State-of-the-Art Review: Eye-Tracking Research in Marketing

Exploration of Ads
Gist Perception   The majority of ads in magazines, in newspapers, on billboards, and on Web sites receive at most a single eye fixation. If exposures occur in the periphery, they also lack visual detail. The question is which meaning, if any, ads can communicate under these exposure conditions. Building on the scene perception literature (Friedman, 1979; Oliva, 2013), Pieters and Wedel (2012) showed that even very coarse images presented for only 100 ms allow consumers to identify, with high accuracy

(over 90%), the ad and the advertised product, and to a lesser extent, the brand. This, however, was the case only for ads with a typical layout (often containing a central diagnostic object). Atypical ads (often containing semantically incongruent objects; see figure 25.1) do much worse under these conditions. Accurate gist perception elicits immediate interest in the ad. Wedel and Pieters (2015) next showed that color buffers gist perception of typical ads during blurred exposures, and that the color of a central diagnostic object matters most.

Attention Capture   Much of the research in marketing that studies exploration of print ads and Web sites is related to scene perception research in psychology (Henderson, 2003; Wu, Wick, & Pomplun, 2014). Marketing stimuli are complex, containing multiple regions of interest (ROIs), including pictures, text, a brand, and a headline. The modality of these regions affects attention capture and engagement. In the first known eye-movement analyses of print ads, Nixon (1924) found that fixations increased with approximately the square root of the size of ads, and the results suggested that pictures guide attention to the text. Whereas little work in psychology has addressed effects of ROI size, a significant amount of work in marketing has done so, building on theories of attention capture and engagement (Loftus, 1983; Posner, 1980). Wedel and Pieters (2000) asked participants to explore print ads appearing in magazines as they would do normally and to identify the brand from a pixelated image of the ad. The brand element received by far the most eye fixations per unit of its size, followed by the text and then the pictorial. Whereas fixations to the brand, and to a lesser extent the pictorial, promoted recognition memory, text fixations did not. Aribarg, Pieters, and Wedel (2010) further investigated this relationship. Attention to the ad predicted ad recognition. However, stated recognition of the ad and its pictorial was biased upward especially when it contained a larger pictorial and a smaller brand. These findings undermine the validity of recognition tests commonly used in advertising practice.

Pieters and Wedel (2004) analyzed data from 33 commercial eye-tracking studies, comprising several thousand participants and more than a thousand ads. They studied the effects of the sizes of brand, pictorial, and text elements. The results showed that each 1% increase in the size of the text element led to a 0.9% increase in gaze duration: a large effect. Increasing the size of the brand and pictorial by 1% led to a 0.3% increase in gaze. Only larger text elements resulted in a longer gaze to the ad as a whole. This result was later confirmed by Wang and Pomplun (2012). Longer gaze on the brand element carried over to longer gaze to the other elements, but there was little transfer in the reverse direction, contrary to earlier claims and beliefs in advertising practice. The average gaze duration to the ads was only 1.7 seconds and increased by 0.8% for every 1% increase in ad size. Thus, fixed exposure durations of 10 seconds or more, as are often used, have limited external validity.

Pieters, Wedel, and Zhang (2007) extended these findings to feature advertising, which provides a unique context because multiple ads are shown simultaneously on the same page and compete for attention. They analyzed commercial eye-tracking data for more than 100 newspaper pages with ads aggregated across close to 10,000 consumers. Competition between the ads on a page affected attention. Competition was operationalized using entropy measures of the similarity of a focal product with distractors, and of the similarities among the distractors, following the theory by Duncan and Humphreys (1989). The optimal sizes of ad elements were derived mathematically. This showed that currently picture and text were 40% and 20% too large, respectively, and the brand was about 75% too small. Using a Bayesian mediation analysis, Zhang, Wedel, and Pieters (2009) linked the eye-tracking data to national sales data for the featured brand, collected in the same weeks as when the ads had appeared in the newspapers. This revealed that the size of feature ads positively affected sales, an effect that was fully mediated by attention to the ads. This study revealed a critical downstream impact of attention on sales.

Pieters, Warlop, and Wedel (2002) investigated the influence that ad originality and familiarity have on consumers' eye fixations on the brand, text, and pictorial elements, and on memory for the advertised brand. They showed that original advertisements (see figure 25.1) draw more attention to the brand and improve brand memory as well. Further, this effect was even stronger for familiar ads. Radach et al. (2003) showed that a specific type of originality, "implicit" advertising, which uses pictures and text that are not related to the advertised product, receive more eye fixations on their ROIs than explicit ads. Pieters, Wedel, and Batra (2010) investigated another way to achieve originality by making ads more complex, building on the work of Berlyne (1958). They distinguished feature complexity (dense visual features) and design complexity (elaborate creative design). It appeared that feature complexity hurts attention to the brand whereas design complexity helps attention to both the pictorial and the advertisement as a whole.

Several studies have dealt with the more challenging topic of attention capture by dynamic stimuli. In a study of televised advertising images during soccer games, d'Ydewalle and Tamsin (1993) found little attention capture for billboards placed around a soccer field. Similar negative findings were obtained by Drèze and Hussherr (2003) for banner ads on Web sites, revealing active avoidance of task-irrelevant stimuli by consumers. d'Ydewalle, Desmet, and Van Rensbergen (1998) studied the effect of transitions between scenes (cuts and edits) on attention to movies. The spatial variance of eye movements increased up to half a second after transitions, providing evidence that these transitions disrupt perception and decrease attentional focus. Teixeira, Wedel, and Pieters (2010) studied attention capture and retention by brands in TV commercials, using a dynamic regression model. Long and uninterrupted presence of the brand increased the spatial variance of eye movements and promoted zapping.

Presenting the brand repeatedly during very short time intervals (pulsing) maximized attention and minimized zapping. Brasel and Gips (2008) found that participants who fast-forwarded through shows on their DVR players paid more attention during commercials because their attention was focused on the center of the screen. Thus, during fast-forwarding, commercials with extensive central brand placement capture more attention to the brand, which also promotes memory.

Scanpaths Individual differences have been a key substantive and methodological focus of research in marketing. Rosbergen, Pieters, and Wedel (1997) first investigated heterogeneity of consumer attention to pictorial, brand, headline, and text elements of a print advertisement, using a finite mixture regression model—which assumes the existence of unobserved groups (segments) of participants that have different parameters characterizing the effect of these ROIs on attention. They found three consumer segments that had, respectively, significant gaze durations on the headline and pictorial; on the headline, pictorial, and brand; and on headline, pictorial, brand, and text. Total ad viewing time increased from the first to the last segment (0.6, 1.0, and 2.7 seconds). These results suggest different scanpaths for different segments (Noton & Stark, 1971).

Following research by Stark and Ellis (1981) and Harris (1993), Pieters, Rosbergen, and Wedel (1999) investigated scanpaths across three repeated exposures to print ads. Fixation frequency decreased by as much as 50% on each subsequent exposure. The authors tested whether the scanpaths on the headline, pictorial, text, and brand were stable across exposures, using Markov models—which assume first-order transition probabilities between these ROIs. They found that scanpaths did not change significantly across repeated exposures. This was later supported in research by Underwood, Foulsham, and Humphrey (2009). However, Foulsham and Kingstone (2013) found no evidence either that scanpaths are stored in memory or that their recapitulation supports it. Thus, the stability of scanpaths observed by Pieters et al. (1999) may be caused by stable bottom-up influences rather than a memory trace of the scanpath.

Interested in the spatial allocation of attention to print ads, Liechty, Pieters, and Wedel (2003) investigated local and global segments of scanpaths. Previous accounts of exploratory eye movements on scenes reported global followed by local exploration (Antes, 1974; Henderson & Hollingworth, 1998; LaBerge, 1998). Using a Hidden Markov model to identify these two modes of exploration—which assumes unobserved states that evolve over time according to a Markov process and that drive eye movements—Liechty et al. (2003) analyzed eye-tracking data of consumers exploring print advertisements in magazines. Participants nearly always started with local exploration in the center of the page, and they switched between local and global exploration about three times during exposure. They spent much longer in local (1.1 second) than in global (0.2 second) exploration. There was no evidence of initial global exploration followed by local exploration. Rather, the problem of interpreting a complex scene seems to be broken up into a sequence of local explorations of the most informative regions, interspersed with global jumps between those regions.

Top-Down Effects Several studies in marketing measured top-down factors such as involvement and familiarity and assessed their impact on attention (Rosbergen, Pieters, & Wedel, 1997; Treistman & Gregg, 1979). Stronger evidence of top-down influences comes from studies in which top-down factors were manipulated through task instructions. Rayner et al. (2001) experimentally manipulated involvement by asking participants to imagine buying one of several products and to try to learn more about them. Participants spent more time looking at the text than the picture of the ads. They read the headlines first, then read the body text, and then attended to the picture. These findings seem counter to those of Rosbergen, Pieters, and Wedel (1997) discussed above, but the latter authors used an exploration task. The Rayner et al. (2001) research showed that attention to ads was influenced by task instructions. Similarly, van Herpen, and van Trijp (2011) induced preference, health, and low-salt goals and showed that especially the latter influences gaze on nutrition labels.

Building on the classic work by Yarbus (1967), Pieters and Wedel (2007) manipulated five goals (exploration, ad learning, brand learning, ad evaluation, and brand evaluation) via task instructions. Text, pictorial, and brand elements all received longer gaze durations under an ad-learning goal. A brand-evaluation goal promoted longer gaze on the text while a brand-learning goal caused longer gaze for the text but shorter for the pictorial. These findings further explain the long gaze on the text in the Rayner et al. (2001) study, which used a brand-learning task. Thus, it appears that participants' lay theories on where information resides (the text), although not necessarily correct, influenced their attention. Wedel, Pieters, and Liechty (2008) showed subsequently that goals influenced the pattern of local and global attention. Goals directed at processing the ad as a whole lead to more switching than goals directed at the brand, which result in more local attention. Evaluation goals led to more global attention than learning goals.

Several later studies also replicated and expanded Yarbus's (1967) findings. Rayner, Miller, and Rotello (2008) found that when participants imagined wanting to buy the brand, they looked more at the text; when they evaluated ads, they looked more at the pictorial. These findings are also discussed by Rayner and Castelhano (2007). DeAngelus and Pelz (2009) replicated Yarbus's research with a larger sample of respondents and self-paced (shorter) exposure durations. Attention was influenced by the task instructions. On the contrary, Greene, Liu, and Wolfe (2012) were not able to predict the task from the observed eye movements. Their studies had fixed exposure durations, however (1 to 60 seconds), which may have resulted in repetitions of the scanpaths (Yarbus,

1967). Self-paced exposures might have shown different results, but even using fixed exposures, Borji and Itti (2014) showed that with powerful classification methods viewers' goals can be reliably inferred from their eye movements.

Shelf Search and Choice

Salience   In marketing, eye-tracking research on visual search is relatively sparse. Janiszewski and Warlop (1993) manipulated segments of commercials to contain a target brand or not, aiming to induce brand learning. They found that when consumers had looked at the brand earlier, attention capture by the brand during subsequent search on displays was enhanced. Janiszewski (1998) investigated the relationship between the space devoted to products on displays and the amount of attention they receive during search. This revealed the bottom-up influence of display layout and object size. The competition for attention created by items surrounding a focal product influenced gaze duration and recall, confirming predictions by Duncan and Humphreys (1989). This research suggested that carefully rearranging the layout of stimuli on a display may maximize attention to them, which was later confirmed by Pieters, Wedel, and Zhang (2007).

Van der Lans, Pieters, and Wedel (2008a, 2008b) investigated target search for brands on shelves, developing an integrated model of the component processes of visual search. Their Hidden Markov model is based on the assumption that participants rely on salience only during localization, not during identification of the search target. Building on prior research in psychology (Itti & Koch, 2001; Wolfe & Horowitz, 2004), they estimated salience maps from both image and eye-movement data. The results revealed large individual differences in salience. Moreover, search goals that were induced by having participants search for different brands had dramatic effects on salience maps: As much as one-third of salience was due to these top-down effects. Salience was found to be highly predictive of search effectiveness. These findings are in line with current views in psychology on the role of salience (Ludwig & Gilchrist, 2002; van Zoest & Donk, 2004).

Central Gaze Bias and Gaze Cascade   Much of the research on consumer decision making has attempted to identify stages in the decision process. Russo and Leclerc (1994), building on earlier work by Russo and Rosen (1975), used video recording of eye movements on shelves and identified three different stages in the choice process. During orientation an overview of the product display was obtained. In the evaluation stage, direct comparisons between two or three products were made. The verification stage involved final examination of a chosen brand. Building on prior work on the gaze cascade—the phenomenon that gaze becomes increasingly biased toward a stimulus that ends up being chosen (Shimojo, Simion, Shimojo, & Scheier, 2003)— Atalay, Bodur, and Rasolofoarison (2012) found a central gaze cascade effect:

progressively increasing attention on the central option prior to a decision. Tatler (2007) concludes that the center may be an optimal location for early information processing or it may be a convenient location from which to start exploration of the scene. This "center-of-gravity effect" was already observed by Buswell (1935), but research in marketing has shown that it has downstream effects. Interested in the effects of the visual layout of shelves on attention and its effects on consideration and choice, Chandon, Hutchinson, Bradlow, and Young (2009) experimentally manipulated shelf layouts. They found that more space devoted to a brand increased attention, in line with Janiszewski (1998), and that top- and middle-shelf positions gained more attention, in line with Atalay et al. (2012). Further, they showed that the impact of these spatial factors on consideration and choice is fully mediated by visual attention. Shi, Wedel, and Pieters (2013) used a hierarchical Hidden Markov model—a model with two interacting layers of unobserved states that evolve over time according to Markov processes—to study processing stages during choice on comparison Web sites. These Web sites display products in the rows and descriptions of their attributes in the columns. Participants switched a little less than once every second between information acquisition by row and by column, and attended to about two to three attributes or products before switching. Eye movements showed a strong left-to-right tendency and were concentrated on contiguous elements of the display, predispositions which persisted even when the row/column orientation of the Web site was changed. The gaze cascade was observed during the last moments of the decision, but mostly when information was acquired by product. Whether the gaze cascade corresponds to a verification stage, corresponds to a motor preparation response, or is caused by neural convergence is still an open question (Orquin & Loose, 2013).

Integrating Search and Choice   Understanding consumer choice of products is central to marketing, and research has long recognized limited information, bounded rationality, and the use of heuristics in decision making (Payne, Bettman, & Johnson, 1993; Orquin & Loose, 2013). Stewart, Pickering, and Sturt (2004) showed that regressions on a brand name during reading indicate the acceptation of brand extensions. Milosavljevic et al. (2012) documented the relative impacts of salience and preference in a two-alternative forced-choice task. After short exposures, up to 200 ms, salience influenced choices more than preferences. After 200 ms the effect reversed, but this threshold increased to up to 500 ms under high cognitive load.

Following the recognition by Russo and Leclerc (1994) that eye movements provide major insights into decision-making processes, Reutskaja, Nagel, Camerer, and Rangel (2011) conducted an eye-tracking choice experiment to compare three computational models: an optimal search model, a satisficing search model in which a participant stops searching once a satisfactory alternative is encountered, and a hybrid search model that incorporates aspects of both. The results provided strong support for the hybrid model. Toubia, de Jong, Stieger, and Füller (2012) demonstrated that by making tasks incentive compatible—aligning the incentives of the participant and the researcher—participants utilized up to 20% more of the information presented to them. Their results are in line with earlier findings by Pieters and Warlop (1999), who studied the opposing effects of time pressure and motivation on decision making.

In natural conditions, search and choice often occur simultaneously, and the integration of eye tracking into models of choice has great potential to improve the understanding of human decision making (Orquin & Loose, 2013). For example, Krajbich, Armel, and Rangel (2010) first showed how to include eye-movement data in the Drift Diffusion Model—which is based on the assumption that stochastic evidence on choice alternatives accumulates over time until a decision threshold is reached—to account for how the relative value of choice options affects choice outcomes and decision time. Relatedly, Satomura, Wedel, and Pieters (2014) extend the Drift Diffusion model by including image characteristics and unobserved individual differences to describe perceptual decisions on copycat brands. In the first study that develops an integrated statistical model of visual search and choice, Stüttgen, Boatwright, and Monroe (2012) used eye-tracking data collected during a conjoint choice experiment. Their model assumes "satisficing" decisions—in which the choice alternatives are evaluated sequentially rather than simultaneously. Search was modeled following van der Lans et al. (2008a).

Integration and Implications

Attention research in marketing has extensively built on psychological theories. The main concepts studied, with key references from psychology and marketing, are summarized in table 25.1. The present chapter has shown that marketing research has confirmed predictions from extant attention theories in natural settings. In addition, the marketing literature has extended theories, and formalized them through Bayesian models, which has enabled assessment of their predictions. Facilitated by these psychological theories and statistical models, the attention research reviewed here has direct implications for marketing practice. For example, for advertising it has revealed that during short exposures atypical ads (award-winning ads are often atypical) perform much worse than typical ads; that advertisers design ads with pictorial elements that are typically too large, while text, price, and brand elements are too small; that designing ads to be original works because it promotes attention and stimulates memory for the ad; that achieving originality by designing ads to be more complex improves attention and memory only if it does not result in cluttered low-level features; that in order to optimize attention to pages with many feature ads, one needs to reduce clutter in the layout of the page and the individual ads; that a central brand placement

Table 25.1

Theoretical Constructs Studied in Marketing, with Sample References

Concepts	References in Psychology	References in Marketing
Attention capture and engagement	Nixon, 1924; Loftus, 1983; Duncan & Humphreys, 1992; Wang & Pomplun, 2012	Janiszewski, 1998; Pieters & Wedel, 2004; Pieters, Wedel, & Zhang, 2007
Salience maps based on the local distribution of visual features	Itti & Koch, 2001; Koch & Ullman, 1985; Wolfe & Horowitz, 2004	van der Lans, Pieters, & Wedel, 2008a, 2008b
Gist perception from the global distribution of visual features	Friedman, 1979; Oliva, 2013	Pieters & Wedel, 2012; Wedel & Pieters, 2015
Stability of scanpaths across repeated exposures	Noton & Stark, 1971; Foulsham & Kingstone, 2013	Pieters, Rosbergen, & Wedel, 1999
Central fixation tendencies and their downstream effects	Buswell, 1935; Mannan, Ruddock, & Wooding, 1995; Tatler, 2007	Atalay, Bodur, & Rasolofoarison, 2012; Chandon, Hutchinson, Bradlow, & Young, 2009
Individual differences in attention	Rayner, 2009; Rutishauser & Koch, 2007; Torralba, Oliva, Castelhano, & Henderson 2006	Rosbergen, Pieters, & Wedel, 1997; Wedel & Pieters, 2000
The gaze cascade	Shimojo, Simion, Shimojo, & Scheier, 2003	Atalay, Bodur, & Rasolofoarison, 2012; Shi, Wedel, & Pieters, 2013
Competition for attention by targets and distractors	Desimone & Duncan, 1995; Duncan & Humphreys, 1992; Folk, Remington, & Johnston, 1992	Janiszewski, 1998; Pieters, Wedel, & Zhang, 2007
The "what and where" streams of visual processing	Bullier, Schall, & Morel, 1996; Ungerleider & Mishkin, 1982	Liechty, Pieters, & Wedel, 2003; van der Lans, Pieters, & Wedel, 2008a
Top-down influences on salience of locations	Einhäuser, Rutishauser, & Koch, 2008; Hayhoe & Ballard, 2005	van der Lans, Pieters, & Wedel, 2008b
Top-down influences on informativeness of objects	Yarbus, 1967; Navalpakkam & Itti, 2005; Spratling & Johnson, 2004	Pieters & Wedel, 2007; Wedel, Pieters, & Liechty, 2008
Effects of attention on memory and search performance	Williams, Reingold, Moscovitch, & Behrmann, 1997; Zelinsky & Sheinberg, 1997	Wedel & Pieters, 2000; Van der Lans, Pieters, & Wedel, 2008b
The role of visual attention in consumer decision making	Russo & Rosen, 1975	Russo & Leclerc, 1994; Stüttgen, Boatwright, & Monroe, 2012

in TV commercials helps during fast-forwarding; and that pulsing the brand in short repeated time intervals in TV commercials retains attention and prevents zapping. It has also revealed challenges for ads to capture attention in complex and dynamic environments such as Web and video content. For retailing, the research has shown that increasing brands' shelf facings promotes attention and consideration; that central and top positions stimulate attention and promote choice; and that brands can be made to pop out on the shelf by careful selection of basic features and communicating them in advertising. It was shown that improving attention to ads through better design affects memory for ads and attitudes toward advertised brands; accurate gist perception and sustained attention to ads was revealed to lead to more positive attitudes toward the brand; and eye movements on shelves were found to predict search performance, brand consideration, impulse buying, and brand choice. Crucially, attention to ads was shown to have an immediate impact on sales.

Future Directions

Some of the very early interest of psychologists in marketing problems (Adams, 1917; Nixon, 1924; Strong, 1925) has provided stepping-stones for a better understanding of consumers' attention, memory, and decision making and has had major impetus on subsequent research in both marketing and psychology. While for a period some of that initial interest had declined, marketing provides an ever more fruitful testing ground for attention research, with its complex pictorial and textual, static, dynamic and interactive, and multisensory stimuli. The fundamental study of attention to these complex stimuli in natural settings will undoubtedly give rise to new theoretical questions and fundamental insights and will lead to the resolution of pressing practical problems.

Box 25.1
Key Points

- In marketing, Bayesian modeling has been gainfully employed to disentangle the influences of multiple cognitive mechanisms that operate simultaneously on eye-movement data in real-life settings.
- Research in marketing has confirmed and extended predictions from theories of attention, scene perception, search, and decision making for complex stimuli, such as ads, shelves, packages, and Web sites, in natural settings.

> **Box 25.2**
> **Outstanding Issues**
>
> - Understanding consumers' visual attention to complex multimodal and interactive stimuli in natural contexts.
> - Understanding the constructive role of attention in decision making and its effects on downstream measures in marketing.

Acknowledgments

The author is grateful to Jacob Orquin and the editors and reviewers for their comments on the manuscript and to Seoungwoo Lee for his help in collecting and processing the eye-movement data for the illustrations.

References

Adams, H. F. (1917). The memory value of mixed sizes of advertisements. *Journal of Experimental Psychology, 2*, 448–465.

Antes, J. R. (1974). The time course of picture viewing. *Journal of Experimental Psychology, 103*, 62–70.

Aribarg, A., Pieters, R., & Wedel, M. (2010). Raising the BAR: Bias adjustment of recognition tests in advertising. *JMR, Journal of Marketing Research, 47*, 387–400.

Atalay, A. S., Bodur, H. O., & Rasolofoarison, D. (2012). Shining in the center: Central gaze cascade effect on product choice. *Journal of Consumer Research, 39*, 848–866.

Berlyne, D. E. (1958). The influence of complexity and novelty in visual figures on orienting responses. *Journal of Experimental Psychology, 55*, 289–296.

Borji, A., & Itti, L. (2014). Defending Yarbus: Eye movements reveal observers' task. *Journal of Vision, 14*(29), 1–21.

Brasel, S. A., & Gips, J. (2008). Breaking through fast-forwarding: Brand information and visual attention. *Journal of Marketing, 72*(6), 31–48.

Bullier, J., Schall, J. D., & Morel, A. (1996). Functional streams in occipito-frontal connections in the monkey. *Behavioural Brain Research, 76*, 89–97.

Buswell, G. T. (1935). *How people look at pictures: A study of the psychology and perception in art*. Chicago: University of Chicago Press.

Chandon, P., Hutchinson, J. W., Bradlow, E. T., & Young, S. H. (2009). Does in-store marketing work? Effects of the number and position of shelf facings on brand attention and evaluation at the point of purchase. *Journal of Marketing, 73*(6), 1–17.

d'Ydewalle, G., Desmet, G., & Van Rensbergen, J. (1998). Film perception: The processing of film cuts. In G. Underwood (Ed.), *Eye guidance in reading and scene perception* (pp. 357–367). Amsterdam: Elsevier.

d'Ydewalle, G., & Tamsin, F. (1993). On the visual processing and memory of incidental information: Advertising panels in soccer games. In D. Brogan, A. Gale, & K. Carr (Eds.), *Visual search 2* (pp. 401–408). London: Taylor & Francis.

DeAngelus, M., & Pelz, J. B. (2009). Top-down control of eye movements: Yarbus revisited. *Visual Cognition, 17*(6–7), 790–811.

Desimone, R., & Duncan, J. (1995). Neural mechanisms of selective visual attention. *Annual Review of Neuroscience, 18*, 193–222.

Drèze, X., & Hussherr, F.-X. (2003). Internet advertising: Is anybody watching? *Journal of Interactive Marketing, 17*(4), 8–23.

Duncan, J., & Humphreys, G. W. (1989). Visual search and stimulus similarity. *Psychological Review, 96*, 433–458.

Duncan, J., & Humphreys, G. W. (1992). Beyond the search surface: Visual search and attentional engagement. *Journal of Experimental Psychology: Human Perception and Performance, 18*, 578–588.

Einhäuser, W., Rutishauser, U., & Koch, C. (2008). Task-demands can immediately reverse the effects of sensory-driven saliency in complex visual stimuli. *Journal of Vision, 8*(2), 2, 1–19.

Fletcher, J. E., Krugman, D. M., Fox, R. J., Fischer, P. M., & Rojas, T. (1995). Masked recall and eye-tracking of adolescents exposed to cautionary notices in magazine ads. In P. S. Ellen & P. J. Kaufman (Eds.), *Marketing and Public Policy Conference Proceedings* (Vol. 5, pp. 128–135). Atlanta: Georgia State University.

Folk, C. L., Remington, R. W., & Johnston, J. C. (1992). Involuntary covert orienting is contingent on attentional control settings. *Journal of Experimental Psychology: Human Perception and Performance, 18*, 1030–1044.

Foulsham, T., & Kingstone, A. (2013). Fixation-dependent memory for natural scenes: An experimental test of scanpath theory. *Journal of Experimental Psychology. General, 142*, 41–56.

Fox, R. J., Krugman, D. M., Fletcher, J. E., & Fischer, P. M. (1998). Adolescents' attention to beer and cigarette print ads and associated product warnings. *Journal of Advertising, 27*(3), 57–68.

Friedman, A. (1979). Framing pictures: The role of knowledge in automatized encoding and memory for gist. *Journal of Experimental Psychology. General, 108*, 316–355.

Greene, M. R., Liu, T., & Wolfe, J. M. (2012). Reconsidering Yarbus: A failure to predict observers' task from eye movement patterns. *Vision Research, 62*, 1–8.

Harris, C. M. (1993). On the reversibility of Markov scanning in free viewing. In D. Brogan, A. Gale, & K. Carr (Eds.), *Visual search 2* (pp. 123–135). London: Taylor & Francis.

Hayhoe, M., & Ballard, D. (2005). Eye movements in natural behavior. *Trends in Cognitive Sciences, 9*, 188–194.

Henderson, J. M. (2003). Human gaze control during real-world scene perception. *Trends in Cognitive Sciences, 7*, 498–504.

Henderson, J. M. & Hollingworth, A. (1998). Eye movements during scene viewing: An overview. In Underwood, G. (Ed.), *Eye guidance in reading and scene perception* (pp. 269–293). Amsterdam: Elsevier.

Itti, L., & Koch, C. (2001). Computational modeling of visual attention. *Nature Reviews. Neuroscience, 2*, 194–203.

Janiszewski, C. (1998). The influence of display characteristics on visual exploratory search behavior. *Journal of Consumer Research, 25*, 290–301.

Janiszewski, C., & Warlop, L. (1993). The influence of classical conditioning procedures on subsequent attention to the conditioned brand. *Journal of Consumer Research, 20*, 171–189.

Karslake, J. S. (1940). The Purdue eye-camera: A practical apparatus for studying the attention value of advertisements. *Journal of Applied Psychology, 24*, 417–440.

Kersten, D. (1999). High-level vision as statistical inference. In M. Gazzaniga (Ed.), *The new cognitive neurosciences* (pp. 353–363). Cambridge, MA: MIT Press.

Kingstone, A., Smilek, D., Ristic, J., Friesen, C. K., & Eastwood, J. D. (2003). Attention, researchers! It is time to take a look at the real world. *Current Directions in Psychological Science, 12*, 176–180.

Koch, C., & Ullman, S. (1985). Shifts in selective visual attention: Towards the underlying neural circuitry. *Human Neurobiology, 4*, 219–227.

Krajbich, I., Armel, C., & Rangel, A. (2010). Visual fixations and the computation and comparison of value in goal-directed choice. *Nature Neuroscience, 13*, 1292–1298.

Kroeber-Riel, W. (1979). Activation research: Psychobiological approaches in consumer research. *Journal of Consumer Research, 5*, 240–250.

Krugman, D. M., Fox, R. J., Fletcher, J. E., Fischer, P. M., & Rojas, T. H. (1994). Do adolescents attend to warnings in cigarette advertising? An eye-tracking approach. *Journal of Advertising Research, 34*, 39–52.

Kruschke, J. K. (2013). Bayesian estimation supersedes the t-test. *Journal of Experimental Psychology. General, 142*, 573–603.

LaBerge, D. (1998). Attentional emphasis in visual orienting and resolving. In R. D. Wright (Ed.), *Visual attention* (pp. 417–454). New York: Oxford University Press.

Lee, M. D. (2011). How cognitive modeling can benefit from hierarchical Bayesian models. *Journal of Mathematical Psychology, 55*, 1–7.

Liechty, J., Pieters, R., & Wedel, M. (2003). Global and local covert visual attention: Evidence from a Bayesian hidden Markov model. *Psychometrika, 68*, 519–541.

Loftus, G. R. (1983). Eye fixations on text and scenes. In R. A. Monty & J. W. Senders (Eds.), *Eye movements in reading: Perceptual and language processes* (pp. 359–376). Hillsdale, NJ: Erlbaum.

Lohse, G. L. (1997). Consumer eye movement patterns on yellow pages advertising. *Journal of Advertising, 26*(1), 61–73.

Ludwig, C. J., & Gilchrist, I. D. (2002). Stimulus-driven and goal-driven control over visual selection. *Journal of Experimental Psychology: Human Perception and Performance, 28*, 902–912.

Mannan, S., Ruddock, K., & Wooding, D. (1995). Automatic control of saccadic eye movements made in visual inspection of briefly presented 2-D images. *Spatial Vision, 9*, 363–386.

Milosavljevic, M., Navalpakkam, V., Koch, C., & Rangel, A. (2012). Relative visual saliency differences induce sizable bias in consumer choice. *Journal of Consumer Psychology, 22*(1), 67–74.

Navalpakkam, V., & Itti, L. (2005). Modeling the influence of task on attention. *Vision Research, 45*, 205–231.

Nixon, H. K. (1924). Attention and interest in advertising. *Archives de Psychologie, 72*, 5–67.

Noton, D., & Stark, L. (1971). Eye movements and visual perception. *Scientific American, 224*, 34–43.

Oliva, A. (2013). Scene perception. In J. S. Werner & M. Chalupa (Eds.), *The new visual neurosciences* (pp. 725–732). Cambridge, MA: MIT Press.

Orquin, J. L., & Loose, S. M. (2013). Attention and choice: A review on eye movements in decision making. *Acta Psychologica, 144*, 190–206.

Payne, J. W., Bettman, J. R., & Johnson, E. J. (1993). *The adaptive decision maker*. Cambridge, UK: Cambridge University Press.

Pieters, R., Rosbergen, E., & Wedel, M. (1999). Visual attention to repeated print advertising: A test of scanpath theory. *JMR, Journal of Marketing Research, 36*, 424–438.

Pieters, R., & Warlop, L. (1999). Visual attention during brand choice: The impact of time pressure and task motivation. *International Journal of Research in Marketing, 16*(1), 1–16.

Pieters, R., Warlop, L., & Wedel, M. (2002). Breaking through the clutter: Benefits of advertisement originality and familiarity for brand attention and memory. *Management Science, 48*, 765–781.

Pieters, R., & Wedel, M. (2004). Attention capture and transfer in advertising: Brand, pictorial and text size effects. *Journal of Marketing, 68*(2), 36–50.

Pieters, R., & Wedel, M. (2007). Goal control of attention to advertising: The Yarbus implication. *Journal of Consumer Research, 34*, 224–233.

Pieters, R., & Wedel, M. (2012). Ad gist: Ad communication in a single eye fixation. *Marketing Science, 31*, 59–73.

Pieters, R., Wedel, M., & Batra, R. (2010). The stopping power of advertising: Measures and effects of visual complexity. *Journal of Marketing*, *74*(5), 48–60.

Pieters, R., Wedel, M., & Zhang, J. (2007). Optimal feature advertising design under competitive clutter. *Management Science*, *53*, 1815–1828.

Poffenberger, A. T. (1925). *Psychology in advertising*. Chicago: Shaw.

Posner, M. I. (1980). Orienting of attention. *Quarterly Journal of Experimental Psychology*, *32*, 3–25.

Radach, R., Lemmer, S., Vorstius, C., Heller, D., & Radach, K. (2003). Eye movements in the processing of print advertisements. In J. Hyönä, R. Radach, & H. Deubel (Eds.), *The mind's eye: Cognitive and applied aspects of eye movement research* (pp. 9–42). Oxford, UK: Elsevier Science.

Rayner, K. (2009). The 35th Sir Frederick Bartlett Lecture: Eye movements and attention in reading, scene perception, and visual search. *Quarterly Journal of Experimental Psychology*, *62*, 1457–1506.

Rayner, K., & Castelhano, M. S. (2007). Eye movements during reading, scene perception, visual search, and while looking at print advertisements. In M. Wedel & R. Pieters (Eds.), *Visual marketing* (pp. 9–42). New York: Erlbaum.

Rayner, K., Miller, B., & Rotello, C. M. (2008). Eye movements when looking at print advertisements: The goal of the viewer matters. *Applied Cognitive Psychology*, *22*, 697–707.

Rayner, K., Rotello, C. M., Stewart, A. J., Keir, J., & Duffy, S. A. (2001). Integrating text and pictorial information: Eye movements when looking at print advertisements. *Journal of Experimental Psychology. Applied*, *7*, 219–226.

Reutskaja, E., Nagel, R., Camerer, C. F., & Rangel, A. (2011). Search dynamics in consumer choice under time pressure: An eye-tracking study. *American Economic Review*, *101*, 900–926.

Rosbergen, E., Pieters, R., & Wedel, M. (1997). Visual attention to advertising: A segment-level analysis. *Journal of Consumer Research*, *24*, 305–314.

Russo, J. E., & Leclerc, F. (1994). An eye-fixation analysis of choice processes for consumer nondurables. *Journal of Consumer Research*, *21*, 274–290.

Russo, J. E., & Rosen, L. D. (1975). An eye fixation analysis of multialternative choice. *Memory & Cognition*, *3*, 267–276.

Rutishauser, U., & Koch, C. (2007). Probabilistic modeling of eye movement data during conjunction search via feature-based attention. *Journal of Vision*, *7*(6), 1–20.

Satomura, T., Wedel, M., & Pieters, F. R. (2014). Copy alert: A method and metric to detect visual copycat brands. *JMR, Journal of Marketing Research*, *51*, 1–13.

Shi, S. W., Wedel, M., & Pieters, F. R. (2013). Information acquisition during online decision making: A model-based exploration using eye-tracking data. *Management Science*, *59*, 1009–1026.

Shimojo, S., Simion, C., Shimojo, E., & Scheier, C. (2003). Gaze bias both reflects and influences preference. *Nature Neuroscience, 6,* 1317–1322.

Spratling, M. W., & Johnson, M. H. (2004). A feedback model of visual attention. *Journal of Cognitive Neuroscience, 16,* 219–237.

Starch, D. (1923). *Principles of advertising.* Chicago: Shaw.

Stark, L. W., & Ellis, S. R. (1981). Scanpaths revisited: Cognitive models direct active looking. In D. F. Fisher, R. A. Monty, & J. W. Senders (Eds.), *Eye movements: Cognition and visual perception* (pp. 193–226). Hillsdale, NJ: Erlbaum.

Stewart, A. J., Pickering, M. J., & Sturt, P. (2004). Using eye movements during reading as an implicit measure of the acceptability of brand extensions. *Applied Cognitive Psychology, 18,* 697–709.

Strong, E. K. (1925). *The psychology of selling and advertising.* New York: McGraw-Hill.

Stüttgen, P., Boatwright, P., & Monroe, R. T. (2012). A satisficing choice model. *Marketing Science, 31,* 878–899.

Tatler, B. W. (2007). The central fixation bias in scene viewing: Selecting an optimal viewing position independently of motor biases and image feature distributions. *Journal of Vision, 7*(14), 4, 1–17.

Teixeira, T. S., Wedel, M., & Pieters, R. (2010). Moment-to-moment optimal branding in TV commercials: Preventing avoidance by pulsing. *Marketing Science, 29,* 783–804.

Torralba, A., Oliva, A., Castelhano, M. S., & Henderson, J. M. (2006). Contextual guidance of eye movements and attention in real-world scenes: The role of global features in object search. *Psychological Review, 113,* 766–786.

Toubia, O., de Jong, M. G., Stieger, D., & Füller, J. (2012). Measuring consumer preferences using conjoint poker. *Marketing Science, 31,* 138–156.

Treistman, J., & Gregg, J. P. (1979). Visual, verbal, and sales responses to print ads. *Journal of Advertising Research, 19,* 41–47.

Underwood, G., Foulsham, T., & Humphrey, K. (2009). Saliency and scan patterns in the inspection of real-world scenes: Eye movements during encoding and recognition. *Visual Cognition, 17,* 812–834.

Ungerleider, L. G., & Mishkin, M. (1982). Two cortical visual systems. In D. Ingle, R. J. W. Mansfeld, & M. S. Goodale (Eds.), *The analysis of visual behavior* (pp. 549–586). Cambridge, MA: MIT Press.

van der Lans, R., Pieters, R., & Wedel, M. (2008a). Competitive brand salience. *Marketing Science, 27,* 922–931.

van der Lans, R., Pieters, R., & Wedel, M. (2008b). Eye-movement analysis of search effectiveness. *Journal of the American Statistical Association, 103,* 452–461.

van Herpen, E., & van Trijp, H. C. M. (2011). Front-of-pack nutrition labels: Their effect on attention and choices when consumers have varying goals and time constraints. *Appetite, 57*(1), 148–160.

van Raaij, F. W. (1977). Consumer information processing for different information structures and formats. In W. D. Perreault, Jr., (Ed.), *Advances in consumer research* (Vol. 4, pp. 176–184). Atlanta, GA: Association for Consumer Research.

Van Zoest, W., & Donk, M. (2004). Bottom-up and top-down control in visual search. *Perception, 33*, 927–937.

Wang, H.-C., & Pomplun, M. (2012). The attraction of visual attention to texts in real-world scenes. *Journal of Vision, 12*, 1–17.

Wedel, M., & Pieters, R. (2000). Eye fixations on advertisements and memory for brands: A model and findings. *Marketing Science, 19*, 297–312.

Wedel, M., & Pieters, R. (2008). Eye-tracking for visual marketing. *Foundations and Trends in Marketing, 1*(4), 231–320.

Wedel, M., & Pieters, R. (2015). The buffer effect: The role of color when advertising exposures are brief and blurred. *Marketing Science, 34*, 134–143.

Wedel, M., Pieters, R., & Liechty, J. (2008). Attention switching during scene perception: How goals influence the time course of eye movements across advertisements. *Journal of Experimental Psychology. Applied, 14*, 129–138.

Williams, D. E., Reingold, E. M., Moscovitch, M., & Behrmann, M. (1997). Patterns of eye movements during parallel and serial visual search tasks. *Canadian Journal of Experimental Psychology, 51*, 151–164.

Wolfe, J. M., & Horowitz, T. S. (2004). What attributes guide the deployment of visual attention and how do they do it? *Nature Reviews. Neuroscience, 5*, 495–501.

Wu, C. C., Wick, F. A., & Pomplun, M. (2014). Guidance of visual attention by semantic information in real-world scenes. *Frontiers in Psychology, 5*, 54.

Yarbus, A. L. (1967). *Eye movements and vision.* New York: Plenum Press.

Zelinsky, G. J. (2008). A theory of eye movements during target acquisition. *Psychological Review, 115*, 787–835.

Zelinsky, G. J., & Sheinberg, D. L. (1997). Eye movements during parallel–serial visual search. *Journal of Experimental Psychology: Human Perception and Performance, 23*, 244–262.

Zhang, J., Wedel, M., & Pieters, R. (2009). Sales effects of attention to feature advertisements: A Bayesian mediation analysis. *JMR, Journal of Marketing Research, 46*, 669–681.

26

Exploring Attention in the "Reel" World: Visual and Auditory Influences on Reactions to Wins and Near Misses in Multiline Slot Machine Play

Mike J. Dixon, Jeff Templeton, Karen Collins, Lisa Wojtowicz, Kevin A. Harrigan, Jonathan A. Fugelsang, and Vivian Siu

Historical Context and State-of-the-Art Review

In research on attention, the wheel has come full circle. The need to understand attention was initially prompted by a real-world problem—namely, the attention failures of World War II pilots, who missed processing key verbal directives conveyed over their headsets. Instead of investigating attention in situ, researchers typically explored attention and attention failures in the cognition laboratory where experimenters ran a tight ship. They carefully crafted sets of displays containing single targets embedded among predetermined numbers of distractors. By measuring how quickly and accurately research participants could find these targets, researchers sought to reveal the factors that impacted search efficiency and the mechanisms by which attention operated. Recently however, attention research has moved back to the real world. Rather than search for specific letters embedded among other letters, researchers now investigate problems like searching for weapons hidden in luggage at airports (e.g., Menneer, Cave, & Donnelly, 2009) or how to effectively cue distracted drivers that danger is imminent (see chapter 6 in the current volume).

Here we investigate how attention operates in an unusual, highly complex, real-world scenario, namely, slot machine play. In the old-school, three-reel, one-armed bandits, players spun the mechanical reels by pulling on a lever. The reels would spin and stop in sequence, and the player hoped that when all three reels had come to a stop, three of the same symbols (e.g., red sevens) would lie beneath the pay line. In these games the visual displays were relatively simple, and it was easy to tell if one had won or lost. In modern multiline slot machines there are more reels (typically five), there are more visible symbols (typically 15), and players can and do bet on many lines at once (20 lines is common, but machines will vary from 5 playable lines to over 1,000!). In multiline play gamblers hope that a winning combination will occur on one or more of the wagered lines. When players gamble on many lines, some of these lines will include complex zigzag combinations across the symbol array. An example of a player wagering on nine lines is shown in figure 26.1. Because of the complexity of the

Figure 26.1
In the Magic Melody multiline simulator game, credit gains occur when the same symbol falls along any of the nine depicted played lines.

displays, players will often rely on the machines to indicate whether they won or lost on a given spin. When players spin and lose their entire wager, the machine goes into a state of quiet in both the visual and the auditory domain. When players spin and win more than they wagered, they are reinforced by animations highlighting the winning symbols and by catchy "winning songs" whose duration is titrated to the size of the win. When, however, players bet on multiple lines (e.g., bet 10 cents on each of nine lines for a spin wager of 90 cents) and they win back *less* than their spin wager (e.g., the player bets 90 cents and wins back 20 cents), the machine still highlights the symbols responsible for the "win" and still plays "winning jingles," known in the industry as the "credit rack," despite the fact that the player has lost money on this spin. We call these outcomes losses disguised as wins (LDWs; Dixon, Harrigan, Sandhu, Collins, & Fugelsang, 2010). We have shown that when players are asked to estimate the number of spins on which they won more than they wagered, players tend to overestimate the number of true wins. In our view attention plays a key role in this win-overestimation effect. The celebratory feedback serves to draw attention to these outcomes, making them more memorable. Because of the similarity between the feedback for LDWs and actual wins, they get conflated in memory (Jensen, Dixon, Harrigan, Sheepy, Fugelsang, & Jarick, 2013).

The sounds of slot machines have a marked impact on players. In a recent study, we (i.e., Dixon, Harrigan, et al., 2014) compared the reactions of players during standard play when credit gains were accompanied by winning sounds, compared to play where all winning sounds were removed—only the visual animations highlighted any credit gains. We found that the vast majority of players preferred the game with the winning sounds—they found it more exciting. Because both games contained LDWs, players overestimated the number of times that they won during play. Importantly, compared to the game where celebratory sounds were removed, gamblers playing the standard game with celebratory sounds for wins and LDWs more markedly overestimated the number of times they won during slots play. In this study, the link between attention and sound was key. Sound appeared to draw attention to the LDWs, making them more memorable and more likely to be confused with actual wins.

In a follow-up study, we (i.e., Dixon, Collins, Harrigan, Graydon, & Fugelsang, 2015) showed that sound could be used to draw attention to the fact that LDWs were outcomes that cost rather than gained the player money. One group of participants played the standard game and showed the standard LDW-triggered win overestimation effect. They recalled winning far more often than they did in reality. Another group of participants played exactly the same game except that "negative" sounds (i.e., "raspberry"-type sounds) accompanied all losses (including LDWs). Pairing LDWs with negative sounds drew attention to the fact that LDWs were in fact negative outcomes, and players in this condition gave high-fidelity win estimates, eliminating the LDW-triggered win-overestimation effect.

The impact of sound on slot machine play can be considered an instance of cross-modal attention in the real (reel?) world. In the sound-off condition, the reactions to slot machines are triggered purely by visual inputs (the sights of the spinning reels and the eye-catching animations during rewarding outcomes). In the standard condition where celebratory sounds accompany credit gains, the sounds appeared to further draw attention to the LDWs, resulting in an exacerbation of the LDW-win-overestimation effect.

In the following series of new experiments we sought to further explore the role of visual and auditory attention in slot machine play. As can be seen in the display in figure 26.1, some of the symbols appear to "capture attention" more readily than others. For example, the stereo amplifier is the only symbol with straight edges—a feature that sets it apart from the other more roundish symbols. We wondered if such low-level visual features might draw attention to these symbols during play and cause players to react differently to key outcomes involving these symbols compared to other symbols.

As gambling researchers, we paid special attention to a specific class of outcomes called near misses. In slot machines, near misses are outcomes that resemble jackpot wins but fall just short (Reid, 1986). If the jackpot involves three red 7s falling on the pay line, a classic near miss would be two red 7s on the pay line and the third red 7 just above or just below the pay line. Near misses are an important feature of slot machine games because they influence both the psychology of the players and their overt behavior. Near misses evoke high levels of self-perceived, subjective arousal (Griffiths, 1990, 2003; Parke & Griffiths, 2004) and trigger the urge to continue gambling (Clark, Lawrence, Astley-Jones, & Gray, 2009; Clark, Crooks, Clarke, Aitken, & Dunn, 2011). They also can promote risky and impulsive betting behavior (Luo, Wang, & Qu, 2011). Finally, near misses have been shown to extend play on slot machine simulators (Kassinove & Schare, 2001) and on actual slot machines (Cote, Caron, Aubert, & Ladouceur, 2003). Since extended play times, on average, translate to greater losses incurred by players, near misses may have profound implications for problem gambling.

Near-miss effects have been explained in various ways. For some researchers, the surface visual similarity of a near miss (two red 7s on the pay line) to a big win (three red 7s on the pay line) is key. Near-miss effects have been couched in terms of Pavlovian generalization—three red 7s triggers a big win and a spike in arousal due to the rewarding properties associated with winning money. After extensive play, outcomes that closely resemble winning outcomes (near misses) will begin to trigger reward responses. Thus, players begin to psychologically respond to near misses as though they are wins. One problem with this account is that in real-world slots play jackpot wins are exceedingly rare. Conditioning (and Pavlovian generalization) relies on the consistent pairing of specific stimuli (three red 7s) with reward (winning large amounts of money). In typical slots play gamblers would be exposed to so few instances of jackpot wins (one in approximately 240,000 spins) that it is difficult to see how a subset of red 7s (i.e., two red 7s) would become sufficiently conditioned to trigger reward responses. The other problem is how players react to near misses versus wins. Whereas wins are accompanied by elation, near misses are greeted by disappointment, negative affect, and frustration (Bossuyt, Moors, & De Houwer, 2014).

Frustration elicits marked psychophysical changes that can be documented using skin conductance responses (SCRs; Lobbestael, Arntz, & Wiers, 2008). Clark, Crooks, Clarke, Aitken, and Dunn (2011) used a simplified two-reel slot simulator to show that near misses triggered higher SCRs than regular losses. Dixon et al. (2011) recorded the SCRs of slots players for losses, wins, and near misses, as gamblers played a realistic three-reel slots simulator. Like Clark et al., they showed that SCRs for near misses were larger than regular losses, but also larger than small wins. Dixon et al. proposed that these arousal patterns were due to the frustration of "just missing" a big win.

In a subsequent study, Dixon et al. (2013) replicated the SCR findings and also measured the time taken between outcome delivery and the initiation of the next spin—the so-called post-reinforcement pause (PRP; Peters et al., 2010). They showed that the larger the win, the longer the PRP. Crucially, the PRP after near misses was significantly smaller than all wins, as well as significantly smaller than regular losses. This finding is crucial because it suggests that the combination of measuring SCRs and PRPs together allows one to discriminate between reactions to frustration and reward. While reward triggers large SCRs and long PRPs, frustration triggers large SCRs but very short PRPs.

Dixon et al. reasoned that the frustration of "just missing" the jackpot causes players to spin quickly to get out of this frustrated state. Although we typically tend to avoid situations which induce frustration, when there is a clear attainable goal (e.g., a jackpot), just missing this goal can lead people to persist in the behaviors related to goal pursuit (Aarts et al., 2010). Thus, games with near misses, although frustrating, cause players to play longer and spend more money. A study by Bossuyt, Moors, and De Houwer (2014) highlights the interaction of frustration, persistence in betting behavior, and what they call the tendency to repair goal incongruence (i.e., just missing the jackpot is incongruent with the desired outcome). They had players play a slot machine simulator and polled their reactions to the various outcomes. Consistent with Dixon et al. (2013) and Clark et al. (2009), they showed that near misses triggered more negative affect than regular losses. Participants rated these outcomes as more disappointing, frustrating, and anger inducing. In their version of a slot machine, after any outcome players could invoke a "second chance" option in which they could essentially win back the money they had lost on that spin by conducting a form of "side bet" against the computer. Crucially, players opted to take the second chance significantly more often following near misses than following regular losses. In sum, although the particular emotional reactions (frustration, disappointment, anger) to near misses remains a matter of debate, previous research has shown that near misses have powerful psychophysiological and psychological effects on slots players, and these effects can be measured during slot machine play.

To date, however, near-miss research has been conducted using relatively simple three-reel, single pay-line games. In modern casinos some of the most heavily used games are the aforementioned highly complex multiline games. At first glance, the complexity of the symbol arrays (straight-line wins, diagonal wins, zigzag wins) would seemingly make it hard, if not impossible, for players to detect near misses were they to occur. Here we show how, by judicious use of specific visual and auditory cues that draw attention to specific symbols, slot machine designers could potentially overcome the inherent complexity of multiline games and render near misses salient to the player. If near misses are indeed apparent to players, then it should be possible to demonstrate their characteristic psychophysical and behavioral responses (i.e., large SCRs and short PRPs) that we have shown in the three-reel games (Dixon et al., 2011; Dixon et al., 2013). This line of research was prompted by certain commercially available games, which appear to take steps to render certain key symbols more salient than others. We have emulated these practices in our slot machine simulator but have for copyright purposes created our own symbols, our own pay tables, and our own sounds and have empirically controlled the outcomes that gamblers playing our simulator will encounter. The study can be viewed as a "proof of concept" designed to show that it is possible to trigger near-miss effects in complex multiline games. Ultimately, we show how low-level visual principles and cross-modal attention capture can be used to create near-miss effects in this highly complex, real-world situation.

Increasing the Salience of Near Misses Using High- and Low-Level Features

Figure 26.2 shows four symbol arrays from our multiline slot machine simulator. For each of these outcomes, the five reels would stop in sequence from left to right, settling on the outcome shown. In the top left panel, three gramophones are aligned from left to right, leading to a win of 100 credits. The top right panel (the three-part stereo), shows a special kind of slot machine win that we call a horizontal triplet. Horizontal triplet wins are different from other wins (such as the gramophone win shown in the top left panel) in that (1) the triplet starts on reel 3 whereas regular wins typically start on reel 1; (2) the horizontal triplet is typically worth more than other symbol wins (in commercially available games such triplets trigger high paying bonuses); (3) the horizontal triplet forms a three-part connected object whereas other wins are replications of the same symbols; (4) in commercially available games horizontal triplets are described on a separate page of the pay table, whereas regular symbols are grouped together; and (5) horizontal triplet wins (as the name suggests) can only occur when the three component symbols are *horizontally* aligned (no diagonal or zigzag alignments lead to wins for these special triplets). The bottom left panel of figure 26.2 shows a gramophone near miss, and the bottom right panel shows a near miss involving the horizontal triplet. Both panels show identical near miss configurations, but certain elements of the stereo near miss appear to render it more noticeable.

In the following experiments we empirically assessed whether specific high- and low-level features of the horizontal triplets may make both wins and near misses stand out from comparison wins and near misses (involving the gramophone). In terms of "high-level" features, the fact that the stereo win is presented last, and on its own page in the game's pay table (description of the symbols and their worth), may serve to sensitize players to look for this high-paying outcome during play. By presenting this type of win last in the pay table, designers can capitalize on the recency effects in memory (e.g., Neath, 1993). By having this outcome be worth more credits than most of the other symbols, one can further sensitize players to these special outcomes.

The second set of features are low-level visual features. A robust literature has demonstrated that humans are capable of discerning basic elements in a visual scene in a rapid fashion independent of attentional focus (see Wolfe, 1994, and chapter 2 in

Wins

Gramophone win (pays 100 credits)   Stereo win (pays 100 credits)

Near-misses

Gramophone near-miss (0 credits)   Stereo near-miss (0 credits)

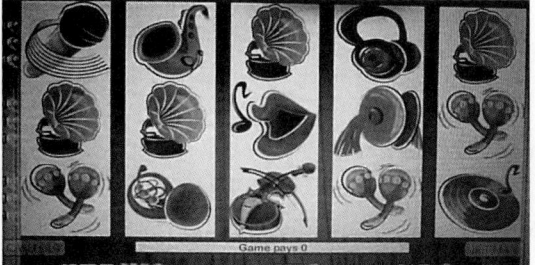
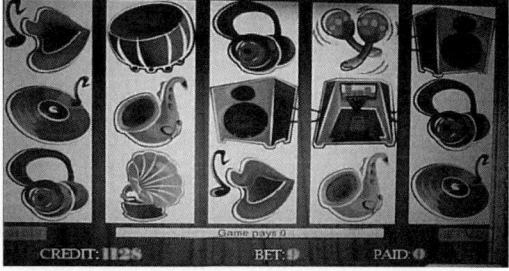

Figure 26.2
Gramophone wins (top left), and near misses (bottom left), along with stereo wins (top right) and stereo near misses (bottom right).

the current volume for a review). Among the component elements that can be readily distinguished in this manner are edges (Biederman & Ju, 1988), curved versus straight lines (Neisser, 1963; Biederman, 2001), color (Treisman & Gelade, 1980), size and luminance (Wolfe, 1994), and texture segmentation (Biederman & Ju, 1988; Bergen & Julesz, 1983). If a target is unique on one of these attributes compared to other items in the visual field, observers can readily find the target without using attentional resources (Wolfe, 1994; Treisman & Gelade, 1980). Furthermore, "even with inhomogeneous distractors, feature search remains efficient if the target is categorically unique" (Wolfe, 1994, p. 232).

Figure 26.3 shows all of the different symbols in our slots simulator. Although depicted in gray scale, the symbols shown actually bore different colors (the stereo pieces, the gramophone, saxophone, and maracas were all predominantly yellow; the

Figure 26.3
The stereo triplet is shown on the top left, followed by the circular (gramophone), as well as other round symbols that served as distractors in our visual search experiments.

kettledrum, headphones, mouth, and LP record were predominantly red; and the violin, microphone, tambourine, and French horn were predominantly purple). Thus, any given symbol shared the same color with three or more other symbols in the stimulus set. As can be seen in figures 26.2 and 26.3, the different stereo pieces have connecting "wires" that appear to join the individual stereo components. Such wires make use of the gestalt principle of element connectedness (Wagemans et al., 2012). Kellman and Shipley (1991), for example, have shown that line segments will be perceptually "filled in" to complete objects like the stereo shown in figures 26.2 and 26.3. Finally, whereas most of the symbols shown in figure 26.3 are composed of curved lines, the symbols forming the stereo are composed of straight edges that coterminate in vertices. Numerous studies show that straight edges, when embedded among curved stimuli, can be found with a high degree of efficiency in visual search (Biederman, 2001; Neisser, 1963; Wolfe & Horowitz, 2004). We propose that the combination of high-level and low-level features may make both the stereo wins and stereo near misses highly salient—more salient than comparison stimuli (the gramophones). If so, then players should react differently to these outcomes than other outcomes during slots play. Specifically, we proposed that if the high-level features of the horizontal triplet cause players to look for these objects, and the low-level features of the stereo make these objects particularly easy to find, then, despite the complexity of multiline slots games, we could still show that stereo wins lead to greater arousal responses than other symbols (e.g., gramophone wins that are worth the same amount), and that the stereo symbols will trigger classic near miss effects that will emerge during slots play (e.g., larger SCRs than regular losses).

Our approach was to start with standard laboratory visual search techniques, gradually increase the ecological validity, and culminate in experiments emulating typical slots play. At step 1, we recreated the types of symbols used in commercially available slot machines, including horizontal triplets and other regular symbols. In creating these symbols, we used commercially available machines as only a rough guideline (incorporating features such as the connecting wires of the triplet shown in the top right panel of figure 26.2). Our symbols were an artist's renderings of images belonging to a general musical theme. The exact low-level features (straight edges, color schemes comprising the triplet and regular symbols) were of our own devising and only roughly approximated the exact features used in commercially available slots. (Here we emphasize that these experiments are a proof of concept concerning how it may be possible to use certain features to make specific symbols more salient and influence players' reactions to these symbols.) Next we used classic visual search techniques to show that the low-level features of the objects we assigned to the horizontal triplet (the speakers and amplifier of the stereo) could indeed draw attention to these components, allowing them to be easily found when embedded among the other symbols. Next we had participants search for slots outcomes formed by the horizontal triplets, and comparison (round) shapes. Specifically, we had participants search for alignments of these symbols that during slots play would form a winning alignment, a near miss alignment, or a losing alignment of these symbols. Finally, we recruited players at a nearby casino to play our simulator to see if indeed they reacted differently to wins and near misses involving these horizontal triplets compared to regular symbols whose win amounts were equal in value to the triplets.

Visual Search Experiments

As previously mentioned, objects possessing certain shape features (vertical lines) will "pop out" when they are presented among symbols with different shape features (curved lines; Neisser, 1963; Wolfe & Horowitz, 2004).

In the search task we asked 50 participants to search either for a straight-edged, relatively square target (one of the stereo pieces) or for a generally round target (the gramophone symbol). For both searches, targets were embedded in arrays of round distractors (the saxophone and the other round symbols depicted on the bottom two rows of figure 26.3). The experiment consisted of three blocks of 180 trials each. Each block contained three types of arrays displaying 5, 10, or 15 symbols, respectively. On half of the trials, a predefined target was present; on the other half, only distractor symbols

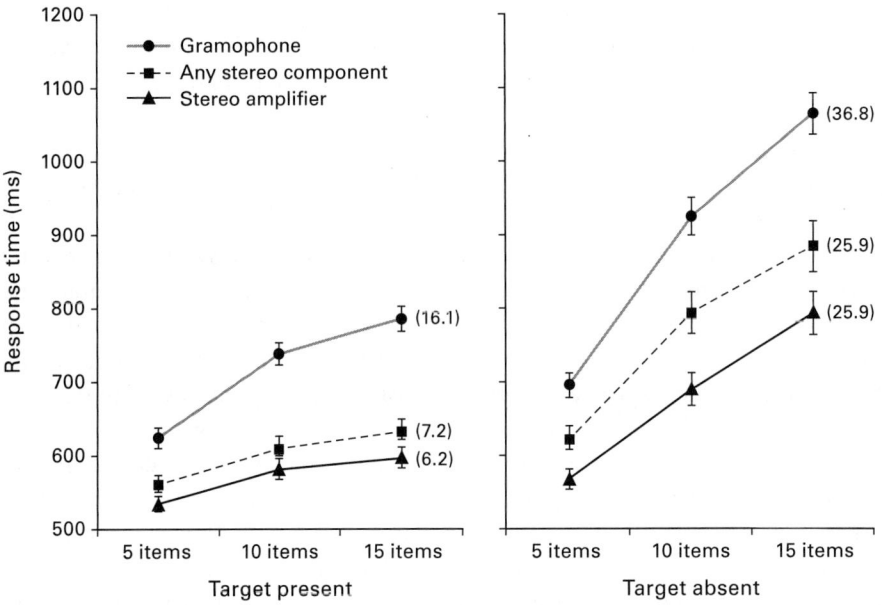

Figure 26.4
Target-present and -absent average search times (search slopes in milliseconds per item are in parentheses) when seeking a round gramophone shape among round distractors, versus searching for square stereo shapes among round distractors. Error bars reflect standard errors of the mean.

were shown. Participants indicated whether the target symbol was present or absent by pressing an appropriate button. Participants were told to press the correct button as quickly as possible without making errors. Symbols were randomly assigned to any of the 15 locations. The symbols constituting a given array were also randomly selected with the provision that on both present and absent trials at least one of the distractors had the same color scheme as the target. In one block of trials, the target was the square amplifier of the stereo triplet. The distractors were always round symbols (one of the distractors was always the gramophone). In a second block of trials, the target was the round gramophone embedded within mostly round distractors (one of the distractors was also the stereo amplifier). In the third block of trials, the target was any one of the square triplet shapes embedded within the round distractors. Block order was counterbalanced across participants. Arrays remained on screen until a response was made. Figure 26.4 shows the mean search times for target-present and target-absent trials for 49 participants (one was removed due to technical problems).

As can be seen in figure 26.4, searching for the square targets is more efficient than searching for the round targets in an array of round distractors. An analysis of variance (ANOVA) on the slopes confirmed these conclusions. When targets were present, there was a main effect of target type, $F(2, 96) = 46.04$, $p < .001$. Bonferroni-corrected post hocs revealed that searching for the gramophone was significantly less efficient than searching for any (square) stereo piece ($p < .001$) or searching for the amplifier of the stereo ($p < .001$). Finding any stereo piece (three targets in the template) was just as efficient as finding the amplifier (one target in the template; $p = .829$). For target-absent trials, there was a main effect of target type, $F(2, 96) = 38.5$, $p < .001$. Bonferroni-corrected post hocs revealed that declaring that the gramophone was absent was significantly less efficient than declaring that there were no stereo pieces ($p < .001$) or declaring that there was no stereo amplifier ($p < .001$).

These findings are in accord with the findings and theory of Duncan and Humphreys (1989), who proposed that search efficiency was determined by how similar targets are to nontargets (high similarity leads to inefficient search), and how similar non-targets are to one another (low similarity among nontargets can further decrease search efficiency when targets are similar to nontargets). Searching for gramophones among other round shapes constitutes a search in which the target–nontarget similarity is high. Consequently, the search for these targets is relatively inefficient (the gray lines in figure 26.4). By contrast, searching for any stereo piece or just the amplifier led to relatively efficient search. Here, all of the potential targets have straight edges whereas all of the distractors have curved edges. In terms of Duncan and Humphrey's model, this would lead to low target–nontarget similarity and, hence, efficient search. The fact that there was no difference in the efficiency of search with three possible targets (any stereo piece) versus one possible target (the stereo amplifier) is in accord with Menneer, Cave, and Donnelly (2009), who showed that multitarget search could be as efficient as single-target search so long as the multiple targets shared a common feature (in this case straight edges) that differentiated targets from distractors.

In sum, Experiment 1 mirrors a traditional search experiment and shows that the square-shaped stereo pieces were more easily detected than the round gramophone pieces. Our next empirical move was to see if this increased search efficiency would translate to a context that more closely resembled slots play where players look not for single symbols but for the patterns formed by sets (in this case triads) of symbols. We tested a different set of 50 participants. They were presented with three blocks of 180 search trials. Prior to each block, participants were shown symbol arrays that in a slots game would constitute (1) a losing configuration (where any related symbols occupied nonadjacent locations), (2) a winning configuration (where somewhere within the array three related symbols were horizontally aligned), or (3) a near miss configuration (somewhere in the array there were two related symbols horizontally aligned, with the third related symbol either just above or below that horizontal alignment). On each trial, players were shown a 15-symbol array (like those in figure 26.2 which show the slot machine symbols after the last reel has stopped). Participants were to indicate whether the array contained a "Win," "Near Miss," or "Loss" by pressing the appropriately designated button. Participants were instructed to respond as quickly as possible without making errors. Crucially, in one block participants were told that wins or near miss trials would be formed by the alignment of the gramophone symbol (e.g., players looked for three gramophones in a row on wins). In another block players were told that wins or near misses would be formed by the alignment of the square shapes that comprised the stereo triplet (e.g., players looked for the stereo with the left speaker, amplifier, and right speaker all in a row). Wins or near misses involving these stimuli could occur on any row, and the leftmost symbol of the win or near miss could appear on reel 1, 2, or 3. Loss trials always presented one stereo symbol and one gramophone symbol in the array. Importantly, during these first two blocks players searched the arrays with a predefined target group (i.e., the three gramophones in one block, the three stereo components in another block). The order in which these two blocks were presented was counterbalanced. In the third block, participants searched arrays without out a specific target symbol per se. Instead, they were told that wins and near misses could be triggered by either the stereo components or the gramophones. They were told to report via a button press whether the arrays contained a winning, losing, or near miss outcome formed by either the stereo pieces *or* the gramophones.

The results, shown in figure 26.5, are intriguing. Across all types of search, participants were faster to detect wins than near misses and were faster to detect near misses than losses. Search times were slowest in the complex template condition, where wins, near misses, and losses could be formed by either the stereo pieces or the gramophones. In this condition there were no differences between search times for these two classes of objects. In the simple template condition (where players knew in advance what symbols to look for) participants were faster to identify wins, near misses, and losses formed by the straight-edged and connected stereo pieces than when these same outcomes were formed by the round gramophone pieces.

These impressions were confirmed by an outcome (win, near miss, loss) by template complexity (single class of object, two classes of object) by object (stereo, gramophone) repeated-measures ANOVA. In brief, all main effects and all two-way and three-way interactions were significant (all p values <.001). To make sense of the three-way interaction, we analyzed the two levels of template complexity separately. When participants were required to hold in mind a complex template involving the stereo pieces *or* the gramophone pieces, there was a main effect of outcome, $F(2, 98) = 91.42$, $p < .001$, but no main effect of objects, $F(1, 49) = 0.059$, *ns*, and no outcome by object interaction, $F(2, 98) = .406$, *ns*. (see the two overlapping lines at the top of figure 26.5). When participants could form a simple template involving only a single set of objects (either the stereo components or the gramophones), there was both a main effect of outcome, $F(2, 98) = 88.34$, $p < .001$; a main effect of object, $F(1, 49) = 25.75$, $p < .001$; and an outcome by object interaction, $F(2, 98) = 42.32$, $p < .001$. Although response times were

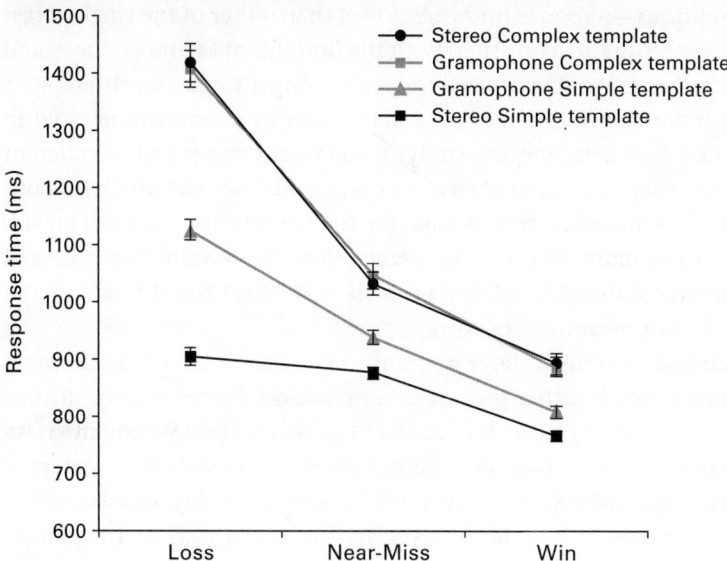

Figure 26.5
Average response times to detect wins, near misses, or losses. The top two lines reflect when play-ers searched the arrays using a complex template (outcomes on any given trial could be formed by either the stereo components or the gramophones). The bottom two lines reflect the use of simple templates—in one block players searched for wins, near misses, and losses formed by the gramophones, and in a different block players searched for wins, near misses, and losses formed by the stereo pieces.

slower for gramophones than stereos for all outcomes, the interaction was caused by an exacerbation of this slowing for gramophone losses.

In sum, when players searched with a specific target in mind, they were significantly faster at detecting wins, near misses, and losses involving the straight-edged stereo targets than they were with round targets. Presumably, this superior performance for the stereo condition is due to the ease with which the straight-edged symbols can be extracted from the round-shaped distractors (as in Experiment 1) when participants know to look for such straight-edged components. To explain this pattern of data, we drew upon Duncan and Humphrey's notions concerning target–nontarget similarity. When participants are searching for the outcomes formed by stereo pieces (the black line in the bottom of figure 26.5), the straight edges of the target serve to effectively reduce target–nontarget similarity, leading to faster search times than when search-ing for gramophones, where target–nontarget similarity is greater. Importantly, when players are searching for two classes of targets (the outcomes could be determined by either the stereo or the gramophone), there is no difference between the stereo and the gramophone conditions—response times are slower than either of the single-target group conditions. These results are consistent with the findings of Menneer, Cave, and Donnelly (2009), who showed that compared to search using a single target template, search that required holding two different potential targets in a template resulted in large search costs when (as in the present study) targets were comprised of different visual feature sets where there was no single feature that could guide the search for both target categories. We note, however, that in this task the complexity was even greater than in search tasks using multiple potential targets. Here, there were two potential object classes (stereos and gramophones), but each class of object could be aligned in one of three outcomes (loss, near miss, or win).

In the visual search tasks described above we transitioned from a classic visual search experiment in which a target is either present or absent, and the number of distractors is varied, to an atypical kind of search experiment in which the 15-element arrays approximate what would be seen when the final reel stops during slots play, and players must look for wins, near misses, or losses in the presented display. Our transition to actual slots play is far from complete however. In our search task all 15 symbols appeared at once whereas in slots play the symbols appear in their final positions one reel (three symbols) at a time. Also, in the search task the stereo wins and near misses could span across any of the reels (e.g., across reels 1, 2, and 3) whereas in the slot machine stereo wins could only occur on reels 3, 4, 5, and gramophone wins would always span across reels 1, 2, and 3. Despite such differences, our search results still could have implications for whether or not players would show near miss effects during actual slots play. The visual search experiments indicate that when participants hold a template that includes only the stereo pieces, searching for stereo components is easier than searching for the gramophones. Recall that in the slots game there are certain (high-level) game features that may actually bias players to preferentially look for the objects that make up the horizontal triplet. For example, the horizontal triplet is (unlike the regular symbols) described on a separate page at the very end of the pay table and is described as a special kind of relatively large win that follows different rules than the other symbols in the game. If these high-level features serve to sufficiently bias participants to preferentially look for stereo pieces (i.e., form a template involving only the stereo pieces), then one would predict that noticing the stereo within the complex 15-symbol array as it unfolds should be relatively easy. If the stereo pieces are sufficiently salient, then attention will be drawn to these pieces, and when the stereo pieces fall in a winning alignment, they should trigger greater reactions than wins involving the gramophone pieces. Similarly, if the stereo pieces are more salient than the gramophone pieces, then near misses involving the stereo pieces should be sufficiently salient as well. If near misses involving the horizontal triplet are attended to, then we should see the characteristic arousal signature. Specifically, we should show higher SCRs for stereo near misses than for gramophone near misses or for regular losses. If, however, the high-level features associated with the stereo are insufficient to bias players to specifically look for these outcomes during slots play, then in terms of psychophysiological measures such as SCRs, the wins and the near misses triggered by the gramophones or the horizontal triplet might look identical. If the displays are so complex in terms of the myriad ways of winning (zigzag, diagonal wins, etc.), it may be that no near miss effects are evident whatsoever, and near misses will look equivalent to regular losses for both the gramophones and the horizontal triplets.

Subjective and Physiological Reactions to Wins, Losses, and Near Misses during Multiline Slots Play

To see whether actual gamblers playing a multiline slot machine simulator would react differently to wins and near misses involving the horizontal triplets, we recruited 120 players from a local casino. In exchange for participation they received a $25 Walmart gift card. Players were told that $10.00 had been loaded into the machine. They were told that after 250 spins they could "keep their winnings" (i.e., they could keep the remaining balance in the slot machine at the end of play). Participants played 250 spins, wagering 1 credit per line on each of nine play lines for a total spin wager of 9 credits. The play session was composed of 89 regular losses (credit gains of zero), 66 credit gains of 2–8 credits, and 35 wins between 9 and 58 credits. There were also 6 gramophone wins worth 100 credits each and 6 stereo wins worth 100 credits each. For both of these winning outcomes, the three symbols "animated" by enlarging and contracting, and the middle pay line (on which both types of wins fell) illuminated. Finally there were 30 gramophone near misses (0 credits) and 30 stereo near misses (0 credits). Like regular losses, there were no animations or winning jingles accompanying these outcomes. Near misses were always of the "classic" variety—for example, two gramophones in the middle of reel 1 and 2, respectively, followed by a third gramophone on reel 3, just above or just below this horizontally aligned pair (see the gramophone near miss depiction in figure 26.2). For the stereo horizontal triplet, near misses always comprised the left speaker and the amplifier in the middle row of reels 3 and 4 (see figure 26.2) with the right speaker just failing to complete the figure. The right speaker appeared on reel 5, either just above the first two components (as in figure 26.2) or just below the first two stereo components. Following the fifth reel's stopping, there was a 4-second period in which the spin button was disabled (this study measured phasic changes in heart rate as well as SCRs). Upon initiating a spin, an animation showed all reels spinning. The reels stopped in sequence with the elapsed time between one reel's stopping and the next being approximately 450 ms. Reel 1 stopped after 464 ms; reel 2 stopped 445 ms later, reel 3 stopped 447 ms later, reel 4 stopped 440 ms later, and reel 5 stopped 450 ms after reel 4.

To assess differences in the various symbols' ability to capture attention, we also asked a number of subjective questions following slots play. Participants were shown three symbols: the gramophone, the left speaker of the stereo, and a depiction of a record album. Beside each symbol was the following text: "When playing, some people claim that certain symbols were more noticeable than others (they attract your attention). Please rate how much the following symbol attracted your attention while playing the game." Participants were asked to click on a number from 1 to 7 and told that 1 = "not grabbing attention" and 7 = "very attention grabbing." Next they were shown three winning outcomes (all of which they encountered during slots play): the stereo horizontal triplet, three adjacent record albums, and three adjacent gramophones. They were asked how many credits the depicted win was worth, and to answer by clicking on a number. Their choices were: 10, 20, 50, 100, 200, 500, and "don't remember." Next they were shown screen shots of six different outcomes: a gramophone win (three adjacent gramophones), a gramophone near miss, a loss (with neither gramophones nor stereo components), a stereo near miss, another loss (with both gramophones, and the left speaker of the stereo—not analyzed), and a stereo horizontal triplet win. For each image they were asked to select the phrase that best described their reaction to this outcome during the game. Players clicked on one of the following phrases: "Mildly Bored," "Very Bored," "Mildly Excited," "Very Excited," "Mildly Frustrated," or "Very Frustrated."

Results

Following slots play, when participants were shown depictions of the gramophone, the LP record, and the left speaker of the stereo and asked to rate how attention-grabbing these symbols were on a 7-point Likert scale, the stereo piece received the highest average ratings (mean = 5.8), followed by the gramophone (mean = 4.20). The LP record received the lowest attention-grabbing ratings (mean = 3.70). These attention-grabbing ratings were analyzed using a repeated-measures ANOVA. This analysis revealed a main effect of symbol, $F(2, 234) = 41.05$, $p < .001$. Post hoc analyses revealed that all three outcomes were significantly different from each other. Crucially, the stereo speaker received significantly higher attention-grabbing ratings than the gramophone ($p < .001$).

Following slots play, participants were shown depictions of the horizontal triplet win, a gramophone win, and an LP win, and they were asked to recall how much each win was worth. These data were analyzed by categorizing these estimates as correct if they gave the correct value (i.e., 10 credits for the LP, 100 credits for the gramophone and stereo), incorrect if they misremembered the win's worth, or "didn't know" if they checked the "I do not remember button." These responses are shown in table 26.1.

Table 26.1

Number of Participants giving Correct, Incorrect, and Unsure Estimates of Credit Values for Three Slot Machine Symbols

Estimations of Credit Worth by Participants	LP	Gramophone	Stereo
Correct number of credits	35	65	102
Incorrect number of credits	53	30	7
"Didn't know"	32	25	11

Note: LP, depiction of a record album.

Table 26.2

Number of Participants Applying Ratings of Frustration for Losses, Gramophone Near Misses, and Stereo Near Misses (NM)

Rating	Loss	Gramophone NM	Stereo NM
"Mildly frustrated"	8	14	18
"Very frustrated"	5	4	10
Nonfrustration response (very bored, mildly bored, mildly excited, or very excited)	107	102	92

Note: NM, near miss.

Chi-square analyses revealed that the value of the gramophone win (100 credits) was better recalled than the value of the LP record win (10 credits), $\chi^2(2) = 20.79$, $p < .01$. Crucially, despite being worth the same amount, the value of a stereo win (100 credits) was better recalled than the value of a gramophone win (also 100 credits), $\chi^2(2) = 27.49$, $p < .01$.

To see whether players rated stereo near misses as preferentially frustrating, we observed the frequencies with which players rated depictions of losses, gramophone near misses, and stereo near misses as mildly frustrating, extremely frustrating, or nonfrustrating (very boring, mildly boring, mildly exciting, extremely exciting). The observed frequencies for each type of rating are shown in table 26.2.

When gramophone near misses were compared to losses, no differences were noted, $\chi^2(2) = 1.86$, *ns*. When stereo near misses were compared to losses, significantly more players indicated that stereo near misses were frustrating, $\chi^2(2) = 6.64$, $p < .05$. However, when stereo near misses were compared to gramophone near misses, there was no difference between conditions, $\chi^2(2) = 3.59$, *ns*.

As can be seen in table 26.2, the majority of participants did not endorse near misses as frustrating. Instead they rated these outcomes as being very boring, mildly boring, mildly exciting, or very exciting. To assess whether players gave different

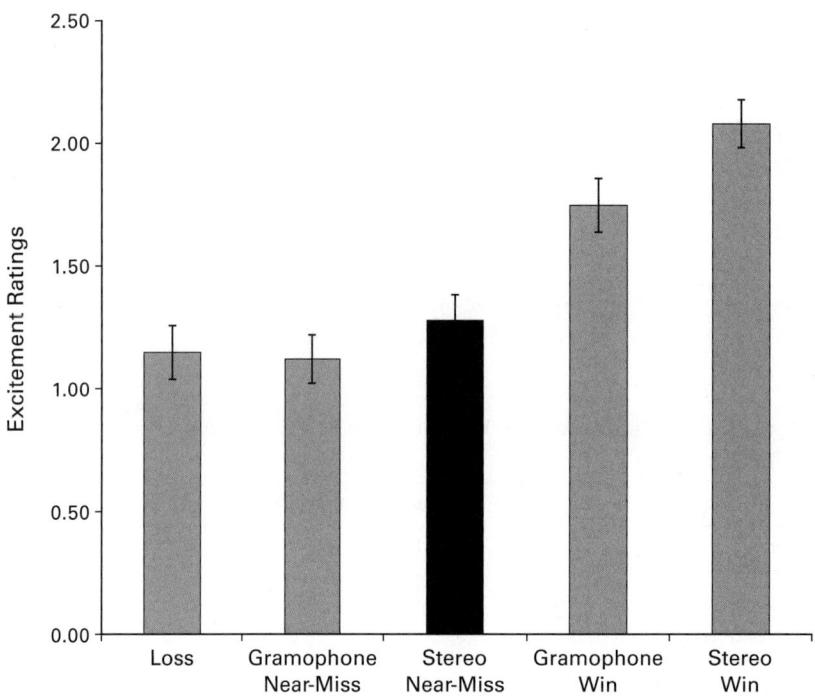

Figure 26.6
Average excitement ratings for losses, gramophone near misses, stereo near misses, gramophone wins, and stereo wins. Error bars reflect standard errors of the mean.

excitement ratings to losses, gramophone near misses, stereo near misses, gramophone wins, or stereo wins, we eliminated any participants who gave frustration ratings and then translated the very bored, mildly bored, mildly excited, very excited ratings to an excitement scale that ranged from 0 to 3. A total of 75 participants had complete excitement ratings for all five outcomes. The excitement data were analyzed using a repeated-measures ANOVA with outcome as the repeated factor. The repeated-measures ANOVA yielded a highly significant main effect of outcome, $F(4, 296) = 28.995$, $p < .001$. Bonferroni-corrected post hoc pairwise comparisons indicated gramophone wins were rated as more exciting than all types of losses ($p < .005$). Losses were rated equivalently to both gramophone near misses and stereo near misses, which did not differ. Crucially, the stereo wins were rated as being significantly more exciting than any other type of outcome, including gramophone wins ($p < .03$). The excitement ratings for these outcomes are shown in figure 26.6.

To summarize the collective results presented thus far, the subjective ratings indicated that stereo pieces were more "attention capturing" than other symbols.

Furthermore, when the stereo pieces were aligned as wins, they were better remembered (in terms of their worth) and were rated as being more exciting than gramophone wins of equivalent value. If during play they were indeed more exciting, one would expect that they would trigger preferentially large SCR responses.

The SCRs for losses, gramophone near misses, stereo near misses, gramophone wins, and stereo wins were calculated by first defining a 2-second window that occurred 1 second after outcome delivery (the final reel's stopping). Brief 2-second windows were used, following the recommendation of Dawson, Schell, and Filion (2000) to minimize the chances of nonspecific SCRs from contaminating the data, and to replicate the procedures used in our previous research measuring psychophysiological responses to various slots outcomes during slots play (Dixon et al., 2011; Dixon et al., 2013). To calculate the SCR, the skin conductance level at the beginning of the window was subtracted from the peak skin conductance level within the window. To reduce the potential skew of the SCRs, a square-root transformation was applied to these difference scores (Dawson, Schell, & Filion, 2000). For each participant, five mean SCRs were calculated based on the outlier-free averages of that participant's SCR amplitudes for that outcome. Since the number of observations within each outcome was very different (e.g., there were 89 regular losses but only 6 stereo-triplet wins) prior to calculating the means, outliers were eliminated using the procedures of Van Selst and Jolicoeur (1994), which uses a sliding rejection criterion based on the number of observations on which the mean is based.

The SCRs were analyzed using a repeated-measures ANOVA with outcome (loss, gramophone near miss, stereo near miss, gramophone win, stereo win) as the repeated variable. This analysis showed a main effect of outcome, $F(4, 476) = 28.706$, $p < .001$, which is depicted in figure 26.7. Bonferroni-corrected post hoc analyses revealed no significant differences between losses, gramophone near misses, and stereo near misses. Gramophone wins were significantly larger than stereo near misses ($p = .03$). Crucially, SCRs for stereo wins were significantly larger than all other outcomes, including gramophone wins ($p < .001$).

The SCRs indicated that even though stereo wins and gramophone wins were worth exactly the same amount, players reacted far more strongly to the stereo wins. One interpretation of these data is that these results are simply an artifact of how we measured SCRs. Recall that both gramophone wins and stereo wins were measured using a 2-second window beginning 1 second after the last reel stopped. One could argue that since the gramophone win occurred after the third reel stopped, any SCR for that win might have dissipated by the time the 2-second window emerged. We think that this interpretation is unlikely because even though the gramophone win occurred when the third reel stopped, the celebratory, rewarding feedback for both types of wins occurs only after the final reel has stopped. It is this exciting feedback that likely triggers the SCRs. Converging evidence that players did indeed

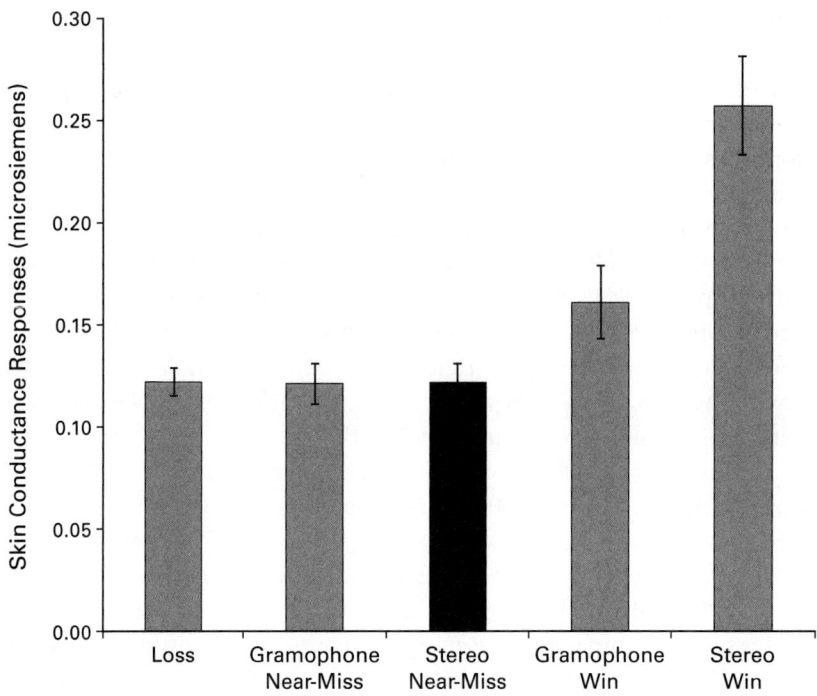

Figure 26.7
Average skin conductance responses for losses, gramophone near misses, stereo near misses, gramophone wins, and stereo wins. Error bars reflect standard errors of the mean.

find the stereo wins more rewarding comes from their excitement ratings. Players rated stereo wins as significantly more exciting than gramophone wins—indeed the SCR ratings in figure 26.7 and the excitement ratings in figure 26.6 closely mirror one another.

Although participants clearly responded to the stereo wins with more arousal during play and with higher excitement ratings following play, there was no hint of the predicted near miss effect. Neither the gramophone near misses nor the stereo near misses showed elevated SCRs relative to the regular losses. While there was some support from the subjective frustration ratings that stereo near misses were more frustrating than regular losses, overall few players rated these outcomes as frustrating, and we were unable to show that stereo near misses were any more frustrating than gramophone near misses. The SCR data in figure 26.7 shows an absence of the near miss effect in this multiline slot game—an effect that we have twice replicated (Dixon et al., 2011; Dixon et al., 2013) using simpler three-reel machines.

Using Multisensory Inputs to Signal the Presence of Near Misses

The failure to find any compelling near miss effects prompted us to look (and listen) closely to the commercially available game upon which we patterned our simulator. While we had adequately emulated the *visual* features of the horizontal triplet in our simulator, we noticed that in the commercially available game, within the auditory domain, the slot machine signaled the appearance of the triplet's components on the reels with highly salient, sudden-onset ascending tones. Bally composer Peter Inouye calls these "anticipations" that "ramp up" on the second tone: "Since you need the third [symbol] to hit, once the second one hits, you'll kind of hear a ramp up to kind of build anticipation for the third one to hit ... if you only get the first two, you really want that third one to hit, so when that third one hits it's kind of rewarding and makes you feel a lot better when it actually does hit" (Inouye, 2013).

Recall that in multiline games when players spin the reels, the reels stop in succession. In the commercially available game, when the third reel stopped, and the leftmost component of the horizontal triplet was visible on this reel, the appearance of this leftmost component was signaled by a highly salient tone (as though to signal to the player that the first part of the horizontal triplet was in a winning position). The third reel in the right panels of figure 26.8 shows the pairing of the visual appearance of left speaker with the signaling sound. If reel 4 stopped so that the second component of the triplet landed adjacent ("connected to") the first component (i.e., the amplifier was connected to the left speaker), the second component was also accompanied by another highly salient tone, at a slightly higher pitch than the first component's tone (see reel 4, right panels of figure 26.8). If the final component (on the fifth reel) landed beside this pair (completing the horizontal triplet), this symbol was accompanied by a tone at an even higher pitch, followed by the winning animation. Thus, in the auditory domain, on horizontal triplet wins, the joining of the three symbols was accompanied by three tones rising in frequency. The lower right panel depicts how the last tone was conspicuously absent on horizontal triplet near miss outcomes.

Particular tones in video game soundtracks are often used to create a sense of conclusion and resolution (known as a cadence). A strong cadence leads to a strong subjective sense of conclusion or finality. Leaving a cadence open is used in songs to build energy and foster the urge to complete whatever task is ongoing. The lower right panel of figure 26.8 shows the uncompleted sound sequence that accompanies near misses in the commercially available game (and in our simulator). By creating a near miss with an open sequence, the sounds may lead to a sense of an unfinished and unresolved outcome. Since we have been conditioned through our music listening to expect a closing cadence, open cadences like those employed in slot machine near misses may add to their salience. Given the profound influence of sound on other aspects of multiline

Wins

Gramophone win

No sounds signal appearance
of gramophone symbols

Stereo win

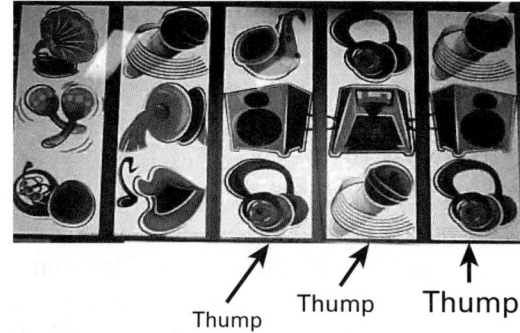

Thump   Thump   Thump

Near-misses

Gramophone near-miss

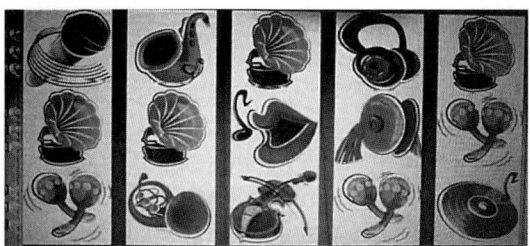

No sounds signal appearance
of gramophone symbols

Stereo near-miss

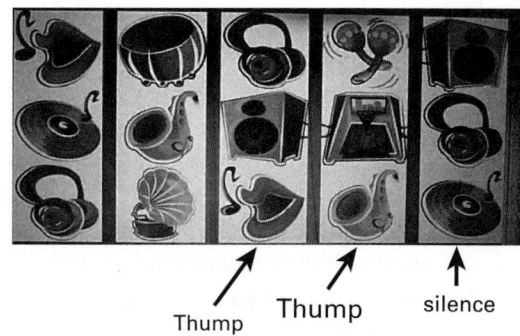

Thump   Thump   silence

Figure 26.8
For gramophone wins no signaling sounds were present. On stereo wins the sounds rose in pitch, concluding with a final closing cadence upon appearance of the third completing symbol. On stereo near misses this closing cadence was noticeably absent.

games (Dixon, Harrigan, et al., 2014; Dixon et al., 2015) we proposed that adding signaling sounds to our simulator could heighten the salience of near misses for the stereo horizontal triplet. If so, we should then see the characteristic near miss arousal effect on SCRs that was so conspicuously absent in the previous experiment.

In addition to the sounds, two other minor changes were made to the simulator. In the previous experiment, we enforced a 4-second delay between outcome delivery and the initiation of the next spin (allowing phasic changes in heart rate to be measured). Experienced players dislike this artificial slowing of game play (Loba et al., 2001). Here we allowed players to play more naturally, spinning as quickly as they wished. Removing this enforced delay allowed us to measure the PRP—the delay between outcome delivery and the initiation of the next spin. Finally we introduced a force transducer into the button that participants used to initiate spins, so we could measure how hard they pressed to initiate each spin.

We predicted that if adding signaling sounds to the horizontal triplet heightens sensitivity to horizontal triplet near misses, then stereo near misses should trigger a larger SCR than either regular losses or gramophone near misses (which had no accompanying sounds). If stereo near misses are particularly frustrating, the force players apply to the button to initiate the next spin should be greater following stereo near misses than regular losses and gramophone near misses. Finally if stereo near misses are particularly frustrating losses, then following these outcomes, players should initiate the next spin quite quickly to get out of this frustrated state. In short, if the addition of these signaling sounds draws attention to the components of the horizontal triplet, and if these cross-modal inputs make near misses more salient, then we should see the typical near miss arousal signatures (in SCRs) that have previously been shown in simple three-reel games.

A total of 58 gamblers (30 males) were recruited from a different local slots venue. All procedures were identical to the previous casino experiment except for the force transducer and the addition of signaling sounds accompanying the stereo component symbols. The force transducer was inserted into a custom-made button casing, similar to a mouse, that translated how hard participants pressed the button into a millivolt signal. The addition of signaling sounds (demonstrated in figure 26.8) used frequency intervals between the sounds that were in a root, fifth, octave pattern, from the third to the fifth reel sound, respectively.

Subjective Results

As in the previous experiment, subjective questions were posed to this group of players. Players were shown depictions of the gramophone, the LP record, and the left speaker of the stereo and asked to rate how attention grabbing these symbols were on a 7-point Likert scale, where higher numbers meant more attention grabbing. A total of 56 players had complete data for all ratings. These responses were analyzed using a repeated-measures ANOVA. This analysis revealed a main effect of symbol, $F(2, 110) = 22.884$, $p < .001$. Bonferroni-corrected post hoc analyses revealed that all three outcomes were significantly different from each other. The gramophone was more attention-grabbing than the LP ($p < .003$). Crucially, the stereo speaker received higher attention-grabbing ratings (mean = 5.11) than either the gramophone (mean = 4.21, $p = .002$) or the LP (mean = 3.31, $p < .001$).

After the session, players were also asked how much various symbols were worth when they were aligned in a winning outcome. Table 26.3 shows the number of participants who gave correct, incorrect, or "don't know" estimates for each of the symbols.

Table 26.3

Number of Participants giving Correct, Incorrect, and Unsure Estimates of Credit Values for Three
Slot Machine Symbols

	LP	Gramophone	Stereo
Correct number of credits	12	35	46
Incorrect number of credits	24	12	2
"I don't know"	22	11	10

Note: LP, depiction of a record album.

Chi-square analyses revealed that the value of the gramophone win was better recalled than the value of the LP record win, $\chi^2(2) = 26.92$, $p < .01$. Crucially, the value of a stereo win was better recalled than the value of a gramophone win, $\chi^2(2) = 7.36$, $p < .05$.

As in the previous experiment, following play, gamblers were asked to indicate their reactions to the various outcomes. In this sample there were very few frustration ratings (only 9 of 58 players indicated that stereo near misses were mildly or very frustrating). There were no differences in the propensity to rate stereo near misses as any more frustrating than losses or gramophone near misses.

To assess excitement ratings, we once again eliminated those few participants who gave frustration ratings, and we translated the "very bored," "mildly bored," "mildly excited," "very excited" ratings to an excitement scale that ranged from 0 to 3. A total of 42 participants had complete excitement ratings for all five outcomes. A repeated-measures ANOVA yielded a significant main effect of outcome, $F(4, 164) = 24.276$, $p < .001$. Bonferroni-corrected post hoc pairwise comparisons indicated that gramophone wins were rated as more exciting than losses ($p < .003$) and gramophone near misses ($p < .007$). Crucially, stereo wins were rated as significantly more exciting than any other type of outcome, including gramophone wins (all p values $< .001$). These excitement ratings are shown in figure 26.9.

As in the previous experiment, the subjective reports of the participants indicated that stereo pieces were more attention grabbing, the worth of stereo wins was better remembered, and stereo wins were rated as being more exciting than all other outcomes. What remained to be seen was whether the addition of signaling sounds to the stereo pieces would allow us to see the characteristic psychophysical near miss responses during slots play.

Based on the last experiment, we made several a priori predictions: that stereo near misses would trigger larger SCRs than gramophone near misses (or regular losses) and that stereo wins would trigger greater SCRs than gramophone wins. Stereo near misses would also trigger smaller PRPs than gramophone near misses and regular losses. Stereo wins would trigger larger PRPs than gramophone wins. Finally, stereo

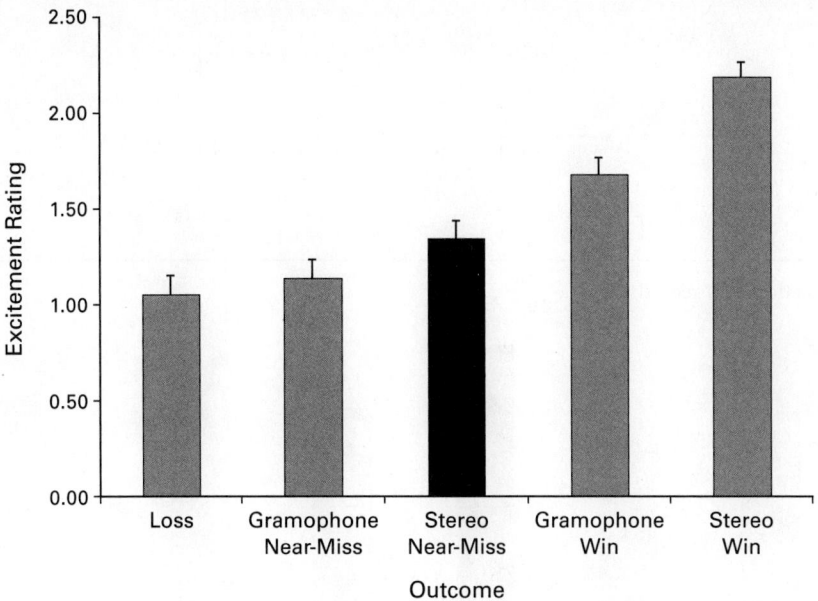

Figure 26.9
Average excitement ratings for losses, gramophone near misses, stereo near misses, gramophone wins, and stereo wins. Error bars reflect standard errors of the mean.

near misses would trigger greater force to initiate the next spin than either gramophone near misses or regular losses (since the force predictions centered around frustration, no predictions were made concerning wins). These planned comparisons justified one-tailed tests.

As can be seen in figure 26.10, stereo wins triggered larger SCRs than gramophone wins (replicating the results of our previous experiment). Unlike the previous experiment, figure 26.10 indicates a clear near miss effect for the stereo pieces. These impressions were verified by a repeated-measures ANOVA on the SCRs for each of the five outcomes. This analysis revealed a main effect of outcome, $F(4, 228) = 14.10$, $p < .001$. For the planned comparisons (denoted by the asterisks in figure 26.10), stereo wins triggered significantly greater SCRs than gramophone wins ($p < .001$). Stereo near misses triggered significantly greater SCRs than gramophone near misses ($p < .001$, one tailed) and were significantly greater than regular losses ($p = .002$, one tailed). Unexpectedly the gramophone wins did not differ from the regular losses ($p = .90$).

Force Measurements
Two participants were removed because they applied excessive force to the spin button on every single trial (beyond the range of the force transducer). For the remaining

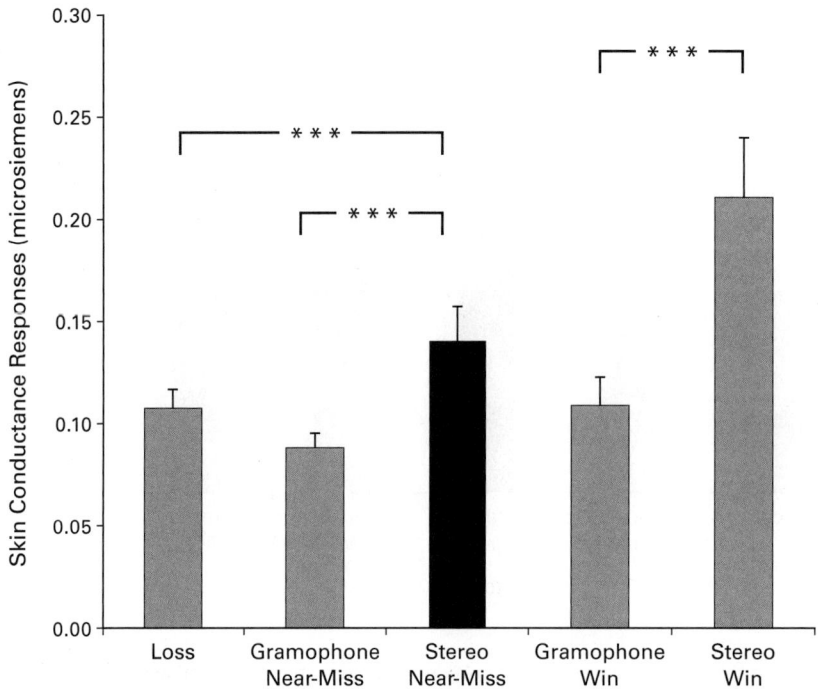

Figure 26.10
Average skin conductance response amplitudes for the key outcomes. Planned contrasts denoted by the asterisks were significant at $p < .002$. Error bars reflect standard errors of the mean.

56 participants, a repeated-measures ANOVA revealed a main effect of outcome, $F(4, 220) = 3.231$, $p < .05$. Bonferroni-corrected post hoc tests revealed that gramophone wins triggered significantly more force than regular losses ($p = .05$). In terms of the planned contrasts, although stereo near misses were associated with somewhat greater force than gramophone near misses, this planned contrast fell short of significance ($p = .063$, one tailed). Crucially, though, stereo near misses were associated with greater force than regular losses ($p = .005$, one tailed). The force data are shown in figure 26.11.

PRP results are shown in figure 26.12. They indicate long pauses for the two actual wins but relatively short pauses for all types of losses. A repeated-measures ANOVA verified these impressions. There was a large main effect of outcome, $F(4, 228) = 312.39$, $p < .001$. PRP's following gramophone wins were significantly longer than any type of loss (all p values < 0.01). Planned comparisons indicated that stereo wins had longer PRPs than gramophone wins ($p < .001$). The stereo near miss PRPs were equivalent to losses and gramophone PRPs.

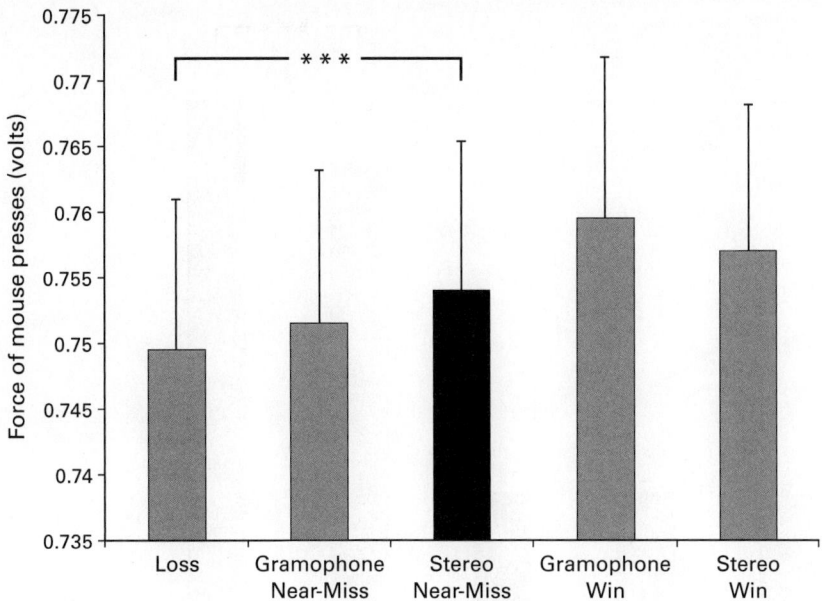

Figure 26.11

Average force applied to mouse to initiate subsequent spin following the five key outcomes. The planned contrast denoted by the asterisks was significant at $p < .01$. Error bars reflect standard errors of the mean.

Integration

Our predictions were that the addition of signaling sounds to a multiline horizontal triplet would render near misses on these outcomes more salient than in the experiment without signaling sounds. The SCR measures during play suggest that players now reacted to near misses involving this horizontal triplet with the classic SCR arousal response. Support for the contention that these SCR responses were due to frustration comes from the combination of the force and PRPs. In terms of force, players hit the spin button significantly harder following stereo near misses than following regular losses. The link between force and frustration has been shown in classic work on children. In a study by Haner and Brown (1955), children were asked to drop marbles into a rack to win a prize. As they neared completion, the experimenter released all of the marbles and a buzzer played that children could stop by pressing a plunger that measured force. The closer they were to completion, the more force the children applied to the plunger.

Although not predicted, wins also triggered large force—a finding that can be interpreted in terms of excitement and arousal—the thrill of winning a large amount may

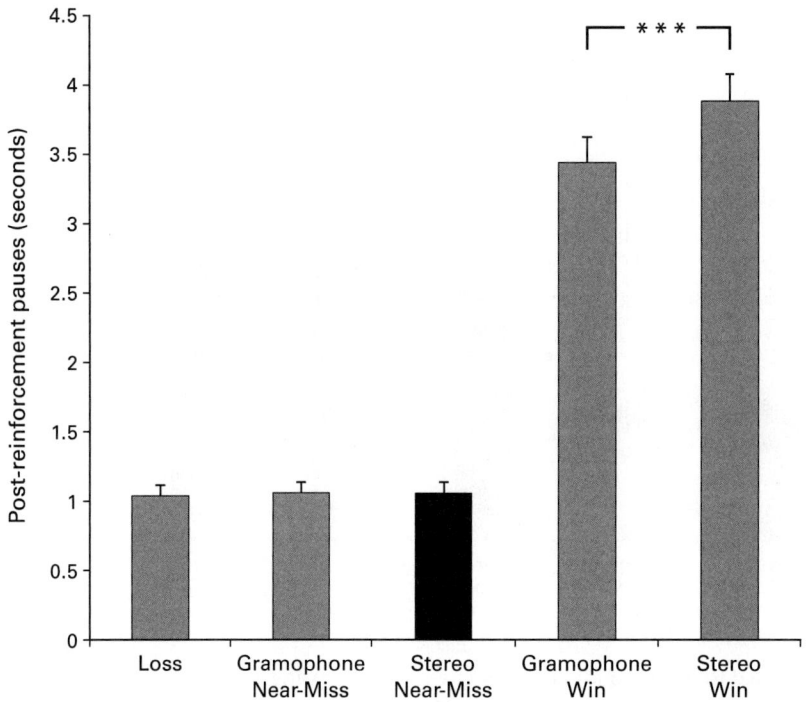

Figure 26.12
Average postreinforcement pauses for the key outcomes. Error bars reflect standard errors of the mean.

have increased arousal, lengthened PRPs, and increased force due to excitement. Thus, force increases, like SCR increases, can arise from either rewarding events (wins) or frustrating events. The PRP results are therefore crucial in constraining our interpretation of the near miss data. Some researchers suggest that near misses trigger arousal responses because they are misinterpreted as a kind of win. For example, Peters, Hunt, and Harper (2010) suggest that near misses may acquire win-like sequalae via Pavlovian generalization. That is, since they share many visual features of a win, players react to these outcomes as though they were wins. Indeed, both the SCR data and the force data could be interpreted using this kind of cognitive miscategorization view. The PRP data, however, suggests that wins and near misses are perceived very differently. The near miss PRPs are very much smaller than the win PRPs, and equivalent in size to the loss PRPs. When the three measures (SCRs, force, and PRPs) are considered together, the data support the interpretation that near misses are especially frustrating losses.

Unexpectedly, however, after game play, only a small subset of players explicitly rated the near misses as frustrating. One possibility is that there are differences between the retrospective evaluations of near misses using static depictions of the outcomes (a screenshot of a near miss), compared to the frustration that accrues "online" during actual game play. Alternatively, it may be that some players view these outcomes as disappointing rather than actively frustrating (Bossuyt, Moors, & De Houwer, 2014). Although the exact nature of players responses to near misses remains a matter of debate, for the purposes of this chapter, it is clear that adding cross-modal inputs to near miss stimuli in the context of a multiline game triggers the characteristic SCR arousal responses seen in simpler three-reel games. In sum, the two casino experiments suggest that while the visual features of the horizontal triplets served to draw attention to wins involving the stereo, these visual features alone were not enough to draw attention to triplet near misses. Once signaling sounds were added to the horizontal triplets, strong near miss effects could now be seen in our measurements of SCR and force.

These findings highlight a real-world instance of how cross-modal pairings of visual stimuli and auditory stimuli can dramatically impact observers. Although they lack the laboratory precision of showing, for example, "superadditivity" of sensitivity to the presented stimuli (for a review, see chapter 6 in the current volume), the categorical changes in reactions to near misses in this highly complex situation stands as an example of multisensory integration in a real-world scenario. The cadential progression of sounds used, and their multisensory alignment with the three-part object in the slots game, is an example of the complexity of the types of sounds used in video/slots games. As noted by Spence and Ho in their chapter, such complex meaningful sounds stand in stark contrast to the single-tone stimuli employed in most laboratory assays of cross-modal attention. Our findings also present another instance of just how important sound is in influencing the players of slot machines (Dixon, Harrigan, et al., 2014; Dixon et al., 2015).

This research stands as a proof of concept that near misses can be presented in multiline slots games, and that when signaling sounds accompany horizontal triplets, players significantly react to these outcomes. This is an important finding since near misses have been shown to increase the time players spend on a given slot machine. Since, over the long haul, more time on the machine translates to greater losses, such a finding could have implications for problem gambling.

Future Directions

In this chapter we showed how cross-modal inputs can be used to draw attention to highly specific aspects of slots play. Other, more global, aspects of attention have proven to provide important insights into gambling behavior. A significant number of people with gambling problems gamble to escape the cares of everyday life (Blaszczynski & Nower, 2002). Thus, slots gambling can be seen as a form of distraction. Some players talk of entering "the zone" where they forget about everything else but the game they are playing (Dow Schull, 2012). Researchers have linked escape gambling and dissociation (Ledgerwood & Petry, 2006) as being integral parts of players' experiences. Jacobs (1988; Kuley & Jacobs, 1988) has shown that those with gambling problems score higher on scales that tap feelings of dissociation, including amnestic and fugue-like states while gambling. Griffiths, Parke, Wood, and Parke (2006) conclude that problem gamblers either seek out games that will induce a dissociative experience or are more likely to experience dissociation while playing. Either way, slot machines, where a single player interacts with a single machine, may be a form of gambling that is preferentially linked to dissociative states.

It is possible that certain types of slots games may be particularly appealing to those who seek dissociative states. In single-line games, players typically will encounter long chains of losses interrupted by occasional wins. Such long losing streaks interleaved with periodic wins can lead to a "jagged" play experience. In modern multiline games where players can bet on multiple lines, losing streaks are far shorter. Indeed, in some games players will encounter reinforcing stimuli almost every other spin (Harrigan, Dixon, MacLaren, Collins, & Fugelsang, 2011). Most of these reinforced outcomes, however, will be LDWs (bet 20, win back 5, net loss = 15). Despite the fact that these outcomes are monetary losses, the machine celebrates these outcomes with eye-catching animations and winning jingles. Such outcomes enable the machine to dish out potentially rewarding sights and sounds, at high frequencies, but at no cost to the machine owner. This high celebratory event frequency can lead to a smoother playing experience that might possibly increase the propensity of players to enter in, and stay in, the zone. Indeed in a recent study (Dixon, Graydon et al., 2014) we have shown that problem gamblers are more likely to forget everything around them and become absorbed by slots play, but only when they are playing multiple lines.

The experiments provided above show a proof of concept regarding how it is possible for slots designers to incorporate the power of the near miss in these complex multiline games. Future research could explore the complex interactions between game features that serve to keep players in the zone (high celebratory event frequency) with those that foster the illusion that players may be "close to winning" big prizes (i.e., near misses). The current results also show how looking at slot machine gambling provides a fruitful means of investigating the role of attention in a compelling real-world setting.

Box 26.1
Key Points

- Manipulating players' attention can have profound affects on slot machine play.
- Game designers have the potential to bias players to preferentially look for certain symbols using descriptions in pay tables and render certain symbols more salient by using a combination of low-level visual properties and cross-modal acoustic signaling.

Box 26.2

Outstanding Issues

- Future research should investigate how certain game features (win frequency) can manipulate more global levels of attention (absorption, forgetting everything around oneself but the game).

References

Aarts, H., Ruys, K. I., Veling, H., Renes, R. A., de Groot, J. H. B., van Nunen, A. M., & Geertjes, S. (2010). The art of anger: Reward context turns avoidance responses to anger-related objects into approach. *Psychological Science*, *21*, 1406–1410.

Bergen, J. R., & Julesz, B. (1983). Parallel versus serial processing in rapid pattern discrimination. *Nature*, *303*, 696–698.

Biederman, I. (2001). Recognition-by-components: A theory of human image understanding. In S. Yantis (Ed.), *Visual perception: Essential readings* (pp. 320–340). Philadelphia: Psychology Press.

Biederman, I., & Ju, G. (1988). Surface versus edge-based determinants of visual recognition. *Cognitive Psychology*, *20*, 38–64.

Blaszczynski, A., & Nower, L. (2002). A pathways model of problem and pathological gambling. *Addiction*, *97*, 487–499.

Bossuyt, E., Moors, A., & De Houwer, J. (2014). Unexpected and just missed: The separate influence of the appraisals of expectancy and proximity on negative emotions. *Emotion*, *14*, 284–300.

Clark, L., Crooks, B., Clarke, R., Aitken, M. R. F., & Dunn, B. D. (2011). Physiological responses to near miss outcomes and personal control during simulated gambling. *Journal of Gambling Studies*, *28*, 123–137.

Clark, L., Lawrence, A. J., Astley-Jones, F., & Gray, N. (2009). Gambling near-misses enhance motivation to gamble and recruit win-related brain circuitry. *Neuron*, *61*, 481–490.

Cote, D., Caron, A., Aubert, J., & Ladouceur, R. (2003). Near wins prolong gambling on a video lottery terminal. *Journal of Gambling Studies*, *19*, 380–407.

Dawson, M. E., Schell, A. M., & Filion, D. L. (2000). The electrodermal system. In J. T. Cacioppo, L. G., Tassinary, & G. G. Bernston (Eds.), *Handbook of psychophysiology* (2nd ed., pp. 200–223). New York: Cambridge University Press.

Dixon, M. J., Harrigan, K. A., Santesso, D. L., Graydon, C., Fugelsang, J. A., & Collins, K. (2014). The impact of sound in modern multiline video slot machine play. *Journal of Gambling Studies*. doi:10.1007/s10899-013-9391-8.

Dixon, M. J., Harrigan, K. A., Jarick, M., MacLaren, V. V., Fugelsang, J. A., & Sheepy, E. (2011). Psychophysical arousal signatures of near-misses in slot machine play. *International Gambling Studies*, *11*, 393–407.

Dixon, M. J., Harrigan, K. A., Sandhu, R., Collins, K., & Fugelsang, J. A. (2010). Losses disguised as wins in multiline video slot machines. *Addiction*, *105*, 1819–1824.

Dixon, M. J., MacLaren, V., Jarick, M., Fugelsang, J. A., & Harrigan, K. A. (2013. The frustrating effects of just missing the jackpot: Slot machine near-misses trigger large skin conductance responses, but no post-reinforcement pauses. *Journal of Gambling Studies*, *29*, 661–674.

Dow Schull, N. (2012). *Addiction by design: Machine gambling in Las Vegas*. Princeton, NJ: Princeton University Press.

Duncan, J., & Humphreys, G. W. (1989). Visual search and stimulus similarity. *Psychological Review*, *96*, 433–458.

Griffiths, M. D. (1990). The cognitive psychology of gambling. *Journal of Gambling Studies*, *6*, 31–42.

Griffiths, M. (2003). Internet gambling: Issues, concerns, and recommendations. *Cyberpsychology & Behavior*, *6*, 557–568. doi:10.1089/109493103322725333.

Griffiths, M., Parke, A., Wood, R., & Parke, J. (2006). Internet gambling: An overview of psychological impacts. *UNLV Gaming Research & Review Journal*, *10*, 27–39.

Haner, C. F., & Brown, P. A. (1955). Clarification of the instigation to action concept in the frustration–aggression hypothesis. *Journal of Abnormal and Social Psychology*, *51*, 204–206.

Harrigan, K., Dixon, M., MacLaren, V., Collins, K., & Fugelsang, J. (2011). The maximum rewards at the minimum price: Reinforcement rates and payback percentages in multi-line slot machines. *Journal of Gambling Issues*, *26*, 11–29.

Inouye, P. (2013). Beyond cha-ching! Music for slot machines. Presentation at The Game Developers' Conference, San Francisco, CA. March 25–29, 2013. [Audio Recording] Available at http://www.gdcvault.com/search.php#&category=free&firstfocus=&keyword=peter+inouye&conference_id=/.

Jacobs, D. F. (1988). Evidence for common dissociative-like reaction among addicts. *Journal of Gambling Behavior*, *4*(1), 27–37.

Jensen, C., Dixon, M., Harrigan, K., Sheepy, E., Fugelsang, J., & Jarick, M. (2013). Misinterpreting "winning" in multiline slot machine games. *International Gambling Studies*, *13*, 112–126. doi:10.1080/14459795.2012.717635.

Kassinove, H. I., & Schare, M. L. (2001). Effects of the near-miss and the big win on persistence at slot machine gambling. *Psychology of Addictive Behaviors*, *15*, 155–158.

Kellman, P. J., & Shipley, T. F. (1991). A theory of visual interpolation in object perception. *Cognitive Psychology*, *23*, 141–221.

Kuley, N. B., & Jacobs, D. F. (1988). The relationship between dissociative-like experiences and sensation seeking among social and problem gamblers. *Journal of Gambling Behavior, 4*(3), 197–207.

Ledgerwood, D. M., & Petry, N. M. (2006). Psychological experience of gambling and subtypes of pathological gamblers. *Psychiatry Research, 19*, 411–416.

Loba, P., Stewart, S. H., Klein, R. M., & Blackburn, J. R. (2001). Manipulations of the features of standard video lottery terminal (VLT) games: Effects in pathological and non-pathological gamblers. *Journal of Gambling Studies, 17*, 297–320.

Lobbestael, J., Arntz, A., & Wiers, R. W. (2008). How to push someone's buttons. *Cognition and Emotion, 22*, 353–373.

Luo, Q., Wang, Y., & Qu, C. (2011). The near-miss effect in slot-machine gambling: Modulation of feedback-related negativity by subjective value. *Neuroreport, 22*, 989–993. doi:10.1097/WNR.0b013e32834da8ae.

Menneer, T., Cave, K. R., & Donnelly, N. (2009). The cost of search for multiple targets: Effects of practice and target similarity. *Journal of Experimental Psychology. Applied, 15*, 125–139.

Neath, I. (1993). Contextual and distinctive processes and the serial position function. *Journal of Memory and Language, 32*, 820–840.

Neisser, U. (1963). Decision-time without reaction time: Experiments in visual scanning. *American Journal of Psychology, 76*, 376–385.

Parke, J., & Griffiths, M. (2004). Gambling addiction and the evolution of the "near-miss." *Addiction Research and Theory, 12*, 407–411.

Peters, H., Hunt, M., & Harper, D. (2010). An animal model of slot machine gambling: The effect of structural characteristics on response latency and persistence. *Journal of Gambling Studies, 26*, 521–531.

Reid, R. L. (1986). The psychology of the near miss. *Journal of Gambling Studies, 2*, 32–39.

Treisman, A., & Gelade, G. (1980). A feature-intergration theory of attention. *Cognitive Psychology, 12*, 97–136.

Van Selst, M., & Jolicoeur, P. (1994). A solution to the effect of sample size on outlier elimination. *Quarterly Journal of Experimental Psychology, A, 47*, 631–650.

Wagemans, J., Elder, J. H., Kubovy, M., Palmer, S. E., Peterson, M. A., Singh, M., von der Heydt, R. (2012). A century of Gestalt psychology in visual perception: I. Perceptual grouping and figure–ground organization. *Psychological Bulletin, 138*, 1172–1217. doi: 10.1037/a0029333.

Wolfe, J. M. (1994). Guided search 2.0: A revised model of visual search. *Psychonomic Bulletin & Review, 1*, 202–238.

Wolfe, J. M., & Horowitz, T. S. (2004). What attributes guide the deployment of visual attention and how do they do it? *Nature Reviews Neuroscience, 5*, 495–501.

27

Attention in Educational Contexts: The Role of the Learning Task in Guiding Attention

Andrew M. Olney, Evan F. Risko, Sidney K. D'Mello, and Arthur C. Graesser

Historical Context

Attention is crucial for effective learning. Unfortunately, students do not always pay attention. Over the past century, educators and researchers have tried to maximize the time students pay attention, in various ways, in order to optimize learning (see Berliner, 1990, for a review). These efforts have largely focused on the construct *level of attention*. The level of a student's attention, sometimes referred to as engagement,[1] may be described as absent, passive, partial, or active (Currie, 1861). Efforts to measure students' level of attention in educational contexts has ranged from various types of probed self-report (e.g., mind wandering probes; Risko, Anderson, Sarwal, Engelhardt, & Kingstone, 2012), amount of note-taking (Scerbo et al., 1992), fidgeting (Farley, Risko, & Kingstone, 2013), heart rate (Bligh, 2000), and more generally performance. Because a student's level of attention is not directly observable, direct observation can be misleading. For example, a student may have his or her gaze directed toward the teacher but in fact might be mind wandering (Risko, Anderson, Sarwal, Engelhardt, & Kingstone, 2012).

Much work has demonstrated that level of attention is predictive of student achievement in authentic classroom environments (see Denham & Lieberman, 1980, for a review), and recent studies have confirmed these effects with increasing methodological precision. For example, inattention during reading and lectures in the form of mind wandering has been shown to be negatively associated with memory for the source material (Lindquist & McLean, 2011; Risko et al., 2012; Smallwood et al., 2008; Szpunar, Khan, & Schacter, 2013). One emerging finding in these more recent studies is that changing the task structure can improve a student's level of attention, leading to increased learning (Szpunar et al., 2013). These results suggest that a deeper look at how task type might impact student attention is warranted. We attempt to provide such a perspective in the present chapter by unifying work on student attention with work on optimal learning activities (Chi, 2009; Menekse, Stump, Krause, & Chi, 2013).

Theories of optimal learning activities typically emphasize the role of task type and deemphasize the role of attention (Chi, 2009; Menekse, Stump, Krause, & Chi, 2013). For example, the Interactive–Constructive–Active–Passive (ICAP) hypothesis predicts that task type (as defined by overt behaviors) will largely determine learning outcomes and rank orders the effectiveness of these activities as $I \geq C \geq A \geq P$ (Chi, 2009). Examples of these activities include dialogue as part of a constructive activity (interactive), summarizing by adding new ideas or reorganizing old ideas (constructive), taking notes without adding new ideas or organization (active), and viewing a lecture or video with no other overt behavior (passive). In a reanalysis of 15 studies, Chi (2009) consistently found the $I \geq C \geq A \geq P$ pattern, and follow-up studies in both controlled laboratory settings and classroom conditions provide additional evidence for ICAP (Menekse et al., 2013). Here we explore the idea that task structure can improve students' attention. Thus, the effect of task type on learning may be partially mediated by the influence of task type on attention.

The following sections review and reanalyze studies of attention in educational settings with respect to the task types in ICAP. This analysis is then integrated with theoretical accounts of mind wandering, sequential action, and monitoring. The resulting elaborated theory of mind wandering explains how a learning task structure can enhance a student's level of attention within the learning task. Furthermore, it motivates a new hypothesis we refer to as Interactive–Constructive–Active–Passive–Attention (ICAP-A), which predicts that attention level will follow the $I \geq C \geq A \geq P$ pattern. We also explore how ICAP-A could be used to suggest ways of changing task structure to optimize learning performance via improved attention.

State-of-the-Art Review

A growing body of work has investigated attention during lectures (Bunce et al., 2010; Lindquist & McLean, 2011; Risko et al., 2012; Szpunar et al., 2013; Young et al., 2009) and, critically for the present proposal, across different types of activities within lectures. These various studies provide an opportunity to investigate the relationship between the task structure, as defined by an ICAP learning activity, and student attention. In a recent classroom study of mind wandering, Lindquist and McLean (2011) found evidence consistent with the idea that students who engaged in a more passive or minimally active task during a lecture tend to mind wander more than students engaged in a more active or constructive task. At various times within a live, in-person lecture, students received a mind-wandering probe. In addition, various measures of classroom activities and performance in the class were collected. Critically for the present purposes, the study found a significant negative association between mind wandering and note-taking and between mind wandering and exam grades. In other words, students who took more notes during the lecture mind wandered less than individuals who did not take notes. Within ICAP, the act of taking notes would have changed the task type from passive/active to active/constructive, depending on whether the notes were verbatim or elaborative in nature. Thus, the effect of note-taking can be interpreted as a shift in task type and the associated reduction in mind wandering evidence for the posited relation between the task type and student attention. While consistent with this hypothesis, it is important to note that the Lindquist and McLean (2011) research was correlational. Stronger evidence for a link between extent of note-taking and mind wandering would require an experimental design.

In a classroom study, Young et al. (2009) investigated student attention across four different lecture types/class activities using a measure of the vigilance decrement and found evidence consistent with the idea that more interactive activities could improve student attention. The vigilance decrement refers to performance costs in passive monitoring tasks (Mackworth, 1948) and is typically attributed to limits on human ability to attend for extended periods of time in those tasks. Risko et al. (2012) demonstrated the existence of a vigilance decrement in lectures using mind wandering as the measure. Young et al. (2009) used a measure of subjective workload (i.e., NASA-TLX) which previous research had demonstrated yields a "signature" pattern (i.e., relatively high contribution of mental demand and frustration to overall perceived workload) associated with the vigilance decrement. Students, in an actual class, completed the NASA-TLX during four different types of lecture: (1) standard lecture, (2) guest lecture, (3) lecture + small group discussion, and (4) lecture + multimedia case studies. Critically, the signature of the vigilance decrement was (statistically) present only in the standard lecture. The absence of vigilance decrement in the guest lecture may be due to a novelty effect. In addition, while no formal statistical comparison was provided, Young et al.'s (2009) table 1 suggests that the NASA-TLX pattern that least resembles the signature pattern occurred in the lecture + small group discussion condition—a condition that might be considered interactive in ICAP. Thus, Young et al.'s (2009) data support the hypothesized relation between task type and student attention.

Evidence that constructive activities included in lectures increase student attention was reported by Bunce et al. (2010) in a study investigating self-reported attention lapses across three chemistry classes over 6 weeks. Critically, Bunce et al. (2010) included self-reported measures of attention lapses both during the constructive activities and during the nonconstructive parts of lectures that were preceded by constructive activities, thus allowing an assessment of potential "carryover" of the attentional benefits of constructive activities. Individual response devices ("clickers") were used to indicate when students' attention had lapsed, and the frequency of lapses was assessed during periods of standard lecturing and two other activities—questions answered by clicker (constructive) and demonstrations (passive/active). Bunce et al. (2010) found a reduced number of attention lapses during both clicker question periods and demonstrations relative to standard lecture periods. Thus, again, the constructive task was associated with greater attention than a more passive task. Interestingly, Bunce et al. (2010) also found evidence that attention, during the standard lectures, increased following constructive tasks, suggesting some carryover.

Lastly, a recent laboratory study by Szpunar et al. (2013) provides further evidence for a link between task type and student attention. Szpunar et al. (2013) investigated lecture viewing with note-taking across three conditions of varying lecture-related activities. Participants watched a video lecture on statistics in four segments of approximately 5.5 minutes each. After viewing a lecture segment, all participants solved unrelated math problems for approximately 1 minute. In addition, in a test condition participants received test questions on the material just covered whereas in a study condition participants received the same test questions accompanied by answers, and participants were told to study and not solve them. Participants in a no-test condition only did unrelated math problems (for the same amount of time as the other individuals completed their tasks) after viewing each lecture segment. In response to mind wandering probes during the lecture, participants in the test condition reported the lowest level of mind wandering (19%), followed by the study (39%) and the no-test (41%) conditions. In addition, those in the test condition also scored higher on the final test (89%) than either those in the study condition (65%) or those in the no-test condition (70%). One interpretation of these results (similar to that offered in Bunce et al., 2010) is that interpolated testing shifted the participants' (spontaneously adopted) mode of interacting with the lecture (i.e., during the lecture segments). Consistent with this idea, and the Lindquist and McLean (2011) work above, Szpunar et al. (2013) reported that participants in the test condition took more notes (24%), suggesting more active processing, than participants in the study (9%) and no-test (7%) conditions. Thus, not only did test participants report half the level of mind wandering but they also took approximately three times as many notes as participants in the other conditions.

While lecture is certainly a common instructional format, most students spend an equal or greater amount of time reading instructional materials. Within ICAP, simple reading would likely be classified as active and certainly, on the whole, more active than a standard lecture. Accordingly, student attention should be greater during reading than listening to a lecture. A brief survey of mind wandering rates across studies involving lectures and reading seems to support this notion (see Varao Sousa, Carriere, & Smilek, 2013, p. 2). Varao Sousa et al. (2013) also provided a direct test of this general idea by comparing mind wandering rates across three conditions—reading a text aloud, reading silently, and listening to the text being read. In each condition participants responded to mind wandering probes during the task. Critically, mind wandering was greatest in the listening condition, followed by silent reading; the least amount of mind wandering was reported in the reading-aloud condition. Thus, Varao Sousa et al. (2013) seems to follow the ICAP pattern closely. Specifically, as the task shifts from passive to more active (i.e., listening—silently reading—reading aloud), student inattention decreases. Critically, these results were mirrored by differences in performance across these conditions. In other words, participants' memory for the material was greater in the two reading conditions than in the listening condition.

A study by Moss et al. (2013) provides evidence that constructive activities improve student attention relative to active activities during reading. The study investigated three reading strategies: rereading (active), paraphrasing (active), and self-explanation (constructive). After they were trained to perform these strategies, participants read texts aloud with instructions to use a particular strategy and rated their degree of mind wandering after each paragraph. Mind wandering ratings were lower for self-explanation than for paraphrasing or rereading, following the ICAP pattern. Moreover, the self-explanation strategy led to greater learning gains than the other conditions. These results complement and extend those of Varao Sousa et al. (2013) by showing how shifting the task type to constructive can further reduce mind wandering and increase learning during reading.

Educators have long suggested that the nature of the standard lecture encourages a passive form of activity relative to other pedagogical techniques (e.g., discussion, problem solving; Young et al., 2009). Consistent with this idea, in a study of a standard class lecture, Cameron and Giuntoli (1972) found that students reported not listening or superficially listening (passive) 38% of the time, closely following (active) 37% of the time, actively meeting the speaker's mind or engaging in episodic recall in response to what the speaker is saying (constructive) 21% of the time, and wanting to speak (interactive) 5% of the time. Thus, a broad base of students were engaged in what would be considered passive and active activities, and increasingly fewer were engaged in constructive or interactive activities. Presumably during any learning task, a group of students will naturally manifest a distribution of attention where the shape and center of the distribution depends on task type. Indeed a single student may shift in task type over a learning session and adopt a more or less demanding mode of processing. In the studies reviewed above, the overall pattern of student attention shifted as a function of ICAP task type (see table 27.1). It is important to note that the pairwise comparisons

Table 27.1

Pairwise Comparisons of Attention according to Interactive–Constructive–Active–Passive Task Type

	Active	Constructive	Interactive
Passive	Lindquist & McLean (2011); Varao Sousa et al. (2013)	Bunce et al. (2010); Szpunar et al. (2013)	Young et al. (2009)
Active	—	Moss et al. (2013)	—
Constructive	—	—	—

presented here should be taken with a grain of salt (e.g., a few rely on different measures of attention, and all use different types of content). For the most part, research comparing student attention across all the different ICAP task types is currently unavailable. We hope the theoretical framework offered here will inspire such efforts.

Integration

The previous section reviewed and reanalyzed studies of attention in educational settings with respect to the task types in the ICAP hypothesis. The trend across these studies suggests an ICAP ordering of attention relative to task type. Specifically, student attention may be greatest in interactive tasks followed by constructive, active, and lastly passive tasks. The current section offers a theoretical account of why an ICAP ordering of attention might be expected given a popular theoretical account of mind wandering (i.e., Control Failures × Concerns model; McVay & Kane, 2010; Kane & McVay, 2012), elaborated with novel conceptualizations of proactive and reactive control. The resulting elaborated theory licenses the hypothesis that learning task structure, such as ICAP, can enhance student attention within the learning task. We refer to this hypothesis as ICAP-A. Thus, the goal of the current section is to situate the reviewed work in a theoretical framework that can inform our basic understanding of task by attention interactions and provide guidance on how this knowledge could be translated into learning gains in the classroom.

A Model of Mind Wandering

Mind wandering arguably represents the quintessential representation of attentional disengagement in educational contexts and, as such, provides a useful basis for an account of attention relative to task type in that context. A prominent theory of mind wandering presents it as an interaction between executive failure of control and current concerns, sometimes called the Control Failures × Concerns model (McVay & Kane, 2010; Kane & McVay, 2012). This theory integrates two theoretical frameworks (Watkins, 2008; Klinger, 2009) to argue that mind wandering results from automatically generated thoughts (current concerns; see Klinger, 2009) that the executive control system fails to suppress by not maintaining the appropriate level of construal (elaborated control theory; see Watkins, 2008). Klinger defines current concerns as the set of established but unattained goals. Because neuroimaging studies have also tied mind wandering to the activity of the brain at "rest" via the brain's so-called default network (Mason et al., 2007), there is reason to believe that current concerns reflect a kind of ever-present "background noise" in the brain. Thus, mind wandering about "what's for dinner" (i.e., a current concern) in, for example, a one-on-one tutoring session, would represent a failure to suppress this "noise" and sustain attention on the educational goal.

To explain the relation between current concerns and executive control, the Control Failures × Concerns model relies on the elaborated control theory of Watkins (2008). The elaborated control theory is based upon the idea of construal. Watkins (2008) defines construal in terms of action identification theory (Vallacher & Wegner, 1987), which explains the connection between cognition and action as a cyclical process: Intentions generate actions, and actions are interpreted to infer intentions. For example, one may decide that it would be nice to ride a bike around the neighborhood (intention generating action). On encountering a hill, one's progress becomes effortful, and this may be interpreted as exercise (action interpreted to infer intention). In both cases the concrete action is the same, that of riding a bicycle. However this concrete action is consistent with the two more abstract goals of riding for pleasure and riding for exercise. Critically, the new interpretation of getting exercise can trigger a shift in the higher level goal away from riding for pleasure, even though the concrete behavior is unchanged.

Elaborated control theory applies this notion of cyclical feedback to repetitive thought and further specifies that construal may be either abstract or concrete. Abstract construal ("Why am I riding this bike?") promotes trait-based action interpretation ("I enjoy riding") and outcome-based goal intention ("Riding will be fun"). Concrete construal ("How do I ride this bike?") promotes state-based action interpretation ("I'm swerving to avoid a pothole") and means-based goal intention ("I need to brake and turn to avoid the pothole"). Elaborated control theory proposes that the level of construal depends on the difficulty of the current situation and follows a U-shaped curve. Specifically, low difficulty affords abstract construal, intermediate difficulty requires concrete construal, and high difficulty again affords abstract construal as goals are severely blocked.

The Control Failures × Concerns model holds that to prevent mind wandering, executive control must maintain appropriate construal relative to task difficulty (i.e., low difficulty = abstract; intermediate difficulty = concrete; high difficulty = abstract). For example, failure to maintain concrete construal on difficult tasks enables mind wandering because abstract construal activates current concerns, increasing the probability of mind wandering. Control in the Control Failures × Concerns model can be either proactive or reactive, and (as will be detailed below) tasks can vary in the extent to which they engage these mechanisms. Critically, it is this variation in proactive and reactive control across tasks that we suggest is helpful in explaining the variation in student attention across the ICAP task types. Specifically, the greater the engagement of proactive and reactive control in response to the task, the greater the student attention to the task. It will be argued that the ICAP ordering engages greater to less control as individuals shift from interactive to passive tasks. To support this argument, the Control Failures × Concerns model must be elaborated with mechanisms of proactive control and reactively initiated control. We briefly describe each in turn.

Proactive Control

On way to think of proactive control is in terms of models of sequential action. Cooper, Ruh, and Mareschal (2014) have recently proposed the goal circuit (GC) model of sequential action, which shares a heritage of ideas with action identification theory (e.g., Norman & Shallice, 1986). The GC model explicitly models three kinds of influences on action—namely, environmental affordances, task-specific ordering constraints, and top-down control. All three may be considered proactive in the sense that they exert an influence that is strictly forward in time, as opposed to a reactive system that implements feedback. Environmental affordances specify the preconditions of particular actions. Obviously, one cannot drive a car to the store if no car is present. Perhaps less obvious is the fact that simultaneously present environmental cues can facilitate or trigger particular actions. For example, seeing toast, butter, and a knife on a counter is sufficient to infer that someone is making toast. Task-specific ordering constraints operate at a more abstract level that might be thought of as a subtask level. For example, to make toast, bread must be obtained, the bread must be placed into a heating device, and then the heating device must be activated. If the heating device is activated before the toast is placed there, the operation of making toast will fail. Finally, top-down control, also known as the supervisory system, allows executive control during nonroutine action sequences.

Each of these influences is represented by a path in a recurrent network, as shown in figure 27.1 (plate 6). It should be noted that both input and output layers use a localist encoding, that is, each node corresponds to a particular environmental affordance, action, or goal, represented below in italics. Environmental affordances are represented by input nodes that feed into a hidden layer and link to action nodes, for example, *pencil is present*. Action nodes represent changes to the environment, and these changes are reflected in the environmental affordance nodes during the next time step, for example, *pick up pencil*. Top-down control is likewise represented by input goal nodes that feed into a hidden layer and link to predicted goals, for example, *write an answer*. These predicted goals become the goal inputs during the next time step. Task-specific ordering constraints are represented by recurrent connections on the hidden layer. The hidden layer receives both environmental affordance input and top-down control input and links to the analogous nodes in the output layer, so the task-specific ordering constraint pathway effectively merges environmental affordances and goals into a single context, for example, *write next letter of a word*. In the case of goal nodes, all superordinate goals of the current goal are simultaneously active such that a hierarchy of goals are active at any given time. For example, the nodes for *write answer*, *write word*, *and write letter* might all be simultaneously active in a goal layer to represent the hierarchical goal structure of writing an answer.

With respect to the current discussion, the GC model illustrates how proactive control may stem from multiple sources: environmental affordances, task ordering

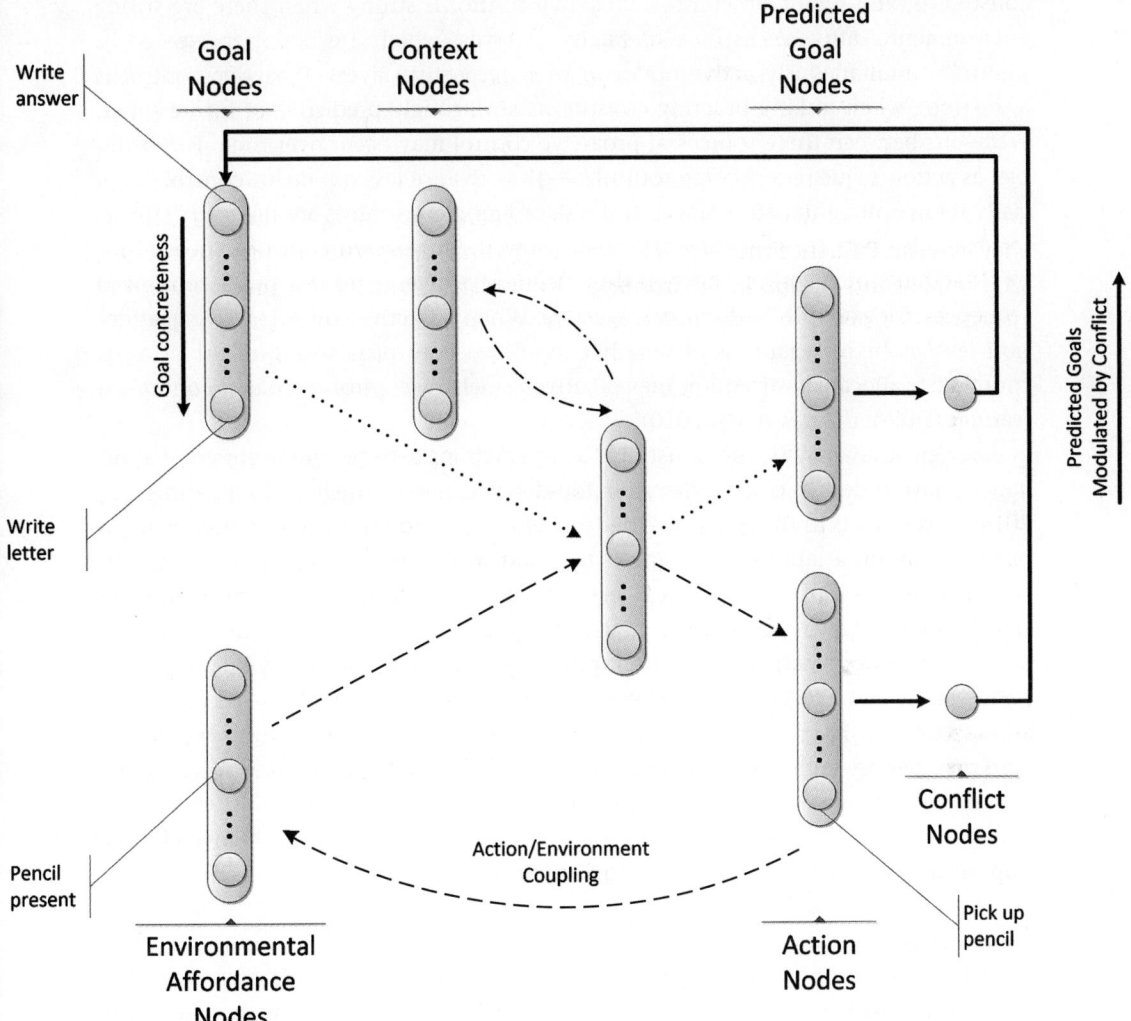

Figure 27.1 (plate 6)
An elaborated version of the goal circuit model with conflict nodes at the outputs. Layers are fully connected, with lines indicating the following mechanisms: the environmental affordance loop (dash), the goal circuit loop (dot), the task ordering loop (dash-dot), and the modulation of conflict nodes on top-down control (solid). Selected nodes are annotated according to the *writing* example given in the text.

constraints, and top-down control. Proactive control is strong when there are strong environmental affordances or a hierarchy of goals, which are both represented as multiple simultaneously active nodes in their respective layers. Proactive control is also strong when tasking ordering constraints are strongly predictive of future states. Trade-offs between these sources of proactive control may occur over time. For example, as action sequences become routinized, they require less top-down control, especially if environmental affordances and task ordering constraints are relatively strong, for example, *write the next letter of a word*. However, top-down control is needed for novel action sequences, especially if these sequences diverge from strongly routinized sequences, for example, *write the next sentence*. When proactive control is strong, attention level is high, regardless of whether top-down control is strong or attention is consciously allocated—attention may be driven solely by environmental or contextual features (Dijksterhuis & Aarts, 2010).

A recent study may further illustrate the interactions between environmental affordances, task ordering constraints, and top-down control (Mueller & Oppenheimer, 2014). Participants in the study watched a video lecture and took notes either by hand or by typing on a laptop. Notes taken by hand were significantly shorter and were significantly less verbatim with the lectures; moreover, hand note-takers performed significantly better on conceptual questions at posttest. In terms of the GC model, we may consider both handwriting and typing to be highly automatized in the participants, who were college students. However, typing is faster than writing—so fast that it allows verbatim note-taking that could not be accomplished by writing because of working memory constraints. Instead, the task of writing forces a strategic top-down control with the goals of condensing the information (as seen in the shorter and less verbatim effects) and prioritizing important information (as may be evidenced by the improved performance on conceptual questions at posttest).

Reactive Control

The GC model provides a mechanistic explanation of proactively initiated control that is congruent with the Control Failures × Concerns model, but it does not explicitly model reactively initiated control. This is because the GC model specifies the top-down control at the goal node layer without specifying what drives the presence or absence of goals at those nodes. Botvinick, Braver, Barch, Carter, and Cohen (2001) describe a model of reactively initiated control that accounts for behavior and corresponding brain activation on a diverse set of tasks. The model consists of a single conflict monitoring node that is connected to multiple output nodes of a connectionist network, as shown in figure 27.1 (plate 6). If multiple output nodes in the connectionist network are active, then the conflict node is active. In the GC model above, an active conflict node would mean that multiple goal output nodes or action nodes were simultaneously active. It is important to note that given the nature of the GC model, multiple active output nodes are likely to occur either when the action sequence is being learned (so multiple actions are equally possible) or when the environmental affordances or task ordering are relatively weak (so multiple actions or goals are equally possible).

The conflict node, by itself, does not provide reactively initiated control. Reactively initiated control is provided by using the conflict node to influence top-down control nodes. Thus, periods of high conflict (e.g., a spelling error of *ghost* as *goast* where nodes for *h*, *o*, and *a* are highly active) lead to a greater amount of top-down control activation, and periods of low conflict lead to a lower amount of top-down control activation (e.g., the production of each letter with no competing letters active). This feedback mechanism allows the model to account for a range of behavioral data where online control appears to adapt to error and trial-type frequency. In the GC model, such control would be applied proportionally more during the learning of novel action sequences, because in novel sequences various related and unrelated goals and actions will be active. However, control based on conflict monitoring would slowly fade if those sequences were routinized and task-irrelevant goals were suppressed.

In summary, reactively initiated control consistent with the Control Failures × Concerns model can be implemented by augmenting the GC connectionist model with conflict nodes at the action output and predicted goal output layers and using conflict node activation to modulate top-down control via the goal node input layer. This elaborated version of the Control Failures × Concerns model implies that novelty should increase attention because of the conflict and error generated in novel tasks. However, as the task is better learned, attention driven by this reactive mechanism should shift to proactive mechanisms as conflict diminishes and learning enhances proactive control.

Elaborating the Control Failures × Concerns model with the GC model and conflict nodes also provides an account of difficulty in terms of the need for control. As the GC model learns action sequences, it makes fewer errors: As the sequences become routinized they become less difficult. Learning is accelerated by top-down control, strong environmental affordances, and strong task ordering constraints. In general, conflict monitoring declines as action sequences are learned. However, monitoring is enhanced in novel situations or when the environmental affordances or task ordering constraints are relatively weak. For example, composing a new action sequence out of well-known action subsequences will enhance conflict monitoring because ordering constraints between those subsequences will be weak "at the joints." Situations like this that require relatively more monitoring can be described as more difficult. This depiction of difficulty does not preclude additional considerations of difficulty like relational complexity (Halford, Wilson, & Phillips, 1998); however, a more complete account of difficulty in educational contexts is outside the scope of the present discussion.

ICAP-A

The preceding discussion has placed the notions of control and difficulty in the Control Failures × Concerns model into a more mechanistic framework. Specifically, the above accounts describe proactive control in terms of sequential action, reactively initiated control in terms of monitoring, and difficulty in terms of the need for control. Elaborated with these accounts, the Control Failures × Concerns model licenses the hypothesis that learning task structure, such as ICAP, can enhance student attention within the learning task. The ICAP-A hypothesis asserts that because proactive control and reactive control each increase as task type advances from passive to interactive, attention should likewise increase from passive to interactive. The remainder of this section describes how the three dimensions increase across the ICAP task types in order to motivate the ICAP-A hypothesis.

Passive tasks, by definition, involve no overt activity. Proactive control via sequential action is therefore nonexistent in passive tasks. Likewise, there can be no monitoring of actions because there is no sequential action. It should be noted that ICAP focuses on overt activities and that a student could be engaging in a constructive task covertly. In the description of other task types that follow this focus on overt activities will be preserved; however, to the extent that the GC model could also apply to covert, or mental, action sequences, the same rationale for an ordering of attention can be applied to both overt and covert task types.

Active tasks involve actions that do not require the integration of new ideas with the learning materials. For example, underlining and verbatim repetition are examples of active tasks. As discussed previously, proactive control may stem from environmental affordances, task ordering constraints, and top-down control. In active tasks, the environmental affordances and task ordering constraints are relatively strong. Underlining, for example, provides continuous visual and motor feedback to guide the production of a straight line that stops at clause boundaries. Repetition similarly provides a tight coupling of perception and action. Presumably, tasks like these are highly practiced and fairly well routinized, meaning they require less top-down control. Top-down control may be further weakened in these tasks by the relatively weak conflict detection produced by monitoring: These tasks have limited variation in action, so little conflict is expected to occur. Accordingly, in active tasks, student attention is limited by weaker top-down control and simpler task complexity.

Constructive tasks involve overt behaviors that integrate new ideas with the learning materials. Examples of constructive activities are problem solving and making analogies. Constructive tasks may involve routinized subsequences of action but would require combining these subsequences in novel ways. Therefore, unlike active tasks which are likely to have weak top-down control resulting from routinized action sequences, constructive tasks are likely to have relatively stronger top-down control while having comparable environmental affordances and task ordering constraints. For example, solving a math problem may involve basic operations like long division that are highly routinized, but the problem as a whole consists of a novel sequence of such operations. This novel sequence requires closer attention to the goal structure of the problem in order to correctly chain basic operations together. Likewise, the need for monitoring increases for constructive tasks because they are novel and multiple actions may seem appropriate at any given point. This increased need for monitoring should serve to increase top-down control in constructive tasks. Constructive tasks, as a result, have higher top-down control affordances than active tasks, which should result in a correspondingly higher level of student attention.

Interactive tasks are co-constructive. That is, in addition to individually being engaged in a constructive task, the participants are collaboratively assisting each other in a constructive task. Interactive tasks share proactive control and reactive control attributes with constructive tasks. However, in addition to these shared attributes, there are additional influences for control that come from the other partner in the interaction via redundant systems for proactive and reactive control. Both participants may reinforce their proactive control by observing each other's actions and inferring relevant goal structure from those actions. Likewise, both participants reinforce reactive control by monitoring each other's actions and providing top-down control, that is, correction, when needed. In other words, in interactive tasks, participants are not only monitoring their progress on the task but are also monitoring each other. Increased conflict from interpersonal monitoring would further serve to strengthen top-down control. Therefore interactive tasks should promote even greater levels of attention than constructive tasks, but only if redundancies in proactive and reactive control are exploited.

A summary of the ICAP-A hypothesis is presented in table 27.2. Although the elaborated Control Failures × Concerns model makes more nuanced predictions, including the novelty and variability of a task and the amount of previous practice on a task, table 27.2 presents the ICAP-A hypothesis for the generic case of routinized action subsequences that may require novel recombination for certain problems. Under these conditions, students should manifest no action control for passive tasks, low control for active tasks, high control for constructive tasks, and redundant control for interactive tasks. As discussed earlier, the hypothesis is primarily concerned with overt student behaviors, and it is quite likely that even in a passive task like lecture viewing some students are engaging in covert activities like self-explanation. Accordingly, the

Table 27.2

Pairwise Comparisons of Attention according to Interactive–Constructive–Active–Passive Task Type

	Passive	Active	Constructive	Interactive
Top-down control	None	Low	High	Redundant
Reactive control	None	Low	High	Redundant
Examples	Lecture viewing; being read to	Underlining; reading aloud	Problem solving; self-explaining aloud	Joint problem solving; tutoring

hypothesis does not assert that task type uniquely determines student attention at an individual level but rather that task type shifts the distribution of attention at the group level. ICAP-A suggests that attention for a group of students may be improved by increasing the opportunities for top-down and reactive control.

Future Directions

Emerging evidence from research on attention in education suggests that task structure can support attention. The hypothesis presented in this chapter, ICAP-A, proposes a framework for explaining how attention may be improved by changing task structure. These mechanisms, rooted in theoretical accounts of mind wandering, sequential action, and monitoring are well-known and have been individually validated. However, the ICAP-A hypothesis has not been directly tested, and much of the research reviewed here addresses pairwise contrasts (e.g., active vs. constructive); moreover, as shown in table 27.1, several pairwise contrasts remain unaddressed empirically, particularly for interactive tasks. Direct experimental evidence is an important direction for future research. One possible approach for future research is to augment designs that have evaluated ICAP in a four-condition experiment (Menekse et al., 2013) with mind-wandering probes that query thought contents (Schoen, 1970).

An important implication of ICAP-A is that enhanced learning performance will be achieved in interactive learning tasks by virtue of improved attention. While this prediction is no different from ICAP, the mechanisms described in this chapter offer a more detailed explanation of the conditions under which this prediction will be true. ICAP-A predicts that redundant proactive control and reactive control each contribute to optimal attention. One avenue for future work is to ablate these mechanisms in an interactive task and compare to a constructive task control and measure whether attention decreases. For example, in a mathematics problem-solving task, a worked example which makes the abstract problem-solving structure clear (a strong goal hierarchy) could be contrasted with a worked example without this proactive support. Likewise, feedback regarding errors made could be manipulated to provide or remove the error signal needed for reactive control.

The mechanisms behind ICAP-A also suggest future directions for the enhancement of instructional design for optimal learning performance. Both environmental affordances and task ordering constraints are frequently discussed in the learning sciences (Chandler & Sweller, 1991; Pavlik & Anderson, 2005); however, goal hierarchies may be an underutilized avenue for optimizing instructional design. Considerations of goal hierarchies have previously been proposed in the learning sciences (Farrell, Anderson, Reiser, & Boyle, 1987; Collins, Brown, & Holum, 1991; Anderson et al., 1995; Olney, in press), and the GC model provides an account whereby top-down control provided by a strong goal hierarchy results in faster learning. However, the mechanisms behind ICAP-A provide an additional dimension to this account in terms of attention—namely, that strong practice with weak goal hierarchies will actually be harmful to performance because it will increase mind wandering. Future work could pursue this avenue by training participants either with or without strong goal hierarchies on the same task. For example, teaching solutions to mathematics problems could be done completely procedurally (weak goal hierarchy) or by training participants to solve problems using the abstract problem-solving structure of the task (strong goal hiearchy). If mind wandering does indeed occur more with weak hierarchies, even though the task in both cases is putatively constructive, then this provides yet another reason for instructional designers to strengthen goal hierarchies by making thinking visible (cf. Collins et al., 1991).

Finally, there are still many effects unaccounted for in the elaborated Control Failures × Concerns model. Perhaps the most prominent omissions are a model of deliberate decisions to go off-task and the modulating influence of affect (or emotion) on attention, and all of cognition for that matter (Clore & Huntsinger, 2007; Dalgleish & Power, 1999; Fielder, 2001). The elaborated model provides an account for on-task behavior and a nondeliberate loss of control through weak proactive and reactive control. However, it does not model deliberate behavior by which a goal hierarchy is largely replaced in a top-down manner. Such action would occur outside the framework of the GC model and provide inputs to the goal node layer. The GC model assumes as given the initial values of the goal nodes and any later deliberate change. Likewise, the model does not elaborate on the link between attention and affect. This omission may be significant given that negative moods appear to lead to an increase in mind wandering (Smallwood, Fitzgerald, Miles, & Phillips, 2009), presumably due to a shift toward current concerns and away from the task. Affect has also been shown to influence monitoring, such that enhanced negative affect in response to errors results in a more sustained response to conflict (Ichikawa et al., 2011). These studies suggest that affect may influence both proactive and reactive control and therefore be a nontrivial influence on attention in educational settings.

Box 27.1
Key Points

- Attention may be increased by changing the structure of the learning task, enhancing learning.
- An elaborated theory of mind wandering may explain how the learning task influences attention through mechanisms of proactive and reactive control.
- The ICAP-A hypothesis claims that interactive, constructive, active, and passive learning tasks differ in their needs and opportunities for proactive and reactive control, suggesting ways of changing task structure to optimize learning performance via improved attention.

Box 27.2
Outstanding Issues

- Future research should study the stratification of attention across learning tasks types using methods that measure both the occurrence and the contents of mind wandering.

- Attention researchers should investigate the contributions of proactive and reactive control to attention, particularly for interactive tasks which are hypothesized to be the most beneficial for learning.

- Researchers in education, attention, and affect should consider the role of deliberate decisions to go off-task and of affect on attention, as evidence suggests that these factors are inextricably linked to attention in real-world educational settings.

Acknowledgments

This research was supported by the National Science Foundation (1235958 and 1352207), Institute of Education Sciences (R305C120001 and R305A130030), Office of Naval Research (N00014–12-C-0643), and Army Research Laboratory (W911NF-12-2-0030). Any opinions, findings and conclusions, or recommendations expressed in this chapter are those of the author and do not represent the views of these organizations.

Note

1. Although attention and engagement are sometimes used synonymously, we consider engagement to be a multidimensional construct that includes both attention and affect (cf. D'Mello, 2013).

References

Anderson, J. R., Corbett, A. T., Koedinger, K. R., & Pelletier, R. (1995). Cognitive tutors: lessons learned. *Journal of the Learning Sciences*, *4*, 167–207.

Berliner, D. C. (1990). What's all the fuss about instructional time. In M. Ben-Peretz & R. Bromme (Eds.), *The nature of time in schools: Theoretical concepts, practitioner perceptions* (pp. 3–35). New York: Teachers College Press.

Bligh, D. (2000). *What's the use of lectures?* San Francisco: Jossey-Bass.

Botvinick, M. M., Braver, T. S., Barch, D. M., Carter, C. S., & Cohen, J. D. (2001). Conflict monitoring and cognitive control. *Psychological Review*, *108*, 624–652.

Bunce, D. M., Flens, E. A., & Neiles, K. Y. (2010). How long can students pay attention in class? A study of student attention decline using clickers. *Journal of Chemical Education*, *87*, 1438–1443.

Cameron, P., & Giuntoli, D. (1972). Consciousness sampling in the college classroom or is anybody listening? *Intellect, 101*(2343), 63–64.

Chandler, P., & Sweller, J. (1991). Cognitive load theory and the format of instruction. *Cognition and Instruction, 8*, 293–332. doi:10.1207/s1532690xci0804_2.

Chi, M. T. H. (2009). Active–constructive–interactive: A conceptual framework for differentiating learning activities. *Topics in Cognitive Science, 1*, 73–105. doi:10.1111/j.1756-8765.2008.01005.x.

Clore, G. L., & Huntsinger, J. R. (2007). How emotions inform judgment and regulate thought. *Trends in Cognitive Sciences, 11*, 393–399. doi:10.1016/j.tics.2007.08.005.

Collins, A., Brown, J. S., & Holum, A. (1991). Cognitive apprenticeship: Making thinking visible. *American Educator, 15*, 38–46.

Cooper, R. P., Ruh, N., & Mareschal, D. (2014). The goal circuit model: A hierarchical multi-route model of the acquisition and control of routine sequential action in humans. *Cognitive Science, 38*, 244–274. doi:10.1111/cogs.12067.

Currie, J. (1861). *The principles and practice of common-school education*. Edinburgh, UK: James Gordon.

Dalgleish, T., & Power, M. (Eds.). (1999). *Handbook of cognition and emotion*. Sussex, UK: Wiley.

Denham, C., & Lieberman, A. (1980, May). *Time to learn: A review of the beginning teacher evaluation study*. (RIE No. ED192454). Washington, DC: National Institute of Education.

Dijksterhuis, A., & Aarts, H. (2010). Goals, attention, and (un)consciousness. *Annual Review of Psychology, 61*, 467–490.

D'Mello, S. K. (2013). A selective meta-analysis on the relative incidence of discrete affective states during learning with technology. *Journal of Educational Psychology, 105*, 1082–1099.

Farley, J., Risko, E. F., & Kingstone, A. (2013). Everyday attention and lecture retention: The effects of time, fidgeting, and mind wandering. *Frontiers in Psychology, 4*, 619.

Farrell, R. G., Anderson, J. R., Reiser, B. J., & Boyle, C. F. (1987). Cognitive principles in the design of computer tutors. In P. Morris (Ed.), *Modelling cognition* (pp. 93–133). Chichester, UK: Wiley.

Fielder, K. (2001). Affective states trigger processes of assimilation and accommodation. In K. Martin & G. Clore (Eds.), *Theories of mood and cognition: A user's guidebook* (pp. 85–98). Mahwah, NJ: Erlbaum.

Halford, G. S., Wilson, W. H., & Phillips, S. (1998). Processing capacity defined by relational complexity: Implications for comparative, developmental, and cognitive psychology. *Behavioral and Brain Sciences, 21*, 803–831.

Ichikawa, N., Siegle, G. J., Jones, N. P., Kamishima, K., Thompson, W. K., Gross, J. J., et al. (2011). Feeling bad about screwing up: Emotion regulation and action monitoring in the anterior cingulate cortex. *Cognitive, Affective & Behavioral Neuroscience, 11*, 354–371. doi:10.3758/s13415-011-0028-z.

Kane, M. J., & McVay, J. C. (2012). What mind wandering reveals about executive-control abilities and failures [abstract]. *Current Directions in Psychological Science, 21*, 348–354.

Klinger, E. (2009). Daydreaming and fantasizing: Thought flow and motivation. In K. D. Markman, W. M. P. Klein, & J. A. Suhr (Eds.), *Handbook of imagination and mental simulation* (pp. 225–239). New York: Psychology Press.

Lindquist, S. I., & McLean, J. P. (2011). Daydreaming and its correlates in an educational environment. *Learning and Individual Differences, 21*, 158–167.

Mackworth, N. H. (1948). The breakdown of vigilance during prolonged visual search. *Quarterly Journal of Experimental Psychology, 1*, 6–21.

Mason, M. F., Norton, M. I., Van Horn, J. D., Wegner, D. M., Grafton, S. T., & Macrae, C. N. (2007). Wandering minds: The default network and stimulus-independent thought [abstract]. *Science, 315*, 393–395.

McVay, J. C., & Kane, M. J. (2010). Does mind wandering reflect executive function or executive failure? Comment on Smallwood and Schooler (2006) and Watkins (2008). *Psychological Bulletin, 136*, 188–197.

Menekse, M., Stump, G. S., Krause, S., & Chi, M. T. H. (2013). Differentiated overt learning activities for effective instruction in engineering classrooms. *Journal of Engineering Education, 102*, 346–374. doi:10.1002/jee.20021.

Moss, J., Schunn, C. D., Schneider, W., & McNamara, D. S. (2013). The nature of mind wandering during reading varies with the cognitive control demands of the reading strategy. *Brain Research, 1539*, 48–60. http://doi.org/10.1016/j.brainres.2013.09.047.

Mueller, P. A., & Oppenheimer, D. M. (2014). The pen is mightier than the keyboard: Advantages of longhand over laptop note taking. *Psychological Science, 25*, 1159–1168.

Norman, D., & Shallice, T. (1986). Attention to action. In R. J. Davidson, G. E. Schwartz, & D. Shapiro (Eds.), *Consciousness and self-regulation* (pp. 1–18). New York: Springer. doi:10.1007/978-1-4757-0629-1_1.

Olney, A. M. (2014). Scaffolding made visible. In R. Sottilare, A. Graesser, X. Hu, & B. Goldberg (Eds.), *Design recommendations for adaptive intelligent tutoring systems: Volume 2 - Instructional management* (pp. 327–340). Orlando, FL: U.S. Army Research Laboratory.

Pavlik, P. I., & Anderson, J. R. (2005). Practice and forgetting effects on vocabulary memory: An activation-based model of the spacing effect. *Cognitive Science, 29*, 559–586.

Risko, E. F., Anderson, N., Sarwal, A., Engelhardt, M., & Kingstone, A. (2012). Everyday attention: Variation in mind wandering and memory in a lecture. *Applied Cognitive Psychology, 26*, 234–242.

Scerbo, M. W., Warm, J. S., Dember, W. N., & Grasha, A. F. (1992). The role of time and cuing in a college lecture. *Contemporary Educational Psychology, 17*, 312–328.

Schoen, J. R. (1970). Use of consciousness sampling to study teaching methods. *Journal of Educational Research, 63*, 387–390.

Smallwood, J. M., Fitzgerald, A., Miles, L. K., & Phillips, L. H. (2009). Shifting moods, wandering minds: Negative moods lead the mind to wander. *Emotion (Washington, D.C.)*, *9*, 271–276.

Smallwood, J. M., McSpadden, M., & Schooler, J. W. (2008). When attention matters: The curious incident of the wandering mind. *Memory & Cognition*, *36*, 1144–1150.

Szpunar, K. K., Khan, N. Y., & Schacter, D. L. (2013). Interpolated memory tests reduce mind wandering and improve learning of online lectures. *Proceedings of the National Academy of Sciences of the United States of America*, *110*, 6313–6317.

Varao Sousa, T. L. Carriere, J. S., & Smilek, D. (2013). The way we encounter reading material influences how frequently we mind wander. *Frontiers in Psychology*, *4*, 892.

Vallacher, R. R., & Wegner, D. M. (1987). What do people think they're doing? Action identification and human behavior. *Psychological Review*, *94*, 3–15.

Watkins, E. R. (2008). Constructive and unconstructive repetitive thought. *Psychological Bulletin*, *134*, 163–206.

Young, M. S., Robinson, S., & Alberts, P. (2009). Students pay attention! Combating the vigilance decrement to improve learning during lectures. *Active Learning in Higher Education*, *10*, 41–55.

28
Visual Attention in Sports

Daniel Memmert

Introduction and Historical Context

Visual attention could be defined as "focusing on specific features, objects or locations or on certain thoughts or activities" (Goldstein, 2011, p. 391). The field of cognitive science has always placed great importance on the mechanisms of visual attention, which is why behavior that triggers attentional processes has been analyzed for nearly 100 years now (Styles, 2008). Since attention has a limited capacity (Broadbent, 1958; Cowan, 1995), events that are relevant for behavior should be in the focus of attention. At the same time, this implies that certain internal or external information in our environment needs to be blanked out or attenuated, so that people can work more effectively with other pieces of information. While a number of different metaphors of attention have been and still are discussed (cf. Fernandez-Duque & Johnson, 1999; Moran, 2012), generally, visual attention is likened to a spotlight (the spotlight metaphor; Eriksen & St. James, 1986), whose size (diameter) varies.

Attention is also a crucial factor in sport-related performance (Abernethy, Maxwell, Masters, van der Kamp, & Jackson, 2007; Memmert, 2009; Moran, 1996, 2011; Wulf, 2007). This is especially true in team and racket sports, where the orientation and direction of attention play an important role. In order to act successfully in any complex situations of any kind of sport, the respective players need to rely on attentional skills during the execution of different simultaneous tasks. In field hockey, for example, one has to lead the ball and shield it from the opponent while being on the lookout for open teammates.

Assistant referees in a football match need to maintain a remarkable level of attentional performance in order to correctly identify offside decisions. In order to recognize an offside position correctly, for example, assistant referees need to keep the ball and many moving players in their attentional focus: the attacking player, the ball, the defender, and the potentially offside player in question. Simultaneous attention to two events (e.g., players, ball) cannot be achieved by shifting the eyes from one object to the other, given the time-consuming nature of saccades (Maruenda, 2004).

Furthermore, in this last example, it needs to be considered that one half of the pitch already consists of a huge area, so the linesmen perceive many actions only peripherally. Therefore, they will always have a problem, when a long pass is played and they need to decide if the attacker is standing offside at the precise moment when the ball was played. To make correct decisions about offside situations, assistant referees would benefit from the ability to perceive two spatially distinct and far apart actions peripherally (the football player performing the pass and the last player in the defense).

State-of-the-Art Review

Athletes and referees should be able to control the direction of their attention to such an extent that they can distinguish between relevant and irrelevant information in complex environments. Based on neurological and psychological findings (see Brown & Friston, 2013; Coull, 1998; Knudsen, 2007), attention research in the contexts of sports (cf. Memmert, 2009, for a review in the area of sport psychology) has investigated the processes involved in orienting attention (attentional orienting), selecting stimuli for preferential processing (selective attention), sustaining attention on a target for an extended period of time (sustained attention), and dividing attention between multiple inputs (divided attention):

1. *Attentional orienting* Attentional orienting is the "log-in" (i.e., focusing) and "log-out" (i.e., defocusing) of attention on a given stimulus—for example, a teammate on the other side of the pitch signals through lifting his or her arm that he or she is ready to take the ball, whereupon the other player directs his or her attention to this teammate to play a long pass.

2. *Selective attention* Selective attention entails choosing between competing stimuli within a given time. Selective attention is closely related to attentional orienting but differs from a neurophysiological viewpoint, as selective attention involves choosing among different stimuli, whereas orienting then directs attention toward this particular stimulus—for example, a player ignoring the "hail of catcalls" coming from the audience of the opposing team before playing a corner ball, to focus on playing the ball to the strong heading players.

3. *Sustained attention* Sustained attention entails maintaining attention toward a specific stimulus for a certain time span—for example, fixating the ball during a 60-meter pass to control the ball from the air in a playable manner.

4. *Divided attention* Divided attention involves simultaneous distribution of attention to different stimuli—for example, secure shielding of the ball while scanning the field for unmarked teammates.

To better understand attention in the world of sports and make them accessible for training purposes (see also Moran, 2014), these attention processes are described in more detail below. The emphasis of this review is on divided attention because this process is regarded as highly important in a variety of different team and racket sports (for a review, see Memmert, 2009).

Attentional Orienting

Attentional orienting serves to direct attention toward task-relevant stimuli or distinctive features of a stimulus. With this mechanism, the focus of attention can be directed to different locations. Tenenbaum and Bar-Eli (1995) also speak of attention's being "logged-in" or being "logged-out" (i.e., one focuses or de-focuses on a specific area). According to Posner (1980), attentional orienting moves information into the focus of (covert) attention. Accordingly, the processing of information outside the focus of attention becomes more difficult. Evidence for this assumption has been accumulated in research using the "cuing paradigm" (see Posner, 1980, for more details) wherein performance in signal detection tasks is enhanced by precuing the location where the target stimulus is likely to occur (for a review, see Nougier & Rossi, 1999).

Experts in team sports like hockey (Enns & Richards, 1997), football (Lum, Enns, & Pratt, 2002), or volleyball (Castiello & Umiltà, 1992) appear to possess greater attentional flexibility regarding attentional orienting in the visual field compared to novices. For example, male volleyball and football players showed greater modulation of the automatic and voluntary orienting of visual attention (Lum et al., 2002). However, this cognitive ability is not as refined in experts in individual sports like swimming (Nougier, Rossi, Alain, & Taddei, 1996). Only experts who carry out sports in which a high degree of attentional workload is necessary perform better in orienting than novices (Nougier, Ripoll, & Stein, 1989; Nougier, Stein, & Bonnel, 1991). Further research has shown that experts can modify their attentional resources according to specific requirements of a particular task (Castiello & Umiltà, 1992; Nougier et al., 1989). For example, expert volleyball players can better modulate the effective size of their attentional focus to accommodate different task constraints than nonexperts (Pesce-Anzeneder & Bösel, 1998).

In addition to orienting to different locations in the external environment, based on the work of Wulf and Prinz (2001), an internal attentional focus can also be distinguished from an external focus during the execution of movement tasks (e.g., Beilock, Bertenthal, McCoy, & Carr, 2004; Rowe & McKenna, 2001). With respect to an internal focus of attention, the athlete attends to the movement itself (e.g., the snap of the wrist during a free throw in basketball). With respect to an external focus of attention, the athlete instead attends to the movement effect (i.e., the movement goal; e.g., the back rim of a basketball hoop during a free throw). A recent series of studies has concentrated on motor expertise regarding the internal versus external attentional-focus paradigm (for an overview, see Wulf, 2013). Within this paradigm, attentional focus is connected with acquiring motor skills in racket sports, as well as with motor control over complex movement (cf. Wulf, 2007). The central question within this paradigm is this: What is the difference between an internal and an external attentional focus? Current study results derived with a dual-task paradigm (Gray, 2011) suggest that an external attentional focus is usually superior to an internal, movement-related focus for many different movement skills, skill levels, and target groups (Wulf, Shea, & Park, 2001; Wulf, 2007). An external attentional focus leads to greater achievement in the process of acquiring and learning new skills not only in novices but also in experts (for an alternative position, see Gray, 2004; Wulf, McConnel, Gärtner, & Schwarz, 2002). In fact, Beilock and colleagues have shown that experts' performance suffers when attention is oriented internally (i.e., is skill focused) while executing their highly automatized movement actions (Beilock, Carr, MacMahon, & Starkes, 2002).

Selective Attention

Selective attention directs attention toward specific events, locations, or fixed points at a given time or within given time frames (Coull, 1998; Posner & Boies, 1971). Selective attention is strongly connected to attentional orienting as both subprocesses are involved in attentional direction and control. However, the critical difference is that selective attention is associated with situations in which specific stimuli are preferred over others, as opposed to simply orienting attention to single locations.

With the exception of attentional orienting, selective attention has historically been the most investigated attentional topic in the context of sport sciences (Memmert, 2009). One reason for this could be the fact that the ability to selectively attend is critical in sport-specific situations, which are typically comprised of many stimuli competing for attention. Another reason could be the fact that there are various paradigms to test selective attention methodologically. The question of interest is which visual cues do experts look at in competitions and matches (visual search strategies) in order to react quickly and appropriately, and importantly how do individuals learn to selectively attend to this information. In this case, Magill (1998) talks about "information-rich areas" (IRAs), which include visual characteristics that can be used to anticipate movement effects, such as the trajectory of a ball after a strike. In badminton, for example, it is helpful to direct attention not only to regions far away from the body center (i.e., the racket of the opponent or the trajectory of the ball) but also to regions close to the body center (i.e., the arm and upper body of the opponent) in order to receive important information necessary to predict the direction of the shuttlecock (Abernethy & Russell, 1987).

One active area of research in sport science involves investigating how visual or verbal cues provided in the form of training can direct attention toward areas that are important and rich in information. Verbal instructions offer a meaningful possibility for a coach to direct attention toward IRAs (i.e., learning cues; for an overview, see Jackson & Farrow, 2005). Normally, sport-specific IRAs are taught via video-based simulations and training sessions with live demonstrations (e.g., squash: Abernethy, Wood, & Parks, 1999; field hockey: Williams, Ward, & Chapman, 2003). Sport-specific laboratory tests following these training sessions showed that the applied attention-directing training programs led to significant improvements in anticipatory perfor-mance. Hagemann and Memmert (2006) were able to show, for example, that verbal instructions training participants to utilize IRAs in badminton are just as effective in an actual training setting as in video-based laboratory training sessions. In addition to explicit verbal instructions, studies seem to suggest that implicit and guided discovery learning are also promising strategies in this context. For example, a number of stud-ies show that these methods are similarly effective regarding the directing of attention in an efficient manner (e.g., Farrow & Abernethy, 2002; Smeeton, Ward, & Williams, 2004; Williams et al., 2003; Williams, Ward, Knowles, & Smeeton, 2002). However, recent studies have demonstrated important boundary conditions on effective training techniques. For example, sometimes simple instructions (e.g., focus on *an* opponent player) lead to a reduced focus of selective attention, and therefore important cues in a specific situation (e.g., an unmarked teammate) may not be integrated into the process of decision making (Memmert & Furley, 2007). Specifically, in a 6-month longitudi-nal study, Memmert (2007) demonstrated, in real-life training scenarios, that children who were given concrete instructions and cues for IRAs in team sports and all forms of practice and play did *not* deliver any learning progress regarding the production of surprising, novel, or flexible results during the tactical tasks. More research in this area is clearly needed.

Sustained Attention

According to Coull (1998), the processes responsible for maintaining attentional focus on a specific stimulus or place for an ongoing time are called sustained atten-tion or concentration. This term is often used interchangeably with the word "vigi-lance." However, vigilance refers to long-lasting attentional processes between minutes and hours, instead of seconds and minutes, as it is the case in sustained attention. Some studies highlight that vigilance is a rather time-relevant process while sustained attention is a rather spatial process and that both processes operate independently (Fernandez-Duque & Posner, 1997). There is literature proposing ways in which ath-letes can improve their concentration in team sports (see Moran, 2011). For example, coaches can use "trigger words" as verbal cues to support the players in focusing on important movement sequences or tactical actions (Moran, 2009).

Divided Attention

According to Coull (1998), divided attention refers to the act of concentrating on two or more sources of information simultaneously. A multitude of sport-practical require-ments depend on this ability. Within the visual domain, dividing attention can be thought of as maintaining different "windows of attention." Thus, the critical question concerns determining if and how people perceive two or more stimuli simultaneously (peripherally). Divided attention can be analyzed within a dual-task-paradigm (the participant is asked to concentrate on two tasks simultaneously). This approach affords a number of different analyses, for example, (1) analysis of a distraction effect caused by a secondary task on the performance of the primary task; (2) analysis of the influence of instructions that set up an attentional focus on the performance of the primary task; and (3) examination of performance within the secondary task while performing the primary task. Thus, the dual-task paradigm can be used flexibly to address issues related to divided attention in sports contexts. In sport-relevant attention research, dual tasks were recently applied within the inattentional-blindness paradigm (for a general review, see chapter 9 of the current volume). This paradigm demonstrates that observers whose attention is directed toward a specific object or event often do not perceive unexpected objects, even though these objects are placed in clear view (Mack & Rock, 1998; Most, Scholl, Clifford, & Simons, 2005). This scenario occurs often in sports where unanticipated objects, such as uncovered teammates, appear quite often. In complex game situations, the best solution often is to notice these players and pass the ball to them. This, of course, does not always happen. For example, occasionally, teammates or coaches blame players who were in possession of the ball in a crucial situation because they had not seen another free-standing team player and failed to play the ball toward him or her. Later, the criticized player would reject all of the accusations and assure the accusers that he had simply not seen the player who was directly in front of him or her and in a better position.

Memmert (2006) demonstrated significant differences in inattentional blindness between expert basketball players and novices of the same age groups when tested on the famous "gorilla task" used by Simons and Chabris (1999): Although participants with basketball-specific expertise were no better at counting the passes (the primary task), they were more likely to perceive the unexpected object (here, the gorilla). This result implies that experts in a specific sport can better divide their attention, which in theory would allow them to focus their attention on other stimuli that may seem irrelevant at first. Furthermore, it is known that attentional focusing should receive greater consideration in sports, especially in beginners' training. There is proof, for example, that 8-year-old children are significantly less able to perceive unexpected objects than teenagers and adults (Memmert, 2006). The gorilla task used by Memmert (2006) is somewhat divorced from the complexity of "real" sports. Memmert and Furley (2007) improved on this design by transferring the inattentional-blindness paradigm to sport-related tasks by developing a simple decision-making test in handball and a more complex decision-making test in basketball (see figure 28.1).

Figure 28.1
Presentation of two sport-specific inattentional-blindness test scenarios with different levels of complexity (left: low complexity, handball, four against four; Memmert & Furley, 2007; right: medium complexity, basketball, five against five; Furley, Memmert, & Heller, 2010). The respective free-standing team-player who represents the best pass solution, according to expert opinion, is circled.

In this task, athletes have to solve two problems after watching a handball-specific video sequence, in which four offensive players act against four defensive players, and a completely free-standing teammate is visible in the closing decision-making situation (see figure 28.1, top). In the attention-demanding *primary task* the participants have to judge how far the opponent is positioned from them (far/near; see figure 28.1, top; the opponent is wearing a white T-shirt, the participants have to put themselves into the position of the player with the number 10). For the *secondary task* (dual task) the participants are supposed to specify the best solution for the presented situation within 2 seconds.

In this more realistic task, Memmert and Furley (2007) found inattentional blindness occurred in experienced teenage handball players (i.e., players often failed to notice a free teammate if their attention was directed toward the primary opposing players; cf. figure 28.2). Additionally, these findings were confirmed in more realistic, ecologically valid and challenging decision-making tasks with more complex primary tasks (see figure 28.1, bottom). Basketball experts similarly failed to notice the obvious free-standing team players, who would have represented the best solution in each of the game situations (Furley et al., 2010). These results were also replicated in more realistic contexts with motor response actions, as well as with a more field-relevant first task (triple choice task; see figure 28.2).

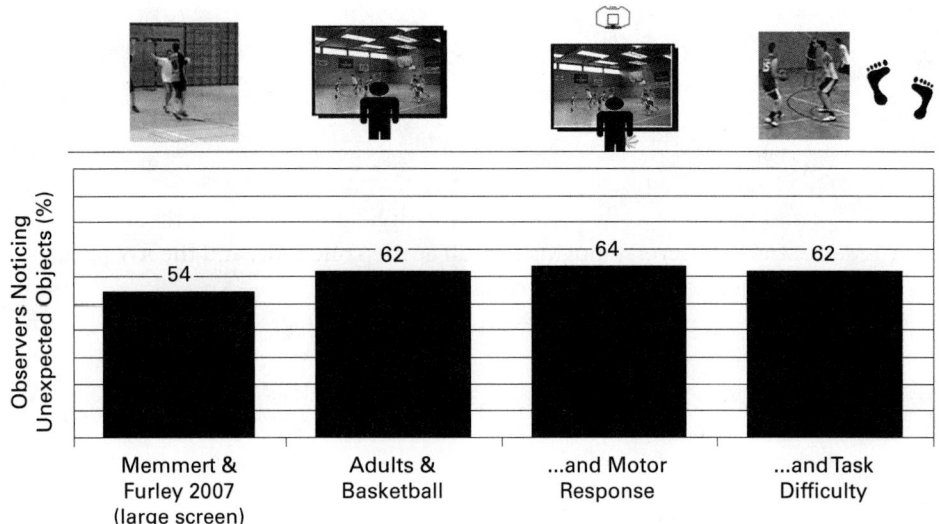

Figure 28.2
Results regarding inattentional blindness depend on the sport and motor action solutions, as well as the first task, which draws on the participants' attention (Furley et al., 2010). In the first experiment teenage handball players were tested, in the second experiment basketball experts had to complete a basketball-specific decision test (verbal naming of the answer), in the third experiment the same target group as in the second experiment had to perform their tactical solutions motorically, and in the fourth experiment the first task was made more difficult (in Experiments 1–3, the distance to the opponent had to be specified: near/far), because the basketball players had to correctly name the foot position of the opponents (left foot in front/right foot in front/ same level).

Integration: The Attention-Window Paradigm

Following the state-of-the-art review, this section will discuss how more traditional laboratory research is relevant to the analysis of attention in athletes or elderly people in everyday contexts, using the example of the Attention-Window paradigm. As discussed in the last section, the particular challenge in team and racket sport lies in perceiving several stimuli at the same time. The striker in soccer, for example, has to keep an eye on the opposing defenders, the goalkeeper, the teammates, and the ball simultaneously. Or the soccer referee has to observe both the position of the goalkeeper and the players in the other half of the playing field.

Because of the importance of attentional breadth for various everyday situations and complex sport settings, different test procedures for the identification of attentional breadth have been developed over the last few years (see figure 28.3). Tests include the Useful Field of View (UFOV; Green & Bavelier, 2003), the Multiple Object Tracking (MOT) test (Cavanagh & Alvarez, 2005; Jans, Peters, & De Weerd, 2010; Pylyshyn, 2007; Scholl, 2001), and the Attention-Window paradigm (AW paradigm; Hüttermann et al., 2013). The aim of the respective tests is to determine an individual's attentional breadth whenever two or more stimuli (one of them foveal, the other peripheral, or both/several peripheral) are to be perceived simultaneously. The UFOV is about fixating on a central stimulus while perceiving a peripheral stimulus simultaneously, the MOT requires tracking several moving stimuli at the same time, and the AW paradigm requires paying attention to two peripheral stimuli.

Regarding this knowledge and based on the testing procedures used up until now (i.e., UFOV, MOT), Hüttermann and colleagues (Hüttermann et al., 2012, 2013; Hüttermann, Simons, & Memmert, 2014) developed a new measure of attention, the above-mentioned AW paradigm. Using this paradigm, one can define the so-called visual "Attention-Window" for any given individual exactly. This window represents the maximum width of the attentional field that is consciously perceived with one glance. This field is defined by the maximum of horizontal, vertical, and diagonal orientation of attention, within which a minimum of two stimuli can be correctly identified with an accuracy of 75%. Concretely speaking, with this task it is possible to exactly define up to which distance people can still perceive two distinct stimuli that are situated at different borders of their visual field. Prior to the actual testing, a control test, in which one single stimulus is supposed to be perceived peripherally across the whole test area, is carried out to exclude the possibility that the visual Attention-Window is reduced because of limitations in peripheral visual performance. The maximum Attention-Window appears in the form of an ellipse (Hüttermann et al., 2013). The pictorial imagination of the attentional focus through the zoom-lens model (Eriksen & St. James, 1986) can now be completed for the first time (cf. Hüttermann, 2014) through valid conclusions and precisely quantified data on the size and form of the visual focus. Results show that the attentional focus can be widened up to a visual angle of an average of about 29°, averaged across all axes, with greater expansion along the horizontal compared to the diagonal and vertical axes.

The following section will describe how the AW paradigm, based on the more traditional laboratory research, is relevant to attention in everyday contexts such as athletic training, sports penalty situations, and aging.

Athletes

Hüttermann, Simons, and Memmert (2014) demonstrated that athletes have up to 25% wider Attention-Windows than nonathletes (Hüttermann et al., 2014; cf. figure 28.4). Test results, however, did not clearly answer the question as to whether the athletes were better than the nonathletes as a result of their long-term experience or whether

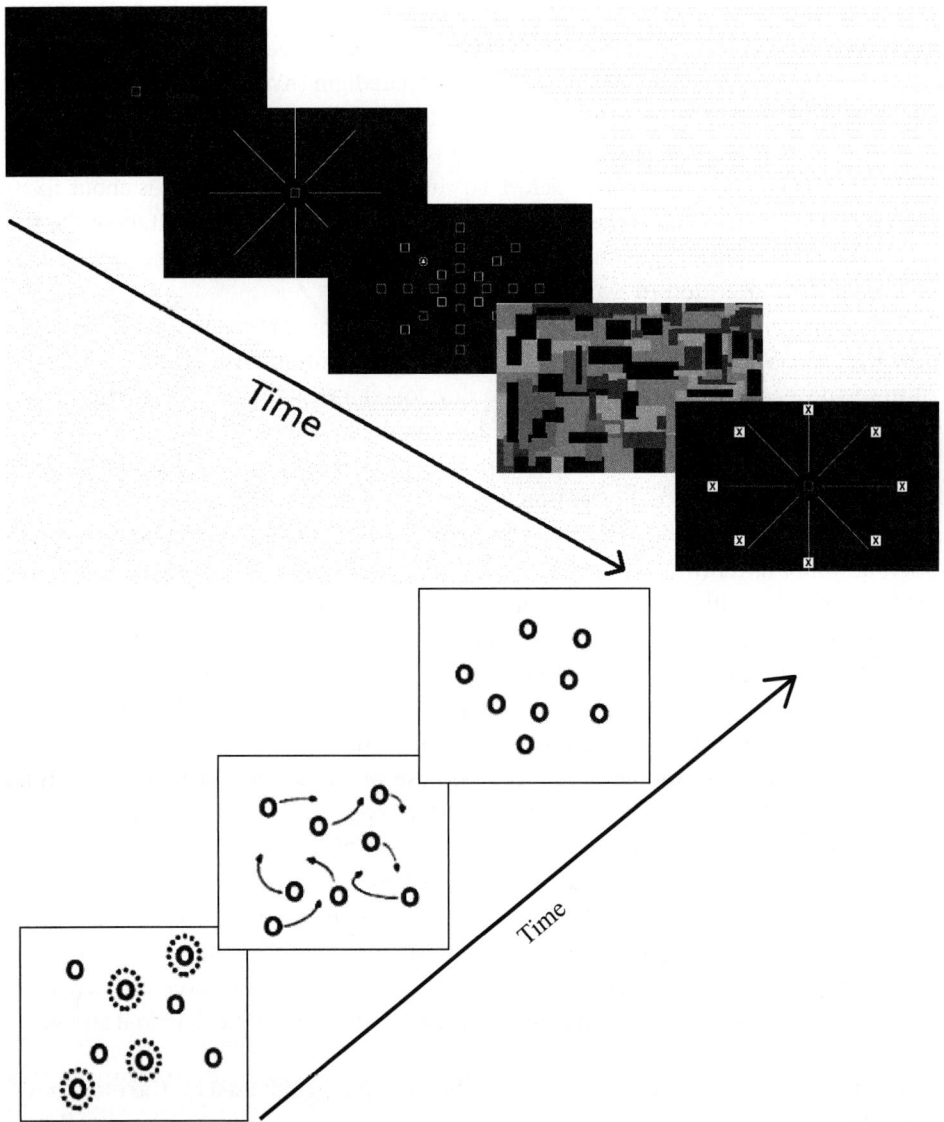

Figure 28.3
Presentation of three different attention tests capturing attentional breadth. (Top) Useful Field of View Task, Green and Bavelier (2003): The task is to discover a stimulus in the periphery while focusing on a central stimulus. (Middle) Multiple Object Tracking Task, Pylyshyn (2007): The task is to trace and perceive several moving objects simultaneously. (Botttom) Attention-Window paradigm (AW paradigm), Hüttermann et al. (2013): This test is about simultaneously perceiving two stimuli peripherally. The maximum, still perceivable, distance between both stimuli in a horizontal, vertical, and diagonal direction is measured, and a maximum attention window is determined.

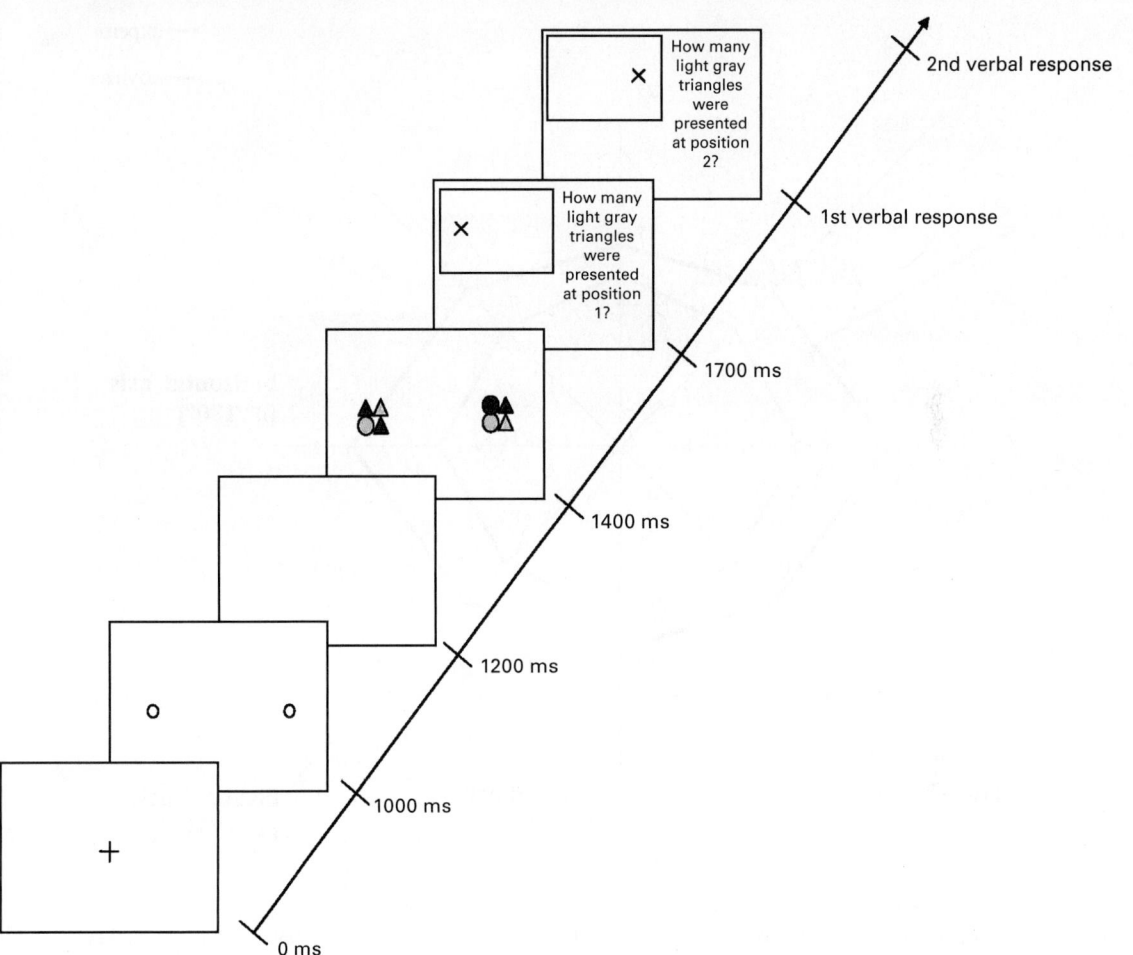

Figure 28.3 (continued)

their wider Attention-Windows were due to innate differences. Nevertheless, the AW paradigm could be utilized as a prospective selection criterion for junior athletes. In addition, in team and racket sports the question is often raised as to which of two eye-movement strategies is more effective when two objects require nearly the same amount of attention simultaneously: Athletes can either focus on one object (the fixate-target strategy) or focus between the objects (the fixate-center strategy). The AW paradigm can also be used to answer this question. Hüttermann et al. (2013) were able to show a higher success rate regarding the identification of the simultaneously presented stimuli when objects were both viewed with the peripheral orientation (fixate-center

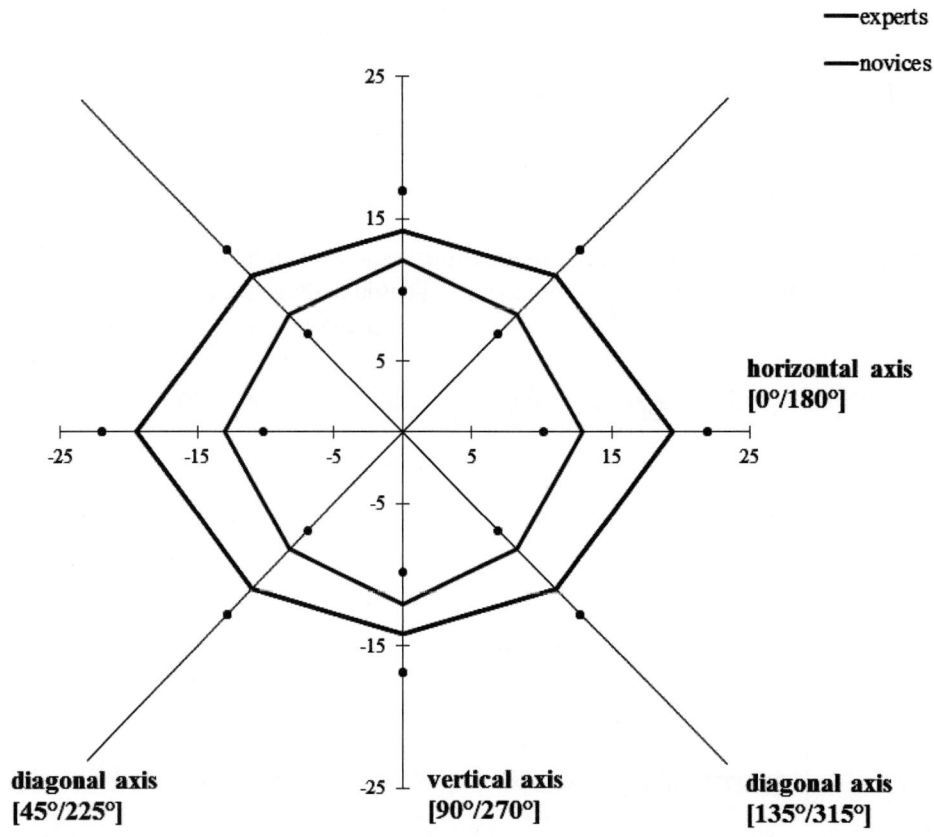

Figure 28.4
Size of the maximum Attention-Window (in degrees) for experts and novices in team and racket sports, within which two stimuli can be perceived simultaneously, defined according to a 75% success criterion. As each stimulus pair was presented symmetrically across the center, the data for each axis (horizontal, diagonal, and vertical) were also displayed symmetrically. The outer, bold Attention-Window presents the average expert values, the inner window the average novice values. From Hüttermann, Simons, & Memmert, 2014, Study 1, reprinted with permission.

strategy: 89.99 ± 5.13%), as opposed to fixating one object while perceiving the other one peripherally (fixate-target strategy: 80.29 ± 10.26%). Experts in team and racket sports performed better than novices in both the fixate-target strategy (84.95 ± 6.25% for experts, 75.12 ± 11.66% for novices) and the fixate-center strategy (92.34 ± 3.66% for experts, 87.38 ± 5.45% for novices).

Penalty Situation

On the basis of previous research highlighting best detection of two stimuli when applying the fixate-center strategy in laboratory-based designs (see the Athletes subsection), Hüttermann, Memmert, and Liesner (2014) conducted a football penalty experiment in which players had to detect the goalkeeper's movement during the run-up. Compared with a free-gaze strategy (foveal gaze on either the ball or the goalkeeper, or saccades) the fixate-center strategy (perceiving an imaginary point between ball and goalkeeper, or rather both stimuli, ball and goalkeeper, peripherally) was superior when it came to detecting goalkeeper movements. Future research will investigate whether appropriate training scenarios applying the fixate-center strategy will lead to a higher number of scored goals.

Elderly People

Hüttermann, Bock, and Memmert (2012) analyzed the Attention-Window in elderly people by leaving it up to the participants where to direct their gaze (the free-gaze condition). The data (averaged across the free-gaze and a fixate-center condition) indicates that attention breadth decreases by 27% from the age of early 20s to the age of late 60s (cf. figure 28.5). This result generalizes previous findings about age-related attention deficits to settings that are relevant in everyday contexts.

Future Directions

In the course of evolution, the human brain was optimized to consciously select and process those kinds of stimuli that are highly relevant for action. Sports-related attention research has often focused on four subprocesses of attention: attentional orienting, selective attention, sustained attention, and divided attention. Additionally, it was shown that the AW paradigm is useful as a measure of attentional breadth in complex environments or varied populations such as athletic training, penalty situations, and elderly people. In past years, different testing procedures in various racket sports were developed regarding measurements of all of the four dimensions of attention, and numerous strategies for optimal attentional direction and control were tested. This has direct consequences for the development of new training programs for players and referees.

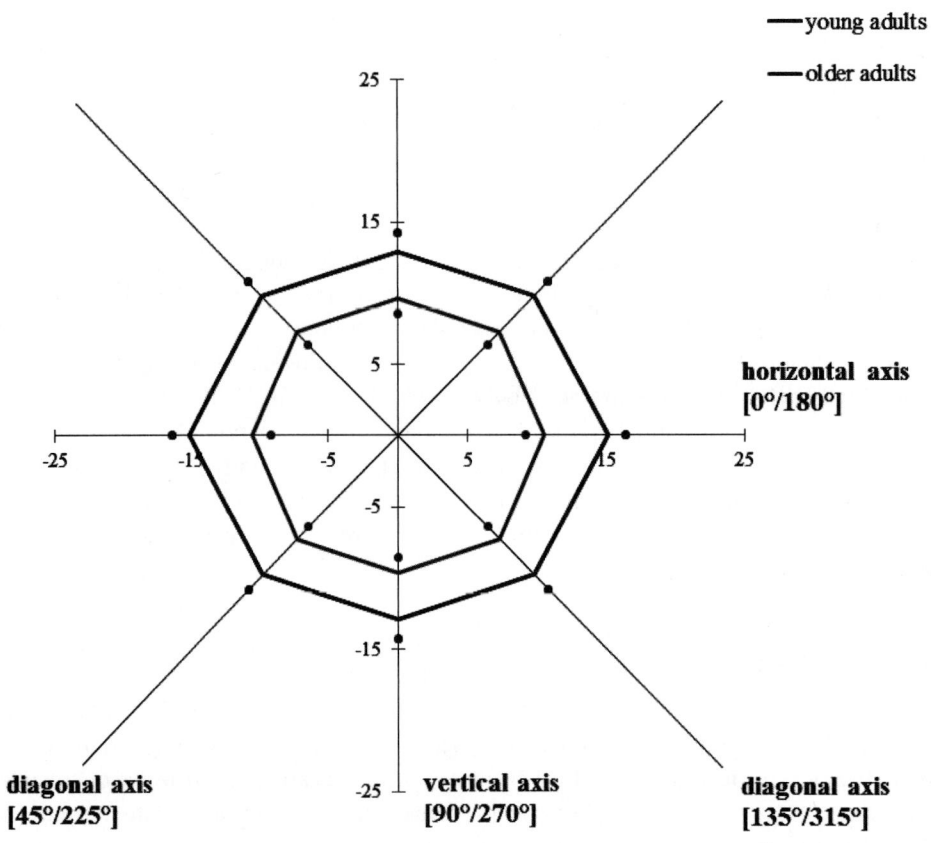

Figure 28.5
The size of the Attention-Window for young and older adults as a function of stimulus separation (free-gaze condition). Since each stimulus pair was presented symmetrically across the center, the data for each axis (horizontal, diagonal, and vertical) was also displayed symmetrically. Symbols represent across-subject means. From Hüttermann et al., 2012.

Further theoretical and methodological avenues (developmental issues, event-related potentials [ERPs]), as well as differential factors (i.e., motivation, working memory, or creativity), need to be added to already existing facets of attention research, especially in complex contexts such as sports. In this respect, recent advances in the sport attention literature have indicated that working memory can be regarded as a central concept in understanding attention in sports (Furley, 2012; Furley & Memmert, 2012, 2013). A series of studies demonstrated that the activated contents of working memory guide an athlete's focus of attention (Furley & Memmert, 2013) by directing attention toward objects in the visual field that are related to the activated contents of working memory. In addition, working memory has been shown to be predictive of controlling attention in a goal-directed manner and avoiding distraction and interference among athletes (Furley & Memmert, 2012). Further research along the lines of the work reviewed here should yield new insights about the role of attention in sports performance.

Box 28.1
Key Points

- Four subprocesses of attention are of central importance in sport practice and therefore have been the primary object of investigation in sport science: attentional orienting, selective attention, divided attention, and sustained attention.

- An external attentional focus of attention seems to be superior to an internal, movement-related focus across many different movement skills, skill levels, and target groups.

- Ignoring (perhaps at the present moment) unimportant information can be understood as an advantage and as a necessity for making decisions in complex game situations. However, it needs to be recognized that inattentional blindness can put people at a disadvantage with respect to their ability to consider important changes in their surroundings or perceive unexpected, free-standing teammates.

- Because of the tremendous importance of a wide Attention-Window in various team and racket sports, test procedures for its measurement have been developed, and different strategies in favor of an optimal attentional direction have been tested.

Box 28.2
Outstanding Issues

- To meet the demands of complex surroundings in sport, further theoretical and methodological avenues (developmental issues, ERP) as well as individual differences (e.g., motivation, working memory, or creativity) need to be integrated more intensively with already existing studies on the facets of attention.

- The study of attentional achievements, not only in competitive sports but also in early training for children and youths, needs to receive more detailed observation in research, because various attention-related resources have not yet been sufficiently developed during this age.

References

Abernethy, B., & Russell, D. G. (1987). Expert–novice differences in an applied selective attention task. *Journal of Sport Psychology, 9,* 326–345.

Abernethy, B., Maxwell, J. P., Masters, R. S. W., van der Kamp, J., & Jackson, R. C. (2007). Attentional processes in skill learning and expert performance. In G. Tenenbaum & R. C. Eklund (Eds.), *Handbook of sport psychology* (3rd ed., pp. 245–263). New York: Wiley.

Abernethy, B., Wood, J. M., & Parks, S. (1999). Can the anticipatory skills of experts be learned by novices? *Research Quarterly for Exercise and Sport, 70,* 313–318.

Beilock, S. L., Bertenthal, B. I., McCoy, A. M., & Carr, T. H. (2004). Haste does not always make waste: Expertise, direction of attention, and speed versus accuracy in performing sensorimotor skills. *Psychonomic Bulletin & Review, 11,* 373–379.

Beilock, S. L., Carr, T. H., MacMahon, C., & Starkes, J. L. (2002). When paying attention becomes counterproductive: Impact of divided versus skill-focused attention of novice and experienced performance of sensorimotor skills. *Journal of Experimental Psychology. Applied, 8,* 6–16.

Broadbent, D. E. (1958). *Perception and communication.* London: Pergamon Press.

Brown, H. R., & Friston, K. J. (2013). The functional anatomy of attention: A DCM study. *Frontiers in Human Neuroscience, 7*(784). doi:10.3389/fnhum.2013.00784.

Castiello, U., & Umiltà, C. (1992). Orienting of attention in volleyball players. *International Journal of Sport Psychology, 23,* 301–310.

Cavanagh, P., & Alvarez, G. A. (2005). Tracking multiple targets with multifocal attention. *Trends in Cognitive Sciences, 9,* 349–354.

Coull, J. T. (1998). Neural correlates of attention and arousal: Insights from electrophysiology, functional neuroimaging and psychopharmacology. *Progress in Neurobiology, 55,* 343–361.

Cowan, N. (1995). *Attention and memory: An integrated framework.* New York: Oxford University Press.

Enns, J., & Richards, J. (1997). Visual attentional orienting in developing hockey players. *Journal of Experimental Child Psychology, 64,* 255–275.

Eriksen, C. W., & St. James, J. D. (1986). Visual attention within and around the field of focal attention: A zoom lens model. *Perception & Psychophysics, 40,* 225–240.

Farrow, D., & Abernethy, B. (2002). Can anticipatory skills be learned through implicit video-based perceptual training? *Journal of Sports Sciences, 20,* 471–485.

Fernandez-Duque, D., & Johnson, M. L. (1999). Attention metaphors: How metaphors guide the cognitive psychology of attention. *Cognitive Science, 23,* 83–116.

Fernandez-Duque, D., & Posner, M. I. (1997). Relating the mechanisms of orienting and alerting. *Neuropsychologia, 35,* 477–486.

Furley, P. (2012). *Working memory and the control of attention in sport: From general mechanisms to individual differences* (Doctoral thesis). Cologne, Germany: German Sport University Cologne.

Furley, P., & Memmert, D. (2012). Working memory capacity as controlled attention in tactical decision making. *Journal of Sport & Exercise Psychology, 34*, 322–344.

Furley, P., & Memmert, D. (2013). "Whom should I pass to?" The more options the more attentional guidance from working. *PLoS ONE, 8*(5), e62278. doi:10.1371/journal.pone.0062278.

Furley, P., Memmert, D., & Heller, C. (2010). The dark side of visual awareness in sport—Inattentional blindness in a real-world basketball task. *Attention, Perception & Psychophysics, 72*, 1327–1337.

Gray, R. (2004). Attending to the execution of a complex sensorimotor skill: Expertise differences, choking, and slumps. *Journal of Experimental Psychology. Applied, 10*, 42–54.

Gray, R. (2011). Links between attention, performance pressure, and movement in skilled motor action. *Current Directions in Psychological Science, 20*, 301–306.

Green, C. S., & Bavelier, D. (2003). Action video game modifies visual selective attention. *Nature, 42*, 534–537.

Goldstein, B. (2011). *Cognitive psychology: Connecting mind, research and, everyday experience* (3rd ed.). Belmont, CA: Wadsworth.

Hagemann, N., & Memmert, D. (2006). Coaching anticipatory skill in badminton: Laboratory-versus field-based perceptual training? *Journal of Human Movement Studies, 50*, 381–398.

Hüttermann, S. (2014). *Das "Attention-Window"-Modell: Eine Exploration verschiedener Einflussfaktoren auf die Größe und Form des visuellen Aufmerksamkeitsfokus im Sport* [*The Attention-Window-Model: An exploration of different influential factors on the size and shape of the visual focus of attention in sports*] (Doctoral thesis). Cologne, Germany: German Sport University Cologne.

Hüttermann, S., Bock, O., & Memmert, D. (2012). The breadth of attention in old age. *Ageing Research, 10*(4), 67–70.

Hüttermann, S., Memmert, D., & Liesner, F. (2014). Finding the happy medium: An analysis of gaze behavior strategies in a representative task design of soccer penalties. *Journal of Applied Sport Psychology, 26*, 172–181.

Hüttermann, S., Memmert, D., Simons, D. J., & Bock, O. (2013). Fixation strategy influences the ability to focus attention on two spatially separate objects. *PLoS ONE, 8*(6), e65673.

Hüttermann, S., Simons, D., & Memmert, D. (2014). The size and shape of the attentional "spotlight" varies with differences in sports expertise. *Journal of Experimental Psychology. Applied, 20*, 147–157.

Jackson, R. C., & Farrow, D. (2005). Implicit perceptual training: How, when and why? *Human Movement Science, 24*, 308–325.

Jans, B., Peters, J. C., & De Weerd, P. (2010). Visual spatial attention to multiple locations at once: The jury is still out. *Psychological Review, 117*, 637–682.

Knudsen, E. (2007). Fundamental components of attention. *Annual Review of Neuroscience, 30*, 57–78.

Lum, J., Enns, J. T., & Pratt, J. (2002). Visual orienting in college athletes: Explorations of athlete type and gender. *Research Quarterly for Exercise and Sport, 73*, 156–167.

Mack, A., & Rock, I. (1998). *Inattentional blindness*. Cambridge, MA: MIT Press.

Magill, R. A. (1998). Knowledge is more than we can talk about: Implicit learning in motor skill acquisition. *Research Quarterly for Exercise and Sport, 69*, 104–110.

Maruenda, F. B. (2004). Can the human eye detect an offside position during a football match? *British Medical Journal, 329*, 1470–1472.

Memmert, D. (2006). The effects of eye movements, age, and expertise on inattentional blindness. *Consciousness and Cognition, 15*, 620–627.

Memmert, D. (2007). Can creativity be improved by an attention-broadening training program?—An exploratory study focusing on team sports. *Creativity Research Journal, 19*, 281–292.

Memmert, D. (2009). Pay attention! A review of attentional expertise in sport. *International Review of Sport & Exercise Psychology, 2*, 119–138.

Memmert, D., & Furley, P. (2007). "I spy with my little eye!"—Breadth of attention, inattentional blindness, and tactical decision making in team sports. *Journal of Sport & Exercise Psychology, 29*, 365–381.

Moran, A. (1996). *The psychology of concentration in sport performers: A cognitive analysis*. Hove, UK: Psychology Press.

Moran, A. (2009). Attention, concentration, and thought management. In B. Brewer (Ed.), *The Olympic handbook of sports medicine and science: Sport psychology* (pp. 18–29). Oxford, UK: Wiley-Blackwell.

Moran, A. (2011). Concentration. In T. Morris & P. Terry (Eds.), *The new sport and exercise psychology companion* (pp. 309–326). Morgantown, WV: Fitness Information Technology.

Moran, A. (2012). Concentration: Attention and performance. In S. M. Murpy (Ed.), *The Oxford handbook of sport and performance psychology* (pp. 117–130). Oxford, UK: Oxford University Press.

Moran, A. (2014). Situational awareness. In R. C. Eklund & G. Tenenbaum (Eds.), *Encyclopedia of sport and exercise psychology* (pp. 678–679). London: Sage.

Most, S. B., Scholl, B. J., Clifford, E. R., & Simons, D. J. (2005). What you see is what you set: Sustained inattentional blindness and the capture of awareness. *Psychological Review, 112*, 217–242.

Nougier, V., Ripoll, H., & Stein, J. F. (1989). Orienting of attention with highly skilled athletes. *International Journal of Sport Psychology, 20*, 205–223.

Nougier, V., & Rossi, B. (1999). The development of expertise in the orienting of attention. *International Journal of Sport Psychology, 30*, 246–260.

Nougier, V., Rossi, B., Alain, C., & Taddei, F. (1996). Evidence for strategic effects in the modulation of the orienting of attention. *Ergonomics, 39*, 1119–1133.

Nougier, V., Stein, J. F., & Bonnel, A. M. (1991). Information processing in sport and "Orienting of attention" *International Journal of Sport Psychology, 22*, 307–327.

Pesce-Anzeneder, C., & Bösel, R. (1998). Modulation of the spatial extent of the attentional focus in high-level volleyball players. *European Journal of Cognitive Psychology, 10*, 247–267.

Posner, M. I. (1980). Orienting of attention. *Quarterly Journal of Experimental Psychology, 32*, 3–25.

Posner, M. I., & Boies, S. J. (1971). Components of attention. *Psychological Review, 78*, 391–408.

Pylyshyn, Z. W. (2007). *Things and places: How the mind connects with the perceptual world (2004 Jean Nicod Lectures)*. Cambridge, MA: MIT Press.

Rowe, R. M., & McKenna, F. P. (2001). Skilled anticipation in real-world tasks: Measurement of attentional demands in the domain of tennis. *Journal of Experimental Psychology. Applied, 7*, 60–67.

Scholl, B. J. (2001). Objects and attention: The state of the art. *Cognition, 80*, 1–46.

Simons, D. J., & Chabris, C. F. (1999). Gorillas in our midst: Sustained inattentional blindness for dynamic events. *Perception, 28*, 1059–1074.

Smeeton, N., Ward, P., & Williams, A. M. (2004). Transfer of perceptual skill in sport. *Journal of Sports Sciences, 19*, 3–9.

Styles, E. A. (2008). *The psychology of attention*. Hove, UK: Psychology Press.

Tenenbaum, G., & Bar-Eli, M. (1995). Contemporary issues in exercise and sport psychology research. In S. J. H. Biddle (Ed.), *European perspectives on sport and exercise psychology* (pp. 292–323). Champaign, IL: Human Kinetics.

Williams, A. M., Ward, P., & Chapman, C. (2003). Training perceptual skill in field hockey: Is there transfer from the laboratory to the field? *Research Quarterly for Exercise and Sport, 74*, 98–103.

Williams, A. M., Ward, P., Knowles, J. M., & Smeeton, N. J. (2002). Anticipation skill in a real-world task: Measurement, training, and transfer in tennis. *Journal of Experimental Psychology. Applied, 8*, 259–270.

Wulf, G. (2007). *Attention and motor skill learning*. Champaign, IL: Human Kinetics.

Wulf, G. (2013). Attentional focus and motor learning: A review of 15 years. *International Review of Sport and Exercise Psychology, 6*, 77–104.

Wulf, G., McConnel, N., Gärtner, M., & Schwarz, A. (2002). Feedback and attentional focus: Enhancing the learning of sport skills through external-focus feedback. *Journal of Motor Behavior, 34*, 171–182.

Wulf, G., & Prinz, W. (2001). Directing attention to movement effects enhances learning: A review. *Psychonomic Bulletin & Review, 8*, 648–660.

Wulf, G., Shea, C. H., & Park, J.-H. (2001). Attention in motor learning: Preferences for and advantages of an external focus. *Research Quarterly for Exercise and Sport, 72*, 335–344.

Contributors

Richard A. Abrams, Washington University, Department of Psychology, St. Louis, Missouri

Lewis Baker, Vanderbilt University, Department of Psychology and Human Development, Nashville, Tennessee

Daphne Bavelier, University of Geneva, Faculty of Psychology and Education Sciences, Geneva, Switzerland; Brain and Cognitive Sciences, University of Rochester, Rochester, New York

Virginia Best, Boston University, Department of Speech, Language and Hearing Sciences, Boston, Massachusetts

Adam B. Blake, University of California, Los Angeles, Department of Psychology, Los Angeles, California

Paul W. Burgess, UCL (University College London), Institute of Cognitive Neuroscience, London, United Kingdom

Alan D. Castel, University of California, Los Angeles, Department of Psychology, Los Angeles, California

Karen Collins, University of Waterloo, Department of Psychology, Waterloo, Ontario, Canada

Mike J. Dixon, University of Waterloo, Department of Psychology, Waterloo, Ontario, Canada

Sidney K. D'Mello, University of Notre Dame, Department of Psychology and Department of Computer Science, Notre Dame, Indiana

Julia Föcker, University of Geneva, Faculty of Psychology and Education Sciences, Geneva, Switzerland

Charles L. Folk, Villanova University, Department of Psychology, Villanova, Pennsylvania

Tom Foulsham, University of Essex, Department of Psychology, Colchester, Essex, United Kingdom

Jonathan A. Fugelsang, University of Waterloo, Department of Psychology, Waterloo, Ontario, Canada

Bradley S. Gibson, University of Notre Dame, Department of Psychology, Notre Dame, Indiana

Matthias S. Gobel, University College London, Department of Experimental Psychology, London, United Kingdom

Davood G. Gozli, University of Toronto, Department of Psychology, Toronto, Ontario, Canada

Arthur C. Graesser, University of Memphis, Department of Psychology, Memphis, Tennessee

Peter A. Hancock, University of Central Florida, Department of Psychology; and Institute for Simulation and Training, Orlando, Florida

Kevin A. Harrigan, University of Waterloo, Department of Psychology, Waterloo, Ontario, Canada

Simone G. Heideman, University of Oxford, Oxford Centre for Human Brain Activity, University Department of Psychiatry, Oxford, United Kingdom

Cristy Ho, University of Oxford, Department of Experimental Psychology, Oxford, United Kingdom

Roxane J. Itier, University of Waterloo, Department of Psychology, Waterloo, Ontario, Canada

Gustav Kuhn, Goldsmiths, University of London, Department of Psychology, London, United Kingdom

Michael F. Land, School of Life Sciences, University of Sussex, Brighton, United Kingdom

Mallorie Leinenger, University of California, San Diego, Department of Psychology, La Jolla, California

Daniel Levin, Vanderbilt University, Department of Psychology and Human Development, Nashville, Tennessee

Steven J. Luck, University of California, Davis, Center for Mind and Brain and Department of Psychology, Davis, California

Gerald Matthews, University of Central Florida, Institute for Simulation and Training, Orlando, Florida

Daniel Memmert, German Sport University Cologne, Institute of Cognitive and Team/Racket Sport Research, Cologne, Germany

Stephen Monsell, University of Exeter, Psychology, College of Life and Environmental Sciences, Exeter, United Kingdom

Meenely Nazarian, University of California, Los Angeles, Department of Psychology, Los Angeles, California

Anna C. Nobre, University of Oxford, Oxford Centre for Human Brain Activity and Department of Experimental Psychology, Oxford, United Kingdom

Andrew M. Olney, University of Memphis, Institute for Intelligent Systems, Memphis, Tennessee

Kerri L. Pickel, Ball State University, Department of Psychological Science, Muncie, Indiana

Jay Pratt, University of Toronto, Department of Psychology, Toronto, Ontario, Canada

Keith Rayner, University of California, San Diego, Department of Psychology, La Jolla, California

Daniel C. Richardson, University College London, Department of Experimental Psychology, London, United Kingdom

Evan F. Risko, University of Waterloo, Department of Psychology, Waterloo, Ontario, Canada

Barbara Shinn-Cunningham, Boston University, Department of Biomedical Engineering, Boston, Massachusetts

Vivian Siu, University of Waterloo, Department of Psychology, Waterloo, Ontario, Canada

Jonathan Smallwood, University of York, Department of Psychology, York, United Kingdom

Charles Spence, University of Oxford, Department of Experimental Psychology, Oxford, United Kingdom

David L. Strayer, University of Utah, Department of Psychology, Salt Lake City, Utah

Pedro Sztybel, University of Notre Dame, Department of Psychology, Notre Dame, Indiana

Benjamin W. Tatler, School of Psychology, University of Aberdeen, Aberdeen, United Kingdom

J. Eric T. Taylor, University of Toronto, Department of Psychology, Toronto, Ontario, Canada

Jeff Templeton, University of Waterloo, Department of Psychology, Waterloo, Ontario, Canada

Robert Teszka, Goldsmiths, University of London, Department of Psychology, London, United Kingdom

Michel Wedel, University of Maryland, Robert H. Smith School of Business, College Park, Maryland

Blaire J. Weidler, Washington University, Department of Psychology, St. Louis, Missouri

Lisa Wojtowicz, University of Waterloo, Department of Psychology, Waterloo, Ontario, Canada

Jeremy M. Wolfe, Departments of Ophthalmology and Radiology, Harvard Medical School; Director, Visual Attention Lab, Brigham and Women's Hospital, Cambridge, Massachusetts

Geoffrey F. Woodman, Vanderbilt University, Department of Psychology; Vanderbilt Vision Research Center, Center for Integrative and Cognitive Neuroscience, Nashville, Tennessee

Reviewers

Naseem Al-Aidroos

Philip Allen

Walter Boot

Gene Brewer

Jennifer Campos

Alan Castel

Monica Castelhano

Kyle Cave

Craig Chapman

J. Allan Cheyne

Joseph Chisholm

Luke Clark

Michael Dodd

Judy Fan

Charles Folk

Tom Foulsham

Andrew Gallup

Michelle Green

Norbert Hagemann

Jari Hietanen

Lorraine Hope

Ira Hyman

Jukka Hyönä

Yoko Ishigami

Andrea Kiesel

Wilfried Kunde

Michael Lawrence

Adrian Lee

Daniel Levin

Stephen Macknik

Jennifer McVay

Nathan Medeiros-Ward

Neil Mennie

Jyoti Mishra

Aidan Moran

Jeff Pelz

Franziska Plessow

Jay Pratt

Bruno Rossion

Leonhard Schilbach

Darryl Schneider

Daniel Simons

Daniel Smilek

Grayden Solman

Salvador Soto-Faraco

Ian Spence

Aikaterini Stefanidi

Karl Szpunar

David Thomson

John Tsotsos

Melissa Vo

Jean Vroomen

Michel Wedel

Theodore Zanto

Index